BARRON'S

FE

FUNDAMENTALS OF ENGINEERING EXAM

3RD EDITION

Masoud Olia, Ph.D., PE
Wentworth Institute of Technology
Boston, Massachusetts

Contributors

Professor Armen Casparian, M.S.
Wentworth Institute of Technology
Boston, Massachusetts

Professor Nader Jalili, Ph.D.
Northeastern University
Boston, Massachusetts

Professor Hamid Nayeb-Hashemi, Ph.D.
Northeastern University
Boston, Massachusetts

Professor Frederick Driscoll, M.S.
Wentworth Institute of Technology
Boston, Massachusetts

Professor Taha Marhaba, Ph.D., PE
New Jersey Institute of Technology
Newark, New Jersey

Farzan Parsinejad, Ph.D.
Chevron Corp.
Richmond, California

Professor Mark Frisina, M.S., PE
Wentworth Institute of Technology
Boston, Massachusetts

Professor Hameed Metghalchi, Sc.D.
Northeastern University
Boston, Massachusetts

Professor Thomas J. Pfaff, Ph.D.
Ithaca College
Ithaca, New York

Professor Thomas Hulbert, M.S., PE
Northeastern University
Boston, Massachusetts

Professor Richard Murphy, Ph.D.
Northeastern University
Boston, Massachusetts

Professor David E. Stevens, M.S.
Wentworth Institute of Technology
Boston, Massachusetts

BARRON'S

© Copyright 2015, 2008 by Barron's Educational Series, Inc.
Previous edition © copyright 2000 by Barron's Educational Series, Inc., under the
title *How to Prepare for the Fundamentals of Engineering (FE/EIT) Exam.*

All inquiries should be addressed to:
Barron's Educational Series, Inc.
250 Wireless Boulevard
Hauppauge, New York 11788
www.barronseduc.com

Library of Congress Control No.: 2014951474

ISBN: 978-1-4380-0310-8

PRINTED IN THE UNITED STATES OF AMERICA
9 8 7 6 5 4 3 2 1

**10%
POST-CONSUMER
WASTE**
Paper contains a minimum
of 10% post-consumer
waste (PCW). Paper used
in this book was derived
from certified, sustainable
forestlands.

Contents

Preface

Whether you are a working, experienced engineer or a soon-to-be engineering graduate, *FE Fundamentals of Engineering Exam* was written to help you pass the Fundamentals of Engineering Exam. Use this book in conjunction with the National Council of Engineering Examiners and Surveyors (NCEES)-supplied handbook of equations and formulas, and you will be well prepared to pass the FE exam.

Until now, preparing for the FE exam has been a time-consuming and sometimes costly proposition. *FE Fundamentals of Engineering Exam* is the most efficient and effective way to study for the morning exam that is required of all examinees. This book also includes a review of the multiconcept, multipart questions found in the afternoon General Exam Option. While examinees can elect to take a discipline-specific exam (in either chemical, civil, electrical, industrial, or mechanical engineering), the majority choose the General Exam Option.

We have included a subject review and sample problems for each of the fourteen subject areas covered by the FE exam. Each chapter includes exam-like problems as well as theoretical questions. Not only are the answers to the questions given but solutions to the problems are provided as well. Finally, a practice exam is included with answers and a scoring analysis.

This book is a collaborative work. How best to review each of the fourteen exam subjects covered by the FE exam requires specialized knowledge. Experts weren't just "consulted" for certain chapters. A team of teaching professors, each an expert in his field and an experienced FE review course instructor, worked together to produce this book.

I am extremely thankful for the contributions made by the coauthors of this book, who are listed below. Their hard work and dedication made this project possible.

Professor Armen S. Casparian, M.S.	*Wentworth Institute of Technology*
Professor Frederick F. Driscoll, M.S.	*Wentworth Institute of Technology*
Professor Mark Frisina, M.S., PE	*Wentworth Institute of Technology*
Professor Thomas E. Hulbert, M.S., PE	*Northeastern University*
Professor Nader Jalili, Ph.D.	*Northeastern University*
Professor Taha Marhaba, Ph.D., PE	*New Jersey Institute of Technology*
Professor Hameed Metghalchi, Sc.D.	*Northeastern University*
Professor Richard J. Murphy, Ph.D.	*Northeastern University*
Professor Hamid Nayeb-Hashemi	*Northeastern University*
Farzan Parsinejad, Ph.D.	*Chevron Corp.*
Professor Thomas J. Pfaff, Ph.D.	*Ithaca College*
Professor David E. Stevens, M.S.	*Wentworth Institute of Technology*

Finally, I would like to thank my wife, Maria, for her support during the period I worked on this project. Without her encouragement and helpful input, completion of this book would not have been possible.

Passing the FE exam is your first step toward becoming a registered engineer. *FE Fundamentals of Engineering* provides the information review and the practice problems needed to give you the confidence to pass the FE exam. We wish you continued success in your engineering career.

Masoud Olia, Ph.D., PE
Wentworth Institute of Technology
Boston, Massachusetts

Introduction

What Is the FE Exam?

Passing the FE exam is the first step in the licensure process to become a Registered, or Professional, Engineer (PE). The FE exam covers a comprehensive range of subject matter taught in an EAC/ABET-accredited baccalaureate program. Becoming a registered engineer is necessary to work as a consulting engineer and, in many industries, to advance to an engineering management position.

Registration requirements vary by state. In general, though, a registered engineering candidate must graduate from an ABET-accredited engineering program, pass the FE exam, and, after several years as a practicing engineer, pass the PE exam in his or her specialty.

How Do I Register/Apply?

As of January 2014, the FE exam has changed from the paper and pencil format to the new computer-based (CB) format.

The new computer-based FE exam is offered in testing windows throughout the year during the following months:

- January, February
- April, May
- July, August
- October, November

Exams are not administered during March, June, September, or December. NCEES policy allows examinees to attempt a particular NCEES exam one time per testing window and no more than three times in a 12-month approval period, which begins with the examinee's first attempt. Some licensing boards have a more restrictive policy. You should visit *www.ncees. org/boards* to read the policy of the licensing board you selected during the registration process.

The FE Exam Format

The FE exam for each discipline consists of 110 questions. You have approximately 5 hours and 20 minutes to complete the entire exam. Therefore, you should manage your time accordingly.

The approximate breakdown of questions by subject for each discipline is shown in the chart on page 2.

The most current version of the appropriate NCEES-supplied reference handbook will be supplied onscreen as a searchable pdf file. You will use a 24-inch monitor while taking the test. This will allow you sufficient space to display both the exam questions and the reference handbook. You can download the current version of the handbook free of charge from *ncees.org/cbt*.

You will be provided with a copy of the NCEES CBT exam rules to review once you arrive. You must indicate your agreement to comply with these rules by providing a digital signature before testing begins.

A representative will confirm that the only items in your possession are ones allowed into the testing area. These items include your ID, an *NCEES-approved calculator*, the key to your locker (if applicable), eyeglasses, a light jacket or sweater, and any other approved comfort items.

After completing approximately 55 questions, you will be prompted to review those questions and then submit them. You will no longer have access to those questions after you have submitted them. You will then have the option of taking a scheduled break. Manage your time accordingly to ensure that you have plenty of time to complete the remaining questions.

Test-Taking Strategies

A major goal of the review process is to improve recall, which is the ability to answer questions quickly. During the approximately 5 hours and 20 minutes of testing, you will have to answer 110 questions. Generally, the questions are unrelated to each other.Each question must be answered, on average, in less than 3 minutes.

Mechanical Discipline		Other Discipline	
Subject	Approximate Number of Problems	Subject	Approximate Number of Problems
Mathematics	6	Mathematics	14
Probability and Statistics	4	Probability and Statistics	7
Computational Tools	3	Chemistry	7
Ethics and Professional Practices	3	Ethics and Professional Practices	3
Engineering Economics	3	Engineering Economics	7
Electricity and Magnetism	4	Electricity and Magnetism	8
Statics	9	Statics	9
Dynamics, Kinematics, and Vibrations	10	Dynamics, Kinematics, and Vibrations	8
Mechanics of Materials	9	Mechanics of Materials	9
Material Properties and Processing	9	Material Properties and Processing	6
Fluid Mechanics	10	Fluid Mechanics of Liquids and Gases	14
Thermodynamics	14	Heat Transfer	10
Heat Transfer	10	Instrumentation and Data Acquisitions	4
Measurement, Instrumentation, and Controls	6	Safety, Health, and Environment	4
Mechanical Design and Analysis	10		
Total	**110**	**Total**	**110**

Remember that you do not want to spend too much time on any one question. If you are unsure of an answer, mark the question and make your best guess. Guessing is a valid test-taking strategy. You are not penalized for answering incorrectly. Answer each question; do not leave any answers blank. Your recall speed will be greatly improved as you practice the problems in *FE Exam*. Remember that you will have access to an electronic copy of the NCEES *FE Reference Handbook* at the exam site. As you work the problems in *FE Exam*, mark the equations and formulas that you use most frequently in your own copy of this handbook so that you will be familiar with their locations during the actual test. Similarly, become well acquainted with your calculator and its functions before exam day.

How to Use This Book

Since the FE exam is comprehensive, the best test-preparation strategy is to study all the exam subjects. To get the most benefit from your study time, however, concentrate on the subjects that have the most questions, as indicated in the breakdown of questions by subject presented here.

FE Exam is designed to give an overview of the concepts and equations used in the FE exam and to provide you with practice using problems like those on the FE exam. This book reflects the relative emphasis given to the various FE exam subjects. Thus Chapter 1, which is about mathematics, has substantially more problems than Chapter 10, which is about material properties and processing. Do not skip any chapters, however. Read each chapter review, and work all the sample problems in this book. It is only through actual practice with the problems that you will become familiar with the FE exam format.

Give yourself sufficient time to prepare. You should allow 8 to 10 weeks of study time. If you have been out of school for several years, you may need more time to prepare for the FE exam. One week before the FE exam, take the practice exams included in this book, using the NCEES *FE Reference Handbook* as your only reference. If the results indicate you need additional review, identify the areas where you did poorly. Concentrate your study on those topics.

Leave yourself a couple of days for a brief overview of all the subject areas.

Do not study the night before the exam. Instead, get a good night's sleep! By that point, you will have done all you can academically to prepare for the FE exam. By your understanding of basic engineering principles and by demonstrating your ability to apply these principles, you will pass the FE exam.

Mini-Diagnostic Exams

The purpose of the mini-diagnostic exams are to give direction to your study. The questions on the diagnostic exams are similar to those found on the FE Mechanical and Other Discipline exams. Taking these exams before you begin your FE study program will help you assess which areas you need to study and how best to allocate your time.

Each mini-diagnostic exam includes 40 questions. Be sure to read the section on test-taking strategies before taking the mini-diagnostic exams. Allow yourself 110 minutes to take both exams. Simulate conditions you will experience while taking the FE exam. Work in a quiet place. Take the diagnostic exams in one sitting. Use your calculator and your copy of the NCEES *FE Reference Handbook* of equations and formulas.

Correct your mini-diagnostic exams and evaluate your results. If you scored more than one answer wrong in any one subject area, you may need to increase your study time in that area. If you were unable to complete the 40 questions in the time allotted for each exam, you may need to work on your problem-solving speed and/or recall. After taking the mini-diagnostic exams and evaluating your results, begin your subject review with Chapter 1, "Mathematics."

ANSWER SHEET
Mini-Diagnostic Exam—Mechanical Discipline

1. Ⓐ Ⓑ Ⓒ Ⓓ
2. Ⓐ Ⓑ Ⓒ Ⓓ
3. Ⓐ Ⓑ Ⓒ Ⓓ
4. Ⓐ Ⓑ Ⓒ Ⓓ
5. Ⓐ Ⓑ Ⓒ Ⓓ
6. Ⓐ Ⓑ Ⓒ Ⓓ
7. Ⓐ Ⓑ Ⓒ Ⓓ
8. Ⓐ Ⓑ Ⓒ Ⓓ
9. Ⓐ Ⓑ Ⓒ Ⓓ
10. Ⓐ Ⓑ Ⓒ Ⓓ

11. Ⓐ Ⓑ Ⓒ Ⓓ
12. Ⓐ Ⓑ Ⓒ Ⓓ
13. Ⓐ Ⓑ Ⓒ Ⓓ
14. Ⓐ Ⓑ Ⓒ Ⓓ
15. Ⓐ Ⓑ Ⓒ Ⓓ
16. Ⓐ Ⓑ Ⓒ Ⓓ
17. Ⓐ Ⓑ Ⓒ Ⓓ
18. Ⓐ Ⓑ Ⓒ Ⓓ
19. Ⓐ Ⓑ Ⓒ Ⓓ
20. Ⓐ Ⓑ Ⓒ Ⓓ

21. Ⓐ Ⓑ Ⓒ Ⓓ
22. Ⓐ Ⓑ Ⓒ Ⓓ
23. Ⓐ Ⓑ Ⓒ Ⓓ
24. Ⓐ Ⓑ Ⓒ Ⓓ
25. Ⓐ Ⓑ Ⓒ Ⓓ
26. Ⓐ Ⓑ Ⓒ Ⓓ
27. Ⓐ Ⓑ Ⓒ Ⓓ
28. Ⓐ Ⓑ Ⓒ Ⓓ
29. Ⓐ Ⓑ Ⓒ Ⓓ
30. Ⓐ Ⓑ Ⓒ Ⓓ

31. Ⓐ Ⓑ Ⓒ Ⓓ
32. Ⓐ Ⓑ Ⓒ Ⓓ
33. Ⓐ Ⓑ Ⓒ Ⓓ
34. Ⓐ Ⓑ Ⓒ Ⓓ
35. Ⓐ Ⓑ Ⓒ Ⓓ
36. Ⓐ Ⓑ Ⓒ Ⓓ
37. Ⓐ Ⓑ Ⓒ Ⓓ
38. Ⓐ Ⓑ Ⓒ Ⓓ
39. Ⓐ Ⓑ Ⓒ Ⓓ
40. Ⓐ Ⓑ Ⓒ Ⓓ

40 Questions

Directions: For each incomplete statement or question, select the answer that best completes the statement or answers the question. Then fill in the corresponding space on the answer sheet.

1. A weight attached to a spring moves up and down, so that the equation of motion is $\frac{d^2x}{dt^2} + 64x = 0$, where x is the displacement in feet of the weight from its equilibrium position at any time t. Suppose that at $t = 0$ second the displacement is 3 inches $[x(0) = \frac{1}{4}]$ and the velocity is 1 foot per second $[x'(0) = 1]$. The amplitude of motion is
   (A) $\frac{1}{4}$ ft
   (B) $\frac{1}{3}$ ft
   (C) $\frac{\sqrt{3}}{2}$ ft
   (D) $\frac{\sqrt{5}}{8}$ ft

2. A function f is defined by a determinant as follows:
   $$f(x) = \begin{vmatrix} 2x^2 & x \\ 8 & 2 \end{vmatrix}$$
   What is the minimum value of this function?
   (A) −4
   (B) −2
   (C) 1
   (D) 0

3. An area in Quadrant I is bounded by $y = x^3, x = 2$, and the x-axis. The volume generated by revolving this area about the y-axis is
   (A) 8π cubic units
   (B) 14π cubic units
   (C) $\frac{64\pi}{5}$ cubic units
   (D) $\frac{8\pi}{3}$ cubic units

4. An experimental drug is being tested for side effects. The producers of the drug claim that only 5% of users will experience side effects. If 20 people are tested, what is the probability that 2 or more will experience side effects?
   (A) 0.0025
   (B) 0.1887
   (C) 0.2641
   (D) 0.7359

5. In a quality control check for a machine that produces bearings, a sample of 23 bearings yields a sample mean of 21 mm and a standard deviation of 2 mm. Find a 95% confidence interval for the mean diameter of a bearing produced by this machine.
   (A) (17.000, 25.000)
   (B) (20.135, 21.865)
   (C) (20.183, 21.817)
   (D) (20.284, 21.716)

6. How many individual cells are in a spreadsheet in the range B3:I12?
   (A) 8
   (B) 10
   (C) 78
   (D) 80

7. An engineer working for a consulting firm has decided to go into private practice. One of her first steps should be to
   (A) submit her letter of resignation
   (B) tell colleagues about her plans
   (C) send an e-mail to the firm regarding her plans
   (D) discuss her plans with her supervisor

8. E. Jones, PE, has taken a loan of $100,000 at an interest rate of 10% compounded annually. He plans to pay it back in 20 equal installments, one each year to partially fund his startup Materials Testing Lab. What should be the annual payment?
   (A) $12,000
   (B) $11,750
   (C) $13,200
   (D) $11,200

Questions 9 and 10 are based on the diagram shown below.

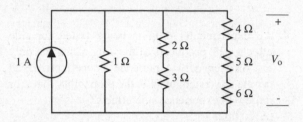

9. What is the value of the output voltage, V_o, in this circuit?
   (A) 0.789 V
   (B) 1.256 V
   (C) 4.321 V
   (D) 0.543 V

10. What is the power dissipated by the 2-kiloohm resistor in this circuit?
    (A) 1.22 W
    (B) 0.31 W
    (C) 0.24 W
    (D) 0.05 W

11. Based on the diagram, what is the force P needed for equilibrium?

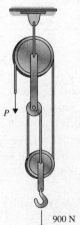

    (A) 225 N
    (B) 300 N
    (C) 900 N
    (D) 600 N

12. What is the tension in the cable shown in the diagram? Neglect the size of the pulley.

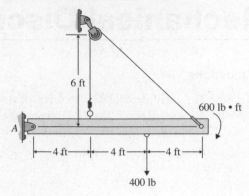

    (A) 950 lb
    (B) 285.7 lb
    (C) 900 lb
    (D) 339.3 lb

13. What is the horizontal force (P) that must be applied to the handle to maintain equilibrium in the diagram?

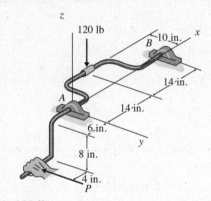

    (A) 150 lb
    (B) 200 lb
    (C) 120 lb
    (D) 240 lb

8

14.

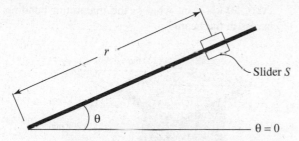

A rod/slider block mechanism rotates in a horizontal plane as shown in the diagram above. The motion is governed by the equations $r(t) = 2.0t^3 - 0.5t^2$ and $\theta(t) = t^2 + 8t$, where the units for r, t, and θ are meters, seconds, and radians, respectively. The mass of the slider block is 2.0 kilograms, and there is negligible friction between the rod and the slider block. The magnitude of the slider block's angular momentum when angular speed $\omega = 20.0$ radians per second is

(A) 6.85×10^6 kg·m²/s
(B) 1.20×10^3 kg·m²/s
(C) 5.18×10^4 kg·m²/s
(D) 4.06×10^5 kg·m²/s

15.

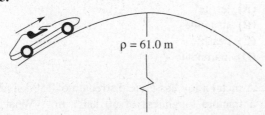

A 1,455-kilogram automobile (including the mass of the driver) is traveling along a hill that has a parabolic shape. The section of the hill near the crest has a radius of curvature of 61.0 meters, as shown in the above diagram. Wind resistance is considered negligible. If the speed of the automobile is 18.3 meters per second at the crest of the hill, the apparent weight of the system at this location should be

(A) 7,988 N
(B) 22,261 N
(C) 6,285 N
(D) 641 N

16.

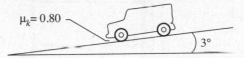

A police officer measures skid marks left by all four wheels of a 1,300-kilogram car to be 20.0 meters along a road surface that has a 3° slope from the horizontal as illustrated in the above diagram. It is known from prevailing conditions that the coefficient of kinetic friction between the tires and the pavement was approximately 0.80 and that the car had a speed of 40.0 kilometers per hour at the end of the skid. The amount of energy dissipated by friction due to the entire skid is

(A) 203,760 J
(B) 13,350 J
(C) 1,380 J
(D) 101,880 J

17.

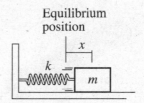

The spring-mass system shown in the above diagram consists of a 1.8-kilogram block and a linear spring with a stiffness of 0.26 kilonewton per meter. The initial displacement of the block is 2.5 centimeters to the right of the equilibrium position, and the block is released from rest. Friction is considered negligible between the block and the horizontal surface. The period of this vibration is necessarily

(A) 1.81 s
(B) 6.28 s
(C) 0.08 s
(D) 0.52 s

18.

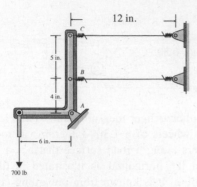

The two wires at *B* and *C* are made of the same material and have the same cross-sectional area. If the bracket is rigid, what is the tension in the wire at *B* due to the applied 700 lb load?

(A) 173.2 lb
(B) 67.6 lb
(C) 390.0 lb
(D) 197.0 lb

19.

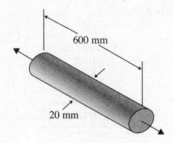

For the solid rod shown, if the new length of the rod due to the applied load is 601.4 mm and its diameter becomes 19.9837 mm, what is the Poisson's ratio (v)?

(A) 0.55
(B) 0.349
(C) 2.865
(D) 0.21

20. If the beam is subjected to a bending moment $M = 50$ kN · m, what is the maximum bending stress in the beam?

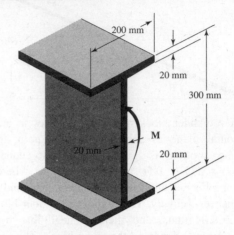

(A) 40.2 MPa
(B) 25.6 MPa
(C) 80.4 MPa
(D) 32.16 MPa

21. When carbon is dissolved in iron to a concentration of 0.5 weight percent carbon and the alloy is heated to 900°C, what phase is present?

(A) ferrite
(B) austenite
(C) pearlite
(D) martensite

22. A metal alloy has a yield strength of 150 ksi and a fracture toughness of 50 ksi · in$^{1/2}$. What is the largest surface crack size that can be applied without fracture in the elastic region under an applied static tension service load?

(A) 0.02 in.
(B) 0.08 in.
(C) 0.2 in.
(D) 0.5 in.

23. Consider an alloy containing 40 weight percent copper and 60 weight percent nickel, and use the copper-nickel equilibrium phase diagram shown below.

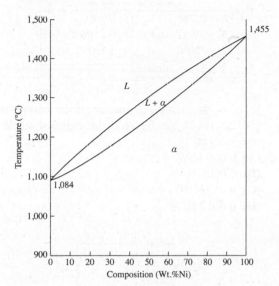

What is the percentage of liquid present in the alloy at 1,300°C?

(A) 11
(B) 88
(C) 37
(D) 63

Questions 24 and 25 refer to the following diagram:

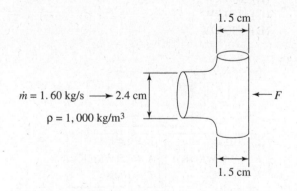

Water flows into the tee shown.

24. What is the average velocity of the water entering the tee?

(A) 4.63 m/s
(B) 3.95 m/s
(C) 3.54 m/s
(D) 0.35 m/s

25. What is the magnitude of the force (F) required to keep the tee stationary?

(A) 5.66 N
(B) 2.21 N
(C) 6.32 N
(D) 7.41 N

26.

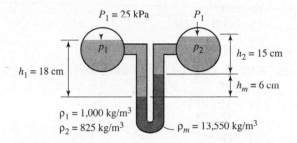

For the U-tube manometer shown above, what is the pressure in vessel 2 (p_2)?

(A) 32.4 kPa
(B) 33.5 kPa
(C) 18.2 kPa
(D) 17.6 kPa

27.

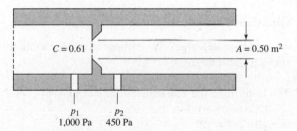

In the above diagram where C is the discharge coefficient, what is the volumetric flow rate through the orifice?

(A) 0.32 m³/s
(B) 0.28 m³/s
(C) 1.28 m³/s
(D) 0.65 m³/s

Questions 28–30 refer to the following:

Air is compressed from an initial state of 100-kilopascal pressure and 25°C temperature to a final pressure of 500 kilopascals in a polytropic process ($n = 1.5$) in a piston and cylinder device.

28. What is the final temperature of the air?
- **(A)** 35°C
- **(B)** 43°C
- **(C)** 199°C
- **(D)** 237°C

29. What is the value of the work interaction in this process?
- **(A)** −122 kJ/kg
- **(B)** −10 kJ/kg
- **(C)** 10 kJ/kg
- **(D)** 120 kJ/kg

30. What is the amount of energy transfer via heat interaction?
- **(A)** −272 kJ/kg
- **(B)** −32 kJ/kg
- **(C)** 31 kJ/kg
- **(D)** 272 kJ/kg

Questions 31–32 refer to the following:

At steady-state, a 25 W curling iron has an outer surface temperature of 80°C.

31. Determine the rate of heat transfer in $\frac{kWh}{s}$.
- **(A)** 0.5×10^{-5}
- **(B)** 0.7×10^{-5}
- **(C)** 2×10^{-5}
- **(D)** 5×10^{-5}

32. What is the rate of entropy production in $\frac{kJ}{s \cdot k}$?
- **(A)** 0.05
- **(B)** 0.07
- **(C)** 0.2
- **(D)** 0.5

33.

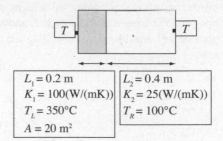

What is the thermal resistance of the composite wall shown in the figure if the exposed surface area is 20 m²?
- **(A)** 0.009 K/W
- **(B)** 0.0009 K/W
- **(C)** 0.012 K/W
- **(D)** 0.0012 K/W

34.

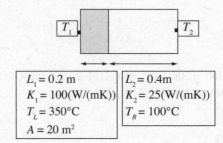

What is the heat transfer rate for the system shown in the figure?
- **(A)** 100 kW
- **(B)** 200 kW
- **(C)** 280 kW
- **(D)** 320 kW

35.

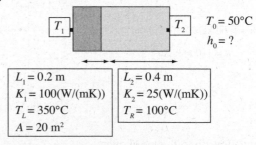

What is the convective heat transfer coefficient, h_o, of the system in the figure?
- **(A)** $100 \ \frac{W}{m^2 K}$
- **(B)** $200 \ \frac{W}{m^2 K}$
- **(C)** $280 \ \frac{W}{m^2 K}$
- **(D)** $320 \ \frac{W}{m^2 K}$

36. For each of the following systems, what parameter is invariant?

I. $T_1(s) = \dfrac{12}{s^2 + 8s + 12}$ II. $T_2(s) = \dfrac{16}{s^2 + 8s + 16}$

(A) settling time
(B) overshoot
(C) damping ratio
(D) damped natural frequency

37. The open-loop Bode diagram of a feedback control system is shown below. What are the gain margin (in dB) and phase margin (in degrees), respectively?

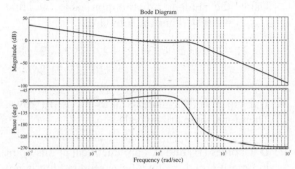

(A) 5 dB, 90°
(B) 10 dB, 180°
(C) 20 dB, 105°
(D) 10 dB, 105°

38.

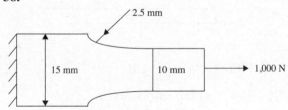

Find the maximum stress in a step shaft shown in the figure if it is subjected to 1,000 N.

(A) 10 MPa
(B) 15.1 MPa
(C) 19.09 MPa
(D) 21.07 MPa

39.

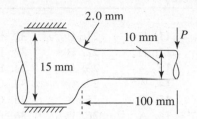

A cantilever rotating step shaft shown in the figure is machined from AISI 1040 CD steel with an ultimate tensile strength of 590 MPa. What is the maximum load that can be applied to the shaft in order for it to have infinite life?

(A) 151.2 N
(B) 178.4 N
(C) 200.5 N
(D) 251.7 N

40.

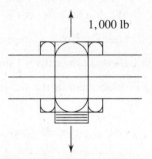

Two components are fastened together by using a ½ UNC bolt grade 5. If the assembly is subjected to a force of 1,000 lb, what is the factor of safety against the bolt failure? Assume the stiffness of the part is twice the stiffness of the bolt. The system is designed for replacement parts.

(A) 3
(B) 6
(C) 9
(D) 12

Answer Key

1. D	11. B	21. B	31. B
2. A	12. D	22. B	32. B
3. C	13. A	23. C	33. B
4. C	14. A	24. C	34. C
5. B	15. C	25. A	35. C
6. D	16. A	26. D	36. B
7. D	17. D	27. A	37. D
8. B	18. A	28. D	38. C
9. A	19. B	29. A	39. B
10. D	20. A	30. C	40. B

Answers Explained

1. D The solutions of the auxiliary equation $m^2 + 64 = 0$ are $\pm 8i$. Hence,

$$x = c_1 \cos 8t + c_2 \sin 8t \text{ and}$$
$$x' = -8c_1 \sin 8t + 8c_2 \cos 8t$$

The initial condition $x(0) = \dfrac{1}{4}$ implies $c_1 = \dfrac{1}{4}$, and the initial condition $x'(0) = 1$ implies $c_2 = \dfrac{1}{8}$. Thus, the displacement of the weight at any time t is given by

$$x = \frac{1}{4} \cos 8t + \frac{1}{8} \sin 8t$$

The amplitude of motion is $\sqrt{c_1^2 + c_2^2}$

$$= \sqrt{\left(\frac{1}{4}\right)^2 + \left(\frac{1}{8}\right)^2} = \frac{\sqrt{5}}{8} \text{ ft.}$$

2. A For $f(x) = 4x^2 - 8x$:

$$f'(x) = 8x - 8 \quad \text{and} \quad f''(x) = 8$$

Solving yields $f'(x) = 0$, which implies $x = 1$. Since $f''(1) = 8 > 0$, there is a minimum at $x = 1$, and this minimum value is $f(1) = -4$.

3. C Take the volume of the outer cylinder, and subtract from it the volume of the hollowed-out region:

$$\text{Volume} = \pi(2)^2(8) - \pi \int_0^8 \left(\sqrt[3]{y}\right)^2 dy$$

$$= 32\pi - \frac{3\pi}{5} y^{5/3} \Big|_0^8 = \frac{64\pi}{5} \text{ cubic units.}$$

4. C Let X represent the number of people tested that experience side effects. Note that X is a binomial random variable with $n = 20$ and $p = 0.05$. Now

$$P(X \geq 2) = 1 - P(X = 0) - P(X = 1)$$
$$= 1 - C(20, 0)(0.05)^0(1 - 0.05)^{20-0} - C(20, 1)(0.05)^1(1 - 0.05)^{20-1}$$
$$= 1 - 0.3585 - 0.3774$$
$$= 0.2641$$

5. B Since the population standard deviation is unknown, we use a t-interval which is

$$\bar{x} \pm t_{\alpha/2, n-1} \frac{s}{\sqrt{n}} \, .$$

Using the student t-distribution table, we find $t_{0.025, 22} = 2.074$. Thus the interval is

$$\left(21 - 2.074\left(2 / \sqrt{23}\right), 21 + 2.074\left(2 / \sqrt{23}\right)\right)$$
$$= (20.135, 21.865)$$

6. D B to I is 8 columns; 3 to 12 is 10 rows. Therefore, there are 80 cells.

7. D Choices (B) and (C) are unethical at this point in time. Formal resignation is required but is not the first step the engineer should take. A discussion with her supervisor is the first step in the process.

8. B

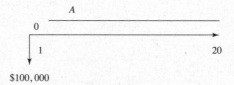

$$A = \$100,000 \, (A/P, 10\%, 20)$$
$$A = \$100,000 \, (\Phi.1175) = \$11,750$$
$$\text{Annual Payment} = \$11,750$$

9. A Determine the total equivalent resistance "seen" by the current source. First, find the equivalent resistance of each branch.

Branch 1: $1 \, \Omega$
Branch 2: $2 \, \Omega + 3 \, \Omega = 5 \, \Omega$
Branch 3: $4 \, \Omega + 5 \, \Omega + 6 \, \Omega = 15 \, \Omega$
Apply eq. 12.17(b):

$$R_T = \frac{1}{\dfrac{1}{1 \, \Omega} + \dfrac{1}{5 \, \Omega} + \dfrac{1}{15 \, \Omega}} = 0.789 \, \Omega$$

Voltage $V_o = (1 \text{ A})(0.789 \, \Omega) = 0.789 \text{ V}$

10. D From Problem 9, the voltage across each branch is known to be 0.789 V. Therefore, the current through the center branch can be determined as follows:

$$I = \frac{0.789 \text{ V}}{2\,\Omega + 3\,\Omega} = 0.1578 \text{ A}$$

Apply eq. 12.9:

$$P = I^2 \times R = (0.1578 \text{ A})^2 (2\,\Omega) = 0.05 \text{ W}$$

11. B

Draw a free-body diagram for the lower pulley.

$3P = 900$

$P = 300 \text{ N}$

$P \quad P \quad P$

900 N

12. D Draw a free-body diagram for the beam.

Take the moment about point A:

$$\sum M_A = T(4) + \frac{6}{10}\,T(12) - 400(8) - 600 = 0$$

$T = 339.3 \text{ lb}$

13. A Take the moment about the x-axis:

$$\sum M_x = P(8) - 120(10) = 0 \quad P = 150 \text{ lb}$$

14. A Given: $r(t) = 2.0t^3 - 0.5t^2$ and $\theta(t) = t^2 + 8t$.

Then,

$$\omega = \dot{\theta} = \frac{d\theta}{dt} = 2t + 8 = 20.0 \text{ rad/s}$$

$$t = \frac{20 - 8}{2} = 6.0 \text{ s}$$

Angular momentum is

$$H = mr^2\dot{\theta} = (2.0 \text{ kg})[2(6)^3 - 0.5(6)^2]^2(20.0 \text{ rad/s})$$
$$= 6.85 \times 10^6 \text{ kg·m}^2/\text{s}$$

15. C Apparent weight is reaction N_R. Draw the FBD of the car, and identify weight and normal and friction forces.

Normal direction is toward center of curvature.

$$\Sigma F_n - \frac{mV^2}{R} = \frac{(1{,}455 \text{ kg})(18.3 \text{ m/s})^2}{61.0 \text{ m}}$$

$$= 7{,}988.0 \text{ N}$$
$$= W_c - N_R = 7{,}988.0 \text{ N}$$
$$N_R = (1{,}455 \text{ kg})(9.81 \text{ m/s}^2) - 7{,}988$$
$$= 6{,}285 \text{ N}$$

16. A The equation for work of a force is $W_{1\rightarrow 2} = \int F \cdot dr$. For a constant-friction force, $W_{1\rightarrow 2} = (F_F)(20.0 \text{ m})$ since the friction force and displacement are in the same direction. $F_F = 0.8N_R$, where N_R is the normal reaction between the tires and pavement, given by $N_R = W \cos 3°$ (from equilibrium in the direction perpendicular to incline). Then,

$$N_R = (1{,}300)(9.81)\cos 3° = 12{,}735.5 \text{ N}$$

Energy dissipated is friction work, so,

$$W_{1\rightarrow 2} = (0.8 \times 12{,}735.5\text{N})(20.0 \text{ m})$$
$$= 203{,}768.4 \text{ N·m}$$
$$W_{1\rightarrow 2} \approx 203{,}760 \text{ J}$$

17. D Period of vibration τ is independent of the initial displacement.

$$\tau = \frac{1}{f_n} \quad \text{and} \quad f_n = \frac{\omega_n}{2\pi} = \frac{1}{2\pi}\sqrt{\frac{k}{m}}$$

Then,

$$\tau = 2\pi\sqrt{\frac{m}{k}}$$

where $k = 0.260 \text{ kN/m} = 260.0 \text{ N/m}$. Substitute known values:

$$\tau = 2\pi\sqrt{\frac{1.8 \text{ kg}}{260.0 \text{ N/m}}} = 2\pi\sqrt{0.00688 \text{ s}^2}$$

$$= 6.28(0.083 \text{ s}) = 0.523 \text{ s} = 0.52 \text{ s}$$

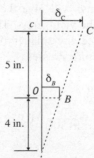

18. A This problem is statistically indeterminate, which means that there are more unknowns than equilibrium equations. There are four unknowns (x- and y-components of the reaction at A and the tension in wire B and C). To solve the problem, first take the moment about the pin reaction at A:

$$\sum M_A = F_B(4) + F_C(9) - 700(6) = 0$$

Now we need an additional equation, which can be obtained from the geometry of deformation (rotation of the bracket):

$$\frac{\delta_B}{4} = \frac{\delta_C}{9}$$

$$\frac{F_B(L)}{4AE} = \frac{F_C(L)}{9AE}$$

$$F_C = \frac{9}{4} F_B$$

Solve for F_B: $F_B = 173.2$ lb

19. B Poisson's ratio is defined as:

$$V = -\frac{\varepsilon_{lateral}}{\varepsilon_{axiel}} = -\frac{\frac{19.9837 - 20}{20}}{\frac{600 - 601.4}{600}} = 0.349$$

20. A

$$\sigma_{max} = \frac{M\,c}{I} = -\frac{(50,000)(0.15)}{\frac{1}{12}\left[(0.2)(0.3^3) - (0.18)(0.26^3)\right]}$$

$$= 40.24 \text{ MPa}$$

21. B Refer to Figure 10.8, an iron-carbon phase diagram.

22. B In order to avoid fracture in the elastic region, the applied tension stress must not exceed the yield stress. In the fracture stress/crack size relationship and using the yield stress as the fracture stress, solving for a will give the crack size to cause fracture at the yield stress. That is the largest crack size since any smaller size cracks will require larger fracture stresses.

$$K_{IC} = \sigma\sqrt{\pi a}$$

$$a = \frac{K_{IC}^2}{\pi\sigma^2} = \frac{50^2}{\pi 150^2} = 0.079 \text{ in.} = 0.08 \text{ in.}$$

23. C $f_L = \dfrac{C_s - C_o}{C_s - C_L} = \dfrac{67 - 60}{67 - 48} = 0.37$

24. C $\bar{V} = \dfrac{Q}{A}$

$$Q = \frac{\dot{m}}{\rho} \qquad\qquad A = \frac{\pi D^2}{4}$$

$$= \frac{1.60\,\frac{\text{kg}}{\text{s}}}{1,000\,\frac{\text{kg}}{\text{m}^3}} \qquad\qquad = \frac{\pi(0.024\text{ m})^2}{4}$$

$$= 1.60 \times 10^{-3}\,\frac{\text{m}^3}{\text{s}} \qquad = 4.52 \times 10^{-4}\text{ m}^2$$

$$\bar{V} = \frac{1.60 \times 10^{-3}\,\frac{\text{m}^3}{\text{s}}}{4.52 \times 10^{-4}\text{ m}^2}$$

$$= 3.54\,\frac{\text{m}}{\text{s}}$$

25. A $\sum \mathbf{F} = (\dot{m}\mathbf{V})_{in} - (\dot{m}\mathbf{V})_{out}$

This equation can be separated into its horizontal and vertical components:

$$\sum \mathbf{F}_x = (\dot{m}\mathbf{V}_x)_{in} - (\dot{m}\mathbf{V}_x)_{out}$$

and

$$\sum \mathbf{F}_y = (\dot{m}\mathbf{V}_y)_{in} - (\dot{m}\mathbf{V}_y)_{out}$$

Both the horizontal component of the outlet velocity, $(\mathbf{V}_x)_{out}$, and the vertical component of the inlet velocity, $(\mathbf{V}_y)_{in}$, are zero. Also, because the tee is symmetrical and both outlet ports have the same diameter, the momentum leaving each port will be of equal magnitude and opposite direction. Therefore,

$$\sum \mathbf{F}_y = 0$$

and

$$\sum \mathbf{F}_x = (\dot{m}\mathbf{V}_x)_{in} = \mathbf{F}$$

$$\mathbf{F} = \left(1.60\,\frac{\text{kg}}{\text{s}}\right)\left(3.54\,\frac{\text{m}}{\text{s}}\right)\left(\frac{\text{N}}{\frac{\text{kg}\cdot\text{m}}{\text{s}^2}}\right)$$

$$= 5.66 \text{ N}$$

26. D

$$p_2 = p_1 + \rho_1 g h_1 - \rho_m g h_m - \rho_2 g h_2$$

$$= 25 \text{ kPa} + \left(1,000 \frac{\text{kg}}{\text{m}^3}\right)\left(9.81 \frac{\text{m}}{\text{s}^2}\right)(0.18 \text{ m})\left(\frac{\text{N}}{\frac{\text{kg·m}}{\text{s}^2}}\right)$$

$$\times \left(\frac{\text{kPa}}{1,000 \frac{\text{N}}{\text{m}^2}}\right) - \left(13,550 \frac{\text{kg}}{\text{m}^3}\right)\left(9.81 \frac{\text{m}}{\text{s}^2}\right)(0.06 \text{ m})$$

$$\times \left(\frac{\text{N}}{\frac{\text{kg·m}}{\text{s}^2}}\right)\left(\frac{\text{kPa}}{1,000 \frac{\text{N}}{\text{m}^2}}\right) - \left(825 \frac{\text{kg}}{\text{m}^3}\right)\left(9.81 \frac{\text{m}}{\text{s}^2}\right)$$

$$\times (0.15 \text{ m})\left(\frac{\text{N}}{\frac{\text{kg·m}}{\text{s}^2}}\right)\left(\frac{\text{kPa}}{1,000 \frac{\text{N}}{\text{m}^2}}\right)$$

$$= 25 \text{ kPa} + 1.76 \text{ kPa} - 7.98 \text{ kPa} - 1.21 \text{ kPa}$$

$$= 17.57 \text{ kPa}$$

$$= 17.6 \text{ kPa}$$

27. A The flow rate of fluid through an orifice flow meter is given by the equation

$$Q = CA\sqrt{2g\left[\left(\frac{P_1}{\gamma} + z_1\right) - \left(\frac{P_2}{\gamma} + z_2\right)\right]}$$

Because z_1 is equal to z_2, this equation can be simplified to

$$Q = CA\sqrt{2g\left(\frac{P_1 - P_2}{\gamma}\right)}$$

Substitute known values:

$$Q = (0.61)(0.50 \text{ m}^2)$$

$$\times \sqrt{2\left(9.81 \frac{\text{m}}{\text{s}^2}\right)\left(\frac{1,000 \text{ Pa} - 450 \text{ Pa}}{9,810 \frac{\text{N}}{\text{m}^3}}\right)\left(\frac{\frac{\text{N}}{\text{m}^2}}{\text{Pa}}\right)}$$

$$= 0.32 \text{ m}^3/\text{s}$$

28. D

$$\frac{T_2}{T_1} = \left(\frac{p_2}{p_1}\right)^{\frac{n-1}{n}}$$

$$T_2 = T_1\left(\frac{p_2}{p_1}\right)^{\frac{n-1}{n}}$$

$$= (298 \text{ K})\left(\frac{500 \text{ kPa}}{100 \text{ kPa}}\right)^{\left(\frac{1.5-1}{1.5}\right)}$$

$$= 510 \text{ K}$$

$$= 237°\text{C}$$

29. A $w = \int p \, dv$

$$= \frac{p_2 v_2 - p_1 v_1}{1 - n}$$

$$= \frac{R(T_2 - T_1)}{1 - n}$$

$$= \frac{\overline{R}(T_2 - T_1)}{M(1 - n)}$$

$$= \frac{\left(8.314 \frac{\text{kJ}}{\text{kmol K}}\right)(510 \text{ K} - 298 \text{ K})}{\left(29 \frac{\text{kg}}{\text{kmol}}\right)(1 - 1.5)}$$

$$= -122 \text{ kJ/kg}$$

30. C Energy balance:

$$\frac{dU}{dt} = \dot{Q} - \dot{W} + \sum \dot{m}_i h_i - \sum \dot{m}_e h_e$$

$$= \dot{Q} - \dot{W}$$

For a finite period of time:

$$\Delta U = Q - W$$
$$\Delta u = q - w$$
$$q = \Delta u + w$$
$$= c_v(T_2 - T_1) + w$$
$$= \left(0.718 \frac{\text{kJ}}{\text{kg·K}}\right)(510 \text{ K} - 298 \text{ K}) - 122 \frac{\text{kJ}}{\text{kg}}$$

$$= 31 \text{ kJ/kg}$$

31. B

$$\frac{dU}{dt} = \dot{Q} - \dot{W}$$

$$\dot{Q} = \dot{W} = -25\ W = -25\ W\ \frac{kW}{10^3 W}\ \frac{h}{3600\,s}$$

$$= 0.69 \times 10^{-5}\ \frac{kWh}{s}$$

$$= 0.70 \times 10^{-5}\ \frac{kWh}{s}$$

32. B

$$\frac{dS}{dt} = \frac{\dot{Q}}{T} + \sum \dot{m}_i s_i - \sum \dot{m}_e s_e + \dot{\sigma}$$

$$\dot{\sigma} = -\frac{\dot{Q}}{T} = -\frac{\dot{W}}{T} = -\frac{-25\ W}{353\ K} = 0.071\ \frac{kJ}{s \cdot K}$$

$$= 0.07\ \frac{kJ}{s \cdot K}$$

33. B

$$R_{tot} = \sum R = \frac{L_1}{K_1 A} + \frac{L_2}{K_2 A} = \frac{1}{A}\left(\frac{L_1}{K_1} + \frac{L_2}{K_2}\right) = 0.0009$$

$$= K/W$$

34. C

$$\dot{Q} = \frac{\Delta T}{\sum R} = (T_1 - T_2) / \left(\frac{L_1}{K_1 A} + \frac{L_2}{K_2 A}\right) = 2.8 \times 10^5\ W$$

$$= 280\ W$$

35. C

$$\dot{Q} = \frac{T_1 - T_2}{R_1 + R_2} = \frac{(T_1 - T_o)}{R_o} : \frac{T_1 - T_2}{\dfrac{L_1}{K_1 A} + \dfrac{L_2}{K_2 A}} = \frac{T_2 - T_o}{\dfrac{1}{h_o A}} \rightarrow h_o$$

$$= 280\ \frac{W}{m^2 K}$$

36. B The roots of the characteristic equation are
$s_{1,2} = -4 \pm \sqrt{16-12} = -2, -6$ and $s_{1,2} = -4 \pm \sqrt{16-12} = -4, -4$ for system T_1 and T_2, respectively. Hence, the only parameter that is not changing is the overshoot as these systems are either critically damped or overdamped with no overshoot (i.e., OV = 0).

37. D The system is stable since the gain margin (GM) is negative (below 0 dB) and the phase margin (PM) is above −180°. From the diagram, PM = −75 − (−180) = 105°. Also from the diagram, GM = 10 dB.

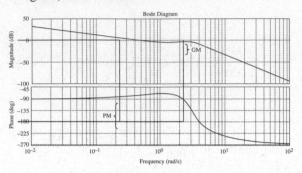

38. C

$$\frac{r}{d} = \frac{2.5}{10} = 0.25$$

$$\frac{D}{d} = \frac{15}{10} = 1.5$$

From the stress concentration charts, $K_t = 1.5$

$$\sigma_{max} = K_t \frac{P}{A} = 1.5 \cdot \frac{1,000}{\dfrac{\pi}{4}(10)^2} = 19.09\ MPa$$

39. B First find the fatigue stress concentration,

$$\frac{r}{d} = \frac{2.0}{10} = 0.2$$

$$\frac{D}{d} = \frac{15}{10} = 1.5$$

From the stress concentration charts, $K_t = 1.38$
The notch sensitivity factor is $q = 0.8$

$$K_f = 1 + 0.8(1.38 - 1) = 1.304$$

Now find the endurance limit:

$$S_e = k_a k_b k_c k_d k_e k_g (0.5 S_{ut})$$
$$k_a = a(S_{ut})^b = 4.51(590)^{-0.265} = 0.83$$
$$k_b = 1.24 d^{-0.107} = 1.24(10)^{-0.107} = 0.969$$
$$k_c = k_d = k_d = k_g = 1$$
$$S_e = 0.83 \cdot 0.969 \cdot 1 \cdot 1 \cdot 1 \cdot (0.5 \cdot 590) = 237\ MPa$$

$$\sigma_a = K_f \frac{M_c}{I} = 1.304 \frac{(P \cdot 100) \cdot 5}{\dfrac{\pi}{64} 10^4} = 237$$

$$P = 178.4\ N$$

40. B

$$K_m = 2K_b$$

$$c = \frac{K_b}{K_b + K_m} = \frac{1}{3}$$

$$A_t = 0.1419 \text{ in}^2, \ S_p = 85 \text{ ksi}$$

$$F_i = 0.75 \ A_t \ S_p = 0.75 \cdot 0.1419 \cdot 85 = 9.04 \text{ kip}$$

$$n_o = \frac{A_t S_p - F_i}{cP} = \frac{0.1419 \cdot 85 - 9.04}{0.333 \cdot 1} = 9.07 \sim 9$$

Diagnostic Chart: Mechanical Discipline

Subject Area	Total Number of Questions	Number Correct	Number Incorrect	Reason for Incorrect Answer			
				Lack of Knowledge	Misread Problem	Careless or Mathematical Error	Wrong Guess
Mathematics (1–3)	3						
Probability and Statistics (4–5)	2						
Computational Tools (6)	1						
Ethics and Professional Practices (7)	1						
Engineering Economics (8)	1						
Electricity and Magnetism (9–10)	2						
Statics (11–13)	3						
Dynamics, Kinematics, and Vibrations (14–17)	4						
Mechanics of Materials (18–20)	3						
Materials Properties and Processing (21–23)	3						
Fluid Mechanics (24–27)	4						
Thermodynamics (28–32)	5						
Heat Transfer (33–35)	3						
Measurement, Instrumentations, and Controls (36–37)	2						
Mechanical Design and Analysis (38–40)	3						
Total	40						

ANSWER SHEET
Mini-Diagnostic Exam—Other Discipline

1. Ⓐ Ⓑ Ⓒ Ⓓ   11. Ⓐ Ⓑ Ⓒ Ⓓ   21. Ⓐ Ⓑ Ⓒ Ⓓ   31. Ⓐ Ⓑ Ⓒ Ⓓ

2. Ⓐ Ⓑ Ⓒ Ⓓ   12. Ⓐ Ⓑ Ⓒ Ⓓ   22. Ⓐ Ⓑ Ⓒ Ⓓ   32. Ⓐ Ⓑ Ⓒ Ⓓ

3. Ⓐ Ⓑ Ⓒ Ⓓ   13. Ⓐ Ⓑ Ⓒ Ⓓ   23. Ⓐ Ⓑ Ⓒ Ⓓ   33. Ⓐ Ⓑ Ⓒ Ⓓ

4. Ⓐ Ⓑ Ⓒ Ⓓ   14. Ⓐ Ⓑ Ⓒ Ⓓ   24. Ⓐ Ⓑ Ⓒ Ⓓ   34. Ⓐ Ⓑ Ⓒ Ⓓ

5. Ⓐ Ⓑ Ⓒ Ⓓ   15. Ⓐ Ⓑ Ⓒ Ⓓ   25. Ⓐ Ⓑ Ⓒ Ⓓ   35. Ⓐ Ⓑ Ⓒ Ⓓ

6. Ⓐ Ⓑ Ⓒ Ⓓ   16. Ⓐ Ⓑ Ⓒ Ⓓ   26. Ⓐ Ⓑ Ⓒ Ⓓ   36. Ⓐ Ⓑ Ⓒ Ⓓ

7. Ⓐ Ⓑ Ⓒ Ⓓ   17. Ⓐ Ⓑ Ⓒ Ⓓ   27. Ⓐ Ⓑ Ⓒ Ⓓ   37. Ⓐ Ⓑ Ⓒ Ⓓ

8. Ⓐ Ⓑ Ⓒ Ⓓ   18. Ⓐ Ⓑ Ⓒ Ⓓ   28. Ⓐ Ⓑ Ⓒ Ⓓ   38. Ⓐ Ⓑ Ⓒ Ⓓ

9. Ⓐ Ⓑ Ⓒ Ⓓ   19. Ⓐ Ⓑ Ⓒ Ⓓ   29. Ⓐ Ⓑ Ⓒ Ⓓ   39. Ⓐ Ⓑ Ⓒ Ⓓ

10. Ⓐ Ⓑ Ⓒ Ⓓ   20. Ⓐ Ⓑ Ⓒ Ⓓ   30. Ⓐ Ⓑ Ⓒ Ⓓ   40. Ⓐ Ⓑ Ⓒ Ⓓ

Mini-Diagnostic Exam— Other Discipline

40 Questions

Directions: For each incomplete statement or question, select the answer that best completes the statement or answers the question. Then fill in the corresponding space on the answer sheet.

1. The distance between points (2,3) and (5,7) is

    (A) $\sqrt{149}$

    (B) 7

    (C) $\sqrt{7}$

    (D) 5

2. If $0 \leq x < \pi/2$, then $\tan x \cos x =$
    (A) $\sec x$
    (B) $\csc x$
    (C) $\cot x$
    (D) $\sin x$

3. For $b > a > 0$, the value of $\lim\limits_{x \to 0} \dfrac{8x^2}{\cos x - 1}$ is

    (A) 0
    (B) 8
    (C) −16
    (D) −∞

4. The value of $\int_a^b \dfrac{1}{x}\, dx$ is

    (A) $b^2 - a^2$
    (B) $\ln (b - a)$
    (C) $\dfrac{\ln b}{\ln a}$
    (D) $\ln \dfrac{b}{a}$

5. The solution of the differential equation $\dfrac{dy}{dt} = ky$, with initial condition $y = 1$ when $t = 0$, is
    (A) $y = \ln kt$
    (B) $y = e^{kt}$
    (C) $y = \sin kt$
    (D) $y^2 = 2t + k$

6. An area in quadrant I is bounded by $y = x^3$, $x = 2$, and the x-axis. The area of this region is
    (A) 6 square units
    (B) 8 square units
    (C) 4 square units
    (D) 2 square units

7. For a certain city on June 1, the mean high temperature is $\mu = 73$ degrees with a standard deviation of $\sigma = 3.1$ degrees. Assuming that daily high temperature is normally distributed, what is the probability of the high temperature exceeding 80 degrees on June 1?
    (A) 0.0107
    (B) 0.0250
    (C) 0.2420
    (D) 0.9893

8. A gardener is testing a new type of carrot. From a sample of 15 carrots, she calculates a mean length of 7.2 inches and a standard deviation of 1.23 inches. Find a 90% confidence interval for the mean length of this new type of carrot.

    (A) (6.519, 7.881)
    (B) (6.557, 7.822)
    (C) (6.641, 7.759)
    (D) (6.577, 7.822)

9. In the reaction below, how many grams of water are produced from the combustion of 300 grams of ethane, C_2H_6?

$$2C_2H_6 + 7O_2 \rightarrow 4CO_2 + 6H_2O$$

    (A) 540 g
    (B) 180 g
    (C) 1,080 g
    (D) 740 g

10. What happens to the volume of an ideal gas when the pressure is halved and the absolute temperature is doubled?
    (A) The volume is doubled.
    (B) The volume is halved.
    (C) The volume is quadrupled.
    (D) The volume is unchanged.

11. After leaving a company and setting up your own engineering practice, a former competitor contacts you to solve a problem similar to one you solved while working at the company. Your previous design seems to be a solution to this former competitor's problem. Can you accept this assignment?
    (A) No. Information from the original design would be required.
    (B) Yes. Knowledge transfer happens.
    (C) No. One should not accept money twice for the same job.
    (D) Yes. No stated nondisclosure statement was signed.

12. What amount of money would have to be deposited in a bank at the end of each year to accumulate a college fund of $50,000 at the end of 18 years? Interest on the account is 6%.
    (A) $1,520
    (B) $1,560
    (C) $1,580
    (D) $1,620

13. A company can manufacture a product with two different machines. Machine A has a $4.00 manufacturing cost per unit and a fixed cost of $3,000 for tools. Machine B costs $45,000 to purchase and has a $0.50 manufacturing cost per unit. With an annual anticipated volume of 7,000 units, the break-even point, in years, is most nearly
    (A) 1.5 yr
    (B) 1.7 yr
    (C) 1.9 yr
    (D) 2.1 yr

14. An automated measurement system has an initial cost of $36,000, and annual maintenance is $2,700. After 3 years, the salvage value is $9,000. If the interest rate is 10%, the equivalent uniform annual cost is most nearly
    (A) $14,457
    (B) $14,658
    (C) $14,860
    (D) $15,020

15.

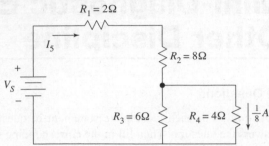

What is the total power delivered to the circuit shown above?
    (A) 1.52 W
    (B) 0.54 W
    (C) 2.14 W
    (D) 0.86 W

16.

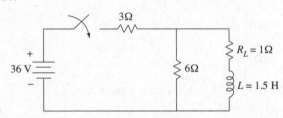

If the initial current in the inductor shown above is 0 ampere, what is the current through the inductor 0.25 second after the switch is closed?
    (A) 2.67 A
    (B) 1.95 A
    (C) 5.24 A
    (D) 3.15 A

17.

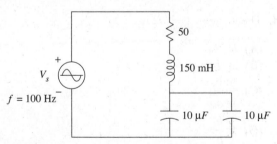

What is the input impedance for the circuit shown above?
    (A) $52.1\angle16.3°$ kΩ
    (B) $4.52\angle24.8°$ kΩ
    (C) $7.62\angle-62.1°$ kΩ
    (D) $3.24\angle-38.7°$ kΩ

18.

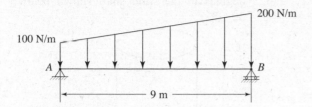

What is the reaction at the roller support (point *B*) in the above diagram?
(A) 750 N
(B) 600 N
(C) 500 N
(D) 525 N

19.

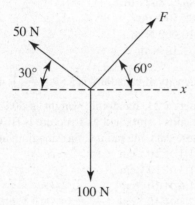

If the resultant of the forces shown in the above diagram is vertical, what is force **F**?
(A) 50 N
(B) 100 N
(C) 86.6 N
(D) 60 N

20. What is the angle between force **F** and the *x*-axis, where $F = 30i + 50j - 20k$ newtons?
(A) 30°
(B) 65.8°
(C) 60.9°
(D) 12.8°

21. A long-range artillery rifle located at point *A* fires a shell at an angle of 45° with respect to the horizontal, as illustrated below.

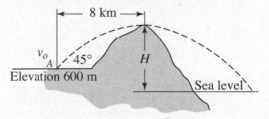

If the shell just barely clears the mountain peak at the top of its trajectory, the height *H* of the mountain peak above sea level is necessarily
(A) 7,998.0 m
(B) 4,600.0 m
(C) 4,000.0 m
(D) 12,600.0 m

22. A 60.0-kilogram crate at rest on a horizontal surface is suddenly subjected to a constant towing force as illustrated below.

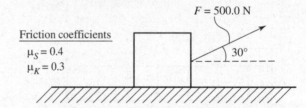

On the basis of the given values of the friction coefficients at the interface between the horizontal surface and the lower surface of the crate, the velocity of the crate at the instant when the towing force has been applied for 12.0 seconds is
(A) 66.2 m/s
(B) 108.3 m/s
(C) 497.4 m/s
(D) 82.9 m/s

23. A weight attached to a spring moves up and down, so that the equation of motion is $\dfrac{d^2x}{dt^2} + 64x = 0$ where *x* is the displacement in feet of the weight from its equilibrium position at any time *t*. Suppose that at *t* = 0 second the displacement is 3 inches [$x(0) - \dfrac{1}{4}$] and the velocity is 1 foot per second [$x'(0) = 1$]. The displacement *x* of the weight at any time *t* is given by:
(A) $x = -\sin 8t$
(B) $x = \dfrac{1}{4}\cos 8t + \dfrac{1}{8}\sin 8t$
(C) $x = 2\cos 8t$
(D) $x = \cos 8t - \sin 8t$

25

24. What is the average shear stress in the pin at A (double shear) in the diagram below? The diameter of the pin at A is 2 centimeters.

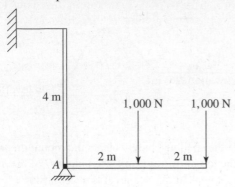

(A) 7.96 MPa
(B) 3.98 MPa
(C) 6.36 MPa
(D) 12.72 MPa

25. The maximum bending stress in the 6-centimeter-diameter shaft beam shown below is

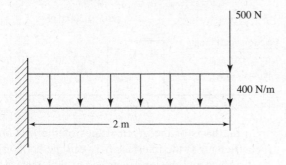

(A) 84.88 MPa
(B) 66.02 MPa
(C) 112.6 MPa
(D) 32.44 MPa

26. What is the torque T causing an angle of twist of 3 degrees in the solid shaft shown below? ($G = 30$ GPa)

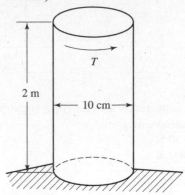

(A) 441,786 N·m
(B) 8,550 N·m
(C) 10,000 N·m
(D) 7,710 N·m

27. The density of vanadium is 5.89 $\left(\dfrac{g}{cm^3}\right)$, its atomic number is 23, its atomic weight is 50.94 atomic mass units (amu), and its structure is BCC. Based on these data, the radius of a vanadium atom is
(A) 1.1×10^{-8} cm
(B) 1.3×10^{-8} cm
(C) 1.6×10^{-8} cm
(D) 2.1×10^{-8} cm

28. The strong magnetic behavior exhibited by permanent magnets is called
(A) paramagnetism
(B) diamagnetism
(C) ferromagnetism
(D) superconductivity

Questions 29–30 refer to the following diagram:

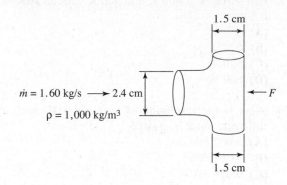

Water flows into the tee shown.

29. What is the average velocity of the water entering the tee?
    (A) 4.63 m/s
    (B) 3.95 m/s
    (C) 3.54 m/s
    (D) 0.35 m/s

30. What is the magnitude of the force (F) required to keep the tee stationary?
    (A) 5.66 N
    (B) 2.21 N
    (C) 6.32 N
    (D) 7.41 N

31.

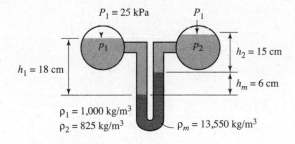

For the U-tube manometer shown above, what is the pressure in vessel 2 (p_2)?
    (A) 32.4 kPa
    (B) 33.5 kPa
    (C) 18.2 kPa
    (D) 17.6 kPa

32. The Reynolds number is
    (A) a measure of the rate of fluid flow
    (B) the ratio of inertial forces to viscous forces
    (C) the ratio of pressure forces to viscous forces
    (D) rarely important in fluid flow analysis

33. Air at 500 K and 20 m/s enters a nozzle and leaves at 350 K. What is the exit velocity of air? ($c_p = 1$ kJ/kg·K)

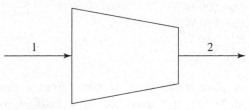

    (A) 26 m/s
    (B) 225 m/s
    (C) 450 m/s
    (D) 550 m/s

34. Air enters a compressor at 1 bar pressure and a temperature of 300 K. The pressure ratio across the compressor is 8. Assuming isentropic compression and a specific heat ratio of 1.4, calculate the air exit temperature.

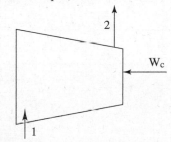

    (A) 240 K
    (B) 540 K
    (C) 800 K
    (D) 2,400 K

35. What is the flux rate for a slab (thermal conductivity $K = 20$ W/m · K, thickness $L = 0.4$ m) if two side temperatures are 400 K and 300 K?
    (A) 400 W/m^2
    (B) 600 W/m^2
    (C) 800 W/m^2
    (D) 1,000 W/m^2

36. A window is made of 2 pieces of glass 5 mm thick that enclose an air space 5 mm thick. The window separates room air at 25°C from outside ambient air at −12°C. The convection coefficient outside is 60 W/m² · K and inside is 10 W/m² · K. What is the rate of energy loss through a window that is 1 m long by 0.8 m wide? (The thermal conductivity of glass is 1.4 W/m · K and for the stagnant air between two glasses is 0.03 W/m · K.)
(A) 500 W
(B) 100 W
(C) 150 W
(D) 200 W

37. A plate 2 cm thick is taken from a furnace at 800°C and quenched in a bath of oil at 30°C. Calculate the time required to cool the plate to 100°C (convective heat transfer coefficient $h = 400$ W/m² · K, thermal conductivity $k = 50$ W/mK, density is 8,000 kg/m³, and specific heat is 500 J/kg · K).
(A) 240 s
(B) 200 s
(C) 280 s
(D) 320 s

38. Calculate the rate of blackbody radiation between two surfaces at 400 K and 300 K with an area of 0.5 m².
(A) 300 W
(B) 400 W
(C) 500 W
(D) 600 W

39. What is the resolution of a 16-bit ADC with a full-scale input range of 5 V?
(A) 0.067 mV
(B) 0.05 mV
(C) 0.076 mV
(D) 0.01 mV

40. The overall risk of a hazard is 21 on a 0 – 100 scale. If the seriousness of hazard score is estimated to be 7 on a 0 – 10 scale, then compute the likelihood of the occurrence on a 0 – 10 scale.
(A) 0.3
(B) 3
(C) 14
(D) 147

Answer Key

1. D	11. C	21. B	31. D
2. D	12. D	22. A	32. B
3. C	13. B	23. B	33. D
4. D	14. A	24. B	34. B
5. B	15. B	25. A	35. C
6. C	16. D	26. D	36. B
7. A	17. A	27. B	37. A
8. C	18. A	28. C	38. C
9. A	19. C	29. C	39. C
10. C	20. C	30. A	40. B

Answers Explained

1. D Distance $= \sqrt{(7-3)^2 + (5-2)^2}$

$$= \sqrt{16+9} = 5$$

2. D $\tan x \cos x = \dfrac{\sin x}{\cos x} \cdot \cos x = \sin x$, provided that $\cos x \neq 0$.

3. C By L'Hôpital's rule

$$\lim_{x \to 0}\left(\frac{8x^2}{\cos x - 1}\right) = \lim_{x \to 0}\left(\frac{16x}{-\sin x}\right)$$

$$= \lim_{x \to 0}\left(\frac{16}{-\cos x}\right) = -16$$

4. D $\displaystyle\int_a^b \frac{1}{x}\,dx = \ln x\big|_a^b = \ln b - \ln a = \ln\frac{b}{a}$,

provided that $b > a > 0$.

5. B The differential equation $\dfrac{dy}{dt} - ky = 0$ is linear

in y with integrating factor $e^{\int -k\,dt} = e^{-kt}$, where k is a constant. Hence, the general solution is

$$ye^{-kt} = c \quad \text{or} \quad y = ce^{kt}$$

If $y = 1$ when $t = 0$, then $c = 1$. Therefore, $y = e^{kt}$.

6. C Area $= \displaystyle\int_0^2 x^3 dx = \frac{x^4}{4} = 4$ square units.

7. A Let X be the high temperature on June 1. By converting to the standard normal and using the unit normal distribution table, we have

$$P(X > 80) = P\,\frac{X-73}{3.1} > \frac{80-73}{3.1} = P(Z > 2.3)$$

$$= 0.0107$$

8. C Since the population standard deviation is unknown, use a *t*-interval that is

$$\bar{x} \pm t_{a/2,n-1}\frac{s}{\sqrt{n}}\,.$$

Use the student *t*-distribution table to find $t_{-0.05,14}$ = 1.761. Thus, the interval is $(7.2 - 1.761(1.23/\sqrt{15}), 7.2 + 1.761(1.23/\sqrt{15}) = (6.641, 7.759)$.

9. A The reaction is already balanced. Assume O_2 is in excess, and use the factor-label method.

$$300 \text{ g } C_2H_6\,\frac{(1.0 \text{ mol } C_2H_6)}{(30.0 \text{ g } C_2H_6)}$$

$$\times \frac{(6.0 \text{ mol } H_2O)}{(2.0 \text{ mol } C_2H_6)}\,\frac{(18.0 \text{ g } H_2O)}{(1.0 \text{ mol } H_2O)} = 540 \text{ g}$$

10. C Check the general gas law, Equation 3.3 in Chapter 3, and follow through with numbers.

11. C

12. D

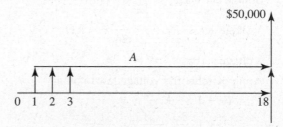

$A = \$50,000(A/F,6,18)$
$= \$50,000(0.0324) = \$1,620$

13. B Break-even analysis:

Set costs of Machine A = costs of Machine B

Let x = break-even quantity. Then

$$4.00x + \$3,000 = 0.50x + \$45,000$$
$$3.5x = \$42,000$$
$$x = 12,000 \text{ units}$$

At 12,000 units B/E and usage of 7,000 units per year:

$$\frac{12,000}{7,000} = 1.7 \text{ yr}$$

14. A

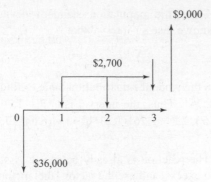

$9,000

$2,700

0   1   2   3

$36,000

Annual cost = $36,000(A/P,10,3) + $2,700 − $9,000(A/F,10,3)

= $36,000(0.4021) + $2,700 − $9,000(0.3021)

= $14,476 + $2,700 − $2,719

= $14,457

15. B Voltage across R_4: $V_4 = 4\,\Omega \times \frac{1}{8}A = \frac{1}{2}V$

Voltage across R_3: $^{1}/_{2}$ V because R_3 and R_4 are in parallel

Current through R_3: $I_3 = \dfrac{\frac{1}{2}V}{6\,\Omega} = \frac{1}{12}A$

Source current: $I_s = \frac{1}{12}A + \frac{1}{8}A = \frac{5}{24}A$

Voltage across R_2: $V_2 = \frac{5}{24}A \times 8\,\Omega = \frac{5}{3}V$

Voltage across R_1: $V_1 = \frac{5}{24}A \times 2\,\Omega = \frac{5}{12}V$

Apply Kirchhoff's voltage law to obtain

$V_s = V_1 + V_2 + V_3$

$= \frac{5}{12}V + \frac{5}{3}V + \frac{1}{2}V = \frac{31}{12}V$

Total power delivered to the system is

$P_T = V_s I_s = \frac{31}{12}V \times \frac{5}{24}A \cong 0.54\ W$

16. D Consider R_L and L as a load and obtain a Thevenin equivalent circuit of R_1, R_2, and the source voltage.

$E_{TH} = \dfrac{6\,\Omega}{3\,\Omega + 6\,\Omega} \times 36\ V = 24\ V$

and

$R_{TH} = \dfrac{3\,\Omega \times 6\,\Omega}{3\,\Omega + 6\,\Omega} = 2\,\Omega$

The time constant of the circuit depends on L and $R_L + R_{TH}$ or

$\tau = \dfrac{L}{R} = \dfrac{1.5\ H}{3\,\Omega} = 0.5\ s$

The maximum current that can flow depends on E_{TH} and $R_L + R_{TH}$:

$I_{max} = \dfrac{24\ V}{1\,\Omega + 2\,\Omega} = 8\ A$

The inductor current is

$I_L = 8\ A \times \left(1 - e^{-\frac{0.25\ s}{0.5\ s}}\right) = 3.15\ A$

17. A Combine the capacitors into a single equivalent capacitor:

$C_{eq} = 10\ \mu F + 10\ \mu F = 20\ \mu F$

The capacitive reactance is

$X_C = \dfrac{1}{2 \times \pi \times 100\ Hz \times 20 \times 10^{-6}\ F}$

$= 79.5\,\Omega$

The inductive reactance is

$X_L = 2 \times \pi \times 100\ Hz \times 150\ mH = 94.2\,\Omega$

The input impedance is

$Z_T = (50 + j94.2 - j79.6)\Omega$

$= (50 + j14.6)\Omega = 52.1\angle 16.3°\ k\Omega$

18. A $\sum M_A = 0$

$B(9) - 900(4.5) - 450(6) = 0$

$B = 750\ N$

19. C Since the resultant is vertical, $\Sigma F_x = 0$. Then

$F \cos 60 - 50 \cos 30 = 0$

$F = 86.6\ N$

20. C $\mathbf{F} = 30\mathbf{i} + 50\mathbf{j} - 20\mathbf{k}$

$\cos\theta_x = \dfrac{F_x}{\sqrt{F_x^2 + F_y^2 + F_z^2}}$

$= \dfrac{30}{\sqrt{30^2 + 50^2 + (-20)^2}}$

$\theta = 60.9°$

21. B Projectile motion without air friction:

Elevation equation: $y(t) = y_o + v_{yo}t + \frac{1}{2}gt^2$

At position A, $y_o = 600$ m

At top of trajectory, $y = y_{max} = H$ at time t_M

$x(t_M) = 8,000.0$ m given

Note that at y_{max}, $v_y = 0$ (velocity is entirely horizontal, v_x).

$$v_x = v_o \cos \alpha = v_o \cos 45°$$
$$= 0.707 v_o = \text{constant}$$
$$x(t) = v_x t = 0.707 v_o t \rightarrow$$
$$8,000.0 \text{ m} = 0.707 v_o t_M$$

$$t_M = \frac{11,315.4}{v_o}$$

$$y(t_M) = H = 600.0 \text{ m} + v_o \sin 45° \, t_M$$

$$+ \frac{1}{2}(-9.81)t_M^2 = 600.0 + 0.707 v_o t_M - 4.9 t_M^2$$

Since the vertical velocity component is $v_y(t) = v_{yo} - 9.81t$ at t_M, write

$$0 = 0.707 v_o - 9.81 t_M \rightarrow t_M = 0.072 v_o$$

Equate the previous expression for t_M to obtain

$$\frac{11,315.4}{v_o} = 0.072 v_o \rightarrow v_o^2 = 157,158.0 \quad \text{and}$$

$$v_o = 396.4 \text{ m/s}$$
$$t_M = 0.072(396.4) = 28.5 \text{ s}$$

Substitute into the elevation equation above to obtain

$$H = 600.0 + 0.707(396.4)(28.5) - 4.9(28.5)^2$$
$$= 4,607.1 \text{ m}$$
$$\approx 4,600 \text{ m}$$

22. A Use the impulse/momentum principle (IMP):

$$mv_1 + \text{IMP}_{1 \to 2} = mv_2$$

Given $v_1 = 0$, obtain v_2 as follows:

Draw an FBD of the crate, and identify external forces: kinetic friction F_{FK}, normal reaction N_R, weight W, and towing F_T.

$$\text{IMP}_{1 \to 2} = \sum F_x(\Delta t) \quad \text{and}$$
$$\Delta t = 12.0 \text{ s (given)}$$

$$\sum F_x = F_T \cos 30° - F_{FK}$$
$$= (500.0 \text{ N})(0.866) - (0.3)N_R$$

From $\Sigma F_y = 0 \rightarrow F_T \sin 30° - W + N_R = 0$:

$(500.0 \text{ N})(0.5) - (60)(9.81) + N_R = 0 \rightarrow$
$N_R = 588.6 - 250.0 = 338.6 \text{ N}$

Substitute to obtain $\Sigma F_x = 433.0 \text{ N} - 0.3$
$(338.6 \text{ N}) = 331.4 \text{ N}$

From the IMP equation:

$$0 + (331.4 \text{ N})(12.0 \text{ s}) = (60 \text{ kg})v_2$$

$$v_2 = \frac{3,977.0 \text{ N·s}}{60 \text{ kg}} = 66.2 \text{ m/s}$$

23. B The solutions of the auxiliary equation $m^2 + 64 = 0$ are $\pm 8i$. Hence,

$$x = c_1 \cos 8t + c_2 \sin 8t \text{ and}$$
$$x' = -8c_1 \sin 8t + 8c_2 \cos 8t$$

The initial condition $x(0) = \frac{1}{4}$ implies $c_1 = \frac{1}{4}$, and the initial condition $x'(0) = 1$ implies $c_2 = \frac{1}{8}$.

Thus, the displacement of the weight at any time t is given by

$$x = \frac{1}{4} \cos 8t + \frac{1}{8} \sin 8t$$

24. B $\sum M_A = 0 \quad T(4) - 1,000(2) - 1,000(4) = 0$
$$T = 1,500 \text{ N}$$

$$\sum F_x = 0 \quad A_x = 1,500 \text{ N}$$
$$\sum F_y = 0 \quad A_y = 2,000 \text{ N}$$

Resultant pin reaction at A:

$$A = \sqrt{1,500^2 + 2,000^2} = 2,500 \text{ N}$$

Since the pin at A is in double shear, $V = A/2 = 1,250 \text{ N}$. Then

$$\tau = \frac{V}{A_s} = \frac{1,250}{\frac{\pi}{4}(0.02)^2} = 3.98 \text{ MPa}$$

25. A The maximum moment occurs at the fixed support and is equal to

$$M_{max} = 800(1) + (500)(2) = 1,800 \text{ N·m}$$

Then

$$\sigma_{max} = \frac{M_{max}c}{I} = \frac{1,800(0.03)}{\frac{\pi}{64}(0.06)^4} = 84.88 \text{ MPa}$$

26. D $\phi = \dfrac{T\,l}{G\,J}$

Solve for torque, T:

$$T = \frac{\phi G J}{l} = \frac{\left(3\dfrac{\pi}{180}\right)(30 \times 10^9)\left(\dfrac{\pi}{32}(0.1)^4\right)}{2}$$

$$= 7{,}710 \text{ N·m}$$

Note that the unit for the angle of twist must be radians.

27. B The structure is BCC, so there are 2 atoms in a unit cell. Obtain the radius by finding the lattice parameter from the volume. The radius is,

for BCC, $4r = a\sqrt{3}$.

$$\text{Volume} = \frac{\text{mass}}{\rho} = \frac{2 \cdot 50.94}{6.023 \times 10^{23} \cdot 5.89}$$

$$= 2.87 \times 10^{-23} \text{ cm}^3$$

Then

$$r = \frac{\sqrt{3}}{4}(\text{volume})^{\frac{1}{3}} = \frac{\sqrt{3}\cdot(2.87\times10^{-23})^{\frac{1}{3}}}{4}$$

$$= 1.32 \times 10^{-8} \text{ cm}$$

$$= 1.3 \times 10^{-8} \text{ cm}$$

28. C

29. C $\bar{V} = \dfrac{Q}{A}$

$Q = \dfrac{\dot{m}}{\rho}$ $\qquad A = \dfrac{\pi D^2}{4}$

$$= \frac{1.60\,\dfrac{\text{kg}}{\text{s}}}{1{,}000\,\dfrac{\text{kg}}{\text{m}^3}} \qquad = \frac{\pi(0.024 \text{ m})^2}{4}$$

$$= 1.60 \times 10^{-3}\,\frac{\text{m}^3}{\text{s}} \qquad = 4.52 \times 10^{-4} \text{ m}^2$$

$$\bar{V} = \frac{1.60 \times 10^{-3}\,\dfrac{\text{m}^3}{\text{s}}}{4.52 \times 10^{-4} \text{ m}^2}$$

$$= 3.54\,\frac{\text{m}}{\text{s}}$$

30. A $\sum \mathbf{F} = (\dot{m}\mathbf{V})_{\text{in}} - (\dot{m}\mathbf{V})_{\text{out}}$

This equation can be separated into its horizontal and vertical components:

$$\sum \mathbf{F}_x = (\dot{m}\mathbf{V}_x)_{\text{in}} - (\dot{m}\mathbf{V}_x)_{\text{out}}$$

and

$$\sum \mathbf{F}_y = (\dot{m}\mathbf{V}_y)_{\text{in}} - (\dot{m}\mathbf{V}_y)_{\text{out}}$$

Both the horizontal component of the outlet velocity, $(\mathbf{V}_x)_{\text{out}}$, and the vertical component of the inlet velocity, $(\mathbf{V}_y)_{\text{in}}$, are zero. Also, because the tee is symmetrical and both outlet ports have the same diameter, the momentum leaving each port will be of equal magnitude and opposite direction. Therefore,

$$\sum \mathbf{F}_y = 0$$

and

$$\sum \mathbf{F}_x = (\dot{m}\mathbf{V}_x)_{\text{in}} = \mathbf{F}$$

$$\mathbf{F} = \left(1.60\,\frac{\text{kg}}{\text{s}}\right)\left(3.54\,\frac{\text{m}}{\text{s}}\right)\left(\frac{\text{N}}{\dfrac{\text{kg·m}}{\text{s}^2}}\right)$$

$$= 5.66 \text{ N}$$

31. D $p_2 = p_1 + \rho_1 g h_1 - \rho_m g h_m - \rho_2 g h_2$

$$= 25 \text{ kPa} + \left(1{,}000\,\frac{\text{kg}}{\text{m}^3}\right)\left(9.81\,\frac{\text{m}}{\text{s}^2}\right)(0.18 \text{ m})\left(\frac{\text{N}}{\dfrac{\text{kg·m}}{\text{s}^2}}\right)$$

$$\times \left(\frac{\text{kPa}}{1{,}000\,\dfrac{\text{N}}{\text{m}^2}}\right) - \left(13{,}550\,\frac{\text{kg}}{\text{m}^3}\right)\left(9.81\,\frac{\text{m}}{\text{s}^2}\right)(0.06 \text{ m})$$

$$\times \left(\frac{\text{N}}{\dfrac{\text{kg·m}}{\text{s}^2}}\right)\left(\frac{\text{kPa}}{1{,}000\,\dfrac{\text{N}}{\text{m}^2}}\right) - \left(825\,\frac{\text{kg}}{\text{m}^3}\right)\left(9.81\,\frac{\text{m}}{\text{s}^2}\right)$$

$$\times (0.15 \text{ m})\left(\frac{\text{N}}{\dfrac{\text{kg·m}}{\text{s}^2}}\right)\left(\frac{\text{kPa}}{1{,}000\,\dfrac{\text{N}}{\text{m}^2}}\right)$$

$$= 25 \text{ kPa} + 1.76 \text{ kPa} - 7.98 \text{ kPa} - 1.21 \text{ kPa}$$
$$= 17.57 \text{ kPa}$$
$$= 17.6 \text{ kPa}$$

32. B The Reynolds number for flow in a pipe is given by the equation:

$$\mathrm{Re} = \frac{\rho \overline{V} D}{\mu}$$

At first glance, the numerator of this expression looks like mass flow rate per unit area, so answer (A) seems reasonable. However, if the Reynolds number were purely a measure of the rate of fluid flow, there would be no need of viscosity in the relationship, so it can be inferred that the meaning may be somewhat broader. Examine the other choices for a better fit.

The pressure of the fluid does not appear in the relationship for the Reynolds number, nor is it easy to derive a pressure-force term from the parameters used. Answer (C) seems unlikely.

An inertial force term would be proportional to

$$F_i \approx \rho \, \overline{V}^2$$

And a viscous force would be proportional to

$$F_v \approx \mu \frac{V}{L}$$

where L is a characteristic length for the flow. The ratio of these two forces is as follows:

$$\frac{F_i}{F_v} \approx \frac{\rho \overline{V}^2}{\mu \dfrac{V}{L}} = \frac{\rho \overline{V} L}{\mu} - \mathrm{Re}_L$$

Therefore, the answer is (B).

Choice (D) is incorrect because the Reynolds number is always an important parameter when analyzing a fluid-flow situation. The Reynolds number gives an indication of whether the flow is likely to be laminar or turbulent.

33. D

$$h_1 - h_2 + \frac{1}{2}\left(V_1^2 - V_2^2\right) = 0$$

$$V_2^2 = 2\left(h_1 - h_2\right) + V_1^2$$

$$V_2 = \left[2\left(h_1 - h_2\right) + V_1^2\right]^{1/2}$$

$$V_2 = \left[2c_p\left(T_1 - T_2\right) + V_1^2\right]^{1/2}$$

$$V_2 = \left[2\left|\frac{1.0\,\mathrm{kJ}}{\mathrm{kg\cdot K}}\right|\frac{(500-350)\,\mathrm{K}}{}\left|\frac{10^3\,\mathrm{J}}{\mathrm{kJ}}\right|\frac{\mathrm{N-m}}{\mathrm{J}}\left|\frac{\mathrm{kg-m}}{\mathrm{N-s^2}}\right.\right.$$

$$\left.+ \left(20\frac{\mathrm{m}}{\mathrm{s}}\right)^2\right]^{1/2}$$

$$V_2 = 548\,\frac{\mathrm{m}}{\mathrm{s}} = 550\,\frac{\mathrm{m}}{\mathrm{s}}$$

34. B

$$\frac{T_2}{T_1} = \left(\frac{p_2}{p_1}\right)^{\frac{k-1}{k}}$$

$$T_2 = 300(8)^{\frac{1.4-1}{1.4}} = 543\ \mathrm{K}$$

35. C $Q'' = k\,dT/dx$
$$= (20\ \mathrm{W/mK})(400 - 300)\mathrm{K}/0.4\ \mathrm{m}$$
$$= 800\ \mathrm{W/m}$$

36. B

$$q = \frac{T_{\infty 1} - T_{\infty 2}}{\dfrac{1}{Ah_1} + \dfrac{L}{Ak_g} + \dfrac{L}{Ak_{air}} + \dfrac{L}{Ak_g} + \dfrac{1}{Ah_2}}$$

$$= \frac{25 + 12}{\dfrac{1}{0.8} + \left(\dfrac{1}{10} + \dfrac{0.005}{1.4} + \dfrac{0.005}{0.03} + \dfrac{0.005}{1.4} + \dfrac{0.005}{1.4} + \dfrac{1}{60}\right)}$$

$$= 100.66\ \mathrm{W}$$
$$= 100\ \mathrm{W}$$

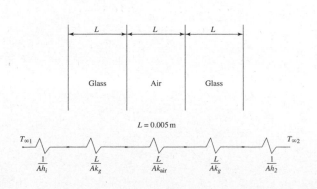

37. A The Biot number should be calculated first to see if the lumped thermal capacity approximation is valid.

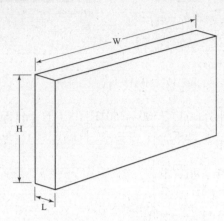

$$B_i = \frac{hL_c}{k}$$

$$L_c = \text{Characteristic length} = \frac{V}{A}$$

$$L_c = \frac{W \cdot H \cdot L}{2 \cdot W \cdot H} = \frac{L}{2}$$

Surface area of the edges has been neglected

$$B_i = \frac{(400 \text{ W} / \text{m}^2 \cdot \text{K}) \cdot (0.01 \text{ m})}{50 \text{ W} / \text{m}^2 \cdot \text{K}} = 0.08 < 0.1,$$

so the lumped thermal capacity is applicable:

$$\frac{T - T_\infty}{T_i - T_\infty} = \exp(-(\frac{hA_s}{\rho V_c})t) = \exp(-\frac{t}{t_c})$$

$$t = -t_c \ln T - T_{inf} T_i - T_{inf}$$
$$t = -100 \ln 100 - 30{,}800 - 30 = 239.8 \text{ s}$$
$$= 240 \text{ s}$$

38. C $\dot{Q} = \varepsilon \sigma \cdot A(T_1^4 - T_2^4)$

$$\varepsilon = 1 \text{ (blackbody)}$$

$$\dot{Q} = \left(5.67 \times 10^{-8} \frac{\text{W}}{\text{m}^2 \cdot \text{K}^4}\right) \cdot (0.5 \text{ m}^2)[(400)^4 - (300)^4]$$

$$\dot{Q} = 496 \text{ W} = 500 \text{ W}$$

39. C The resolution is $R/2^n = 5/2^{16} = 0.000076$ or 0.076 mV.

40. B Equation 16.5: Risk = (Likelihood of Occurrence score) * (Seriousness of Hazard score)

Risk = 21 (0 – 100 scale)

Seriousness of Hazard score = 7 (0 – 10 scale)

Likelihood of Occurrence score = ?

21 = (Likelihood of Occurrence score)(7)

Likelihood of Occurrence = 3 (0 – 10 scale)

Diagnostic Chart: Other Discipline

Subject Area	Total Number of Questions	Number Correct	Number Incorrect	Reason for Incorrect Answer			
				Lack of Knowledge	Misread Problem	Careless or Mathematical Error	Wrong Guess
Mathematics (1–6)	6						
Engineering Probability and Statistics (7–8)	2						
Chemistry (9–10)	2						
Ethics and Business Practices (11)	1						
Engineering Economics (12–14)	3						
Electricity and Magnetism (15–17)	3						
Engineering Mechanics: Statics (18–20)	3						
Engineering Mechanics: Dynamics (21–23)	3						
Strength of Materials (24–26)	3						
Materials Properties (27–28)	2						
Fluid Mechanics (29–34)	6						
Heat Transfer (35–38)	4						
Instruments and Data Acquisitions (39)	1						
Safety, Health, and Environment (40)	1						
Total	**40**						

Mathematics

David E. Stevens

Introduction

The intent of this chapter is to review some of the important mathematical topics that an engineer may encounter in the design and analysis of a physical system. The major topics include algebra, trigonometry, complex numbers, analytic geometry, linear algebra, vectors, sequences and series, differential and integral calculus, and differential equations with Laplace transforms.

1.1
Straight Lines

Assigned to each point P in the coordinate plane is an ordered pair (a,b), called the **rectangular coordinates of the point.** The x-coordinate a is the directed distance from the y-axis to point P, and the y-coordinate b is the directed distance from the x-axis to point P, as shown in Figure 1.1.

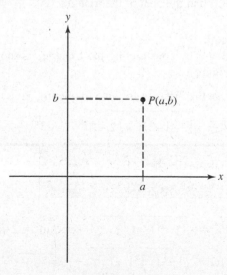

Figure 1.1

The distance between two points $P(x_1,y_1)$ and $Q(x_2,y_2)$ in the coordinate plane is given by the **distance formula:**

$$PQ = \sqrt{(x_1 - x_2)^2 + (y_1 - y_2)^2}$$

The rectangular coordinates of the midpoint M of a line segment joining points $P(x_1,y_1)$ and $Q(x_2,y_2)$ in the coordinate plane are given by the **midpoint formula:**

$$M\left(\frac{x_1 + x_2}{2}, \frac{y_1 + y_2}{2}\right)$$

The **slope** m of a nonvertical line through distinct points $P(x_1,y_1)$ and $Q(x_2,y_2)$, where $x_1 \neq x_2$, is

$$m = \frac{y_2 - y_1}{x_2 - x_1}$$

If $x_1 = x_2$, the line is vertical and its slope is undefined. There are three important forms for the equation of a straight line:

1. *Slope-intercept form:* $y = mx + b$, where m is the slope and b is the y-intercept.
2. *General form:* $Ax + By = C$, where A, B, and C are real numbers.
3. *Point-slope form:* $y - y_1 = m(x - x_1)$, where m is the slope and (x_1,y_1) is a point on the line.

Two distinct lines with slopes m_1 and m_2 are **parallel** if and only if $m_1 = m_2$, and are **perpendicular** if and only if $m_1 = -\dfrac{1}{m_2}$.

Example 1.1

A triangle has vertices at points $A(-2,2)$, $B(3,2)$, $C(1,-2)$ in the coordinate plane.

A. Find the perimeter of the triangle.
B. Find the area of the triangle.

Solution:

Begin by constructing the triangle, as shown in Figure 1.2.

Use the distance formula to obtain

$$AB = \sqrt{[3-(-2)]^2 + (2-2)^2} = \sqrt{25} = 5$$

> Write radicals in simplified radical form:
> $\sqrt{20} = \sqrt{4 \cdot 5} = \sqrt{4} \cdot \sqrt{5} = 2\sqrt{5}.$

$$BC = \sqrt{(1-3)^2 + (-2-2)^2} = \sqrt{20} = 2\sqrt{5}$$

$$AC = \sqrt{[1-(-2)]^2 + (-2-2)^2} = \sqrt{25} = 5$$

A. The perimeter P of the triangle is the sum of the lengths of its sides:

$$P = 5 + 2\sqrt{5} + 5 = 10 + 2\sqrt{5} \approx 14.5 \text{ units}$$

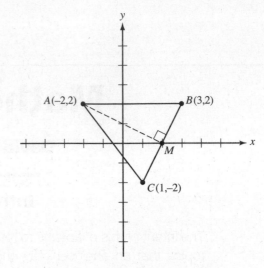

Figure 1.2

B. Since $AB = AC$, this triangle is isosceles with $\overline{BC}$ the nonequal side. To find the area of this triangle, recall from geometry that the altitude to the nonequal side of an isosceles triangle bisects that side. Thus, if $\overline{AM}$ is the altitude to base $\overline{BC}$, then M is the midpoint of $\overline{BC}$. By the midpoint formula, the coordinates of midpoint M are

$$M\left(\frac{1+3}{2}, \frac{-2+2}{2}\right) = M(2,0)$$

and, by the distance formula, the length of altitude $\overline{AM}$ is

$$AM = \sqrt{[1-(-2)]^2 + (-2-2)^2} = \sqrt{20} = 2\sqrt{5}$$

The area of triangle ABC is half the product of its base and altitude:

$$\text{Area} = \frac{1}{2}(BC)(AM) = \frac{1}{2}(2\sqrt{5})(2\sqrt{5}) = 10 \text{ square units}$$

As an alternative method for finding the area of this triangle, let $\overline{CN}$ be the altitude to base $\overline{AB}$. Since $\overline{AB}$ is a horizontal line segment, $CN = 4$. Hence:

$$\text{Area} = \frac{1}{2}(AB)(CN) = \frac{1}{2}(5)(4) = 10 \text{ square units}$$

Example 1.2

In each case, find the equation of a line that satisfies the given conditions:

A. Passes through points $(-2,-3)$ and $(2,1)$
B. Passes through point $(2,1)$ and is perpendicular to $3x - 5y = 10$

Solution:

A. The slope of the line that passes through points $(-2,-3)$ and $(2,1)$ is

$$m = \frac{1 - (-3)}{2 - (-2)} = \frac{4}{4} = 1$$

Using the point-slope form for the equation of a straight line with $(x_1,y_1) = (-2,-3)$ as the fixed point and $m = 1$ as the slope, find the equation of this line:

$$y - y_1 = m(x - x_1)$$

$$y + 3 = x + 2$$

$$y = x - 1$$

B. Determine the slope of line $3x - 5y = 10$ by writing it in slope-intercept form:

$$y = \frac{3}{5}x - 2$$

Hence, the slope of line $3x - 5y = 10$ is $\frac{3}{5}$. Therefore, the slope of a line perpendicular to this line is

$$m = -\frac{5}{3}$$

Use the point-slope form for the equation of a straight line with fixed point $(x_1,y_1) = (2,1)$ and slope $m = -\frac{5}{3}$ to find the desired equation:

$$y - y_1 = m(x - x_1)$$

$$y - 1 = -\frac{5}{3}(x - 2)$$

Write lines in *slope-intercept form*:
$y = mx + b$.

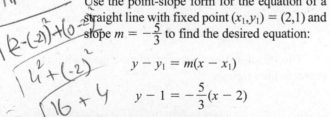

$$y = -\frac{5}{3}x + \frac{13}{3}$$

1.2
Quadratic Equations

An equation of the form $ax^2 + bx + c = 0$, where a, b, and c are real numbers with $a \neq 0$, is called a **quadratic equation in standard form.** The roots of this equation are given by **the quadratic formula:**

$$x = \frac{-b \pm \sqrt{b^2 - 4ac}}{2a}$$

In this formula, the quantity $b^2 - 4ac$ is called the **discriminant.** It indicates the nature of the roots of the quadratic equation:

1. Two distinct real roots if $b^2 - 4ac > 0$
2. Two complex conjugate roots if $b^2 - 4ac < 0$
3. One real root of multiplicity 2 if $b^2 - 4ac = 0$

Example 1.3

Solve each quadratic equation for x.

A. $x^2 + 6x = 18$ **B.** $x^2 + 5 = 2x$

Solution:

A. Before identifying a, b, and c, write $x^2 + 6x = 18$ in standard form:

$$x^2 + 6x - 18 = 0$$

Now, $a = 1$, $b = 6$, and $c = -18$. By the quadratic formula,

$$x = \frac{-6 \pm \sqrt{6^2 - 4(1)(-18)}}{2(1)} = \frac{-6 \pm \sqrt{108}}{2}$$

$$= -3 \pm 3\sqrt{3}$$

Hence, the two real roots of this equation are $-3 + 3\sqrt{3} \approx 2.20$ and $-3 - 3\sqrt{3} \approx -8.20$.

B. Begin by writing $x^2 + 5 = 2x$ in standard form:

$$x^2 - 2x + 5 = 0$$

Therefore, $a = 1$, $b = -2$, and $c = 5$. By the quadratic formula,

$$x = \frac{-(-2) \pm \sqrt{(-2)^2 - 4(1)(5)}}{2(1)}$$

$\sqrt{-1}$ is defined as the *imaginary unit i* (see Section 1.8).

$$= \frac{2 \pm \sqrt{-16}}{2} = 1 \pm 2i$$

Thus, the roots of this equation are the complex conjugates $1 + 2i$ and $1 - 2i$.

1.3
Conic Sections

If the graph of a **general quadratic equation in two unknowns, written as**

$$Ax^2 + Bxy + Cy^2 + Dx + Ey + F = 0$$

exists and is not degenerate, then its graph is

1. A *circle* if $A = C$ and $B = 0$.
2. A *parabola* if the discriminant $B^2 - 4AC = 0$.
3. An *ellipse* (or a circle, see definition 1) if the discriminant $B^2 - 4AC < 0$.
4. A *hyperbola* if the discriminant $B^2 - 4AC > 0$.

A **circle** is the set of all points P in a plane that lie a fixed distance (radius) from a given point (center), as shown in Figure 1.3.

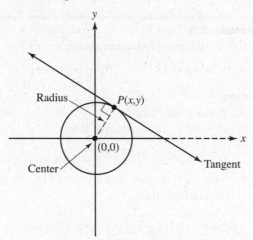

Figure 1.3

A **tangent** to a circle is a line with an endpoint not within the circle that intersects the circle in one and only one point. Every tangent to a circle is perpendicular to a radius drawn to the point of tangency. The standard-form equation of a circle with center at the origin and radius r is

$$x^2 + y^2 = r^2$$

A **parabola** is the set of all points P in a plane that are equidistant from a fixed line (directrix) and a fixed point (focus) not on the line. In Figure 1.4, $PQ = PF$.

The standard-form equation of a parabola with its vertex at the origin and a vertical axis of symmetry is

$$x^2 = 4py$$

The parabola opens upward if $p > 0$, and opens downward if $p < 0$. The standard-form equation of a parabola with its vertex at the origin and a horizontal axis of symmetry is

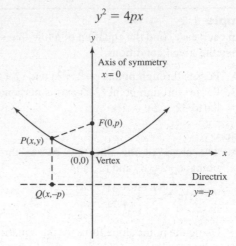

Figure 1.4

The parabola opens right if $p > 0$, and opens left if $p < 0$.

An **ellipse** is the set of all points P in a plane, the sum of whose distances from two fixed points (foci) is a constant that is equal to the length of the major axis of the ellipse. In Figure 1.5, $PF_1 + PF_2 = 2a$.

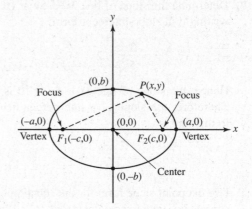

Figure 1.5

The standard-form equation of an ellipse with its center at the origin is

$$\frac{x^2}{a^2} + \frac{y^2}{b^2} = 1$$

If $a > b$, the vertices of the ellipse lie on the x-axis. If $b > a$, the vertices of the ellipse lie on the y-axis. For an ellipse, the center-to-focus distance is given by

$$c = \sqrt{|a^2 - b^2|}$$

A **hyperbola** is the set of all points P in a plane, the difference of whose distances from two fixed points (foci) is a constant that is equal to the length of the transverse axis of the hyperbola. In Figure 1.6, $PF_1 - PF_2 = 2a$.

40

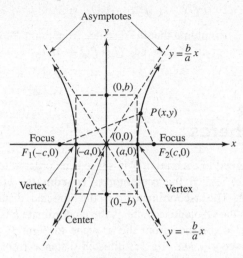

Figure 1.6

The standard-form equation of a hyperbola with its center at the origin is

$$\frac{x^2}{a^2} - \frac{y^2}{b^2} = 1$$

if its vertices are on the x-axis, and

$$\frac{y^2}{a^2} - \frac{x^2}{b^2} = 1$$

if its vertices are on the y-axis. For a hyperbola, the center-to-focus distance is given by

$$c = \sqrt{a^2 - b^2}$$

The graph of the equation $xy = k$ (k is a constant) is also a hyperbola. Its center is at the origin, and its asymptotes are the x- and y-axes.

If the center (or vertex) of a conic section is translated from the origin to another point (h, k) in the coordinate plane, then the standard-form equation of the conic section is adjusted by replacing x with $x - h$ and y with $y - k$.

Example 1.4

Classify the graph of each general quadratic equation in two unknowns as a circle, a parabola, an ellipse, or a hyperbola.

A. $4x^2 - 4xy + y^2 - 5\sqrt{5}x + 5 = 0$
B. $8y^2 + 6xy - 9 = 0$

Solution:

A. In the general quadratic equation $4x^2 - 4xy + y^2 - 5\sqrt{5}x + 5 = 0$, $A = 4$, $B = -4$, and $C = 1$. Since the discriminant $B^2 - 4AC = 0$, the graph of this equation is a parabola.

B. In the general quadratic equation $8y^2 + 6xy - 9 = 0$, $A = 8$, $B = 6$, and $C - 0$. Since the discriminant $B^2 - 4AC = 36 > 0$, the graph of this equation is a hyperbola.

Example 1.5

Determine the coordinates of (1) the vertices and (2) the foci for the conic section defined by each equation.

A. $16x^2 + 9y^2 = 144$ **B.** $16x^2 - 9y^2 = 144$

Solution:

A. Divide both sides of this equation by 144:

$$\frac{x^2}{3^2} + \frac{y^2}{4^2} = 1$$

(1) This is the standard-form equation of an ellipse with its center at the origin and vertices on the y-axis (since $4 > 3$). Hence, the coordinates of the vertices are $(0,4)$ and $(0,-4)$.

(2) The center-to-focus distance is $c = \sqrt{4^2 - 3^2} = \sqrt{7}$. Therefore, the coordinates of the foci are $(0, \sqrt{7}) \approx (0, 2.646)$ and $(0, -\sqrt{7}) \approx (0, -2.646)$.

B. Divide both sides of this equation by 144:

$$\frac{x^2}{3^2} - \frac{y^2}{4^2} = 1$$

(1) This is the standard-form equation of a hyperbola with center at the origin and vertices on the x-axis (since the x^2-term is positive). Hence, the coordinates of the vertices are $(3, 0)$ and $(-3, 0)$.

(2) The center-to-focus distance is $c = \sqrt{3^2 + 4^2} = 5$. Therefore, the coordinates of the foci are $(5, 0)$ and $(-5, 0)$.

Example 1.6

In each case, determine the general quadratic equation of the conic section with the given characteristics.

A. Parabola: vertex at the origin and focus at $(-2, 0)$
B. Hyperbola: asymptotes along the x- and y-axes and passes through point $(3, 2)$

Solution:

A. A parabola with vertex at the origin and focus on the x-axis has the form $y^2 = 4px$, with $(p, 0)$ the coordinates of the focus. Since the x-coordinate of the focus given is -2, the general quadratic equation of this parabola is found by replacing p with -2 in the equation $y^2 = 4px$:

$$y^2 = -8x \quad \text{or} \quad y^2 + 8x = 0$$

B. A hyperbola with asymptotes along the x- and y-axes has the form $xy = k$. Since this hyperbola passes through point $(3, 2)$, $k = xy = 6$. The general quadratic equation of this hyperbola is found by replacing k with 6 in the equation $xy = k$:

$$xy = 6 \quad \text{or} \quad xy - 6 = 0$$

Example 1.7

A tangent to the circle with center $C(1, -2)$ and radius 3 passes through point $P(5, -5)$.

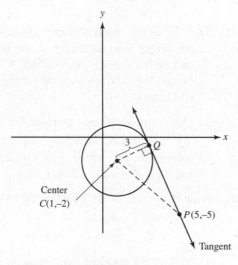

Figure 1.7

A. Find the equation of the circle.
B. Find the distance from point P to a point of tangency Q on the circle.

Solution:

A. The equation of a circle with radius 3 and center at the origin is $x^2 + y^2 = 9$. Hence, by a translation of axis, the equation of a circle with center at $C(1, -2)$ and radius 3 is

$$(x - 1)^2 + (y + 2)^2 = 9$$

B. Since every tangent to a circle is perpendicular to a radius drawn to the point of tangency Q, the triangle formed by points C, P, and Q must be a right triangle with hypotenuse CP, as shown in Figure 1.7.

Since the radius of the circle is 3, $CQ = 3$, and, by the distance formula, $CP = 5$. Hence, by the Pythagorean theorem

$$(PQ)^2 = (CP)^2 - (CQ)^2 = 5^2 - 3^2 = 16$$

implying that the distance from point P to the point of tangency Q is $PQ = 4$.

1.4 Spheres

Assigned to each point P in space is an ordered triple (a,b,c), called the **rectangular coordinates** of the point. The x-coordinate a is the directed distance from the yz-plane to point P, the y-coordinate b is the directed distance from the xz-plane to point P, and the z-coordinate c is the directed distance from the xy-plane to point P, as shown in Figure 1.8.

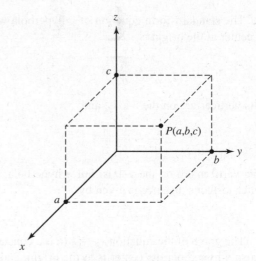

Figure 1.8

The distance between two points $P(x_1, y_1, z_1)$ and $Q(x_2, y_2, z_2)$ in space is given by the **distance formula:**

$$PQ = \sqrt{(x_2 - x_1)^2 + (y_2 - y_1)^2 + (z_2 - z_1)^2}$$

A sphere is a surface in space, each point of which is at a given fixed distance (radius) from a given fixed point (center). The standard-form equation of a sphere with center at (h,k,m) and radius r is

$$(x - h)^2 + (y - k)^2 + (z - m)^2 = r^2$$

The general-form equation of a sphere is

$$x^2 + y^2 + z^2 + Ax + By + Cz + D = 0$$

where A, B, C, and D are real numbers.

Example 1.8

A sphere with center $(1, -3, 2)$ passes through point $(4, 1, 2)$.

A. Find the radius of the sphere.
B. Find the general-form equation of the sphere.

Solution:

A. The radius r of the sphere is the distance from its center to a point on the sphere. By the distance formula,

$$r = \sqrt{(4-1)^2 + \left[1-(-3)\right]^2 + (2-2)^2} = \sqrt{25} = 5$$

B. The standard-form equation of a sphere with center $(1, -3, 2)$ and radius 5 is given by

$$(x - 1)^2 + (y + 3)^2 + (z - 2)^2 = 25$$

Its general-form equation is found by expanding and simplifying its standard-form equation:

> To expand, use $(a + b)^2 = a^2 + 2ab + b^2$.

$$x^2 + y^2 + z^2 - 2x + 6y - 4z - 11 = 0$$

1.5
Logarithms

The **logarithmic function with base b** is defined by

$$y = \log_b x \quad \text{if and only if} \quad x = b^y \text{ for } x > 0$$

A base-10 logarithm, written as $\log x$, is called a **common logarithm.** A base-e ($e \approx 2.718...$) logarithm, written as $\ln x$, is called a **natural logarithm.** To change from one logarithmic base to another, use the **change of base formula:**

$$\log_b x = \frac{\log_a x}{\log_a b}, \, a > 0, \, a \neq 1$$

Four **logarithmic identities** hold for all bases b, with $b > 0$ and $b \neq 1$:

1. $\log_b 1 = 0$ 2. $\log_b b = 1$
3. $b^{\log_b u} = u, u > 0$ 4. $\log_b b^u = u$

If $b > 0$, $b \neq 1$, then, for positive real numbers x and y, there are three **properties of logarithms:**

1. $\log_b xy = \log_b x + \log_b y$ 2. $\log_b \frac{x}{y} = \log_b x - \log_b y$
3. $\log_b x^n = n \log_b x$

Example 1.9

Evaluate $\log_4 24$ using:

A. common logarithms   **B.** natural logarithms

Solution:

A. Use the change of base formula with $a = 10$, $b = 4$, and the common logarithm key on a calculator:

$$\log_4 24 = \frac{\log 24}{\log 4} \approx 2.29$$

B. Use the change of base formula with $a = e$, $b = 4$, and the natural logarithm key on a calculator:

$$\log_4 24 = \frac{\ln 24}{\ln 4} \approx 2.29$$

Example 1.10

Simplify each expression.

A. $e^{2\ln(x+1)}$ **B.** $\log_2 16^{x^2}$

Solution:

A. Apply logarithmic property 3, and use the fact that $b^{\log_b u} = u$, for $u > 0$:

$$e^{2\ln(x+1)} = e^{\ln(x+1)^2} = (x + 1)^2$$

provided that $x > -1$.

B. Begin by applying the properties of exponents and matching the base of the exponential expression to the base of the logarithm:

$$\log_2 16^{x-2} = \log_2 (2^4)^{x-2} = \log_2 2^{4x-8}$$

Now, since $\log_b b^u = u$ for any permissible base b and any real number u,

$$\log_2 16^{x-2} = \log_2 2^{4x-8} = 4x - 8$$

Example 1.11

Use the properties of logarithms to write each expression as a single logarithm.

A. $\log_6 50 + 2 \log_6 4 - 3 \log_6 2$
B. $\ln(x^2 - 1) - 2 \ln(x + 1)$

Solution:

A. Begin with property 3 and then apply the other two logarithmic properties:

$$\log_6 50 + 2\log_6 4 - 3\log_6 2 = \log_6 50 + \log_6 4^2 - \log_6 2^3$$
$$= \log_6 50 + \log_6 16 - \log_6 8$$
$$= \log_6 \frac{(50)(16)}{8}$$
$$= \log_6 100$$

B. Assuming that $x > 1$, begin with property 3 and then apply property 2, reducing the fraction to lowest terms:

$$\ln(x^2 - 1) - 2\ln(x + 1) = \ln(x^2 - 1) - \ln(x + 1)^2$$
$$= \ln \frac{x^2 - 1}{(x+1)^2}$$
$$= \ln \frac{(x-1)(x+1)}{(x+1)^2}$$
$$= \ln \frac{x-1}{x+1}$$

Example 1.12

Solve each equation for x.

A. $4e^{2x-1} = 100$ **B.** $2 - \log x = \log 3$

Solution:

A. Begin by dividing both sides of the equation by 4 to isolate the exponential expression on one side:

$$e^{2x-1} = 25$$

Now, take the natural logarithm of each side, apply log property 3, and solve for x:

$$\ln e^{2x-1} = \ln 25$$
$$(2x - 1)\ln e = \ln 25$$
$$x = \frac{1 + \ln 25}{2} \approx 2.109$$

B. Begin by grouping the logarithms on one side of the equation and then condense them into a single logarithm by applying the properties of logarithms:

$$2 - \log x = \log 3$$
$$\log 3 + \log x = 2$$
$$\log 3x = 2$$

Now, change to exponential form and solve for x:

If no base is written, it is understood to be base 10.

$\log 3x = 2$ if and only if $10^2 = 3x$

implying that $x = \frac{100}{3}$.

1.6 Triangle Trigonometry

The interior angles of a polygon are usually measured in radians or degrees. The relationship between **radian measure** and **degree measure** is as follows:

$$\pi \text{ radians} = 180 \text{ degrees}$$

The **sum of the interior angles of an n-sided polygon** is

$$(n - 2)180° \quad \text{or} \quad (n - 2)\pi$$

An angle whose measure is less than 90° ($\pi/2$ radians) is called an **acute angle,** and an angle whose measure is between 90° and 180° is an **obtuse angle.** If an angle measures 90°, it is called a **right angle.** The Pythagorean theorem states that in any right triangle

$$x^2 + y^2 = r^2$$

as shown in Figure 1.9.

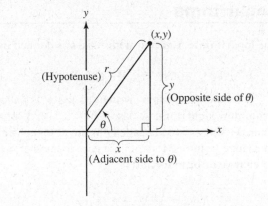

Figure 1.9

For Figure 1.9, there are six **trigonometric functions**:

1. $\sin \theta = \dfrac{y}{r} = \dfrac{\text{opp}}{\text{hyp}}$ 2. $\cos \theta = \dfrac{x}{r} = \dfrac{\text{adj}}{\text{hyp}}$

3. $\tan \theta = \dfrac{y}{x} = \dfrac{\text{opp}}{\text{adj}}$ 4. $\csc \theta = \dfrac{r}{y} = \dfrac{\text{hyp}}{\text{opp}}$

5. $\sec \theta = \dfrac{r}{x} = \dfrac{\text{hyp}}{\text{adj}}$ 6. $\cot \theta = \dfrac{x}{y} = \dfrac{\text{adj}}{\text{opp}}$

Figure 1.10 shows in which quadrants the trigonometric functions are positive:

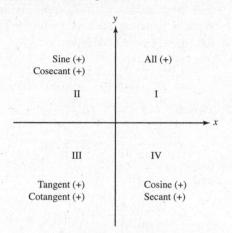

Figure 1.10

The **inverse trigonometric functions** are defined as follows:

1. $\sin^{-1} k$, with $-1 \leq k \leq 1$, denotes the angle between $90°$ and $90°$ whose sine is k.
2. $\cos^{-1} k$, with $-1 \leq k \leq 1$, denotes the angle between $0°$ and $180°$ whose cosine is k.
3. $\tan^{-1} k$ denotes the angle between $-90°$ and $90°$ whose tangent is k.

The **law of sines** states that in a given triangle the ratio of the length of the side to the sine of the angle opposite that side has the same constant value:

$$\frac{a}{\sin A} = \frac{b}{\sin B} = \frac{c}{\sin C}$$

The **law of cosines** states that the square of the length of any side of a triangle is the sum of the squares of the other two sides minus twice the product of those two sides and the cosine of the angle between them:

$$a^2 = b^2 + c^2 - 2bc \cos A$$

If a and b are any two sides of a triangle and θ is the angle between those two sides, then the **area** A of the triangle is

$$A = \frac{ab \sin \theta}{2}$$

Example 1.13

Given that $\sin \theta = \frac{3}{5}$, find $\cos \theta$ if:

A. θ is acute.   **B.** θ is obtuse.

Solution:

If $\sin \theta = \frac{3}{5}$, then, by the definition of sine, $y = 3$ and $r = 5$. Hence, by the Pythagorean theorem, $x^2 + 3^2 = 5^2$, implying that $x = \pm 4$.

A. If θ is acute, then θ is in quadrant I, where $\cos \theta > 0$. Hence,

$$\cos \theta = \frac{x}{r} = \frac{4}{5}$$

B. If θ is obtuse, then θ is in quadrant II, where $\cos \theta < 0$. Hence,

$$\cos \theta = \frac{x}{r} = \frac{-4}{5} = -\frac{4}{5}$$

Example 1.14

Given a right triangle with legs of lengths 5 meters and 12 meters, find:

A. the length of the hypotenuse
B. the angle (in radians) opposite the longer leg

Solution:

A. If z is the length of the hypotenuse, then, by the Pythagorean theorem,

$$z^2 = 5^2 + 12^2$$

implying that the length of the hypotenuse $z = 13$ m.

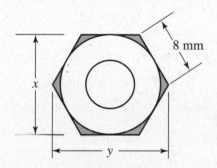

Figure 1.11

B. If θ is the angle opposite the longer leg, then, by the definition of tangent,

$$\tan \theta = \frac{\text{opp}}{\text{adj}} = \frac{12}{5}$$

To find θ, use the inverse tangent key on a calculator set in radian mode:

$$\theta = \tan^{-1} \frac{12}{5} \approx 1.176 \text{ rad}$$

Example 1.15

Each side of a hexagonal nut is 8 mm long, as shown in Figure 1.11.

A. Find distance x.   **B.** Find distance y.

Solution:

A. Since a hexagonal nut has six sides, the sum of its interior angles is $(6 - 2)180° = 720°$. Therefore, each interior angle measures $720°/6 = 120°$, implying that triangle ABC is isosceles with $\theta = 60°$, as shown in Figure 1.12.

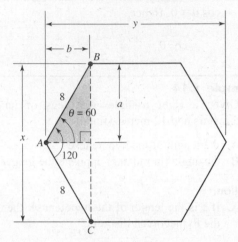

Figure 1.12

In the right triangle in Figure 1.12,

$$\sin 60° = \frac{a}{8}$$

implying that $a = 8 \sin 60° \approx 6.93$ mm. Thus,

$$x = 2a \approx 13.9 \text{ mm}$$

B. Again, in the right triangle in Figure 1.12,

$$\cos 60° = \frac{b}{8}$$

implying that $b = 8 \cos 60° = 4$ mm. Therefore,

$$y = 2b + 8 = 16 \text{ mm}$$

Example 1.16

Two angles of a triangle measure 62° and 48°. The length of the side opposite the unknown angle is 52 cm.

A. Find the measure of the unknown angle.

B. Find the lengths of the unknown sides.

Solution:

A. The sum of the interior angles of a triangle is 180°. Hence, if θ is the unknown angle, then

$$\theta = 180° - (62° + 48°) = 70°$$

B. Let x and y be the sides opposite the angles that measure 48° and 62°, respectively. To find these unknown sides, apply the law of sines by matching each side with the sine of the angle opposite that side:

$$\frac{52}{\sin 70°} = \frac{x}{\sin 48°} = \frac{y}{\sin 62°}$$

Hence:

$$x = \frac{52 \sin 48°}{\sin 70°} \approx 41.1 \text{ cm}$$

and

$$y = \frac{52 \sin 62°}{\sin 70°} \approx 48.9 \text{ cm}$$

Example 1.17

The sides of a triangle have lengths 22.5 m, 16.2 m, and 14.9 m.

A. Determine the measure in degrees of the largest angle.

B. Find the area of the triangle.

Solution:

A. The largest angle in a triangle is always opposite its longest side. Hence, the largest angle is opposite the side of length 22.5 m. If θ is the largest angle, then, by the law of cosines,

$$22.5^2 = 16.2^2 + 14.9^2 - 2(16.2)(14.9) \cos \theta$$

Solving for $\cos \theta$ gives

$$\cos \theta = \frac{22.5^2 - 16.2^2 - 14.9^2}{-2(16.2)(14.9)}$$

The inverse cosine function in conjunction with a calculator set in degree mode shows that θ is obtuse:

$$\theta = \cos^{-1} \left[\frac{22.5^2 - 16.2^2 - 14.9^2}{-2(16.2)(14.9)} \right] \approx 92.6°$$

B. From part **A**, the angle between the sides of lengths 16.2 m and 14.9 m is 92.6°. Thus, the area A of this triangle is

$$A = \frac{ab \sin \theta}{2} = \frac{(16.2)(14.9)\sin 92.6°}{2} \approx 121 \text{ m}^2$$

1.7
Analytic Trigonometry

Below is a list of the most frequently used trigonometric identities. For additional trigonometric identities, refer to the *Fundamentals of Engineering Reference Handbook*.

1. $\sin \theta \csc \theta = 1$
2. $\cos \theta \sec \theta = 1$
3. $\tan \theta \cot \theta = 1$
4. $\tan \theta = \dfrac{\sin \theta}{\cos \theta}$
5. $\cot \theta = \dfrac{\cos \theta}{\sin \theta}$
6. $\sin^2\theta + \cos^2\theta = 1$
7. $1 + \tan^2\theta = \sec^2\theta$
8. $1 + \cot^2\theta = \csc^2\theta$
9. $\cos (\alpha \pm \beta) = \cos \alpha \cos \beta \mp \sin \alpha \sin \beta$
10. $\sin (\alpha \pm \beta) = \sin \alpha \cos \beta \pm \cos \alpha \sin \beta$
11. $\sin 2\theta = 2 \sin \theta \cos \theta$
12. $\cos 2\theta = \cos^2\theta - \sin^2\theta = 2 \cos^2\theta - 1 = 1 - 2 \sin^2\theta$

Example 1.18

Given that $\sin \theta = \frac{3}{5}$ and θ is acute, find:

A. $\sin 2\theta$ **B.** $\cos (\theta + \pi)$

Solution:

From Example 1.13, if $\sin \theta = \frac{3}{5}$ and θ is acute, then $\cos \theta = \frac{4}{5}$.

A. Use trigonometric identity 11; then

$$\sin 2\theta = 2 \sin \theta \cos \theta = 2\left(\frac{3}{5}\right)\left(\frac{4}{5}\right) = \frac{24}{25}$$

B. Use trigonometric identity 9; then

$$\cos (\theta + \pi) = \cos \theta \cos \pi - \sin \theta \sin \pi$$
$$= \left(\frac{4}{5}\right)(-1) - \left(\frac{3}{5}\right)(0) = -\frac{4}{5}$$

Example 1.19

Simplify each expression to a single trigonometric function or a constant.

A. $\cos \theta \tan^2\theta + \cos \theta$ **B.** $\dfrac{\cos^2\theta}{1 - \sin \theta} - \sin \theta$

Solution:

A. Begin by factoring out $\cos \theta$ and then applying trigonometric identities 7 and 2:

As an alternate approach, begin with $\tan^2 \theta = \dfrac{\sin^2 \theta}{\cos^2 \theta}$, then simplify.

$$\cos \theta \tan^2\theta + \cos \theta = \cos \theta (\tan^2\theta + 1)$$
$$= \cos \theta \sec^2\theta$$
$$= (\cos \theta \sec \theta)\sec \theta$$
$$= (1)\sec \theta$$
$$= \sec \theta$$

B. Begin by subtracting fractions and then apply trigonometric identity 6:

$$\frac{\cos^2\theta}{1 - \sin \theta} - \sin \theta = \frac{\cos^2\theta - \sin \theta(1 - \sin \theta)}{1 - \sin \theta}$$
$$= \frac{\cos^2\theta - \sin \theta + \sin^2 \theta}{1 - \sin \theta}$$
$$= \frac{1 - \sin \theta}{1 - \sin \theta}$$
$$= 1$$

provided that $1 - \sin \theta \neq 0$

Example 1.20

Write each expression as a first power of the cosine function.

A. $\cos^2\theta$ **B.** $\sin^4\theta$

Solution:

A. Use identity 12, $\cos 2\theta = 2 \cos^2\theta - 1$, to solve for $\cos^2\theta$:

$$\cos 2\theta = 2 \cos^2\theta - 1$$
$$2 \cos^2\theta = 1 + \cos 2\theta$$
$$\cos^2\theta = \frac{1}{2}(1 + \cos 2\theta)$$

B. Use identity 12, $\cos 2\theta = 1 - 2\sin^2\theta$, to solve for $\sin^2\theta$:

$$\cos 2\theta = 1 - 2 \sin^2\theta$$
$$2 \sin^2\theta = 1 - \cos 2\theta$$
$$\sin^2\theta = \frac{1}{2}(1 - \cos 2\theta)$$

Now, express $\sin^4 \theta$ in terms of the cosine function:

$$\sin^4\theta = \sin^2\theta \sin^2 \theta$$
$$= \frac{1}{2}(1 - \cos 2\theta) \frac{1}{2}(1 - \cos 2\theta)$$
$$= \frac{1}{4}(1 - 2 \cos 2\theta + \cos^2 2\theta)$$

Finally, using the results of part **A**, replace $\cos^2 2\theta$ with $\frac{1}{2}(1 + \cos 4\theta)$:

$$\sin^4\theta = \frac{1}{4}\left[1 - 2\cos 2\theta + \frac{1}{2}(1 + \cos 4\theta)\right]$$

$$= \frac{1}{4}\left(\frac{3}{2} - 2\cos 2\theta + \frac{1}{2}\cos 4\theta\right)$$

$$= \frac{1}{8}(3 - 4\cos 2\theta + \cos 4\theta)$$

$$w_k = r^{1/n}e^{i(\theta+2\pi k)/n} = r^{1/n}\left[\cos\left(\frac{\theta + 2\pi k}{n}\right) + i\sin\left(\frac{\theta + 2\pi k}{n}\right)\right], \text{ for } k = 0, 1, 2, \ldots, n-1$$

1.8
Complex Numbers

The **imaginary unit** i is defined by

$$i = \sqrt{-1}$$

where $i^2 = -1$.

A complex number z in **rectangular form** is written as

$$z = ai + b$$

where a and b are real numbers and i is the imaginary unit, as shown in the complex plane in Figure 1.13.

A complex number z in **polar form** is given by

$$z = re^{i\theta} = r(\cos\theta + i\sin\theta)$$

where $a = r\cos\theta$, $b = r\sin\theta$, $r = \sqrt{a^2 + b^2}$, and $\tan\theta = \frac{b}{a}$, as shown in Figure 1.13.

The **complex conjugate** of $z = a + bi$ is $\bar{z} = a - bi$. The product $z\bar{z}$ is the nonnegative real number $a^2 + b^2$.

DeMoivre's theorem is used to raise a complex number z in polar form to an integer power:

$$z^k = r^k e^{ik\theta} = r^k(\cos k\theta + i\sin k\theta)$$

If $z = re^{i\theta} = r(\cos\theta + i\sin\theta)$, with $z \neq 0$ and n a positive integer, then z has exactly n distinct nth roots $w_0, w_1, w_2, \ldots, w_{n-1}$, which are given by the **nth root formula:**

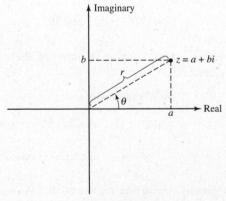

Figure 1.13

Example 1.21

In each case, convert the given complex number to the indicated form.

A. $3 + 3i$ to polar form
B. $4.47e^{5.82i}$ to rectangular form

Solution:

A. For $3 + 3i$, $a = 3$ and $b = 3$. Thus:

$$r = \sqrt{a^2 + b^2} = \sqrt{3^2 + 3^2} = \sqrt{18} = 3\sqrt{2} \approx 4.24$$

Since $3 + 3i$ lies in quadrant I in the complex plane, θ must be a first-quadrant angle that satisfies the equation $\tan\theta = \frac{b}{a} = \frac{3}{3} = 1$. Hence,

$$\theta = \tan^{-1} 1 = \frac{\pi}{4} \approx 0.785$$

Thus, $3 + 3i = 3\sqrt{2}\ e^{(\pi/4)i} \approx 4.24e^{0.785i}$.

B. To express $4.47e^{5.82i}$ in rectangular form, use a calculator set in radian mode:

$$4.47e^{5.82i} = 4.47(\cos 5.82 + i\sin 5.82) \approx 4.00 - 2.00i$$

Example 1.22

In each case, given that $z_1 = 3 + 5i$ and $z_2 = 1 - 5i$, perform the indicated operation.

A. $z_1 z_2$ **B.** $\dfrac{z_1}{z_2}$

Solution:

A. To find the product of two complex numbers in rectangular form, multiply each part of the first complex number by each part of the second complex number:

$$z_1 z_2 = (3 + 5i)(1 - 5i)$$

$$= 3 - 15i + 5i - 25i^2$$

$$= 28 - 10i$$

B. To find the quotient of two complex numbers in rectangular form, multiply by the complex conjugate of the denominator to obtain division by a nonnegative real number:

$$\frac{z_1}{z_2} = \frac{3 + 5i}{1 - 5i} = \frac{(3 + 5i)(1 + 5i)}{(1 - 5i)(1 + 5i)}$$

$$= \frac{-22 + 20i}{26}$$

$$\approx -0.846 + 0.769i$$

Example 1.23

In each case, given that $z_1 = 5e^{0.3i}$ and $z_2 = 3e^{2.5i}$, perform the indicated operation and write the answer in rectangular form:

A. $z_1 z_2$ **B.** $z_1 + z_2$

Solution:

A. To multiply (or divide) complex numbers in polar form, add (or subtract) exponents:

$$z_1 z_2 = (5e^{0.3i})(3e^{2.5i}) = 15e^{(0.3i + 2.5i)} = 15e^{2.8i}$$

Now, convert to rectangular form:

> To evaluate, use the *radian mode* on a calculator.

$$15e^{2.8i} = 15(\cos 2.8 + i \sin 2.8) \approx -14.1 + 5.02i$$

B. To find the sum (or difference) of two complex numbers in polar form, convert the numbers to rectangular form and then add (or subtract) the real and imaginary parts of these numbers, separately. Hence,

$$z_1 + z_2 = 5e^{0.3i} + 3e^{2.5i}$$

$$= 5(\cos 0.3 + i \sin 0.3) + 3(\cos 2.5 + i \sin 2.5)$$

$$= (5 \cos 0.3 + 3 \cos 2.5) + i(5 \sin 0.3 + 3 \sin 2.5)$$

$$\approx 2.37 + 3.27i$$

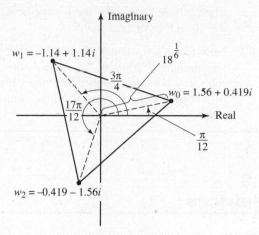

Figure 1.14

Example 1.24

In each case, given that $z = 3 + 3i$, find the indicated power or roots and write the answer in rectangular form.

A. z^4 **B.** the three cube roots of z

Solution:

A. From Example 1.21A, $z = 3 + 3i = 3\sqrt{2}\,e^{(\pi/4)i}$. Hence, by DeMoivre's theorem:

$$z^4 = \left(3\sqrt{2}\,e^{(\pi/4)i}\right)^4 = \left(3\sqrt{2}\right)^4 e^{4(\pi/4)i} = 324e^{\pi i}$$

Now, convert this answer to rectangular form:

$$324e^{\pi i} = 324(\cos \pi + i \sin \pi) = -324 + 0i = -324$$

B. Again, from Example 1.21A, $z = 3 + 3i = 3\sqrt{2}\,e^{(\pi/4)i}$. Hence, by the nth root formula, this complex number has three distinct cube roots given by

$$w_k = \left(3\sqrt{2}\right)^{1/3} e^{i[(\pi/4) + 2\pi k]/3}$$

for $k = 0, 1,$ and 2. Hence, the three cube roots are

$$w_0 = \left(3\sqrt{2}\right)^{1/3} e^{i(\pi/12)} = 18^{1/6}\left(\cos \frac{\pi}{12} + i \sin \frac{\pi}{12}\right)$$
$$\approx 1.56 + 0.419i$$

$$w_1 = \left(3\sqrt{2}\right)^{1/3} e^{i(3\pi/4)} = 18^{1/6}\left(\cos \frac{3\pi}{4} + i \sin \frac{3\pi}{4}\right)$$
$$\approx -1.14 + 1.14i$$

$$w_2 = \left(3\sqrt{2}\right)^{1/3} e^{i(17\pi/12)} = 18^{1/6}\left(\cos \frac{17\pi}{12} + i \sin \frac{17\pi}{12}\right)$$
$$\approx -0.419 - 1.56i$$

Figure 1.14 shows the three cube roots of $3 + 3i$ in the complex plane.

Note that w_0, w_1, and w_2 form the vertices of an *equilateral triangle*. In general, the nth roots of a complex number form the vertices of a regular n-sided polygon in the complex plane.

1.9
Matrices

A **matrix** is a rectangular array of numbers enclosed by a pair of brackets. Each number in the matrix is called an *element*. Consider matrix A with m *rows* and n *columns*:

$$A = \begin{bmatrix} a_{11} & a_{12} & \cdots & a_{1j} & \cdots & a_{1n} \\ a_{21} & a_{22} & \cdots & a_{2j} & \cdots & a_{2n} \\ \vdots & \vdots & \vdots & \vdots & \vdots & \vdots \\ a_{i1} & a_{i2} & \cdots & a_{ij} & \cdots & a_{in} \\ \vdots & \vdots & \vdots & \vdots & \vdots & \vdots \\ a_{m1} & a_{m2} & \cdots & a_{mj} & \cdots & a_{mn} \end{bmatrix}$$

This matrix is of *order $m \times n$* and can be denoted more compactly by writing $A = [a_{ij}]$, where a_{ij} stands for the element in the ith row, jth column. Two matrices $A = [a_{ij}]$ and $B = [b_{ij}]$ are equal if both are of order $m \times n$ and $a_{ij} = b_{ij}$ for all $i = 1, 2, 3, \ldots, m$ and $j = 1, 2, 3, \ldots, n$.

A **square matrix** is a matrix of order $n \times n$. The elements $a_{11}, a_{22}, \ldots, a_{nn}$ are called its *main diagonal elements*. A square matrix of order $n \times n$, with the digit 1 along its main diagonal and zeros elsewhere, is called an **identity matrix** of order $n \times n$. The **inverse** of a square matrix A of order $n \times n$ is denoted by A^{-1}. The product of a square matrix A and its inverse A^{-1} is the identity matrix I:

$$AA^{-1} = A^{-1}A = I$$

The **transpose** of matrix A, denoted by A^T, is formed by switching the rows of matrix A with its columns.

Three rules are used to add, subtract, and multiply matrices (matrix division is undefined):

1. **Matrix addition:** If $A = [a_{ij}]$ and $B = [b_{ij}]$ are matrices of order $m \times n$, then the sum $A + B$ is a matrix of order $m \times n$ defined by $A + B = [a_{ij} + b_{ij}]$.
2. **Scalar multiplication:** If $A = [a_{ij}]$ is a matrix of order $m \times n$ and k is a real number, then the scalar multiple kA is a matrix of order $m \times n$ defined by $kA = [ka_{ij}]$.

3. **Matrix multiplication:** If $A = [a_{ij}]$ is a matrix of order $m \times n$ and $B = [b_{ij}]$ is a matrix of order $n \times p$, then the product AB is a matrix of order $m \times p$ defined by using a row times column procedure; that is, $AB = [c_{ij}]$, where $c_{ij} = a_{i1}b_{1j} + a_{i2}b_{2j} + \ldots + a_{in}b_{nj}$.

Example 1.25

In each case, given the matrices $A = \begin{bmatrix} 2 & 5 \\ -3 & 0 \\ 1 & 4 \end{bmatrix}$ and

$B = \begin{bmatrix} 4 & 6 \\ -2 & -5 \\ 0 & 3 \end{bmatrix}$, perform the indicated operation.

A. $2A - B$ **B.** $A^T + B^T$

Solution:

A. Using the rules for scalar multiplication and matrix addition, proceed as follows:

$$2A - B = 2A + (-1)B = 2\begin{bmatrix} 2 & 5 \\ -3 & 0 \\ 1 & 4 \end{bmatrix} + (-1)\begin{bmatrix} 4 & 6 \\ -2 & -5 \\ 0 & 3 \end{bmatrix}$$

$$= \begin{bmatrix} 4 & 10 \\ -6 & 0 \\ 2 & 8 \end{bmatrix} + \begin{bmatrix} -4 & -6 \\ 2 & 5 \\ 0 & -3 \end{bmatrix} = \begin{bmatrix} 0 & 4 \\ -4 & 5 \\ 2 & 5 \end{bmatrix}$$

B. First, find the transpose matrices A^T and B^T by switching rows and columns:

$$A^T = \begin{bmatrix} 2 & -3 & 1 \\ 5 & 0 & 4 \end{bmatrix}, \text{ and } B^T = \begin{bmatrix} 4 & -2 & 0 \\ 6 & -5 & 3 \end{bmatrix}$$

Now, use the rule for matrix addition to obtain the sum of the transposes:

$$A^T + B^T = \begin{bmatrix} 2 & -3 & 1 \\ 5 & 0 & 4 \end{bmatrix} + \begin{bmatrix} 4 & -2 & 0 \\ 6 & -5 & 3 \end{bmatrix} = \begin{bmatrix} 6 & -5 & 1 \\ 11 & -5 & 7 \end{bmatrix}$$

Example 1.26

Given the matrices $A = \begin{bmatrix} 3 & 4 \\ 2 & 3 \end{bmatrix}$, $B = \begin{bmatrix} 3 & -4 \\ -2 & 3 \end{bmatrix}$,

and $C = \begin{bmatrix} 2 & -1 \\ 3 & 4 \\ 0 & -2 \end{bmatrix}$, find each product.

A. AB **B.** CA

Solution:

A. Since matrices A and B are square matrices of order 2×2, the product AB is defined and is also of order 2×2. Using the rule for matrix multiplication, proceed as follows:

$$AB = \begin{bmatrix} 3 & 4 \\ 2 & 3 \end{bmatrix}\begin{bmatrix} 3 & -4 \\ -2 & 3 \end{bmatrix} =$$

$$\begin{bmatrix} (3)(3) + (4)(-2) & (3)(-4) + (4)(3) \\ (2)(3) + (3)(-2) & (2)(-4) + (3)(3) \end{bmatrix} = \begin{bmatrix} 1 & 0 \\ 0 & 1 \end{bmatrix}$$

Since this product yields the identity matrix I of order 2×2, matrices A and B are inverses of each other.

B. Since matrix C has two columns and matrix A has two rows, the product CA is defined and is of order 3×2, since matrix C has three rows and matrix A has two columns. Using the rule for matrix multiplication, proceed as follows:

$$CA = \begin{bmatrix} 2 & -1 \\ 3 & 4 \\ 0 & -2 \end{bmatrix}\begin{bmatrix} 3 & 4 \\ 2 & 3 \end{bmatrix} =$$

$$\begin{bmatrix} (2)(3) + (-1)(2) & (2)(4) + (-1)(3) \\ (3)(3) + (4)(2) & (3)(4) + (4)(3) \\ (0)(3) + (-2)(2) & (0)(4) + (-2)(3) \end{bmatrix} = \begin{bmatrix} 4 & 5 \\ 17 & 24 \\ -4 & -6 \end{bmatrix}$$

Note that the number of columns in matrix A is *not* the same as the number of rows in matrix C. Hence, the product AC is not defined. Unlike multiplication of real numbers, matrix multiplication, in general, is not commutative; that is, $AC \neq CA$.

1.10
Determinants and Inverses of Matrices

A **determinant** of order n contains n^2 elements arranged in n rows and n columns and enclosed by two vertical bars. The **minor** of an element a_{ij} in a determinant is denoted by M_{ij}, and represents the determinant that remains after deleting the row and column in which the element a_{ij} appears. The **cofactor** of an element a_{ij} in a determinant, denoted by C_{ij}, differs from M_{ij} at most in sign and is given by $C_{ij} = (-1)^{i+j}M_{ij}$.

The value of an nth-order determinant is a sum found by multiplying each element in any row or column by its corresponding cofactor, and then adding the products. This sum is called the **expansion of the determinant** by the specified row or column. The expansion of a second-order determinant by any row or column is

$$\begin{vmatrix} a_{11} & a_{12} \\ a_{21} & a_{22} \end{vmatrix} = a_{11}a_{22} - a_{12}a_{21}$$

The inverse of an $n \times n$ matrix A can be found by using the **inverse formula:**

$$A^{-1} = \frac{\text{adj}(A)}{|A|}$$

where $\text{adj}(A)$ is the *adjoint* of matrix A, obtained by replacing each element in the transpose matrix A^T with the value of its corresponding cofactor. If $|A| = 0$, then matrix A does not have an inverse.

A **system of n linear equations in n variables** $(x_1, x_2, \ldots, x_n)$, with each equation written in the form $a_1x_1 + a_2x_2 + \ldots + a_nx_n = k$, may be represented by the matrix equation

$$AX = K$$

where A is an $n \times n$ matrix containing the coefficients of the variables, X is an $n \times 1$ matrix containing the variables, and K is an $n \times 1$ matrix containing the constants on the right-hand sides of the equations. The **solution** of the $n \times n$ linear system of equations is given by $X = A^{-1}K$, provided that matrix A has an inverse.

Example 1.27

In each case, given the determinant $|A| = \begin{vmatrix} 1 & 3 & -2 \\ 3 & 0 & 1 \\ -1 & 2 & -4 \end{vmatrix}$,

determine the value of (1) the minor and (2) the cofactor of the given element.

A. a_{31} **B.** a_{23}

Solution:

A. (1) To find M_{31}, the minor of element a_{31}, delete the third row and first column of $|A|$ and evaluate the remaining determinant:

$$M_{31} = \begin{vmatrix} 3 & -2 \\ 0 & 1 \end{vmatrix} = (3)(1) - (-2)(0) = 3$$

(2) The cofactor C_{31} of a_{31} is given by

$$C_{31} = (-1)^{3+1}M_{31} = M_{31} = 3$$

51

B. (1) To find M_{23}, the minor of element a_{23}, delete the second row and third column of $|A|$ and evaluate the remaining determinant:

$$M_{23} = \begin{vmatrix} 1 & 3 \\ -1 & 2 \end{vmatrix} = (1)(2) - (3)(-1) = 5$$

(2) The cofactor C_{23} of a_{23} is given by

$$C_{23} = (-1)^{2+3}M_{23} = -M_{23} = -5$$

Example 1.28

Evaluate each determinant.

A. $|A| = \begin{vmatrix} 1 & 3 & -2 \\ 3 & 0 & 1 \\ -1 & 2 & -4 \end{vmatrix}$ **B.** $|B| = \begin{vmatrix} 1 & -2 & 3 & -1 \\ 0 & 2 & 0 & 3 \\ 0 & 4 & 0 & -2 \\ 2 & 0 & 0 & 4 \end{vmatrix}$

Solution:

A. Since element $a_{22} = 0$, it is best to expand about the second row or second column. Choose the second row:

> Try expanding about the *second column* to obtain the same results.

$$|A| = 3C_{21} + 0C_{22} + 1C_{23} = 3C_{21} + 1C_{23}$$

Now,

$$C_{21} = (-1)^{2+1}\begin{vmatrix} 3 & -2 \\ 2 & -4 \end{vmatrix} = 8$$

and

$$C_{23} = (-1)^{2+3}\begin{vmatrix} 1 & 3 \\ -1 & 2 \end{vmatrix} = -5$$

Hence,

$$|A| = 3C_{21} + 1C_{23} = 3(8) + 1(-5) = 19$$

B. Since the third column of this determinant contains three zeros, it is best to expand about this column:

$$|B| = 3C_{13} + 0C_{23} + 0C_{33} + 0C_{43} = 3C_{13}$$

Now,

$$C_{13} = (-1)^{1+3}\begin{vmatrix} 0 & 2 & 3 \\ 0 & 4 & -2 \\ 2 & 0 & 4 \end{vmatrix} = \begin{vmatrix} 0 & 2 & 3 \\ 0 & 4 & -2 \\ 2 & 0 & 4 \end{vmatrix}$$

Since the first column of this third-order determinant contains two zeros, it is best to expand about this column:

$$C_{13} = 2(-1)^{3+1}\begin{vmatrix} 2 & 3 \\ 4 & -2 \end{vmatrix} = -32$$

Hence,

$$|B| = 3C_{13} = 3(-32) = -96$$

Example 1.29

Find the inverse of each matrix.

A. $A = \begin{bmatrix} 3 & 4 \\ 2 & 3 \end{bmatrix}$ **B.** $\begin{bmatrix} 1 & 2 & -3 \\ 2 & 3 & -4 \\ -1 & 0 & 3 \end{bmatrix}$

Solution:

A. The inverse of matrix A is given by $A^{-1} = \frac{\text{adj}(A)}{|A|}$. Begin by finding the determinant:

$$|A| = \begin{vmatrix} 3 & 4 \\ 2 & 3 \end{vmatrix} = (3)(3) - (4)(2) = 1$$

To find $\text{adj}(A)$, form A^T and then replace each element in A^T with its corresponding cofactor:

$$A^T = \begin{bmatrix} 3 & 2 \\ 4 & 3 \end{bmatrix} \quad \text{implies} \quad \text{adj}(A) = \begin{bmatrix} 3 & -4 \\ -2 & 3 \end{bmatrix}$$

> The notation A^{-1} represents the *inverse* of matrix A and does not mean $\frac{1}{A}$. In fact, matrix division is undefined.

Hence,

$$A^{-1} = \frac{\text{adj}(A)}{|A|} = \frac{\begin{bmatrix} 3 & -4 \\ -2 & 3 \end{bmatrix}}{1} = \begin{bmatrix} 3 & -4 \\ -2 & 3 \end{bmatrix}$$

To verify this inverse, show that $AA^{-1} = A^{-1}A = \begin{bmatrix} 1 & 0 \\ 0 & 1 \end{bmatrix}$, the identity matrix of order 2×2.

B. The inverse of matrix B is defined by $B^{-1} = \frac{\text{adj}(B)}{|B|}$. For the determinant, expand about the third row:

$$|B| = -1(-1)^{3+1}\begin{vmatrix} 2 & -3 \\ 3 & -4 \end{vmatrix} + 3(-1)^{3+3}\begin{vmatrix} 1 & 2 \\ 2 & 3 \end{vmatrix} = -4$$

To find $\text{adj}(B)$, form B^T and then replace each element in B^T with its corresponding cofactor:

$$B^T = \begin{bmatrix} 1 & 2 & -1 \\ 2 & 3 & 0 \\ -3 & -4 & 3 \end{bmatrix} \quad \text{implies} \quad \text{adj}(B) = \begin{bmatrix} 9 & -6 & 1 \\ -2 & 0 & -2 \\ 3 & -2 & -1 \end{bmatrix}$$

52

Hence,

For *scalar multiplication*, multiply *all* elements of the matrix by the scalar.

$$B^{-1} = \frac{\text{adj}(B)}{|B|} = -\frac{1}{4}\begin{bmatrix} 9 & -6 & 1 \\ -2 & 0 & -2 \\ 3 & -2 & -1 \end{bmatrix} = \begin{bmatrix} -\frac{9}{4} & \frac{3}{2} & -\frac{1}{4} \\ \frac{1}{2} & 0 & \frac{1}{2} \\ -\frac{3}{4} & \frac{1}{2} & \frac{1}{4} \end{bmatrix}$$

To check, show that $BB^{-1} = B^{-1}B = \begin{bmatrix} 1 & 0 & 0 \\ 0 & 1 & 0 \\ 0 & 0 & 1 \end{bmatrix}$, the identity matrix of order 3×3.

Example 1.30

Solve each system of linear equations by using the inverse method.

A. $3x + 4y = 6$ **B.** $x + 2y - 3z = -1$
$2x + 3y = 5$ $2x + 3y - 4z = 2$
                  $-x + 3z = 5$

Solution:

A. Let

$$A = \begin{bmatrix} 3 & 4 \\ 2 & 3 \end{bmatrix}, \quad X = \begin{bmatrix} x \\ y \end{bmatrix}, \quad \text{and} \quad K = \begin{bmatrix} 6 \\ 5 \end{bmatrix}$$

Then, by the rules for matrix multiplication and equality of matrices, the matrix equation $AX = K$ is equivalent to the given system of linear equations:

$$\begin{bmatrix} 3 & 4 \\ 2 & 3 \end{bmatrix}\begin{bmatrix} x \\ y \end{bmatrix} = \begin{bmatrix} 6 \\ 5 \end{bmatrix}$$

The solution of this matrix equation is $X = A^{-1}K$, provided that A^{-1} exists. Use the results of Example 1.29**A** to find X:

$$\begin{bmatrix} x \\ y \end{bmatrix} = \begin{bmatrix} 3 & 4 \\ 2 & 3 \end{bmatrix}^{-1}\begin{bmatrix} 6 \\ 5 \end{bmatrix} = \begin{bmatrix} 3 & -4 \\ -2 & 3 \end{bmatrix}\begin{bmatrix} 6 \\ 5 \end{bmatrix} = \begin{bmatrix} -2 \\ 3 \end{bmatrix}$$

Hence, the solution of this 2×2 system of linear equations is $x = -2$ and $y = 3$.

B. Let

$$B = \begin{bmatrix} 1 & 2 & -3 \\ 2 & 3 & -4 \\ -1 & 0 & 3 \end{bmatrix}, \quad X = \begin{bmatrix} x \\ y \\ z \end{bmatrix}, \quad \text{and} \quad K = \begin{bmatrix} -1 \\ 2 \\ 5 \end{bmatrix}$$

Then, by the rules for matrix multiplication and equality of matrices, the matrix equation $BX = K$ is equivalent to the given system of linear equations:

$$\begin{bmatrix} 1 & 2 & -3 \\ 2 & 3 & -4 \\ -1 & 0 & 3 \end{bmatrix}\begin{bmatrix} x \\ y \\ z \end{bmatrix} = \begin{bmatrix} -1 \\ 2 \\ 5 \end{bmatrix}$$

The solution of this matrix equation is $X = B^{-1}K$, provided that B^{-1} exists. Use the results of Example 1.29**B** to find X:

$$\begin{bmatrix} x \\ y \\ z \end{bmatrix} = \begin{bmatrix} 1 & 2 & -3 \\ 2 & 3 & -4 \\ -1 & 0 & 3 \end{bmatrix}^{-1}\begin{bmatrix} -1 \\ 2 \\ 5 \end{bmatrix} = \begin{bmatrix} -\frac{9}{4} & \frac{3}{2} & -\frac{1}{4} \\ \frac{1}{2} & 0 & \frac{1}{2} \\ -\frac{3}{4} & \frac{1}{2} & \frac{1}{4} \end{bmatrix}\begin{bmatrix} -1 \\ 2 \\ 5 \end{bmatrix} = \begin{bmatrix} 4 \\ 2 \\ 3 \end{bmatrix}$$

Hence, the solution of this 3×3 system of linear equations is $x = 4$, $y = 2$, and $z = 3$.

1.11
Vectors

A **vector** is a directed line segment in space that represents a force, a velocity, or a displacement. The *basic vectors,* denoted by the boldface letters **i, j,** and **k,** are directed line segments in the rectangular coordinate system from the origin $(0,0,0)$ to points $(1,0,0)$, $(0,1,0)$, and $(0,0,1)$, respectively. The *position vector* of a point in space, denoted by $\overrightarrow{OP} = x\mathbf{i} + y\mathbf{j} + z\mathbf{k}$, is a directed line segment from the origin $O(0,0,0)$ to a point $P(x,y,z)$ in space, as shown in Figure 1.15.

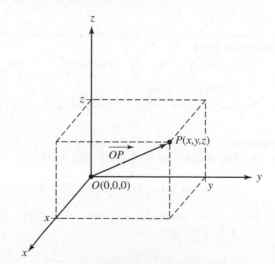

Figure 1.15

The vector $\mathbf{A}$ from $P_1 = (x_1, y_1, z_1)$ to $P_2 = (x_2, y_2, z_2)$ is given by

$$\overrightarrow{P_1P_2} = \mathbf{A} = a_x\mathbf{i} + a_y\mathbf{j} + a_z\mathbf{k}$$

where $a_x = x_2 - x_1$, $a_y = y_2 - y_1$, and $a_z = z_2 - z_1$.

The **magnitude** (or length) of $\mathbf{A} = a_x\mathbf{i} + a_y\mathbf{j} + a_z\mathbf{k}$ is denoted by $|\mathbf{A}|$, and defined by

$$|\mathbf{A}| = \sqrt{\left(a_x\right)^2 + \left(a_y\right)^2 + \left(a_z\right)^2}$$

A vector whose magnitude is 1 is called a **unit vector**. If $\mathbf{A} = a_x\mathbf{i} + a_y\mathbf{j} + a_z\mathbf{k}$, then $\mathbf{A}/|\mathbf{A}|$ is a unit vector in the direction of $\mathbf{A}$.

For the vectors $\mathbf{A} = a_x\mathbf{i} + a_y\mathbf{j} + a_z\mathbf{k}$ and $\mathbf{B} = b_x\mathbf{i} + b_y\mathbf{j} + b_z\mathbf{k}$, and scalar (real number) c, the following operations can be performed:

1. **Addition and subtraction:** $\mathbf{A} \pm \mathbf{B} = (a_x \pm b_x)\mathbf{i} + (a_y \pm b_y)\mathbf{j} + (a_z \pm b_z)\mathbf{k}$
2. **Scalar multiplication:** $c\mathbf{A} = ca_x\mathbf{i} + ca_y\mathbf{j} + ca_z\mathbf{k}$
3. **Dot product:** $\mathbf{A} \cdot \mathbf{B} = a_xb_x + a_yb_y + a_zb_z$
   $= |\mathbf{A}|\,|\mathbf{B}| \cos\theta$,
   where θ is the angle between $\mathbf{A}$ and $\mathbf{B}$. Nonzero vectors $\mathbf{A}$ and $\mathbf{B}$ are perpendicular if and only if $\mathbf{A} \cdot \mathbf{B} = 0$.
4. **Cross product:** $\mathbf{A} \times \mathbf{B} = \begin{vmatrix} \mathbf{i} & \mathbf{j} & \mathbf{k} \\ a_x & a_y & a_z \\ b_x & b_y & b_z \end{vmatrix} = |\mathbf{A}|\,|\mathbf{B}|\,\mathbf{n}\sin\theta$,

where $\mathbf{n}$ is a unit vector perpendicular to the plane formed by $\mathbf{A}$ and $\mathbf{B}$ by the *right-hand rule;* that is, $\mathbf{n}$ points the way your right thumb points when your fingers curl through the angle θ from $\mathbf{A}$ to $\mathbf{B}$. Nonzero vectors $\mathbf{A}$ and $\mathbf{B}$ are parallel if and only if $\mathbf{A} \times \mathbf{B} = 0$. The magnitude of $\mathbf{A} \times \mathbf{B}$ is the area of a parallelogram determined by $\mathbf{A}$ and $\mathbf{B}$.

Example 1.31

Given that $\mathbf{A} = 3\mathbf{i} - 6\mathbf{j} + 2\mathbf{k}$, find:

A. the magnitude of $\mathbf{A}$
B. a unit vector in the direction of $\mathbf{A}$

Solution:

A. The magnitude of $\mathbf{A} = 3\mathbf{i} - 6\mathbf{j} + 2\mathbf{k}$ is

$$|\mathbf{A}| = \sqrt{(3)^2 + (-6)^2 + (2)^2} = \sqrt{49} = 7$$

B. A unit vector in the direction of $\mathbf{A} = 3\mathbf{i} - 6\mathbf{j} + 2\mathbf{k}$ is

$$\frac{\mathbf{A}}{|\mathbf{A}|} = \frac{3\mathbf{i} - 6\mathbf{j} + 2\mathbf{k}}{7} = \frac{3}{7}\mathbf{i} - \frac{6}{7}\mathbf{j} + \frac{2}{7}\mathbf{k}$$

Example 1.32

Given that $\mathbf{A} = 10\mathbf{i} + 11\mathbf{j} - 2\mathbf{k}$ and $\mathbf{B} = 5\mathbf{i} + 12\mathbf{j}$, find:

A. $\mathbf{A} \cdot \mathbf{B}$
B. the angle θ (in degrees) between $\mathbf{A}$ and $\mathbf{B}$

Solution:

A. To find the dot product of two given vectors, multiply their corresponding **i**-, **j**-, and **k**-components and add the results:

$$\mathbf{A} \cdot \mathbf{B} = (10)(5) + (11)(12) + (-2)(0) = 182$$

B. The magnitudes of $\mathbf{A}$ and $\mathbf{B}$ are

$$|\mathbf{A}| = \sqrt{(10)^2 + (11)^2 + (-2)^2} = 15$$

and

$$|\mathbf{B}| = \sqrt{(5)^2 + (12)^2} = 13$$

Now, use the dot product formula $\mathbf{A} \cdot \mathbf{B} = |\mathbf{A}|\,|\mathbf{B}| \cos\theta$ and the results from part A to obtain the angle θ between $\mathbf{A}$ and $\mathbf{B}$:

$$\theta = \cos^{-1}\left(\frac{\mathbf{A} \cdot \mathbf{B}}{|\mathbf{A}|\,|\mathbf{B}|}\right) = \cos^{-1}\left[\frac{182}{(25)(13)}\right] = 21.0°$$

Example 1.33

Given that $\mathbf{A} = -8\mathbf{i} - 2\mathbf{j} - 4\mathbf{k}$ and $\mathbf{B} = 2\mathbf{i} + 2\mathbf{j} + \mathbf{k}$, find:

A. $\mathbf{A} \times \mathbf{B}$ B. $\mathbf{B} \times \mathbf{A}$

Solution:

A. Use the determinant formula for cross products; then

$$\mathbf{A} \times \mathbf{B} = \begin{vmatrix} \mathbf{i} & \mathbf{j} & \mathbf{k} \\ -8 & -2 & -4 \\ 2 & 2 & 1 \end{vmatrix}$$

$$= \mathbf{i}\begin{vmatrix} -2 & -4 \\ 2 & 1 \end{vmatrix} - \mathbf{j}\begin{vmatrix} -8 & -4 \\ 2 & 1 \end{vmatrix} + \mathbf{k}\begin{vmatrix} -8 & -2 \\ 2 & 2 \end{vmatrix}$$

$$= 6\mathbf{i} - 12\mathbf{k}$$

B. Use the determinant formula for cross products; then

$$\mathbf{B} \times \mathbf{A} = \begin{vmatrix} \mathbf{i} & \mathbf{j} & \mathbf{k} \\ 2 & 2 & 1 \\ -8 & -2 & -4 \end{vmatrix}$$

$$= \mathbf{i} \begin{vmatrix} 2 & 1 \\ -2 & -4 \end{vmatrix} - \mathbf{j} \begin{vmatrix} 2 & 1 \\ -8 & -4 \end{vmatrix} + \mathbf{k} \begin{vmatrix} 2 & 2 \\ -8 & -2 \end{vmatrix}$$

$$= -6\mathbf{i} + 12\mathbf{k}$$

Note that the cross-product operation is not commutative; that is, $\mathbf{A} \times \mathbf{B} \neq \mathbf{B} \times \mathbf{A}$. In general, for cross products, $\mathbf{A} \times \mathbf{B} = -(\mathbf{B} \times \mathbf{A})$.

Example 1.34

Given points $P(1,-1,0)$, $Q(2,1,-1)$, and $R(-1,1,2)$, find:

A. a vector perpendicular to the plane formed by P, Q, and R

B. the area of the triangle with vertices P, Q, and R

Solution:

A. By the definition of cross product, $\vec{PQ} \times \vec{PR}$ is a vector perpendicular to the plane formed by P, Q, and R. Begin by writing $\vec{PQ}$ and $\vec{PR}$ in component form:

$$\vec{PQ} = (2-1)\mathbf{i} + [1-(-1)]\mathbf{j} + (-1-0)\mathbf{k} = \mathbf{i} + 2\mathbf{j} - \mathbf{k}$$

$$\vec{PR} = (-1-1)\mathbf{i} + [1-(-1)]\mathbf{j} + (2-0)\mathbf{k} = -2\mathbf{i} + 2\mathbf{j} + 2\mathbf{k}$$

Hence,

$$\vec{PQ} \times \vec{PR} = \begin{vmatrix} \mathbf{i} & \mathbf{j} & \mathbf{k} \\ 1 & 2 & -1 \\ -2 & 2 & 2 \end{vmatrix}$$

$$= \mathbf{i} \begin{vmatrix} 2 & -1 \\ 2 & 2 \end{vmatrix} - \mathbf{j} \begin{vmatrix} 1 & -1 \\ -2 & 2 \end{vmatrix} + \mathbf{k} \begin{vmatrix} 1 & 2 \\ -2 & 2 \end{vmatrix}$$

$$= 6\mathbf{i} + 6\mathbf{k}$$

B. The magnitude of $\vec{PQ} \times \vec{PR}$ is the area of the parallelogram determined by P, Q, and R. Hence, half of this magnitude represents the area A of the triangle formed by P, Q, and R:

$$A = \frac{1}{2}\left|\vec{PQ} \times \vec{PR}\right| = \frac{1}{2}\sqrt{6^2 + 6^2} = 3\sqrt{2} \text{ square units}$$

1.12
Sequences and Series

A **sequence** is an ordered list of numbers: a_1, a_2, a_3, ..., a_n. Each number in the list is called an *element* of the sequence. If a sequence stops after a finite number of elements, it is called a *finite sequence*. If the sequence continues forever, it is called an *infinite sequence*.

A **series** is the indicated sum of the elements in a sequence: $a_1 + a_2 + a_3 + ... + a_n$. Each number in a series is called a *term* of the series. The terms of a series are identical to the elements in its corresponding sequence.

An **arithmetic sequence** is a sequence in which every element after the first is obtained by adding a fixed number d to the preceding element. The constant d is called the *common difference* of the sequence. The nth element a_n of an arithmetic sequence is given by

$$a_n = a_1 + (n-1)d$$

where a_1 is the first element and d is the common difference of the sequence. The sum S_n of an arithmetic series with n terms is given by

$$S_n = n\left(\frac{a_1 + a_n}{2}\right) = \frac{n}{2}[2a_1 + (n-1)d]$$

where a_1 is the first term, a_n is the last term, and d is the common difference of the series.

A **geometric sequence** is a sequence in which every element after the first is obtained by multiplying the preceding element by a fixed nonzero number r. The constant r is called the *common ratio* of the sequence. The nth element a_n of a geometric sequence is given by

$$a_n = a_1 r^{n-1}$$

where a_1 is the first element and r is the common ratio of the sequence. The sum S_n of a geometric series with n terms is given by

$$S_n = \frac{a_1(1 - r^n)}{1 - r} = \frac{a_1 - ra_n}{1 - r}$$

where a_1 is the first term, a_n is the last term, and r is the common ratio ($r \neq 1$).

If $|r| < 1$ in an infinite geometric series, then the series *converges* and has the finite sum

$$S = \frac{a_1}{1 - r}$$

where a_1 is the first term and r is the common ratio. If $|r| \geq 1$ in an infinite geometric series, then the series *diverges* and has no finite sum.

The **Taylor series** generated by a function f at $x = a$ is given by

$$f(x) = f(a) + f'(a)(x - a) + \frac{f''(a)}{2!}(x - a)^2$$

$$+ \ldots + \frac{f^n(a)}{n!}(x - a)^n + \ldots$$

where $f^k(a)$ represents the kth derivative of f at $x = a$. (For a discussion of derivatives, see Section 1.13.) When $a = 0$, the Taylor series is called a **Maclaurin series.**

Example 1.35

In each case, find the 15th element of the given sequence.

A. $23, 17, 11, 5, \ldots$ **B.** $2, -1, \frac{1}{2}, -\frac{1}{4}, \ldots$

Solution:

A. This sequence is an arithmetic sequence with first element $a_1 = 23$ and common difference $d = -6$. Hence, the nth element of this sequence is

$$a_n = a_1 + (n - 1)d = 23 + (n - 1)(-6) = 29 - 6n$$

To find the 15th element a_{15}, replace n with 15:

$$a_{15} = 29 - 6(15) = -61$$

B. This sequence is a geometric sequence with first element $a_1 = 2$ and common ratio $r = -1/2$. Hence, the nth element of this sequence is

$$a_n = a_1 r^{n-1} = 2\left(-\frac{1}{2}\right)^{n-1}$$

To find the 15th element a_{15}, replace n with 15:

$$a_{15} = 2\left(-\frac{1}{2}\right)^{15-1} = 2\left(-\frac{1}{2}\right)^{14} = \frac{1}{8,192}$$

Example 1.36

In each case, find the sum of the given series.

A. $\sum_{i=1}^{40} (4i - 3)$ **B.** $\sum_{i=1}^{10} 5(-2)^{i-1}$

Solution:

A. Since the summation variable i starts at 1 and ends at 40, the number of terms in this series is $n = 40$. Begin by writing this series in expanded form:

$$\sum_{i=1}^{40} (4i - 3) = 1 + 5 + 9 + \ldots + 157$$

This is an arithmetic series with common difference $d = 4$, first term $a_1 = 1$, and last term $a_{40} = 157$. Thus the sum of this series is

$$S_{40} = 40\left(\frac{a_1 + a_{40}}{2}\right) = 40\left(\frac{1 + 157}{2}\right) = 3,160$$

B. Since the summation variable i starts at 1 and ends at 10, the number of terms in this series is $n = 10$. Begin by writing this series in expanded form:

$$\sum_{i=1}^{10} 5(-2)^{i-1} = 5 - 10 + 20 - 40 + \ldots - 2,560$$

This is a geometric series with common ratio $r = -2$ and first term $a_1 = 5$. Thus the sum of this series is

$$S_{10} = \frac{a_1(1 - r^{10})}{1 - r} = \frac{5[1 - (-2)^{10}]}{1 - (-2)} = -1,705$$

Example 1.37

In each case, determine whether the infinite geometric series converges or diverges. If the series converges, find its sum.

A. $\sum_{n=0}^{\infty} \left(\frac{3}{2}\right)^n$ **B.** $\sum_{i=1}^{\infty} (-1)^{i+1}\left(\frac{1}{2}\right)^i$

Solution:

A. Write the series in expanded form:

$$\sum_{n=0}^{\infty} \left(\frac{3}{2}\right)^n = 1 + \frac{3}{2} + \frac{9}{4} + \frac{27}{8} + \ldots$$

This is an infinite geometric series with first term $a_1 = 1$ and common ratio $r = 3/2$. Since $|r| \geq 1$, the series diverges and has no finite sum.

B. Write the series in expanded form:

$$\sum_{i=1}^{\infty} (-1)^{i+1}\left(\frac{1}{2}\right)^i = \frac{1}{2} - \frac{1}{4} + \frac{1}{8} - \frac{1}{16} + \ldots$$

This is an infinite geometric series with $a_1 = \frac{1}{2}$ and common ratio $r = -\frac{1}{2}$. Since $|r| < 1$, the infinite geometric series converges and the sum of the series is

$$S = \frac{a_1}{1-r} = \frac{\frac{1}{2}}{1-\left(-\frac{1}{2}\right)} = \frac{1}{3}$$

$$f(0) + f'(0)x + \frac{f''(0)}{2!}x^2 + \frac{f'''(0)}{3!}x^3 + \dots + \frac{f^{(n)}(0)}{n!}x^n + \dots$$

$$= 1 - \frac{x^2}{2!} + \frac{x^4}{4!} - \frac{x^6}{6!} + \dots + (-1)^n \frac{x^{2n}}{(2n)!} + \dots$$

$$= \sum_{k=0}^{\infty} \frac{(-1)^k x^{2k}}{(2k)!}$$

Example 1.38

In each case, find the Taylor (Maclaurin) series generated by the given function f at $x = 0$.

A. $f(x) = e^x$ **B.** $f(x) = \cos x$

Solution:

A. Begin by finding the derivatives of $f(x) = e^x$:

> From the *derivative formulas* listed in Section 1.13, $D_x(e^x) = e^x$.

$$f'(x) = e^x, \quad f''(x) = e^x, \quad f'''(x) = e^x, \quad \dots,$$
$$f^{(n)}(x) = e^x$$

For $x = 0$, $e^x = 1$. Hence, the Taylor (Maclaurin) series generated by $f(x) = e^x$ at $x = 0$ is

$$f(0) + f'(0)x + \frac{f''(0)}{2!}x^2 + \frac{f'''(0)}{3!}x^3 + \dots + \frac{f^{(n)}(0)}{n!}x^n + \dots$$

$$= 1 + x + \frac{x^2}{2!} + \frac{x^3}{3!} + \dots + \frac{x^n}{n!} + \dots$$

$$= \sum_{k=0}^{\infty} \frac{x^k}{k!}$$

B. The function and its derivatives are as follows:

> From the *derivative formulas* listed in Section 1.13, $D_x(\sin x) = \cos x$ and $D_x(\cos x) = -\sin x$.

$$f(x) = \cos x, \qquad\qquad f'(x) = -\sin x,$$

$$f''(x) = -\cos x, \qquad\qquad f'''(x) = \sin x,$$

$$\vdots \qquad\qquad\qquad\qquad \vdots$$

$$f^{(2n)}(x) = (-1)^n \cos x, \qquad f^{(2n+1)}(x) = (-1)^{n+1}\sin x$$

for $x = 0$, $\cos x = 1$, and $\sin x = 0$. Therefore, the Taylor (Maclaurin) series generated by $f(x) = \cos x$ and $x = 0$ is

1.13 Derivatives

The derivative of $y = f(x)$ with respect to x is defined as the limit of the ratio $\frac{\Delta y}{\Delta x}$ as Δx approaches zero ($\Delta x \to 0$). A physical interpretation of the derivative is the *instantaneous velocity* of a moving particle. A geometric interpretation of the derivative is the *slope of the tangent* to the curve.

The most common notations for the derivative of a function $y = f(x)$ with respect to x are $D_x y$, y', $f'(x)$, and $\frac{dy}{dx}$. The derivative of its derivative is called the second derivative of $y = f(x)$ and is denoted by $D_x^2 y$, y'', f'', or $\frac{d^2 y}{dx^2}$. Similarly, the derivative of the second derivative is called the third derivative, and so on for derivatives of any order.

In the following **rules for differentiation,** c is a constant, x is the independent variable, u and v are functions of x, and n is a rational exponent:

1. Derivative of a constant: $D_x(c) = 0$
2. Derivative of the independent variable: $D_x(x) = 1$
3. Derivative of a power of the independent variable: $D_x(x^n) = nx^{n-1}$
4. Derivative of a constant times a function: $D_x(c\,u) = c\,D_x u$
5. Derivative of a sum or difference of two functions: $D_x(u \pm v) = D_x u \pm D_x v$
6. Derivative of a product of two functions: $D_x(u \cdot v) = u \cdot D_x v + v \cdot D_x u$
7. Derivative of a quotient of two functions:
   $$D_x\left(\frac{u}{v}\right) = \frac{v \cdot D_x u - u \cdot D_x u}{v^2}$$
8. Derivative of a power of a function: $D_x(u^n) = nu^{n-1} \cdot D_x u$

The following is a list of the most frequently used **derivative formulas.** For additional derivative formulas, refer to the *Fundamentals of Engineering Reference Handbook.*

1. $D_x(e^u) = e^u \cdot D_x u$
2. $D_x(\ln u) = \frac{1}{u} \cdot D_x u$
3. $D_x(\sin u) = \cos u \cdot D_x u$

4. $D_x(\cos u) = -\sin u \bullet D_x u$
5. $D_x(\tan u) = \sec^2 u \bullet D_x u$
6. $D_x(\cot u) = -\csc^2 u \bullet D_x u$
7. $D_x(\sec u) = \sec u \tan u \bullet D_x u$
8. $D_x(\csc u) = -\csc u \cot u \bullet D_x u$
9. $D_x(\sin^{-1} u) = \dfrac{1}{\sqrt{1-u^2}} \bullet D_x u$

10. $D_x(\tan^{-1} u) = \dfrac{1}{1-u^2} \bullet D_x u$

Derivatives may be used to evaluate limits in the indeterminate forms 0/0 and ∞/∞ by applying **L'Hôpital's rule:** If $f(x) \to 0$ and $g(x) \to 0$ as $x \to a$, or if $f(x) \to \infty$ and $g(x) \to \infty$ as $x \to a$, then

$$\lim_{x \to a} \frac{f(x)}{g(x)} = \lim_{x \to a} \frac{f'(x)}{g'(x)}$$

provided that the latter limit exists.

If y is defined as an implicit function of x by an equation of the form $F(x, y) = 0$, then $D_x y$ may be found by a method called **implicit differentiation:** Differentiate the equation $F(x, y) = 0$ term by term with respect to x, regarding y as a function of x, and then solve the resulting equation for $D_x y$.

The concept of the derivative of a function of one variable may be extended to the concept of the *partial derivatives* of a function of two or more variables:

1. The **partial derivative of $z = f(x, y)$ with respect to x** is the ordinary derivative of z when y is held constant, and is usually denoted by z_x, $f_x(x, y)$, $\frac{\partial z}{\partial x}$, or $\frac{\partial f}{\partial x}$.
2. The **partial derivative of $z = f(x, y)$ with respect to y** is the ordinary derivative of z when x is held constant, and is usually denoted by z_y, $f_y(x, y)$, $\frac{\partial z}{\partial y}$, or $\frac{\partial f}{\partial y}$.

Example 1.39
For a model rocket rising vertically, the height s (in feet) at time t (in seconds) is given by

$$s = 120t - 16t^2.$$

A. Find the instantaneous velocity v of the rocket at any time t.
B. Find the greatest height the rocket attains.

Solution:
A. The instantaneous velocity v at any time t is the derivative $D_t s$. To find $D_t s$, apply differentiation rules 4 and 5, then rules 2 and 3:

$v = D_t s = 120\, D_t(t) - 16 D_t(t^2)$

$\quad = 120(1) - 16(2t)$

$\quad = 120 - 32t$

B. The rocket attains its greatest height when its velocity $v = 0$. Substituting $v = 0$ into the velocity equation from part **A,** solve for t:

$0 = 120 - 32t \quad \text{implies} \quad t = 3.75 \text{ sec}$

To find the maximum height (s_{max}), substitute $t = 3.75$ into the height equation:

$s_{max} = 120(3.75) - 16(3.75)^2 = 225 \text{ ft}$

Example 1.40
In each case, find the slope of the tangent to the given curve at the point where $x = 1$.

A. $y = e^{-3x} + 2 \sin^3 x$
B. $y = 2x \tan^{-1} x - \ln(x^2 + 1)$

Solution:
A. The required slope is the value of the derivative $D_x y$ at $x = 1$. To find $D_x y$, begin by applying differentiation rules 4 and 5:

$D_x y = D_x(e^{-3x}) + 2\, D_x(\sin^3 x)$

For $D_x(e^{-3x})$, apply derivative formula 1, and for $D_x(\sin^3 x) = D_x(\sin x)^3$, apply differentiation rule 8 with derivative formula 3:

$D_x y = e^{-3x} D_x(-3x) + 2\,[3(\sin x)^2 \bullet D_x(\sin x)]$

$\quad = e^{-3x}(-3) + 2\,[3\,(\sin x)^2 \bullet \cos x \bullet D_x x]$

$\quad = -3e^{-3x} + 6 \sin^2 x \cos x$

Substitute $x = 1$ in this derivative formula, and use a calculator set in radian mode to find the slope of the tangent to the curve at this point:

$-3e^{-3(1)} + 6 \sin^2(1) \cos(1) \approx 2.146$

B. To find $D_x y$ for $y = 2x \tan^{-1} x - \ln(x^2 + 1)$, begin by applying differentiation rules 4 and 5:

$D_x y = 2\, D_x(x \tan^{-1} x) - D_x[\ln(x^2 + 1)]$

For $D_x(x \tan^{-1} x)$, apply differentiation rule 6 and derivative formula 10, and for $D_x[\ln(x^2 + 1)]$, apply derivative formula 2:

$$D_x y = 2\left[x \cdot D_x(\tan^{-1}x) + \tan^{-1}x \cdot D_x(x)\right] - \frac{1}{x^2 + 1}$$
$$\times D_x(x^2 + 1)$$

$$= 2\left(\frac{x}{1 + x^2} + \tan^{-1}x\right) - \frac{2x}{x^2 + 1}$$

$$= 2\tan^{-1}x$$

Substitute $x = 1$ in this derivative formula, and use a calculator set in radian mode to find the slope of the tangent to the curve at this point:

$$2\tan^{-1}(1) \approx 1.571$$

Example 1.41

Use L'Hôpital's rule to evaluate each limit.

A. $\lim\limits_{x \to 0}\left(\dfrac{1 - \cos x}{3x^2}\right)$ **B.** $\lim\limits_{x \to 0}(x \ln x)$

Solution:

A. Since $\cos x \to 1$ as $x \to 0$, the quotient $(1 - \cos x)/(3x^2)$ takes the indeterminate form $0/0$, which is a form for L'Hôpital's rule:

$$\lim_{x \to 0}\left(\frac{1 - \cos x}{3x^2}\right) = \lim_{x \to 0}\left[\frac{D_x(1 - \cos x)}{D_x(3x^2)}\right] = \lim_{x \to 0}\left(\frac{\sin x}{6x}\right)$$

Since $\sin x \to 0$ as $x \to 0$, the quotient $(\sin x)/(6x)$ still takes on the indeterminate form $0/0$. Hence, apply L'Hôpital's rule again:

$$\lim_{x \to 0}\left(\frac{\sin x}{6x}\right) = \lim_{x \to 0}\left[\frac{D_x(\sin x)}{D_x(6x)}\right] = \lim_{x \to 0}\left(\frac{\cos x}{6}\right) = \frac{1}{6}$$

Therefore, $\lim\limits_{x \to 0}\left(\dfrac{1 - \cos x}{3x^2}\right) = \dfrac{1}{6}$.

B. Since $\ln x \to -\infty$ as $x \to 0$, the product $x \ln x$ takes on the indeterminate form $0 \cdot \infty$, which is *not* a form for L'Hôpital's rule. However,

> *L'Hopital's rule* applies only to the forms $\dfrac{0}{0}$ and $\dfrac{\infty}{\infty}$.

rewrite this product as a quotient:

$$\lim_{x \to 0}(x \ln x) = \lim_{x \to 0}\left(\frac{\ln x}{x^{-1}}\right)$$

Now, as $x \to 0$, the quotient $(\ln x)/x^{-1}$ takes on the indeterminate form ∞/∞, which is a form for L'Hôpital's rule:

$$\lim_{x \to 0}\left(\frac{\ln x}{x^{-1}}\right) = \lim_{x \to 0}\left[\frac{D_x(\ln x)}{D_x(x^{-1})}\right] = \lim_{x \to 0}\left(\frac{1/x}{-1/x^2}\right) =$$
$$\lim_{x \to 0}(-x) = 0$$

Therefore, $\lim\limits_{x \to 0}(x \ln x) = 0$.

Example 1.42

For the circle defined by the equation $x^2 + y^2 - 4x + 2y - 20 = 0$, find:

A. $D_x y$ **B.** the equation of the line normal to the circle at $(6,2)$

Solution:

A. Use implicit differentiation, and differentiate each term with respect to x, treating y as a function of x:

$$D_x(x^2) + D_x(y^2) - D_x(4x) + D_x(2y) - D_x(20) = D_x(0)$$

$$2x + 2yD_x y - 4 + 2D_x y = 0$$

Now, solve the last equation for $D_x y$:

$$D_x y = \frac{4 - 2x}{2y + 2} = \frac{2 - x}{y + 1}$$

B. The slope of the tangent to the circle at $(6,2)$ is found by replacing x with 6 and y with 2 in the derivative formula from part **A.** Hence, the slope of the tangent is $-4/3$. The normal to the circle at $(6,2)$ is a line that is *perpendicular* to the tangent at the point of tangency. Hence, the slope of the normal is $3/4$. Use the *point-slope form* of a line to find the equation of the line normal to the circle at $(6,2)$:

> *Perpendicular lines* have negative reciprocal slopes.

$$y - y_1 = m(x - x_1)$$

$$y - 2 = \frac{3}{4}(x - 6)$$

$$y = \frac{3}{4}x - \frac{5}{2}$$

Example 1.43

In each case, find the partial derivatives of the given function with respect to each of the independent variables.

A. $z = x^2 + 5xy + 3y^2$ **B.** $z = e^{x^2y}$

Solution:

A. To find the partial derivative of $z = x^2 + 5xy + 3y^2$ with respect to x, use the basic differentiation rules with the assumption that y is constant:

$$\frac{\partial z}{\partial x} = \frac{\partial(x^2)}{\partial x} - \frac{\partial(5xy)}{\partial x} + \frac{\partial(3y^2)}{\partial x} = 2x + 5y$$

To find the partial derivative of $z = x^2 + 5xy + 3y^2$ with respect to y, use the basic differentiation rules with the assumption that x is constant:

$$\frac{\partial z}{\partial y} = \frac{\partial(x^2)}{\partial y} - \frac{\partial(5xy)}{\partial y} + \frac{\partial(3y^2)}{\partial y} = 5x + 6y$$

B. To find the partial derivative of $z = e^{x^2y}$ with respect to x, use derivative formula 1 with the assumption that y is constant:

$$\frac{\partial z}{\partial x} = e^{x^2y}\frac{\partial(x^2y)}{\partial x} = 2xye^{x^2y}$$

To find the partial derivative of $z = e^{x^2y}$ with respect to y, use derivative formula 1 with the assumption that x is constant:

$$\frac{\partial z}{\partial y} = e^{x^2y}\frac{\partial(x^2y)}{\partial y} = x^2e^{x^2y}$$

1.14
Maxima, Minima, and Points of Inflection

The **maximum value** of a function is the value that is greater than any other value of that function immediately preceding or following. In Figure 1.16, A is a maximum.

The **minimum value** of a function is the value that is less than any other value of that function immediately preceding or following. In Figure 1.16, B is a minimum. A **point of inflection** on a curve is a point at which the curve changes from concave upward to concave downward, or vice versa. In Figure 1.16, C is a point of inflection.

To find the maximum and minimum values of $y = f(x)$, first find the *critical values* of the function by solving the equation $f'(x) = 0$. Then, for each critical value a, determine the sign of $f''(a)$:

1. Maximum at $x = a$ if $f'(a) = 0$ and $f''(a) < 0$
2. Minimum at $x = a$ if $f'(a) = 0$ and $f''(a) > 0$

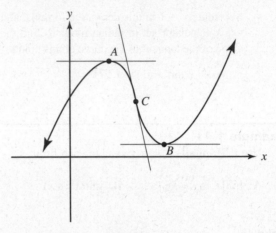

Figure 1.16

To find the points of inflection on the graph of $y = f(x)$, first find the roots of the equation $f''(x) = 0$. If $x = a$ is a root of this equation and if $f''(x)$ changes sign as x increases through a, then $x = a$ is a point of inflection.

Example 1.44

Examine the function $f(x) = \frac{1}{3}x^3 - 2x^2 + 3x + 1$ for

A. maxima and minima   **B.** point of inflection

Solution:

A. Using the rules for differentiation, find $f'(x)$ and $f''(x)$:

$$f'(x) = x^2 - 4x + 3 \quad \text{and} \quad f''(x) = 2x - 4$$

The critical values are found by solving the quadratic equation $f'(x) = 0$:

$$f'(x) = x^2 - 4x + 3 = 0$$

which implies $x = 3$ or $x = 1$

Now, determine the sign of $f''(3)$ and $f''(1)$:

$$f''(3) = 2(3) - 4 = 2 > 0$$

and $f''(1) = 2(1) - 4 = -2 < 0$

Since $f''(3) > 0$, this function has a *minimum value* at $x = 3$ and the minimum value is

$$f(3) = \frac{1}{3}(3)^3 - 2(3)^2 + 3(3) + 1 = 1$$

Since $f''(1) < 0$, this function has a *maximum value* at $x = 1$ and the maximum value is

$$f(1) = \frac{1}{3}(1)^3 - 2(1)^2 + 3(1) + 1 = \frac{7}{3}$$

B. Any point of inflection is found by solving the equation $f''(x) = 0$:

$$2x - 4 = 0 \quad \text{which implies} \quad x = 2$$

As the following table indicates, $f''(x)$ changes sign as x increases through 2.

x	1	2	3
$f''(x)$	-2	0	2

Hence, $x = 2$ is a point of inflection. The graph of this function is shown in Figure 1.17.

This graph is *concave downward* for $x < 2$ and *concave upward* for $x > 2$.

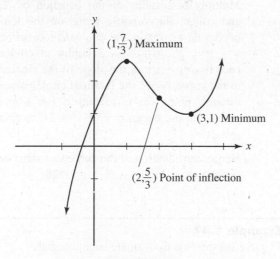

Figure 1.17

Example 1.45

A rectangular beam with dimensions x and y is cut from a circular log of diameter 12 cm, as shown in Figure 1.18. The strength S of the beam is proportional to its width x and to the square of its depth y: $S = kxy^2$.

A. Find the width x of the strongest beam that can be cut from this log.

B. Find the depth y of the strongest beam that can be cut from this log.

Solution:

A. Refer to Figure 1.18, and apply the Pythagorean theorem:

$$y^2 = 12^2 - x^2$$

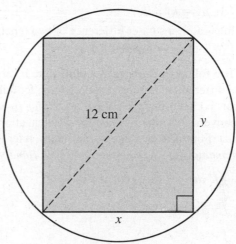

Figure 1.18

Replace y^2 with $12^2 - x^2$ in the strength formula $S = kxy^2$:

$$S = kx(12^2 - x^2) = 144kx - kx^3$$

To find the width x of the beam with the maximum strength (the strongest beam) that can be cut from this log, solve the equation $D_x S = 0$ for x:

$$D_x S = 144k - 3kx^2 = 0 \quad \text{implies} \quad x = 4\sqrt{3} \approx 6.93 \text{ cm}$$

B. To determine the depth y of the strongest beam that can be cut from this log, replace x with $4\sqrt{3}$ in the equation $y^2 = 12^2 - x^2$:

$$y^2 = 12^2 - (4\sqrt{3})^2 = 96 \quad \text{implies} \quad y = 4\sqrt{6} \approx 9.80 \text{ cm}$$

1.15
Indefinite Integrals

The inverse operation of differentiation is called **integration**. The **indefinite integral** of $f(x)$ is denoted by $\int f(x)dx$ and is defined by

$$\int f(x)\, dx = F(x) + C$$

where C is called the *integration constant* and $D_x[F(x)] = f(x)$.

In the following **integration rules,** k is a constant, x is the independent variable, u and v are functions of x, and n is a rational exponent.

1. Integral of a constant: $\int k\,dx = kx + C$
2. Integral of a power of the independent variable:
$\int x^n dx = \frac{x^{n+1}}{n+1} + C$ if $n \neq -1$
3. Integral of a constant times a function:
$\int ku\,dx = k\int u\,dx$
4. Integral of a sum or difference of two functions:
$\int (u \pm v)\,dx = \int u\,dx \pm \int v\,dx$

The following **integral formulas** are inversions of the differentiation rules and derivative formulas in Section 1.13. Formula 2, called *integration by parts,* is the inversion of the formula for differentiation of a product. For additional integral formulas, refer to the *Fundamentals of Engineering Reference Handbook.*

1. $\int u^n\,du = \frac{u^{n+1}}{n+1} + C$ if $n \neq -1$
2. $\int u\,dv = uv - \int v\,du$
3. $\int e^u\,du = e^u + C$
4. $\int \frac{1}{u}\,du = \ln|u| + C$
5. $\int \cos u\,du = \sin u + C$
6. $\int \sin u\,du = -\cos u + C$
7. $\int \sec^2 u\,du = \tan u + C$
8. $\int \csc^2 u\,du = -\cot u + C$
9. $\int \sec u \tan u\,du = \sec u + C$
10. $\int \csc u \cot u\,du = -\csc u + C$
11. $\int \frac{1}{\sqrt{1-u^2}}\,du = \sin^{-1} u + C$
12. $\int \frac{1}{1-u^2}\,du = \tan^{-1} u + C$

Example 1.46

Find the equation of the curve that passes through point (3,4) and whose slope of the tangent to the curve is given by

A. $\frac{dy}{dx} = 2x - 4$ **B.** $\frac{dy}{dx} = -\frac{x}{y}$

Solution:

A. Begin by writing the slope of the tangent to the curve in differential form by treating $\frac{dy}{dx}$ as

a fraction and multiplying both sides by dx:

$$dy = (2x - 4)dx$$

Now, using the rules for integration, integrate each side of this equation and combine the integration constants on the right-hand side:

$$\int dy = \int (2x - 4)\,dx$$

which implies $y = x^2 - 4x + C$

The equation $y = x^2 - 4x + C$ represents a family of parabolas each having $2x - 4$ as the slope of the tangent to the curve. Since the required parabola passes through point (3,4), substitute $x = 3$ and $y = 4$ into the equation $y = x^2 - 4x + C$ to find that $C = 7$.

Hence, the equation of the curve that satisfies the given conditions is $y = x^2 - 4x + 7$.

B. Separating variables, write the slope of the tangent to the curve in differential form:

$$y\,dy = -x\,dx$$

Using the rules for integration, integrate each side of this equation separately:

$\int y\,dy = \int -x\,dx$ which implies $\frac{y^2}{2} = -\frac{x^2}{2} + C$

Multiply both sides of the equation by 2, and collect the variable terms on the left-hand side to obtain the equivalent equation $x^2 + y^2 = 2C$, which is a family of circles each having $-x/y$ as the slope of the tangent to the curve. Since the required circle passes through point (3,4), substitute $x = 3$ and $y = 4$ into the equation $x^2 + y^2 = 2C$ to find that $2C = 25$.

Hence, the equation of the curve that satisfies the given conditions is $x^2 + y^2 = 25$.

Example 1.47

Use substitution to evaluate each integral.

A. $\int \frac{3}{4x - 1}\,dx$ **B.** $\int \frac{1}{4 + x^2}\,dx$

Solution:

A. Let $u = 4x - 1$; then $du = 4dx$, which implies that $dx = (1/4)du$. Substitute these quantities in the given integral:

$$\int \frac{3}{4x - 1}\,dx = 3\int \frac{dx}{4x - 1} = 3\int \frac{(1/4)du}{u} = \frac{3}{4}\int \frac{du}{u}$$

To evaluate this integral, use integral formula 4 with $u = 4x - 1$. Hence,

$$\int \frac{3}{4x - 1}\, dx = \frac{3}{4} \ln|4x - 1| + C$$

B. Factor out 4 from the denominator to obtain

$$\int \frac{1}{4 + x^2}\, dx = \frac{1}{4} \int \frac{dx}{1 + (x^2/4)}$$

Let $u^2 = x^2/4$; then $u = x/2$. Thus, $du = (1/2)\, dx$, which implies that $dx = 2du$. Substitute these quantities in the given integral to obtain

$$\int \frac{1}{4 + x^2}\, dx = \frac{1}{4} \int \frac{dx}{1 + (x^2/4)} = \frac{1}{4} \int \frac{2du}{1 + u^2} = \frac{1}{2} \int \frac{du}{1 + u^2}$$

To evaluate this integral, use integral formula 12 with $u = x/2$. Hence,

$$\int \frac{1}{4 + x^2}\, dx = \frac{1}{2} \tan^{-1} \frac{x}{2} + C$$

Example 1.48

Use integration by parts (integral formula 2) to evaluate each integral.

A. $\int x \sin 3x\, dx$ **B.** $\int \ln 4x^2\, dx$

Solution:

A. Using the integration by parts formula, $\int u\, dv = uv - \int v\, du$, let

$$u = x \quad \text{and} \quad dv = \sin 3x\, dx$$

> The correct choice of u and dv is critical. The formula will not work if these are reversed.

To find du, differentiate both sides of $u = x$, and to find v, integrate both sides of $dv = \sin 3x\, dx$ using integral formula 6:

$$du = dx$$

and

$$v = -\frac{1}{3} \cos 3x$$

Substitute these quantities in the integration by parts formula, and then apply integral formula 5, to evaluate the given integral:

$$\int x \sin 3x\, dx = -\frac{1}{3} x \cos 3x + \frac{1}{3} \int \cos 3x\, dx$$

$$= -\frac{1}{3} x \cos 3x + \frac{1}{9} \sin 3x + C$$

B. Using the integration by parts formula, $\int u\, dv = uv - \int v\, du$, let

$$u = \ln 4x^2$$

and

$$dv = dx$$

To find du, differentiate both sides of $u = \ln 4x^2$ using derivative formula 2, and to find v, integrate both sides of $dv = dx$:

$$du = \frac{1}{4x^2} \cdot 8x\, dx = \frac{2}{x} dx$$

and

$$v = x$$

Substitute these quantities in the integration by parts formula to find that

$$\int \ln 4x^2\, dx = x \ln 4x^2 - 2\int dx = x \ln 4x^2 - 2x + C$$

1.16
Definite Integrals

The **definite integral** of $f(x)$ from a to b is denoted by the symbol $\int_a^b f(x)\, dx$ and is defined by

$$\int_a^b f(x)\, dx = \lim_{\delta \to 0} \sum_{k=1}^n f(\xi_k) \Delta x_k$$

where ξ_k are points in the interval (a,b). The function f is called the *integrand,* and a and b are the *lower limit* and *upper limit,* respectively. A physical interpretation of the definite integral is the *work done* by a variable force. A geometric interpretation of the definite integral is the *area under a curve.* The **fundamental theorem of integral calculus** connects the two concepts of the indefinite integral and the definite integral:

$$\int_a^b f(x)\, dx = F(x)\Big|_a^b = F(b) - F(a)$$

where F is any particular integral of $f(x)$.

Definite integrals are used to find the area between two curves, the volume of a solid of revolution, and the centroid (center of gravity) of an area. The **area between two curves** is found by using either a vertical strip with width dx or a horizontal strip with width dy:

1. By using a vertical strip: $A = \int_a^b$ (length of strip) dx, where the left-hand and right-hand boundaries for the vertical strip are from $x = a$ to $x = b$

2. By using a horizontal strip: $A = \int_c^d$ (length of strip) dy, where the lower and upper boundaries for the horizontal strip are from $y = c$ to $y = d$

The **volume of a solid of revolution** is found by using either a vertical disk with thickness dx or a horizontal disk with thickness dy:

1. By using a vertical disk: $V = \pi \int_a^b$ (radius of disk)^2 dx, where the left-hand and right-hand boundaries for the vertical disk are from $x = a$ to $x = b$

2. By using a horizontal disk: $V = \pi \int_c^d$ (radius of disk)^2 dy, where the lower and upper boundaries for the horizontal disk are from $y = c$ to $y = d$

The coordinates of the **centroid of an area** A are denoted by $(\bar{x}, \bar{y})$ and given by

$$(\bar{x}, \bar{y}) = \left(\frac{M_y}{A}, \frac{M_x}{A} \right)$$

where M_y and M_x are the *first moments* with respect to the y-axis and x-axis, respectively. The first moments are defined as follows:

$$M_y = \int_a^b x (\text{length of vertical strip}) \, dx,$$

where the left-hand and right-hand boundaries for the vertical strip are from $x = a$ to $x = b$ and

$$M_x = \int_c^d y (\text{length of horizontal strip}) \, dy$$

where the lower and upper boundaries for the horizontal strip are from $y = c$ to $y = d$.

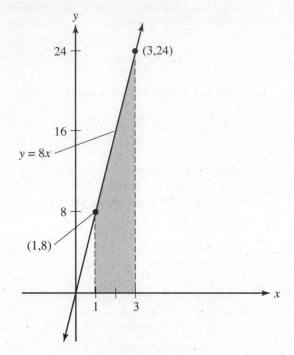

Figure 1.19

Example 1.49

Consider the definite integral $\int_1^3 8x \, dx$.
A. Evaluate this integral.
B. Give a geometric interpretation of this integral.

Solution:

A. Use the fundamental theorem of integral calculus to evaluate the integral:

$$\int_1^3 8x \, dx = 4x^2 \Big|_1^3 = 4(3)^2 - 4(1)^2 = 32$$

B. Geometrically, this definite integral represents the area under the curve $y = 8x$ from $x = 1$ to $x = 3$. A sketch of this area is shown in Figure 1.19.

To verify this answer, use the formula for the area of a trapezoid:

$$A = \frac{1}{2}h(b_1 + b_2) = \frac{1}{2}(2)(8 + 24) = 32 \text{ square units}$$

which agrees with the value of the definite integral found in part **A**.

Example 1.50

Find the area of the shaded region in Figure 1.20:

A. by using a horizontal strip
B. by using a vertical strip

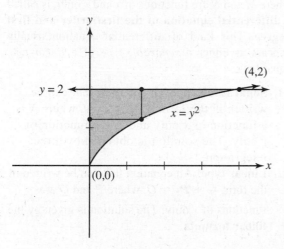

Figure 1.20

Solution:

A. For the horizontal strip in Figure 1.20, the length is the difference between the right-hand x-value ($x = y^2$) and the left-hand x-value ($x = 0$):

$$\text{length of strip} = y^2 - 0 = y^2$$

The lower and upper boundaries for the horizontal strip are from $y = 0$ to $y = 2$. The area of the shaded region is given by the area formula:

$$A = \int_c^d (\text{length of strip})\, dy$$

$$A = \int_0^2 y^2\, dy = \frac{y^3}{3}\Big|_0^2 = \left(\frac{8}{3}\right) - (0) = \frac{8}{3}\ \text{square units}$$

B. For the vertical strip in Figure 1.20, the length is the difference between the upper y-value ($y = 2$) and the lower y-value ($y = \sqrt{x} = x^{1/2}$):

$$\text{length of strip} = 2 - x^{1/2}$$

The left-hand and right-hand boundaries of the vertical strip are from $x = 0$ to $x = 4$. The area of the shaded region is given by the area formula:

$$A = \int_a^b (\text{length of strip})\, dx$$

$$= \int_0^4 (2 - x^{1/2})\, dx = 2x - \frac{2x^{3/2}}{3}\Big|_0^4$$

$$= (8 - \frac{16}{3}) - (0) = \frac{8}{3}\ \text{square units}$$

which agrees with the answer found in part **A.**

Example 1.51

Find the volume generated by revolving the shaded area in Figure 1.20:

A. about the y-axis   **B.** about the x-axis

Solution:

A. If the shaded area in Figure 1.20 is revolved about the y-axis, then the length of the horizontal strip becomes the radius of a horizontal disk. Hence:

$$\text{radius of disk} = y^2 - 0 = y^2$$

The lower and upper boundaries for the horizontal disk are from $y = 0$ to $y = 2$. Thus, the volume generated by revolving the shaded region about the y-axis is given by the volume formula:

$$V = \pi \int_c^d (\text{radius of disk})^2\, dy$$

$$= \pi \int_0^2 (y^2)^2\, dy = \pi \int_0^2 y^4\, dy = \pi\left(\frac{y^5}{5}\right)\Big|_0^2$$

$$= \frac{32\pi}{5}\ \text{cubic units}$$

B. If the shaded area in Figure 1.20 is revolved about the x-axis, then the volume that is generated is a cylinder with a hollowed-out middle. To find this volume, determine the volume of the cylinder and subtract from it the volume of the hollowed-out portion, which can be determined by using a vertical disk:

$$V = \pi r^2 h - \pi \int_a^b (\text{radius of disk})^2\, dx$$

The radius of a vertical disk in the hollowed-out portion is the difference between its upper y-value ($y = \sqrt{x} = x^{1/2}$) and its lower y-value ($y = 0$):

radius of disk $= x^{1/2} - 0 = x^{1/2}$

Thus, the volume of this region is

$$V = \pi(2)^2(4) - \pi \int_0^4 (x^{1/2})^2 \, dx = 16\pi - \pi \int_0^4 x \, dx$$

$$= 8\pi \text{ cubic units}$$

Example 1.52

For the area in Figure 1.20, find

A. the first moments with respect to the y-axis and x-axis
B. the coordinates of the centroid of the area

Solution:

A. The first moment with respect to the y-axis is

$$M_y = \int_0^4 x(2 - x^{1/2}) \, dx = \int_0^4 (2x - x^{3/2})dx$$

$$= x^2 - \frac{2x^{5/2}}{5} \Bigg|_0^4 = \frac{16}{5}$$

and the first moment with respect to the x-axis is

$$M_x = \int_0^2 y(y^2 - 0) \, dy = \int_0^2 y^3 \, dy = \frac{y^4}{4} \Bigg|_0^2 = 4$$

B. From Example 1.50, the area of this region is 8/3 square units. Use this fact and the results from part **A** to find the centroid:

$$(\bar{x}, \bar{y}) = \left(\frac{M_y}{A}, \frac{M_x}{A}\right) = \left(\frac{16/5}{8/3}, \frac{4}{8/3}\right) = \left(\frac{6}{5}, \frac{3}{2}\right)$$

1.17
Differential Equations

An equation involving derivatives (or differentials) of one or more unknown functions is called a **differential equation.** Its *order* is the order of the highest derivative appearing in the equation, and its *degree* is the power to which the derivative of highest order is raised. The *solution* of the equation is a functional relation between the variables involved, not containing derivatives or differentials, that satisfies the differential equation.

Any differential equation that may be written in the derivative form

$$\frac{dy}{dx} = f(x, y)$$

or in the differential form

$$M \, dx + N \, dy = 0$$

where M and N are functions of x and y only, is called **a differential equation of the first order and first degree.** This kind of differential equation usually appears as either a *separable-type* or a *linear-type* equation.

1. **Separable type:** An equation that can be written in the form $X \, dx = Y \, dy$, where X is a function of x only and Y is a function of y only. The solution is obtained by direct integration.
2. **Linear type:** An equation that can be written in the form $\frac{dy}{dx} + Py = Q$, where P and Q are functions of x only. The solution is given by the **linear formula**

$$y \, e^{\int P dx} = \int Q e^{\int P dx} \, dx + C$$

Any differential equation that can be written in the form $y'' + ay' + by = X$, where the coefficients a and b are constants and X is 0 or is a function of x only, is called a **linear differential equation of the second order with constant coefficients.** This kind of differential equation appears as either a *homogeneous-type* or a *nonhomogeneous-type* equation.

1. **Homogeneous type:** An equation in which $X = 0$. The solution of the equation depends on the character of the roots of its *auxiliary equation*, $m^2 + am + b = 0$, as indicated in Table 1.1. The auxiliary equation is formed by replacing the kth derivative with m^k and replacing y with 1.

TABLE 1.1

Roots of Auxiliary Equation	Solution of Differential Equation
Real and distinct roots; m_1, m_2	$y = c_1 e^{m_1 x} + c_2 e^{m_2 x}$
Real repeated root; r	$y = c_1 e^{rx} + c_2 x e^{rx}$
Conjugate imaginary roots; $a \pm bi$	$y = e^{ax}(c_1 \cos bx + c_2 \sin bx)$

TABLE 1.2

For $X =$	Assume $y_p =$
$a_0x^m + a_1x^{m-1} + \ldots + a_m$ ($a_1, a_2, \ldots,$ or a_m may be 0)	$A_0x^m + A_1x^{m-1} + \ldots + A_m$ if 0 is not a root of the auxiliary equation; $x(A_0x^m + A_1x^{m-1} + \ldots + A_m)$ if 0 is a root of the auxiliary equation
ce^{ax}	Ae^{ax} if a is not a root of the auxiliary equation; Axe^{ax} if a is a root of the auxiliary equation; Ax^2e^{ax} if a is a repeated root of the auxiliary equation
$c_1 \sin bx + c_2 \cos bx$ (c_1 or c_2 may be 0)	$A \sin bx + B \cos bx$ if bi is not a root of the auxiliary equation; $x(A \sin bx + B \cos bx)$ if bi is a root of the auxiliary equation

2. **Nonhomogeneous type:** An equation in which $X \neq 0$. The solution is $y = y_c + y_p$, where y_c is called the *complementary solution* or *homogeneous solution* (i.e., the general solution of the corresponding homogeneous equation) and y_p is the *particular solution* of the given nonhomogeneous equation. If X is of the form e^{ax}, $\sin bx$, $\cos bx$, or a polynomial in x, then y_p may be found by the *method of undetermined coefficients*. The procedure in this method is to assume for y_p an expression based on X and its derivatives, and containing certain undetermined coefficients, as indicated in Table 1.2.

To find the values of the undetermined coefficients (A, B, A_0, A_1, ... , A_m), substitute the assumed form of y_p in the differential equation and equate coefficients of corresponding types.

Example 1.53

Solve each differential equation of the first order and first degree.

A. $\dfrac{dy}{dx} = 2xy$ **B.** $\dfrac{dy}{dx} + \dfrac{3y}{x} = 5x$, with $y(2) = 0$

Solution:

A. This equation may be treated as a separable type by writing it in the form

$$\frac{dy}{y} = 2x\,dx$$

or as a linear type, with $P = -2x$ and $Q = 0$, by writing it in the form

$$\frac{dy}{dx} - 2x \cdot y = 0$$

To solve this equation, use either approach.

Separable: Begin by integrating both sides of $(dy)/y = 2x\,dx$:

$$\ln y = x^2 + C$$

Rewriting this expression in exponential form, solve for y:

$$y = e^{x^2+C} = e^{x^2} \cdot e^C = C_1 e^{x^2}$$

Hence, the solution of this differential equation is $y = C_1 e^{x^2}$.

Linear: Begin by evaluating $e^{\int P(x)dx}$ with $P = -2x$:

$$e^{\int -2x\,dx} = e^{-x^2}$$

Now, use the linear formula, with $Q = 0$, to obtain

$$y\,e^{-x^2} = \int 0 \cdot e^{-x^2}\,dx + C = C$$

Multiply both sides of this equation by e^{x^2} to obtain $y = Ce^{x^2}$, which agrees with the answer obtained by using the separable approach.

B. Although this equation is *not* separable, it is linear in y with $P = 3/x$ and $Q = 5x$. Use the properties of logarithms to help evaluate $e^{\int P(x)dx}$:

From the *log identities* (Section 1.5), $e^{\ln f(x)} = f(x)$.

$$e^{\int (3/x)dx} = e^{3\ln x} = e^{\ln x^3} = x^3$$

Now, use the linear formula, with $Q = 5x$, to obtain

$$yx^3 = \int (5x)x^3\, dx + C \quad \text{which implies} \quad yx^3 = x^5 + C$$

Applying the initial conditions, $y(2) = 0$, solve for C:

$$(0)(2)^3 = (2)^5 + C \quad \text{implies} \quad C = -32$$

Hence, the solution of this differential equation is $yx^3 = x^5 - 32$ or

$$y = x^2 - \frac{32}{x^3}$$

Example 1.54

Solve each homogeneous linear differential equation of the second order.

A. $y'' + 8y' + 16 = 0$ **B.** $\dfrac{d^2x}{dt^2} + 4x = 0$

Solution:

A. Find the roots of the corresponding auxiliary equation, $m^2 + 8m + 16 = 0$, by factoring and setting the individual factors equal to 0:

$$(m + 4)(m + 4) = 0$$

implies $m = -4$ is a repeated root

Hence, by Table 1.1, the general solution of this differential equation is

$$y = c_1 e^{-4x} + c_2 x e^{-4x}$$

B. By the quadratic formula, the roots of the corresponding auxiliary equation, $m^2 + 4 = 0$, are $m = 0 \pm 2i$. Hence, by Table 1.1, the general solution of this differential equation is

$$x = e^{0t}(c_1 \cos 2t + c_2 \sin 2t) = c_1 \cos 2t + c_2 \sin 2t$$

Example 1.55

Solve each nonhomogeneous linear differential equation of the second order.

A. $y'' + y' - 2y = 4x^2$ **B.** $y'' - 2y' - 3y = 8e^{3x}$

Solution:

A. To find y_c for this nonhomogeneous differential equation, solve the corresponding homogeneous equation, $y'' + y' - 2y = 0$. Its auxiliary equation, $m^2 + m - 2 = 0$, which may be written as $(m + 2)(m - 1) = 0$, has real distinct roots $m = -2$ and $m = 1$. Hence,

$$y_c = c_1 e^{-2x} + c_2 e^x$$

Referring to Table 1.2, for $X = 4x^2$ with 0 not a root of the auxiliary equation, assume

$$y_p = Ax^2 + Bx + C$$

which implies

$$y'_p = 2Ax + B$$

and

$$y''_p = 2A$$

where A, B, and C are undetermined coefficients. Substitute these values of y''_p, y'_p, and y_p in the given differential equation to obtain

$$2A + (2Ax + B) - 2(Ax^2 + Bx + C) = 4x^2$$

Equating coefficients of the x^2, x, and constant terms on the left-hand and right-hand sides of this equation yields a system of three equations with three unknowns:

$$-2A = 4, \ 2A - 2B = 0, \ \text{and} \ 2A + B - 2C = 0$$

Solve this system of equations to find that $A = -2$, $B = -2$, and $C = -3$. Hence,

$$y_p = -2x^2 - 2x - 3$$

The solution of the nonhomogeneous linear differential equation is $y = y_c + y_p$:

$$y = c_1 e^{-2x} + c_2 e^x - 2x^2 - 2x - 3$$

B. To find y_c for $y'' - 2y' - 3y = 8e^{3x}$, take the corresponding homogeneous equation, $y'' - 2y' - 3y = 0$, form the auxiliary equation, $m^2 - 2m - 3 = 0$, and find its roots. Since this auxiliary equation has real distinct roots $m = 3$ and $m = -1$,

$$y_c = c_1 e^{3x} + c_2 e^{-x}$$

The right-hand member X for this differential equation is $8e^{3x}$ with $a = 3$. Since 3 is also a singular root of the auxiliary equation, assume that

Use the *product rule for differentiation* (Section 1.13) to find y' and y''.

$$y_p = Axe^{3x}$$

where A is an undetermined coefficient, as indicated in Table 1.2. Hence,

$$y'_p = 3Axe^{3x} + Ae^{3x}$$

and

$$y''_p = 9Axe^{3x} + 6Ae^{3x}$$

Substitute these values of y''_p, y'_p, and y_p in the given differential equation to obtain

$$9Axe^{3x} + 6Ae^{3x} - 2(3Axe^{3x} + Ae^{3x}) - 3Axe^{3x} = 8e^{3x}$$

The xe^{3x}-terms drop out, leaving $4Ae^{3x} = 8e^{3x}$, which implies $A = 2$. Hence,

$$y_p = 2xe^{3x}$$

The solution of the nonhomogeneous linear differential equation is $y = y_c + y_p$:

$$y = c_1e^{3x} + c_2e^{-x} + 2xe^{3x}$$

Example 1.56

A weight attached to a spring moves up and down, so that the equation of motion is $\frac{d^2x}{dt^2} + 4x = 0$, where x is the displacement, in feet, of the weight from its equilibrium position at any time t. Suppose that at $t = 0$ sec the displacement is $x = 1$ ft $[x(0) = 1]$ and the velocity is $v = dx/dt = 6$ ft/sec $[x'(0) = 6]$. Find

A. x in terms of t, the equation of motion
B. the amplitude, period, and phase shift of this motion

Solution:

A. From Example 1.54**B**, the general solution of this differential equation is

$$x = c_1 \cos 2t + c_2 \sin 2t$$

Using the initial condition $x(0) = 1$, determine c_1:

$$1 = c_1 \cos 0 + c_2 \sin 0 \quad \text{implies} \quad c_1 = 1$$

To determine c_2 from the given initial conditions, first find the derivative x':

$$x' = -2c_1 \sin 2t + 2c_2 \cos 2t$$

Now, use the initial condition $x'(0) = 6$ with the fact that $c_1 = 1$ to determine c_2:

$$6 = -2(1) \sin 0 + 2c_2 \cos 0 \quad \text{implies} \quad c_2 = 3$$

Hence, the equation of motion is given by

$$x = \cos 2t + 3 \sin 2t$$

B. To determine the amplitude, period, and phase shift from an equation of the form $x = c_1 \cos \omega t + c_2 \sin \omega t$, transform it into the form

$$x = A \sin(\omega t + \phi)$$

where $A = \sqrt{c_1^2 + c_2^2}$ and $\phi = \tan^{-1}\frac{c_1}{c_2}$, with $\sin \phi = \frac{c_1}{A}$ and $\cos \phi = \frac{c_2}{A}$. When the equation of motion is written in this form, the amplitude, period, and phase shift are easily found:

$$\text{amplitude} = A, \quad \text{period} = \frac{2\pi}{\omega}, \quad \text{phase shift} = \frac{\phi}{\omega}$$

For $x = \cos 2t + 3 \sin 2t$, note that $c_1 = 1$, $c_2 = 3$, and $\omega = 2$. Thus,

$$A = \sqrt{1^2 + 3^3} = \sqrt{10} \approx 3.162$$

and

$$\phi = \tan^{-1}\frac{1}{3} \approx 0.3218$$

Hence, the equation of motion is equivalent to

$$x = 3.162 \sin(2t + 0.3218)$$

Thus, the amplitude is 3.162 ft, the period is π sec, and the phase shift is 0.1609 ft, as shown in Figure 1.21.

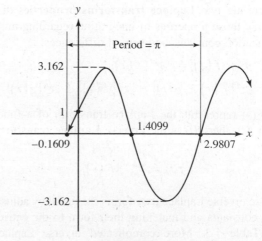

Figure 1.21

1.18
Laplace Transforms

If f is a function defined for $t \geq 0$, then the **Laplace transform** of f, denoted $\mathscr{L}\{f(t)\}$, is defined by the integral

$$\mathscr{L}\{f(t)\} = \int_0^\infty e^{-st} f(t)\,dt = F(s)$$

provided the integral converges. Table 1.3 shows the Laplace transforms of eight basic functions that occur quite frequently. The entries can be verified by evaluating the appropriate integral (see Example 1.57 for some examples.)

Table 1.3

Function	Laplace Transform
1. $f(t) = 1$	$\mathscr{L}\{1\} = \dfrac{1}{s}, \; s > 0$
2. $f(t) = e^{at}$	$\mathscr{L}\{e^{at}\} = \dfrac{1}{s - a}, \; s > a$
3. $f(t) = t^n$	$\mathscr{L}\{t^n\} = \dfrac{n!}{s^{n+1}}, \; s > 0$
4. $f(t) = \sin kt$	$\mathscr{L}\{\sin kt\} = \dfrac{k}{s^2 + k^2}, \; s > 0$
5. $f(t) = \cos kt$	$\mathscr{L}\{\cos kt\} = \dfrac{s}{s^2 + k^2}, \; s > 0$
6. $f(t) = t^n e^{at}$	$\mathscr{L}\{t^n e^{at}\} = \dfrac{n!}{(s - a)^{n+1}}, \; s > a$
7. $f(t) = e^{at}\sin kt$	$\mathscr{L}\{e^{at}\sin kt\} = \dfrac{k}{(s - a)^2 + k^2}, \; s > a$
8. $f(t) = e^{at}\cos kt$	$\mathscr{L}\{e^{at}\cos kt\} = \dfrac{s - a}{(s - a)^2 + k^2}, \; s > a$

There are two **Laplace transforms properties** that mimic those properties of integration regarding multiplication constants and sum or differences:

1. $\mathscr{L}\{cf(t)\} = c\mathscr{L}\{f(t)\}$
2. $\mathscr{L}\{f(t) \pm g(t)\} = \mathscr{L}\{f(t)\} \pm \mathscr{L}\{g(t)\}$

If $F(s)$ represents the Laplace transform of a function $f(t)$, then $f(t)$ is the **inverse Laplace transform** of $F(s)$:

$$f(t) = \mathscr{L}^{-1}\{F(s)\}$$

Basic inverse Laplace transforms are found by adjusting constants and matching their form to the entries in Table 1.3. More complicated inverse Laplace transforms require *partial fraction decomposition* of a rational expression to obtain the entries in Table 1.3. For partial fraction decomposition in this section, there are two cases to consider:

Case 1 – Distinct Linear Factors

To each linear factor, $as + b$, occurring in the denominator of a rational expression, assign a single partial fraction of the form

$$\frac{A}{as + b}$$

where A is a constant to be determined.

Case 2 – Distinct Irreducible Quadratic Factors

To each distinct irreducible quadratic factor, $as^2 + bs + c$, occurring in the denominator of a rational expression, assign a single partial fraction of the form

$$\frac{As + B}{as^2 + bs + c}$$

where A and B are constants to be determined.

Integration by parts (Section 1.15) yields the **Laplace transform of the first and second derivatives** of $f(t)$:

1. $\mathcal{L}\left\{f'(t)\right\} = s\,\mathcal{L}\left\{f(t)\right\} - f(0)$

2. $\mathcal{L}\left\{f''(t)\right\} = s^2\,\mathcal{L}\left\{f(t)\right\} - s\,f(0) - f'(0)$

Because of these results, Laplace transforms are ideally suited to solve differential equations with constant coefficients in which the values of $f(0)$ and $f'(0)$ are known.

Example 1.57

Evaluate each Laplace transform by using the integral definition.

   A. $\mathcal{L}\{1\}$

   B. $\mathcal{L}\{t\}$

Solution:

   A. Using integration formula 3 from Section 1.15:

> If $s < 0$, the Laplace transform *diverges*, since $e^{+\infty} \to \infty$.

$\mathcal{L}\{1\} = \int\limits_0^\infty e^{-st}(1)dt = \left.\frac{-e^{-st}}{s}\right|_0^\infty = \frac{-e^{-\infty}}{s} - \frac{-e^0}{s} = 0 + \frac{1}{s} = \frac{1}{s}, s > 0$

   B. Using integration by parts (formula 2 from Section 1.15) with $u = t$ and $dv = e^{-st}$:

$\mathcal{L}\{t\} = \int\limits_0^\infty e^{-st}(t)dt = \left.\frac{-te^{-st}}{s}\right|_0^\infty + \frac{1}{s}\int\limits_0^\infty e^{-st}dt = 0 + \frac{1}{s}\mathcal{L}\{1\} = \frac{1}{s^2}, s > 0$

Example 1.58

Evaluate each Laplace transform by using Table 1.3.

   A. $\mathcal{L}\left\{5t^2 - 2\sin 3t\right\}$

   B. $\mathcal{L}\left\{\left(1 + e^{2t}\right)^2\right\}$

Solution:

   A. Use the two properties of Laplace transforms to obtain the entries in Table 1.3:

$\mathcal{L}\left\{5t^2 - 2\sin 3t\right\} = 5\mathcal{L}\left\{t^2\right\} - 2\mathcal{L}\left\{\sin 3t\right\}$

$= 5\left(\frac{2!}{s^{2+1}}\right) - 2\left(\frac{3}{s^2 + 9}\right) = \frac{10}{s^3} - \frac{6}{s^2 + 9}$

   B. First, expand to obtain

$\mathcal{L}\left\{(1 + e^{2t})^2\right\} = \mathcal{L}\left\{(1 + 2e^{2t} + e^{4t})\right\}$

Now, use the two properties with Table 1.3:

$\mathcal{L}\left\{(1 + 2e^{2t} + e^{4t})\right\} = \mathcal{L}\{1\} + 2\mathcal{L}\{e^{2t}\} + \mathcal{L}\{e^{4t}\}$

$= \frac{1}{s} + \frac{2}{s - 2} + \frac{1}{s - 4}.$

Example 1.59

Evaluate each inverse Laplace transform.

   A. $\mathcal{L}^{-1}\left\{\frac{4}{s} + \frac{5}{s^4} - \frac{1}{2s + 8}\right\}$

   B. $\mathcal{L}^{-1}\left\{\frac{2s - 5}{s^2 + 9}\right\}$

Solution:

   A. Use the two properties and adjust constants to match the entries in Table 1.3:

$\mathcal{L}^{-1}\left\{\frac{4}{s} + \frac{5}{s^4} - \frac{1}{2s + 8}\right\}$

$= 4\mathcal{L}^{-1}\left\{\frac{1}{s}\right\} + \frac{5}{3!}\mathcal{L}^{-1}\left\{\frac{3!}{s^4}\right\} - \frac{1}{2}\mathcal{L}^{-1}\left\{\frac{1}{s - 4}\right\}$

$= 4 + \frac{5}{6}t^3 - \frac{1}{2}e^{4t}$

   B. Split into two fractions to obtain

$\mathcal{L}^{-1}\left\{\frac{2s - 5}{s^2 + 9}\right\} = \mathcal{L}^{-1}\left\{\frac{2s}{s^2 + 9}\right\} - \mathcal{L}^{-1}\left\{\frac{5}{s^2 + 9}\right\}$

Now, use the properties and adjust constants to match the entries in Table 1.3:

$\mathcal{L}^{-1}\left\{\frac{2s}{s^2 + 9}\right\} - \mathcal{L}^{-1}\left\{\frac{5}{s^2 + 9}\right\} = 2\mathcal{L}^{-1}\left\{\frac{s}{s^2 + 9}\right\}$

$- \frac{5}{3}\mathcal{L}^{-1}\left\{\frac{3}{s^2 + 9}\right\} = 2\cos 3t - \frac{5}{3}\sin 3t$

71

Example 1.60

Solve each differential equation by using Laplace transforms.

A. $y'' - 4y' + 4y = t^3 e^{2t}$, $y(0) = 0$, $y''(0) = 0$

B. $y'' - 6y' + 13y = 26$, $y(0) = 1$, $y'(0) = -3$

Solution:

A. Take the Laplace transform of each of the four terms in the equation:

$$\left[s^2 \mathscr{L}\{y\} - s(0) - 0 \right] - 4\left[s\mathscr{L}\{y\} - 0 \right] + 4\mathscr{L}\{y\} = \frac{3!}{(s-2)^4}$$

Now, solve for $\mathscr{L}\{y\}$:

$$\mathscr{L}\{y\} = \frac{3!}{\left(s^2 - 4s + 4\right)\left(s-2\right)^4} = \frac{3!}{(s-2)^6}.$$

Take the inverve Laplace transform of each side, adjust the constants, and use Table 1.3 to obtain the solution of the differential equation:

$$y = \mathscr{L}^{-1}\left\{ \frac{3!}{(s-2)^6} \right\} = \frac{3!}{5!}\mathscr{L}^{-1}\left\{ \frac{5!}{(s-2)^6} \right\} = \frac{1}{20}t^5 e^{2t}$$

B. Take the Laplace transform of each of the four terms in the equation:

$$\left[s^2 \mathscr{L}\{y\} - s(1) + 3 \right] - 6\left[s\mathscr{L}\{y\} - 1 \right] + 13\mathscr{L}\{y\} = \frac{26}{s}$$

Now, solve for $\mathscr{L}\{y\}$:

$$\mathscr{L}\{y\} = \frac{\dfrac{26}{s} + s - 9}{s^2 - 6s + 13} = \frac{s^2 - 9s + 26}{s\left(s^2 - 6s + 13\right)}$$

Hence,

$$y = \mathscr{L}^{-1}\left\{ \frac{s^2 - 9s + 26}{s\left(s^2 - 6s + 13\right)} \right\}$$

Use partial fraction decomposition to rewrite this rational expression in form

$$\frac{s^2 - 9s + 26}{s\left(s^2 - 6s + 13\right)} = \frac{A}{s} + \frac{Bs + C}{s^2 - 6s + 13}$$

To determine A, B, and C, add the fractions on the right-hand side and set its numerator equal to the numerator on the left-hand side:

$$s^2 - 9s + 26 = A\left(s^2 - 6s + 13\right) + \left(Bs + C\right)s$$

Now, equate the coefficients of like terms on each side:

Constant terms:	$26 = 13A$,	implies $A = 2$
s-terms:	$-9 = -6A + C$,	implies $C = 3$
s^2-terms:	$1 = A + B$,	implies $B = -1$

Hence,

$$y = \mathscr{L}^{-1}\left\{ \frac{s^2 - 9s + 26}{s\left(s^2 - 6s + 13\right)} \right\} = \mathscr{L}^{-1}\left\{ \frac{2}{s} + \frac{-s + 3}{s^2 - 6s + 13} \right\}$$

Finally, rewrite $s^2 - 6s + 13$ in the form $(s-a)^2 + k$, adjust its numerator, and use Table 1.3 to obtain the solution of the differential equation:

> *Complete the square*: add
> $$\left[\frac{1}{2}(-6) \right]^2 = 9 \text{ to } s^2 - 6s, \text{ then subtract}$$
> 9 from 13.

$$y = \mathscr{L}^{-1}\left\{ \frac{2}{s} + \frac{-(s-3)}{(s-3)^2 + 4} \right\} = 2 - e^{3t}\cos 2t$$

1.19 Summary

This chapter has reviewed the essential mathematical skills and concepts required by an engineer to analyze and design a physical system. It is important to note that the mathematics in this chapter forms the foundation for many of the other chapters in this manual. Therefore, it is essential to feel confident with most of this material before continuing. Make notes of any difficulties you may have with the mathematics in the other sections of this manual and refer back to the topic in this section for a review, whenever necessary. If more than a review on any topic is required, refer to one of the many current precalculus or calculus texts that are available.

PRACTICE PROBLEMS

1. Find the approximate perimeter of a triangle whose vertices are points $A(-2,1)$, $B(1,3)$, and $C(4,-3)$ in the coordinate plane.

   (A) 17.5 units
   (B) 14.3 units
   (C) 18.0 units
   (D) 21.3 units

2. Find the equation of a line that passes through points $(4,-1)$ and $(2,5)$ in the coordinate plane.

   (A) $y = -2x + 9$
   (B) $y = -3x + 11$
   (C) $y = -3x + 9$
   (D) $y = -2x + 12$

3. Solve the equation $x(2x - 3) = 5$ for x.

   (A) -1
   (B) $5/2$
   (C) 3
   (D) both (A) and (B)

4. Classify the graph of the equation $x^2 + xy + y^2 - 6 = 0$ as a

   (A) circle
   (B) parabola
   (C) ellipse
   (D) hyperbola

5. Find the coordinates of the foci for the hyperbola whose equation is given by $9y^2 - 4x^2 = 36$.

   (A) $(\pm \sqrt{13}, 0)$
   (B) $(0, \pm \sqrt{13})$
   (C) $(0, \pm 2)$
   (D) $(\pm 3, 0)$

6. Determine the equation of an ellipse with center $(0,0)$, foci $(\pm 2, 0)$, and major axis of length 6.

   (A) $5x^2 + 9y^2 = 45$
   (B) $9x^2 + 5y^2 = 45$
   (C) $x^2 + 9y^2 = 4$
   (D) $5x^2 + y^2 = 20$

7. Find the equation of a circle that is tangent to the y-axis and whose center is at point $(-2,5)$ in the coordinate plane.

   (A) $x^2 + y^2 + 10x - 4y + 16 = 0$
   (B) $x^2 + y^2 - 10y + 25 = 0$
   (C) $x^2 + y^2 + 4x - 10y + 50 = 0$
   (D) $x^2 + y^2 + 4x - 10y + 25 = 0$

8. Find the radius of a sphere whose equation is $x^2 + y^2 + z^2 - 4x + 2y - 4 = 0$.

   (A) 2
   (B) 5
   (C) 3
   (D) 4

9. Evaluate $\log_2 12$ to four significant digits.

   (A) 2.654
   (B) 3.585
   (C) 3.841
   (D) 4.372

10. Simplify the exponential expression $e^{2 \ln x}$.

    (A) $2x$
    (B) 2^x
    (C) x^2
    (D) $x/2$

11. Express $3 \ln x - \ln 2 + 2 \ln 4$ as a single logarithm with a coefficient of 1.

    (A) $\ln 8x^3$
    (B) $\ln 16x$
    (C) $\ln 8x$
    (D) $\ln (x^3/32)$

12. Find the approximate solution of the equation $4e^{-7x} = 15$.

    (A) -0.2981
    (B) -1.672
    (C) 2.570
    (D) -0.1888

13. Given that $\sin \theta = 3/5$ and θ is obtuse, find $\tan \theta$.

    (A) $-4/5$
    (B) $4/5$
    (C) $3/5$
    (D) $-3/4$

14. The two legs of a right triangle measure 12 cm and 18 cm. Find the approximate measure of the smallest angle in the triangle.

    (A) $28.8°$
    (B) $33.7°$
    (C) $31.6°$
    (D) $37.1°$

15. When a helicopter is 1.2 km above one end of an island, the angle of depression to the other end of the island is $22.7°$. Find the approximate length of the island.

    (A) 1.89 km
    (B) 2.87 km
    (C) 2.25 km
    (D) 3.17 km

16. Two angles in an oblique triangle measure 32° and 47°. If the side opposite the unknown angle is 12.5 m, find the approximate length of the smallest side of the triangle.

   (A) 6.75 m
   (B) 6.25 m
   (C) 7.45 m
   (D) 5.85 m

17. Three sides of an oblique triangle measure 196 ft, 212 ft, and 92 ft. Find the approximate measure of the smallest angle.

   (A) 32.1°
   (B) 28.3°
   (C) 21.9°
   (D) 25.7°

18. Given that $\sin \theta = 3/5$ and θ is acute, find $\cos 2\theta$.

   (A) 7/25
   (B) −7/25
   (C) −4/5
   (D) 4/5

19. Express $(\sin x + \cos x)^2 + (\sin x - \cos x)^2$ as a constant.

   (A) 0
   (B) 1
   (C) 2
   (D) 3

20. Express $\cos^2 3x$ in terms of the first power of the cosine function.

   (A) $(1 - \cos 6x)/2$
   (B) $(1 + \cos 6x)/2$
   (C) $(1 + \sin 3x)/2$
   (D) $(1 - \cos 3x)/2$

21. Express the complex number $2.65 + 7.91i$ in polar form.

   (A) $8.34e^{1.25i}$
   (B) $8.34e^{-1.25i}$
   (C) $5.88e^{1.25i}$
   (D) $5.88e^{-1.25i}$

22. Express the quotient $(1 - 2i)/(1 + 2i)$ in rectangular form.

   (A) $0.6 - 0.8i$
   (B) $-0.6 + 0.8i$
   (C) $0.6 + 0.8i$
   (D) $-0.6 - 0.8i$

23. Express the product $(5e^{(17\pi/12)i})(\sqrt{2} + i\sqrt{2})$ in polar form.

   (A) $10e^{(2\pi/3)i}$
   (B) $e^{(5\pi/3)i}$
   (C) $-10e^{(\pi/3)i}$
   (D) $10e^{(5\pi/3)i}$

24. Express the power $(1 + i)^8$ in rectangular form.

   (A) 16
   (B) $8i$
   (C) $1 - i$
   (D) −6

25. Given the matrices $A = \begin{bmatrix} -1 & 1 \\ 1 & -1 \end{bmatrix}$ and $B = \begin{bmatrix} 5 & 5 \\ -2 & 2 \end{bmatrix}$, find $3A - 2B$.

   (A) $\begin{bmatrix} -2 & 8 \\ 3 & -2 \end{bmatrix}$

   (B) $\begin{bmatrix} -13 & -7 \\ 7 & 1 \end{bmatrix}$

   (C) $\begin{bmatrix} -10 & -3 \\ 5 & 1 \end{bmatrix}$

   (D) $\begin{bmatrix} 13 & -7 \\ 7 & 0 \end{bmatrix}$

26. Given the matrices in Problem 25, find BA.

   (A) $\begin{bmatrix} 1 & 0 \\ 0 & 1 \end{bmatrix}$

   (B) $\begin{bmatrix} 0 & 0 \\ 0 & 0 \end{bmatrix}$

   (C) $\begin{bmatrix} 0 & 1 \\ 1 & 0 \end{bmatrix}$

   (D) $\begin{bmatrix} 1 & 1 \\ 1 & 1 \end{bmatrix}$

27. For the matrix $A = \begin{bmatrix} 1 & 0 & 2 \\ 2 & -3 & 4 \\ -3 & 1 & 9 \end{bmatrix}$, find the minor of element a_{11}.

   (A) −31
   (B) 12
   (C) 0
   (D) −45

28. Evaluate the determinant of the matrix in Problem 27.

   (A) −31
   (B) 12
   (C) 0
   (D) −45

29. Find the inverse of the matrix $A = \begin{bmatrix} 1 & 4 \\ 2 & 3 \end{bmatrix}$.

    (A) $\dfrac{1}{2}\begin{bmatrix} -1 & 7 \\ 2 & -1 \end{bmatrix}$

    (B) $\dfrac{1}{3}\begin{bmatrix} -3 & -1 \\ 2 & 4 \end{bmatrix}$

    (C) $\dfrac{1}{5}\begin{bmatrix} -3 & 4 \\ 2 & -1 \end{bmatrix}$

    (D) $\begin{bmatrix} -3 & 3 \\ 2 & 1 \end{bmatrix}$

30. Solve this system of equations:
$$x - y + z = -3$$
$$2x + y = 1$$
$$y - 3z = 7$$

    (A) $x = 0, y = -11, z = 2$
    (B) $x = 0, y = 1, z = -2$
    (C) $x = 1, y = -1, z = 0$
    (D) $x = 0, y = 2, z = -6$

31. Find the magnitude of vector $\mathbf{A} = \mathbf{i} - 2\mathbf{j} + 3\mathbf{k}$.
    (A) $\sqrt{14}$
    (B) $\sqrt{10}$
    (C) $\sqrt{6}$
    (D) 4

32. For the vectors $\mathbf{A} = 2\mathbf{i} + \mathbf{j} + \mathbf{k}$ and $\mathbf{B} = -4\mathbf{i} + 3\mathbf{j} + \mathbf{k}$, find the dot product $\mathbf{A} \cdot \mathbf{B}$.

    (A) 12
    (B) 6
    (C) 4
    (D) -4

33. For the vectors in Problem 32, find the cross product $\mathbf{A} \times \mathbf{B}$.

    (A) $\mathbf{i} + 6\mathbf{j} + 5\mathbf{k}$
    (B) $-2\mathbf{i} + 5\mathbf{j} - 9\mathbf{k}$
    (C) $-2\mathbf{i} - 6\mathbf{j} + 10\mathbf{k}$
    (D) $\mathbf{i} - 6\mathbf{j}$

34. Find the area of a triangle with vertices $P(2,-2,1)$, $Q(3,-1,2)$, and $R(3,-1,1)$.

    (A) 4 square units
    (B) $\sqrt{2}/2$ square units
    (C) $\sqrt{2}$ square units
    (D) 8 square units

35. Find the tenth element of the arithmetic sequence $11, 4, -3, -10, \ldots$.

    (A) -52
    (B) -106
    (C) -99
    (D) -58

36. Find the sum of the arithmetic series $\sum\limits_{i=1}^{58}(7 - 5i)$.
    (A) $-8,149$
    (B) 946
    (C) $-2,369$
    (D) $-9,980$

37. Find the sum of the infinite geometric series $\sum\limits_{i=2}^{\infty}(-0.2)^i$.
    (A) 1
    (B) 1/15
    (C) 1/45
    (D) 1/30

38. Find the Taylor (Maclaurin) series generated by $f(x) = \sin x$ at $x = 0$.

    (A) $x + \dfrac{x^3}{3!} + \dfrac{x^5}{5!} + \ldots + \dfrac{x^{2n-1}}{(2n-1)!}$

    (B) $1 - \dfrac{x^2}{2!} + \dfrac{x^4}{4!} - \ldots + (-1)^{n+1}\dfrac{x^{2n-2}}{(2n-2)!}$

    (C) $x - \dfrac{x^3}{3!} - \dfrac{x^5}{5!} - \ldots - \dfrac{x^{2n-1}}{(2n-1)!}$

    (D) $x - \dfrac{x^3}{3!} + \dfrac{x^5}{5!} - \ldots + (-1)^{n+1}\dfrac{x^{2n-1}}{(2n-1)!}$

39. A rock thrown vertically upward from the surface of the moon reaches a height of $s = 24t - 0.8t^2$ m in t s. Find the velocity v at any time t.

    (A) $v = 16 - 1.4t$ m/s
    (B) $v = 24 - 1.6t$ m/s
    (C) $v = 22 + 1.8t$ m/s
    (D) $v = -13 - 1.9t$ m/s

40. Find the equation of the tangent to the curve $y = x^2 - 3x + 4$ at $x = 2$.

    (A) $y = x + 2$
    (B) $y = 2x - 1$
    (C) $y = x$
    (D) $y = -x$

41. Evaluate $\lim\limits_{x \to 0}\left(\dfrac{\cos x - 1}{\cos 2x - 1}\right)$.

    (A) 1/2
    (B) 1/3
    (C) 2/3
    (D) 1/4

42. Find the equation of the normal to the circle $x^2 + y^2 = 25$ at $(3, -4)$.

    (A) $y = \dfrac{4}{3}x - 6$

    (B) $y = -\dfrac{2}{3}x - 6$

    (C) $y = -\dfrac{4}{3}x + 8$

    (D) $y = -\dfrac{4}{3}x$

43. For $z = x^2 y^2 + e^{2x} y^3$, find $\partial z / \partial x$.

    (A) $xy^2 + 2e^x$
    (B) $2xy^2 + 2e^{2x} y^3$
    (C) $2xy + 2e^x y^3$
    (D) $2x + y^3$

44. Determine a point of inflection for the graph of $y = x^3 + 6x^2$.

    (A) $(-2, 16)$
    (B) $(0, 0)$
    (C) $(-1, 5)$
    (D) $(2, 32)$

45. Find the dimensions of the rectangle with greatest area that can be cut from a semicircle of radius 6 cm.

    (A) 3 cm by 4 cm
    (B) $6\sqrt{2}$ cm by $3\sqrt{2}$ cm
    (C) $\sqrt{2}$ cm by $6\sqrt{2}$ cm
    (D) $6\sqrt{2}$ cm by $6\sqrt{2}$ cm

46. Find the equation of the curve that passes through point (1,2) and whose slope of the tangent to the curve is given by $dy/dx = 4x - 6$.

    (A) $y = x^2 + 1$
    (B) $y = 2x^2 - 6x + 6$
    (C) $y = 2x^2 - 6x$
    (D) $y = 2x^2 + 6x + 1$

47. Evaluate $\int \dfrac{4}{2x - 1} dx$.

    (A) $(2x - 1)^2 + C$
    (B) $x^2 + C$
    (C) $\ln(2x - 1) + C$
    (D) $\ln(2x - 1)^2 + C$

48. Evaluate $\int x \sin 2x \, dx$.

    (A) $x \cos 2x + C$
    (B) $-\dfrac{1}{2} x^2 \sin 2x + C$
    (C) $-\dfrac{1}{2} x \cos 2x + \dfrac{1}{4} \sin 2x + C$
    (D) $x \cos 2x + \cos 2x + C$

49. Evaluate $\int_0^3 \dfrac{5x \, dx}{\sqrt{1 + x^2}}$.

    (A) $5\sqrt{10} - 5$
    (B) 2
    (C) $\sqrt{8} - 3$
    (D) 0

50. Find the area in the first quadrant bounded by $y^2 = 4x$ and $x = 4$.

    (A) 32/3 square units
    (B) 16 square units
    (C) 8/3 square units
    (D) 24 square units

51. Find the volume of the solid generated by revolving about the x-axis the area described in Problem 50.

    (A) 16π cubic units
    (B) 8π cubic units
    (C) 32π cubic units
    (D) 48π cubic units

52. Find the first moment with respect to the x-axis of the area described in Problem 50.

    (A) 2
    (B) 4
    (C) 8
    (D) 16

53. Solve the differential equation $\dfrac{dy}{dx} + \dfrac{y}{x} = 2$.

    (A) $y = cx$
    (B) $y = \dfrac{1}{x} + c$
    (C) $y = 3x + c$
    (D) $y = x + \dfrac{c}{x}$

54. Solve the differential equation $y'' - y' - 6y = 0$.

    (A) $y = c_1 e^{3x} + c_2$
    (B) $y = c_1 e^{3x} + c_2 e^{-2x}$
    (C) $y = c_1 e^{-3x} + c_2 e^{-2x}$
    (D) $y = c_1 e^{-5x} + c_2 e^{-x}$

55. Solve the differential equation $y'' + 4y' = 10 \sin 2x$.

    (A) $y = c_1 + c_2 e^{-4x} + \sin 2x - \cos 2x$
    (B) $y = c_1 + c_2 e^{-4x} - 0.5 \sin 2x - \cos 2x$
    (C) $y = c_1 + c_2 e^{-4x}$
    (D) $y = c_1 + c_2 e^{-4x} + \cos 2x$

56. A weight attached to a spring moves up and down, so that the equation of motion is $\dfrac{d^2 x}{dt^2} + 64x = 0$, where x is the displacement (in feet) of the weight from its equilibrium position at any time t. Suppose that at $t = 0$ sec the displacement is 3 in $[x(0) = 1/4]$ and the velocity is 1 ft/sec $[x'(0) = 1]$. Find the approximate amplitude of the motion.

    (A) 1 ft
    (B) 0.280 ft
    (C) 0.119 ft
    (D) 0.521 ft

57. The Laplace transform of the function f defined

by $f(t) = \begin{cases} 0, & 0 \le t < 3 \\ 2, & t \ge 3 \end{cases}$ is

(A) $\dfrac{2e^{-3s}}{s}, s > 0$

(B) $\dfrac{3e^{-2s}}{s}, s > 0$

(C) $\dfrac{2}{s}, s > 3$

(D) undefined

58. Use Table 1.3 to evaluate $\mathcal{L}\left\{\left(e^{-3t} + e^{3t}\right)^2\right\}$.

(A) $\dfrac{1}{s+9} + \dfrac{1}{s} + \dfrac{1}{s-9}$

(B) $\left(\dfrac{1}{s+3} + \dfrac{1}{s-3}\right)^2$

(C) $\dfrac{1}{s+6} + \dfrac{1}{s} + \dfrac{1}{s-6}$

(D) $\dfrac{1}{s+6} + \dfrac{1}{s-6}$

59. Use Table 1.3 to evaluate $\mathcal{L}^{-1}\left\{\dfrac{3}{s^5} - \dfrac{1}{s^2+9}\right\}$.

(A) $3t^4 - \sin 3t$

(B) $\dfrac{1}{24}t^4 - \dfrac{1}{3}\cos 3t$

(C) $\dfrac{1}{8}t^4 - \dfrac{1}{3}\sin 3t$

(D) $\dfrac{1}{24}t^4 - \cos 3t$

60. Use Laplace transforms to find the solution of the differential equation $y' + 3y = 13\sin 2t$, $y(0) = 6$.

(A) $8e^{-3t} - 2\cos 2t + 3\sin 2t$

(B) $9e^{3t} - 3\cos 2t + 2\sin 2t$

(C) $8e^{-3t} - 2\cos 2t$

(D) $6e^{3t} + \sin 2t$

Answer Key

1. A	13. D	25. B	37. D	49. A
2. B	14. B	26. B	38. D	50. A
3. D	15. B	27. A	39. B	51. C
4. C	16. A	28. D	40. C	52. D
5. B	17. D	29. C	41. D	53. D
6. A	18. A	30. B	42. D	54. B
7. D	19. C	31. A	43. B	55. B
8. C	20. B	32. D	44. A	56. B
9. B	21. A	33. C	45. B	57. A
10. C	22. D	34. B	46. B	58. C
11. A	23. D	35. A	47. D	59. C
12. D	24. A	36. A	48. C	60. A

Answers Explained

1. **A** The length of the sides of the triangle are as follows: $AB = \sqrt{3^2+2^2} = \sqrt{13}$, $BC = \sqrt{3^2+6^2} = \sqrt{45}$, and $AC = \sqrt{6^2+4^2} = \sqrt{52}$. Hence:

Perimeter $P = \sqrt{13} + \sqrt{45} + \sqrt{52} \approx 17.5$ units.

2. **B** Slope $m = \dfrac{-1-5}{4-2} = -3$. Hence,

$$y - 5 = -3(x - 2) \quad \text{or} \quad y = -3x + 11$$

3. **D** The given equation is equivalent to $2x^2 - 3x - 5 = 0$ or $(2x - 5)(x + 1) = 0$. Hence, $x = \dfrac{5}{2}$ or -1.

4. **C** The discriminant $B^2 - 4AC = 1^2 - 4(1)(1) = -3 < 0$. Hence, the graph is an ellipse.

5. **B** The center of this hyperbola is at the origin, and its vertices are on the y-axis with $c = \sqrt{2^2+3^2} = \sqrt{13}$. Hence, the coordinates of the foci are $(0, \pm\sqrt{13})$.

6. **A** Using $c = \sqrt{|a^2-b^2|}$ with $a = 3$ and $c = 2$ implies $b = \pm\sqrt{5}$. Hence, the equation of the ellipse is

$$(x^2/9) + (y^2/5) = 1 \quad \text{or} \quad 5x^2 + 9y^2 = 45$$

7. **D** The equation of the circle is

$$(x + 2)^2 + (y - 5)^2 = 2^2$$
$$\text{or} \quad x^2 + y^2 + 4x - 10y + 25 = 0$$

8. C Complete the squares to rewrite the equation as $(x-2)^2 + (y+1)^2 + (z-0)^2 = 9$. Hence, the radius of the sphere is $\sqrt{9}$ or 3.

9. B To four significant digits, $\log_2 12 = \dfrac{\ln 12}{\ln 2} \approx 3.585$.

10. C Simplify as follows: $e^{2\ln x} = e^{\ln x^2} = x^2$.

11. A Rewrite as

$$\ln x^3 - \ln 2 + \ln 4^2 = \ln\left(\frac{4^2 x^3}{2}\right) = \ln 8x^3$$

12. D The equation $e^{-7x} = \dfrac{15}{4}$ implies

$$-7x = \ln\left(\frac{15}{4}\right)$$

or $\quad x = \dfrac{\ln\left(\frac{15}{4}\right)}{-7} \approx -0.1888$

13. D That θ is obtuse implies

$$\cos\theta = \sqrt{1-\sin^2\theta}$$
$$= -\sqrt{1-\left(\frac{3}{5}\right)^2} = -\frac{4}{5}$$

Hence,

$$\tan\theta = \frac{\sin\theta}{\cos\theta} = \frac{3/5}{-4/5} = -\frac{3}{4}$$

14. B $\tan\theta = \dfrac{12}{18}$ implies $\theta = \tan^{-1}\left(\dfrac{12}{18}\right) \approx 33.7°$.

15. B Since $\tan 22.7° = 1.2/x$, then $x = \dfrac{1.2}{\tan 22.7°} \approx 2.87$ km.

16. A By the law of sines, $\dfrac{12.5}{\sin 101°} = \dfrac{x}{\sin 32°}$. Hence, $x = \dfrac{12.5 \sin 32°}{\sin 101°} \approx 6.75$ m.

17. D By the law of cosines,

$$92^2 = 196^2 + 212^2 - 2(196)(212)\cos\theta$$

Hence,

$$\theta = \cos^{-1}\left(\frac{92^2 - 196^2 - 212^2}{-2(196)(212)}\right) \approx 25.7°$$

18. A That θ is acute implies

$$\cos\theta = \sqrt{1-\sin^2 x} = \sqrt{1-\left(\frac{3}{5}\right)^2} = \frac{4}{5}$$

Hence,

$$\cos 2\theta = \cos^2\theta - \sin^2\theta$$
$$= \left(\frac{4}{5}\right)^2 - \left(\frac{3}{5}\right)^2 = \frac{7}{25}$$

19. C Here,

$$(\sin x + \cos x)^2 = \sin^2 x + 2\sin x\cos x + \cos^2 x$$
$$= 1 + 2\sin x\cos x$$

and

$$(\sin x - \cos x)^2 = \sin^2 x - 2\sin x\cos x + \cos^2 x$$
$$= 1 - 2\sin x\cos x$$

Hence, the sum of these two expressions is 2.

20. B In terms of the first power of the cosine function,

$$\cos^2 3x = \frac{1}{2}(1 + \cos[2(3x)]) = \frac{1+\cos 6x}{2}$$

21. A Since $r = \sqrt{2.65^2 + 7.91^2} \approx 8.34$ and $\theta = \tan^{-1}\left(\dfrac{7.91}{2.65}\right) \approx 1.25$ rad, then $2.65 + 7.91i \approx 8.34e^{1.25i}$.

22. D The rectangular form is as follows:

$$\frac{1-2i}{1+2i} \cdot \frac{1-2i}{1-2i} = \frac{1-4i+4i^2}{1-4i^2}$$
$$= \frac{-3-4i}{5}$$
$$= -0.6 - 0.8i$$

23. D Since $\sqrt{2} + i\sqrt{2} = 2e^{(\pi/4)i}$, then
$$(5e^{(17\pi/12)i})(2e^{(\pi/4)i}) = 10e^{[(17\pi/12)+(\pi/4)]i}$$
$$= 10e^{(5\pi/3)i}$$

24. A In rectangular form,

$$(1 + i)^8 = (\sqrt{2}\, e^{(\pi/4)i})^8 = 16e^{2\pi i} = 16$$

25. B $3A - 2B = \begin{bmatrix} -3 & 3 \\ 3 & -3 \end{bmatrix} + \begin{bmatrix} -10 & -10 \\ 4 & 4 \end{bmatrix}$

$$= \begin{bmatrix} -13 & -7 \\ 7 & 1 \end{bmatrix}$$

26. B $BA = \begin{bmatrix} 5 & 5 \\ -2 & -2 \end{bmatrix}\begin{bmatrix} -1 & 1 \\ 1 & -1 \end{bmatrix}$

$$= \begin{bmatrix} -5 + 5 & 5 - 5 \\ 2 - 2 & -2 + 2 \end{bmatrix} = \begin{bmatrix} 0 & 0 \\ 0 & 0 \end{bmatrix}$$

27. A $\begin{vmatrix} -3 & 4 \\ 1 & 9 \end{vmatrix} = -27 - 4 = -31$

28. D Expand about the first row; then,

$$|A| = 1\begin{vmatrix} -3 & 4 \\ 1 & 9 \end{vmatrix} + 2\begin{vmatrix} 2 & -3 \\ -3 & 1 \end{vmatrix}$$

$$= -31 + 2(-7) = -45$$

29. C $|A| = 3 - 8 = -5$, $A^T = \begin{bmatrix} 1 & 2 \\ 4 & 3 \end{bmatrix}$, and adj($A$)

$$= \begin{bmatrix} 3 & -4 \\ -2 & 1 \end{bmatrix}. \text{ Hence,}$$

$$A^{-1} = \frac{\text{adj}(A)}{|A|} = \frac{1}{5}\begin{bmatrix} -3 & 4 \\ 2 & -1 \end{bmatrix}$$

30. B $\begin{bmatrix} x \\ y \\ z \end{bmatrix} = \begin{bmatrix} 1 & -1 & 1 \\ 2 & 1 & 0 \\ 0 & 1 & -3 \end{bmatrix}^{-1}\begin{bmatrix} -3 \\ 1 \\ 7 \end{bmatrix}$

$$= \frac{1}{7}\begin{bmatrix} 3 & 2 & 1 \\ -6 & 3 & -2 \\ -2 & 1 & -3 \end{bmatrix}\begin{bmatrix} -3 \\ 1 \\ 7 \end{bmatrix} = \begin{bmatrix} 0 \\ 1 \\ -2 \end{bmatrix}$$

Hence, the solution of the system of equations is $x = 0, y = 1$, and $z = -2$.

31. A $|\mathbf{A}| = \sqrt{1^2 + (-2)^2 + 3^2} = \sqrt{14}$

32. D $\mathbf{A}\cdot\mathbf{B} = (2)(-4) + (1)(3) + (1)(1) = -4$

33. C $\mathbf{A} \times \mathbf{B} = \begin{vmatrix} \mathbf{i} & \mathbf{j} & \mathbf{k} \\ 2 & 1 & 1 \\ -4 & 3 & 1 \end{vmatrix}$

$$= \mathbf{i}\begin{vmatrix} 1 & 1 \\ 3 & 1 \end{vmatrix} - \mathbf{j}\begin{vmatrix} 2 & 1 \\ -4 & 1 \end{vmatrix} + \mathbf{k}\begin{vmatrix} 2 & 1 \\ -4 & 3 \end{vmatrix}$$

$$= -2\mathbf{i} - 6\mathbf{j} + 10\mathbf{k}$$

34. B $\overrightarrow{PQ} = \mathbf{i} + \mathbf{j} + \mathbf{k}$ and $\overrightarrow{PR} = \mathbf{i} + \mathbf{j}$. Hence,

$$\overrightarrow{PQ} \times \overrightarrow{PR} = \begin{vmatrix} \mathbf{i} & \mathbf{j} & \mathbf{k} \\ 1 & 1 & 1 \\ 1 & 1 & 0 \end{vmatrix} = -\mathbf{i} + \mathbf{k}$$

Therefore,

$$\text{Area} = \frac{1}{2}|\overrightarrow{PQ} \times \overrightarrow{PR}| = \frac{\sqrt{2}}{2} \text{ square units}$$

35. A In the given sequence,

$$a_{10} = 11 + (10 - 1)(-7) = -52$$

36. A $S_{58} = \dfrac{58}{2}[2(2) + 57(-5)] = -8{,}149$

37. D $S = \dfrac{0.04}{1 - (-0.2)} = \dfrac{1}{30}$

38. D Here, $f(x) = \sin x$, $f'(x) = \cos x$, $f''(x) = -\sin x, f'''(x) = -\cos x$, etc. Hence,

$$\sin x = f(0) + f'(0)x + \frac{f''(0)}{2!}x^2 + \frac{f'''(0)}{3!}x^3 + \cdots$$

$$= x - \frac{x^3}{3!} + \frac{x^5}{5!} - \cdots$$

39. B At any time t, $v = \dfrac{ds}{dt} = 24 - 1.6t$ m/s.

40. C Here, $\dfrac{dy}{dx} = 2x - 3$. At $x = 2$ and $y = 2$,

$\dfrac{dy}{dx} = 1$. Hence,

$$y - 2 = 1(x - 2) \quad \text{or} \quad y = x$$

41. D By L'Hôpital's rule,

$$\lim_{x\to 0}\left(\frac{\cos x - 1}{\cos 2x - 1}\right) = \lim_{x\to 0}\left(\frac{-\sin x}{-2\sin 2x}\right)$$

$$= \lim_{x\to 0}\left(\frac{-\cos x}{-4\cos 2x}\right) = \frac{1}{4}$$

42. D Use implicit differentiation: $2x + 2y\dfrac{dy}{dx} = 0$

implies $\dfrac{dy}{dx} = \dfrac{-x}{y}$. At $(3,-4)$, $\dfrac{dy}{dx} = \dfrac{3}{4}$. Hence,

the slope of the normal is $\dfrac{-4}{3}$. Thus, the equation of the normal is

$$y + 4 = -\frac{4}{3}(x - 3) \quad \text{or} \quad y = -\frac{4}{3}x$$

43. B For the given equation,

$$\frac{\partial z}{\partial x} = (2x)y^2 + (2e^{2x})y^3 = 2xy^2 + 2y^3 e^{2x}$$

44. A Here, $y'' = 6x + 12 = 0$ implies $x = -2$ and $y = 16$. Since y'' changes sign as x increases through -2, $(-2,16)$ is a point of inflection.

45. B The equation of the semicircle is $y = \sqrt{36 - x^2}$. Hence, the area A of the inscribed rectangle is $A = 2x\sqrt{36 - x^2}$. Then,

$$\frac{dA}{dx} = \frac{-2x^2}{(36 - x^2)^{1/2}} + 2(36 - x^2)^{1/2} = 0$$

which implies $x = 3\sqrt{2}$ and $y = 3\sqrt{2}$. Therefore, the dimensions of the rectangle with greatest area are $6\sqrt{2}$ cm by $3\sqrt{2}$ cm.

46. B Here, $dy = (4x - 6)dx$ implies $y = 2x^2 - 6x + C$. If $x = 1$ and $y = 2$, then $C = 6$. Hence, $y = 2x^2 - 6x + 6$.

47. D Here, $\displaystyle\int \frac{4}{2x - 1}dx = 2\ln|2x - 1| + C$

$$= \ln(2x - 1)^2 + C$$

48. C Use integration by parts with $u = x$, $du = dx$, $dv = \sin 2x\, dx$, and $v = -\dfrac{1}{2}\cos 2x$:

$$\int x \sin 2x\, dx = -\frac{1}{2}x\cos 2x + \frac{1}{2}\int \cos 2x\, dx$$

$$= -\frac{1}{2}x\cos 2x + \frac{1}{4}\sin 2x + C$$

49. A Here, $5\displaystyle\int_0^3 x(1 + x^2)^{-1/2}\, dx = \frac{5}{2}\int u^{-1/2}\, du =$

$5(1 + x^2)^{1/2}\big|_0^3 = 5\sqrt{5} - 5.$

50. A Area $= \displaystyle\int_0^4 2x^{1/2}\, dx = \frac{4}{3}x^{3/2}\Big|_0^4 = \frac{32}{3}$ square units.

51. C Volume $= \pi\displaystyle\int_0^4 (2x^{1/2})^2\, dx$

$$= \pi\int_0^4 4x\, dx$$

$$= \pi(2x^2)\Big|_0^4 = 32\pi \text{ cubic units.}$$

52. D $M_x = \displaystyle\int_0^4 y\left(4 - \frac{1}{4}y^2\right)dy = \int_0^4 \left(4y - \frac{1}{4}y^3\right)dy$

$$= 2y^2 - \frac{1}{16}y^4\Big|_0^4 = 16$$

53. D The given equation is linear in y with integrating factor $e^{\int (1/x)\, dx} = e^{\ln x} = x$. Hence,

$$yx = \int 2x\, dx + c \quad \text{or} \quad y = x + \frac{c}{x}$$

54. B The solutions of the auxiliary equation $m^2 - m - 6 = 0$ are 3 and -2. Hence, $y = c_1 e^{3x} + c_2 e^{-2x}$.

55. B The solutions of the auxiliary equation $m^2 + 4m = 0$ are 0 and -4. Hence, $y_c = c_1 + c_2 e^{-4x}$. Substitute $y_p = A\sin 2x + B\cos 2x$ into $y'' + 4y' = 10\sin 2x$ to obtain

$$-4A\sin 2x - 4B\cos 2x$$
$$+ 4(2A\cos 2x - 2B\sin 2x) = 10\sin 2x$$

which implies $A = -\dfrac{1}{2}$ and $B = -1$. Hence,

$$y = c_1 + c_2 e^{-4x} - 0.5\sin 2x - \cos 2x$$

56. B The solutions of the auxiliary equation $m^2 + 64 = 0$ are $\pm 8i$. Hence,

$$x = c_1\cos 8t + c_2\sin 8t$$

with $x(0) = \dfrac{1}{4}$, and $x' = -8c_1\sin 8t + 8c_2\cos 8t$

with $x'(0) = 1$ implies $c_1 = \dfrac{1}{4}$ and $c_2 = \dfrac{1}{8}$.

Therefore, $x = \dfrac{1}{4}\cos 8t + \dfrac{1}{8}\sin 8t$, and amplitude of motion =

$$\sqrt{\left(\frac{1}{4}\right)^2 + \left(\frac{1}{8}\right)^2} \approx 0.280 \text{ ft}$$

57. A Use the integral definition of the Laplace transform:

$$\mathscr{L}\{f(t)\} = \int_0^3 e^{-st}(0)\,dt + \int_3^\infty e^{-st}(2)\,dt$$

$$= 0 - \frac{2}{s}e^{-st}\Big|_3^\infty = \frac{2e^{-3s}}{s}$$

58. C First, expand the expression, then find the Laplace transform of the three terms using Table 1.3:

$$\mathscr{L}\left\{\left(e^{-3t} + e^{3t}\right)^2\right\} = \mathscr{L}\left\{e^{-6t} + 1 + e^{6t}\right\}$$

$$= \frac{1}{s+6} + \frac{1}{s} + \frac{1}{s-6}$$

59. C Adjust the constants to fit the entries in Table 1.3:

$$\mathscr{L}^{-1}\left\{\frac{3}{s^5} - \frac{1}{s^2+9}\right\}$$

$$= \frac{3}{4!}\mathscr{L}^{-1}\left\{\frac{4!}{s^5}\right\} - \frac{1}{3}\mathscr{L}^{-1}\left\{\frac{3}{s^2+9}\right\}$$

$$= \frac{1}{8}t^4 - \frac{1}{3}\sin 3t$$

60. A Take the Laplace transform of the three terms in the equation and solve for $\mathscr{L}\{y\}$:

$$\mathscr{L}\{y\} = \frac{6 + \dfrac{26}{s^2+4}}{s+3} = \frac{6s^2 + 50}{(s+3)(s^2+4)}$$

Hence, $y = \mathscr{L}^{-1}\left\{\dfrac{6s^2+50}{(s+3)(s^2+4)}\right\}$

Use partial fraction decomposition to rewrite this rational expression in the form

$$\frac{6s^2+50}{(s+3)(s^2+4)} = \frac{A}{s+3} + \frac{Bs+C}{s^2+4}$$

Add fractions on the right-hand side, equate numerators, and match like terms to find $A = 8$, $B = -2$, and $C = 6$. Finally, adjust constants and use Table 1.3 to find the solution:

$$y = \mathscr{L}^{-1}\left\{\frac{8}{s+3} + \frac{-2s+6}{s^2+4}\right\}$$

$$= 8\mathscr{L}^{-1}\left\{\frac{1}{s+3}\right\} - 2\mathscr{L}^{-1}\left\{\frac{s}{s^2+4}\right\} + \frac{6}{2}\mathscr{L}^{-1}\left\{\frac{2}{s^2+4}\right\}$$

$$= 8e^{-3t} - 2\cos 2t + 3\sin 2t$$

2

Engineering Probability and Statistics

Thomas J. Pfaff

Introduction

Suppose we have a bag with three red marbles and five blue marbles. Probability is used to answer the type of question that asks, what is the probability of selecting a red marble from the bag? On the other hand, given a bag with some red and blue marbles, statistics is used to try to determine the proportion of red and blue marbles in the bag based on a sample of marbles from the bag. These two topics are linked by the fact that knowledge of probability is used to help make statistically based decisions. For example, probabilities related to normal random variables are used to construct *z*-confidence intervals. This chapter reviews the major topics of both probability and statistics that are typically taught in a probability and statistics course.

- Mode
- Median
- Mean
- Standard Deviation
- Variance
- Combination
- Permutation
- Probability Rules
- Random Variables
- Binomial Random Variable

- Normal Random Variable
- Central Limit Theorem
- Confidence Interval for a Mean and Variance
- Hypothesis Test for a Single Mean
- Design of Experiment-Sample Size
- Goodness-of-Fit Hypothesis Test
- Correlation and Regression Line
- Confidence Interval for Slope and Intercept of Regression Line

2.1
Measures of Central Tendency and Dispersion

It is often difficult to understand the characteristics of large data sets. Hence, we look to summarize certain characteristics of a data set with one or two numerical values. Two such important characteristics are the "middle" of the data set, a measure of central tendency, and the "spread or variability," a measure of dispersion. Measures of central tendency are the mode, median, and mean, while the standard deviation and variance are used to measure dispersion.

The **mode** is the value in the data set that occurs most frequently. There may be more than one mode in a data set. While the mode is easy to calculate, it typically isn't a very good measure of central tendency.

The **median** is the middle value in the data set. To calculate the median, the data set has to be ordered either smallest to largest or largest to smallest value. If there are an odd number of data points, then the median is the middle data point in the sorted list. If there are an even number of data points then the median is the average of the middle two data points. One advantage of the median is that it is resistant to outliers. In other words, a very large or very small data point will not affect the median.

The **mean** or arithmetic mean is the average of the data points. If your data points are a sample of size n from a larger population, then the mean is referred to as the sample mean and given by

$$\bar{x} = \sum_{i=1}^{n} \frac{x_i}{n} \qquad [2.1]$$

The sample mean is an unbiased estimator of the population mean μ; in other words, on average the sample mean will equal the population mean. Unlike the median, the mean is not resistant. One outlier can affect the value of the sample mean. A **weighted mean** is calculated by

The median is resistant to outliers, while the mean is not resistant to outliers.

Note: All tables needed to solve problems in the text and chapter problems can be found in the *Fundamentals of Engineering Supplied Reference Handbook* available through NCEES.

$$\bar{x}_w = \frac{\sum_{i=1}^{n} w_i x_i}{\sum_{i=1}^{n} w_i} \qquad [2.2]$$

where w_i is the weight associated with x_i. Two other means are the geometric mean

$$(x_1 x_2 x_3 \cdots x_n)^{1/n}, \qquad [2.3]$$

with each $x_i > 0$, and the **root-mean-square**

Note that unless all numbers are equal the geometric mean is less than the arithmetic mean.

$$x_{rms} = \sqrt{\frac{\sum_{i=1}^{n} x_i^2}{n}} \qquad [2.4]$$

If we have data for the entire population, then the **population standard deviation** is

$$\sigma = \sqrt{\frac{\sum_{i=1}^{N} (x_i - \mu)^2}{N}} \qquad [2.5]$$

where N is the size of the population. On the other hand, if we have a sample of size n from a population, then the **sample standard deviation** is

$$s = \sqrt{\frac{\sum_{i=1}^{n} (x_i - \bar{x})^2}{n-1}} \qquad [2.6]$$

The sample standard deviation is an unbiased estimator of the population standard deviation. The variance is simply the square of the standard deviation, so the population variance is σ^2 and the sample variance is s^2. To measure the relative variation of a data set with respect to the mean we use the **coefficient of variation**, which is $\frac{s}{\bar{x}}$.

The best single numerical estimate or point estimate for μ is $\bar{x}$ and the point estimate for σ is s.

Example 2.1
The city gas mileage of a sample of six different cars is 19, 23, 23, 17, 18, and 32. Find the mean, median, mode, and standard deviation of this data set.

Solution:
The mean is

$$\bar{x} = \frac{19+23+23+17+18+32}{6} = 22$$

Ordering the data set from smallest to largest 17, 18, 19, 23, 23, 32, the median is the average of 19 and 23. Hence, the median is 21. Finally, the mode, the most frequent value, is 23. The standard deviation is

$$s = \sqrt{\frac{(19-22)^2+(23-22)^2+(23-22)^2+(17-22)^2+(18-22)^2+(32-22)^2}{6-1}} = 5.5$$

Example 2.2
Find the mean, median, and sample standard deviation of the three data sets below and comment on the similarities and differences.

D1: 1, 2, 3, 4, 5, 6, 7
D2: 1, 1, 1, 4, 7, 7, 7
D3: 11, 12, 13, 14, 15, 16, 17

Solution:
The results are

	Mean	Median	Standard Deviation
D1	4	4	2.16
D2	4	4	3.00
D3	14	14	2.16

All three data sets are symmetric, and the mean and median are equal. Data sets D1 and D2 have the same center, but D2 has greater variation or dispersion, so the standard deviation is larger. Data sets D1 and D3 are the same except shifted by 10, so the mean and median are also shifted by 10. On the other hand, the dispersion is the same for the two data sets, so they have the same standard deviation.

2.2
Combinations and Permutations

How many 5-card hands can be dealt from a standard 52-card deck of playing cards? If we have 10 different books, then how many ways can 7 books be selected and lined up on a shelf? Both questions are counting something, but there is a distinct difference. In the case of the card hand, order is unimportant. A card

is in the hand no matter if it is dealt first or last. The card hand is a **combination**, or more specifically a 5-combination of a set of 52 objects. The order of the cards, or the objects, doesn't matter. On the other hand, the books clearly have an order on the shelf from left to right. The arrangement of books is a **permutation,** or more specifically a 7-permutation of a set of 10 objects. It is important to note that the books have a specific order. Recall that $n! = n(n-1) \ldots (2)(1)$. The number of k-combinations of a set of n objects, with $k \le n$, is

$$\binom{n}{k} = C(n,k) = \frac{n!}{k!(n-k)!} \qquad [2.7]$$

> If the set you are counting has no order use $C(n,x)$. If the set should be listed with an order, use $P_n(x)$.

The number of k-permutations of a set of n objects, with $k \le n$, is

$$P_n(k) = \frac{n!}{(n-k)!} \qquad [2.8]$$

Example 2.3
How many 5-card hands can be dealt from a standard 52-card deck of playing cards?

Solution:
Since order is unimportant, we want the number of 5-combinations from a set of 52 objects. Thus, the number of 5-card hands is

$$C(52,5) = \frac{52!}{5!(52-5)!} = 2,598,960$$

Example 2.4
If we have 10 different books, then how many ways can 7 books be selected and lined up on a shelf?

Solution:
Since order matters, we want the number of 7-permuations from a set of 10 objects. Thus, the number of arrangements is

$$P_{10}(7) = \frac{10!}{(10-7)!} = 604,800$$

2.3
General Probability

If we flip a fair coin 10 times, there are a number of probabilistic questions we could ask. For instance, what is the probability of getting at least one tail? In this case the **sample space**, the set of all possible outcomes of a random experiment, is the set of all outcomes of flipping the coin 10 times and denoted by S. We define the **event** E, a subset of the sample space, as getting at least one tail in the 10 coin flips. We use $P(E)$ to represent the probability that the event E occurs. Note that $0 \le P(E) \le 1$, $P(S) = 1$, and the event E^c is when E does not occur. Complements of events can often be calculated by using $P(E^c) = 1 - P(E)$.

> The complement formula is useful when calculating the probability of at least or at most something.

For any two events E and F, the law of total probability is given by $P(E + F) = P(E) + P(F) - P(E,F)$, where $E + F$ is the union and E,F is the intersection of two events. The conditional probability of E given F is $P(E \mid F) = \dfrac{P(E,F)}{P(F)}$, and two events are defined to be independent if $P(E \mid F) = P(E)$. If two events are independent, then $P(E,F) = P(E)P(F)$.

> When possible, independence is assumed in a problem so that we can calculate $P(E,F)$ by using $P(E)P(F)$.

The conditional probability formula can be rewritten as $P(E,F) = P(E/F)P(F)$, and used to calculate $P(E,F)$.

Example 2.5

Let E be the event of getting at least one tail in 10 flips of a fair coin, and let F be the event the first of the 10 coins flipped is a head. Find $P(E)$ and $P(E \mid F)$.

Solution:

Here, the event E^c is getting no tails, which is to say we get all heads, in the 10 coin flips. Since there are 2^{10} possible outcomes of flipping a coin 10 times and only one way to get all heads, $P(E) = 1 - P(E^c) = 1 - 1/2^{10} = 0.9990$.

To find $P(E \mid F)$, we use the formula $P(E \mid F) = \dfrac{P(E,F)}{P(F)}$. Now, $P(F) = 0.5$ since the coin

is a fair coin. To calculate $P(E,F)$ we recognize that for E and F to occur we must get a head on the first coin flip and then at least one tail in the remaining nine coin flips, and so we get $(0.5)(1 - 1/2^9)$. Thus,

$$P(E \mid F) = \frac{P(E,F)}{P(F)} = \frac{(0.5)(1 - 1/2^9)}{0.5} = (1 - 1/2^9) = 0.9980$$

Two points should be noted here. First, the fact that $P(E \mid F) = (1 - 1/2^9)$ is due to the fact that once we know the first coin flip is a head then we really have nine chances to get at least one tail. Second, the events E and F are not independent since $P(E \mid F) \ne P(E)$.

2.4
Probability Distributions

A random variable X is defined as any numerical outcome of a random event. Examples are the number of heads in 10 coin flips or the height of a person selected at random. There are two types of random variables. A **discrete random variable** has a finite (or countable) number of possible outcomes which are often integers. The **expected value of a discrete random variable** X is $E(X) = \sum\limits_{i=1}^{N} x_i p(x_i)$, where each x_i is a possible value of X and $p(x_i) = P(X = x_i)$. More generally, for a function $g(x)$, $E[g(X)] = \sum\limits_{i=1}^{N} g(x)_i p(x_i)$. A **continuous random variable** has a density function $f(x)$ such that $P(a \le X \le b) = \int\limits_a^b f(x)dx$. Two common random variables are binomial, which is discrete, and normal, which is continuous. The **expected value of a continuous random variable** X is $E(X) = \int\limits_{-\infty}^{\infty} xf(x)dx$. Again, for a function $g(x)$, $E[g(X)] = \int\limits_{-\infty}^{\infty} g(x)f(x)dx$. In both cases, the formula for $E[g(X)]$ can be used, for example, to calculate $E(X^2)$ by taking $g(x) = x^2$. One application of the expected value is that if we have a density function which gives the mass of an object, then the expected value is the center of gravity of the object.

> Note that for a density function we must have $f(x) \ge 0$ and $\int\limits_{-\infty}^{\infty} f(x)dx = 1$.

A **binomial random variable** arises when you have an experiment with a fixed number n of observations or trials; each trial results in two possible outcomes, referred to as a success or failure; each trial is independent; and the probability of a success, p, is the same for each observation or trial. If X is the number of successes in the n trials, then X is a binomial random variable with n trials and probability of success p. The possible outcomes for X are $\{0, 1, \ldots n\}$.

> A binomial random variable simply counts the number of successes in n trials.

For any possible outcome a

$$P(X = a) = C(n, a)p^a(1 - p)^{n-a} \qquad [2.9]$$

A **normal random variable** with mean μ and standard deviation σ has a density function

$$f(x) = \frac{1}{\sigma\sqrt{2\pi}} e^{-(x-\mu)^2/2\sigma^2} \qquad [2.10]$$

The normal random variable is associated with a bell-shaped curve because the shape of the graph of the density function is bell-shaped. Both the normal and binomial random variables are regularly used to model real world situations. If we have two independent random variables, X_1 and X_2, each sampled from normal distributions with means μ_1 and μ_2, and standard deviations σ_1 and σ_2, then $\mu_{X_1 \pm X_2} = \mu_1 \pm \mu_2$ and $\sigma^2_{X_1 \pm X_2} = \sigma^2_1 + \sigma^2_2$.

> When calculating the standard deviation of the sum or difference of two independent random variables, we must first add the variances and then take the square root; we cannot just add the standard deviations.

Moreover, the **central limit theorem** states that if $X_1, X_2, \ldots, X_n$ are independent and identically distributed random variables with mean μ and standard deviation σ, then for n large or each X_i normal,

$$\bar{X} = \frac{X_1 + X_2 + \cdots + X_n}{n}$$ is normally distributed with

mean μ and standard deviation $\dfrac{\sigma}{\sqrt{n}}$.

> When applying the central limit theorem don't forget to divide by $\sqrt{n}$ for the standard deviation.

Example 2.6

If a fair coin is flipped 20 times, what is the probability of the coin landing on heads exactly 10 times?

Solution:

If X is the number of heads in 20 coin flips, then X is binomial with $n = 20$ and $p = 0.5$. Thus

$$P(X = 10) = C(20, 10)0.5^{10}(1 - 0.5)^{20-10} = 0.176.$$

Example 2.7

Engineers have provided plans for a car that can accommodate men with sitting heights (from seat to top of head) up to 37.5 inches. If we assume the sitting heights of men are normally distributed with a mean of 36.0 inches and a standard deviation of 1.38 inches, what proportion of men cannot fit in the new car design? What is the probability that the mean height of a sample of 5 men won't fit in the car?

Solution:

Let X be the sitting height of a randomly selected man. Converting to standard normal, mean zero, and standard deviation one, we find

$$P(X > 37.5) = P\left(\frac{X - 36}{1.38} > \frac{37.5 - 36}{1.38}\right)$$
$$= P(Z > 1.1)$$
$$= 0.1357$$

> Use the unit normal distribution table, $R(x)$ column, to calculate this probabilty.

Thus, the proportion of men that cannot fit in the car is 13.6%. On the other hand, by the central limit theorem,

$$P(\bar{X} > 37.5) = P\left(\frac{\bar{X} - 36}{1.38/\sqrt{5}} > \frac{37.5 - 36}{1.38/\sqrt{5}}\right)$$
$$= P(Z > 2.4)$$
$$= 0.0082$$

2.5
Confidence Intervals

A soda dispenser is designed to fill 16 ounce cups of soda. To see if the dispenser is working properly, we sample eight cups filled by the dispenser to obtain the following data set: 15.7, 15.8, 16.1, 16.0, 15.8, 16.2, 16.1, and 15.8 ounces. The best estimate of the average amount of soda per cup μ is given by $\bar{x} = 15.94$ ounces, the **point estimator** or single number from the sample used to estimate the population parameter μ. Based on the 15.94 ounces estimate, it seems that the dispenser is underfilling the cups. Unfortunately, we don't know the accuracy of a point estimator. Constructing a confidence interval addresses this problem.

A level $(1 - \alpha)\%$ **confidence interval** (a,b) for a population mean μ is an interval constructed so that there is a $(1 - \alpha)\%$ chance that it will contain the population mean.

> A confidence interval is constructed so that the likelihood is high of it containing the parameter somewhere. However, the parameter can be anywhere in the confidence interval. The parameter is not necessarily in the middle of the confidence interval.

If the population standard deviation σ is unknown and the sample data are roughly bell shaped then the population mean μ is estimated by a t-confidence interval which is constructed by

$$\left(\bar{x} - t_{\alpha/2,n-1}\frac{s}{\sqrt{n}}, \quad \bar{x} + t_{\alpha/2,n-1}\frac{s}{\sqrt{n}} \right) \quad [2.11]$$

We call this a t-confidence interval, or t-interval, and use $t_{\alpha/2,n-1}$ in the construction of the interval since theory tells us that when σ is unknown the distribution of the sample mean fits into a t distribution with $n - 1$ degrees of freedom. On the other hand, if the population standard deviation σ is known and the sample size n is large (at least 30 is typically acceptable) or the population is normal, then the population mean μ is estimated by a z-confidence interval which is constructed by

$$\left(\bar{x} - z_{\alpha/2}\frac{\sigma}{\sqrt{n}}, \quad \bar{x} + z_{\alpha/2}\frac{\sigma}{\sqrt{n}} \right) \quad [2.12]$$

This interval is called a z-confidence interval, or z-interval, and we use $z_{\alpha/2}$ in the construction since we are using the central limit theorem which shows us that, under these assumptions, the distribution of the sample mean is normal. In this case, if we wish to construct a confidence interval with a $(1 - \alpha)\%$ level and margin of error E, half the width of the interval, then the size of the sample needed is calculated by

$$n \geq (z_{\alpha/2}\sigma / E)^2 \quad [2.13]$$

Similarly, to estimate the difference of two population means $\mu_1 - \mu_2$ from independent populations when the population variances are unknown, we use a two-sample t-interval

$$\left(\begin{array}{l} \bar{x}_1 - \bar{x}_2 - t_{n_1+n_2-2,\alpha/2}\sqrt{s_p^2\left(\frac{1}{n_1}+\frac{1}{n_2}\right)}, \\ \bar{x}_1 - \bar{x}_2 + t_{n_1+n_2-2,\alpha/2}\sqrt{s_p^2\left(\frac{1}{n_1}+\frac{1}{n_2}\right)} \end{array} \right) \quad [2.14]$$

where n_1 and n_2 are the sample sizes for the two samples, s_1^2 and s_2^2 are the variances of each sample, and

$$s_p^2 = \frac{(n_1-1)s_1^2 + (n_2-1)s_2^2}{n_1 + n_2 - 2} \quad [2.15]$$

> Use t when σ is unknown, but use z when σ is known.

If the population variances are known, we use a two-sample z-interval given by

$$\left(\begin{array}{l} \bar{x}_1 - \bar{x}_2 - z_{\alpha/2}\sqrt{\left(\frac{\sigma_1^2}{n_1}+\frac{\sigma_2^2}{n_2}\right)}, \\ \bar{x}_1 - \bar{x}_2 + z_{\alpha/2}\sqrt{\left(\frac{\sigma_1^2}{n_1}+\frac{\sigma_2^2}{n_2}\right)} \end{array} \right) \quad [2.16]$$

A confidence interval for the variance of a normal distribution is given by

$$\frac{(n-1)s^2}{\chi_{\alpha/2,n-1}^2} \leq \sigma^2 \leq \frac{(n-1)s^2}{\chi_{1-\alpha/2,n-1}^2} \quad [2.17]$$

where $\chi_{\alpha/2,n-1}^2$ and $\chi_{1-\alpha/2,n-1}^2$ come from the χ^2 (pronounced chi-square) distribution table similar to $z_{\alpha/2}$ and $t_{\alpha/2,n-1}$.

Example 2.8

Find a 95% confidence interval for the soda dispenser data and decide if there is reason to believe that the machine is underfilling the cups of soda.

Solution:

Since we don't know the population standard deviation, we use the t-interval formula. We know $\bar{x} = 15.94$ ounces and can calculate $s = 0.185$ ounces. Using the t-table to find $t_{0.025,7} = 2.365$, we get

$$\bar{x} - t_{n-1,\alpha/2}\frac{s}{\sqrt{n}} = 15.94 - 2.365(0.185/\sqrt{8}) = 15.79$$

$$\bar{x} + t_{n-1,\alpha/2}\frac{s}{\sqrt{n}} = 15.94 + 2.365(0.185/\sqrt{8}) = 16.09$$

We are 95% certain that $15.79 \le \mu \le 16.09$. Thus we cannot conclude that there is reason to believe that the machine isn't filling each cup with an average of 16 ounces of soda since 16 ounces is a possible value for μ based on the confidence interval.

2.6
Hypothesis Testing for One Mean

Again, using our soda dispenser data we might simply ask the question, Is our soda dispenser filling cups on average with 16 ounces of soda? One way to answer this question is with a hypothesis test. To carry out a **hypothesis test** we need to choose a level of significance α, state the null and alternative hypothesis (H_0 and H_a), calculate a test statistic, find the critical value, and compare the test statistic with the critical value to decide whether or not there is sufficient evidence to reject H_0.

> The null and alternative hypotheses are statements about an unknown parameter such as μ or σ.

As with confidence intervals, the test we choose depends on whether or not we know the population standard deviation. Following the soda dispenser example, we will use level of significance $\alpha = 0.05$, which is common. The null hypothesis is $H_0 : \mu = 16$ ounces, which is read as the soda dispenser fills each cup, on average, with 16 ounces of soda. The alternative hypothesis is $H_a : \mu \ne 16$, which says that the soda dispenser does not fill each cup with 16 oz of soda. Both hypotheses are statements about the population mean, which is what the dispenser actually does, and the point of a hypothesis test is to determine which hypothesis is most likely to be valid. For a t-test, the case when the population standard deviation is unknown, the test statistic is $t = \frac{\bar{x} - \mu_0}{s/\sqrt{n}}$, where μ_0 is the value against which we are testing the population mean. Here,

$$t = \frac{\bar{x} - \mu_0}{s/\sqrt{n}} = \frac{15.94 - 16.00}{0.185/\sqrt{8}} = -0.917.$$

Since this is a two-sided test, the critical value is $\pm t_{\alpha/2,7} = \pm 2.365$. If the test statistic $t = -0.917$ is beyond the critical values, in other words, if $t \le -t_{\alpha/2,7}$ or $t \ge t_{\alpha/2,7}$, then we reject the null hypothesis in favor of the alternative; otherwise we fail to reject the null hypothesis. In this example we don't reject the null hypothesis. If $H_a : \mu < \mu_0$ or $H_a : \mu > \mu_0$, referred to as one-sided tests, then we reject the null hypothesis if $t \le -t_{\alpha,n-1}$ in the first case and if $t \ge t_{\alpha,n-1}$ in the second case.

The only difference for a z-test, the case when the population standard deviation is known, is that the test statistic is $z = \frac{\bar{x} - \mu_0}{\sigma/\sqrt{n}}$ and the critical value is given by $z_{\alpha/2}$ in the two-sided case and $\pm z_\alpha$ in the one-sided case, depending on the inequality in the alternative hypothesis.

2.7
Design of Experiment

Let α be the probability of a type I error, also known as the level of significance, and β be the probability of a type II error. If we want to test $H_1 : \mu \ne \mu_0$ and the population mean is known to be μ_1 then we need a sample size $n \cong \dfrac{(z_{\alpha/2} + z_\beta)^2 \sigma^2}{(\mu_1 - \mu_0)^2}$.

> Always round up when calculating a sample size.

Similarly, if we want to test $H_1 : \mu > \mu_0$ and the population mean is known to be μ_1 then the sample size needed is $n \cong \dfrac{(z_\alpha + z_\beta)^2 \sigma^2}{(\mu_1 - \mu_0)^2}$. It is less likely that μ_1 is known and more likely that a researcher wants to be able to detect a difference in the means, and so a value for $\mu_1 - \mu_0$ is known.

Example 2.9
A researcher is planning to do a hypothesis test with $H_1 : \mu \ne \mu_0$. The researcher would like the probability of type I error to be 0.05 and the probability of a type II error to be 0.20. Moreover, she would like to be able to detect a difference between μ_0 and μ_1 of 5. Assume that the population standard deviation is 17.5. Approximately, what sample size is necessary?

Solution:

We use the formula $n \cong \dfrac{(Z_{\alpha/2} + Z_\beta)^2 \sigma^2}{(\mu_1 - \mu_2)^2}$ $\alpha = 0.05$ and

$\beta = 0.20$. From the unit normal table, we find $Z_{\alpha/2} \cong 1.95$, and $Z_\beta \cong 0.85$. Hence,

$$n \cong \frac{(Z_{\alpha/2} + Z_\beta)^2 \sigma^2}{(\mu_1 - \mu_2)^2} \cong \frac{(1.95 + 0.85)^2 (17.5)^2}{(5)^2} \cong 96.04$$

In this case we need to round up, since increasing the sample size will only improve accuracy, so $n \cong 97$.

2.8
Goodness-of-Fit Hypothesis Test

According to M&M's.com, the colors of plain M&M's are packaged according to the percentages in the table. To test this claim we sample $n = 50$ M&M's and obtain the counts in the observed column of the table. We perform **a goodness-of-fit hypothesis test** with a null hypothesis that the claimed distribution, or multinomial probabilities, given by M&M are correct. The alternative hypothesis is that at least one of the multinomial probabilities does not equal the hypothesized value.

> The table here is a good way to organize the information and do the calculations.

Color	Claim	Observed	Expected	$(O-E)^2/E$
Brown	0.13	8	6.5	0.346
Yellow	0.14	6	7	0.143
Red	0.13	9	6.5	0.962
Blue	0.24	10	12	0.333
Orange	0.20	13	10	0.900
Green	0.16	4	8	2.000
Totals	1.00	50	50	4.684

The observed column is given by our sample. The expected column is calculated by multiplying the sample size, 50 in this case, by the claimed probabilities. The test statistic is given by

$$\chi^2 = \sum \frac{(Observed - Expected)^2}{Expected}$$

In this example, $\chi^2 = 4.684$, which is the value at the bottom right of the table. We reject the null hypothesis if $\chi^2 > \chi^2_{\alpha, k-1}$, where $\chi^2_{\alpha, k-1}$ has $(k-1)$ degrees of freedom, k is the number of categories, and is

found in the chi-square table. With $\alpha = 0.05$, we get $\chi^2_{0.05,5} = 11.0705$ and hence we fail to reject that null hypothesis.

2.9
Regression

For a data set of pairs (x_i, y_i) the sample correlation coefficient $r = \dfrac{S_{xy}}{\sqrt{S_{xx}S_{yy}}}$, where

$$S_{xy} = \sum_{i=1}^{n} x_i y_i - (1/n)\left(\sum_{i=1}^{n} x_i\right)\left(\sum_{i=1}^{n} y_i\right),$$

$$S_{xx} = \sum_{i=1}^{n} x_i^2 - (1/n)\left(\sum_{i=1}^{n} x_i\right)^2, \text{ and}$$

$$S_{yy} = \sum_{i=1}^{n} y_i^2 - (1/n)\left(\sum_{i=1}^{n} y_i\right)^2,$$

measures the linear relationship between the two variables. A value or 1 or −1 means that there is a perfect linear relationship, while a value of 0 says that there is no linear relationship. If there is a linear relationship, then the linear regression line is given by $y = \hat{a} + \hat{b}x$, where $\hat{b} = S_{xy}/S_{xx}$, and $\hat{a} = \bar{y} - \hat{b}\bar{x}$. The **residual** for a particular data point (x_i, y_i) is $y_i - \hat{y}(x_i)$. The regression line is the line for which the sum of the squares of the residuals is minimized. Confidence intervals for the slope and intercept are given by

$$\hat{b} \pm t_{\alpha/2, n-2}\sqrt{\frac{MSE}{S_{xx}}} \text{ and}$$

$$\hat{a} \pm t_{\alpha/2, n-2}\sqrt{\left(\frac{1}{n} + \frac{\bar{x}^2}{S_{xx}}\right)MSE}, \text{ where}$$

$$MSE = S_e^2 = \frac{S_{xx}S_{yy} - S_{xy}^2}{S_{xx}(n-2)}.$$

> When calculating S, make a table with columns x, y, x^2, y^2, and xy, then sum the columns for the values needed.

Example 2.10

The tar and nicotine levels of eight brands of cigarettes are given in the table along with a scatter plot of the data. Find the linear regression line predicting nicotine levels based on tar levels, predict the amount of nicotine in a cigarette with 5 mg of tar, find the residual for the cigarette with 7.8 mg of tar, find a 95% confidence for the slope, find a 95% confidence interval for the intercept, and find r.

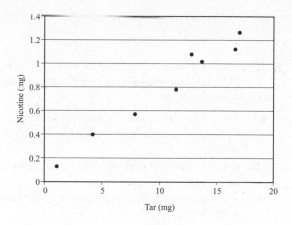

Solution:

Tar (mg)	Nicotine (mg)
1.0	0.13
7.8	0.57
16.6	1.12
12.8	1.08
4.1	0.40
17.0	1.26
13.7	1.01
11.4	0.78

First observe that the scatter plot does suggest a linear relationship. To find the regression line we need

$$S_{xx} = \sum_{i=1}^{n} x_i^2 - (1/n)\left(\sum_{i=1}^{n} x_i\right)^2$$
$$= 1124.7 - (1/8)(84.4)^2 = 234.28, \text{ and}$$

$$S_{xy} = \sum_{i=1}^{n} x_i y_i - (1/n)\left(\sum_{i=1}^{n} x_i\right)\left(\sum_{i=1}^{n} y_i\right)$$
$$= 82.781 - (1/8)(84.4)(6.35) = 15.7885.$$

Now, $\hat{b} = S_{xy}/S_{xx} = 15.7885/234.28 = 0.067$, and $\hat{a} = \bar{y} - \hat{b}\bar{x} = 6.35/8 - (0.067)(84.4/8) = 0.0869$, so that $y = 0.0869 + 0.067x$ (you might sketch this line on the scatter plot above to see how it fits the data). We now predict that a cigarette with 5 mg of tar has $y(5) = 0.0869 + 0.067(5) = 0.4219$ mg of nicotine. The residual for a cigarette with 7.8 mg of tar is $0.57 - y(7.8) = -0.395$.

For the confidence intervals, we first calculate

$$S_{yy} = \sum_{i=1}^{n} y_i^2 - (1/n)\left(\sum_{i=1}^{n} y_i\right)^2$$
$$= 6.1387 - (1/8)(6.35)^2 = 1.098$$

and $MSE = \dfrac{S_{xx}S_{yy} - S_{xy}^2}{S_{xx}(n-2)}$

$$= \frac{(234.28)(1.098) - (15.7885)^2}{234.28(8-2)} = 0.00566.$$

Using the Student's t-distribution table, $t_{0.05/2,6} = 2.447$. Hence,

$$\hat{b} \pm t_{\alpha/2,n-2}\sqrt{\frac{MSE}{S_{xx}}}$$
$$= 0.067 \pm 2.447\sqrt{0.00566/234.28}$$

or $(0.055, 0.079)$, and

$$\hat{a} \pm t_{\alpha/2,n-2}\sqrt{\left(\frac{1}{n} + \frac{\bar{x}^2}{S_{xx}}\right)MSE}$$
$$= 0.0869 \pm 2.447\sqrt{\left(\frac{1}{8} + \frac{10.55^2}{234.28}\right)(0.00566)}$$

or $(-0.0557, 0.2295)$. Lastly, note that

$$r = \frac{S_{xy}}{\sqrt{S_{xx}S_{yy}}} = \frac{15.7885}{\sqrt{(234.28)(1.098)}} = 0.9844.$$

2.10 Summary

Probability and statistics are tools used for decision making. In statistics we are looking to obtain information about an unknown population parameter. We may want to know if a coin is fair, the mean amount of soda a machine puts in a container, or the mean household income in the country. In each case the parameter is unobtainable for one reason or another; we cannot know for certain that a coin is fair, a soda dispenser may be designed to put a certain amount of soda is a container but it could be defective, and there are too many people to sample to know the mean household income. In each case, we have statistical techniques based on probability to estimate the parameter and make a decision. For instance, a binomial random variable shows that the probability of obtaining 9 or 10 tails in 10 flips of a fair coin is 0.01. Hence, if a coin, unknown to be fair or not, is flipped 10 times and 9 or 10 tails occurs we would decide that the coin is unfair, since it is unlikely to observe so many tails with a fair coin. This is the general idea of a hypothesis test, which we use to assess the fairness of our coin or if the soda machine is dispensing the correct amount. We assume the null hypothesis is true and calculate the likelihood of our outcome by comparing a test statistic to a critical value. Similarly, a confidence interval estimates a parameter with 90 or 95% accuracy. Hence, we estimate the mean household income with a sample. With both hypothesis tests and confidence intervals, we can obtain information about an unknown parameter and use this to make a decision.

PRACTICE PROBLEMS

1. Suppose five cards are randomly selected from a standard deck of 52 cards. What is the probability that at least one card is not a heart?

    (A) 0.000495
    (B) 0.010729
    (C) 0.778466
    (D) 0.999505

2. The percentages of adults in each of six northeastern states (ME, MA, VT, CT, NH, NY) that consume fruits and vegetables five or more times a day are 27.0, 29.0, 32.5, 29.8, 28.5, and 25.8. Find the mean and sample standard deviation of the percentage of adults that consume fruits and vegetables five or more times a day in the six northeastern states. (Source: statemaster.com)

    (A) $\bar{x} = 28.77$ $s = 2.33$
    (B) $\bar{x} = 28.77$ $s = 2.12$
    (C) $\bar{x} = 28.75$ $s = 2.33$
    (D) $\bar{x} = 28.75$ $s = 2.12$

3. In a data set with $n = 1292$ it is found that $\bar{x} = \$125,000$ and the median is $\$50,000$. Based on this information,

    (A) we can conclude that the data set is roughly symmetric and neither skewed to the right or left
    (B) we can conclude that the data set is skewed to the left
    (C) we can conclude that the data set is skewed to the right
    (D) we can't make any statements about whether the data set is skewed to the right or left

Questions 4–6

Use the data below, a sample of nine polishing times for bowls and their diameters, for the next three problems.

Diameter (inches)	7.4	5.4	7.5	14.0	7.0	9.0	12.0	5.5	6.0
Time (min)	16.41	12.02	20.21	32.62	17.84	22.82	29.48	15.61	13.25

4. Use the linear regression line to estimate the polishing time of a 10.0 inch diameter bowl.

    (A) 20.80 minutes
    (B) 24.22 minutes
    (C) 23.33 minutes
    (D) 25.25 minutes

5. Find a 95% confidence interval for the slope of the regression line predicting polishing time based on bowl diameter.

    (A) (−0.01, 0.87)
    (B) (1.89, 2.77)
    (C) (1.91, 2.75)
    (D) (1.97, 2.69)

6. Find a 95% confidence interval for the intercept of the regression line predicting polishing time based on bowl diameter.

(A) (−5.50, 6.89)

(B) (−2.72, 4.56)

(C) (−2.89, 4.73)

(D) (−2.24, 4.08)

7. A machine is expected to fill bags of chips with 32 ounces of chips and has a standard deviation of 0.3 ounces. Assuming that the distribution of chips in the bag is normal, what is the probability that a sample of 10 bags will have a mean greater than 32.17 ounces?

(A) 0.0359

(B) 0.4207

(C) 0.5793

(D) 0.9641

8. Let X be a random variable with density function $f(x) = kx^2$ for $1 \leq x \leq 3$ and 0 elsewhere. The expected value of X is

(A) 2.000

(B) 2.308

(C) 2.051

(D) 2.455

9. A researcher is planning to do a hypothesis test with $H_1 : \mu > \mu_0$. The test will be done with a level of significance of 0.05. It is desired that the test be able to detect a difference between μ_0 and μ_1, the actual population mean, of 1.77 and that the probability of a type II error be 0.15. Assume that the population standard deviation is 7.5. Approximately what sample size is necessary?

(A) 126

(B) 127

(C) 157

(D) 158

10. A gardener plants 26 tomato seeds from a packet of seeds that guarantees an 85% germination rate. What is the probability that 24 or fewer of the seeds will germinate?

(A) 0.77

(B) 0.85

(C) 0.92

(D) 0.99

11. A large family owns five chickens. In the past, the five chickens laid an equal amount of eggs. Over the last four weeks the five chickens laid 34, 28, 16, 17, and 25 eggs each. Use a hypothesis test to determine if the chickens are still laying eggs in equal amounts. Based on the hypothesis test,

(A) at the 0.05 and 0.01 level of significance we conclude that the distribution has not changed

(B) at the 0.05 but not at the 0.01 level of significance we conclude that the distribution has changed

(C) at the 0.01 but not at the 0.05 level of significance we conclude that the distribution has changed

(D) at the 0.05 and 0.01 level of significance we conclude that the distribution has changed

12. A student wonders whether or not the grade point average for the junior year in college is different than the grade point average for the first year of college. She finds 10 students to give her their grade point average for their junior year and 10 other students to give her their grade point average for their first year of college. The data is in the table below. What is a 99% confidence interval for the mean difference of first year and junior year grade point averages?

First Year	2.56	2.75	3.16	3.58	3.23	3.89	2.12	3.30	2.43	2.35
Junior Year	2.43	3.52	2.77	3.76	2.71	3.46	3.39	3.64	3.12	3.05

(A) (−0.843, 0.347)

(B) (−0.913, 0.417)

(C) (−0.999, 0.503)

(D) (−1.169, 0.673)

13. To check and see if a random number generator is working properly, twelve values from a unit normal distribution are generated. They are –0.5, –1.3, –0.7, 2.1, 0.1, 1.1, –0.2, 0.5, –0.2. –1.7, –0.9, and 1.6. What is a 95% confidence interval for the variance of this data set?

(A) (0.5440, 2.8825)
(B) (0.5791, 3.3267)
(C) (0.6277, 3.3265)
(D) (0.6683, 3.8391)

14. A couple has three children, and it is known that at least one of the children is a girl. What is the probability that the couple has at least one boy, assuming that having a boy or a girl is equally likely?

(A) 0.875
(B) 0.857
(C) 0.750
(D) 0.500

15. A sample of 23 plates has a mean polishing time of 32.12 minutes, and the process is known to have a population standard deviation of 11.0 minutes. A sample of 37 cookware objects has a mean polishing time of 41.36 minutes, and the process is known to have a population standard deviation of 21.0 minutes. Find a 90% confidence interval for the difference in mean polishing time.

(A) (–17.36, –1.12)
(B) (–16.06, –2.42)
(C) (–11.24, –7.24)
(D) (–10.92, –7.55)

16. In 2006, the number of home runs hit by each of the 16 National League teams is 88, 71, 75, 84, 59, 71, 61, 59, 68, 54, 92, 60, 82, 66, 44, and 99. Find a 99% confidence interval for the mean number of home runs hit by National League teams.

(A) (59.75, 81.87)
(B) (61.14, 80.48)
(C) (63.46, 78.16)
(D) (64.23, 77.39)

17. In 2006, the number of home runs hit by each of the 16 National League teams is 88, 71, 75, 84, 59, 71, 61, 59, 68, 54, 92, 60, 82, 66, 44, and 99. Perform a hypothesis test to see if the average number of team home runs in the National League is different from 64, which is the average for the American League.

(A) At both the $\alpha = 0.05$ and $\alpha = 0.01$ levels of significance there is sufficient evidence that shows that the National League teams hit a different number of home runs.
(B) At the $\alpha = 0.05$ but not at the $\alpha = 0.01$ level of significance there is sufficient evidence that shows that the National League teams hit a different number of home runs.
(C) At the $\alpha = 0.01$ but not at the $\alpha = 0.05$ level of significance there is sufficient evidence that shows that the National League teams hit a different number of home runs.
(D) At both the $\alpha = 0.05$ and $\alpha = 0.01$ levels of significance there is not sufficient evidence that shows that the National League teams hit a different number of home runs.

18. A bag contains 40 red chips and 60 purple chips. If we select 40 chips, with replacement, what is the probability that 30 of these chips will be purple?

(A) 0.01
(B) 0.02
(C) 0.03
(D) 0.04

Questions 19–20

The expense of moving a warehouse is justified only if it can be shown that the daily mean travel distance will be less than 250 miles. In trial runs of 35 delivery trucks, the mean distance is 225 miles. The company assumes that $\sigma = 45$.

19. What is a 96% confidence interval for the mean distance traveled?

(A) (235.09, 264.91)
(B) (234.38, 265.62)
(C) (210.09, 239.91)
(D) (209.38, 240.62)

20. Use a hypothesis test to determine if the distance is less than 250 miles.

    (A) At both the $\alpha = 0.05$ and $\alpha = 0.01$ levels of significance there is sufficient evidence that shows that we should move the warehouse.

    (B) At the $\alpha = 0.05$ but not at the $\alpha = 0.01$ level of significance there is sufficient evidence that shows that we should move the warehouse.

    (C) At the $\alpha = 0.01$ but not at the $\alpha = 0.05$ level of significance there is sufficient evidence that shows that we should move the warehouse.

    (D) At both the $\alpha = 0.05$ and $\alpha = 0.01$ levels of significance there is not sufficient evidence that shows that we should move the warehouse.

21. What is the geometric mean of the numbers 1 through 10?

    (A) 1.49
    (B) 4.53
    (C) 5.00
    (D) 6.20

22. Suppose X is a normal random variable with a mean of 95 and a standard deviation of 3.5. What is $P(X < 97.5)$?

    (A) 0.2420
    (B) 0.3123
    (C) 0.4839
    (D) 0.7580

Answer Key

1. D	**6.** C	**11.** B	**15.** B	**19.** D
2. A	**7.** A	**12.** B	**16.** A	**20.** A
3. C	**8.** B	**13.** D	**17.** D	**21.** B
4. B	**9.** B	**14.** B	**18.** B	**22.** D
5. B	**10.** C			

Answers Explained

1. D We first calculate the probability of all hearts in the hand. The total number of five-card hands is $C(52,5)$, since order doesn't matter. There are 13 cards in the deck that are hearts, so there are $C(13,5)$ hands that are all hearts. Hence, the probability of a five-card hand containing all hearts is

$$\frac{C(13,5)}{C(52,5)} = 0.000495$$

Now,

$$P(\text{At Least One Non-Heart}) = 1 - P(\text{All Hearts})$$
$$= 1 - 0.000495$$
$$= 0.999505$$

2. A The mean is

$$\bar{x} = \frac{27.0 + 29.0 + 32.5 + 29.8 + 28.5 + 25.8}{6} = 28.77$$

The standard deviation is

$$s = \sqrt{\frac{6(27.0^2 + 29.0^2 + 32.5^2 + 29.8^2 + 28.5^2 + 25.8^2) - (27.0 + 29.0 + 32.5 + 29.8 + 28.5 + 25.8)^2}{6(6-1)}}$$
$$= 2.33$$

3. C The fact that the mean is larger than the median tells us that the data set is skewed to the right.

4. B To find the regression line we need

$$S_{xx} = \sum_{i=1}^{n} x_i^2 - (1/n)\left(\sum_{i=1}^{n} x_i\right)^2$$
$$= 676.42 - (1/9)(73.8)^2 = 71.26 \text{ and}$$

$$S_{xy} = \sum_{i=1}^{n} x_i y_i - (1/n)\left(\sum_{i=1}^{n} x_i\right)\left(\sum_{i=1}^{n} y_i\right)$$
$$= 1643.972 - (1/9)(73.8)(180.26) = 165.84$$

Now, $\hat{b} = S_{xy} / S_{xx} = 165.84 / 71.26 = 2.33$, and

$\hat{a} = \bar{y} - \hat{b}\bar{x} = 180.26 / 9 - (2.33)(73.8 / 9) = 0.92$,

so that $y(10) = 0.92 + 2.33(10) = 24.22$ minutes.

5. B To construct the confidence interval we first calculate

$$S_{yy} = \sum_{i=1}^{n} y_i^2 - (1/n)\left(\sum_{i=1}^{n} y_i\right)^2$$
$$= 4013.6 - (1/9)(180.26)^2 = 403.19,$$

and

$$MSE = \frac{S_{xx}S_{yy} - S_{xy}^2}{S_{xx}(n-2)}$$
$$= \frac{(71.26)(403.19) - (165.84)^2}{71.26(9-2)} = 2.46.$$

Using the student's t-distribution table we have $t_{0.05/2.7} = 2.365$. Hence, we have

$$\hat{b} \pm t_{\alpha/2, n-2}\sqrt{\frac{MSE}{S_{xx}}} = 2.33 \pm 2.365\sqrt{2.46 / 71.26}$$

or $(1.89, 2.77)$.

6. C Based on the work already done and noting that $\bar{x} = 8.2$, we have

$$\hat{a} \pm t_{\alpha/2, n-2}\sqrt{\left(\frac{1}{n} + \frac{\bar{x}^2}{S_{xx}}\right)MSE}$$

$$= 0.92 \pm 2.365\sqrt{\left(\frac{1}{9} + \frac{8.2^2}{71.26}\right)(2.46)},$$

or $(-2.89, 4.73)$.

7. A Since we are told that the population is normal, we know that the distribution of the sample mean is normal with a mean of 32 and a standard deviation of $0.3/\sqrt{10} = 0.0949$. Using the unit normal tables, we find that

$$P(\bar{x} > 32.17) = P\left(\frac{\bar{x} - 32}{0.0949} > \frac{32.17 - 32}{0.0949}\right)$$

$$= P(z > 1.8) = 0.0359$$

8. B First we find k by solving $1 = \int_1^3 kx^2 dx$. Since

$\int_1^3 kx^2 dx = k(3^3)/3 - k(1^3)/3$, we find that

$k = 3/26$. To find the expected value we have

$$\int_1^3 \frac{3x^3}{26} dx = \frac{3(3^4)}{26(4)} - \frac{3(1^4)}{26(4)} = 2.308.$$

9. B We use the formula $n \cong \frac{(Z_\alpha + Z_\beta)^2 \sigma^2}{(\mu_1 - \mu_2)^2}$, where

$\alpha = 0.05$ and $\beta = 0.15$. From the unit normal table we find $Z_\alpha \cong 1.65$ and $Z_\beta \cong 1.0$. Hence,

$$n \cong \frac{(Z_\alpha + Z_\beta)^2 \sigma^2}{(\mu_1 - \mu_2)^2} \cong \frac{(1.65 + 1.0)^2 (7.5)^2}{(1.77)^2} \cong 126.086.$$

In this case, we need to round up, since increasing the sample size will only improve accuracy, so $n \cong 127$.

10. C Let X represent the number of seeds that germinate, and note that X is a binomial random variable with $n = 26$ and $p = 0.85$. Now,

$P(X \leq 24) = 1 - P(X = 25) - P(X = 26)$

$= 1 - C(26, 25)(0.85)^{25}(1 - 0.85)^{26-25} - C(26, 26)(0.85)^{26}(1 - 0.85)^{26-26}$

$= 1 - 0.0671 - 0.0146$

$= 0.9183$

11. B We perform a goodness-of-fit hypothesis test. The null hypothesis states that all proportions have stayed the same, i.e., each chicken lays 20% of the eggs, while the alternative is that at least one is different. Creating a table to organize our calculations we have:

Chicken	Claim	Observed	Expected	(O—E)²/E
1	0.20	34	24	4.1667
2	0.20	28	24	0.6667
3	0.20	16	24	2.6667
4	0.20	17	24	2.0417
5	0.20	25	24	0.0417
Totals	1.00	120	120	9.5835

The test statistic is given by

$$\chi^2 = \sum \frac{(Observed - Expected)^2}{Expected}$$

We have $\chi^2 = 9.5835$, which is the value in the bottom right of the table. The critical values of interest are $\chi^2_{0.05,4} = 9.48773$ and $\chi^2_{0.01,4} = 13.2767$. We reject the null hypothesis if $\chi^2 > \chi^2_{\alpha,4}$. Hence, we reject the null hypothesis at the 0.05 but not at the 0.01 level of significance.

12. B Let μ_1 be the mean grade point average of first year students and μ_2 be the mean grade point average of juniors. We construct a t-interval for $\mu_1 - \mu_2$. For a 99% confidence interval with $n_1 + n_2 - 2 = 18$ degrees of freedom, we have $t_{0.01/2,18} = 2.878$. For the confidence interval

$$\sqrt{\frac{(1/n_1 + 1/n_2)\left[(n_1 - 1)s_1^2 + (n_2 - 1)s_2^2\right]}{n_1 + n_2 - 2}}$$

$$= \sqrt{\frac{(1/10 + 1/10)\left[(9)(0.581)^2 + (9)(0.442)^2\right]}{10 + 10 - 2}} = 0.231$$

and so

$$2.937 - 3.185 - 2.878(0.231) \leq \mu_1 - \mu_2$$

$$\mu_1 - \mu_2 \leq 2.937 - 3.185 + 2.878(0.231)$$

and so

$$-0.913 \leq \mu_1 - \mu_2 \leq 0.417$$

13. D The confidence interval for a variance is
$\frac{(n-1)s^2}{\chi^2_{\alpha/2,n-1}} \le \sigma^2 \le \frac{(n-1)s^2}{\chi^2_{1-\alpha/2,n-1}}$. From the χ^2 table we
have $\chi^2_{0.025,11} = 21.9200$ and $\chi^2_{0.975,11} = 3.81575$.
The variance of the data set is $s^2 = 1.33174$, and
so our confidence interval is $(0.6683, 3.8391)$.

14. B The sample space can be listed as $\{bbb, bbg,$
$bgb, gbb, ggg, ggb, gbg, bgg\}$ where g represents
a girl and b represents a boy. Let B be the event
that the couple has at least one boy and G be the
event that the couple has at least one girl. We
now have

$$P(B \mid G) = \frac{P(B, G)}{P(G)}$$
$$= \frac{P(\{bbg, bgb, gbb, ggb, gbg, bgg\})}{P(\{bbg, bgb, gbb, ggb, gbg, bgg, ggg\})}$$
$$= \frac{6/8}{7/8} = 6/7 = 0.857$$

15. B Using the formula for the confidence interval for the difference between two means with
known population standard deviations, we have

$$\bar{x}_1 - \bar{x}_2 \pm z_{\alpha/2}\sqrt{\frac{\sigma_1^2}{n_1} + \frac{\sigma_2^2}{n_2}}$$
$$= 32.12 - 41.36 \pm 1.645\sqrt{\frac{11^2}{23} + \frac{21^2}{37}}.$$

The confidence interval is $(-16.06, -2.42)$.

16. A Since the population standard deviation is
unknown, we use a t-interval, which is
$\bar{x} \pm t_{\alpha/2,n-1}\frac{s}{\sqrt{n}}$. Using the student t-distribution
table we find $t_{0.005,15} = 2.947$. Thus, the interval is
$$\left(70.81 - 2.947(15.01/\sqrt{16}), \quad 70.81 + 2.947(15.01/\sqrt{16})\right)$$
$$= (59.75, 81.87).$$

17. D We perform a hypothesis test with a
null hypothesis of $\mu = 64$ and an alternative
of $\mu \ne 64$. The test statistic is
$t = \frac{\bar{x} - \mu_0}{s/\sqrt{n}} = \frac{70.81 - 64}{15.01/\sqrt{16}} = 1.81$. We would
reject the null hypothesis at a level of sig-
nificance of $\alpha = 0.05$ if the test statistics were
beyond ± 2.131, and at $\alpha = 0.01$ if they were beyond ± 2.947. The null hypothesis is not rejected
in both cases.

18. B Let X represent the number of purple chips selected and note that X is a binomial random variable with $n = 40$ and $p = 0.6$. Thus, $P(X = 30) =$
$C(40,30)(0.6)^{30}(1 - 0.6)^{40-30} = 0.0196$.

19. D Since the population standard deviation is known, we use a z-interval, which is
$\bar{x} \pm z_{\alpha/2}\frac{\sigma}{\sqrt{n}}$. Thus, the interval is
$$\left(225 - 2.0537(45/\sqrt{35}), \quad 225 + 2.0537(45/\sqrt{35})\right)$$
$$= (209.38, 240.62).$$

20. A We perform a hypothesis test with a null hypothesis of $\mu = 250$ and an alternative of $\mu < 250$.
Note that this is a one-tailed test. The test statis-
tic is $z = \frac{\bar{x} - \mu_0}{\sigma/\sqrt{n}} = \frac{225 - 250}{45/\sqrt{35}} = -3.287$. Using
the unit normal distribution table, we find that
the critical value for a level of significance of
$\alpha = 0.05$ is about -1.65, whereas the critical
value for a level of significance of $\alpha = 0.01$ is
about -2.30. Thus, we reject the null hypothesis
at $\alpha = 0.05$ and at $\alpha = 0.01$.

21. B The geometric mean is $\sqrt[10]{(1)(2)\cdots(10)} = 4.53$.

22. D Converting to unit normal and using the table,
we have
$$P(X \le 97.5) = P\left(\frac{X - 95}{3.5} \le \frac{97.5 - 95}{3.5}\right)$$
$$= P(Z \le 0.7) = 0.758$$

Chemistry

Armen S. Casparian

Introduction

Chemistry is the branch of physical science that studies matter and the changes or "molecular rearrangements" it can undergo. These rearrangements, commonly known as chemical reactions, involve the breaking of existing chemical bonds between atoms to form new bonds with other atoms. In the process, different molecules with different properties are formed. Electrons may be gained or lost, and energy changes generally accompany the reaction. Chemical reactions, as distinguished from nuclear reactions, involve the exchange of only electrons — never protons or neutrons. In a sense, chemistry is the recycling of atoms.

Visible evidence of chemical reactions includes

- Bubbling or fizzing, indicating the release of a gas
- Color change
- Temperature change
- Formation of a precipitate
- Dissolving of a solid

Examples of common chemical reactions include

- Rusting of iron or corrosion of any metal
- Generation of current by a battery
- Combustion of fuel to produce energy
- Neutralization of excess stomach acid by an antacid
- Hardening of concrete

Chemistry also studies the structure of matter, including chemical and physical properties, correlating properties on the microscopic scale with behavior observed on the macroscopic scale. Included are such properties as vapor pressure, osmotic presssure, solubility, boiling and melting points, and energy and its transformations. Many of these properties may dictate or influence the outcome of a reaction. [The three transport properties — diffusion, viscosity, and thermal conductivity — are not discussed in this chapter.]

This chapter consists of 12 sections with individual topics discussed under each section. Solved examples reflecting fundamental principles and applications are included in most sections. Additional practice problems with answers are provided at the end of the chapter.

3.1
Basic Concepts, Definitions, Notation, and Units

Symbols and Notation

A. Chemistry is based on the premise that all matter is composed of some combination of the 92 naturally occurring elements. These elements are arranged in a particular order, known as **the Periodic Chart** or **Periodic Table,** in horizontal rows called **periods,** by increasing atomic number, and in vertical columns called **groups or families,** by similar electron configurations, which in turn give rise to similar chemical properties. **Electron configurations** are ordered arrangements of electrons based on specific housing rules. Energy levels range from 1 to 7, and within each energy level or shell are sublevels or orbitals s, p, d, and f. Each of the four orbitals has a distinct shape and population limit: 1 s orbital with a maximum of 2 electrons; 3 p orbitals for a maximum of 6 electrons; 5 d orbitals for a maximum of 10 electrons; 7 f orbitals for a maximum of 14 electrons.

B. Members of the same family are called **congeners.** [Elements after uranium (atomic number 92), that is, elements 93 through 109, are manmade, radioactive, and increasingly unstable, and are not addressed in this chapter.] Some rows of the Periodic Table have names, such as the actinide (row 6) or lanthanide (row 7) series, while columns have such names as alkali metals (column 1A), alkaline-earth metals (column 2A), transition metals, halogens (column 7A), and noble (inert) gases (column 8A).

In general, oxides of nonmetals, when dissolved in water, are acidic. Oxides of metals are alkaline or basic. Thus, substances such as calcium oxide or magnesium oxide are slightly alkaline in the presence of water. Substances such as carbon dioxide or sulfur dioxide, on the other hand, are slightly acidic. Normal rainwater, for example, has a pH of about 5.6 (slightly acidic) due to the presence of carbon dioxide in the atmosphere.

C. Each element is symbolized by either a single capital letter or a capital letter followed by a single lower-case letter. Thus, the number of elements from which a compound is made up can easily be determined by counting the number of capital letters.

D. The **atomic number, Z,** of an element is the number of protons in the nucleus of its atom. The atomic number, which is always an integer and uniquely identifies the element, is written as a *left-hand subscript* to the element's symbol, for example, $_6C$.

E. The **mass number, A,** of an element is the sum of the number of protons and number of neutrons in the nucleus. Protons and neutrons are arbitrarily assigned a mass number of one atomic mass unit (amu) each. Although an element can have only one atomic number, it may have more than one mass number. This fact gives rise to the phenomenon of isotopes. **Isotopes** are atoms with the same atomic number (hence represent the same element) but with different neutron numbers. An element may have several isotopes, one or more of which may be radioactive and hence unstable. For example, oxygen has three naturally occurring isotopes, all of which are stable, while carbon also has three, one of which is radioactive. The mathematical average of all of an element's isotopic mass numbers, weighted by the percent abundances in nature of these isotopes, constitutes the element's average **atomic mass.** It is impossible to predict theoretically how many isotopes an element may have, or how many may be radioactive. The isotopes of a given element have identical chemical properties (reactivities) but different physical properties. The mass number is expressed as a *left-hand superscript* to the element's symbol, for example, ^{12}C.

F. The electronic charge of an element is zero, that is, the number of electrons equals the number of protons in the neutral or uncharged atom in its elemental state. If an atom gains or loses electrons, it becomes charged. A charged atom is called an **ion.** A positively charged ion has lost one or more electrons and is called a **cation,** while a negatively charged ion has gained one or more electrons and is called an **anion.** The charge or oxidation state on the ion is expressed as a *right-hand superscript* to the element's symbol.

Example 3.1

Uranium, the element with atomic number $Z = 92$, has three naturally occurring isotopes: U-234, U-235, and U-238. Only U-235 is fissionable. When it combines with a halogen to form a compound, it commonly forms a cation with a +6 charge.

A. Using proper notation, write the symbol of the U-235 isotope, indicating its electronic charge (also known as the **valence** or **oxidation state**).

B. Determine the number of protons, neutrons, and electrons in one atom of this isotope.

Solution:

A. The symbol, written with proper notation, is

$$^{235}_{92}U^{+6}$$

B. This atom has 92 protons (note left-hand subscript).

It must also have 92 electrons, understood, in its neutral (uncharged) atom. However, since this atom has a charge of +6 (note right-hand superscript), it has *lost* six electrons to become positively charged. Hence, it now has only 86 electrons.

To compute the number of neutrons in this atom, simply subtract the atomic number (left-hand subscript) from the mass number (left-hand superscript): $235 - 92 = 143$ neutrons.

This is the isotope of uranium that is fissionable and produces nuclear energy.

Common Quantities and Units of Measurement

Important, measurable quantities or variables, along with their symbols and common units of measurement, are as follows:

- **Temperature, T,** in degrees centigrade (Celsius) or Fahrenheit or in kelvin units (K).
- **Mass, m,** in grams (g), or kg (SI unit).
- **Number of moles, n,** 6.022×10^{23} atoms, molecules, or formula units.
- **Molar mass, MM,** the mass in grams of 6.022×10^{23} atoms of an element or molecules of a compound. The terms **molecular weight** and **formula weight** are often used interchangeably with molar mass.
- **Avogadro's number** or **constant:** 6.022×10^{23} particles or items per mole.
- **Volume, V,** in cubic centimeters (cm^3) or liters (L).
- **Pressure, P,** in atmospheres (atm) or other suitable units.
- **Density, d,** in grams per cubic centimeter (g/cm^3) or grams per milliliter (g/mL).
- **Concentration** may be expressed in any one of several units, depending on the application.

Very often, amounts of substances either are given or must be calculated. They may be expressed in *grams* (unit of mass), *moles,* or *liters* (unit of volume).

3.2
Basic Calculations

Inorganic Nomenclature and Formula Writing

There are two methods to name inorganic compounds and write their chemical formulas:

1. The crisscross method for ionic compounds.
2. The Greek prefix method for covalent compounds.

In order to decide which method is suitable, you must first decide what class of compound you are dealing with: **ionic** or **covalent.** Recall that ionic compounds are made up of a metal and a nonmetal; salts, such as sodium chloride or potassium nitrate, are ionic compounds. Covalent compounds are made up of two nonmetals; examples are carbon dioxide and phosphorous pentachloride. Ionic compounds are characterized by ionic bonds, in which one or more electrons have been transferred from the metal to the nonmetal atom, while covalent compounds are characterized by covalent bonds, in which electrons are shared between two atoms. If the sharing is unequal, the bond is said to be **polar-covalent,** and one end of the molecule has a partial positive charge while the other has a partial negative charge. Such molecules are referred to as **dipoles** and have dipole moments, which are quantitative measures of their polarity. If the sharing is equal, the bond is said to be nonpolar covalent; the molecule has no charge separation and hence no positive or negative end. In reality, most compounds are *not entirely* ionic or entirely covalent, but are a percentage of each.

If the compound in question is ionic, use the crisscross method of nomenclature. Write the symbols of the metal and nonmetal elements side by side (recall that the nonmetal may be a single anion or a polyatomic oxoanion, such as sulfate or nitrate), along with their respective valences or oxidation states as right-hand superscripts. Then criss-cross the superscripts, that is, interchange them for the two elements, and write them as subscripts, omitting the plus or negative signs. If the two subscripts are the same, drop them both. In nomenclature, the name of the metal element is given first and that of the nonmetal element second, changing it to an "-ide" ending, such as chlor<u>ide</u> or sulf<u>ide</u>, or using the name of the oxoanion, like sulfate or nitrate.

If a compound is covalent, the Greek prefix method is used. Here, knowledge of valences is unnecessary, but you will need to know the first twelve numbers in Greek: mono, di, tri, tetra, penta, hexa, hepta, octa, nona, deca, unideca, and dodeca. The proper Greek prefix simply precedes each element in the formula. As above, the name of the second element takes on the suffix -ide.

You should remember that many compounds have chemical names as well as common names. A compound generally has only one correct chemical name but may have one or more common names. For example, the compound $Ca(OH)_2$ has the chemical name calcium hydroxide; it is also referred to commonly as slaked lime or hydrated lime.

Example 3.2

Write the correct chemical formula for each of the following compounds:

A. Calcium chloride   B. Aluminum sulfate
C. Iron(III) phosphate   D. Sulfur trioxide
E. Diphosphorus pentoxide

Solution:

Compounds **A–C** are ionic, so the crisscross method applies.

Compounds **D** and **E** are covalent, so the Greek prefix method applies.

A. Calcium chloride: Ca^{2+} and Cl^{1-} are the two ions.

Crisscrossing gives $CaCl_2$.

Note that the subscript 1 next to Ca has been omitted because 1's are understood and are not written when specifying numbers of atoms.

B. Aluminum sulfate: Al^{3+} and $(SO_4)^{2-}$ are the two ions.

Crisscrossing gives $Al_2(SO_4)_3$.

Note that parentheses are necessary to express the fact that the subscript 3 refers to and thus multiplies both the sulfur and the oxygen atoms (subscripts) inside the parentheses.

C. Iron(III) phosphate: Fe^{3+} and $(PO_4)^{3-}$ are the two ions.

Note that the iron(III) notation refers to the Fe^{3+} species, to be distinguished from the Fe^{2+} species, written as iron(II). This is the modern way of distinguishing the ferric ion from the ferrous ion. This same system of notation, using Roman numerals in parentheses, is used for all transition metals with multiple oxidation states.

Crisscrossing gives $Fe_3(PO_4)_3$.

Note that fine-tuning is necessary. Since both subscripts are 3, they are omitted. Furthermore, the parentheses then become unnecessary and are dropped as well.

The fine-tuned answer is $FePO_4$.

D. Sulfur trioxide: S and O are the elements.

Sulfur is "mono" or 1 (understood and not written), and oxygen is "tri" or 3. Hence, the chemical formula is SO_3.

E. Diphosphorus pentoxide: P and O are the elements.

Phosphorus is "di" or 2, and oxygen is "pent" or 5. Hence, the chemical formula is P_2O_5.

Example 3.3

Find the total number of atoms in one formula unit (molecule) of compounds **A, B,** and **C** in Example 3.2.

Solution:

In each case, simply add the subscripts next to all the atoms in the formula unit. Where parentheses are used, be sure to multiply subscripts inside parentheses by subscripts outside parentheses.

For $CaCl_2$:   1 Ca atom + 2 Cl atoms = 3 atoms total

For $Al_2(SO_4)_3$:   2 Al atoms + 3 S atoms + 12 O atoms = 17 atoms total

For $FePO_4$:   1 Fe atom + 1 P atom + 4 O atoms = 6 atoms total

Note: The percent oxygen (O) *by number* in $Al_2(SO_4)_3$ is $12/17 \times 100$ or 70.6%, while in $FePO_4$ it is $4/6 \times 100$ or 66.7%. Compare this calculation to the calculation for percent *by mass* in Example 3.5.

Calculating Oxidation States from Formulas

Occasionally, it may be necessary to determine the oxidation state or valence of an ion in a compound from its formula. This situation is often encountered with compounds or polyatomic complexes containing transition metal ions, which may have multiple oxidation states. A simple algebraic equation solves the problem, as illustrated in Example 3.4.

Example 3.4

Find the oxidation state or valence of each underlined atom.

A. $K\underline{Mn}O_4$ **B.** $(\underline{Cr}_2O_7)^{2-}$

Solution:

Here, it is necessary to know the oxidation states assigned to common "fixed" elements. Some guidelines include

- Alkali metals are 1+.
- Alkaline earth metals are 2+.
- Oxygen is generally 2−.
- Halides are generally 1−.

Set up a simple algebraic equation based on the principle that the sum of the known charges of individual atoms in the formula, multiplied by their respective subscripts, must total the net charge on the formula unit of the compound or polyatomic ion.

A. For $KMnO_4$:

$$(1+)(1) + (x)(1) + (2-)(4) = 0$$

Note that x represents the oxidation state of the "unknown" Mn atom. The total is set to zero because the formula unit is electronically neutral or uncharged.

Solve for x: $x = +7$.

Thus, the oxidation state of Mn in $KMnO_4$ is +7. [The Mn ion itself can be expressed as either Mn^{7+} or Mn(VII).]

B. For $(Cr_2O_7)^{2-}$:

$$(x)(2) + (2-)(7) = 2-$$

Note that x represents the oxidation state of the unknown Cr atom. The total is set to 2− because the net charge on the complex is 2−.

Solve for x: $x = 6+$.

Thus, the oxidation state of Cr in $(Cr_2O_7)^{2-}$ is 6+. [The Cr ion itself can be expressed as either Cr^{6+} or Cr(VI).]

Determining the Percent Composition of a Compound

It is often important to be able to compute the percent composition of a compound. This ability is valuable, for example, when comparing compounds such as fertilizers or mineral supplements to determine which fertilizer contains the most nitrogen per unit mass of fertilizer, or which mineral supplement contains the most calcium per unit mass of tablet. This determination requires that you first calculate the molar mass of each of the compounds in question. It is assumed that you are familiar with calculating the molar mass of a compound from the atomic masses of the individual elements in the compound. Example 3.5 illustrates the method.

Example 3.5

Determine the percent composition, by mass, of each of the following compounds:

A. Calcium oxide, CaO (commonly known as lime or quicklime)
B. Calcium carbonate, $CaCO_3$ (commonly known as limestone)

Solution:

A. For CaO, the percent composition must be a two-part answer, because two elements make up CaO. Its molar mass is 56.1 g/mol. Thus:

$$\%Ca = \frac{(\text{atomic mass of Ca})(\text{\# Ca atoms in 1 formula unit})}{\text{molar mass of CaO}} \times 100$$

$$= \frac{(40.1 \text{ g Ca})(1)}{56.1 \text{ g CaO}} \times 100$$

$$= 71.5\%$$

$$\% O = \frac{(\text{atomic mass of O})(\text{\# O atoms in 1 formula unit})}{56.1 \text{ g CaO}} \times 100$$

$$= \frac{(16.0 \text{ g O})(1)}{56.1 \text{ g CaO}} \times 100$$

$$= 28.5\%$$

B. For $CaCO_3$, the percent composition must be a three-part answer, because three elements make up $CaCO_3$. Its molar mass is 100.1 g/mol. In a similar fashion, then,

$$\% Ca = \frac{(40.1 \text{ g Ca})(1)}{100.1 \text{ g CaCO}_3} \times 100$$

$$= 40.0\% \text{ Ca}$$

$$\% C = \frac{(12.0 \text{ g C})(1)}{100.1 \text{ g CaCO}_3} \times 100$$

$$= 12.0\%$$

$$\% O = \frac{(16.0 \text{ g O})(3)}{100.1 \text{ g CaCO}_3} \times 100$$

$$= 48.0\%$$

Note that CaO has a higher percent Ca content by mass than $CaCO_3$.

Determining Empirical and Molecular Formulas from Percent Composition

It is frequently useful to determine the chemical formula of an unknown compound. One method is **combustion analysis.** If the unknown compound is known to contain at least carbon and hydrogen, heating a measured amount in the presence of excess oxygen, a process known as combustion, produces carbon dioxide and water, which can be collected and weighed individually. All of the carbon atoms are contained in the carbon dioxide, and all of the hydrogen atoms are contained in the water. Since 1 mole of carbon dioxide is equivalent to 1 mole of carbon atoms, and 1 mole of water is equivalent to 2 moles of hydrogen atoms, the number of moles of carbon and hydrogen in the unknown compound can be determined. If oxygen is also present in the unknown, it is distributed over the carbon dioxide and water products; its mass can be found by difference and then converted to moles of oxygen in the original or unknown compound. The mass or moles of other elements, such as nitrogen and sulfur, which form known compounds with oxygen, can be determined in the same fashion. **Pyrolysis** is the intense heating of a compound or mixture in the absence of oxygen.

Alternatively, another method is known as **elemental analysis.** A measured amount of the unknown substance is heated until it decomposes into its constituent elements, which are collected and analyzed individually. The percent composition by mass of the compound is obtained.

Example 3.6 illustrates how percent composition data can be used to deduce the chemical formula and ultimately the identity of an unknown substance.

Example 3.6

The percent composition by mass of an unknown compound is 40.9% carbon, 4.57% hydrogen, and 54.5% oxygen. By a separate analysis, its molecular weight or molar mass is found to be 176 g/mol. Find for the unknown compound:

A. the empirical formula
B. the molecular formula

Solution:

A. Recall that the empirical formula represents the smallest group or combination of atoms, in the proper ratio, of which the molecule is composed. Assume a 100-g sample to work with. Then the given percentages can be translated directly into grams as follows:

40.9 g C, 4.57 g H, 54.5 g O

Next, convert these masses to moles by dividing each one by its respective atomic mass:

$$40.9 \text{ g C} \left(\frac{1.00 \text{ mol C}}{12.01 \text{ g C}} \right) = 3.41 \text{ mol C}$$

$$4.57 \text{ g H} \left(\frac{1.00 \text{ mol H}}{1.008 \text{ g H}} \right) = 4.53 \text{ mol H}$$

$$54.5 \text{ g O} \left(\frac{1.00 \text{ mol O}}{16.0 \text{ g O}} \right) = 3.41 \text{ mol O}$$

At this point, the chemistry is done, and you could write the empirical formula in principle, using the calculated numbers as the subscripts: $C_{3.41} H_{4.53} O_{3.41}$.

Chemical formula rules stipulate, however, that these subscripts must not only be in the proper ratio but also must be whole numbers or integers. The problem now is to find a mathematical technique to maintain the ratio but change the subscripts into integers. One way is to divide through by the smallest number:

$$C_{3.41/3.41} H_{4.53/3.41} O_{3.41/3.41} = C_{1.00} H_{1.33} O_{1.00}$$

Only the 1.33 subscript is not an integer. This can be corrected by multiplying through by the factor 3:

$$C_{3.00} H_{4.00} O_{3.00}$$

This, then, is the empirical formula.

B. To find the molecular formula, simply divide the molar mass by the empirical mass. This ratio should always be a whole number. Then multiply each of the subscripts in the empirical formula by this factor to obtain the molecular formula.

The empirical mass of the above chemical formula ($C_3H_4O_3$) is calculated in the same way as a molar mass would be and equals 88.0 g/mol. Thus, the factor is:

$$\frac{\text{Molar mass}}{\text{Empirical mass}} = \frac{176 \text{ g/mol}}{88 \text{ g/mol}} = 2.0$$

Therefore, the correct molecular formula is $C_6 H_8 O_6$.

This is the formula for ascorbic acid, commonly known as vitamin C.

Concentration Units for Aqueous Solutions

Several concentration units exist in chemistry to quantify the strengths of aqueous solutions. The reason for this great variety is that different applications and chemical formulas require different units. The term "concentration" always implies the existence of a **solute,** present as a gas, liquid, or solid, dissolved in a suitable **solvent** to form a **solution.** Unless otherwise specified, the solvent for liquid solutions is understood to be water—hence the designation "aqueous solution," abbreviated as (aq). The common definitions and units are as follows:

- Molarity, M = moles of solute/liter of solution.
- Molality, m = moles of solute/kilogram of solvent.
- Mole fraction, X = number of moles of component i/total number of moles.
- Percent by mass/mass = mass of solute in grams/mass of solution in grams.
- Percent by volume/volume = volume of solute in milliliters/total volume of solution in milliliters.

Conversion from one concentration unit to another may also be quite useful. The "factor label method" is the most efficient way to handle this. For example, to convert the concentration of sulfuric acid, H_2SO_4, solution used in car batteries from g/mL (sometimes also used as a unit of density) into molarity M (in moles/liter), do the following:

Note that normal battery acid concentration is usually about 1.265 g/mL, while pure sulfuric acid solution is 1.800 g/mL, for comparison.

Thus, 1.265 g H_2SO_4/mL soln

$$\left(\frac{1000 \text{ mL soln}}{1.00 \text{ L soln}} \right) \left(\frac{35.0 \text{ g } H_2SO_4}{100 \text{ g soln}} \right) \left(\frac{1.0 \text{ mol } H_2SO_4}{98.0 \text{ g } H_2SO_4} \right)$$

$$= 4.52 \text{ mol/L} = 4.52 \text{ M}$$

3.3
Gas Laws

The Kinetic-Molecular Theory of Gases

The fact that all gases behave similarly with change in temperature or pressure led to the **Kinetic-Molecular Theory** or **Kinetic-Molecular Model**. It can be summarized as follows:

- All matter in the gas phase is composed of discrete particles called molecules.

- In the gaseous state, molecules are relatively far apart.
- The molecules of all substances in the gaseous state are in continuous, rapid motion. This motion is in three dimensions and is often called translation.
- This continuous motion is a measure of the kinetic energy of the system.
- In addition to translation, molecules may also rotate and vibrate, depending on the type of external energy source to which they are subjected. These rotations and vibrations are known as internal modes of energy and are separate from the molecules' kinetic energy.
- Collisions between molecules is assumed to be perfectly elastic. This means that when molecules collide with one another or with the walls of a container, the molecules rebound without any loss in kinetic energy. This assumption explains why, at least in principle, a balloon or tire is not supposed to lose any pressure over time.
- The average kinetic energy of the molecules is directly proportional to its temperature in kelvins.

Gases have the following general properties:

- Gases exert pressure. This pressure can be measured in a variety of units, such as pounds per square inch, atmospheres, pascals (or newtons per square meter), torrs, and so on.
- Gases are highly compressible (unlike liquids and solids).
- Gases diffuse easily.
- Gases expand upon heating, providing the pressure remains constant.
- The pressure exerted by a gas increases with temperature, providing the volume is held constant.

What Is an Ideal Gas?

When discussing an ideal gas, two assumptions are made. First, the gas molecules are assumed to consist of perfect spheres that take up no room or volume themselves when compared with the volume of the container they are in (sometimes referred to as excluded volume). Second, the individual gas molecules do not repel or attract one another. These assumptions lead to the **ideal gas law**, which states:

$$PV = nRT \qquad [3.1]$$

where P is the pressure of the gas system, V is the volume of the container of the system, T is temperature of the gas in kelvins, n is the moles of gas, and R is the universal gas constant. The temperature of the

molecules is assumed to be the same as the temperature of the container. The universal gas constant can be given in several different units.

This law or formula is easy to work with. It accurately predicts the behavior of all of the ideal gases (helium, argon, and so on) as well as some commonly encountered gases such nitrogen and oxygen (which make up air) at or near room temperature.

However, under certain conditions of high pressure or low temperature, gases do not behave ideally. This deviation from ideal behavior occurs for two reasons. First, the molecules themselves have a definite volume and size, so they do occupy a significant fraction of the volume of the container under high pressure. This causes the volume of the gas to be greater than that *calculated* for an ideal gas. Hence, a subtractive correction factor B is necessary. Second, molecules are brought closer together, and hence they attract one another more strongly, especially under conditions of high pressure or low temperature. So, an additive correction factor, A, is necessary. The **van der Waals equation**, given below, represents *real* gas behavior under either of these two conditions, and especially when both conditions prevail. It is mathematically more complicated and therefore more difficult to work with. The constants A and B may be looked up by gas identity in a table in a handbook. There are no less than 10 other equations of state or models of real gas behavior, which are mathematically more complicated but offer little improvement in results.

$$\left[P + \frac{n^2 A}{V^2} \right] [V - n^B] = nRT \qquad [3.2]$$

The General Gas Law

Boyle's law (the pressure of a gas is inversely proportional to its volume) and **Charles's law** (the volume of a gas is directly proportional to its temperature) are generally combined into a more useful form known as the **general gas law,** expressed as

$$\frac{P_1 V_1}{T_1} = \frac{P_2 V_2}{T_2} \qquad [3.3]$$

It is understood that this law is valid only for a closed system in which the total number of gas molecules and hence the mass are constant. The general gas law can be applied to an individual gas (e.g., nitrogen or carbon dioxide) or a mixture of gases (e.g., air). It is assumed that the gas is ideal, that is, follows the ideal gas law (see next section) as its equation of state.

Pressures P and volumes V may be in any units as long as they are consistent. Temperature must be absolute and thus in kelvins. The subscripts 1 and 2 represent the initial and final states or conditions, respectively, for each quantity.

Example 3.7

A fixed quantity of gas occupies a volume of 2.0 L at a temperature of 20°C and a pressure of 1.0 atm. Find the volume that this gas would occupy at 40°C and 1.75 atm pressure.

Solution:

The general gas law (equation 3.3) applies. This is a problem in which the volume of a gas must be calculated at a new or different temperature and pressure. The data given are

$$P_1 = 1.0 \text{ atm} \quad P_2 = 1.75 \text{ atm}$$

$$T_1 = 20°C + 273 \text{ K} = 293 \text{ K}$$
$$T_2 = 40°C + 273 \text{ K} = 313 \text{ K}$$

$$V_1 = 2.0 \text{ L} \quad V_2 = ?$$

Substitute these values in the general gas law and solve for V_2.

$$V_2 = \frac{P_1 V_1 T_2}{P_2 T_1} = \frac{(1.0 \text{ atm})(2.0 \text{ L})(313 \text{ K})}{(1.75 \text{ atm})(293 \text{ K})}$$

$$= 1.22 \text{ L}$$

The Ideal Gas Law

The **ideal gas law,** known also as the **equation of state** for ideal gases, is a very useful and powerful problem-solving tool. It relates the pressure, volume, and temperature of a quantity of a gas. An **ideal gas** is a gas in which every molecule behaves independently of every other molecule (there is an absence of any intermolecular forces) and has no excluded volume. It is represented by the ideal gas law:

$$PV = nRT \qquad [3.4]$$

where P = the pressure in atmospheres, torr, millimeters of mercury, pounds per square inch, pascals, etc.

V = the volume in liters, cubic centimeters, etc.

n = the number of moles of gas

= mass of gas in grams/molar mass in grams per mole

$$= \frac{m}{MM}$$

T = the absolute temperature in kelvin units

R = the universal gas constant in units consistent with the above values, that is, $R = 0.0821$ L · atm/K · mol; 8.31 J/K · mol, etc.

Note that 8.31 J/K is the accepted SI unit. Also, 1.0 J = 1.0 Pa.m^3, so 8.31 Pa.m^3/K is still another possible unit.

The ideal gas law contains four variables and the constant R. Given any three, the fourth can be found. In addition, this law can be used to find the molar mass or molecular weight if the mass of a gas is given.

Note that the ideal gas law is valid for a single, individual gas. If a mixture of gases is present, the ideal gas law is valid for each individual gas. In other words, a separate calculation using this law should be made for each gas present. The pressure of each gas thus calculated represents a partial pressure in the mixture. The total pressure can then be calculated according to Dalton's law (see the next section).

Example 3.8

Exactly 5.75 g of an unknown gas occupies 3.40 L at a temperature of 50°C and a pressure of 0.94 atm. Find the molar mass or molecular weight of the gas.

Solution:

The ideal gas law (equation 3.4) has many applications. One of them is to determine the molar mass of a gas.

$$PV = nRT = \left(\frac{m}{MM}\right) RT$$

In this case, you wish to solve for MM.

The data given in the problem are

$P = 0.94$ atm $\qquad R = 0.0821$ L · atm/K · mol

$V = 3.40$ L $\qquad T = 273 + 50°C = 323$ K

$m = 5.75$ g

Substitute this information in the above equation, and solve for MM:

$$MM = \frac{mRT}{PV}$$

$$= \frac{(5.75 \text{ g})(0.0821 \text{ L} \cdot \text{atm/K} \cdot \text{mole})(323 \text{ K})}{(0.94 \text{ atm})(3.40 \text{ L})}$$

$$= 47.6 \text{ g/mol}$$

Dalton's Law of Partial Pressures

Dalton's law states: The total pressure exerted by a mixture of gases is equal to the sum of the partial pressures of all the gases in the mixture. Each partial pressure is the pressure that the gas would exert if the others gases were not present.

$$P_{total} = \Sigma\, P_i = P_1 + P_2 + P_3 + \ldots + P_n \quad [3.5]$$

This law is particularly useful when gases are collected in vessels above the surface of an aqueous or nonaqueous solution. In an aqueous solution, the partial pressure of water vapor may be subtracted from the total pressure to help determine the partial pressures of the other gases or vapors present. The percent composition of a gaseous mixture may then be calculated, based on Dalton's law. To calculate the composition of the *solution phase* (e.g., in mole percent) **Raoult's law** (see page 114) must be used.

Graham's Law of Effusion

Recall that kinetic molecular theory states that the average speed of molecules in motion is given by the **root-mean-square speed, U_{rms}:**

$$U_{rms} = \sqrt{\frac{3RT}{MM}}$$

Diffusion is the migration or mixing of molecules of different substances as a result of concentration gradients and random molecular motion. **Effusion** is the escape of gas molecules of a single substance through a tiny orifice (pinhole) of a vessel holding the gas. **Graham's law of effusion** states: The rates of effusion of two different gases escaping, A and B, are inversely proportional to the square roots of their molar masses:

$$\frac{\text{Effusion Rate}_A}{\text{Effusion Rate}_B} = \sqrt{\frac{MM_B}{MM_B}} \quad [3.6]$$

Note that, at a fixed temperature, the *effusion time* is inversely proportional to the effusion rate, while the *mean distance traveled,* as well as the *amount of gas* effused, is directly proportional to the effusion rate.

Intermolecular Forces

Chemical bonds between two or more atoms in a molecule are referred to as **intramolecular forces**. They are generally categorized as ionic, covalent, or metallic. These are the bonds that are broken and re-formed during a chemical reaction. Significantly less

in strength are **intermolecular forces**. Intermolecular forces are generally attractive forces that exist between molecules of all sizes, shapes, and masses. Although they are probably easiest to visualize in the gaseous phase, they exist in the liquid and solid phases as well as in the solution phase, where water is acting as the solvent to hydrate the solute.

They are important to understand because they are directly related to macroscopic properties such as melting point, boiling point, vapor pressure and volatility, and the energy needed to overcome forces of attraction between molecules in changes of state. Just as important, they also help determine the solubility of gases, liquids, and solids in various solvents, i.e., whether two substances are soluble or miscible in each other, and in part explain or reflect the "like dissolves like" principle. They are also critical in determining the structure of biologically active molecules such as DNA and proteins.

The categories of intermolecular forces in order of decreasing strength are

- ion–dipole forces, important in salts dissolving in water.
- permanent dipole–permanent dipole forces, important in polar covalent substances.
- hydrogen bonding, important in H–O, H–N, and H–F interactions, really a subcategory of dipole–dipole attractions; water is the most common example.
- dipole-induced–dipole forces, such as oxygen dissolving in water, where a temporary dipole is induced in the diatomic oxygen molecule by water.
- London dispersion forces (also known as van der Waals forces or induced dipole–induced dipole forces), important in nonpolar substances.

For example, polar substances such as acetone (C_3H_6O) will dissolve in other polar substances such as methyl chloride (CH_3Cl) but not in carbon tetrachloride (CCl_4), a nonpolar substance. Polar substances such as ethanol (C_2H_3OH) will dissolve in water in all proportions (is completely miscible) because of hydrogen bonding between the hydrogen atom of the ethanol molecule and the oxygen atom of the water molecule. The same is true of ammonia and hydrofluoric acid, where hydrogen bonding occurs. Note that dimethyl ether (C_2OH_6), an isomer of ethanol, does not undergo hydrogen bonding with water because of its different molecular structure (i.e., an ether versus an alcohol). Hexane (C_6H_{14}) will dissolve in octane (C_6H_{18}) because both substances are characterized by nonpolar bonds. Naturally, size (length) and shape (branching of the molecule) also play a role in solubility considerations, and it is often difficult to determine which factor is more important in predicting solubility between two substances.

The greater the forces of interaction between attractions between molecules in a liquid, the greater the energy that must be supplied to separate them. Hydrogen bonding is a key reason why low molecular weight alcohols have much higher than expected boiling points, i.e., heats of vaporization, in comparison to nonpolar hydrocarbons like hexane.

Special Application

An interesting problem presented itself to an engineer who was trying to spread a thin film of adhesive onto a flat plastic surface to bond with another flat surface. The adhesive used a hexane base. A pressurized air gun was used to spread it as evenly as possible through a spreading nozzle. However, each time the adhesive was spread, tiny bubbles randomly appeared in the adhesive, preventing a tight or perfect seal. The engineer eventually realized that these bubbles were due to the fact that air, which is composed of nitrogen and oxygen, has a finite solubility in hexane since both nitrogen and oxygen are non-polar substances. A rough rule of thumb is that "like dissolves like" for intermolecular forces and solubility. When helium, an inert gas, was substituted for air in the delivery mechanism, the bubbles disappeared and the seal was perfect. Helium has a much lower solubility in hexane.

3.4
Chemical Reactions

Five Categories of Chemical Reactions

The purpose of categorizing a chemical reaction is to help predict the product or products of the reaction. In general, a chemical reaction is a process of molecular rearrangements, in which atoms change partners. Atoms are not destroyed or created. Total mass is conserved, that is, the mass of all of the reactants before the reaction must equal the mass of all of the products after the reaction, although the physical or chemical states of individual substances may change. This mass-conservation requirement explains why all chemical reactions must be properly balanced to be quantitatively valid and useful. Energy changes also accompany reactions—energy is either liberated or absorbed—since existing chemical bonds are broken and new bonds are formed.

With the letters A, B, C, and D used to represent simple elements or polyatomic ions, the five categories of reactions are as follows:

- Combination or synthesis

$$A + B \rightarrow AB$$

- Decompostion

$$AB \rightarrow A + B$$

- Single replacement/displacement

$$A + BC \rightarrow AC + B$$

- Double replacement/displacement

$$AB + CD \rightarrow AD + CB$$

- Combustion (complete or efficient)

$$(CH)_x + O_2 \rightarrow H_2O + CO_2$$

The notation (g), (ℓ), (s), or (aq) immediately following a reactant or product is often employed to designate the chemical state of the substance.

Another class of reactions that cuts across all other reaction categories consists of **oxidation-reduction reactions**. Oxidation-reduction ("red-ox") reactions are reactions in which one substance is oxidized while another is simultaneously reduced. The processes of oxidation and reduction can be defined as follows. A substance is oxidized or reduced, respectively, if any *one* of the following conditions is met:

Oxidation	Reduction
• The substance loses electrons.	• The substance gains electrons.
• The substance gains oxygen atoms.	• The substance loses oxygen atoms.
• The substance loses hydrogen atoms.	• The substance gains hydrogen atoms.

Red-ox reactions are actually formed by the addition of two half-reactions—an oxidation reaction and a reduction reaction.

For example, each of the following shows an *oxidation reaction:*

$$Zn \rightarrow Zn^{2+} + 2e^- \qquad [3.7]$$
[Zn has lost 2 electrons.]

$$C + O_2 \rightarrow CO_2 \qquad [3.8]$$
[C has gained 2 O atoms.]

And each of the following shows a *reduction reaction:*

$$Cu^{2+} + 2e^- \rightarrow Cu \qquad [3.9]$$
[Cu^{2+} has gained 2 electrons.]

$$C + 2H_2 \rightarrow CH_4 \qquad [3.10]$$
[C has gained 4 H atoms.]

Each reaction also has an associated electromotive potential, measured in electron volts (eV) and found in electromotive potential tables. This topic is discussed more fully in Section 3.11.

Writing and Balancing Chemical Reactions

A chemical reaction consists of a reactant side (the left) and a product side (the right) separated by an arrow, which indicates *yields* or *produces.* Reactants and products are present as elements and/or compounds and are represented by appropriate symbols, as discussed in Section 3.1. Reactions must be balanced before they can be used in calculations to provide quantitative information. Again, a symbol placed immediately after the element or compound designates its chemical state: (g) for gas or vapor; (l) for liquid; (s) for solid, powder, precipitate, or crystal; (aq) for solution, where water is understood to be the solvent for the indicated solute.

The simplest balancing method is called *balancing by inspection.* Although this method is often interpreted to mean balancing by trial and error, three simple rules regarding the order of balancing must be followed:

1. Balance metal atoms or atoms present in the greatest number first.
2. Balance nonmetal atoms second.
3. Balance hydrogen and oxygen atoms last.

Balancing means inserting integers, known as **stoichiometric coefficients,** in front of elements or compounds to ensure the same number of like atoms on both sides of the reaction. Balancing requires that you keep track of every kind of atom that appears in the reaction. Subscripts of atoms in compounds are fixed by nature and cannot be altered to achieve balancing.

The following are examples of **balanced reactions:**

(A) $2Al(s) + 6HCl(aq) \rightarrow 2AlCl_3(aq) + 3H_2(g)$
(B) $3CaCl_2(aq) + 2Na_3PO_4(aq) \rightarrow Ca_3(PO_4)_2(s) + 6NaCl(aq)$
(C) $2C_4H_{10}(g) + 13O_2(g) \rightarrow 8CO_2(g) + 10H_2O(g)$
(D) $CaCO_3(s) \rightarrow CaO(s) + CO_2(g)$

The following points regarding these four reactions are noteworthy:

1. (A) is an example of a single-replacement reaction.
   (B) is an example of a double-replacement reaction.
   (C) is an example of a combustion reaction.
   (D) is an example of a decomposition reaction.

2. Only (C) is an example of a **homogeneous reaction,** since all of the reactants and products are present in the same physical state—gaseous. The other three are **heterogeneous reactions,** since the reactants and products are present in more than one state.

3. (A) and (C) are also examples of oxidation-reduction reactions. In (A), Al is oxidized while HCl is reduced; HCl is the **oxidizing agent** while Al is the reducing agent. In (C), C_4H_{10} is oxidized while O_2 is reduced; O_2 is the oxidizing agent, while C_4H_{10} is the **reducing agent.**

4. There are two, and only two, levels of interpretation of the stoichiometric coefficients that balance these reactions. Consider (C). On the microscopic or invisible level, the reaction states that 2 molecules of C_4H_{10} react with 13 molecules of O_2 to produce 8 molecules of CO_2 and 10 molecules of H_2O. On the macroscopic or visible level, the reaction states that 2 moles of C_4H_{10} [2 times the molar mass of C_4H_{10} = (2 mol)(58 g/mol) = **116 g**] react with 13 moles of O_2 [13 times the molar mass of oxygen = (13 mol)(32 g/mol) = **416 g**] to produce 8 moles of CO_2 [8 times the molar mass of CO_2 = (8 mol)(44 g/mol) = **352 g**] and 10 moles of H_2O [10 times the molar mass of H_2O = (10 mol)(18 g/mol) = **180 g**]. In other words, the only two correct units of stoichiometric coefficients are molecules and moles. The unit grams can be obtained by the use of molar masses, as shown, but are not directly readable.

5. Note, from point 4 above, that in (C) mass is automatically conserved. The total mass of reactants is 116 grams + 416 grams = 532 grams. The total mass of products is 352 grams + 180 grams = 532 grams. The **conservation of mass principle** requires that this equality always occurs.

Simple Stoichiometry

The term "stoichiometry" refers to the mass relationships between two reactants, two products, or, more commonly, a reactant and a product. **Stoichiometry** depends on a balanced reaction, which is essential in predicting the amount of product generated by a given amount of reactant or, conversely, the amount of reactant required to generate a given amount of product. The method recommended in Example 3.9 is the **Factor-Label Method;** alternatively, the method of proportions is also used.

Example 3.9

Consider reaction (C), the complete combustion of butane with oxygen to produce carbon dioxide and water, presented in the previous section. Calculate the mass of carbon dioxide that can be produced from 816 g of butane, assuming oxygen is in abundant or unlimited supply.

Solution:

Since the reaction is balanced, you may proceed by setting up three conversion factors, written in parentheses and arranged from left to right, multiplying the given mass of butane. Next, draw a connecting "tie" line between the 2 in front of the butane and the 8 in front of the carbon dioxide, since these are the only two substances of interest. This will form the basis of the middle conversion factor below. The first and third conversion factors are simply the molar masses of butane and carbon dioxide, respectively. Thus:

$$②C_4H_{10}(g) + 13O_2(g) \rightarrow ⑧CO_2(g) + 10\,H_2O(g)$$

$$816\text{ g }C_4H_{10} \times \frac{1.0\text{ mol }C_4H_{10}}{58.0\text{ g }C_4H_{10}}$$

$$\times \frac{8.0\text{ mol }CO_2}{2.0\text{ mol }C_4H_{10}}$$

$$\times \frac{44.0\text{ g }CO_2}{1.0\text{ mol }CO_2}$$

$$= 2{,}480\text{ g }CO_2$$

A Special Application of a Decomposition Reaction— AN EXPLOSION

Liquid nitroglycerin, $C_3H_5N_3O_9$, is a well-known, powerful explosive. It falls under the category of a decomposition reaction. Examining its products, the number of molecules, the physical states of the reactant and products, the amount of heat energy, ΔH, released, explains its tremendous destructive power. Consider the balanced reaction for this decomposition below:

$$4C_3H_5N_3O_9(l) \rightarrow 6N_2(g) + 12CO_2(g)$$
$$+ 10H_2O(g) + O_2(g)$$

$$\Delta H° = 5{,}678\text{ kJ/mol}$$

Note the following three points:

1. A liquid reactant changes into all gaseous products, thereby requiring more volume.

2. Four moles of reactant decompose into 29 moles of products, thereby requiring more volume.

3. A tremendous amount of heat energy, 5,678 kJ/mol, is released. This release is indicated by the negative sign of ΔH. This released energy heats the already gaseous products, forcing them to expand, and thereby requiring more volume, i.e., increased temperature causes increased pressure at constant volume.

The decomposition of liquid nitroglycerin underscores the importance of understanding several individual chemical principles at work simultaneously.

Limiting Reagents

In many reactions, the reactants are not present in stoichiometric ratios. One of the reactants, called the **limiting reagent**, is present in short supply. This reactant determines the outcome of the reaction, that is, the maximum amount of any product of interest that can be generated. The other reactant or reactants are thus present **in excess.** In any reaction where two or more reactants are present and their respective amounts are given, the limiting reagent must be identified before the maximum amount of any product can be calculated.

Example 3.10

In an experiment, 50 g of hydrogen gas and 50 g of oxygen gas are ignited to form water. Calculate:
A. The mass of water formed.
B. The mass of any hydrogen or oxygen that is left over or unreacted.

Solution:

First, write the balanced chemical reaction as follows:

$$2H_2 + O_2 \rightarrow 2H_2O$$

Since the amounts of both reactants are given, one of the reactants may be the limiting reagent. To find which reactant is the limiting reagent, convert each mass given in grams into moles, and compare. Thus:

$$50 \text{ g H}_2 \times \frac{1.0 \text{ mol H}_2}{2.016 \text{ g H}_2} = 24.8 \text{ mol H}_2$$

$$50 \text{ g O}_2 \times \frac{1.0 \text{ mol O}_2}{32 \text{ g O}_2} = 1.56 \text{ mol O}_2$$

From the balanced reaction, it is clear that 2 mol of H_2 require 1 mol of O_2. Thus, 24.8 mol of H_2 would require 12.4 mol of O_2. However, only 1.56 mol of O_2 are actuallly present or available for reaction. Thus, O_2 is in short supply and is the limiting reagent. Also, H_2 is present in excess and will be left over after the reaction.

A. To find the mass of water produced, again apply the factor-label method, using the molar amount of O_2 present as the starting point.

$$1.56 \text{ mol O}_2 \times \frac{2 \text{ mol H}_2O}{1 \text{ mol O}_2} \times \frac{18.0 \text{ g H}_2O}{1.0 \text{ mol H}_2O}$$

$$= 56.2 \text{ g H}_2O$$

B. To find the mass of H_2 left over, subtract the amount of H_2 reacted with O_2 from the total mass of H_2 initially present and available in moles, since O_2 is the limiting reagent.

24.80 mol H_2 initally present
− 3.12 mol H_2 reacted (equivalent to 2 × 1.56 mol O_2)

21.68 mol H_2 left over or unreacted $\times \dfrac{2.016 \text{ g H}_2}{1.0 \text{ mol H}_2}$

$$= 43.7 \text{ g H}_2$$

Percent Yield

The result obtained in Example 3.10A is the theoretically predicted maximum amount of water. In reality, the amount of water actually collected may, for at least two reasons, be somewhat less. First, some reactions may not go to completion but instead may reach an equilibrium condition. Second, other reactions may have more than one pathway and may produce secondary or tertiary products that compete with the main route. In a sense, the percent yield of a reaction is a measure of its efficiency.

In either case, a **percent yield** may be computed as follows:

$$\% \text{ Yield} = \frac{\text{actual yield (g)}}{\text{theoretical yield (g)}} \times 100 \quad [3.11]$$

Example 3.11

Suppose that, in Example 3.10, the actual mass of water collected or measured is 48.0 g. Compute the percent yield of the reaction.

Solution:

$$\% \text{ Yield} = \frac{48.0 \text{ g}}{56.2 \text{ g}} \times 100 = 90.2\%$$

Consecutive and Simultaneous Reactions

Reactions that are carried out one after another in sequence to yield a final product are called **consecutive reactions.** In **simultaneous reactions,** two or more products react independently of each other in separate reactions at the same time.

An example of a consecutive reaction involves the purification of titanium dioxide, TiO_2, the most widely used white pigment for paints. To free TiO_2 of unwanted colored impurities, it must first be converted into $TiCl_4$, and then reconverted into TiO_2.

$$2 \text{ TiO}_2(s) \text{ (impure)} + 3C(s) + 4Cl_2(g) \rightarrow 2TiCl_4(g) + CO_2(g) + 2CO(g)$$

$$TiCl_4(g) + O_2(g) \rightarrow TiO_2(s) \text{ (pure)} + 2 \text{ Cl}_2(g)$$

Note that the $TiCl_4$ product in the first reaction becomes a reactant in the second. This is the connecting link. For example, one could ask how many grams of carbon are required to produce 1.0 kilogram of pure TiO_2. First, it must be realized that 1 mole of TiO_2 (pure) requires 1 mole of $TiCl_4$. Then 2 moles of $TiCl_4$ requires 3 moles of carbon. In this way, using the factor-label method, a calculation with conversion factors can be set up, connecting $TiO_2(s)$ with $C(s)$.

> Exothermic reactions are reactions that produce or generate heat to the environment or surroundings. By convention, their ΔH values are always negative. Endothermic reactions, on the other hand, absorb heat from the environment or surroundings. Their ΔH values are always positive. Units of heat energy are usually in either kilojoules per mol (kJ/mole) or kilocalories per mole (kcal/mole).

112

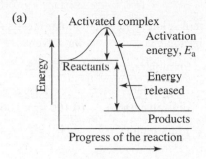

(a)

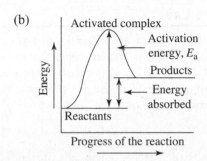

(b)

Figure 3.1 Change in potential energy for (a) exothermic and (b) endothermic reactions

3.5
Colligative Properties of Solutions

Basic Concept

There are four colligative properties of solutions:

- Boiling-point elevation.
- Freezing-point depression.
- Vapor-pressure lowering.
- Osmotic pressure.

The term "colligative" means having to do with the *collection* or *number* of particles and/or ions dissolved in solution. It is an oversimplification, since size and polarity also play a role, albeit a less important one, in the chemistry of solutions. Colligative properties depend on the concentration of the solute (molality) and its "i" (van't Hoff factor), i.e., the total number of dissolved particles. They do not depend on the specific chemical formula of the solute per se. Thus, the freezing point depression or lowering of 1.0 m NaI and 1.0 m KCl are the same.

Before describing each of the four colligative properties or effects in detail, it is useful to review the concept of a solution. A **solution** is a physical state wherein a solute has been dissolved completely and uniformly in a solvent, forming a solution of uniform composition. Unless specified otherwise, the solvent

is assumed to be water, whose normal boiling point, freezing point, and density are well known. In the equations for each of the four properties, the concentration of the solution appears, although concentration units vary. It is critically important, when doing calculations with solutions, to distinguish among units of mass, volume, and density for solute, solvent, and solution, respectively.

For solutions, there are three types of solutes:

- Strong electrolytes.
- Nonelectrolytes.
- Weak electrolytes.

Strong electrolytes are generally ionic compounds, such as salts and strong acids and bases, that dissociate and ionize *completely*, that is, 100%, in water. The solutions they form with water become strong conductors of electric currents. Examples include NaCl, KNO_3, $CaCl_2$, HCl, and NaOH. Thus, 1 mole of NaCl produces 2 moles of ions or particles; 1 mole of $CaCl_2$ produces 3 moles of ions or particles (1 mole of Ca^{2+} ions and 2 moles of Cl^- ions). The **van't Hoff factor** is the total number of ions or particles per formula unit or mole: 2.0 for NaCl, 3.0 for $CaCl_2$, and so on. Experimentally measured van't Hoff factors are generally less than theoretically predicted ones for a given solute.

Nonelectrolytes are generally alcohols and sugars, such as ethanol (C_2H_5OH) and glucose ($C_6H_{12}O_6$), or other organic substances that dissolve in water but do not ionize. One mole of a nonelectrolyte always produces 1 mole of particles. Thus the van't Hoff factor is 1.0.

A **weak electrolyte,** such as acetic acid, is a substance that is only partially dissociated and ionized, typically 10% or less, leaving 90% or more undissociated and thereby behaving much like a nonelectrolyte. Its van't Hoff factor is greater than 1 but less than the total number of ions available for dissociation.

Boiling-Point Elevation

When a nonvolatile solute is dissolved in water or another suitable solvent, the normal boiling point of the solvent is always raised. The amount of elevation in temperature, ΔT, is given by the equation

$$\Delta T = iK_b m \qquad [3.12]$$

where i = the van't Hoff factor, discussed above,

　　m = the molality of the solution, that is, the number of moles of solute per kilogram of water or other solvent,

　　K_b = the ebullioscopic or boiling-point-elevation constant, characteristic of water (or other solvent) in °C · kg/mol.

Example 3.12

What is the boiling point of a solution made by dissolving 70 g of NaCl in 300 g of water? The K_b for water is 0.512°C · kg/mol.

Solution:

Use equation 3.12: $\Delta T = iK_bm$, and note that $i = 2$ for NaCl. Then:

$$\text{molality, m} = \frac{\text{mol NaCl}}{\text{kg H}_2\text{O}}$$

$$= \frac{70 \text{ g NaCl}/58.5 \text{ g/mol}}{0.300 \text{ kg H}_2\text{O}} = 3.99$$

$$\Delta T = (2)(0.512)(3.99 \text{ mol/kg}) = 4.08°\text{C}$$

$$T_{\text{new}} = T_{\text{norm}} + \Delta T = 100°\text{C} + 4.08°\text{C}$$
$$= 104.1°\text{C}$$

Freezing-Point Depression

When a nonvolatile solute is dissolved in water or another suitable solvent, the normal freezing point of the solvent is always lowered. The amount of depression, ΔT, is given by the equation

$$\Delta T = iK_fm \qquad [3.13]$$

where I and m have the same definitions as for equation 3.10, and K_f = the cryoscopic or freeezing-point depression constant in °C · kg/mol.

Example 3.13

The freezing-point depression equation is often used to find the molar mass of an unknown solute. Find the molar mass (also known as the molecular weight) of a substance when 1.14 g of it are dissolved in 100.0 g of liquid camphor, whose freezing point is lowered by 2.48°C. K_f for camphor is 37.9°C · kg/mol.

Solution:

Use equation 3.13: $\Delta T = iK_fm$. Assume that the unknown substance is an organic nonelectrolyte, so $i = 1$. Then,

$$\Delta T = 2.48°\text{C} \quad \text{and} \quad K_f = 37.9°\text{C} \cdot \text{kg/mol}$$

First, solve for the molality, m:

$$m = \frac{\Delta T}{i K_f} = \frac{2.48°\text{C}}{(1)(37.9)} = 0.0654 \text{ mol/kg}$$

But

$$m = \frac{\text{mol solute}}{\text{kg solvent}} = \frac{\text{g solute/molar mass solute}}{\text{kg solvent}}$$

Solve for molar mass, *MM*, of solute:

$$MM = \frac{\text{g solute}}{(\text{kg solvent})(m)} = \frac{1.14 \text{ g}}{(0.100 \text{ kg})(0.0654)}$$

$$= 174.3 \text{ g/mol}$$

Henry's Law

Henry's law concerns the solubility of a gas in a liquid, most often water. The law states that the solubility, *S*, of a gas in a liquid is directly proportional to the partial pressure, *P*, of the gas above the liquid; it can be written as follows:

$$S = k_HP \qquad [3.14]$$

where k_H, called Henry's constant, depends on the gas, the solvent, and the temperature. The law implies that, at constant temperature, doubling the partial pressure of a gas doubles its solubility.

Henry's law helps explain such phenomena as the bends experienced by deep-sea divers who surface too quickly and the carbonation of beverages in tightly capped bottles.

Raoult's Law for Vapor-Pressure Lowering

Raoult's law states that, when a nonvolatile solute is added to water or another solvent, the vapor pressure of the solvent is depressed. In other words, the vapor pressure of the solution is lower than the vapor pressure of the pure solvent. The vapor pressure of the solution is directly proportional to the mole fraction of the solvent multiplied by the vapor pressure of the pure solvent:

$$P_{\text{soln}} = X_{\text{solvent}} P°_{\text{solvent}} \qquad [3.15]$$

Where P_{soln} = the vapor pressure of the solution,

X_{solvent} = the mole fraction of the solvent

$$= \frac{\text{mol solvent}}{\text{mol solvent} + \text{mol solute}}$$

$P°_{\text{solvent}}$ = the vapor pressure of the pure solvent.

Vapor pressures are a measure of volatility. They may be expressed in any suitable units, such as atmospheres, millimeters of mercury, torr, and pounds per square inch. Note that Raoult's law applies to all ideal solutions (where solute and solvent molecules experience similar intermolecular forces, as in a benzene-toluene solution) and to all real but dilute solutions (where $X_{solvent} > 0.85$). For a binary system with only one solute and one solvent, Raoult's law may alternatively be expressed as follows:

$$P_{soln} - P_{solvent} = \Delta P = X_{solute} P_{solvent}° \quad [3.16]$$

since

$$X_{solvent} + X_{solute} = 1 \quad [3.17]$$

Example 3.14

Calculate the vapor pressure at 100°C of a solution prepared by dissolving 10 g of sucrose, $C_{12}H_{22}O_{11}$, in 100 g of water.

Solution:

This is a real solution that is dilute, so Raoult's law should apply.

$$P_{soln} = X_{solvent} P_{solvent}° = X_{H_2O} P_{H_2O}°$$

where $P_{H_2O}° = 760$ torr since the normal boiling point of water is 100°C, and its vapor pressure at that temperature must be equal to atmospheric pressure. On the other hand, the vapor pressures of pure solvents can always be looked up in tables.

To find X_{H_2O}, compute the number of moles of sucrose *and* the number of moles of water.

moles of sucrose = 10 g $C_{12}H_{22}O_{11}$

$$\times \frac{1.0 \text{ mol } C_{12}H_{22}O_{11}}{342.3 \text{ g } C_{12}H_{22}O_{11}}$$

$$= 0.0292 \text{ mol}$$

moles of water = 100 g $H_2O \times \frac{1.0 \text{ mol } H_2O}{18.0 \text{ g } H_2O}$

$$= 5.55 \text{ mol}$$

Thus,

$$X_{H_2O} = \frac{5.55 \text{ mol}}{5.55 \text{ mol} + 0.0292 \text{ mol}}$$

$$= 0.9948$$

Finally,

$$P_{soln} = (0.9948)(760 \text{ torr}) = 756 \text{ torr}$$

Additionally,

$$\Delta P = 760 \text{ torr} - 756 \text{ torr} = 4.0 \text{ torr}$$

Vapor pressure is a function of temperature. The dependence of the vapor pressure of a pure solvent on temperature is given by the **Clausius-Clapeyron equation**:

$$\ln\frac{P_2}{P_1} = \frac{\Delta H_{vap}°}{R}\left(\frac{1}{T_1} - \frac{1}{T_2}\right) \quad [3.18]$$

where P_1 and P_2 are the vapor pressures at *absolute* temperatures T_1 and T_2, respectively, and must be in kelvins, and $\Delta H_{vap}°$ is the molar heat of vaporization of the pure solvent at its *normal* or *standard* boiling point.

A frequently useful, empirical approximation for many liquids, when *hydrogen bonding is unimportant,* is **Trouton's rule**:

$$\frac{\Delta H_{vap}°}{T_b} = 85 \quad [3.19]$$

This ratio, expressed in joules per kelvin, allows the calculation of either the standard molar heat of vaporization, $\Delta H_{vap}°$, or the boiling point, T_b, when one or the other is known.

Osmotic Pressure

Consider a U-tube in which a pure solvent such as water in the left arm is brought into contact with an aqueous solution made with a nonvolatile solute such as a sugar or protein in the right arm, and the two arms are separated by a semipermeable membrane (e.g., cellophane), permeable only to solvent molecules. The two sides or arms of the U-tube are initially at the same height. It is observed that a pressure develops across the membrane that drives solvent molecules from the left arm into the right arm in an apparent attempt to dilute the solution. This pressure is called the **osmotic pressure**, Π, and the process is known as osmosis. It is given by the equation:

$$\Pi = iMRT \quad [3.20]$$

where i = the van't Hoff factor,

M = the molarity of the solution,

R = the universial gas constant, 0.0821 L · atm/K · mol,

T = the absolute temperature in kelvins.

Some time later, the system in the U-tube reaches equilibrium, and a height differential, Δh, between the solution levels of the two arms is observed. Because of this height differential, a hydrostatic pressure, ΔP, exists and is given by the equation

$$\Delta P = \rho g \Delta h \quad [3.21]$$

where ρ = the density of the solution, and g = the acceleration constant due to gravity.

The final hydrostatic pressure is equal to the initial osmotic pressure.

Osmometry can also be used to find the molar mass of an unknown substance dissolved in solution through its molarity, M, when the other quantities are measured experimentally.

A good example of osmotic pressure is the following example:

A very dilute sugar solution of concentration 0.010 M in water is separated from pure water by an osmotic (i.e., semipermeable) membrane. Find the osmotic pressure that develops at 25°C (298 K).

Use $\pi = iMRT$ where $i = 1$ for sugar and all other nonelectrolytes.

$= (1)(0.010\ M)(0.0821\ L\text{-atm/K-mol})$ (298 K) = 0.24 atm

This can also be converted into units of torr since 1.0 atm = 760 torr.

Thus, 0.24 atm = 180 torr.

3.6
Acids, Bases, and Salts

Basic Concept

Substances known as acids and bases are commonly encountered and have important chemical properties. Therefore, this group of substances deserves separate treatment.

An **acid** is any substance that, when dissolved in water, produces or causes to be produced hydrogen ions, H^+. A **base** is any substance that, when dissolved in water, produces or causes to be produced hydroxide ions, OH^-. The **Brönsted-Lowry model** stresses the fact that acids are *proton donors* and bases are *proton acceptors,* as well as the notion of *conjugate* (meaning "related") acids and bases. The **Lewis model** states that acids are substances that are electron-pair acceptors and bases are substances that are electron-pair donors. This concept broadens the meaning of acids and bases considerably to include many salts that would otherwise be excluded. Realistically, since a hydrogen ion is a bare proton, it is too reactive to exist as a stable species by itself. Hence, the real acid species is the hydronium ion, H_3O^+, a protonated water molecule, that is, $[H^+(H_2O)]$. Acids are powerful dehyrating agents.

When an acid and a base are added together, a neutralization (a double-replacement reaction) ensues, with water and a salt as the neutralization products. Quantitatively, 1 mole of acid exactly neutralizes 1 mole of base.

This is true for monoprotic acids and monobasic bases. Monoprotic acids contain one H^+ ion; diprotic acids contain two H^+ ions, and thus require two OH^- ions to neutralize them.

Acids and bases may be strong or weak. This property is determined by molecular structure and is different from concentration. A strong acid is one that dissociates and ionzies completely—100%—into its hydrogen/hydronium ions and anions, known as its conjugate base. Similarly, a strong base is one that dissociates and ionizes completely into hydroxide ions and its cations, known as its conjugate acid. Weak acids and bases, in contrast, dissociate and ionize less than 100%, typically less than 10%, into their respective ions. Recall that these substances are weak electrolytes, while strong acids and bases are strong electrolytes.

There are four common strong acids. All others can be assumed, by exclusion, to be weak.

- Hydrochloric acid, HCl   [HBr and HI are also strong but uncommon; HF is weak.]
- Nitric acid, HNO_3 [This is a *monoprotic acid* since it has only one hydrogen ion.]
- Sulfuric acid, H_2SO_4 [This is a *diprotic acid* since it has two hydrogen ions.]
- Perchloric acid, $HClO_4$

Examples of weak acids include acetic acid, boric acid, hydrofluoric acid, ascorbic acid, carbonic acid, and phosphoric acid.

Examples of strong bases are sodium hydroxide (NaOH) and potassium hydroxide (KOH). Examples of weak bases are ammonia (NH_3), which, when dissolved in water becomes ammonium hydroxide (NH_4OH), and organic amines.

When dissolved in water, **oxides of metals** (e.g., CaO, K_2O, MgO) are generally basic, while **oxides of nonmetals** (e.g., CO_2, SO_2, SO_3) are generally acidic.

Concentration Units and the pH Scale

Two units are routinely used to express concentrations of acids and bases:

- Molarity, M
- pH scale

Molarity is defined as the number of moles of H^+ ion for acids or of OH^- ion for bases per liter of solution. Concentration is symbolized by the use of brackets; that is, [ ] means moles per liter of the bracketed quantity.

The **pH**, representing the power of hydrogen ions, can be calculated from molarity. This value is useful for solutions whose acidic concentrations are less than 1.0 M, for example, 3.5×10^{-3} M.

$$pH = -\log[H^+] \qquad [3.22]$$

Equation 3.20 may also be written as follows:

$$[H^+] = 10^{-pH} \qquad [3.23]$$

The pH scale runs from 0 to 14, because of the autoionization of water. Recall that K_w for water is 1.0×10^{-14} at 25°C, since $K_w = [H^+][OH^-] = 1.0 \times 10^{-14}$.

The pOH may be analogously computed for basic solutions:

$$pOH = -\log[OH^-] \qquad [3.24]$$

In addition, since

$$pH + pOH = 14 \qquad [3.25]$$

you can always find pOH from pH, or vice versa.

For the more commonly used pH, the **acid range** of solutions is **0 to 7**, while the **basic (or alkaline) range** is **7 to 14.** A pH of 7.0 represents a neutral solution such as distilled water. The pH values of common substances are shown in Figure 3.2.

Properties of acids include:

- They taste sour.
- They neutalize bases to produce water and a salt.
- They react with metals to yield H_2 gas.

Properties of bases include:

- They taste bitter.
- They feel slippery.
- They react with fats, oils, and greases.
- They neutralize acids to produce water and salt.

Concentration of Hydrogen Ions Compared to Distilled Water		Examples of Solutions at This pH
10, 000, 000	pH = 0	Battery acid, strong hydrofluoric acid
1, 000, 000	pH = 1	Hydrochloric acid secreted by stomach lining
100, 000	pH = 2	Lemon juice, gastric acid, vinegar
10, 000	pH = 3	Grapefruit, orange juice, soda
1, 000	pH = 4	Tomato juice, acid rain
100	pH = 5	Soft drinking water, black coffee
10	pH = 6	Urine, saliva
1	pH = 7	"Pure" water Neutral
1/10	pH = 8	Sea water
1/100	pH = 9	Baking soda
1/1, 000	pH = 10	Great Salt Lake, Milk of Magnesia
1/10, 000	pH = 11	Ammonia solution
1/100, 000	pH = 12	Soapy water
1/1, 000, 000	pH = 13	Bleaches, oven cleaner
1/10, 000, 000	pH = 14	Liquid drain cleaner

Figure 3.2 The pH values of common substances

Calculations for Strong Acids and Bases

Examples 3.15 to 3.22 represent routinely encountered problems and calculations for strong acids and bases. In general, 1 mole of acid (H^+ ion) neutralizes 1 mole of base (OH^- ion) to produce 1 mole of water (H_2O).

Example 3.15

An aqueous solution of hydrochloric acid is 0.034 M. Find:

A. $[H^+]$
B. $[Cl^-]$

Solution:

Since HCl is a strong acid, it dissociates and ionizes completely (100%) into its component ions. Thus,

A. $[H^+] = 0.034$ M
B. $[Cl^-] = 0.034$ M

It is safe to assume that [HCl] in terms of molecules is 0.

Example 3.16

For a 0.035 M HNO_3 solution find:
A. $[H^+]$
B. pH

Solution:

A. HNO_3 is a strong acid. Thus, a 0.035 M solution of HNO_3 yields 0.035 M H^+ ions (and 0.035 M NO_3^- ions), since there is complete dissociation and ionization.
$[H^+] = 0.035$ M
B. pH $= -\log[H^+] = -\log(0.035$ M$) = 1.46$

Example 3.17

Find $[OH^-]$ for the solution in Example 3.16.

Solution:

Since this is an aqueous solution,

$$[H^+][OH^-] = K_w = 1.0 \times 10^{-14}$$

Thus,

$$[OH^-] = \frac{K_w}{[H^+]} = \frac{1.0 \times 10^{-14}}{0.035 \text{ M}} = 2.86 \times 10^{-13} \text{ M}$$

This shows a very small OH^- ion concentration, but not zero, and illustrates the constant and reciprocal nature of H^+ ions and OH^- ions.

Example 3.18

A sample of lemon juice has a pH of 2.4. Find its $[H^+]$.

Solution:

Recall that $[H^+] = 10^{-pH}$. Thus:

$$[H^+] = 10^{-2.4} = 4.0 \times 10^{-3} \text{ M}$$

Example 3.19

Find, for a 0.014 M solution of slaked lime, $Ca(OH)_2$:
A. $[OH^-]$
B. pH

Solution:

A. The subscript 2 next to (OH^-) indicates that 1 mol of $Ca(OH)_2$ produces 2 mol of OH^- ions (along with 1 mol of Ca^{2+} ions).

It is assumed that $Ca(OH)_2$ is a strong base. Thus,

$$[OH^-] = 2(0.014 \text{ M}) = 0.028 \text{ M}$$

B. Furthermore, pOH $= -\log [OH^-] = -\log (0.028$ M$) = 1.55$. Then,

$$\text{pH} = 14.0 - \text{pOH} = 14.0 - 1.55 = 12.45$$

Example 3.20 Acid-Base Reaction

A laboratory technician mixes 400 mL of a 0.125 M NaOH solution with 600 mL of a 0.100 M HCl solution. Find the pH of the resulting solution.

Solution:

First compute the number of moles of acid and base to determine whether they are equal or, if not, which one is present in excess.

moles of H^+ = (volume HCl)(molarity HCl)
$\qquad = 0.060$ mol
moles of OH^- = (volume NaOH)(molarity NaOH)
$\qquad = 0.050$ mol
Since 0.06 mol of $H^+ > 0.05$ mol of OH^-, the final solution will be acidic.

Furthermore, since 1 mol of acid reacts exactly with 1 mol of base, subtract moles of base from moles of acid: 0.060 mol $-$ 0.050 mol = **0.010 mol of H^+ left over or unreacted** and present after mixing. This amount is in a total or combined volume of 1000 mL or 1.00 L. Thus,

$$[H^+] = \frac{0.01 \text{ mol } H^+}{1.0 \text{ L}} = 0.01 \text{ M}$$

$$\text{pH} = 2.0$$

Example 3.21 Acid-Base Titration/ Neutralization

A laboratory technician wishes to find the concentration of an unknown base. He performs a titration in which 42.50 mL of 0.150 M HCl exactly neutralizes 25.00 mL of the base. Determine the concentration of the unknown base.

Solution:

Since moles of acid (A) equal moles of base (B) at the **endpoint** (point of neutralization), use the relationship

$$V_A M_A = V_B M_B$$

Solve for M_B:

$$M_B = \frac{V_A M_A}{V_B} = \frac{(42.50 \text{ mL})(0.150 \text{ M})}{25.00 \text{ mL}}$$

$$= 0.255 \text{ M}$$

Example 3.22 Dilution of a Solution

A laboratory technician is asked to prepare 500 mL of a 0.750 M solution of HCl. The stock solution of HCl that she has is labeled 6.0 M. How much (what volume) should she take from the stock solution bottle?

Solution:

The total number of moles of HCl ultimately desired in solution is

$$\text{moles of HCl} = V_{\text{HCl}} M_{\text{HCl}} = (0.500 \text{ L})(0.750 \text{ mol/L})$$

$$= 0.0375 \text{ mol}$$

This is the amount that must come from the 6.0 M solution. Since this amount is simply being redistributed from a concentrated solution (M_1 and V_1) to a more dilute one (M_2 and V_2), do the calculation directly as follows:

$$M_1 V_1 = M_2 V_2$$

Solve for V_1, substituting the values:

$$V_1 = \frac{(0.750 \text{ M})(0.500 \text{ L})}{6.0 \text{ M}}$$

$$= 0.0625 \text{ L} = 62.5 \text{ mL of stock solution}$$

This same strategy can be applied to *any* dilution problem.

Calculations for Weak Acids and Bases

In contrast to strong acids and bases, weak acids and bases dissociate and ionize into H^+ ions and OH^- ions, respectively, but to a very limited extent—normally less than 10% and more commonly less than 5%. Thus, for every 100 molecules of acid initially present, typically fewer than 5 molecules dissociate and ionize into hydrogen ions. Weak acids produce few hydrogen ions, while weak bases produce few hydroxide ions. Their reactions are governed by an equilibrium acid (a) or base (b) constant: K_a or K_b.

For example, acetic acid (found in vinegar) is a weak acid. Its equilibrium is represented as follows:

$$CH_3COOH(aq) + H_2O(l)$$
$$\rightleftharpoons H_3O^+(aq) + CH_3COO^-(aq) \quad [3.26]$$

Here, H_3O^+ is the active acid species, and CH_3COO^- is called the **conjugate base** of acetic acid. Note that the double arrows indicate an equilibrium process, meaning that this reaction does not go to completion and that all four chemical species in the reaction are present at any given time in varying concentrations. The equilibrium concept is reviewed in Section 3.7.

The equilibrium acid constant expression, K_a, is as follows:

$$K_a = \frac{[H_3O^+][CH_3COO^-]}{[CH_3COOH]} \quad [3.27]$$

The K_a value for acetic acid at room temperature is 1.75×10^{-5}.

The average percent dissociation and ionization, depending on initial concentration of the parent acid and temperature, is about 3 to 5%.

Table 3.1 gives K_a and K_b values for weak acids and bases, respectively, at 25°C.

There are two ways to compute $[H^+]$ or $[H_3O^+]$ for a solution of a weak acid. If the percent dissociation/ionization is given, simply multiply this by the initial concentration of the acid to obtain $[H^+]$. See Example 3.23. If the K_a value is given along with the initial concentration of the acid, equation 3.23 for K_a can be rearranged and solved for $[H_3O^+]$. See Example 3.24.

Example 3.23

A solution of 0.14 M nitrous acid, HNO_2, is 5.7% ionized. Calculate $[H^+]$.

Solution:

Nitrous acid is obviously a weak acid since its percent ionization is given as less than 10%, and certainly much less than 100%. Thus,

$$[H^+] = 5.7\% \times 0.14 \text{ M} = 7.98 \times 10^{-3} \text{ M}$$

Table 3.1 Weak Acid and Base Constants at 25°C

Monoprotic Acids			K_a
$HC_2O_2Cl_3$	Trichloroacetic acid	(Cl_3CCO_2H)	2.2×10^{-1}
HIO_3	Iodic acid		1.69×10^{-1}
$HC_2HO_2Cl_2$	Dichloracetic acid	(Cl_2CHCO_2H)	5.0×10^{-2}
$HC_2H_2O_2Cl$	Chloroacetic acid	(ClH_2CCO_2H)	1.36×10^{-3}
HNO_2	Nitrous acid		7.1×10^{-4}
HF	Hydrofluoric acid		6.8×10^{-4}
$HOCN$	Cyanic acid		3.5×10^{-4}
$HCHO_2$	Formic acid	(HCO_2H)	1.8×10^{-4}
$HC_3H_5O_3$	Lactic acid	$(CH_3CH(OH)CO_2H)$	1.38×10^{-4}
$HC_4H_3N_2O_3$	Barbituric acid		9.8×10^{-5}
$HC_7H_5O_2$	Benzoic acid	$(C_6H_5CO_2H)$	6.28×10^{-5}
$HC_4H_7O_2$	Butanoic acid	$(CH_3CH_2CH_2CO_2H)$	1.52×10^{-5}
HN_3	Hydrazoic acid		1.8×10^{-5}
$HC_2H_3O_2$	Acetic acid	(CH_3CO_2H)	1.8×10^{-5}
$HC_3H_5O_2$	Propanoic acid	$(CH_3CH_2CO_2H)$	1.34×10^{-5}
$HOCl$	Hypochlorous acid		3.0×10^{-8}
$HOBr$	Hypobromous acid		2.1×10^{-9}
HCN	Hydrocyanic acid		6.2×10^{-10}
HC_6H_5O	Phenol		1.3×10^{-10}
HOI	Hypoiodous acid		2.3×10^{-11}
H_2O_2	Hydrogen peroxide		1.8×10^{-12}

Example 3.24

K_a for formic acid, HCOOH, contained in bee stings, is 1.8×10^{-4}. Find the pH of a solution that initally contains 0.20 mol of HCOOH in 500 mL of solution.

Solution:

Since formic acid has a K_a value, it must be a weak acid. Its equilibrium equation and K_a expression are as follows:

$$HCOOH(aq) + H_2O(l) \rightleftharpoons H_3O^+(aq) + HCOO^-(aq)$$

$$K_a = \frac{[H_3O^+][HCOO^-]}{[HCOOH]} = 1.8 \times 10^{-4}$$

The strategy used in solving this type of problem is one that is commonly used in other equilibrium problems discussed in Section 3.7. The above expression for K_a equated to its value is a useful equation; it can be solved for $[H_3O^+]$, which can then be converted into pH.

The *initial concentration* of HCOOH is given:

$$[HCOOH] = \frac{0.20 \text{ mol}}{0.500 \text{ L}} = 0.400 \text{ M}$$

However, all concentrations inserted into the K_a expression must be equilibrium or final values. Therefore, let $x =$ the number of moles per liter of HCOOH that dissociate and ionize into products. Since 1 mol of HCOOH produces 1 mol of H_3O^+ *and* 1 mol of $HCOO^-$, and all products come from only one reactant, *at equilibrium,* then,

$$[H_3O^+] = x, \quad [HCOO^-] = x, \quad [HCOOH] = 0.40 - x$$

Substitute these values in the K_a expression and equation:

$$\frac{(x)(x)}{0.40 - x} = 1.8 \times 10^{-4}$$

This is a quadratic equation in x. A simplifying assumption may be made. Since $x \ll 0.40$, it may be neglected in the denominator, reducing this to an easier-to-solve abridged quadratic equation. Then,

$$x^2 = (1.8 \times 10^{-4})(0.40)$$

$$x = 8.49 \times 10^{-3} \text{ M}$$

120

Polyprotic Acids		K_{a1}	K_{a2}	K_{a3}
H_2SO_4	Sulfuric acid	Large	1.0×10^{-2}	
H_2CrO_4	Chromic acid	5.0	1.5×10^{-6}	
$H_2C_2O_4$	Oxalic acid	5.6×10^{-2}	5.4×10^{-5}	
H_3PO_3	Phosphorous acid	3×10^{-2}	1.6×10^{-7}	
H_2SO_3	Sulfurous acid	1.2×10^{-2}	6.6×10^{-8}	
H_2SeO_3	Selenous acid	4.5×10^{-3}	1.1×10^{-8}	
H_2TeO_3	Tellurous acid	3.3×10^{-3}	2.0×10^{-8}	
$H_2C_3H_2O_4$	Malonic acid	1.4×10^{-3}	2.0×10^{-6}	
	$(HO_2CCH_2CO_2H)$			
$H_2C_8H_4O_4$	Phthalic acid	1.1×10^{-3}	3.9×10^{-6}	
$H_2C_4H_4O_6$	Tartaric acid	9.2×10^{-4}	4.3×10^{-5}	
$H_2C_6H_6O_6$	Ascorbic acid	7.9×10^{-5}	1.6×10^{-12}	
H_2CO_3	Carbonic acid	4.5×10^{-7}	4.7×10^{-11}	
H_3PO_4	Phosphoric acid	7.1×10^{-3}	6.3×10^{-8}	4.5×10^{-13}
H_3AsO_4	Arsenic acid	5.6×10^{-3}	1.7×10^{-7}	4.0×10^{-12}
$H_3C_6H_5O_7$	Citric acid	7.1×10^{-4}	1.7×10^{-5}	6.3×10^{-6}

Weak Bases		K_b
$(CH_3)_2NH$	Dimethylamine	9.6×10^{-4}
CH_3NH_2	Methylamine	4.4×10^{-4}
$CH_3CH_2NH_2$	Ethylamine	4.3×10^{-4}
$(CH_3)_3N$	Trimethylamine	7.4×10^{-5}
NH_3	Ammonia	1.8×10^{-5}
N_2H_4	Hydrazine	9.6×10^{-7}
NH_2OH	Hydroxylamine	6.6×10^{-9}
C_5H_5N	Pyridine	1.5×10^{-9}
$C_6H_5NH_2$	Aniline	4.1×10^{-10}
PH_3	Phosphine	10^{-28}

Thus,

$$[H_3O^+] = 8.49 \times 10^{-3} \text{ M}$$

Also:

$$pH = -\log[H_3O^+] = -\log(8.49 \times 10^{-3} \text{ M})$$

$$= 2.07$$

Note: A good rule of thumb to follow to determine whether the simplifying assumption is justified is as follows: Let M_a = initial molarity of acid; then,

If $M_a/K_a > 100$, $M_a - x \cong M_a$ is a good approximation.

The error incurred is usually only 1 or 2%, which is generally acceptable. The M_a/K_a ratio in the example above produces a value of about 2,000.

Hydrolysis

Some compounds that appear neutral actually have acidic or alkaline properties. For example, when sodium acetate, $NaCH_3COO$, is dissolved in water, it reacts with water to form acetic acid (a weak acid) and hydroxide ion. This process is called **hydrolysis**, and the resulting solution is slightly alkaline. Remember that the acetate ion is the conjugate base of acetic acid. Other basic salts in this category include sodium cyanide, potassium nitrite, and lithium carbonate.

Similarly, ammonium chloride, NH_4Cl, produces a slightly acidic solution when dissolved in water, again because of hydrolysis. Ammonium ion is the **conjugate acid** of ammonium hydroxide, and/or ammonia, NH_3, is the conjugate base of NH_4^+. Another salt in this category is aluminum chloride, $AlCl_3$. Gases such as CO_2 and SO_2 also undergo hydrolysis and produce H^+ ions and HCO_3^- and HSO_3^- ions, respectively.

Example 3.25

Calculate the pH of a 0.20 M solution of ammonium chloride, NH_4Cl.

Solution:

NH_4Cl is the salt of a weak base, NH_3, and hence hydrolyzes when dissolved in water to form a slightly acidic solution.

$$NH_4^+(aq) + H_2O(l) \rightleftharpoons H_3O^+(aq) + NH_3(aq)$$

The hydrolysis constant expression, K_h, can be expressed as follows:

$$K_h = \frac{[H_3O^+][NH_3]}{[NH_4^+]}$$

The value for K_h is not given but can be found by applying the relationship

$$K_h = \frac{K_w}{K_b} = \frac{1.0 \times 10^{-14}}{1.8 \times 10^{-5}} = 5.56 \times 10^{-10}$$

where K_b is the value of the equilibrium base constant for NH_3 in H_2O. It is analogous to K_a:

$$NH_3 + H_2O \rightleftharpoons NH_4^+ + OH^-$$

The same strategy is now used as in Example 3.24. *At equilibrium,* then,

$$[H_3O^+] = x, \quad [NH_3] = x, \quad [NH_4^+] = 0.20 - x$$

Therefore,

$$5.56 \times 10^{-10} = \frac{(x)(x)}{0.20 - x}$$

Use the same simplifying assumption as above, assume $x \ll 0.20$ M, and solve for x:

$$x^2 = (0.20 \text{ M})(5.56 \times 10^{-10})$$

$$x = 1.1 \times 10^{-5}$$

Thus,

$$[H_3O^+] = 1.1 \times 10^{-5}$$

and

$$pH = -\log[H_3O^+] = -\log(1.1 \times 10^{-5} \text{ M})$$

$$= 4.96$$

The pH is less than 7.0, reflecting an acidic solution, as expected.

Note: For the hydrolysis of an ion producing a *basic* solution, $K_h = K_w/K_a$.

Buffer Solutions and the Henderson-Hasselbalch Equation

Buffer solutions are solutions that protect against large shifts in pH in the event of a shock due to the sudden addition of a strong acid or base. The pH of a buffer solution does, however, change slightly. Such a solution can be made with a combination of either

- A weak acid and its salt (conjugate base).
- A weak base and its salt (conjugate acid).

Thus, acetic acid plus sodium acetate constitutes a buffer system, as does ammonium hydroxide plus ammonium chloride.

To determine the pH of a buffer solution, the **Henderson-Hasselbalch equation** may be used. In the case of a weak acid (e.g., CH_3COOH) and its conjugate base salt (e.g., $NaCH_3COO$):

$$pH = pK_a + \log \frac{[\text{conjugate base}]}{[\text{weak acid}]} \qquad [3.28]$$

The buffering capacity of a system refers to the amount of acid or base it can absorb before its pH changes. Buffering capacity is generally at a maximum when [weak acid] = [conjugate base], so that $pH = pK_a$.

Amphoterism

Amphoterism refers to the property of certain compounds to act as both acids and bases. For example, the compound $Al(OH)_3$, which is categorically a base, may react with acid to form $[Al(H_2O)_6]^{3+}$ *and* with excess base to form $[Al(OH)_4]^{-1}$. Both of these species are termed **complex ions** and are water soluble. They are governed by complex-ion equilibrium constants, as explained in Section 3.7. Aluminum oxide, Al_2O_3, is also amphoteric. The hydroxides, as well as the oxides, of zinc(II) and chromium(III) are similarly amphoteric.

3.7
Chemical Equilibrium

Basic Concept

In reality, most reactions do not go to completion even if the reactants are present in stoichiometric amounts or ratios. Rather, they reach a condition of equilibrium, denoted by double arrows in the reaction equation. **Equilibrium** means that there is a balance between the reactant side and the product side, or simply between the reactants and the products. The equilibrium condition is dynamic, not static, allowing microscopic changes in reactant and product concentrations, such that no *net* change in reactant or product concentrations occurs, to take place, provided that no external stresses are applied. At any given time, all species in the reaction equation—reactants and products—are present at equilibrium in varying amounts. These varying amounts, however, are related and connected to one another in a mathematical formula known as the **equilibrium constant expression, K_c.**

For an equilibrium reaction, which can be generally represented as

$$a\text{A} + b\text{B} \rightleftharpoons g\text{G} + h\text{H} \qquad [3.29]$$

the K_c expression is

$$K_c = \frac{[\text{G}]^g[\text{H}]^h}{[\text{A}]^a[\text{B}]^b} \qquad [3.30]$$

where a, b, g, and h represent the stoichiometric coefficients in the balanced reaction, and the brackets [] stipulate molar concentrations.

Four *points of caution* regarding this expression are noteworthy:

1. Only concentrations in units of molarity are permitted.
2. Only species that have concentrations can appear. Thus, species in the solution state (aq) or in the gaseous state (g) are included, but species that are pure solids (s) or pure liquids (l) are not.
3. K_c is a function of temperature (van't Hoff equation; see page 126).
4. The reaction must be at equilibrium.

A reaction in which all species are present in the same physical state (e.g., all gases or all solutions) is called a **homogeneous reaction,** while a reaction in which species are present in more than one physical state (i.e., mixed physical states) is called a **heterogeneous reaction.** If all species in a reaction are present in the gaseous state, an alternative form of the equilibrium constant expression can be written as K_p:

$$K_p = \frac{P_\text{G}{}^g P_\text{H}{}^h}{P_\text{A}{}^a P_\text{B}{}^b} \qquad [3.31]$$

Here, P represents the partial pressure of each gas present at equilibrium. Any pressure unit is acceptable as long as it is consistent with the others and with the units of K_P, but atmospheres and torr (1 torr = 1 mm Hg) are common.

A formula to convert K_c to K_p, and vice versa, is as follows:

$$K_p = K_c\,(RT)^{\Delta n} \qquad [3.32]$$

where T = the absolute temperature in kelvins,

R = the gas constant in units consistent with partial pressures,

Δn = the total moles of product − the total moles of reactant (in the balanced reaction).

Example 3.26

Write the equilibrium constant expression K_c for the following reaction at 25°C:

$$2\text{N}_2\text{O}_5(g) \rightleftharpoons 4\text{NO}_2(g) + \text{O}_2(g)$$

Solution:

Note that this is a homogeneous, gas-phase reaction. All three substances can have concentrations in units of molarity and should appear in the K_c expression:

$$K_c = \frac{[\text{NO}_2]^4[\text{O}_2]}{[\text{N}_2\text{O}_5]^2}$$

If you were given a value for K_c and were asked to convert it to K_P, then, since all species are in the gaseous state, you could write:

$$K_p = K_c\,(RT)^{\Delta n}$$
$$= K_c\,(0.0821\ \text{L} \cdot \text{atm/K} \cdot \text{mol} \times 298\ \text{K})^3$$

since $\Delta n = (4 + 1) - 2 = 3$.

Example 3.27

For the reaction

$$2\text{SO}_2(g) + \text{O}_2(g) \rightleftharpoons 2\text{SO}_3(g)$$

at 827°C, the value for $K_c = 37.1$. What is the value for the reverse reaction?

Solution:

$$K_c(\text{reverse}) = \frac{1}{K_c(\text{forward})} = \frac{1}{37.1} = 0.0269$$

The Meaning of K_c

Reactions with K_c values significantly greater than 1 (e.g., on the order of 10^3 or larger) approach completion, and their equilibria lie far to the right or product side. In fact, reactions that go to completion have exceptionally large or even undefined K_c values. In contrast, reactions that have K_c values much less than 1 (e.g., on the order of 10^{-4} or smaller), as in the case of the K_a value for acetic acid in Section 3.6, do not undergo significant change, and their equilibria lie far to the left or reactant side—very little product forms. In the case of acids, the smaller the value of K_a, the weaker the acid.

Calculations for K_c

K_c values for a given reaction at a given temperature are always constant, regardless of which of the three approaches or pathways to equilibrium is followed:

- **Only** reactants present initially.
- **Only** products present initially.
- **Both** reactants **and** products present initially.

This concept, together with knowledge of the approach or pathway to equilibrium, can prove very valuable in developing a strategy to solve equilibrium problems, as shown in Example 3.28.

Example 3.28

The reaction below is carried out in a 5-L vessel at 600 K.

$$CO(g) + H_2O(g) \rightleftharpoons CO_2(g) + H_2(g)$$

At equilibrium, it is found that 0.020 mol of CO, 0.0215 mol of H_2O, 0.070 mol of CO_2, and 2.00 mol of H_2 are present. For this reaction, calculate:
A. K_c
B. K_p

Solution:

A. The reaction is balanced as it stands. All four substances are in the gaseous state and thus have concentrations that can be inserted into the K_c expression.

Since the *molar* amounts are given *at equilibrium,* all you need do is to divide each of them by the volume to convert it into molarity, and then substitute each in the K_c expression. Thus,

$$K_c = \frac{[CO_2][H_2]}{[CO][H_2O]}$$

$$= \frac{(0.070 \text{ mol/5.0 L})(2.00 \text{ mol/5.0 L})}{(0.020 \text{ mol/5.0 L})(0.0215 \text{ mol/5.0 L})}$$

$$= 326$$

B. To convert to K_p, simply use the conversion formula given in Equation 3.30:

$$K_p = K_c(RT)^{\Delta n}$$

Note that in this case, since

$$\Delta n = (1 + 1) - (1 + 1) = 0$$
$$K_p = K_c$$

In fact, $K_p = K_c$ whenever the product side and the reactant side contain identical numbers of total moles.

Example 3.29

Ammonia gas, $NH_3(g)$, is introduced into a previously evacuated reaction vessel in such a way that its initial concentration is 0.500 M. The ammonia decomposes into nitrogen and hydrogen gases, according to the reaction given below, and eventually reaches equilibrium. The equilibrium concentration of nitrogen is found to be 0.116 M. Determine the value of K_c.

$$2NH_3(g) \rightleftharpoons N_2(g) + 3H_2(g)$$

Solution:

Knowing the approach or pathway to equilibrium is vital to solving this problem. It is clear in the statement that only reactant is present initially; therefore, all of the products are formed from the decomposition N_2 of NH_3. The equilibrium concentration of N_2 is given as 0.116 M. Because of the 3 to 1 stoichiometric ratio between H_2 and N_2, 3(0.116 M) = 0.348 M of H_2 must have also formed and is present at equilibrium.

Also, because of the stoichiometric ratio between NH_3 and N_2, 2(0.116 M) = 0.232 M of NH_3 must have decomposed. Thus, the concentration of NH_3 remaining and present at equilibrium is 0.500 M − 0.232 M = 0.268 M. Therefore, the values for the three species *at equilibrium* are as follows:

$[NH_3]$ = 0.268 M $[N_2]$ = 0.116 M $[H_2]$ = 0.348 M

Substitute these values in the K_c expression:

$$K_c = \frac{[N_2][H_2]^3}{[NH_3]^2}$$

$$= \frac{(0.116 \text{ M})(0.348 \text{ M})^3}{(0.268 \text{ M})^2}$$

$$= 0.0681 \text{ M}^2$$

Predicting Equilibrium Using a Test Quotient

It is often useful to know whether a reaction reached equilibrium and, if so, when. This information can be obtained by measuring the concentrations of all species in the reaction and then determining a test quotient, Q. Q is identical in form to K_c except that the values substituted in the expression may or may not be equilibrium concentrations. The criteria for equilibrium are as follows:

- If $Q = K_c$, the reaction is at equilibrium.
- If $Q < K_c$, the reaction is not at equilibrium and needs to proceed to the right.
- If $Q > K_c$, the reaction is not at equilibrium and needs to proceed to the left.

Example 3.30

Consider the reaction given in Example 3.27 along with the following concentrations:

$$[SO_2] = 0.054 \text{ M} \quad [O_2] = 0.020 \text{ M}$$
$$[SO_3] = 0.012 \text{ M}$$

A. Is this reaction at equilibrium?
B. If not, in which direction will it proceed?

Solution:

A. $Q = \dfrac{[SO_3]^2}{[SO_2]^2[O_2]} = \dfrac{(0.012 \text{ M})^2}{(0.054 \text{ M})^2(0.020 \text{ M})}$

$\quad = 2.47$

Since $K_c = 37.1$, Q does not equal K_c.

Thus, the reaction is not at equilibrium.

B. Since $Q < K_c$, the reaction must proceed to the right to reach equilibrium.

Stresses and Le Châtelier's Principle

Once a reaction has reached equilibrium, any external stress that is imposed results in a shift of the equilibrium and a subsequent change in the concentrations of the reactants and products involved. The direction of the shift—right (product side) or left (reactant side)—is in accordance with **Le Châtelier's principle,** which states: If a stress is imposed on a reaction at equilibrium, the reaction will shift either right or left in the way or direction that accommodates or relieves the stress.

There are five categories of stress:

- The reactant concentration or product concentration is increased or decreased.
- The temperature of the reaction is increased or decreased.
- The volume of the reaction vessel is increased or decreased.
- The pressure of the reaction is increased or decreased (in any one of three different ways).
- A catalyst is added to the reaction.

Example 3.31

Consider the following reaction already at equilibrium.

$$CS_2(g) + 3Cl_2(g) \rightleftharpoons S_2Cl_2(g) + CCl_4(g)$$

The enthalpy or heat of reaction is $\Delta H^\circ = -84.3$ kJ.

Predict what will happen—increase, decrease, or no change—to the equilibrium concentration of CCl_4 when each of the following stresses is imposed:

A. Cl_2 is added.
B. S_2Cl_2 is removed.
C. The temperature is increased.
D. The volume of the reaction vessel is decreased.
E. The pressure is decreased by increasing the volume of the reaction vessel.
F. A catalyst is added.

Solution:
Use Le Châtelier's principle to answer each part.

A. Adding Cl_2 will cause the reaction to shift to the right, increasing $[CCl_4]$.
B. Removing S_2Cl_2 will cause the reaction to shift to the right, increasing $[CCl_4]$.
C. To answer this part, the meaning of ΔH° must be understood. This is the amount of heat energy either liberated or absorbed in the reaction and may be expressed in either kilojoules (kJ) or kilocalories (kcal). If heat energy is liberated, the reaction is **exothermic**, and the sign of ΔH is by convention negative. If heat energy is absorbed, the reaction is **endothermic**, and the sign of ΔH is by convention positive. Here, the sign is negative, so heat is produced. Hence, since raising the temperature adds heat and heat behaves as a product, increasing the temperature will shift the reaction to the left, decreasing $[CCl_4]$.

D. In the balanced reaction, 4 molecules of reactant are reacting to form 2 molecules of product. Increasing the volume of the vessel favors the reactant or left side of the reaction, while decreasing the volume favors the product side. Hence, decreasing the volume will cause the reaction to shift to the right, increasing [CCl$_4$]. In the special case where the two sides of a reaction contain equal numbers of molecules, changing the volume has no effect.

E. If the pressure is decreased by increasing the volume, the effect can be evaluated as a volume stress. As discussed in part **D**, increasing the volume will cause the reaction to shift to the left and decrease [CCl$_4$].

Note, however, that, if the pressure is increased or decreased by the addition of a reactant or product, the effect should be evaluated as a stress discussed in part **A** or **B**.

Also, note that, if the pressure is increased by the addition of an inert or unreactive gas, there is no effect on the equilibrium.

F. A catalyst speeds up the rate of reaction. Since the reaction has already reached equilibrium, adding a catalyst will have no effect.

Dependence of K$_c$ on Temperature

The exact dependence of K_c (or K_p) on temperature is given by the **van't Hoff equation**:

$$\ln \frac{K_c(2)}{K_c(1)} = \frac{\Delta H_{RX}}{R}\left(\frac{1}{T_1} - \frac{1}{T_2}\right) \quad [3.33]$$

where $K_c(1)$ is the equilibrium constant value at absolute temperature T_1 and $K_c(2)$ is the value at T_2, ΔH_{RX} is the enthalpy or heat of reaction, and R is the gas constant in units of either joules per mole or calories per mole, consistent with the units of ΔH_{RX}. The similarity of the van't Hoff equation to the Clausius-Clapeyron equation given in Section 3.5 is noteworthy.

3.8
Solubility Product Constants

Basic Concept

Like the acids and bases discussed in Section 3.6, another group of substances comprised generally of ionic compounds that have very limited solubility in water deserves separate treatment. Examples include magnesium hydroxide (commonly known as milk of magnesia, an antacid), barium sulfate (used in enemas to diagnose colonic tumors), and silver chloride (used in the photographic process). These substances, generally referred to as "insoluble" or "slightly soluble," are governed by a solubility equilibrium expression and a solubility product constant known as K_{sp}, identical in concept to the principle of equilibrium discussed in Section 3.7.

The solubility of any salt depends on temperature. The solubility of most salts increases with increasing temperature. When a salt has dissolved to the maximum extent in a given volume of water at a given temperature, the solution is said to be **saturated**. Until that point is reached, the solution is **unsaturated**. If a saturated solution is heated above room temperature so that more salt can be dissolved and be held in solution, the solution is **supersaturated.** The solubility of a salt is commonly measured in units of moles per liter, grams per liter, or milligrams per liter (often referred to as parts per million or ppm).

If two liquids, such as ethanol and water, dissolve freely in each other, they are said to be **miscible** in all proportions. Oil and water are immiscible with each other.

Definition of Solubility Product Constant K$_{sp}$

Consider, for example, the slightly soluble salt $CaF_2(s)$ in equilibrium with water. Its limited dissociation and ionization in water can be expressed as follows:

$$CaF_2(s) \rightleftharpoons Ca^{2+}(aq) + 2\,F^-(aq) \quad [3.34]$$

and its **solubility product constant, K$_{sp}$,** as

$$K_{sp} = [Ca^{2+}][F^-]^2 \quad [3.35]$$

Note that [CaF$_2$(s)] does not appear in the K_{sp} expression since it represents the undissolved portion and is a pure solid.

Also note that [Ca^{2+}] represents the concentration of Ca^{2+} ion actually dissolved in the water, while [F$^-$] represents the F$^-$ ion concentration actually dissolved in the water.

Note also that

$$[Ca^{2+}(aq)] = 2\,[F^-(aq)] = [CaF_2(aq)]$$

CaF$_2$(aq) is dissolved.

The smaller the K_{sp} value, the lower the solubility of the salt.

Clearly, the K_{sp} value of a salt can be calculated by determining the solubility of each ion and then using equation 3.35. K_{sp} values, like other equilibrium constants, are functions of temperature.

Calculating Molar Solubility from K_{sp}

The molar solubility of any slightly soluble salt, along with the concentration of any of its ions, can be calculated from its K_{sp} value. The formula of the salt must be known, however, so that its dissociation and ionization can be written.

It is important to note that simply comparing the K_{sp} values of two salts to determine which has a higher or lower molar solubility can often be misleading. This comparison of numerical values is valid only when the two salts have the same stoichiometry, that is, the subscripts in the formulas are identical.

Table 3.2 lists K_{sp} values for many salts at 25°C.

Example 3.32

Calculate

A. the molar solubility of calcium flouride, CaF_2, and

B. the concentration of the flouride ion, in solution at 25°C. The K_{sp} of CaF_2 is 4.0×10^{-11} at 25°C.

Solution:

A. The equilibrium for this salt is as follows:

$$CaF_2(s) \rightleftharpoons Ca^{2+}(aq) + 2F^-(aq)$$

The K_{sp} expression is

$$K_{sp} = [Ca^{2+}][F^-]^2 = 4.0 \times 10^{-11} \, M^3$$

Let S = the solubility of the Ca^{2+} ion in moles per liter.

Then $2S$ = the solubility of the F^- ion.

Substitute S in the K_{sp} expression:

$$(S)(2S)^2 = 4.0 \times 10^{-11}$$
$$4S^3 = 4.0 \times 10^{-11}$$
$$S = 2.15 \times 10^{-4} \, M$$

Thus, the molar solubility of CaF_2 is 2.15×10^{-4} M. This is *also* $[Ca^{2+}]$.

B. $[F^-] = 2 \times 2.15 \times 10^{-4} \, M = 4.30 \times 10^{-4} \, M$

Example 3.33

The molar solubility of tin iodide, SnI_2, is 1.28×10^{-2} mol/L. What is K_{sp} for this compound?

Solution:

The solubility equilibrium for SnI_2 is

$$SnI_2(s) \rightleftharpoons Sn^{2+}(aq) + 2I^-(aq)$$

The K_{sp} expression is

$$K_{sp} = [Sn^{2+}][I^-]^2$$

Note that 1.0 mol of SnI_2 produces 1.0 mol of Sn^{2+}, but 2.0 mol of I^-.

$$[Sn^{2+}] = 1.28 \times 10^{-2} \, M$$

$$[I^-] = (2) \times 1.28 \times 10^{-2} \, M = 2.56 \times 10^{-2} \, M$$

Substitute these values in the K_{sp} expression:

$$K_{sp} = (1.28 \times 10^{-2} \, M)(2.56 \times 10^{-2} \, M)^2$$
$$= 8.4 \times 10^{-6} \, M^2$$

Common and Uncommon Ion and pH Effects

Occasionally, the solvent in which a salt is to be dissolved is not pure water but rather contains an ion in common with the salt to be dissolved. The common ion may be a cation or an anion, such as Ca^{2+} or F^-. The common ion may also be the H^+ ion or OH^- ion, which manifests itself simply in a pH value less than or greater than 7, respectively. This is the **common ion effect,** and the net result is to reduce significantly the solubility of the salt in question.

The presence of a background salt with no ions in common with the salt to be dissolved is known as the **uncommon ion** or **salt effect.** In contrast, this effect slightly increases the solubility of the salt in question.

In general, most salts are more soluble in hot water than in cold; that is, the solubilities of most salts increase with increasing temperature. In the temperature range of 0 to 100°C, many common salts demonstrate a nearly linear relationship between solubility and temperature, with varying, positive slopes. Some notable exceptions include KNO_3, which shows an exponential relationship, and Li_2SO_4, which shows a linear but decreasing solubility with temperature.

Example 3.34

Magnesium hydroxide, $Mg(OH)_2$, is a common antacid and has a $K_{sp} = 1.8 \times 10^{-11}$. Find the molar solubility of $Mg(OH)_2$ in a solution whose pH is 13.12.

Solution:

To find the molar solubility of $Mg(OH)_2$ in distilled water, the method outlined in Example 3.32 would be followed. In this case, however, the solution is basic with an excess of OH^- ions. Since these OH^- ions are in common with the OH^- ions contained in $Mg(OH)_2$, the common ion effect must be taken into account because it will reduce the solubility of $Mg(OH)_2$.

First determine $[OH^-]$ as follows:

$$pOH = 14.00 - pH = 14.00 - 13.12 = 0.88$$

$$[OH^-] = 10^{-pOH} = 10^{-0.88} = 1.32 \times 10^{-1} \text{ M}$$

The K_{sp} expression for $Mg(OH)_2$ is:

$$K_{sp} = [Mg^{2+}][OH^-]^2$$

Use the strategy outlined in Example 3.32.

Let $[Mg^{2+}] = S$

and $[OH^-] = 2S + 1.32 \times 10^{-1}$.

Substitute these values, along with the K_{sp} value, in the K_{sp} expression:

$$1.8 \times 10^{-11} = (S)(2S + 1.32 \times 10^{-1})^2$$

This is a cubic equation in S. To simplify the calculation, assume that $2S$ is negligible compared with 1.32×10^{-1} as an additive term. Then:

$$1.8 \times 10^{-11} = (S)(1.32 \times 10^{-1})^2$$

$$S = 1.0 \times 10^{-9} \text{ M}$$

This result represents the solubility of $Mg(OH)_2$.

Predicting Precipitation

When two solutions are mixed, it is possible that an insoluble precipitate will form and settle to the bottom. Whether this happens depends on the identities of the ions present in the solutions and their respective concentrations. If the ions for a potentially insoluble precipitate are present, all that needs to be done to predict precipitation is to calculate a hypothetical solubility product constant, Q, in the same manner as K_{sp} is calculated. The Q value is then compared to the K_{sp} value, subject to the following criteria or conditions:

- If $Q \leq K_{sp}$, no precipitate forms.
- If $Q > K_{sp}$, a precipitate forms.

In general, conditions that favor **completeness of precipitation** are

- A very small value of K_{sp}
- High initial ion concentrations
- A concentration of common ion that considerably exceeds the concentration of the target ion to be precipitated

Example 3.35

Will a precipitate form if 50.0 mL of 1.2×10^{-3} M $Pb(NO_3)_2$ are added to 50.0 mL of 2.0×10^{-4} M Na_2SO_4?

Solution:

The addition of the two solutions will result in a double-replacement reaction. Two products, $NaNO_3$ and $PbSO_4$, will form and are thus candidates as insoluble salts. A check of Table 3.2 reveals that only $PbSO_4$ has a K_{sp} value (1.6×10^{-8}) and is therefore a possible precipitate. To confirm whether a precipitate actually forms, evaluate Q as follows:

$$Q = [Pb^{2+}(aq)][SO_4^{2-}(aq)]$$

$$[Pb^{2+}(aq)] = \frac{(0.050 \text{ L})(1.2 \times 10^{-3} \text{ M})}{0.100 \text{ L}}$$

$$= 6.0 \times 10^{-4} \text{ M}$$

$$[SO_4^{2-}(aq)] = \frac{(0.050 \text{ L})(2.0 \times 10^{-4} \text{ M})}{0.100 \text{ L}}$$

$$= 1.0 \times 10^{-4} \text{ M}$$

Substitute these values for concentrations into the expression for Q:

$$Q = (6.0 \times 10^{-4} \text{ M})(1.0 \times 10^{-4} \text{ M})$$
$$= 6.0 \times 10^{-8}$$

But $K_{sp} = 1.6 \times 10^{-8}$.

Since $Q > K_{sp}$, a precipitate forms!

3.9
Chemical Kinetics

Basic Concept

Chemical kinetics is the study of the rates and mechanisms of chemical reactions. The reaction rate is a kinetic property and depends on the mechanism. In contrast, thermodynamic properties (discussed in Section 3.10) are independent of mechanism. Thermodynamics can tell us whether or not a reaction is spontaneous, but kinetics tells us how fast the reaction occurs and if it happens fast enough to be of any interest or value.

The most common way to measure the rate of reaction is by the disappearance of a reactant, that is,

Table 3.2 Solubility Product Constants at 25°C

Salt	Solubility Equilibrium	K_{sp}
Fluorides		
MgF_2	$MgF_2(s) \rightleftharpoons Mg^{2+}(aq) + 2F^-(aq)$	6.6×10^{-9}
CaF_2	$CaF_2(s) \rightleftharpoons Ca^{2+}(aq) + 2F^-(aq)$	3.9×10^{-11}
SrF_2	$SrF_2(s) \rightleftharpoons Sr^{2+}(aq) + 2F^-(aq)$	2.9×10^{-9}
BaF_2	$BaF_2(s) \rightleftharpoons Ba^{2+}(aq) + 2F^-(aq)$	1.7×10^{-6}
LiF	$LiF(s) \rightleftharpoons Li^+(aq) + F^-(aq)$	1.7×10^{-3}
PbF_2	$PbF_2(s) \rightleftharpoons Pb^{2+}(aq) + 2F^-(aq)$	3.6×10^{-8}
Chlorides		
$CuCl$	$CuCl(s) \rightleftharpoons Cu^+(aq) + Cl^-(aq)$	1.9×10^{-7}
$AgCl$	$AgCl(s) \rightleftharpoons Ag^+(aq) + Cl^-(aq)$	1.8×10^{-10}
Hg_2Cl_2	$Hg_2Cl_2(s) \rightleftharpoons Hg_2^{2+}(aq) + 2Cl^-(aq)$	1.2×10^{-18}
$TlCl$	$TlCl(s) \rightleftharpoons Tl^+(aq) + Cl^-(aq)$	1.8×10^{-4}
$PbCl_2$	$PbCl_2(s) \rightleftharpoons Pb^{2+}(aq) + 2Cl^-(aq)$	1.7×10^{-5}
$AuCl_3$	$AuCl_3(s) \rightleftharpoons Au^{3+}(aq) + 3Cl^-(aq)$	3.2×10^{-25}
Bromides		
$CuBr$	$CuBr(s) \rightleftharpoons Cu^+(aq) + Br^-(aq)$	5×10^{-9}
$AgBr$	$AgBr(s) \rightleftharpoons Ag^+(aq) + Br^-(aq)$	5.0×10^{-13}
Hg_2Br_2	$Hg_2^{2+}(s) \rightleftharpoons Hg_2^{2+}(aq) + 2Br^-(aq)$	5.6×10^{-23}
$HgBr_2$	$HgBr_2(s) \rightleftharpoons Hg^{2+}(aq) + 2Br^-(aq)$	1.3×10^{-19}
$PbBr_2$	$PbBr_2(s) \rightleftharpoons Pb^{2+}(aq) + 2Br^-(aq)$	2.1×10^{-6}
Iodides		
CuI	$CuI(s) \rightleftharpoons Cu^+(aq) + I^-(aq)$	1×10^{-12}
AgI	$AgI(s) \rightleftharpoons Ag^+(aq) + I^-(aq)$	8.3×10^{-17}
Hg_2I_2	$Hg_2I_2(s) \rightleftharpoons Hg_2^{2+}(aq) + 2I^-(aq)$	4.7×10^{-29}
HgI_2	$HgI_2(s) \rightleftharpoons Hg^{2+}(aq) + 2I^-(aq)$	1.1×10^{-28}
PbI_2	$PbI_2(s) \rightleftharpoons Pb^{2+}(aq) + 2I^-(aq)$	7.9×10^{-9}
Hydroxides		
$Mg(OH)_2$	$Mg(OH)_2(s) \rightleftharpoons Mg^{2+}(aq) + 2OH^-(aq)$	7.1×10^{-12}
$Ca(OH)_2$	$Ca(OH)_2(s) \rightleftharpoons Ca^{2+}(aq) + 2OH^-(aq)$	6.5×10^{-6}
$Mn(OH)_2$	$Mn(OH)_2(s) \rightleftharpoons Mn^{2+}(aq) + 2OH^-(aq)$	1.6×10^{-13}
$Fe(OH)_2$	$Fe(OH)_2(s) \rightleftharpoons Fe^{2+}(aq) + 2OH^-(aq)$	7.9×10^{-16}
$Fe(OH)_3$	$Fe(OH)_3(s) \rightleftharpoons Fe^{3+}(aq) + 3OH^-(aq)$	1.6×10^{-39}
$Co(OH)_2$	$Co(OH)_2(s) \rightleftharpoons Co^{2+}(aq) + 2OH^-(aq)$	1×10^{-15}
$Co(OH)_3$	$Co(OH)_3(s) \rightleftharpoons Co^{3+}(aq) + 3OH^-(aq)$	3×10^{-45}
$Ni(OH)_2$	$Ni(OH)_2(s) \rightleftharpoons Ni^{2+}(aq) + 2OH^-(aq)$	6×10^{-16}
$Cu(OH)_2$	$Cu(OH)_2(s) \rightleftharpoons Cu^{2+}(aq) + 2OH^-(aq)$	4.8×10^{-20}
$V(OH)_3$	$V(OH)_3(s) \rightleftharpoons V^{3+}(aq) + 3OH^-(aq)$	4×10^{-35}
$Cr(OH)_3$	$Cr(OH)_3(s) \rightleftharpoons Cr^{3+}(aq) + 3OH^-(aq)$	2×10^{-30}
Ag_2O	$Ag_2O(s) + H_2O \rightleftharpoons 2Ag^+(aq) + 2OH^-(aq)$	1.9×10^{-8}
$Zn(OH)_2$	$Zn(OH)_2(s) \rightleftharpoons Zn^{2+}(aq) + 2OH^-(aq)$	3.0×10^{-16}
$Cd(OH)_2$	$Cd(OH)_2(s) \rightleftharpoons Cd^{2+}(aq) + 2OH^-(aq)$	5.0×10^{-15}
$Al(OH)_3$ (alpha form)	$Al(OH)_3(s) \rightleftharpoons Al^{3+}(aq) + 3OH^-(aq)$	3×10^{-34}
Cyanides		
$AgCN$	$AgCN(s) \rightleftharpoons Ag^+(aq) + CN^-(aq)$	2.2×10^{-16}
$Zn(CN)_2$	$Zn(CN)_2(s) \rightleftharpoons Zn^{2+}(aq) + 2CN^-(aq)$	3×10^{-16}
Sulfites		
$CaSO_3$	$CaSO_3(s) \rightleftharpoons Ca^{2+}(aq) + SO_3^{2-}(aq)$	3×10^{-7}
Ag_2SO_3	$Ag_2SO_3(s) \rightleftharpoons 2Ag^+(aq) + SO_3^{2-}(aq)$	1.5×10^{-14}
$BaSO_3$	$BaSO_3(s) \rightleftharpoons Ba^{2+}(aq) + SO_3^{2-}(aq)$	8×10^{-7}

Table 3.2 Solubility Product Constants at 25°C (Continued)

Salt	Solubility Equilibrium	K_{sp}
Sulfates		
$CaSO_4$	$CaSO_4(s) \rightleftharpoons Ca^{2+}(aq) + SO_4^{2-}(aq)$	2.4×10^{-5}
$SrSO_4$	$SrSO_4(s) \rightleftharpoons Sr^{2+}(aq) + SO_4^{2-}(aq)$	3.2×10^{-7}
$BaSO_4$	$BaSO_4(s) \rightleftharpoons Ba^{2+}(aq) + SO_4^{2-}(aq)$	1.1×10^{-10}
$RaSO_4$	$RaSO_4(s) \rightleftharpoons Ra^{2+}(aq) + SO_4^{2-}(aq)$	4.3×10^{-11}
Ag_2SO_4	$Ag_2SO_4(s) \rightleftharpoons 2Ag^+(aq) + SO_4^{2-}(aq)$	1.5×10^{-5}
Hg_2SO_4	$Hg_2SO_4(s) \rightleftharpoons Hg_2^{2+}(aq) + SO_4^{2-}(aq)$	7.4×10^{-7}
$PbSO_4$	$PbSO_4(s) \rightleftharpoons Pb^{2+}(aq) + SO_4^{2-}(aq)$	1.6×10^{-8}
Chromates		
$BaCrO_4$	$BaCrO_4(s) \rightleftharpoons Ba^{2+}(aq) + CrO_4^{2-}(aq)$	2.1×10^{-10}
$CuCrO_4$	$CuCrO_4(s) \rightleftharpoons Ba^{2+}(aq) + CrO_4^{2-}(aq)$	3.6×10^{-6}
Ag_2CrO_4	$Ag_2CrO_4(s) \rightleftharpoons 2Ag^+(aq) + CrO_4^{2-}(aq)$	1.2×10^{-12}
Hg_2CrO_4	$Hg_2CrO_4(s) \rightleftharpoons Hg_2^{2+}(aq) + CrO_4^{2-}(aq)$	2.0×10^{-9}
$CaCrO_4$	$CaCrO_4(s) \rightleftharpoons Ca^{2+}(aq) + CrO_4^{2-}(aq)$	7.1×10^{-4}
$PbCrO_4$	$PbCrO_4(s) \rightleftharpoons Pb^{2+}(aq) + CrO_4^{2-}(aq)$	1.8×10^{-14}
Carbonates		
$MgCO_3$	$MgCO_3(s) \rightleftharpoons Mg^{2+}(aq) + CO_3^{2-}(aq)$	3.5×10^{-8}
$CaCO_3$	$CaCO_3(s) \rightleftharpoons Ca^{2+}(aq) + CO_3^{2-}(aq)$	4.5×10^{-9}
$SrCO_3$	$SrCO_3(s) \rightleftharpoons Sr^{2+}(aq) + CO_3^{2-}(aq)$	9.3×10^{-10}
$BaCO_3$	$BaCO_3(s) \rightleftharpoons Ba^{2+}(aq) + CO_3^{2-}(aq)$	5.0×10^{-9}
$MnCO_3$	$MnCO_3(s) \rightleftharpoons Mn^{2+}(aq) + CO_3^{2-}(aq)$	5.0×10^{-10}
$FeCO_3$	$FeCO_3(s) \rightleftharpoons Fe^{2+}(aq) + CO_3^{2-}(aq)$	2.1×10^{-11}
$CoCO_3$	$CoCO_3(s) \rightleftharpoons Co^{2+}(aq) + CO_3^{2-}(aq)$	1.0×10^{-10}
$NiCO_3$	$NiCO_3(s) \rightleftharpoons Ni^{2+}(aq) + CO_3^{2-}(aq)$	1.3×10^{-7}
$CuCO_3$	$CuCO_3(s) \rightleftharpoons Cu^{2+}(aq) + CO_3^{2-}(aq)$	2.3×10^{-10}
Ag_2CO_3	$Ag_2CO_3(s) \rightleftharpoons 2Ag^+(aq) + CO_3^{2-}(aq)$	8.1×10^{-12}
Hg_2CO_3	$Hg_2CO_3(s) \rightleftharpoons 2Hg^+(aq) + CO_3^{2-}(aq)$	8.9×10^{-17}
$ZnCO_3$	$ZnCO_3(s) \rightleftharpoons Zn^{2+}(aq) + CO_3^{2-}(aq)$	1.0×10^{-10}
$CdCO_3$	$CdCO_3(s) \rightleftharpoons Cd^{2+}(aq) + CO_3^{2-}(aq)$	1.8×10^{-14}
$PbCO_3$	$PbCO_3(s) \rightleftharpoons Pb^{2+}(aq) + CO_3^{2-}(aq)$	7.4×10^{-14}
Phosphates		
$Mg_3(PO_4)_2$	$Mg_3(PO_4)_2(s) \rightleftharpoons 3Mg^{2+}(aq) + 2PO_4^{3-}(aq)$	6.3×10^{-26}
$SrHPO_4$	$SrHPO_4(s) \rightleftharpoons Sr^{2+}(aq) + HPO_4^{2-}(aq)$	1.2×10^{-7}
$BaHPO_4$	$BaHPO_4(s) \rightleftharpoons Ba^{2+}(aq) + HPO_4^{2-}(aq)$	4.0×10^{-8}
$LaPO_4$	$LaPO_4(s) \rightleftharpoons La^{3+}(aq) + PO_4^{3-}(aq)$	3.7×10^{-23}
$Fe_3(PO_4)_2$	$Fe_3(PO_4)_2(s) \rightleftharpoons 3Fe^{2+}(aq) + 2PO_4^{3-}(aq)$	1×10^{-36}
Ag_3PO_4	$Ag_3PO_4(s) \rightleftharpoons 3Ag^+(aq) + PO_4^{3-}(aq)$	2.8×10^{-18}
$FePO_4$	$FePO_4(s) \rightleftharpoons Fe^{3+}(aq) + PO_4^{3-}(aq)$	4.0×10^{-27}
$Zn_3(PO_4)_2$	$Zn_3(PO_4)_2(s) \rightleftharpoons 3Zn^{2+}(aq) + 2PO_4^{3-}(aq)$	5×10^{-36}
$Pb_3(PO_4)_2$	$Pb_3(PO_4)_2(s) \rightleftharpoons 3Pb^{2+}(aq) + 2PO_4^{3-}(aq)$	3.0×10^{-44}
$Ba_3(PO_4)_2$	$Ba_3(PO_4)_2(s) \rightleftharpoons 3Ba^{2+}(aq) + 2PO_4^{3-}(aq)$	5.8×10^{-38}
Ferrocyanides		
$Zn_2[Fe(CN)_6]$	$Zn_2[Fe(CN)_6](s) \rightleftharpoons 2Zn^{2+}(aq) + Fe(CN)_6^{4-}(aq)$	2.1×10^{-16}
$Cd_2[Fe(CN)_6]$	$Cd_2[Fe(CN)_6](s) \rightleftharpoons 2Cd^{2+}(aq) + Fe(CN)_6^{4-}(aq)$	4.2×10^{-18}
$Pb_2[Fe(CN)_6]$	$Pb_2[Fe(CN)_6](s) \rightleftharpoons 2Pb^{2+}(aq) + Fe(CN)_6^{4-}(aq)$	9.5×10^{-19}

the change in concentration of the reactant with time. Thus,

$$\text{Reaction rate} = \frac{-\Delta C}{\Delta t} = \frac{-(C_2 - C_1)}{t_2 - t_1} \quad [3.36]$$

where C = the change in the concentration of the reactant, and t = the elapsed time. Subscripts 1 and 2 refer to concentrations and corresponding times, respectively. The reactant concentration decreases with time, making $(C_2 - C_1)$ a negative quantity. The minus sign is used to make it a positive quantity.

The reaction rate can be calculated just as well from the change in concentration with time of a product, in which case $\Delta C/\Delta t$ is a positive quantity. It is important to remember that the reaction rate is constantly changing, so $\Delta C/\Delta t$ represents an average rate. The instantaneous rate is given by the first derivative, dC/dt.

Reaction Rate Laws, Orders, and Constants

For the general reaction:

$$a\text{A} + b\text{B} \longrightarrow g\text{G} + h\text{H} \quad [3.37]$$

the **rate law or equation** is expressed as

$$\text{Rate} = k[\text{A}]^x[\text{B}]^y \quad [3.38]$$

where k = the **rate constant** (in reciprocal time units), [A] and [B] represent the molar concentrations of reactants A and B, and x and y are exponents denoting the **order** of the reaction for the reactant species. The exponents x and y must be determined experimentally. The sum of the exponents $(x + y)$ is called the **overall order** of the reaction. In the present case, it is said that the reaction is x order in A, y order in B, and $(x + y)$ order overall.

Consider, for example, the following reaction:

$$\text{NO}_2(g) + \text{CO}(g) \rightleftharpoons \text{NO}(g) + \text{CO}_2(g) \quad [3.39]$$

Experimentally, it is determined that the rate law for this reaction is

$$\text{Rate} = k[\text{NO}_2]^2 \quad [3.40]$$

This means that the rate is second order in NO_2, zeroth order in CO, and second order overall.

Consider, as another example, the decomposition of dinitrogen pentoxide, N_2O_5:

$$2\text{N}_2\text{O}_5(g) \rightleftharpoons 4\text{NO}_2(g) + \text{O}_2(g) \quad [3.41]$$

Here, it is experimentally determined that the rate law for this reaction is

$$\text{Rate} = k[\text{N}_2\text{O}_5] \quad [3.42]$$

Thus, the rate of decomposition is first order in N_2O_5. The rate constant k is also experimentally measured:

$$k = 5.2 \times 10^{-3} \text{ at } 65°\text{C}.$$

First- and Second-Order Reactions

For any reaction that is known to be first order in a particular species, as in the N_2O_5 decomposition discussed above, the rate can be written as

$$\text{Rate} = \frac{-d[\text{A}]}{dt} = \frac{-\Delta[\text{A}]}{\Delta t} = k[\text{A}] \quad [3.43]$$

For a **first-order reaction,** this means that doubling the concentration of A doubles the reaction rate. This equation may be integrated to give a more useful result:

$$\ln \frac{[\text{A}]_0}{[\text{A}]} = k\,\Delta t$$

This can also be written as

$$\ln[\text{A}] = \ln[\text{A}]_0 - k\,\Delta t \quad [3.44]$$

In similar fashion, the integrated rate equation for a **second-order reaction** is

$$\frac{1}{[\text{A}]} = \frac{1}{[\text{A}]_0} + k\,\Delta t \quad [3.45]$$

Half-Life of a Reaction

A quantity defined as the half-life of the reaction can also be determined. The **half-life** is the time required for the concentration of the reactant to reach half of its initial value. For a *first-order reaction,* the half-life is independent of the initial concentration and is given as

$$t_{1/2} = \frac{0.693}{k} \quad [3.46]$$

For a *second-order reaction,* the half-life depends on initial concentration and is given as

$$t_{1/2} = \frac{1}{k[\text{A}]_0} \quad [3.47]$$

Example 3.36

Consider the reaction between peroxydisulfate ion, $S_2O_8^{2-}$, and iodide ion, I^-, shown below and the experimental data about the concentrations and reaction rates at 25°C given in the table that follows.

$$\text{S}_2\text{O}_8^{2-}(aq) + 3\text{I}^-(aq) \longrightarrow 2\text{SO}_4^{2-}(aq) + \text{I}_3^-(aq)$$

Experiment	$[S_2O_8^{2-}]$	$[I^-]$	Initial Rate
1	0.080 M	0.034 M	2.2×10^{-4} M/sec
2	0.080 M	0.017 M	1.1×10^{-4} M/sec
3	0.16 M	0.017 M	2.2×10^{-4} M/sec

A. Write the rate law expression and determine the overall order, that is, $(x + y)$, for this reaction.

B. Determine the rate constant k for this reaction at 25°C.

Solution:

A. Examine the data in the preceding table. Comparing experiment 1 with 2, it is clear that doubling the iodide ion concentration, while holding the peroxydisulfate ion concentration constant, doubles the reaction rate. Similarly, comparing experiment 2 with 3 reveals that doubling the peroxydisulfate ion concentration, while holding the iodide ion concentration constant, also doubles the reaction rate.

Thus, this reaction is first order (linear) in iodide ion concentration, and first order (linear) in peroxydisulfate ion concentration. The rate law can be written as follows:

$$\text{Rate} = k[S_2O_8^{2-}]^x[I^-]^y$$

where $x = 1$ and $y = 1$.

Hence,

$$\text{Rate} = k[S_2O_8^{2-}]^1[I^-]^1$$

The overall order is $(x + y) = (1 + 1) = 2$, indicating second order overall.

B. The rate constant k can be determined using the data from any of the three experiments. Take the data from experiment 1, and use the rate law expression determined in part **A**. Thus,

$$2.2 \times 10^{-4} \text{ M/sec} = k \, (0.034 \text{ M})^1(0.080 \text{ M})^1$$

$$k = 8.09 \times 10^{-2} \, (\text{M} \cdot \text{sec})^{-1}$$

Example 3.37

The thermal decomposition of phosphine, PH_3, is known to be a first-order reaction:

$$4PH_3(g) \rightarrow P_4(g) + 6H_2(g)$$

The half-life, $t_{1/2}$, is 35.0 sec at 680°C.

A. Compute the rate constant k for this reaction.

B. Find the time required for 75% of the initial concentration of PH_3 to decompose.

Solution:

A. For a first-order reaction, $t_{1/2} = 0.693/k$.

Hence,

$$k = \frac{0.693}{t_{1/2}} = \frac{0.693}{35.0 \text{ sec}}$$

$$= 0.0198 \text{ sec}^{-1}$$

B. To find the time, use the integrated, first-order rate law, equation 3.42, and solve for Δt.

If 75% of PH_3 is to decompose, 25% must remain! Thus,

$$[A]_0 = 100\% = 1.00$$

$$[A] = 25\% = 0.25$$

Substitute these values and solve for Δt:

$$\Delta t = 70.0 \text{ sec}$$

Dependence on Temperature: The Arrhenius Equation

The Arrhenius model for rates of reactions assumes that molecules must undergo collisions before they can react. The number of collisions per unit time is A, the **frequency factor**. This model also assumes that not all collisions will result in or lead to a reaction. Rather, only those with sufficient energy, called the **activation energy, E_a**, to allow molecules to reach an **activated complex** will achieve this result.

The rate constant k is a function of temperature and can be expressed theoretically as the **Arrhenius equation**:

$$k = Ae^{-E_a/RT} \qquad [3.48]$$

where A = the frequency factor, a measure of the number of collisions per second,

E_a = the activation energy in joules or calories per mole,

R = the universal gas constant in units consistent with those for E_a,

T = the absolute temperature in kelvins.

A more useful form of the equation can be written as

$$\ln k = \ln A - \frac{E_a}{RT} \qquad [3.49]$$

The rate constants k_1 and k_2 at two different temperatures, T_1 and T_2, are related as follows:

$$\ln \frac{k_1}{k_2} = \frac{E_a}{R}\left(\frac{1}{T_2} - \frac{1}{T_1}\right) \qquad [3.50]$$

Example 3.38

The reaction below is important in the chemistry of air pollution:

$$NO(g) + O_3(g) \rightarrow NO_2(g) + O_2(g)$$

The frequency factor $A = 8.7 \times 10^{12}$ sec^{-1} and the rate constant $k = 300$ sec^{-1} at 75°C.

A. Find the activation energy, E_a, in joules per mole for this reaction.

B. Find the rate constant k of this reaction at 0°C, assuming E_a to be constant.

Solution:

A. Use the Arrhenius equation and solve for E_a.

$$\ln k = \ln A - \frac{E_a}{RT}$$

Use $R = 8.31$ J/K · mole and $T = 273 + 75°C = 348$ K. Then,

$$\ln 300 = \ln 8.7 \times 10^{12} - \frac{E_a}{(8.31)(348)}$$

$$E_a = 69,700 \text{ J/mol or } 69.7 \text{ kJ/mol}$$

B. Denote the rate constant given at 75°C as k_1, and the rate constant at 0°C as k_2, and use

$$\ln \frac{k_1}{k_2} = \frac{E_a}{R}\left(\frac{1}{T_2} - \frac{1}{T_1}\right)$$

Substitute the above information, and solve for k_2.

$$\ln k_2 = \ln 300 - \frac{69,700}{8.31}\left(\frac{1}{273} - \frac{1}{348}\right)$$

$$= 5.7038 - 6.626 = -0.9222$$

$$k_2 = 0.3976 \text{ sec}^{-1}$$

Catalysis

Substances that increase the rate of reaction without themselves being consumed in the reaction are called **catalysts.** They work by lowering the activation energy, E_a, or the barrier required for the reaction to proceed. A catalyst may be in the same physical state as the other species in the reaction (homogeneous catalysis) or in a different physical state (heterogeneous catalysis).

3.10
Chemical Thermodynamics and Thermochemistry

Basic Concept

Thermodynamics is the study of energy transformations. **Thermochemistry** is the study of the thermal energy changes that accompany chemical and physical changes. **Heat** is the energy transferred between objects or systems and is measured by temperature changes. Energy is a property of a substance or system; it cannot be measured directly.

Any process occurring within a system that produces or releases heat into the surroundings is called **exothermic**, and its value is given a negative sign by convention. Any process that absorbs heat from the surroundings into the system is called **endothermic,** and its value has a positive sign. Note that an exothermic reaction is not necessarily a spontaneous reaction, and its negative sign should *not* be interpreted as such. This type of interpretation and criterion is reserved for ΔG, the Gibbs free energy change, discussed on page 136.

Enthalpy and Hess's Law

Since chemical reactions involve the reforming of bonds, energy changes generally accompany reactions. To initiate a reaction, an activation energy, E_a, must first be supplied for the reactants to reach an activated complex, whether or not the reaction is spontaneous. When a reaction occurs, a change in **enthalpy,** that is, heat content, known as the **heat of reaction,** ΔH_{RX}, is associated with the process. This is the amount of heat exchanged (liberated or absorbed) with the surroundings under constant external pressure and may be measured experimentally (directly) or calculated theoretically (indirectly) and expressed in either kilojoules (kJ) or kilocalories (kcal).

The enthalpy change known as the **standard heat of formation, $\Delta H_f°$**, is defined as the heat of reaction for a product derived only from its elements, the superscript ° meaning measured under normal conditions (usually NTP as opposed to STP).

Regardless of the reaction, **Hess's law** states: The enthalpy change of a chemical reaction is the same regardless of the number of steps or separate reactions. If the reaction of interest is the sum of several other reactions, the individual heats of reaction must be added algebraically. In other words,

$$\Delta H_{RX} = \Sigma \Delta H_{products} - \Sigma \Delta H_{reactants} \quad [3.51]$$

Example 3.39

The standard heats of formation, $\Delta H_f°$, for NO_2 and N_2O_4, are as follows: $\Delta H_f°[NO_2] = 33.9$ kJ/mole, $\Delta H_f°[N_2O_4] = -19.5$ kJ/mol. Predict or calculate the heat of reaction, ΔH_{RX}, for the following reaction at standard conditions.

$$2NO_2(g) \rightleftharpoons N_2O_4(g)$$

Solution:

Use Hess's law.

$$\Delta H_{RX}° = \Sigma \Delta H_{products} - \Sigma \Delta H_{reactants}$$

$$= \Delta H_f°[N_2O_4] - 2\Delta H_f°[NO_2]$$

$$= (1.00 \text{ mol})(-19.5 \text{ kJ/mol})$$
$$- (2 \text{ mol})(33.9 \text{ kJ/mol})$$
$$= -87.3 \text{ kJ}$$

Note that the negative sign means that the reaction is exothermic.

Example 3.40

Given the following reactions and data for $\Delta H°$:

$$N_2O_4(g) \rightarrow 2NO_2(g) \ \Delta H°(I) = 87.30 \text{ kJ}$$

$$2NO(g) + O_2(g) \rightarrow 2NO_2(g) \ \Delta H°(II) = -114.14 \text{ kJ}$$

Compute $\Delta H_{RX}°$ for the following reaction:

$$2NO(g) + O_2(g) \rightarrow N_2O_4(g)$$

Solution:

Recognize that, if the first given reaction is reversed and then added to the second reaction, the desired reaction is obtained, since the $2NO_2$ terms on opposite sides of the reaction cancel. Change

the sign of $\Delta H°(I)$, so that the value is -87.30 kJ, and then add the two $\Delta H°$ values. Thus:

$$\Delta H_{RX}° = \Delta H°(I) + \Delta H°(II)$$

$$= -87.30 \text{ kJ} + (-114.14 \text{ kJ}) = -201.44 \text{ kJ}$$

Another way to solve this problem in principle is to use Hess's law and set the solution up as follows:

$$\Delta H_{RX}° = \Sigma \Delta H_{products} - \Sigma \Delta H_{reactants}$$

$$= \Delta H_f°[N_2O_4] - \{2\Delta H_f°[NO] + \Delta H_f°[O2]\}$$

The problem with this approach is that, while $\Delta H_f°[O_2] = 0$ and the value for $\Delta H_f°[N_2O_4]$ is given in Example 3.39, the value for $\Delta H_f°[NO]$ is unknown.

The First Law and the Conservation of Energy

In understanding the laws of thermodynamics, it is useful to define *system* and distinguish it from *surroundings* very carefully. In general, there are three types of systems:

1. **Closed system**—a system that no mass enters or leaves.
2. **Isolated system**—a system that no mass or heat enters or leaves.
3. **Open system**—a system that mass or heat can enter or leave.

Furthermore, there are two kinds of *boundaries/walls* or processes:

1. **Diathermal walls**—walls that permit the transfer of heat between system and surroundings.
2. **Adiabatic walls**—walls that do not permit the transfer of heat between system and surroundings (e.g., good insulators).

The **first law of thermodynamics** can be stated in several ways. For example:

- The energy of the universe is constant. The total possible energy change is equal to the sum of the energy change of the system plus the energy change of the surroundings.
- During a chemical or physical change in any defined system, energy can neither be created nor destroyed, but only changes form.

Expressed in mathematical terms:

$$\Delta E = q - w \quad [3.52]$$

where ΔE = the total internal energy change of the system,

q = the heat absorbed *by* the system *from* the surroundings,

w = the work done *by* the system *on* the surroundings.

For adiabatic processes, $q = 0$, so $\Delta E = -w$.

For isothermal changes in ideal gases, $\Delta E = 0$, since E is a function of temperature only.

For pressure-volume work of an ideal gas (i.e., expansion or contraction):

$$w = \int P\, dV = P\,\Delta V \text{ if } P \text{ is constant} \quad [3.53]$$

$$= \int P\, dV = \int \frac{nRT}{V} dV$$

$$= nRT \ln \frac{V_2}{V_1} \quad \text{if } P \text{ is not constant}$$
$$[3.54]$$

ΔE and ΔH are related by the equation $\Delta H = \Delta E + P\,\Delta V$ at constant P.

Other types of thermodynamic work may include electrical work and magnetic work.

Example 3.41

Ten moles of an ideal gas in a piston and cylinder assembly absorb 2,500 J of heat. The gas expands from 2.0 L to 8.5 L against a constant external pressure of 2.5 atms. What is the internal energy change of this system?

Solution:

$\Delta E = q - w = q - P\,\Delta V$

$\quad = 2{,}500 \text{ J} - (2.5 \text{ atm})(8.5 \text{ L} - 2.0 \text{ L})$

$\quad = 2{,}500 \text{ J} - 13.75 \text{ L} \cdot \text{atm} \times (101.3 \text{ J/L} \cdot \text{atm})$

$\quad = 2{,}500 \text{ J} - 1{,}390 \text{ J} = 1{,}110 \text{ J}$

Since the answer is positive, this amount of energy is *gained* by the system.

The Second Law and Entropy

The **second law of thermodynamics** may be expressed in several ways. For example:

- The **entropy**, S, that is, the randomness or disorder, of the universe either stays the same or increases, but never decreases.

- During a physical change or chemical reaction, the *total* **entropy change, ΔS,** of the system plus the surroundings is either zero or positive, never negative. It may be calculated as the reversible heat q, either absorbed or liberated by a system, divided by the absolute temperature T:

$$\Delta S = \frac{q}{T} \quad [3.55]$$

- The construction of a perpetual motion machine is impossible.
- It is impossible to build a heat engine with 100% efficiency.

A corollary of the second law, often referred to as the **zeroth law,** states: Heat always flows from regions of hot to regions of cold *spontaneously*.

Entropy is also a measure of the statistical disorder or randomness of a system. For all real or natural changes, which are spontaneous and irreversible, the entropy of a system increases, so the entropy *change, ΔS,* is positive. For idealized changes, which are reversible or infinitely slow, ΔS is equal to zero. Mathematically, this may be expressed as follows:

$$\Delta S_{\text{total}} = \Delta S_{\text{system}} + \Delta S_{\text{surroundings}}$$
$$= \Delta S_{\text{universe}} \geq 0 \quad [3.56]$$

For phase transitions, ΔS_{trans}

$$= \frac{q_{\text{trans}}}{T} = \frac{\Delta H_{\text{trans}}}{T} \quad [3.57]$$

where ΔH_{trans} may be ΔH_{vap} or ΔH_{fus}.

For the heating of any substance, without a phase transition, the heat gained or lost, q, may be expressed as

$$q = mC\,\Delta T \quad [3.58]$$

where m = the mass of the substance in grams or moles,

C = the specific heat or heat capacity in suitable units,

ΔT = the change in temperature in degrees Celsius or kelvins.

For other processes involving changes in temperature and/or volume:

$$\Delta S = C_v \ln\frac{T_2}{T_1} + nR \ln\frac{V_2}{V_1} \quad [3.59]$$

Example 3.42

One mole of an ideal gas expands reversibly and isothermally from an initial volume of 2 L to a final volume of 20 L. Calculate the entropy change, in calories/kelvin, of the system and the surroundings.

Solution:

Since this process is isothermal, $T_1 = T_2$, use $\Delta S = nR \ln V_2/V_1$. Then:

$$\Delta S_{sys} = (1.00 \text{ mol})(1.99 \text{ cal/K} \cdot \text{mol}) \ln \frac{20 \text{ L}}{2.0 \text{ L}}$$

$$= 4.57 \text{ cal/K or } 4.57 \text{ eu (“eu” stands for entropy unit)}$$

Since this is a reversible process:

$$\Delta S_{sys} = -\Delta S_{surr} = -4.57 \text{ cal/K}$$

Example 3.43

Calculate the entropy change when 100 g of water vaporizes at 100°C.

Solution:

This is a phase transition. Use $\Delta S = \Delta H_{vap}/T$ (equation 3.55), where $\Delta H_{vap} = 540$ cal/g and $T = 373$ K. Then:

$$\Delta S = \frac{(540 \text{ cal/g})(100 \text{ g})}{373 \text{ K}}$$

$$= 145 \text{ cal/K}$$

The Third Law and Absolute Zero

The **third law of thermodynamics** states that it is impossible to reach absolute zero in a finite number of steps. Another version states that the absolute entropy of a perfect crystal is zero at absolute zero, since every atom is perfectly in place and has no kinetic energy.

Gibbs Free Energy and the Spontaneity of a Reaction

The Gibbs free energy change, ΔG, is defined as:

$$\Delta G = \Delta H - T\Delta S \qquad [3.60]$$

Its chief value is that it is an unequivocal measure and indication of a reaction's spontaneity. It includes both an enthalpy term, ΔH, and an entropy term, ΔS.

Every reaction wants simultaneously to reach a state of minimum energy and a state of maximum entropy. Thus, if the sign of ΔG is negative, the reaction proceeds spontaneously as written; if positive, the *reverse* reaction proceeds spontaneously.

For equilibrium reactions, ΔG can be expressed as

$$\Delta G = -RT \ln K_c \quad \text{in general} \qquad [3.61]$$

$$\Delta G = -RT \ln K_p \quad \text{for gas-phase reactions} \quad [3.62]$$

Example 3.44

For this reaction:

$$H_2(g) + S(s) \rightarrow H_2S(g)$$

$\Delta H_f° = -20.2$ kJ/mol and $\Delta S_f° = +43.1$ J/K · mol.
A. Compute $\Delta G°$ for the reaction.
B. Tell whether the reaction is spontaneous.

Solution:

A. Use $\Delta G = \Delta H - T\Delta S$. Note that the $\Delta G°$ that will be computed is really $\Delta G_f°$. At standard conditions, $T = 298$ K, and all quantities are 1.0 mol. Then,

$$\Delta G° = -20.2 \times 10^3 \text{ J} - (298 \text{ K})(+43.1 \text{ J/K})$$

$$= -33,908 \text{ J} = -33.9 \text{ kJ}$$

B. Since the sign of ΔG is negative, this reaction is spontaneous as written.

Example 3.45

At 1,500°C,

$$CO(g) + 2H_2(g) \rightleftharpoons CH_3OH(g)$$

and $K_p = 1.4 \times 10^{-7}$.
A. Compute ΔG for this reaction.
B. State whether the reaction is spontaneous.

Solution:

A. Use $\Delta G = -RT \ln K_p$. Then,

$$\Delta G = -(8.31 \text{ J/K} \cdot \text{mol})(1773 \text{ K}) \ln (1.4 \times 10^{-7})$$

$$= 233 \text{ kJ/mol}$$

B. Since the sign of ΔG is positive, this reaction is *not* spontaneous as written.

3.11
Electrochemistry

Basic Concept

One category of chemical reactions described briefly in Section 3.4 is oxidation-reduction reactions, or "redox" reactions for short. Recall the following criteria for each:

Oxidation	Reduction
• Loss of electrons	• Gain of electrons
• Gain of oxygen atoms	• Loss of oxygen atoms
• Loss of hydrogen atoms	• Gain of hydrogen atoms

If two different metals are placed in physical contact with their respective ionic solutions and are then connected by a suitable conductor but separated by a salt bridge, forming a closed circuit, a potential difference or voltage and current flow are observed (see Figure 3.3 on page 138). This arrangement is known as an **electrochemical cell,** and the underlying electrochemical reaction involves simultaneous oxidation and reduction. Electrochemical cells are the basis of operation of all batteries. If the reaction and current flow are spontaneous, as in any battery, the cell is called a **voltaic** or **galvanic cell.** If, instead, electrical energy must be supplied, the cell is called an **electrolytic cell,** and the process is known as **electrolysis.**

An electrochemical cell is comprised of two **half-cells.** In one, called the **anode, oxidation** occurs; in the other, called the **cathode, reduction** occurs. Each reaction is called a **half-reaction.** The sum of the two half-reactions is the overall or net reaction. The rusting of iron or the corrosion of any metal involves an oxidation-reduction process, with a net electrolytic potential.

Units of measurement commonly used in electrochemistry include the volt (V), the electron volt (eV), the coulomb (C), and the ampere (A).

The Nernst Equation

The law describing the relationship between the concentrations of ionic solutions and the voltage produced in the net reaction is called the **Nernst equation** and is expressed as follows:

$$E_{net} = E_{net}° - \frac{RT}{nF} \ln Q \qquad [3.63]$$

where R = the gas constant = 8.31 J/K · mol,

T = the absolute temperature in kelvins,

F = the Faraday constant = 96,485 C/mol electrons,

n = the number of moles of electrons transferred per unit reaction,

Q = the ratio of molar concentrations of products to molar concentrations of reactants, for all (aq) or (g) species in solution [(s) or (l) species are excluded].

For the condition where $T = 25°C = 298$ K, and $\ln Q = 2.303 \log Q$, the Nernst equation can be simplified to:

$$E_{net} = E_{net}° - \frac{0.059}{n} \log Q \qquad [3.64]$$

Since 1 volt = 1 joule per coulomb and 1 ampere = 1 coulomb per second,

(current in amperes)(time in seconds) = total charge transferred in coulombs

Also,

(potential in volts)(charge in coulombs) = total energy produced or used in joules

Tables exist that present the reduction reactions of metals, along with selected nonmetals, and their corresponding potentials, relative to the standard hydrogen electrode, whose reduction potential is taken as zero. More extensive tables include reduction potentials organized separately, according to acidic and alkaline conditions. It is important to note that reduction reactions and potentials depend on the pH of solution, as well as the presence of any complexing agents. This listing is often referred to as the **electromotive series,** and the reduction potentials given are the $E°$ values in volts. If the E^0 value listed is positive, the reaction is spontaneous as written. If negative, the reverse reaction is spontaneous. The superscript zero refers to standard conditions, that is, a temperature of 25°C, a concentration of 1.00 M for any ionic solution, and a partial pressure of 1.00 atmosphere for any gas present.

The listing is also referred to as the **activity series** of metals, since they indicate which metal is more likely, relative to another, to release its electrons, oxidize, and react. For example, the metal lithium has the highest *negative* $E°$ value, −3.045

volt, while magnesium has an $E°$ value of -2.363 volt. Therefore, elemental lithium will oxidize and react more easily than elemental magnesium. The nonmetal fluorine has the highest *positive* $E°$ value, $+2.87$ volts, is easiest to reduce, and is the most electronegative element. In other words, a large negative $E°$ value means that the reactant is a good reducing agent, and a large positive $E°$ value indicates a good oxidizing agent.

Table 3.3 lists standard reduction potentials at 25°C for a number of half-cell reactions.

A Simple Electrochemical Cell

Consider the copper/zinc electrochemical cell in Figure 3.3.

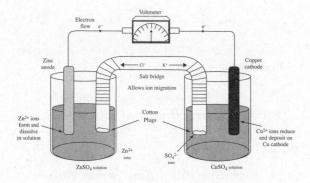

Figure 3.3. Electrochemical cell

In the right half-cell, the **cathode,** Cu^{2+} ions in solution are being reduced to Cu^0, indicating copper metal or solid and deposit on the Cu cathode. In the left half-cell, the **anode,** Zn^0 is being oxidized to Zn^{2+} ions, which dissolve in solution. The anions—SO_4^{2-}(aq)—are spectator ions and do not participate in the reaction. The salt bridge is necessary to maintain electrical neutrality and retard polarization.

The short-hand notation for this cell is as follows:

$$Zn(s)|ZnSO_4(aq)(1.00\ M)|| \qquad CuSO_4(aq)(1.00\ M)|Cu(s) \qquad [3.65]$$

It is understood that the anode cell, where oxidation occurs, is written first and is separated by a double vertical line from the cathode cell, where reduction occurs. Single vertical lines separate the solid electrode from the aqueous solution into which it is immersed. Concentrations of the solutions are expressed in moles per liter in parentheses.

The two half-reactions, followed by the **net (algebraic sum) reaction**, can be written as follows:

At Cathode

$$Cu^{2+}(aq) + 2e^- \rightarrow Cu^0(s) \quad E° = 0.337\ V \qquad [3.66]$$

At Cathode

$$Zn^0(s) \rightarrow Zn^{2+}(aq) + 2e^- \quad E° = 0.763\ V \qquad [3.67]$$

Net reaction

$$Cu^{2+}(aq) + Zn^0(s) \rightarrow Cu^0(s) + Zn^{2+}(aq)$$
$$E°_{net} = 1.100\ V \qquad [3.68]$$

Note that the electrons exactly cancel each other in the two half-reactions, and do not appear in the net reaction. If this does not happen automatically, one or both of the half-reactions must be multiplied by a suitable coefficient (integer) or coefficients to obtain exact cancellation of electrons. Note also that the positive value (understood) for $E_{net}°$ indicates that this reaction proceeds **spontaneously**, and the cell produces 1.100 volts under standard conditions.

Useful Quantitative Relationships

1 mole of electrons (e^-) = 96,485 coulombs (C) [3.69]

charge (c) = current (C/sec) $\times$ time (sec) [3.70]

1 ampere (A) = 1 coulomb/1 second [3.71]

1 volt (V) = 1 joule (J)/1 coulomb [3.72]

Faraday constant F = 96,485 coulombs/mole e^- [3.73]

Gas constant R = 8.31 joules/kelvin $\cdot$ mole [3.74]

Number of moles of e^- = current $\times$ time/F [3.75]

Example 3.46

Consider the electrochemical cell with the following net reaction, which is observed to proceed spontaneously. All species are in their standard states, and $T = 25°C$.

$$Mg(s) + Sn^{2+}(aq) \rightarrow Mg^{2+}(aq) + Sn(s)$$

A. Write the two half-reactions with their respective $E°$ values.

B. Indicate which species is oxidized and which is reduced.

C. Identify the anode and cathode.

D. Compute $E_{net}°$.

E. Confirm that the reaction proceeds spontaneously.

Solution:

A. $Mg(s) \rightarrow Mg^{2+}(aq) + 2e^- \quad E° = 2.370\ V$

$Sn^{2+}(aq) + 2e^- \rightarrow Sn(s) \quad E° = -0.140\ V$

B. Mg(s) is oxidized to Mg^{2+}, while Sn^{2+} is reduced to Sn(s).

C. The anode is the electrode where oxidation takes place, while the cathode is the electrode where reduction takes place. Since Mg is oxidized, the half-cell containing the Mg electrode and Mg^{2+}(aq) solution must be the anode (whatever the anion may be). Similarly, the cathode is the half-cell containing the Sn electrode dipping into Sn^{2+}(aq) solution.

D. Adding the two half-cell potentials determined in part **A** gives $E_{net}° = 2.230$ V.

E. Since $E_{net}°$ is a positive number, the net reaction must proceed spontaneously.

Table 3.3 Standard Reduction Potentials at 25°C

E° (volts)	Half-Cell Reaction
+2.87	$F_2(g) + 2e^- \rightleftharpoons 2F^-(aq)$
+2.08	$O_3(g) + 2H^+(aq) + 2e^- \rightleftharpoons O_2(g) + H_2O$
+2.05	$S_2O_8^{2-}(aq) + 2e^- \rightleftharpoons 2SO_4^{2-}(aq)$
+1.82	$Co^{3+}(aq) + e^- \rightleftharpoons Co^{2+}(aq)$
+1.77	$H_2O_2(aq) + 2H^+(aq) + 2e^- \rightleftharpoons 2H_2O$
+1.695	$MnO_4^-(aq) + 4H^+(aq) + 3e^- \rightleftharpoons MnO_2(s) + 2H_2O$
+1.69	$PbO_2(s) + SO_4^{2-}(aq) + 4H^+(aq) + 2e^- \rightleftharpoons PbSO_4(s) + 2H_2O$
+1.63	$2HOCl(aq) + 2H^+(aq) + 2e^- \rightleftharpoons Cl_2(g) + 2H_2O$
+1.51	$Mn^{3+}(aq) + e^- \rightleftharpoons Mn^{2+}(aq)$
+1.49	$MnO_4^-(aq) + 8H^+(aq) + 5e^- \rightleftharpoons Mn^{2+}(aq) + 4H_2O$
+1.46	$PbO_2(s) + 4H^+(aq) + 2e^- \rightleftharpoons Pb^{2+}(aq) + 2H_2O$
+1.44	$BrO_3^-(aq) + 6H^+(aq) + 6e^- \rightleftharpoons Br^-(aq) + 3H_2O$
+1.42	$Au^{3+}(aq) + 3e^- \rightleftharpoons Au(s)$
+1.36	$Cl_2(g) + 2e^- \rightleftharpoons 2Cl^-(aq)$
+1.33	$Cr_2O_7^{2-}(aq) + 14H^+(aq) + 6e^- \rightleftharpoons 2Cr^{3+}(aq) + 7H_2O$
+1.24	$O_3(g) + H_2O + 2e^- \rightleftharpoons O_2(g) + 2OH^-(aq)$
+1.23	$MnO_2(s) + 4H^+(aq) + 2e^- \rightleftharpoons Mn^{2+}(aq) + 2H_2O$
+1.23	$O_2(g) + 4H^+(aq) + 4e^- \rightleftharpoons 2H_2O$
+1.20	$Pt^{2+}(aq) + 2e^- \rightleftharpoons Pt(s)$
+1.07	$Br_2(aq) + 2e^- \rightleftharpoons 2Br^-(aq)$
+0.96	$NO_3^-(aq) + 4H^+(aq) + 3e^- \rightleftharpoons NO(g) + 2H_2O$
+0.94	$NO_3^-(aq) + 3H^+(aq) + 2e^- \rightleftharpoons HNO_2(aq) + H_2O$
+0.91	$2Hg^{2+}(aq) + 2e^- \rightleftharpoons Hg_2^{2+}(aq)$
+0.87	$HO_2^-(aq) + H_2O + 2e^- \rightleftharpoons 3OH^-(aq)$
+0.80	$NO_3^-(aq) + 4H^+(aq) + 2e^- \rightleftharpoons 2NO_2(g) + 2H_2O$
+0.80	$Ag^+(aq) + e^- \rightleftharpoons Ag(s)$
+0.77	$Fe^{3+}(aq) + e^- \rightleftharpoons Fe^{2+}(aq)$
+0.69	$O_2(g) + 2H^+(aq) + 2e^- \rightleftharpoons H_2O_2(aq)$
+0.54	$I_2(s) + 2e^- \rightleftharpoons 2I^-(aq)$
+0.49	$NiO_2(s) + 2H_2O + 2e^- \rightleftharpoons Ni(OH)_2(s) + 2OH^-(aq)$
+0.45	$SO_2(aq) + 4H^+(aq) + 4e^- \rightleftharpoons S(s) + 2H_2O$
+0.401	$O_2(g) + 2H_2O + 4e^- \rightleftharpoons 4OH^-(aq)$
+0.34	$Cu^{2+}(aq) + 2e^- \rightleftharpoons Cu(s)$
+0.27	$Hg_2Cl_2(s) + 2e^- \rightleftharpoons 2Hg(l) + 2Cl^-(aq)$
+0.25	$PbO_2(s) + H_2O + 2e^- \rightleftharpoons PbO(s) + 2OH^-(aq)$
+0.2223	$AgCl(s) + e^- \rightleftharpoons Ag(s) + Cl^-(aq)$
+0.172	$SO_4^{2-}(aq) + 4H^+(aq) + 2e^- \rightleftharpoons H_2SO_3(aq) + H_2O$
+0.169	$S_4O_6^{2-}(aq) + 2e^- \rightleftharpoons 2S_2O_3^{2-}(aq)$
+0.16	$Cu^{2+}(aq) + e^- \rightleftharpoons Cu^+(aq)$
+0.15	$Sn^{4+}(aq) + 2e^- \rightleftharpoons Sn^{2+}(aq)$
+0.14	$S(s) + 2H^+(aq) + 2e^- \rightleftharpoons H_2S(g)$
+0.07	$AgBr(s) + e^- \rightleftharpoons Ag(s) + Br^-(aq)$

Table 3.3 Standard Reduction Potentials at 25°C (Continued)

E° (volts)	Half-Cell Reaction
0.00	$2H^+(aq) + 2e^- \rightleftharpoons H_2(g)$
−0.13	$Pb^{2+}(aq) + 2e^- \rightleftharpoons Pb(s)$
−0.14	$Sn^{2+}(aq) + 2e^- \rightleftharpoons Sn(s)$
−0.15	$AgI(s) + e^- \rightleftharpoons Ag(s) + I^-(aq)$
−0.25	$Ni^{2+}(aq) + 2e^- \rightleftharpoons Ni(s)$
−0.28	$Co^{2+}(aq) + 2e^- \rightleftharpoons Co(s)$
−0.34	$In^{3+}(aq) + 3e^- \rightleftharpoons In(s)$
−0.34	$Tl^+(aq) + e^- \rightleftharpoons Tl(s)$
−0.36	$PbSO_4(s) + 2e^- \rightleftharpoons Pb(s) + SO_4^{2-}(aq)$
−0.40	$Cd^{2+}(aq) + 2e^- \rightleftharpoons Cd(s)$
−0.44	$Fe^{2+}(aq) + 2e^- \rightleftharpoons Fe(s)$
−0.56	$Ga^{3+}(aq) + 3e^- \rightleftharpoons Ga(s)$
−0.58	$PbO(s) + H_2O + 2e^- \rightleftharpoons Pb(s) + 2OH^-(aq)$
−0.74	$Cr^{3+}(aq) + 3e^- \rightleftharpoons Cr(s)$
−0.76	$Zn^{2+}(aq) + 2e^- \rightleftharpoons Zn(s)$
−0.81	$Cd(OH)_2(s) + 2e^- \rightleftharpoons Cd(s) + 2OH^-(aq)$
−0.83	$2H_2O + 2e^- \rightleftharpoons H_2(g) + 2OH^-(aq)$
−0.88	$Fe(OH)_2(s) + 2e^- \rightleftharpoons Fe(s) + 2OH^-(aq)$
−0.91	$Cr^{2+}(aq) + e^- \rightleftharpoons Cr(s)$
−1.16	$N_2(g) + 4H_2O + 4e^- \rightleftharpoons N_2O_4(aq) + 4OH^-(aq)$
−1.18	$V^{2+}(aq) + 2e^- \rightleftharpoons V(s)$
−1.216	$ZnO_2^-(aq) + 2H_2O + 2e^- \rightleftharpoons Zn(s) + 4OH^-(aq)$
−1.63	$Ti^{2+}(aq) + 2e^- \rightleftharpoons Ti(s)$
−1.66	$Al^{3+}(aq) + 3e^- \rightleftharpoons Al(s)$
−1.79	$U^{3+}(aq) + 3e^- \rightleftharpoons U(s)$
−2.02	$Sc^{3+}(aq) + 3e^- \rightleftharpoons Sc(s)$
−2.36	$La^{3+}(aq) + 3e^- \rightleftharpoons La(s)$
−2.37	$Y^{3+}(aq) + 3e^- \rightleftharpoons Y(s)$
−2.37	$Mg^{2+}(aq) + 2e^- \rightleftharpoons Mg(s)$
−2.71	$Na^+(aq) + e^- \rightleftharpoons Na(s)$
−2.76	$Ca^{2+}(aq) + 2e^- \rightleftharpoons Ca(s)$
−2.89	$Sr^{2+}(aq) + 2e^- \rightleftharpoons Sr(s)$
−2.90	$Ba^{2+}(aq) + 2e^- \rightleftharpoons Ba(s)$
−2.92	$Cs^+(aq) + e^- \rightleftharpoons Cs(s)$
−2.92	$K^+(aq) + e^- \rightleftharpoons K(s)$
−2.93	$Rb^+(aq) + e^- \rightleftharpoons Rb(s)$
−3.05	$Li^+(aq) + e^- \rightleftharpoons Li(s)$

Example 3.47

Consider the reaction in Example 3.46. Instead of standard conditions of 1.00 M concentrations for each solution, assume now that the $Mg^{2+}(aq)$ solution is 0.85 M and that the $Sn^{2+}(aq)$ solution is 0.015 M. Find the $E_{net}°$ under these conditions, assuming $T = 25°C$.

Solution:

Since nonstandard conditions exist, the Nernst equation applies. However, use the simplified version of this equation, equation 3.62, since $T = 25°C$.

$$E_{net} = E_{net}° - \frac{0.059}{n}\log Q$$

In this case, $n = 2$, since 2 mol of electrons are exchanged in the *net* reaction.

Also,

$$Q = \frac{[Mg^{2+}(aq)]}{[Sn^{2+}(aq)]} = \frac{0.85 \text{ M}}{0.015 \text{ M}}$$

by checking the *net* reaction

140

$E_{net}°$ was calculated in part **D** of Example 3.46 as 2.230 V. Substituting gives

$$E_{net} = 2.230 \text{ V} - \frac{0.059}{2} \log \frac{0.85 \text{ M}}{0.015 \text{ M}}$$

$$= 2.178 \text{ V}$$

Example 3.48

Compute the Gibbs free energy change, ΔG, for the cell in Example 3.47.

Solution:

$\Delta G = -nFE_{net}$

$= -(2 \text{ mol e}^-)(96{,}485 \text{ C/mol e}^-)(2.178 \text{ V})$

$= -420{,}290 \text{ J or } -420.3 \text{ kJ}$

Example 3.49

Consider the following cell reaction, in which all species are at standard-state conditions:

$$Cu^{2+}(aq) + H_2(g) \rightarrow Cu(s) + 2H^+(aq)$$

A. Predict the effect on the emf (electromotive force) of this cell of adding NaOH solution to the hydrogen half-cell until pH = 7.

B. Compute the number of coulombs required to deposit 4.20 g of Cu(s) in the copper half-cell.

C. How long in seconds will this deposition take if the measured current is 4.0 A?

Solution:

A. Refer to the simplified Nernst equation, (equation 3.62). The emf is the E_{net}.

$$E_{net} = E_{net}° - \frac{0.059}{2} \log \frac{[H^+]^2}{[Cu^{2+}]}$$

If NaOH(aq) is added, H^+ ions (aq) will be neutralized, thereby raising the pH. [H^+(aq)] will decrease, reducing the magnitude of the log term, which is subtracted from $E_{net}°$. This, in turn, will increase the E_{net} or the emf of the cell.

B. First, compute the number of electrons that must be transferred to deposit 4.20 g of Cu(s). Then, recall that 1.00 mol of Cu^{2+}(aq) is deposited as Cu(s) for every 2.0 mol of electrons used, and that the Faraday constant $F = 96{,}485 \text{ C}/1.00 \text{ mol electrons}$. Thus,

$$4.20 \text{ g Cu(s)} \left(\frac{1.00 \text{ mol Cu}}{63.5 \text{ g Cu}} \right) = 0.06614 \text{ mol Cu(s)}$$
$$\text{deposited}$$

$$0.06614 \text{ mol Cu(s)} \left(\frac{2 \text{ mol e}^-}{1.00 \text{ mol Cu(s)}} \right) = 0.1323 \text{ mol e}^-$$

$$0.1323 \text{ mol e}^- \left(\frac{96{,}485}{1.00 \text{ mol e}^-} \right) = 1.28 \times 10^4 \text{ C}$$

C. time, t (sec) $= \dfrac{\text{charge (C)}}{\text{current (C/sec)}}$

$$= \frac{1.28 \times 10^4 \text{ C}}{4.0 \text{ A}}$$

$$= 3.2 \times 10^3 \text{ sec}$$

This is about 10 hr.

3.12
Organic Chemistry

Basic Concept

Organic chemistry is the chemistry of carbon and its compounds. Carbon has the ability to bond to itself as well as to other elements including hydrogen, oxygen, nitrogen, sulfur, fluorine, chlorine, bromine, and iodine. The number of catalogued organic compounds (natural plus synthetic) is in the millions and rapidly expanding. In this section, the rules for the structures of simple organic compounds and the associated rules for nomenclature are outlined.

Hydrocarbons

Hydrocarbons are the simplest class of organic compounds and contain only carbon and hydrogen. Their classical names form the basis for naming more complex organic compounds. Carbon normally makes four covalent bonds. Compounds for which all possible carbon bonds contain hydrogen are called **alkanes** or **saturated hydrocarbons. Unsaturated hydrocarbons** contain one or more carbon–carbon double bonds or triple bonds, or **aromatic** (benzene-based) bonds. In general, unsaturated hydrocarbons are more reactive (by saturating or reducing agents) than their saturated counterparts.

The solubility of alkanes is related to the nonpolar nature of their molecules. Alkanes are water insoluble. Water molecules are polar and are bonded to one another by strong **hydrogen bonds.** Alkane molecules are nonpolar and are bonded by weak intermolecular forces known as **London dispersion** forces. Therefore, alkane molecules are not readily attracted by water molecules and cannot become part of the water structure (an irregular but generally hexagonal network). This fact demonstrates the general solubility principle of **like dissolves like**. Because most alkanes have densities less than the density of water, these compounds usually float on top.

The structure of normal (n) or straight-chain alkanes is shown below. The chemical formula for alkanes is C_nH_{2n+2}.

$$H-\underset{\underset{H}{|}}{\overset{\overset{H}{|}}{C}}-H \qquad \text{Methane, } CH_4$$

$$H-\underset{\underset{H}{|}\;\underset{H}{|}}{\overset{\overset{H}{|}\;\overset{H}{|}}{C-C}}-H \qquad \text{Ethane, } C_2H_6$$

$$H-\underset{\underset{H}{|}}{\overset{\overset{H}{|}}{C}}\cdots\underset{\underset{H}{|}}{\overset{\overset{H}{|}}{C}}-H$$

A general alkane structure

Table 3.4 lists formulas and properties for the first 10 hydrocarbons, known as the **homologous series**, along with the C_{20} and C_{30} hydrocarbons.

Alkanes can also be branched-chain hydrocarbons, which can form numerous isomers. **Isomers** are substances with the same chemical formula but different structural formulas, giving rise to different chemical and physical properties. To determine whether two compounds are isomers of each other, count the number of carbon atoms, hydrogen atoms, and any other type of atom present in each compound. If the numbers of each type of atom in the two compounds are equal, the compounds are isomers. The C_{20} hydrocarbon (eicosane), for example, has over 366,000 isomers.

The analogous series consists of hydrocarbons with one double bond; these are called **alkenes** and have the general formula C_nH_{2n}. Hydrocarbons with one triple bond are called **alkynes** and have the general formula C_nH_{2n-2}.

$$\overset{H}{\underset{H}{>}}C=C\overset{H}{\underset{H}{<}} \qquad \begin{array}{l}\text{Ethylene or ethene,}\\ C_2H_4\end{array}$$

$$H-C\equiv C-H \qquad \begin{array}{l}\text{Acetylene or ethyne,}\\ C_2H_2\end{array}$$

Many organic compounds have common names that were assigned before the formal system of nomenclature, called the **IUPAC system**, was developed in 1960. The following rules now apply to the naming of organic compounds:

1. Select the longest continuous chain of carbon atoms.
2. Number the carbon atoms in this chain, beginning with the end nearest a branch.
3. Assign the names and (position) numbers that indicate the branches of substituents (i.e., functional groups, including "-ene" for double bond and "-yne" for triple bond). If more than one of the same substituent appears, use Greek prefixes (di, tri, tetra, etc.). See Table 3.6, page 146, for the names of specific functional groups.
4. Number the longest continuous chain in such a way so that the substituents have the lowest possible numbers.

Example 3.50

Using the IUPAC system, name the three isomers of pentane and draw their structures.

Solution:

The three isomers of pentane are drawn below and labeled with both their common and IUPAC names.

$$H-\underset{\underset{H}{|}}{\overset{\overset{H}{|}}{C}}-\underset{\underset{H}{|}}{\overset{\overset{H}{|}}{C}}-\underset{\underset{H}{|}}{\overset{\overset{H}{|}}{C}}-\underset{\underset{H}{|}}{\overset{\overset{H}{|}}{C}}-\underset{\underset{H}{|}}{\overset{\overset{H}{|}}{C}}-H$$

Normal pentane (C_5H_{12})

Isopentane or 2-Methylbutane (C_5H_{12})

Neopentane or 2,2-Dimethylpropane (C_5H_{12})

Cyclic Hydrocarbons

Beginning with propane, C_3H_8, cyclic structures of hydrocarbons are possible. Hence, we have cyclopropane (C_3H_6), cyclobutane, (C_4H_8), and so on as shown in Table 3.5. Ringed hydrocarbons are generally more reactive (because of ring tension) than their straight or branched-chain counterparts.

Aromatic Hydrocarbons

Compounds containing a benzene ring, C_6H_6, are classified as aromatic. The benzene ring, shown in Figure 3.4, has a system of alternating single and double bonds. Benzene has interesting properties. Even with three double bonds, benzene undergoes, not addition, but rather substitution, because it is relatively unreactive. It forms compounds such as toluene and monochloro-, dichloro-, and trichlorobenzene. The six carbon–carbon bonds are all equal and are hybrids of single and double bonds. They are longer than normal double bonds but shorter than single bonds. This type of ring system exhibits **resonance,** which gives it its characteristic stability.

For nomenclature purposes each of the six carbon atoms in the benzene ring is assigned a number, 1–6, in clockwise fashion. Common aromatic compounds—benzene derivatives—with their common names and IUPAC names are shown in Figure 3.4. When substituent groups are added two at a time, the positions they occupy on the ring may be 1,2 (also known as **ortho** or *o*); 1,3 (also known as **meta** or *m*); and 1,4 (also known as **para** or *p*).

Functional Groups

When hydrogen atoms are replaced by certain other atoms, functional groups are formed. These functional groups give rise to discrete chemical and physical properties. The names of these functional groups, along with their structures, are listed in Table 3.6.

Example 3.51

Name, according to IUPAC rules, the four compounds shown below.

A. CH_3—CO—CH_3

B. $CH_3CH=CHCH_2CH_3$

C. $CH_3CH=CCHCH_3$ with CI substituents

D. $CH_3CHOHCH_3$

Solution:

A. Dimethyl ketone (also acetone, the common name)

B. 2-Pentene

C. 3,4 Dichloro-2-pentene

D. 2-Propanol (also isopropanol, the common name)

3.13 Summary

In conclusion, the study of chemistry is useful in two principal ways. First, it helps to predict whether a chemical reaction can occur spontaneously between two or more substances (chemical thermodynamics), what the products of the reaction will be, whether energy will be absorbed or released (thermochemistry), and how fast the reaction will occur (chemical kinetics). Second, material properties of solids and properties of solutions, such as solubility, boiling and melting points, vapor pressures, and osmotic pressures, can be predicted and measured. Knowledge of these properties is essential in designing and preparing chemical systems used in all phases of engineering problems, from environmental to biomedical applications.

The Periodic Chart or Periodic Table tells us the number of individual, naturally occurring elements is 92 and that these elements are a combination of metals, nonmetals, and metalloids. However, the number of compounds, both natural as well as artificially synthesized, runs over 35 million. Many of these are polymers. Compounds may be metallic, ionic, or covalent in bonding character. More often, they are a combination.

Matter occurs in four states—gas, liquid, solid, and solution. Mixtures may be homogeneous. For example, the composition of solutions is uniform throughout and only one phase is present. Mixtures can also be heterogeneous, where two of more phases are present. Colloids, or colloidal dispersions, are an intermediate state between homogenous and heterogeneous mixtures. Specific examples of colloids include milk, blood, mayonnaise, and butter. They do not separate upon standing and exhibit the Tyndall effect. Aerosols, emulsions, and foams are general examples. Suspensions, such as paints or cough medicines that need to be shaken or stirred before use, are a sub-category of colloids.

Chemistry is also helpful in examining the chemical composition of solutions and mixtures. It can help elucidate the structure of matter by correlating molecular structure on the microscopic scale with observable properties on the macroscopic scale.

Table 3.4 The Homologous Series for Alkanes

	Alkanes and Their Properties				
Name	Molecular Formula	Condensed Formula	Melting Point (°C)	Boiling Point (°C)	State at 25°C
Methane	CH_4	CH_4	−183	−162	Gas
Ethane	C_2H_6	CH_3CH_3	−172	−88	Gas
Propane	C_3H_8	$CH_3CH_2CH_3$	−188	−42	Gas
Butane	C_4H_{10}	$CH_3(CH_2)_2CH_3$*	−138	−0.5	Gas
Pentane	C_5H_{12}	$CH_3(CH_2)_3CH_3$	−130	36	Liquid
Hexane	C_6H_{14}	$CH_3(CH_2)_4CH_3$	−94	69	Liquid
Heptane	C_7H_{16}	$CH_3(CH_2)_5CH_3$	−91	98	Liquid
Octane	C_8H_{18}	$CH_3(CH_2)_6CH_3$	−57	126	Liquid
Nonane	C_9H_{20}	$CH_3(CH_2)_7CH_3$	−54	151	Liquid
Decane	$C_{10}H_{22}$	$CH_3(CH_2)_8CH_3$	−30	174	Liquid
Eicosane	$C_{20}H_{42}$	$CH_3(CH_2)_{18}CH_3$	36	343	Solid
Triacontane	$C_{30}H_{62}$	$CH_3(CH_2)_{28}CH_3$	66	446	Solid

*A condensed way to write the formulas of long alkane chains is to enclose CH_2 in parentheses and indicate the number of CH_2 units that occur in the chain as a subscript to the parentheses.

Table 3.5 Cyclic Hydrocarbons

	Structural Formulas and Symbols for Common Cycloalkanes	
Name	Structural Formula	Symbol
Cyclopropane		△
Cyclobutane		□
Cyclopentane		⬠
Cyclohexane		⬡
Cycloheptane		⬡
Cyclooctane		⯃

Benzene exhibits resonance.

In a toluene molecule, a methyl group replaces a hydrogen atom on the benzene ring. Any alkyl group can bond to the ring. If an ethyl group substitutes for a hydrogen atom in benzene, then ethylbenzene results.

Toluene   Ethylbenzene

Many aromatic compounds contain a functional group substituted for a hydrogen atom on the benzene ring. Consider each of the following:

Chlorobenzene   Bromobenzene   Nitrobenzene

Phenol   Aniline   Benzoic acid   Benzaldehyde

Disubstituted benzenes can have three structural isomers. For example, the three isomers of dichlorobenzene are

1,2-Dichlorobenzene   1,3-Dichlorobenzene   1,4-Dichlorobenzene

In 1,2-dichlorobenzene, the two chlorine atoms are bonded to adjacent carbon atoms. In 1,3-dichlorobenzene, the two chlorine atoms are bonded to carbon atoms separated by one carbon atom, and in 1,4-dichlorobenzene the two chlorine atoms are separated by two carbon atoms.

Three prefixes, ortho, meta, and para, may be used to designate the position of groups. Ortho, or just *o*, means 1,2 disubstituted; meta, or *m*, means 1,3 disubstituted; and para, or *p*, means 1,4 disubstituted. Consider the following:

1,2-Dinitrobenzene   1,3-Dinitrobenzene   1,4-Dinitrobenzene
Orthodinitrobenzene   Metadinitrobenzene   Paradinitrobenzene
o-Dinitrobenzene   *m*-Dinitrobenzene   *p*-Dinitrobenzene

On occasion, it is necessary to consider the benzene ring as a substituent group bonded to a long chain or ring [R]. If a hydrogen atom is removed from benzene, a phenyl group, C_6H_5—, results.

Benzene, C_6H_6 Phenyl group, C_6H_5—[R]

An example of a compound that contains a phenyl group is 3-phenylhexane.

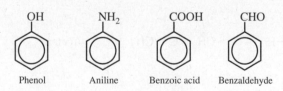

3-Phenylhexane

Figure 3.4   Aromatic hydrocarbons

145

Table 3.6 Functional Groups

	Classes and Functional Groups of Organic Compounds			
Class	Functional Group	Example of Expanded Structural Formula	Example of Condensed Structural Formula	Common Name
Alkane	—	$H-\underset{\underset{H}{\mid}}{\overset{\overset{H}{\mid}}{C}}-\underset{\underset{H}{\mid}}{\overset{\overset{H}{\mid}}{C}}-H$	CH_3CH_3	Ethane
Alkene	$\text{C}=\text{C}$	$\underset{H}{\overset{H}{}}C=C\underset{H}{\overset{H}{}}$	$H_2C=CH_2$	Ethylene
Alkyne	$-C\equiv C-$	$H-C\equiv C-H$	$HC\equiv CH$	Acetylene
Aromatic	(ring structure)	(ring structure)	(benzene ring)	Benzene
Alcohol	$-O-H$	$H-\underset{\underset{H}{\mid}}{\overset{\overset{H}{\mid}}{C}}-\underset{\underset{H}{\mid}}{\overset{\overset{H}{\mid}}{C}}-O-H$	CH_3CH_2-OH	Ethyl alcohol
Ether	$-\overset{\mid}{C}-O-\overset{\mid}{C}-$	$H-\underset{\underset{H}{\mid}}{\overset{\overset{H}{\mid}}{C}}-O-\underset{\underset{H}{\mid}}{\overset{\overset{H}{\mid}}{C}}-H$	CH_3-O-CH_3	Dimethyl ether
Amine	$-N-H$ with H above	$H-\underset{\underset{H}{\mid}}{\overset{\overset{H}{\mid}}{C}}-N-H$ with H above	CH_3-NH_2	Methylamine
Aldehyde	$-\overset{\overset{O}{\parallel}}{C}-H$	$H-\underset{\underset{H}{\mid}}{\overset{\overset{H}{\mid}}{C}}-\overset{\overset{O}{\parallel}}{C}-H$	CH_3-CH with O above	Acetaldehyde
Ketone	$-\overset{\mid}{C}-\overset{\overset{O}{\parallel}}{C}-\overset{\mid}{C}-$	$H-\underset{\underset{H}{\mid}}{\overset{\overset{H}{\mid}}{C}}-\overset{\overset{O}{\parallel}}{C}-\underset{\underset{H}{\mid}}{\overset{\overset{H}{\mid}}{C}}-H$	$CH_3-\overset{\overset{O}{\parallel}}{C}-CH_3$	Acetone
Carboxylic acid	$-\overset{\overset{O}{\parallel}}{C}-O-H$	$H-\underset{\underset{H}{\mid}}{\overset{\overset{H}{\mid}}{C}}-\overset{\overset{O}{\parallel}}{C}-O-H$	$CH_3-\overset{\overset{O}{\parallel}}{C}-OH$	Acetic acid
Ester	$-\overset{\overset{O}{\parallel}}{C}-O-\overset{\mid}{C}-$	$H-\underset{\underset{H}{\mid}}{\overset{\overset{H}{\mid}}{C}}-\overset{\overset{O}{\parallel}}{C}-O-\underset{\underset{H}{\mid}}{\overset{\overset{H}{\mid}}{C}}-H$	$CH_3-\overset{\overset{O}{\parallel}}{C}-O-CH_3$	Methyl acetate
Amide	$-\overset{\overset{O}{\parallel}}{C}-N-H$ with H above N	$H-\underset{\underset{H}{\mid}}{\overset{\overset{H}{\mid}}{C}}-\overset{\overset{O}{\parallel}}{C}-N-H$ with H above N	$CH_3-\overset{\overset{O}{\parallel}}{C}-NH_2$	Acetamide
Halide	$-X$ $(X = F, Cl, Br, I)$	$H-\underset{\underset{H}{\mid}}{\overset{\overset{H}{\mid}}{C}}-\underset{\underset{H}{\mid}}{\overset{\overset{H}{\mid}}{C}}-Br$	CH_3CH_2-Br	Ethyl bromide

Many analytical techniques and modern instruments, such as atomic absorption spectroscopy (AAS), ion-coupled plasma (ICP) spectroscopy, gas chromatography (GC), high-performance liquid chromatography (HPLC), and detectors such as mass spectroscopy (MS) or flamed induced ionization (FID) can identify elements and compounds in solutions as well as quantify the amounts, some as low as parts per trillion (ppt). These methods are frequently used to detect contaminants or impurities in complex, multiphase systems.

PRACTICE PROBLEMS

1. An element has the electronic configuration $1s^2 2s^2 2p^4$.
   a. How many electrons and protons does one atom of this element have?
   b. Does this atom most likely form a stable cation or anion?

2. Each of the species Cl^-, Ar, K^+, and Ca^{2+} has the same number of
   (A) protons
   (B) electrons
   (C) neutrons
   (D) isotopes

3. The correct chemical formula for chromium(III) oxide is
   (A) CrO
   (B) Cr_3O_2
   (C) Cr_2O_3
   (D) Cr_3O_3

4. What is the correct chemical name for MnO_2?

5. Determine for urea, $CO(NH_2)_2$,
   a. the molar mass
   b. the percent nitrogen by mass

6. Determine the empirical formula of mustard gas, a chemical warfare agent, which has the following percent composition: 30.2% carbon, 5.07% hydrogen, 44.58% chlorine, and 20.16% sulfur.

7. How many atoms are represented in one formula unit of $Ca_3(PO_4)_2$, the main component of bone mass?
   (A) 16
   (B) 13
   (C) 9
   (D) 5

8. The oxidation number or valence of Cl in ClO_3^- is
   (A) +5
   (B) +3
   (C) +1
   (D) −1

9. In this reaction:

   $$Fe_2O_3(s) + 3CO(g) \rightarrow 2Fe(s) + 3CO_2(g)$$

   what is the limiting reagent when 45.3 g of CO react with 79.8 g of Fe_2O_3?

10. In excess acid, how much hydrogen gas will be produced with 50 g of Al? The reaction is

    $$2\ Al(s) + 6\ HCl(aq) \rightarrow Al_2Cl_6(aq) + 3\ H_2(g)$$

    (A) 1.85 g
    (B) 3.7 g
    (C) 5.6 g
    (D) 11 g

11. In the reaction of problem 10, if only 4.5 g of H_2 (g) are actually produced, what is the percent yield of the reaction?
    (A) 20%
    (B) 40%
    (C) 60%
    (D) 80%

12. Which aqueous solution has the lowest freezing point, assuming that all are 1.0 molal?
    (A) NaCl (table salt)
    (B) C_2H_5OH (ethanol or grain alcohol)
    (C) $C_{12}H_{22}O_{11}$ (sucrose or table sugar)
    (D) $KAl(SO_4)_2$ (alum)

13. What is the freezing point of a solution of 92 g of ethanol, C_2H_5OH, in 500 g of H_2O?
    (A) −3.72°C
    (B) −7.44°C
    (C) −11.2°C
    (D) −15.0°C

14. How many atoms are represented by 1.0 mg of Mg?
    (A) 2.48×10^{13}
    (B) 1.24×10^{13}
    (C) 6.022×10^{13}
    (D) 6.022×10^{23}

15. Which **substance** is oxidized and which substance is reduced in the following reaction?

    $$3Cu + 8HNO_3 \rightarrow Cu(NO_3)_2 + 2NO + 4H_2O$$

147

16. The equilibrium constant at 427°C for the reaction for the formation of ammonia:

$$N_2(g) + 3H_2(g) \rightleftharpoons 2NH_3(g)$$

is $K_p = 9.4 \times 10^{-5}$.
a. What is ΔG for this reaction?
b. Is the reaction spontaneous?

17. Ozone, O_3, in the atmosphere is subject to reaction with nitric oxide, NO:

$$O_3(g) + NO(g) \rightarrow NO_2(g) + O_2(g)$$

a. Calculate $\Delta G°$ for the following reaction at 25°C from the given data.
b. Tell whether the reaction is spontaneous.

$$\Delta H° = -199 \text{ kJ}, \qquad \Delta S° = -4.1 \text{ J/K}$$

18. Which of the following has the greatest entropy, S?
(A) Crystalline NaCl at 10 K
(B) A liquid solution of Ag in Cu at 500 K
(C) Gaseous CH_4 at 500 K
(D) Liquid He at 4 K

19. What is the pH of a 0.014 M $Ca(OH)_2$ solution?
(A) 1.55
(B) 1.85
(C) 12.15
(D) 12.45

20. A laboratory technician mixes 400 mL of 0.125 M HCl with 800 mL of 0.050 M HCl. What is the pH of the resulting solution?
(A) 1.12
(B) 2.50
(C) 10.5
(D) 12.2

21. Calculate the molar solubility of the salt PbI_2 in
a. distilled water
b. a solution that is 0.150 M NaI(aq)

22. Calculate $\Delta H_{RX}°$ for the reaction below from the standard heats of formation given in the table that follows:

$$Ca_3P_2(s) + 6HCl(g) \rightarrow 3CaCl_2(s) + 2PH_3(g)$$

Compound	$\Delta H°_f$ (kJ/mol)
$Ca_3P_2(s)$	−504
$PH_3(g)$	+9
$HCl(g)$	−92
$CaCl_2(s)$	−794

23. a. Calculate $E_{net}°$ for a silver-aluminum cell in which the electrochemical reaction is as follows:

$$Al(s) + 3Ag^+(aq) \rightarrow Al^{3+}(aq) + Ag(s)$$

b. Would increasing $[Al^{3+}]$ increase or decrease the cell emf?

24. At 160°C, $K_p = 0.750$ for the following decomposition:

$$PCl_5(g) \rightleftharpoons PCl_3(g) + Cl_2(g)$$

When a sample of PCl_5 (g) is introduced into a previously evacuated vessel at an initial partial pressure of 1.00 atm and allowed to reach equilibrium, what will be the partial pressure of PCl_3 (g)?
(A) 0.0569 atm
(B) 0.285 atm
(C) 0.569 atm
(D) 1.00 atm

25. The enzyme aspartase catalyzes the conversion of L-aspartate to fumarate plus ammonium ion. The equilibrium constant $K_c = 7.4 \times 10^{-3}$ at 29°C. ΔH_{RX} for the reaction is 14.40 kcal/mol.
a. Find K_c at the physiologically interesting temperature of 39°C (normal body temperature).
b. Determine ΔG for this reaction at 39°C.
c. Determine ΔS for this reaction at 39°C.

26. K_a for lactic acid, the acid responsible for dental caries, is 1.30×10^{-4}. Calculate, for a 0.200 M solution of lactic acid,
a. the pH
b. the percent ionization

27. Given that K_a for HCN is 4.0×10^{-10}, find the pH of a 0.50 M NaCN solution. [*Hint*: NaCN undergoes hydrolysis, since HCN is a weak acid, and the resulting solution is slightly basic.]
(A) 7.0
(B) 9.4
(C) 10.5
(D) 11.5

28. Will a precipitate form if 1.00 mg of Na_2CrO_4 is added to 225 mL of a 0.00015 M $AgNO_3(aq)$ solution? K_{sp} for Ag_2CrO_4 is 1.2×10^{-12}.

29. a. Name the following four compounds using the IUPAC system:

$$C_6H_{14} \quad CH_3—O—CH_3$$
$$C_2H_5OH \quad CH_2{=}CHCH_2CH_3$$

b. Which are structural isomers?
c. Which is an unsaturated hydrocarbon?

30. Draw the structures for the following compounds:
  a. paradichlorobenzene
  b. 4,5-dibromo-2-octene
  c. cyclohexanol

Answer Key

1. **a.** 8 electrons and 8 protons
   b. anion
2. B
3. C
4. Manganese(IV) oxide
5. **a.** 60 g/mol
   b. 46.6%
6. $C_4H_8C_2S$
7. B
8. A
9. $Fe_2O_3(s)$
10. C
11. D
12. D
13. B
14. A
15. Cu is oxidized and N (in HNO_3) is reduced
16. **a.** +53.9 kJ/mol
    b. No
17. **a.** −198 kJ
    b. The reaction is spontaneous.
18. C
19. D
20. A
21. **a.** $S = 0.0013$ moles/liter
    b. $S = 3.5 \times 10^{-7}$ M.
22. −1,308 kJ/mol
23. **a.** 2.46 V
    b. Decrease
24. C
25. **a.** 1.60×10^{-2}
    b. 2.56×10^3 cal/mol
    c. 37.9 cal/K or 37.9 eu
26. **a.** 2.29
    b. 2.54%
27. D
28. No, $Q < K_{sp}$
29. **a.** n-Hexane, dimethyl ether, ethanol or ethyl alcohol, 1-butene
    b. Dimethyl ether and ethanol
    c. 1-Butene
30. **a.**

b.

c.

Answers Explained

1. **a.** Add the superscripts to get the total number of electrons: 8. Since the atom is neutral, 8 also represents the number of protons, and therefore the atomic number. Hence, the atom represented is oxygen.
   b. Oxygen normally forms a stable anion, since it needs two more electrons to complete its second energy level and meet the octet rule.

2. **B** Cl^- has 17 protons and 18 electrons; Ar has 18 protons and 18 electrons; K^+ has 19 protons and 18 electrons; Ca^{2+} has 20 protons and 18 electrons. The number of neutrons or isotopes cannot be determined without the mass number. Thus, the only quantity in common is the number of electrons.

3. **C** Since chromium(III) oxide is ionic, use the criss-cross method with Cr^{3+} and O^{2-}. Thus, Cr_2O_3 is the correct chemical formula.

4. Since MnO_2 is an ionic compound, and manganese has multiple oxidation states, manganese(IV) oxide is the correct chemical name.

5. **a.** Molar mass is the sum of the atomic masses or weights, multiplied by the appropriate subscripts appearing in the formula, here $CO(NH_2)_2$. Thus:

$$MM = 12.01 \text{ g}(1) + 16.00 \text{ g}(1) + 14.01 \text{ g}(2) + 1.008 \text{ g}(4) = 60.0 \text{ g/mol}$$

   b. Percent N (by mass) $= \dfrac{14.01 \text{ g}(2)}{60.0 \text{ g}} \times 100$

$$= 46.6\%$$

6. See Example 3.6. To convert percentages into grams, assume a 100.0-g sample. Then, convert grams of each element into moles using the atomic masses of carbon, hydrogen, chlorine, and sulfur. The result is $C_{2.517}H_{5.030}Cl_{1.255}S_{0.630}$.

Divide the number of moles of each element, that is, each subscript, by the smallest number, which is 0.630 mol, to obtain the empirical formula of mustard gas: $C_4H_8Cl_2S$.

7. **B** Add the subscripts in the chemical formula to obtain the total number of atoms: 13.

8. **A** See Example 3.4. Set up a simple algebraic equation as follows, letting x represent the unknown oxidation number of chlorine; -2 is the oxidation number of oxygen, and the sum is equal to the net charge on the oxoanion, -1:

$$x(1) + -2(3) = -1$$

Solve for x to obtain $x = +5$.

9. The limiting reagent is the reactant in short supply. To identify it, convert each reactant's mass, given in grams, into moles:

$$45.3 \text{ g CO} \frac{(1.0 \text{ mol CO})}{(28.0 \text{ g CO})} = 1.62 \text{ mol CO}$$

$$79.8 \text{ g Fe}_2\text{O}_3 \frac{(1.0 \text{ mol Fe}_2\text{O}_3)}{(159.7 \text{ g Fe}_2\text{O}_3)} = 0.500 \text{ mol Fe}_2\text{O}_3$$

From the stoichiometry of the balanced reaction, it is clear that 1.0 mol Fe_2O_3 requires 3 mol CO, and 0.500 mol Fe_2O_3 would require 1.50 mol CO. Thus, CO is present in excess and Fe_2O_3 is the limiting reagent.

10. **C** In the given reaction, HCl is present in excess. Thus, Al is the limiting reagent, which determines the maximum yield of H_2 gas. Use the factor-label method:

$$50 \text{ g Al} \frac{(1.0 \text{ mol Al})}{(26.98 \text{ g Al})} \frac{(3 \text{ mol H}_2)}{(2 \text{ mol Al})} \frac{(2.016 \text{ g H}_2)}{(1.0 \text{ mol H}_2)}$$
$$= 5.6 \text{ g H}_2$$

11. **D** See Example 3.12. Percent yield $= \frac{4.5 \text{ g}}{5.6 \text{ g}} \times 100 = 80\%$

12. **D** Refer to equation 3.11. Since all concentrations are the same, look for the solute that has the highest van't Hoff or i factor—the number of ions or particles per mole. For NaCl, $i = 2$; for C_2H_5OH, $i = 1$; for $C_{12}H_{22}O_{22}$, $i = 1$; for $KAl(SO_4)_2$, $i = 4$. Thus, alum produces the lowest freezing point.

13. **B** Use equation 3.13, and solve for ΔT. Molality m is given by the equation

$$m = \frac{\dfrac{92 \text{ g C}_2\text{H}_5\text{OH}}{46.0 \text{ g/mol C}_2\text{H}_5\text{OH}}}{0.500 \text{ kg H}_2\text{O}}$$

Also, $K_f(H_2O) = 1.86$, and $i = 1$ for C_2H_5OH.

Thus, $\Delta T = 7.44\,°C$, and $T_f = -7.44\,°C$.

14. **A** The given quantity, 1.0 mg, of magnesium $= 1.0 \times 10^{-9}$ g Mg. Use the factor-label method, the atomic mass of magnesium, and Avogadro's number:

$$1.0 \times 10^{-9} \text{ g Mg}$$
$$\frac{(1.0 \text{ mol Mg})}{(24.31 \text{ g Mg})} \frac{(6.022 \times 10^{23} \text{ atoms Mg})}{(1.0 \text{ mol Mg})}$$
$$= 2.48 \times 10^{13} \text{ atoms Mg}$$

15. In the given reaction, copper goes from an oxidation state of 0 to $+2$ in $Cu(NO_3)_2$, losing 2 electrons. Nitrogen goes from an oxidation state of $+5$ in HNO_3 to $+2$ in NO, gaining 3 electrons. Therefore, copper is oxidized, while nitrogen is reduced.

16. **a.** Use equation 3.62. The K_p value is given. Change 427°C to 700 K, and use $R = 8.31$ J/K·mol. Then, $\Delta G = +53,900$ J/mol $= +53.9$ kJ/mol.

    b. Since the sign for ΔG is positive, the reaction is *not* spontaneous.

17. **a.** Use equation 3.60, and substitute the values given. Convert 25°C to 298 K. Then, $\Delta G° = -198$ kJ.

    b. Since the sign of ΔG is negative, the reaction is spontaneous.

18. **C** Gas-phase systems have more kinetic energy and random motion and therefore greater entropy.

19. **D** Note that a 0.014 M $Ca(OH)_2$ solution is a strong base with 2 mol of OH^- per mole of $Ca(OH)_2$. First, find $[OH^-]$: $[OH^-] = 0.014$ M(2) = 0.028 M.

    Then use equation 3.22, and substitute 0.028 M to calculate pOH: pOH = 1.55. Finally, use equation 1.23 to calculate pH: pH = 12.45.

20. **A** This problem involves mixing two solutions of the same acid but with different concentrations. The strategy here is to find the number of moles of H^+ in the first solution and the number of moles of H^+ in the second solution, then add the numbers together, and divide the sum by the total volume (assume volumes are additive). The pH can then be found using equation 3.20.

    Moles $H_1^+ = V_1M_1 = (0.400 \text{ L})(0.125 \text{ M})$
    $= 0.050$
    Moles $H_2^+ = V_2M_2 = (0.800 \text{ L})(0.050 \text{ M})$
    $= 0.040$

150

Total moles H^+ = 0.090

Total volume = 1200 mL = 1.2 L

$$[H^+] = \frac{\text{total moles } H^+}{\text{total volume}}$$

$$= \frac{0.090 \text{ mol}}{1.2 \text{ L}} = 0.075 \text{ M}$$

$$pH = -\log[H^+] = -\log(0.0750 \text{ M}) = 1.12$$

21. a. Let S = the solubility (molar concentration) of the Pb^{2+} ion in distilled water. Then $2S$ = the solubility of the I^{1-} ion. From Table 3.2, the solubility product constant or K_{sp} of the PbI_2 is given as 7.9×10^{-9}. Since there are no other common or interfering ions present, we can write the solubility product constant expresssion as:

$$K_{sp} = [Pb^{2+}][I^{1-}]^2 = (S)(2S)^2 = 7.9 \times 10^{-9}$$

$$4S^3 = 7.9 \times 10^{-9}$$

S = 0.00125 M or 0.00125 moles/liter or 0.0013 moles/liter (rounded off)

S represents the concentration of Pb^{2+} ion in solution, and hence the molarity solubility of PbI_2, since the Pb^{2+} and PbI_2 are in a one-to-one ratio. Twice this amount, however, would represent the I^{1-} concentration. To determine the concentration of lead ion in grams/liter, one would only have to multiply the above answer (in moles/liter) by the atomic mass of lead, which is 207 grams/mol.

b. This problem illustrates the common ion effect. The common ion is I^-, and the background concentration of I^- is given as 0.15 M, since NaI is a strong electrolyte. From Table 3.2, the K_{sp} of PbI_2 is 7.9×10^{-9}.

Let S represent the solubility of Pb^{2+}; then $2S$ represents the solubility of I^-. Set up the K_{sp} expression as follows:

$$K_{sp} = [Pb^{2+}][I^-]^2 = (S)(2S + 0.15 \text{ M})^2$$

Solve for S: $S = 3.5 \times 10^{-7}$ M.

This value also represents the molar solubility of PbI_2.

22. Use Hess's law, see Example 3.40, and use the heats of formation given in the table. Then

$$\Delta H°_{RX} = [2(9) + 3(-794)] - [1(-504) + 6(-92)] = -1,308 \text{ kJ/mol}$$

23. a. See Example 3.47. Assume all species are in their standard states. Find the $E°$ values for aluminum and silver from Table 3.3, Standard Reduction Potentials, and add. Note that Al is oxidized while Ag^+ is reduced. Therefore,

$$E_{net}° = 1.66 + 0.80 = 2.46 \text{ V}$$

b. See the Nernst equation, equation 3.61. Here, $Q = \frac{[Al^{3+}]}{[Ag^+]^3}$. Therefore, increasing $[Al^{3+}]$ would increase $\log Q$ but decrease the cell emf.

24. C Use equation 3.31. Set up the K_p expression as follows:

$$K_p = \frac{[PCl_3][Cl_2]}{[PCl_5]} = \frac{(x)(x)}{1.00 \text{ atm} - x} = 0.750$$

Solve for x: $x = 0.569$ atm. This value represents the partial pressure of $PCl_3(g)$.

25. a. Use the van't Hoff equation, equation 3.33. $K_c(2) = 1.37 \times 10^{-2}$. Be sure to use kelvins for temperature.

b. Use equation 3.59. $\Delta G = 2.65 \times 10^3$ cal/mol.

c. Use equation 3.58 and solve for ΔS. $\Delta S = 37.9$ cal/K or 37.9 eu.

26. a. Follow the strategy outlined in Example 3.25. Lactic acid, $HC_3H_5O_3$, is a monoprotic acid like formic acid and acetic acid. The pH of lactic acid is 2.29.

b. The percent ionization, %I, is defined as $\frac{[H_3O^+]}{[HC_3H_5O_3]} \times 100$ and equals 2.54%.

27. D Follow the strategy outlined in Example 3.25, and use the hint given. The only difference is that NaCN produces a basic instead of an acidic solution. The pH of a 0.50 M NaCN solution is 11.5.

28. Assume that the total volume is 225 mL. This is also the volume for both Na_2CrO_4 and $AgNO_3$. Convert 1.0 mg of Na_2CrO_4 into moles, then into molarity. The molarity of $AgNO_3$ is given. Then follow the procedure of Example 3.36 to determine that a precipitate will *not* form.

29. a. See Example 3.51, Table 3.5 for homologous series for alkanes, and Table 3.8 for functional groups. The names of the four compounds are *n*-hexane, dimethyl ether, ethanol or ethyl alcohol, and 1-butane, respectively.

b. Dimethyl ether and ethanol have the same chemical formula, although they have different structural formulas.

c. Only 1-butene has a double bond, which makes it unsaturated.

30. Follow the IUPAC rules for interpreting names and writing structures.

Computational Tools

Thomas E. Hulbert, PE

Introduction

The FE exam requires knowledge of computer organization and terminology and the ability to analyze program segments written in pseudocode and generic programming code. Other questions are based on spreadsheets used to tabulate, analyze, and display data. This chapter reviews basic computer terminology and the organization of computers. It also reviews generic programming and number systems and the use of spreadsheets using Microsoft Excel. Sample problems are solved in the chapter.

4.1 Computer Organization

A computer is a programmable machine that has instructions in a well-defined manner and will execute the prerecorded list of instructions or program. Computers are electronic and digital. The hardware consists of wires, circuits, and transistors. The instructions are called software, and data are stored. All general-purpose computers have the following components:

- Input Device: a keyboard and mouse, through which data and instructions enter a computer.
- Central Processing Unit (CPU): the component that actually executes instructions.

Computer Organization and Terms

- Memory: a device that stores, at least temporarily, data and programs.
- Mass Storage Device: the component that permanently retains large amounts of data in devices such as disk drives and tape drives.
- Output Device: a display screen, printer, or other device that lets you see what the computer has done.

In addition to these components, many others make it possible for the basic components to work together efficiently. For example, every computer requires a *bus* or cable that transmits data from one part of the computer to another.

153

Computers can be generally classified by size and power as follows:

- Personal Computer (PC): a single-user computer based on a microprocessor that includes a keyboard for data entry, a monitor to display information, and a device for storing data.
- Workstation: a single-user computer, similar to a personal computer, with a more powerful microprocessor and a higher-quality monitor.
- Minicomputer: a multiuser computer capable of supporting many users at a time.
- Mainframe: a powerful multiuser computer capable of supporting many hundreds or thousands of users simultaneously.
- Supercomputer: a fast computer that can perform many millions of instructions per second

4.2
Computer Terms and Definitions

The **central processing unit (CPU)** is the brain of the machine and is often referred to as the central processor. The CPU is where most calculations take place and thus determines the computing power of the system. CPUs have one or more printed circuit boards. On smaller systems like the PC, the CPU is a single chip called a microprocessor. Typical components of a CPU are the arithmetic logic unit (ALU), which performs arithmetic and logical operations and the control unit (CU), which takes instructions from memory, decodes, and executes them using ALU as required.

Baud is the number of signaling elements that occur each second. It was named after J.M.E. Baudot who invented the Baudot telegraph code. One bit of information (signaling element) is encoded in each electrical change. The baud indicates the number of bits per second that are transmitted; *300 baud* is 300 bits transmitted each second (abbreviated 300 bps). It is possible to encode multiple bits in each electrical change. (4,800 baud will allow 9,600 bits each second operating at 2,400 baud.)

The **Internet** is a decentralized network connecting millions of computers in over 100 countries which exchanges data and information. Each host connection is independent, enabling its users to choose services to use. The Internet can be accessed via America Online or a commercial Internet service provider (ISP).

WWW is the World Wide Web which is a subset of the Internet. The Internet is the actual network of networks where all the information resides. Pages can be accessed using a Web browser. Hypertext Transfer Protocol (HTTP) is the method used to transfer Web pages to your computer, and to link your computer to other Web sites.

Multitasking or multiprogramming is the allocation of the main memory of the computer among several users so that multiple tasks can be performed at the same time by these users.

Time-sharing is a procedure where each user, controlled by the operating system, takes turns for a very short period of time and uses the entire memory of the computer. The memory used is then stored in a reserve area for that user and the next user's memory is loaded. Since the swapping is so rapid, users feel that their work is operating in real time.

Teleprocessing is accessing the computer from a remote location on a local or wide area network (LAN or WAN) via telephone, microwave links, satellite, coaxial cable, or fiber optic cable. An analog signal is transmitted. A modem is used to convert the signal to digital for processing by the computer.

Additional Web Sites for Computer Terminology

The following Web Sites have more computer definitions and additional information on computers:

www.JustAnswer.com
www.techterms.org
www.whatistechtarget.com
www.zerocut.com
www.folduc.org
www.helpwithpcs.com
www.right-track.com

4.3
Coding

Coding refers to the manner in which alphanumeric data and control characters are handled by series of bits. Characters that can be displayed or printed, including numerals and symbols (&, >, etc.), are alphanumeric data. Control characters (tab, carriage return, etc.) are keystrokes and are not recorded. Computers use binary numbers only; thus, all symbolic data must be expressed in binary codes.

The Extended Binary Coded Decimal Interchange Code (EBCDIC) is widely used on mainframe computers. It uses a byte (eight bits) for each character, allowing a maximum of 256 different characters. Since strings of bits are difficult to read, the hexadecimal or octal format is used to simplify working with EBCDIC data. Each byte is converted into two strings of four bits each. The two strings are then converted to hexadecimal (base −16) numbers. Since $(1111)_2 = (15)_{10} = (F)_{16}$, the largest possible EBCDIC character is coded FF in hexadecimal.

Number Systems

The American Standard Code for Information Interchange (ASCII) is a seven-bit code allowing 128 combinations. It is commonly used in microcomputers, although a higher order bit structure employing a 16-bit Unicode character format has been developed. ASCII-coded magnetic tape and disk files are used to transfer data and documents between computers that normally are unable to share data.

Unicode provides a unique number for every character; no matter the platform, the program, or the language. It is a univeral character encoding that is maintained and controlled by the Unicode Consortium. Unicode has been adopted by most major industry leaders and is required in most standards and operating systems. Building Unicode into client server or multitiered applications and Web sites result in cost savings over previous systems. It minimizes reengineering and allows data to be moved through systems without corruption. It covers all of the characters for all modern and ancient writing systems worldwide, including technical symbols, punctuations, and special characters. In summary, Unicode is not a software program or a font. It is a character encoding system similar to ASCII used by developers throughout the world.

Examples of number system conversions are included in the sample problems.

4.4
Program Design

A program is a sequence of instructions that performs a function in the computer. The program is designed to accomplish an algorithm. An **algorithm** is a problem-solving procedure that consists of a finite number of well-defined steps. Each step in the algorithm includes one or more instructions (such as READ and OPEN). These instructions are known as source code statements and are readable English phrases.

A computer will not normally understand source code statements. Source code is translated into machine-readable object code and absolute memory locations. Finally, an executable program is produced.

Software is an executable program kept on disk or tape.

Firmware is a program placed in ROM (read only memory) or EPROM (Erasable programmable read-only memory).

Hardware is the computer.

4.5
Computer Languages

Machine language instructions are understood by the computer's CPU (its native language). To make an instruction compatible, it consists of two parts:

the operation to be performed (op-code).
the operand expressed as a storage location.

Each instruction is expressed as a series of bits, a form known as intrinsic machine code. Octal and hexadecimal coding is a more convenient alternative. In either case, coding a machine language program is tedious and time-consuming and is seldom done by hand.

Assembly language is one step away from machine language and more symbolic than machine language. Mnemonic codes are used to specify the operations, and the operands are referred to by variable names rather than by addresses. Blocks of code that are to be repeated at multiple locations in the program are known as macros. Macro instructions are written once and are referred to by symbolic names in the source code.

Languages

Assembly language code is translated into machine language by an assembler or macroassembler. After assembly, portions of other programs or function libraries are combined by a linker. In order to run, the program is placed in the computer's memory by a loader. Assembly language is preferred for highly efficient programs. However, the coding is difficult and cumbersome.

The following table compares several languages and the corresponding instructions.

Language	Instruction
FORTRAN	+
Assembly language	AR
Machine language	1A
Intrinsic machine code	1111 0001

Because the instructions for high-level languages resemble English, these languages are easier to use than low-level languages. An interpreter or a compiler, either of which is considered software, translates high-level statements into machine language. A compiler, when activated, performs the checking and conversion functions on all instructions, and a true stand-alone, executable program is created. An interpreter, however, checks the instructions and converts them line by line into machine code during execution, but produces no stand-alone program capable of use without the interpreter.

Interpreters can check syntax as each statement is entered by the programmer. For some languages and implementations of other languages there is little distinction between an interpreter and a compiler; in that case, the conversion software is known as a pseudo- or incremental compiler.

4.6
Microsoft Excel

To start Microsoft Excel for Windows, double-click the Microsoft Excel icon on the Microsoft Office Shortcut Bar or press Start and proceed through the program's pull-down menu to Excel and its icon. The Microsoft Excel window appears. If necessary, enlarge the window by clicking the Maximize button in the upper-right corner of the window.

Excel

To start Microsoft Excel for Macintosh, double-click the Microsoft Excel icon in the Finder. The Microsoft Excel window appears. If necessary, enlarge the window by clicking the zoom box in the upper-right corner of the window.

When Microsoft Excel starts, it creates a new, empty workbook. Workbooks can be reopened when Microsoft Excel is restarted by naming and saving them the first time in the startup directory or folder.

Opening a Workbook

The workbook is the normal document or file type in Microsoft Excel and is the electronic equivalent of a three-ring binder. Inside workbooks are sheets, such as worksheets and chart sheets. Each sheet's name appears on a tab at the bottom of the workbook. Sheets can be moved or copied between workbooks and can be rearranged within a workbook.

Open workbooks by going to my computer and clicking on folders or clicking on Finder even if Excel has not been started. Also, open files in other formats such as Lotus 1-2-3; then create new workbooks in Microsoft Excel. Search your drive for a specific workbook, even if its name is not known, and create template workbooks that can be used a number of times.

Creating Worksheets

When a workbook is created or opened, Microsoft Excel displays it in a window. Several workbook windows can be open at the same time. In Microsoft Excel for Macintosh, the screen is slightly different.

Most of the work in Microsoft Excel will be on a worksheet, which is a grid of rows and columns. Each cell is the intersection of a row and a column and has a unique address, or reference. For example, the cell where column B and row 5 intersect is cell B5. It will be referenced in the white slot to the left and above the worksheet. Cell references are used when formulas are written or specific cells must be cited.

Generally, first select the cell or cells to work with, and then enter data or choose a command. Selected cells appear highlighted on the screen. The active cell is the cell in which data are entered when typing is started. Only one cell, indicated by a heavy border, is active at a time.

To change the active cell, move the mouse pointer into cell B4 and click. Notice that the cell's reference (B4) and its values (region) appear in the formula bar.

To scroll through the worksheet, proceed as follows:

1. Move the pointer to the down arrow in the vertical scroll bar on the right edge of the window, and then click. The worksheet scrolls down one row.
2. Click the scroll bar area below the scroll box. The worksheet scrolls down one screen.
3. Move the pointer into the scroll box. Hold the mouse button down, move the pointer back to the top of the vertical scroll bar, and then release the mouse button. This procedure is called dragging. The worksheet scrolls back up to the first row.
   To scroll any distance, drag the scroll box.
   To scroll down one page, click the scroll bar.
   To scroll down one row, click the down arrow.
   To enter the data, select the cell and fill in alpha or numeric information.

Microsoft Excel fills in the label based on the initial selection. This operation is called AutoFill, and it works with several types of data. For example, the days of the week and series of numbers can be filled by clicking and dragging on the appropriate cells.

Columns and rows can be totaled or averaged, and numerous formulas and functions are available. Data such as numbers, days, or other lists can also be entered with AutoFill. To move data around on the worksheet, drag it or use the Cut, Copy, and Paste commands on the Edit menu.

Building Formulas to Calculate Values

A formula can be used to do simple things, such as adding the values in two cells, or to perform much more complex operations. Here is how Microsoft Excel uses the SUM worksheet function to add two values together:

1. On a worksheet, write two numbers in A1 and A2, say 1 and 2.
2. Move the mouse to A3, and click.
3. Move the cursor to the tool bar and hit Σ. SUM(A1:A2) will appear in A3.
4. Hit Enter, and the sum will appear in A3.

A formula in Microsoft Excel can always be recognized because it starts with an equal sign (=). You can see the formula itself only when editing the cell that contains the formula. Otherwise, the value produced by the formula is displayed.

A formula can be created that adds the amounts in a column, and that formula can be copied into the other columns. To total the column, select the cell where the total will be placed, and click the AutoSum button. Microsoft Excel looks at the data around the selected cell and guesses that you want to add the column of numbers above this cell. It writes the formula, suggesting the range of numbers to sum. Press Enter to accept the proposed formula.

Formulas can be copied to different locations on the worksheet by dragging the fill handle or using Copy. The Function Wizard, located on the tool bar on the fx key, can simplify formula writing and also name cell references or ranges to create more understandable formulas.

When the Function Wizard opens, the function categories appear in the left box. Click on a category to bring up function names in the right box. The function name is explained in bold type below the category box. After highlighting category and function, click OK to finish and the formula will appear in the bold box on the worksheet and in the formula line above the cells. Editing can be done only in the formula line and acknowledged with the symbols to the left.

Absolute Versus Relative Cells

Absolute or relative cells are used when a formula requires reference to a variable cell or a particular range and are copied from one cell to another. Examples and illustrations follow.

Format	Relative/Absolute	Meaning
A6	Absolute/Absolute	Places data in a specific cell
$A6	Absolute/ Relative	Places data in a specific column but a relative row
A$6	Relative/Absolute	Places data in a specific row but a relative column

How to

Select contiguous cells	Select first cell; press and hold shift; select last cell.
Select noncontiguous cells	Select and hold CTRL; select individual cells.
Open contiguous workbooks	Select file open and select 1st file; press and hold last; click open.
Open noncontiguous workbooks	Select file open and select 1st file; press and hold CTRL and select other files to be opened; click open.
Arrange windows	Select windows-arrange; make selections and click OK.
Select a workbook	Select name of window file to become active.
Rename worksheets	Double click on sheet tab; type new name; press Enter.
Copy a worksheet	Open the workbooks; select sheet to be copied and select edit, move, copy sheet; select workbook from the "to book" drop down list; select

157

location from "before sheet" list box; select "create a copy" and click OK.

Format numbers

Select format-cells and click number tab; select desired format options and click OK or use toolbar and select cells; click the appropriate format style.

The basics for using Excel have been summarized above. If you are not familiar with a spreadsheet software package, open Excel and navigate through the basics. More advanced functions are explained through the highlighting on the various symbols or by referring to the Help functions.

4.7
Sample Problems

Examples 4.1 and 4.2 illustrate the appropriate approach to solving problems. Following these examples is a series of problems. An answer key and answer explanations are provided.

The first sample problem is a spreadsheet example.

Example 4.1

In a spreadsheet, the number in cell A3 is set to 5. Then A4 is set to A3 + A3, where $ indicates absolute cell address. This formula is copied into cells A5 and A6. The number shown in cell A6 is most nearly

A. 10
B. 20
C. 25
D. 35

Solution:

First,

	A	
3	5	
4		Insert A3 + A3.
5		
6		

Then,

	A	
3	5	
4	10	
5		COPY A4 + 3.
6		COPY A5 + 3.

	A
3	5
4	10
5	15
6	20

The answer is choice (B).

The second sample problem involves a program segment from a BASIC program.

Example 4.2

The program segment

```
INPUT X, M
S = 1
T = 1
FOR K = 1 TO M
T = T*X/K
S = S + T
NEXT K
```

calculates the sum

A. $S = 1 + XT + 2XT + 3XT + ... + MXT$
B. $S = 1 + XT + (1/2)XT + (1/3)XT + ... + (1/M)XT$
C. $S = 1 + X/1 + X^2/2 + X^3/3 + ... + X^M/M$
D. $S = 1 + M/1! + M^2/2! + M^3/3! + ... + X^M/M!$

Solution:

As the program steps from one instruction to the next, the contents remain from the preceding iteration. Therefore, as $T = T*X/K$ passes the first time, it is $X/1!$ $S = S + T$ becomes $S = 1 + X/1!$.

At K = 2, $S = 1 + X/1 + X^2/1*2$.

At K = 3, $S = 1 + X/1! + X^2/2! + X^3/1*2*3$.

The answer is choice (D).

4.8
Summary

This chapter provides brief highlights of rapidly changing technologies: computer programming and spreadsheet software. It is strongly suggested that your preparations for the exam include referencing additional material on generic programming and practice in using a spreadsheet package.

PRACTICE PROBLEMS

Number Systems (Refer to the NCEES *FE Reference Handbook*, "Electrical and Computer Engineering" section.)

1. Binary number 0101110 is what number in base 10?
   (A) 32
   (B) 46
   (C) 48
   (D) 56

2. Base-10 number 135 is equivalent to what binary number?
   (A) 011000011
   (B) 10000111
   (C) 010001010
   (D) 10000011

3. Base-5 number $(213144)_5$ converts to what binary number?
   (A) 0 0111 0110 1101
   (B) 1 0011 0111 1100
   (C) 0 1101 1001 0101
   (D) 1 1100 1000 0011

Computers

4. Which of the following best defines a compiler?
   (A) hardware that is used to translate high-level language to assembly language
   (B) hardware that collects and stores executable commands in a program
   (C) software that processes and stores executable commands in a program
   (D) software that is used to translate high-level language into machine code

5. How many times will the line labeled "START" execute in the following program segment?

   ```
         I = 2
         J = 1
   START  J = J + I
         I = J ^ 2
         IF J < 100 THEN GO TO START
         ELSE GO TO FINISH
   FINISH   PRINT J
   ```

   (A) 3
   (B) 4
   (C) 5
   (D) 6

6. Based on the following program segment, what is the DECISION?

   ```
   A = 1
   B = 2
   C = 3
   D = 4
   DECISION = I
   IF (A > B) OR (C < D) THEN DECISION = G
   IF (A < B) OR (C > D) THEN DECISION = H
   IF (A > B) OR (C > D) THEN A = 5
   IF (A < B) AND (C < D) THEN DECISION = J
   ```

   (A) G
   (B) H
   (C) I
   (D) J

Spreadsheets

7. In the following portion of a spreadsheet, the value of D1 is set to (A1 + B1 + C1)/3. This formula is copied into the range of cells D2:D3. The value of D4 is set to SUM(D1:D3)/4. What is the number in D4?

	A	B	C	D
1	1	2	3	
2	4	5	6	
3	7	8	9	
4				
5				

   (A) 8
   (B) 3.75
   (C) 15
   (D) 30

Questions 8–9

A segment of a spreadsheet is shown below. Use the numbers in the cells to answer Questions 8 and 9.

	A	B	C	D
1	20	21	22	23
2	5	A2^2	B2*A$1	
3	6	A3^2	B3*B$1	
4	7	A4^2	B4*C$1	
5	8	A5^2	B5*D$1	

8. What will be the top to bottom values in column B?
   (A) 25, 30, 35, 40
   (B) 25, 36, 49, 64
   (C) 40, 42, 44, 46
   (D) 5, 6, 7, 8

9. What will be the top to bottom values in column C?
   (A) 25, 36, 49, 64
   (B) 100, 126, 154, 184
   (C) 500, 756, 1078, 1472
   (D) 125, 196, 34

10. A partial spreadsheet is shown below.

	A	B	C	D
1	2	$A1		
2	4	B$1		
3	0	A2+B2		
4	–5	A$2*B2		??

The contents of column B are copied and pasted into columns C and D. What number will be the result in cell D4?
   (A) 8
   (B) 16
   (C) 0
   (D) 4

Answer Key

1. B		**6.** D	
2. B		**7.** B	
3. D		**8.** B	
4. D		**9.** C	
5. A		**10.** D	

Answers Explained

1. B $2^6*0 = 0$
$2^5*1 = 32$
$2^4*0 = 0$
$2^3*1 = 8$
$2^2*1 = 4$
$2^1*1 = 2$
$2^0*0 = 0$
    46

2. B $2^7*1 = 128$
$2^6*0 =$
$2^5*0 =$
$2^4*0 =$
$2^3*0 =$
$2^2*1 = 4$
$2^1*1 = 2$
$2^0*1 = 1$
    135

Base-2 equivalent is 10000111.

3. D Convert base 5 to base 10.

$(213144)_5$ $2*5^5 = 6250$
            $1*5^4 = 625$
            $3*5^3 = 375$
            $1*5^2 = 25$
            $4*5^1 = 20$
            $4*5^0 = 4$

$(7299)_{10}$ $2^{12}*1 =$
           $2^{11}*1 =$
           $2^{10}*1 =$
           $2^9*0 =$
           $2^8*0 =$
           $2^7*1 =$
           $2^6*0 =$
           $2^5*0 =$
           $2^4*0 =$
           $2^3*0 =$
           $2^2*0 =$
           $2^1*1 =$
           $2^0*1 =$

Base-2 equivalent is 1 1100 1000 0011.

4. D Compilers are software packages that convert programs. They take a high-level language and convert it into a machine code that the computer can understand (low-level language).

5. A Set up a table of values for I and J during each loop. J is indexed before I, so the latest or new value of J is used to calculate I.

Loop	J	I
0(initial)	1	2
1	3	9
2	12	144
3	158	24964
4		

After 3 loops, J > 100, so the program continues and prints.

6. D The program must be followed step by step from the first statement. The first and second IF statements are G and H. The third IF statement is not 5; however, the fourth IF statement is J. Therefore, DECISION is J.

7. B

	A	B	C	D	
1	1	2	3	**2**	
2	4	5	6	**5**	
3	7	8	9	**8**	Copy from D2 to D3.
4				**3.75**	D4 is SUM(D1:D3)/4 or 15/4 = 3.75

8. B

	A	B	C	D	
1	20	21	22	23	
2	5	**25***			* A2^2=5*5=25
3	6	**36**			
4	7	**49**			
5	8	**64**			

9. C

	A	B	C	D	
1	20	21	22	23	
2	5	**25**	**500**		
3	6	**36**	**756****		**B3*B$1=756
4	7	**49**	**1078**		
5	8	**64**	**1472**		

10. D

Copy and paste column B to columns C and D

	A	B	C	D
1	2	$A1	$A1	$A1
2	4	B$1	C$1	D$1
3	0	A2+B2	B2+C2	C2+D2
4	−5	A$2*B2	B$2*C2	C$2*D2

Calculate the values in the cells

	A	B	C	D
1	2	2	2	2
2	4	2	2	2
3	0	A2+B2	4	4
4	−5	A$2*B2	4	4

5

Ethics and Professional Practices

Thomas E. Hulbert, PE

Introduction

Ethics, as defined by *Webster's New World Dictionary*, is the system or code of morals of a particular person, religion, group, or profession. Engineering, being a profession, has its own code of ethics adopted by the National Society of Professional Engineers (NSPE). All Professional Engineers are bound by this code. This chapter includes a strategy for solving ethics questions on the exam with example problems. It also describes the requirements for contracts.

5.1 Engineering and Ethics

All of the occupations known as professions require advanced education and training and also involve intellectual skills. Traditionally, the professions have included law, medicine, theology, teaching, architecture, and engineering. For the most part, members of these professions are held in high esteem by the general public; they also have special responsibilities to the community. These professionals must keep their knowledge and skills current through continuing education, must participate in learned societies, must exercise professional judgment in their work, and must adhere to a professional code of ethics.

> Engineers and ethical dilemmas

Experienced members of a profession are best qualified to determine what training and what experience are necessary for new practitioners to enter their specific profession and its subspecialties. In this way, the professions are self-regulating. Professional boards and societies control the training and evaluation criteria by which new members are admitted and certified to practice. Once credentialed, however, professionals have independence in the workplace and are expected to exercise educated judgment.

Like all other professionals, engineers are highly trained and highly skilled. They are bound by the licensing and ethical rules of their society and exercise autonomy in the workplace. In addition, because the quality of engineering work can vitally affect public safety, engineers are often regulated by state and local statutes, which may include additional codes of ethics.

The professional services of engineers are used by all levels of government, by independent agencies, by the developers of commercial projects, and by individuals. The work of engineers is crucial to the public welfare. Engineers influence public safety through their choices of siting, design, materials specification, and construction methods. To serve the public effectively, engineers must maintain a high level of current technical competence. They must follow the literature, attend lectures and classes, and participate in seminars to follow advances in their fields.

Technical expertise, however, does not guarantee appropriate service. Strict compliance with ethical guidelines is essential to public welfare. Unethical behavior is tantamount to professional incompetence. Engineers must, therefore, be honest and ethical in all their judgments and dealings.

5.2
NCEES Model Rules of Professional Conduct

A code of ethics is a set of guidelines, not a list of regulations. Behaving in an ethical manner is a matter of "doing what is right," not of rigidly following a set of specific rules. Thus, the engineering ethics code spells out the parameters of proper behavior for engineers when acting in their professional capacities. Basically, engineering ethics is universal ethics with specific reference to the demands on and the responsibilities of the professional engineer.

The National Council of Examiners for Engineering and Surveying (NCEES) has promulgated a set of "Model Rules of Professional Conduct." All engineers must know this code, understand it, and follow the guidance that it provides. You will find the code reproduced in the NCEES *FE Reference Handbook* at the exam. However, for purposes of the exam, as well as for your practice of engineering, knowledge of the code is not enough. You must be able to apply the principles of the code to practical problems and to unusual situations on the exam and in your professional life.

The Preamble

The preamble, that is, the introduction, states the purpose of the code and briefly describes the model rules and the ways in which they must be applied.

- The "Model Rules of Professional Conduct" were developed to comply with the main purpose of state laws: to safeguard life, health, and property and to promote the public welfare. The rules simply translate this purpose of state laws into a code for engineers. The rules are binding on all registered engineers and surveyors, who must maintain high standards of ethical and moral conduct.
- The practice of professional engineering is not a matter of right but is a privilege that must be earned. Once an engineer has, through training and registration, earned the privilege to practice the profession, he or she must perform services only in areas of his or her current competence and to the highest standards. The engineer must beware of false or boastful advertising and must be open, objective, and truthful in representation to the public.
- Conflict of interest is unethical.
- Unfair competition is unethical.

Obligations to Three Constituencies

According to the Model Rules of Professional Conduct, licensed professional engineers (the designating term is "registrants") have major obligations to three discrete constituencies in this order of importance: first to society; second to employers and clients; and third to other registrants.

> Engineers' responsibility to society, employers, and fellow engineers

I. Registrants' Obligations to Society
This section makes clear that public safety and public welfare are higher considerations than money making or meeting deadlines for clients or employers. It further requires registrants to be forthright in all their dealings with governmental authorities and with the public and to limit their professional activities to their areas of competence. It stresses that registrants must avoid conflicts of interest or the appearance of such conflicts.

a. Registrants bear responsibility for the public welfare during and as an outcome of the performance of their services. Even when they are engaged by clients, employers, or customers, first consideration is to public safety and the public welfare.

b. Registrants must think beyond the requirements of a specific project toward the long-term health and welfare of the public. For example, when any number of materials might meet the needs of a particular enterprise and even exceed safety standards for the job, the engineer should be mindful of long-term welfare and should give adequate consideration to using recycled or recyclable materials if available and equally applicable.

c. If a client or employer overrules an engineer's professional judgment in a manner that adversely affects the public's life, health, welfare, or property, the engineer must report the action to the authorities. The registrant must safeguard the public even if such action displeases the hiring body. Safety takes precedence over cost cutting. If the engineer's professional judgment cannot prevail, he or she must "blow the whistle."

d. Registrants often find that they must submit professional reports or statements or give testimony at civil trials. These reports, statements, and testimonies must be complete, objective, and honest, including all relevant and pertinent information of which the registrant is aware.

e. Engineers are often asked for their professional opinions with regard to various proposals, projects, and actions. They must refrain from expressing public opinions unless they have adequate knowledge of all the facts, are competent in the particular field under question, and have completely evaluated the subject matter.

f. "Registrants shall issue no statements, criticisms, or arguments on technical matters which are inspired or paid for by interested parties, unless they explicitly identify the interested parties on whose behalf they are speaking and reveal any interest they have in the matters." The effect of this rule is to prohibit the giving of "paid endorsements" unless the engineer clearly identifies the sponsor.

g. Registrants must deal with only honest, ethical individuals and firms. Engineers must be careful to avoid associating their names with persons or businesses of unsavory or even dubious reputation.

h. If a registrant has reason to suspect another engineer or firm of violation of any one of these rules, the registrant is required to notify the state board and, if requested, to assist the board in making its determination of violation.

II. Registrants' Obligations to Employers and Clients

This section emphasizes that registrants should provide services and advice only in areas in which they are completely competent. A second theme in this section is a requirement of confidentiality on the part of registrants. Finally, the section returns to the need for avoiding conflicts of interest or even the appearance of such conflicts and urges full disclosure of all contacts that may imply conflict of interest.

a. The first statement in this section is so obvious that it needs no explanation. It reads in its entirety: "Registrants shall undertake assignments only when qualified by education or experience in the specific technical fields of engineering or land surveying involved."

b. Beyond not taking on assignments for which they are not fully qualified, registrants must not approve any plans or documents that deal with subject matter beyond their areas of competence. Furthermore, they must not sign or seal any plans or documents not prepared under their direct control and personal supervision, even if they are fully competent to do so. Registrants must sign and seal only work for which they can honestly assume responsibility.

c. Although registrants may not approve plans drawn by engineers not under their supervision, they may assume coordination of entire projects in which each design segment is signed and sealed by a fully competent engineer who has been responsible for preparation of that design segment.

d. Like all other professionals, engineers must maintain confidentiality in their relationships with clients. Information given about a project is privileged information. A registrant must obtain prior consent from a client or employer before disclosing any data or other information about a project.

e. Obviously, registrants should not accept bribes. Furthermore, engineers must not accept gifts, money, or promises of future benefits from contractors or contractors' agents or from anyone else in a position to influence purchases, awards, or special treatment from these individuals. Even if the registrant would not be influenced by such gifts, the appearance of impropriety or conflict of interest would damage confidence.

f. "Full prior disclosure" is a phrase to keep in mind for the purpose of avoiding misunderstandings and misinterpretations. If there is any possibility of a potential conflict of interest or any circumstance that might appear to influence a registrant's judgment or quality of service, the registrant should tell the employer or client in advance. Full prior disclosure is a sure means for maintaining trust.

g. A general rule of thumb is that one cannot serve two masters. Application of the rule is that a registrant should not be paid by two parties for services with regard to the same project unless all parties know of and agree to the arrangement.

h. Although it is possible for a registrant to work for more than one private party to the same project if all parties agree (as provided in rule **g**), this arrangement is absolutely prohibited if one of the parties is a governmental body. Because this rule is strictly enforced, you must familiarize yourself with its precise language. In its entirety, the rule states, "Registrants shall not solicit or accept a professional contract from a governmental body on which a principal or officer of their organization serves as a member. Conversely, registrants serving as members, advisors, or employees of a governmental body or department, who are the principals or employees of a private concern, shall not participate in decisions with respect to professional services offered or provided by said concern to the governmental body which they serve."

III. Registrants' Obligations to Other Registrants

This section prohibits false advertising, bribery, and influence peddling and also prohibits damaging the reputation of other registrants.

a. Registrants must not brag or inflate their importance when soliciting or bidding for contracts. They must describe their qualifications accurately and state all aspects of previous experience and assignments clearly. In competing for commissions, registrants must not overstate their credentials to the detriment of the competition.

b. Gaining assignments through bribery is also a form of unfair competition. The rule is very specific in prohibiting political contributions for the purpose of influencing the granting of contracts.

c. Finally, innuendo or slander to denigrate a competitor is unprofessional and unethical. This final rule brings us back full circle to our original statement that engineering ethics is ethics with specific reference to engineering situations. You should recognize this rule as "Thou shalt not bear false witness against thy neighbor."

Note the reference to Sustainability at the end of the "Model Rules" in the NCEES *FE Reference Handbook*.

5.3
Resources for Engineering Ethics

You will find that the World Wide Web is an excellent resource in your preparation for ethics problems on your exam. The rules are easy to understand. Their application to real situations, however, may be ambiguous. The NSPE's Board of Ethical Review maintains a website at *www.nspe.org*. This website features numerous ethics case studies published by the board. The "ethics" button at the NSPE website offers links to the websites of Case Western Reserve University's Ethics Center, the National Institute for Engineering Ethics, and Texas A&M University ethics case studies. As you study previous cases and their determinations, you will develop the ability to analyze and solve problems on your own.

Additional references are available in several texts on engineering ethics. The Cambridge University Press has three books on ethics. The first is *Engineering, Ethics, and the Environment* by Vesilind and Gunn. It is the only book devoted to environmental ethics for engineers. The second is *Ethics in Engineering Practice and Research* by Whitbeck. It includes real-world case studies and example problems on engineering ethics applications and issues. The third is *Engineering Ethics* by Pinkus, Shuman, Hummon, and Wolfe. The cases are centered on the space shuttle and show examples of balancing costs, schedule, deadlines, and risk. A fourth book is published by McGraw-Hill and titled *Ethics in Engineering* by Martin. Current issues are presented in many case studies covering moral as well as technological conflicts. These references are useful for engineers with limited exposure to ethical dilemmas and issues and how to resolve them.

5.4
Problem-Solving Strategy for the FE Exam

The problems on the FE Exam reflect the problems an engineer faces in real life. On the surface, the rules of ethics seem clear. In practice, however, you will often find that circumstances do not fit neatly into a category, that words can be interpreted in more than one way, or that the facts of a case are not complete or may even be in conflict. In addition, identifying the greatest good is not always easy.

Solving ethical problems

The exam problems seek to determine how well you might analyze and solve actual problems that could arise in your work as an engineer.

Here is how you should approach the exam problems:

- Read the entire problem. Reread. Be sure you understand the facts and the question.
- Eliminate any choices that are obviously ridiculous or clearly unethical.
- Pay particular attention to the following: conflict of interest issues, contract or agreement disputes as governed by the code, and accuracy and proper representation of individual and professional information.

Now try the questions presented in Examples 5.1 and 5.2. These are typical questions.

Example 5.1

An engineer who enjoys a fine reputation for his many years of work in modernizing and reconfiguring large public buildings and who is currently engaged in the conversion of an old school into a condominium is asked by a neighbor to renovate her 80-year-old, 12-room colonial. What should the engineer do?

(A) Hire workers experienced in home renovation, and sign a contract with the neighbor to complete the job.

(B) Sign a contract with the neighbor, explaining that his experience is with larger projects.

(C) Turn down the job.

(D) Sign a contract only if he is certain that the neighbor understands that he must devote most of his time to the larger project.

Solution:

Consider the choices:

(A) is not permitted. An engineer must be personally competent to oversee the job. Workers are not licensed to sign off on plans or product.

(B) is not a good choice. Full disclosure of incompetence is not an option under the code.

(C) is what the engineer must do. He must not accept an assignment for which he is not competent.

(D) introduces a twist on conflict of interest—in this case, a conflict of time. The engineer has first responsibility to the client for whom he is already working.

Example 5.2

You are the engineer responsible for the design and installation of plumbing in a new public elementary school under construction. The project is running behind schedule and above budget. The latest catch is that the supplier has not delivered the specified backflow preventers. Although a strike at the manufacturer's plant has been settled, the back order list is moving slowly. The engineer serving as coordinator of the project orders you to obtain less expensive backflow preventers that are readily available from another manufacturer. These backflow preventers are of lesser quality than those specified in the plans, have a history of occasional failure, and are likely to need earlier replacement. What should you do?

(A) Order and install the available backflow preventers but refuse to sign off on this portion of the job.

(B) Order and install the available backflow preventers, informing the school board that this equipment may not last as long but will help the project stay on schedule and closer to budget.

(C) Order and install the available backflow preventers, and say nothing to the school board.

(D) Wait for the specified backflow preventers and explain your action to the school board.

Solution:

Again consider the choices:

(A) may seem like a good course of action but is not. If you are in charge of this segment of the job, you are responsible. If you order and install the backflow preventers, you must sign off. However, the alternative backflow preventers may fail and may contaminate the school's drinking water. Obviously, the health and safety of the schoolchildren are more important than the schedule or the budget.

(B) is a cop-out. Full disclosure that the health and safety of the public may be jeopardized by your action is not an option; it is an ethical requirement.

(C) is an obviously unethical action.

(D) is the ethical course. Your coordinating engineer may be hard on you for a while. Health and safety, however, are your first considerations. The school board should appreciate your responsible choice.

5.5
Contracts

Contracts are involved and required on most engineering projects. They are developed for the protection of two or more parties. One side offers and the other accepts the terms and conditions. Contracts may be oral or written. However, to resolve disputes in court or other resolution processes, contracts must be written and properly executed. They must contain the following elements: mutual agreement, lawful subject matter, valid consideration, legally competent parties, and compliance with provisions of the law. Finally, formal contracts must be in a legal format. In contrast, informal contracts do not depend or rely on formality.

Contracts have become more prevalent and required in today's society because of many issues.

Environmental issues: laws exist but the project can still have adverse effects on the environment. Emerging "green" laws and programs must be considered in all negotiations.

Sustainability: Ethical responsibility may go beyond the legal requirements in order to prevent environmental damage.

Alternative energy sources: They must be taken into consideration as they become more cost effective now and in the future.

5.6
Summary

The three major sections of the Model Rules of Professional Conduct address the three primary obligations of the engineer:

1. Obligation to society.
2. Obligation to employers and clients.
3. Obligation to other registrants.

The principles articulated in these sections define the appropriate behavior of professional engineers. Although the meaning of the code is clear, questions may arise concerning its application to specific situations.

In real life, you may have to make judgment calls. "Does this token gift from a prospective contractor constitute the appearance of undue influence?" "Is my expertise on this particular building material adequate for me to serve as a witness in a liability case?" "How do I weigh the risks against the benefits of a certain course of action?" "For this situation, is the safety versus economics tradeoff an ethical one?"

On the exam, you will be asked questions to prove that you know the provisions of the code and that you can solve ethical problems that arise in the workplace.

PRACTICE PROBLEMS

1. What is the best word to complete the following sentence: "Engineers are to uphold the health, safety, and public _____"?
   (A) trust
   (B) good
   (C) confidence
   (D) welfare

2. The order of ranking of ethical responsibility of the professional engineer, from highest to lowest importance, is
   (A) society as a whole, the client, the profession, oneself
   (B) oneself, the client, society as a whole, the profession
   (C) oneself, the client, the profession, society as a whole
   (D) the client, society as a whole, the profession, oneself

3. An engineer employed at a large manufacturing company has been involved in developing a new production technique. After several years, she decides to leave and to establish her own consulting practice. A competitor of her former employer contacts her and requests that she solve a similar manufacturing problem. All indications are that the original design she developed for her previous employer is also the best solution for the competitor. Can the engineer accept the assignment?
   (A) Yes, provided that she left her previous employer with no written nondisclosure agreement.
   (B) No, information from the original employer would have to be used in the new design.
   (C) Yes, consulting engineers work for competing companies and transfer of knowledge naturally occurs.
   (D) No, this would constitute accepting money more than once for the same project.

4. An engineer working for a large consulting firm has decided to establish his own business. Although the new business is several months from start-up, the engineer should
   (A) approach current colleagues to tell them of his new plans
   (B) discuss his plans with his current supervisor
   (C) resign immediately
   (D) start developing promotional literature while he has use of his current employer's computers

5. An engineer employed by a Big Three automaker is responsible for redesigning the steering system. While working on the new design, she discovers a defect in the steering system currently in production. After a thorough statistical analysis, she concludes that the defect might contribute to fatalities. She should
   (A) first inform the company of the defect and then notify the National Transportation Safety Board to initiate an inquiry
   (B) send an anonymous tip to the consumer "hot line" on a national radio network
   (C) continue to follow the company policy of selecting the most cost-effective design
   (D) inform her supervisor and resign from the company if it does not implement a new design

6. The principal in an 18-month-old engineering firm has developed a promotional brochure for the firm. The brochure contains a "list of clients," implying that companies on the list are clients of the firm, and a "list of projects of the firm," implying that these projects were performed by the firm. All of the engineers have come from other firms. In fact, the client list is actually a list of the companies at which the engineers were previously employed. Similarly, the project list is a list of the projects performed by the firm's engineers at their previous places of employment. It is
   (A) unethical to produce a promotional brochure in which the new firm claims for itself projects done by its engineers at their former firms and claims their former firms as clients
   (B) acceptable to publish such lists since long-established business relationships exist among engineering firms, engineers, and their clients
   (C) ethical to take credit for projects performed by another firm or group of engineers as long as the engineers are now members of the new firm
   (D) not misleading to refer to "projects of the firm" in the new firm's brochure because they were, after all, projects of the firm's members

7. An engineering firm is hired to develop plans for an office/retail complex. Prior to the bid date, several contractors ask the engineers for details about the plans and the project. The firm
   (A) cannot reveal information about the project that might prejudice any contractor against bidding on the job
   (B) should provide information to bidders in whom it has confidence to complete the contract
   (C) cannot reveal data or information, without consent of the client, unless specifically required by law to do so
   (D) can supply the requested information as long as it is accurate

8. A licensed professional engineer and land surveyor works 35 hours per week on a flex-time basis for a state governmental agency. In addition, he is associated with XYZ Engineering and Surveying as the PE in charge. He spends about 20 hours per week supervising engineering services at the firm plus an additional 12 hours of work on the weekends and is available for consultation 24 hours a day. XYZ has granted him 10% of the shares of the stock in the firm, and he receives 5% of the gross billings for his seal as extra compensation for taking charge of engineering. This agreement is contingent on the understanding that, if any of the three principals of XYZ becomes licensed as a PE, the agreement will become void and the engineer will return the stock. Both organizations—the state governmental agency and XYZ—are aware that he is a dual employee and have no objections. If this arrangement comes to the attention of the Board of Ethical Review, the board will determine that
   (A) a flexible schedule is not sufficient grounds for dual employment
   (B) the engineer is stretching the role of responsible charge and cannot assume this role for two employers
   (C) it is acceptable and ethical to be involved with both the state governmental agency and XYZ Engineering and Surveying in the manner described
   (D) the engineer cannot work for a government agency and a private organization at the same time

Answer Key

1.	D	5.	A
2.	A	6.	A
3.	B	7.	C
4.	B	8.	C

Answers Explained

1. **D** The public good and the public welfare are practically synonymous. Since "public welfare" is the term used in the Model Rules, it is the better choice.

2. **A** The code stresses that the order of responsibility is to society as a whole, the client or customer, and then to professional colleagues. The code makes no mention of self-interest. Obviously, then, that must come last.

3. **B** The code is very specific: "Registrants shall not reveal facts, data, or information obtained in a professional capacity… ." Even without a written nondisclosure agreement, the engineer is bound to maintain confidentiality.

4. **B** The code requires full disclosure and frankness with regard to possible future conflicts of interest. Current colleagues may be the second to know. There is no reason for the engineer to leave his current job before the new business is in place as long as he does not use his current position as a springboard for lining up his own clients. Choice (D) represents a dishonest use of the employer's facilities and a form of unfair competition.

5. **A** Safety must come first. The engineer is required to report unsafe practices to the authorities. No company wants to expose itself to product liability lawsuits. When an expensive recall might be involved, though, the company might well attempt to discount an alert engineer's fears. The engineer must first alert the company to the problem and cooperate in trying to find a solution consistent with maintaining the public welfare. However, if the company resists taking appropriate measures, the engineer must "blow the whistle."

6. **A** Despite the mitigating language in some other choices, it should be obvious that (A) is the correct choice. The brochure constitutes false advertising at its worst.

7. **C** Choice (D) looks good, but (C) meets the confidentiality requirement. Both (A) and (B) additionally violate the requirement for fair dealing among contractors.

8. **C** The problem does not say that XYZ Engineering and Surveying is doing any work for the state governmental agency, so there is no question of conflict of interest here. The engineer has two part-time jobs, which do not involve the same projects, and both employers have agreed that this situation is acceptable.

Engineering Economics

Thomas E. Hulbert, PE

Introduction

Engineering designs are subject to many levels of review. The design must meet the physical needs of the project, must utilize appropriate materials, must result in a structurally sound product, must be aesthetically pleasing, and must be economically feasible. If an engineering review indicates that the project meets all other criteria, the project must then be evaluated in financial terms. Beyond simply understanding the costs of materials, the engineer must have a firm grasp of labor, maintenance, insurance, interest, and depreciation costs or charges.

The material in this chapter should serve as a general review for the Fundamentals of Engineering examination. Since you will be taking this examination with an open copy of the "National Council of Examiners for Engineering and Surveying (NCEES FE) Reference Handbook," the topics of this chapter follow the book. Be sure to review this material before you attempt to solve the problems at the end of the chapter. If you feel that you need more in-depth instruction, refer to a textbook on the subject of engineering economics. Five helpful references are listed in Section 6.14.

6.1
Fundamental Concepts

Engineering economics is based on two fundamental concepts: equivalence and money increase in value over time, that is, future value. **Equivalence** refers to

Equivalence

the fact that various cost-benefit statements and differing cash-flow patterns may actually be equivalent, though not necessarily equal. **Future value** refers to the fact that money earns interest and thus increases its value over time. In engineering economics calculations, time is assumed to be one year, unless otherwise specified, and calculations are considered to have been made at the end of the year unless otherwise stated. Interest is the difference between the ending amount of money (future value) and the initial amount of money (present value). The interest rate for a given period would be the ratio of this difference divided by the present value.

For a simple illustration of equivalence, consider the following two alternatives:

Alternative A: Receive $100 today.
Alternative B: Receive $110 one year from today.

The choice between A and B is not clearly evident. If you choose B, you indicate that your minimum acceptable rate of return is 10% per year. If you choose A, your rate of return is indefinite—it may be greater or less than 10%. Some individuals may be indifferent toward the choice between A and B. For them, A and B are equally attractive or equivalent even though the two alternatives involve different amounts of money at different times and may not, in the final analysis, prove to be equal.

Another example of equivalence is the exchange of U.S. dollars into other currencies. In the exchange, only the exchange rate itself is considered, not the buying power of either currency. The amount of foreign currency received is equivalent to the amount of U.S. dollars, but it is not necessarily equal in purchasing power.

In summary, cash flow is the monies generated from or invested in a project. Costs are the imput in dollars to the project, and the output is the benefits derived from the project. Time value of money is the worth of money as it "moves back" to the present. Time value is defined as the changing value of money over time caused by changes in the purchas-

ing power or from real earning potential of alternative investments over time. Monies increase in value over time for many reasons. These include inflation, risk, and cost of money. The cost of money (interest rate) is somewhat predictable and is the primary component used in economic analysis. Cost of money is the amount paid for use of money or the return on investment.

6.2
Nomenclature and Definitions

The following symbols are listed in the NCEES *FE Reference Handbook* and are those used in engineering economics problems and texts. The explanations of some symbols have been expanded for greater clarity.

A Uniform amount per interest period

B Benefit or savings derived from public projects

BV Book value, Remaining value after the depreciation charge for the year

C Initial value or present worth of all costs; the sum of costs in public works projects (see B above)

d Combined interest rate per interest period

D_j Depreciation in year j

F Future worth, value, or amount

f General inflation rate per interest period (treated same as i)

G Uniform gradient amount per interest period

i Interest rate per interest period (years unless otherwise stated)

i_e Effective interest rate (annual)—used in engineering economics studies, unless otherwise stated

m Number of compounding periods per year

n Number of compounding periods or expected life of asset (usually years)

P Present worth, value, or amount

r Nominal annual interest rate or rate per annum

S_n Expected salvage value in year n

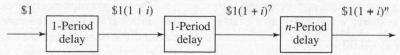

Note: Delays are normally expressed in years.

6.3
Nominal and Effective (Nonannual Compounding) Interest Rates

The basis of engineering economics calculations is the same as that used by financial institutions in their calculation of interest.

Refer to the figure at the top of this page.

At the end of the first year there will be $1(1 + i)$. At the end of the second year, there will be $1(1 + i)^2$. Therefore, at an interest rate of i, $1 today is equivalent to $1(1 + i)^n$ received n years from today or at the end of n periods.

In many financial transactions, an annual nominal interest rate is specified, but the compounding process actually occurs several times throughout the year. For example, if a nominal interest rate of 12% per year is compounded semiannually, the effect is to pay a 6% interest rate per 6 months for two 6-month periods. If $1,000 is invested under these conditions, it will become $1,000 \times 1.06^2 = \$1,123.60$. The effective interest rate is $(1.06^2 - 1) = 0.12360$ or 12.36%.

If the nominal interest rate is r per year compounded m times per year, the effective annual interest rate is

$$i_e = \left(1 + \frac{r}{m}\right)^m - 1$$

As a second example, suppose a finance company charges 1.5% per month compounded monthly. The nominal interest rate is $1.5\% \times 12 = 18\%$, but the effective interest rate is actually $(1.015)^{12} - 1 = 0.1956$ or 19.56%.

6.4
Cash-Flow Diagrams

The flow of money is shown on a diagram known as a cash-flow diagram (Figure 6.1). Money at present, or time = 0, has a present value, which is designated as P. Uniform annual receipts or disbursements are designated as A; and money in the future has a future value, which is designated as F.

Figure 6.2 illustrates the gradient—a linearly increasing amount of money starting at time = 0 with the gradient line at 0. The cash-flow diagrams are represented as follows:

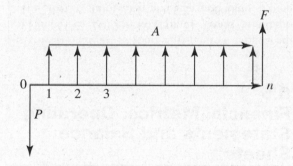

Figure 6.1 Cash-flow diagram

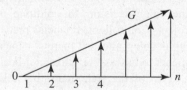

Figure 6.2 Gradient

6.5
Interest Formulas and Tables

The NCEES *FE Reference Handbook* includes the eight commonly used factors along with their original names, the current expression (symbol) for each factor, and the algebraic formula for each. You can solve problems either by searching a table or by using a calculator with the formulas. A complete set of the factor tables provided for the exams is included at the end of this chapter.

A portion of a factor table for 8% interest is shown below. The left column represents the time period, generally in years. The four factors and their respective values are shown in the matrix of the table.

Factor Table for *i* = 8.00%

n	P/F	P/A	P/G	F/P
1	0.9259	0.9259	0.0000	1.0800
2	0.8573	1.7833	0.8573	1.1664
3	0.7938	2.5771	2.4450	1.2597
4	0.7350	3.3121	4.6501	1.3605
5	0.6806	3.9927	7.3724	1.4693
6	0.6302	4.6229	10.5233	1.5869
.	.	.	.	.
.	.	.	.	.
.	.	.	.	.
30	0.0994	11.2578	103.4558	10.0627
40	0.0460	11.9246	126.0422	21.7245
50	0.0213	12.2335	139.5928	46.9016
60	0.0099	12.3766	147.3000	101.2571
100	0.0005	12.4943	155.6107	2,199.7613

6.6
Financial Metrics: Operating Statements and Balance Sheets

In evaluating engineering designs, financial evaluations have to be assessed and considered before making the final decision to continue and complete a project. Controllers, financial managers, and accountants develop and use many financial reports and statements to assess the "financial health" of an enterprise. Two of the statements that relate to the evaluation of engineering designs after traditional engineering economic analysis is completed are the operating statement and the balance sheet.

Accounting and Engineering Economics

The Operating (Profit and Loss) Statement is an itemized list of income or sales, usually for a month. Also, a corresponding list of expenses is tabulated. To calculate gross profit, subtract costs (labor, material, selling and administrative expenses, depreciation (see below) from the sales for the period. After deducting state and federal taxes, the result is Profit after Tax. Net Profits are then distributed to stockholders, employees, and retained earnings (see Balance Sheet). The portion of the profit going to retained earnings provides funds for acquiring assets such as the purchase of capital equipment.

The depreciation charge (expense) is a systematic write-off of the asset over time. The system is passed by Congress and administered by the IRS. The capital asset is purchased and paid for and then charged, usually monthly, on P & L as an expense using the Modified Accelerated Cost Recovery System (MACRS). See Section 6.7.

The retained earnings show on the Balance Sheet. The Balance Sheet is an itemized listing of all of the net assets of the company. This is what the company owns. The other side is all of the liabilities, what is owed by the company. The retained earnings are a portion of the owners' equity. The fundamental accounting equation is:

ASSETS = LIABILITIES + OWNERS' EQUITY

Increases in retained earnings can result in opportunities for capital investment, thus increasing assets and liabilities as well as owners' equity while maintaining a balance in the above equation.

These two statements plus other financial data are used to measure the financial growth and stability of a company through the calculation and comparison of a number of ratios. These can be found in accounting books but are beyond the knowledge required for the exam.

6.7
Depreciation Methods

The value of the client's existing assets—heavy equipment, machinery, and tools—is one of many factors entering into your analysis of the economic feasibility of a project. It is likely that some of these assets were placed in service years ago and will have already been partially, or even fully, depreciated. You must review the depreciation method used by your client, because in a challenger-defender analysis the depreciation method employed at the start must be continued.

In the accounting world, depreciation is an expense. As previously stated, it is a systematic write-off of an asset over time. The Internal Revenue Service Code specifies the available systems of depreciation and the lifetimes allotted to various types of physical assets. The taxpayer may choose from among a limited number of optional methods. The after-tax analysis reduces the rate of return. Since depreciation charges are not linear (except in straight-line depreciation), you must calculate their impact on the rate of return.

<table>
<tr><td style="background:gray;">Calculation of depreciation charges</td></tr>
</table>

You must be familiar with two depreciation methods in order to tackle problems on the FE exam.

The first is **straight-line depreciation** (*SL*)

$$SL = \frac{FC - SV}{n}, \text{ same for every year}$$

where FC = first cost,
SV = salvage value,
n = total number of years.

The modified ACRS has been in effect since 1986 and is the second method used. See the table below.

MACRS Factors Table

Year	Recovery Period (years) 3	5	7	10
	Recovery Rate (years)			
1	33.3	20.0	14.3	10.0
2	44.5	32.0	24.5	18.0
3	14.8	19.2	17.5	14.4
4	7.4	11.5	12.5	11.5
5		11.5	8.9	9.2
6		5.8	8.9	7.4
7			8.9	7.4
8			4.5	6.6
9				6.6
10				6.5
11				3.3

The above is a part of the table for MACRS depreciation. Note that there is a depreciation factor for one year following the number of years write-off. For example, a 5-year asset is written off over 6 years. MACRS assumes the midyear convention where assets are charged off beginning July 1 resulting in a one-half year residual charge for the $n + 1$ year (in this example, year 6 depreciation is provided for the 5-year asset which is one-half the 5th year percentage).

Examples of the type of equipment or asset written off for different number of years are listed below:

5-year property—Automobiles, computers and peripheral equipment, property used in research and experimentation

7-year property—Office furniture and fixtures

15-year property—Land improvements, roads, pipelines, power production, and phone distribution equipment

27.5-year property—Residential rental property

39-year property—Nonresidential real estate, including home offices (Note: the value of land may not be depreciated.)

There is an alternative MACRS depreciation system (known as ADS), where depreciation is deducted over generally longer periods than under the normal MACRS, using the straight-line method described above.

6.8
Bonds

Bonds are similar to stocks in that their values or prices fluctuate up and down with the market. Bonds, however, are somewhat more secure in that they have a par or purchase price that is guaranteed if the bonds are held to maturity. In return for the security, the interest is set at an unvarying rate, which the bond-owner earns as long as he or she holds the bonds.

Once a bond has been issued, its price is free to fluctuate; bond yield is influenced by bond price. The bond yield is computed on the basis of comparison of annual interest to current cost. You should set up cash flow diagrams to solve problems concerning bond yields.

Bonds are often issued to finance construction projects. The interest paid out to bondholders must be figured into the cost of the project.

6.9
Benefit–Cost Analysis

In a simple benefit–cost (*B/C*) analysis, the benefits (*B*) of a project should exceed the estimated costs (*C*). This calculation is based on a ratio of benefits to costs or of incremental benefits to the corresponding incremental costs. The higher the ratio (> 1.0) or the larger the positive difference, the more attractive the project will be to the government or nonprofit agency funding the project. Benefit–cost analysis is used only in situations where no taxes are involved and where a rough estimate is adequate.

6.10
Capitalized Costs

Capitalized costs are the annualized expenditures, such as maintenance, replacement, and operating costs, necessary to keep an asset in service in perpetuity, or forever. To calculate capitalized costs, determine the annualized cost and divide by the interest rate:

$$\text{Capitalized costs} = P = \frac{A}{i}$$

where P = present sum
    A = annual costs
    i = interest rate

6.11
Inflation

In economic analysis you can treat inflation as simply another interest rate. To account for inflation, deflate future dollars by the general inflation rate. Alternatively, you can use the inflated interest rate, which includes both the influence of inflation and an inflation-free interest rate, in your calculations.

6.12
Methods of Analysis

- Present worth analysis (when comparing alternatives that have the same lives).

    For each alternative, start with the initial investment at time = 0 as a negative value. Then add the discounted values of all revenues for that alternative and subtract the discounted values of all costs over the lifetime. Select the alternative having the highest Net Present Worth.
- Annual worth analysis (when alternatives have different lives).

    Convert all revenues and costs to their equivalent annual values. Add the annual values of the revenues (positive values) and costs (negative values) and select the alternative having the highest Net Annual Cost. For situations where costs dominate, one could determine the Annual Cost Analysis (costs assumed positive and revenues negative) and select the alternative having the lowest Net Annual Cost

- Rate-of-return analysis (when interest rate is unknown)

    The rate of return on an investment is the interest rate that makes the total annual worth of all revenues equal the total annual worth of all costs. Convert all cash flows to annual costs, and create an expression for revenues minus costs. Plug in values for interest (i) by trial and error. When the value switches signs (+ to −), linearly interpolate, using successive factor tables to determine the interest rate or rate of return.
- Rate of return on incremental investment

    Consider alternative designs or production methods to solve the problem. For example, suppose that a new machine must be purchased for production of a particular part and that three machines are under consideration. To calculate the return on the incremental or added investment:

    1. List the alternatives in hierarchical order of capital cost (investment), beginning with the least costly alternative.
    2. Given revenues and costs for each, calculate a rate of return (or interest rate) for each alternative so as to make the present worth of that alternative equal to zero, using the factor P/A.
    3. Compare the rate of return for each alternative with the minimum acceptable rate of return for the enterprise.
    4. Eliminate from consideration any alternative that does not meet the minimum rate of return.
    5. Of the remaining alternatives, compare the one with the lowest investment to the one with the next higher investment by considering the incremental (or added) investments and the incremental revenues. For each alternative, subtract the incremental investment from the incremental cost and calculate the rate of return.
    6. Select the alternative that yields the highest return on added investment.

- Break-even analysis

    On the same graph, plot the fixed and variable costs to produce a product and the revenues generated from the sale of that product. The intersection of the total cost (sum of fixed costs and variable A) and revenue lines is the break-even point. Profits are generated above the break-even point; losses below. See the example in Figure 6.3.

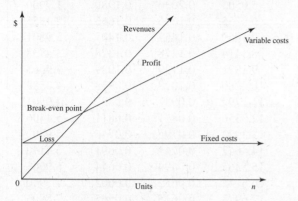

Figure 6.3 Break-even analysis

6.13
The Factors and
the Factor Tables

The algebraic formulas for factors below are given in the reference manual.

> Eight factors at various interest rates

Factors

(P/F) Present Worth given Future Worth

(P/A) Present Worth given Annual Cost/ Revenue

(F/A) Future Worth given Annual Cost/ Revenue

(A/P) Annual Cost/Revenue given Present Worth

(P/G) Present Worth given Gradient Value

(A/F) Annual Cost/Revenue given Future Worth

(F/P) Future Worth given Present Worth

(A/G) Annual Cost/Revenue given Gradient Value

Tables

$i = 0.50\%$

n	P/F	P/A	P/G	F/P	F/A	A/P	A/F	A/G
1	0.9950	0.9950	0.0000	1.0050	1.0000	1.0050	1.0000	0.0000
2	0.9901	1.9851	0.9901	1.0100	2.0050	0.5038	0.4988	0.4988
3	0.9851	2.9702	2.9604	1.0151	3.0150	0.3367	0.3317	0.9967
4	0.9802	3.9505	5.9011	1.0202	4.0301	0.2531	0.2481	1.4938
5	0.9754	4.9259	9.8026	1.0253	5.0503	0.2030	0.1980	1.9900
6	0.9705	5.8964	14.6552	1.0304	6.0755	0.1696	0.1646	2.4855
7	0.9657	6.8621	20.4493	1.0355	7.1059	0.1457	0.1407	2.9801
8	0.9609	7.8230	27.1755	1.0407	8.1414	0.1278	0.1228	3.4738
9	0.9561	8.7791	34.8244	1.0459	9.1821	0.1139	0.1089	3.9668
10	0.9513	9.7304	43.3865	1.0511	10.2280	0.1028	0.0978	4.4589
11	0.9466	10.6770	52.8526	1.0564	11.2792	0.0937	0.0887	4.9501
12	0.9419	11.6189	63.2136	1.0617	12.3356	0.0861	0.0811	5.4406
13	0.9372	12.5562	74.4602	1.0670	13.3972	0.0796	0.0746	5.9302
14	0.9326	13.4887	86.5835	1.0723	14.4642	0.0741	0.0691	6.4190
15	0.9279	14.4166	99.5743	1.0777	15.5365	0.0694	0.0644	6.9069
16	0.9233	15.3399	113.4238	1.0831	16.6142	0.0652	0.0602	7.3940
17	0.9187	16.2586	128.1231	1.0885	17.6973	0.0615	0.0565	7.8803
18	0.9141	17.1728	143.6634	1.0939	18.7858	0.0582	0.0532	8.3658
19	0.9096	18.0824	160.0360	1.0994	19.8797	0.0553	0.0503	8.8504
20	0.9051	18.9874	177.2322	1.1049	20.9791	0.0527	0.0477	9.3342
21	0.9006	19.8880	195.2434	1.1104	22.0840	0.0503	0.0453	9.8172
22	0.8961	20.7841	214.0611	1.1160	23.1944	0.0481	0.0431	10.2993
23	0.8916	21.6757	233.6768	1.1216	24.3104	0.0461	0.0411	10.7806
24	0.8872	22.5629	254.0820	1.1272	25.4320	0.0443	0.0393	11.2611
25	0.8828	23.4456	275.2686	1.1328	26.5591	0.0427	0.0377	11.7407
30	0.8610	27.7941	392.6324	1.1614	32.2800	0.0360	0.0310	14.1265
40	0.8191	36.1722	681.3347	1.2208	44.1588	0.0276	0.0226	18.8359
50	0.7793	44.1428	1,035.6966	1.2832	56.6452	0.0227	0.0177	23.4624
60	0.7414	51.7256	1,448.6458	1.3489	69.7700	0.0193	0.0143	28.0064
100	0.6073	78.5426	3,562.7934	1.6467	129.3337	0.0127	0.0077	45.3613

i − 1.00%

n	P/F	P/A	P/G	F/P	F/A	A/P	A/F	A/G
1	0.9901	0.9901	0.0000	1.0100	1.0000	1.0100	1.0000	0.0000
2	0.9803	1.9704	0.9803	1.0201	2.0100	0.5075	0.4975	0.4975
3	0.9706	2.9410	2.9215	1.0303	3.0301	0.3400	0.3300	0.9934
4	0.9610	3.9020	5.8044	1.0406	4.0604	0.2563	0.2463	1.4876
5	0.9515	4.8534	9.6103	1.0510	5.1010	0.2060	0.1960	1.9801
6	0.9420	5.7955	14.3205	1.0615	6.1520	0.1725	0.1625	2.4710
7	0.9327	6.7282	19.9168	1.0721	7.2135	0.1486	0.1386	2.9602
8	0.9235	7.6517	26.3812	1.0829	8.2857	0.1307	0.1207	3.4478
9	0.9143	8.5650	33.6959	1.0937	9.3685	0.1167	0.1067	3.9337
10	0.9053	9.4713	41.8435	1.1046	10.4622	0.1056	0.0956	4.4179
11	0.8963	10.3676	50.8067	1.1157	11.5668	0.0965	0.0865	4.9005
12	0.8874	11.2551	60.5687	1.1268	12.6825	0.0888	0.0788	5.3815
13	0.8787	12.1337	71.1126	1.1381	13.8093	0.0824	0.0724	5.8607
14	0.8700	13.0037	82.4221	1.1495	14.9474	0.0769	0.0669	6.3384
15	0.8613	13.8651	94.4810	1.1610	16.0969	0.0721	0.0621	6.8143
16	0.8528	14.7179	107.2734	1.1726	17.2579	0.0679	0.0579	7.2886
17	0.8444	15.5623	120.7834	1.1843	18.4304	0.0643	0.0543	7.7613
18	0.8360	16.3983	134.9957	1.1961	19.6147	0.0610	0.0510	8.2323
19	0.8277	17.2260	149.8950	1.2081	20.8109	0.0581	0.0481	8.7017
20	0.8195	18.0456	165.4664	1.2202	22.0190	0.0554	0.0454	9.1694
21	0.8114	18.8570	181.6950	1.2324	23.2392	0.0530	0.0430	9.6354
22	0.8034	19.6604	198.5663	1.2447	24.4716	0.0509	0.0409	10.0998
23	0.7954	20.4558	216.0660	1.2572	25.7163	0.0489	0.0389	10.5626
24	0.7876	21.2434	234.1800	1.2697	26.9735	0.0471	0.0371	11.0237
25	0.7798	22.0232	252.8945	1.2824	28.2432	0.0454	0.0354	11.4831
30	0.7419	25.8077	355.0021	1.3478	34.7849	0.0387	0.0277	13.7557
40	0.6717	32.8347	596.8561	1.4889	48.8864	0.0305	0.0205	18.1776
50	0.6080	39.1961	879.4176	1.6446	64.4632	0.0255	0.0155	22.4363
60	0.5504	44.9550	1,192.8061	1.8167	81.6697	0.0222	0.0122	26.5333
100	0.3697	63.0289	2,605.7758	2.7048	170.4814	0.0159	0.0059	41.3426

i = 1.50%

n	P/F	P/A	P/G	F/P	F/A	A/P	A/F	A/G
1	0.9852	0.9852	0.0000	1.0150	1.0000	1.0150	1.0000	0.0000
2	0.9707	1.9559	0.9707	1.0302	2.0150	0.5113	0.4963	0.4963
3	0.9563	2.9122	2.8833	1.0457	3.0452	0.3434	0.3284	0.9901
4	0.9422	3.8544	5.7098	1.0614	4.0909	0.2594	0.2444	1.4814
5	0.9283	4.7826	9.4229	1.0773	5.1523	0.2091	0.1941	1.9702
6	0.9145	5.6972	13.9956	1.0934	6.2296	0.1755	0.1605	2.4566
7	0.9010	6.5982	19.4018	1.1098	7.3230	0.1516	0.1366	2.9405
8	0.8877	7.4859	26.6157	1.1265	8.4328	0.1336	0.1186	3.4219
9	0.8746	8.3605	32.6125	1.1434	9.5593	0.1196	0.1046	3.9008
10	0.8617	9.2222	40.3675	1.1605	10.7027	0.1084	0.0934	4.3772
11	0.8489	10.0711	48.8568	1.1779	11.8633	0.0993	0.0843	4.8512
12	0.8364	10.9075	58.0571	1.1956	13.0412	0.0917	0.0767	5.3227
13	0.8240	11.7315	67.9454	1.2136	14.2368	0.0852	0.0702	5.7917
14	0.8118	12.5434	78.4994	1.2318	15.4504	0.0797	0.0647	6.2582
15	0.7999	13.3432	89.6974	1.2502	16.6821	0.0749	0.0599	6.7223
16	0.7880	14.1313	101.5178	1.2690	17.9324	0.0708	0.0558	7.1839
17	0.7764	14.9076	113.9400	1.2880	19.2014	0.0671	0.0521	7.6431
18	0.7649	15.6726	126.9435	1.3073	20.4894	0.0638	0.0488	8.0997
19	0.7536	16.4262	140.5084	1.3270	21.7967	0.0609	0.0459	8.5539
20	0.7425	17.1686	154.6154	1.3469	23.1237	0.0582	0.0432	9.0057
21	0.7315	17.9001	169.2453	1.3671	24.4705	0.0559	0.0409	9.4550
22	0.7207	18.6208	184.3798	1.3876	25.8376	0.0537	0.0387	9.9018
23	0.7100	19.3309	200.0006	1.4084	27.2251	0.0517	0.0367	10.3462
24	0.6995	20.0304	216.0901	1.4295	28.6335	0.0499	0.0349	10.7881
25	0.6892	20.7196	232.6310	1.4509	30.0630	0.0483	0.0333	11.2276
30	0.6398	24.0158	321.5310	1.5631	37.5387	0.0416	0.0266	13.3883
40	0.5513	29.9158	524.3568	1.8140	54.2679	0.0334	0.0184	17.5277
50	0.4750	34.9997	749.9636	2.1052	73.6828	0.0286	0.0136	21.4277
60	0.4093	39.3803	988.1674	2.4432	96.2147	0.0254	0.0104	25.0930
100	0.2256	51.6247	1,937.4506	4.4320	228.8030	0.0194	0.0044	37.5295

$i = 2.00\%$

n	P/F	P/A	P/G	F/P	F/A	A/P	A/F	A/G
1	0.9804	0.9804	0.0000	1.0200	1.0000	1.0200	1.0000	0.0000
2	0.9612	1.9416	0.9612	1.0404	2.0200	0.5150	0.4950	0.4950
3	0.9423	2.8839	2.8458	1.0612	3.0604	0.3468	0.3268	0.9868
4	0.9238	3.8077	5.6173	1.0824	4.1216	0.2626	0.2426	1.4752
5	0.9057	4.7135	9.2403	1.1041	5.2040	0.2122	0.1922	1.9604
6	0.8880	5.6014	13.6801	1.1262	6.3081	0.1785	0.1585	2.4423
7	0.8706	6.4720	18.9035	1.1487	7.4343	0.1545	0.1345	2.9208
8	0.8535	7.3255	24.8779	1.1717	8.5830	0.1365	0.1165	3.3961
9	0.8368	8.1622	31.5720	1.1951	9.7546	0.1225	0.1025	3.8681
10	0.8203	8.9826	38.9551	1.2190	10.9497	0.1113	0.0913	4.3367
11	0.8043	9.7868	46.9977	1.2434	12.1687	0.1022	0.0822	4.8021
12	0.7885	10.5753	55.6712	1.2682	13.4121	0.0946	0.0746	5.2642
13	0.7730	11.3484	64.9475	1.2936	14.6803	0.0881	0.0681	5.7231
14	0.7579	12.1062	74.7999	1.3195	15.9739	0.0826	0.0626	6.1786
15	0.7430	12.8493	85.2021	1.3459	17.2934	0.0778	0.0578	6.6309
16	0.7284	13.5777	96.1288	1.3728	18.6393	0.0737	0.0537	7.0799
17	0.7142	14.2919	107.5554	1.4002	20.0121	0.0700	0.0500	7.5251
18	0.7002	14.9920	119.4581	1.4282	21.4123	0.0667	0.0467	7.9681
19	0.6864	15.6785	131.8139	1.4568	22.8406	0.0638	0.0438	8.4073
20	0.6730	16.3514	144.6003	1.4859	24.2974	0.0612	0.0412	8.8433
21	0.6598	17.0112	157.7959	1.5157	25.7833	0.0588	0.0388	9.2760
22	0.6468	17.6580	171.3795	1.5460	27.2990	0.0566	0.0366	9.7055
23	0.6342	18.2922	185.3309	1.5769	28.8450	0.0547	0.0347	10.1317
24	0.6217	18.9139	199.6305	1.6084	30.4219	0.0529	0.0329	10.5547
25	0.6095	19.5235	214.2592	1.6406	32.0303	0.0512	0.0312	10.9745
30	0.5521	22.3965	291.7164	1.8114	40.5681	0.0446	0.0246	13.0251
40	0.4529	27.3555	461.9931	2.2080	60.4020	0.0366	0.0166	16.8811
50	0.3715	31.4236	642.3606	2.6916	84.5794	0.0318	0.0118	20.4420
60	0.3048	34.7609	823.6975	3.2810	114.0515	0.0288	0.0088	23.6961
100	0.1380	43.0984	1,464.7527	7.2446	312.2323	0.0232	0.0032	33.9863

i = **4.00%**

n	P/F	P/A	P/G	F/P	F/A	A/P	A/F	A/G
1	0.9615	0.9615	0.0000	1.0400	1.0000	1.0400	1.0000	0.0000
2	0.9246	1.8861	0.9246	1.0816	2.0400	0.5302	0.4902	0.4902
3	0.8890	2.7751	2.7025	1.1249	3.1216	0.3603	0.3203	0.9739
4	0.8548	3.6299	5.2670	1.1699	4.2465	0.2755	0.2355	1.4510
5	0.8219	4.4518	8.5547	1.2167	5.4163	0.2246	0.1846	1.9216
6	0.7903	5.2421	12.5062	1.2653	6.6330	0.1908	0.1508	2.3857
7	0.7599	6.0021	17.0657	1.3159	7.8983	0.1666	0.1266	2.8433
8	0.7307	6.7327	22.1806	1.3686	9.2142	0.1485	0.1085	3.2944
9	0.7026	7.4353	27.8013	1.4233	10.5828	0.1345	0.0945	3.7391
10	0.6756	8.1109	33.8814	1.4802	12.0061	0.1233	0.0833	4.1773
11	0.6496	8.7605	40.3772	1.5395	13.4864	0.1141	0.0741	4.6090
12	0.6246	9.3851	47.2477	1.6010	15.0258	0.1066	0.0666	5.0343
13	0.6006	9.9856	54.4546	1.6651	16.6268	0.1001	0.0601	5.4533
14	0.5775	10.5631	61.9618	1.7317	18.2919	0.0947	0.0547	5.8659
15	0.5553	11.1184	69.7355	1.8009	20.0236	0.0899	0.0499	6.2721
16	0.5339	11.6523	77.7441	1.8730	21.8245	0.0858	0.0458	6.6720
17	0.5134	12.1657	85.9581	1.9479	23.6975	0.0822	0.0422	7.0656
18	0.4936	12.6593	94.3498	2.0258	25.6454	0.0790	0.0390	7.4530
19	0.4746	13.1339	102.8933	2.1068	27.6712	0.0761	0.0361	7.8342
20	0.4564	13.5903	111.5647	2.1911	29.7781	0.0736	0.0336	8.2091
21	0.4388	14.0292	120.3414	2.2788	31.9690	0.0713	0.0313	8.5779
22	0.4220	14.4511	129.2024	2.3699	34.2480	0.0692	0.0292	8.9407
23	0.4057	14.8568	138.1284	2.4647	36.6179	0.0673	0.0273	9.2973
24	0.3901	15.2470	147.1012	2.5633	39.0826	0.0656	0.0256	9.6479
25	0.3751	15.6221	156.1040	2.6658	41.6459	0.0640	0.0240	9.9925
30	0.3083	17.2920	201.0618	3.2434	56.0849	0.0578	0.0178	11.6274
40	0.2083	19.7928	286.5303	4.8010	95.0255	0.0505	0.0105	14.4765
50	0.1407	21.4822	361.1638	7.1067	152.6671	0.0466	0.0066	16.8122
60	0.0951	22.6235	422.9966	10.5196	237.9907	0.0442	0.0042	18.6972
100	0.0198	24.5050	563.1249	50.5049	1,237.6237	0.0408	0.0008	22.9800

$i = 6.00\%$

n	P/F	P/A	P/G	F/P	F/A	A/P	A/F	A/G
1	0.9434	0.9434	0.0000	1.0600	1.0000	1.0600	1.0000	0.0000
2	0.8900	1.8334	0.8900	1.1236	2.0600	0.5454	0.4854	0.4854
3	0.8396	2.6730	2.5692	1.1910	3.1836	0.3741	0.3141	0.9612
4	0.7921	3.4651	4.9455	1.2625	4.3746	0.2886	0.2286	1.4272
5	0.7473	4.2124	7.9345	1.3382	5.6371	0.2374	0.1774	1.8836
6	0.7050	4.9173	11.4594	1.4185	6.9753	0.2034	0.1434	2.3304
7	0.6651	5.5824	15.4497	1.5036	8.3938	0.1791	0.1191	2.7676
8	0.6274	6.2098	19.8416	1.5938	9.8975	0.1610	0.1010	3.1952
9	0.5919	6.8017	24.5768	1.6895	11.4913	0.1470	0.0870	3.6133
10	0.5584	7.3601	29.6023	1.7908	13.1808	0.1359	0.0759	4.0220
11	0.5268	7.8869	34.8702	1.8983	14.9716	0.1268	0.0668	4.4213
12	0.4970	8.3838	40.3369	2.0122	16.8699	0.1193	0.0593	4.8113
13	0.4688	8.8527	45.9629	2.1329	18.8821	0.1130	0.0530	5.1920
14	0.4423	9.2950	51.7128	2.2609	21.0151	0.1076	0.0476	5.5635
15	0.4173	9.7122	57.5546	2.3966	23.2760	0.1030	0.0430	5.9260
16	0.3936	10.1059	63.4592	2.5404	25.6725	0.0990	0.0390	6.2794
17	0.3714	10.4773	69.4011	2.6928	28.2129	0.0954	0.0354	6.6240
18	0.3505	10.8276	75.3569	2.8543	30.9057	0.0924	0.0324	6.9597
19	0.3305	11.1581	81.3062	3.0256	33.7600	0.0896	0.0296	7.2867
20	0.3118	11.4699	87.2304	3.2071	36.7856	0.0872	0.0272	7.6051
21	0.2942	11.7641	93.1136	3.3996	39.9927	0.0850	0.0250	7.9151
22	0.2775	12.0416	98.9412	3.6035	43.3923	0.0830	0.0230	8.2166
23	0.2618	12.3034	104.7007	3.8197	46.9958	0.0813	0.0213	8.5099
24	0.2470	12.5504	110.3812	4.0489	50.8156	0.0797	0.0197	8.7951
25	0.2330	12.7834	115.9732	4.2919	54.8645	0.0782	0.0182	9.0722
30	0.1741	13.7648	142.3588	5.7435	79.0582	0.0726	0.0126	10.3422
40	0.0972	15.0463	185.9568	10.2857	154.7620	0.0665	0.0065	12.3590
50	0.0543	15.7619	217.4574	18.4202	290.3359	0.0634	0.0034	13.7964
60	0.0303	16.1614	239.0428	32.9877	533.1282	0.0619	0.0019	14.7909
100	0.0029	16.6175	272.0471	339.3021	5,638.3681	0.0602	0.0002	16.3711

i = 8.00%

n	P/F	P/A	P/G	F/P	F/A	A/P	A/F	A/G
1	0.9259	0.9259	0.0000	1.0800	1.0000	1.0800	1.0000	0.0000
2	0.8573	1.7833	0.8573	1.1664	2.0800	0.5608	0.4808	0.4808
3	0.7938	2.5771	2.4450	1.2597	3.2464	0.3880	0.3080	0.9487
4	0.7350	3.3121	4.6501	1.3605	4.5061	0.3019	0.2219	1.4040
5	0.6806	3.9927	7.3724	1.4693	5.8666	0.2505	0.1705	1.8465
6	0.6302	4.6229	10.5233	1.5869	7.3359	0.2163	0.1363	2.2763
7	0.5835	5.2064	14.0242	1.7138	8.9228	0.1921	0.1121	2.6937
8	0.5403	5.7466	17.8061	1.8509	10.6366	0.1740	0.0940	3.0985
9	0.5002	6.2469	21.8081	1.9990	12.4876	0.1601	0.0801	3.4910
10	0.4632	6.7101	25.9768	2.1589	14.4866	0.1490	0.0690	3.8713
11	0.4289	7.1390	30.2657	2.3316	16.6455	0.1401	0.0601	4.2395
12	0.3971	7.5361	34.6339	2.5182	18.9771	0.1327	0.0527	4.5957
13	0.3677	7.9038	39.0463	2.7196	21.4953	0.1265	0.0465	4.9402
14	0.3405	8.2442	43.4723	2.9372	24.2149	0.1213	0.0413	5.2731
15	0.3152	8.5595	47.8857	3.1722	27.1521	0.1168	0.0368	5.5945
16	0.2919	8.8514	52.2640	3.4259	30.3243	0.1130	0.0330	5.9046
17	0.2703	9.1216	56.5883	3.7000	33.7502	0.1096	0.0296	6.2037
18	0.2502	9.3719	60.8426	3.9960	37.4502	0.1067	0.0267	6.4920
19	0.2317	9.6036	65.0134	4.3157	41.4463	0.1041	0.0241	6.7697
20	0.2145	9.8181	69.0898	4.6610	45.7620	0.1019	0.0219	7.0369
21	0.1987	10.0168	73.0629	5.0338	50.4229	0.0998	0.0198	7.2940
22	0.1839	10.2007	76.9257	5.4365	55.4568	0.0980	0.0180	7.5412
23	0.1703	10.3711	80.6726	5.8715	60.8933	0.0964	0.0164	7.7786
24	0.1577	10.5288	84.2997	6.3412	66.7648	0.0950	0.0150	8.0066
25	0.1460	10.6748	87.8041	6.8485	73.1059	0.0937	0.0137	8.2254
30	0.0994	11.2578	103.4558	10.0627	113.2832	0.0888	0.0088	9.1897
40	0.0460	11.9246	126.0422	21.7245	259.0565	0.0839	0.0039	10.5699
50	0.0213	12.2335	139.5928	46.9016	573.7702	0.0817	0.0017	11.4707
60	0.0099	12.3766	147.3000	101.2571	1,253.2133	0.0808	0.0008	11.9015
100	0.0005	12.4943	155.6107	2,199.7613	27,484.5157	0.0800	0.0000	12.4545

i = 10.00%

n	P/F	P/A	P/G	F/P	F/A	A/P	A/F	A/G
1	0.9091	0.9091	0.0000	1.1000	1.0000	1.1000	1.0000	0.0000
2	0.8264	1.7355	0.8264	1.2100	2.1000	0.5762	0.4762	0.4762
3	0.7513	2.4869	2.3291	1.3310	3.3100	0.4021	0.3021	0.9366
4	0.6830	3.1699	4.3781	1.4641	4.6410	0.3155	0.2155	1.3812
5	0.6209	3.7908	6.8618	1.6105	6.1051	0.2638	0.1638	1.8101
6	0.5645	4.3553	9.6842	1.7716	7.7156	0.2296	0.1296	2.2236
7	0.5132	4.8684	12.7631	1.9487	9.4872	0.2054	0.1054	2.6216
8	0.4665	5.3349	16.0287	2.1436	11.4359	0.1874	0.0874	3.0045
9	0.4241	5.7590	19.4215	2.3579	13.5735	0.1736	0.0736	3.3724
10	0.3855	6.1446	22.8913	2.5937	15.9374	0.1627	0.0627	3.7255
11	0.3505	6.4951	26.3962	2.8531	18.5312	0.1540	0.0540	4.0641
12	0.3186	6.8137	29.9012	3.1384	21.3843	0.1468	0.0468	4.3884
13	0.2897	7.1034	33.3772	3.4523	24.5227	0.1408	0.0408	4.6988
14	0.2633	7.3667	36.8005	3.7975	27.9750	0.1357	0.0357	4.9955
15	0.2394	7.6061	40.1520	4.1772	31.7725	0.1315	0.0315	5.2789
16	0.2176	7.8237	43.4164	4.5950	35.9497	0.1278	0.0278	5.5493
17	0.1978	8.0216	46.5819	5.5045	40.5447	0.1247	0.0247	5.8071
18	0.1799	8.2014	49.6395	5.5599	45.5992	0.1219	0.0219	6.0526
19	0.1635	8.3649	52.5827	6.1159	51.1591	0.1195	0.0195	6.2861
20	0.1486	8.5136	55.4069	6.7275	57.2750	0.1175	0.0175	6.5081
21	0.1351	8.6487	58.1095	7.4002	64.0025	0.1156	0.0156	6.7189
22	0.1228	8.7715	60.6893	8.1403	71.4027	0.1140	0.0140	6.9189
23	0.1117	8.8832	63.1462	8.9543	79.5430	0.1126	0.0126	7.1085
24	0.1015	8.9847	65.4813	9.8497	88.4973	0.1113	0.0113	7.2881
25	0.0923	9.0770	67.6964	10.8347	98.3471	0.1102	0.0102	7.4580
30	0.0573	9.4269	77.0766	17.4494	164.4940	0.1061	0.0061	8.1762
40	0.0221	9.7791	88.9525	45.2593	442.5926	0.1023	0.0023	9.0962
50	0.0085	9.9148	94.8889	117.3909	1,163.9085	0.1009	0.0009	9.5704
60	0.0033	9.9672	97.7010	304.4816	3,034.8164	0.1003	0.0003	9.8023
100	0.0001	9.9993	99.9202	13,780.6123	137,796.1234	0.1000	0.0000	9.9927

i = **12.00%**

n	P/F	P/A	P/G	F/P	F/A	A/P	A/F	A/G
1	0.8929	0.8929	0.0000	1.1200	1.0000	1.1200	1.0000	0.0000
2	0.7972	1.6901	0.7972	1.2544	2.1200	0.5917	0.4717	0.4717
3	0.7118	2.4018	2.2208	1.4049	3.3744	0.4163	0.2963	0.9246
4	0.6355	3.0373	4.1273	1.5735	4.7793	0.3292	0.2092	1.3589
5	0.5674	3.6048	6.3970	1.7623	6.3528	0.2774	0.1574	1.7746
6	0.5066	4.1114	8.9302	1.9738	8.1152	0.2432	0.1232	2.1720
7	0.4523	4.5638	11.6443	2.2107	10.0890	0.2191	0.0991	2.5515
8	0.4039	4.9676	14.4714	2.4760	12.2997	0.2013	0.0813	2.9131
9	0.3606	5.3282	17.3563	2.7731	14.7757	0.1877	0.0677	3.2574
10	0.3220	5.6502	20.2541	3.1058	17.5487	0.1770	0.0570	3.5847
11	0.2875	5.9377	23.1288	3.4785	20.6546	0.1684	0.0484	3.8953
12	0.2567	6.1944	25.9523	3.8960	24.1331	0.1614	0.0414	4.1897
13	0.2292	6.4235	28.7024	4.3635	28.0291	0.1557	0.0357	4.4683
14	0.2046	6.6282	31.3624	4.8871	32.3926	0.1509	0.0309	4.7317
15	0.1827	6.8109	33.9202	5.4736	37.2797	0.1468	0.0268	4.9803
16	0.1631	6.9740	36.3670	6.1304	42.7533	0.1434	0.0234	5.2147
17	0.1456	7.1196	38.6973	6.8660	48.8837	0.1405	0.0205	5.4353
18	0.1300	7.2497	40.9080	7.6900	55.7497	0.1379	0.0179	5.6427
19	0.1161	7.3658	42.9979	8.6128	63.4397	0.1358	0.0158	5.8375
20	0.1037	7.4694	44.9676	9.6463	72.0524	0.1339	0.0139	6.0202
21	0.0926	7.5620	46.8188	10.8038	81.6987	0.1322	0.0122	6.1913
22	0.0826	7.6446	48.5543	12.1003	92.5026	0.1308	0.0108	6.3514
23	0.0738	7.7184	50.1776	13.5523	104.6029	0.1296	0.0096	6.5010
24	0.0659	7.7843	51.6929	15.1786	118.1552	0.1285	0.0085	6.6406
25	0.0588	7.8431	53.1046	17.0001	133.3339	0.1275	0.0075	6.7708
30	0.0334	8.0552	58.7821	29.9599	241.3327	0.1241	0.0041	7.2974
40	0.0107	8.2438	65.1159	93.0510	767.0914	0.1213	0.0013	7.8988
50	0.0035	8.3045	67.7624	289.0022	2,400.0182	0.1204	0.0004	8.1597
60	0.0011	8.3240	68.8100	897.5969	7,471.6411	0.1201	0.0001	8.2664
100	0.0000	8.3332	69.4336	83,522.2657	696,010.5477	0.1200	0.0000	8.3321

i = **18.00%**

n	P/F	P/A	P/G	F/P	F/A	A/P	A/F	A/G
1	0.8475	0.8475	0.0000	1.1800	1.0000	1.1800	1.0000	0.0000
2	0.7182	1.5656	0.7182	1.3924	2.1800	0.6387	0.4587	0.4587
3	0.6086	2.1743	1.9354	1.6430	3.5724	0.4599	0.2799	0.8902
4	0.5158	2.6901	3.4828	1.9388	5.2154	0.3717	0.1917	1.2947
5	0.4371	3.1272	5.2312	2.2878	7.1542	0.3198	0.1398	1.6728
6	0.3704	3.4976	7.0834	2.6996	9.4423	0.2859	0.1059	2.0252
7	0.3139	3.8115	8.9670	3.1855	12.1415	0.2624	0.0824	2.3526
8	0.2660	4.0776	10.8292	3.7589	15.3270	0.2452	0.0652	2.6558
9	0.2255	4.3030	12.6329	4.4355	19.0859	0.2324	0.0524	2.9358
10	0.1911	4.4941	14.3525	5.2338	23.5213	0.2225	0.0425	3.1936
11	0.1619	4.6560	15.9716	6.1759	28.7551	0.2148	0.0348	3.4303
12	0.1372	4.7932	17.4811	7.2876	34.9311	0.2086	0.0286	3.6470
13	0.1163	4.9095	18.8765	8.5994	42.2187	0.2037	0.0237	3.8449
14	0.0985	5.0081	20.1576	10.1472	50.8180	0.1997	0.0197	4.0250
15	0.0835	5.0916	21.3269	11.9737	60.9653	0.1964	0.0164	4.1887
16	0.0708	5.1624	22.3885	14.1290	72.9390	0.1937	0.0137	4.3369
17	0.0600	5.2223	23.3482	16.6722	87.0680	0.1915	0.0115	4.4708
18	0.0508	5.2732	24.2123	19.6731	103.7403	0.1896	0.0096	4.5916
19	0.0431	5.3162	24.9877	23.2144	123.4135	0.1881	0.0081	4.7003
20	0.0365	5.3527	25.6813	27.3930	146.6280	0.1868	0.0068	4.7978
21	0.0309	5.3837	26.3000	32.3238	174.0210	0.1857	0.0057	4.8851
22	0.0262	5.4099	26.8506	38.1421	206.3448	0.1848	0.0048	4.9632
23	0.0222	5.4321	27.3394	45.0076	244.4868	0.1841	0.0041	5.0329
24	0.0188	5.4509	27.7725	53.1090	289.4944	0.1835	0.0035	5.0950
25	0.0159	5.4669	28.1555	62.6686	342.6035	0.1829	0.0029	5.1502
30	0.0070	5.5168	29.4864	43.3706	790.9480	0.1813	0.0013	5.3448
40	**0.0013**	**5.5482**	**30.5269**	**750.3783**	**4,163.2130**	**0.1802**	**0.0002**	**5.5022**
50	0.0003	5.5541	30.7856	3,927.3569	1,813.0937	0.1800	0.0000	5.5428
60	0.0001	5.5553	30.8465	20,555.1400	114,189.6665	0.1800	0.0000	5.5526
100	0.0000	5.5556	30.8642	15,424,131.91	85,689,616.17	0.1800	0.0000	5.5555

6.14 References

The material in this chapter provides a brief overview of engineering economics in preparation for the FE exam. Candidates who have not had a formal course in this subject may wish to review in more detail by referring to one of the following textbooks:

Engineering Economy, 16th ed., Sullivan, Wicks, and Koelling. Prentice-Hall, 2015. ISBN 0-13-3439274.

Principles of Engineering Economic Analysis, 6th ed., White, Case, and Pratt. John Wiley, 2012. ISBN 978-1-118-43342-3.

Principles of Engineering Economy, 8th ed., Grant, Leavenworth, and Ireson. John Wiley, 1990. ISBN 0-471-63526-X.

Engineering Economic Analysis, 9th ed., Newnan, Eschenbach, and Lavelle. Oxford University Press, 2004. ISBN 0-19-516807-0.

Engineering Economy, 7th ed., Blank and Tarquin. McGraw-Hill, 2011. ISBN 978-0073376301.

PRACTICE PROBLEMS

1. About how many years will be required for $10,000 invested at 5% per year compounded annually to double in value?
   - **(A)** 5
   - **(B)** 12
   - **(C)** 15
   - **(D)** 18

2. Mrs. DeMarco deposits $200 in a savings account at the beginning of each year. The account draws interest at 6% per year compounded annually. At the end of 15 years the value of the account will be
   - **(A)** $4,230
   - **(B)** $4,560
   - **(C)** $4,982
   - **(D)** $5,135

3. It is estimated that 50 acres of land will be required for future expansion 20 years after the construction of presently needed facilities. The purchase price of the land is $200 per acre at the present time. Uniform annual taxes are 2% of the purchase price. If money is worth 6% per year, what is the total future value in 20 years of the land and the tax payments?
   - **(A)** $750 per acre
   - **(B)** $788 per acre
   - **(C)** $812 per acre
   - **(D)** $852 per acre

4. A payment of $500 is made at the end of each of the next 12 years. What is the present worth of the payments at 6% interest compounded annually?
   - **(A)** $3,973
   - **(B)** $4,192
   - **(C)** $4,281
   - **(D)** $5,420

5. Let S be the future worth, P the principal invested, i the effective interest per compounding period, and n the number of compounding periods. In which of the following formulas are these quantities correctly related?
   - **(A)** $S = P(1 + ni)$
   - **(B)** $S = P(1 + ni)^{n-1}$
   - **(C)** $S = P(1 + i)^n$
   - **(D)** $S = P(1 + i)^{n-1}$

6. A company can manufacture a product with off-the-shelf hand tools. Fixed costs will be $1,000 for tools and $1.50 manufacturing cost per unit. As an alternative, an automated system will cost $15,000 with a $0.50 manufacturing cost per unit. With an annual anticipated volume of 5,000 units, the break-even point in years is
   - **(A)** 2.0
   - **(B)** 2.8
   - **(C)** 3.6
   - **(D)** 15.0

7. A sum of $400,000 is borrowed under an agreement to repay at the rate of $10,000 per month with interest on the unpaid balance compounding at 1% per month. If the first payment occurs at the end of the first month, the number of monthly repayments will be
   - **(A)** 34
   - **(B)** 40
   - **(C)** 48
   - **(D)** 52

8. Suppose that $10,000 is borrowed now at 15% interest per annum compounded annually. The first repayment of $3,000 is made 4 years from now. The amount that will then remain to be paid is
   - **(A)** $7,000
   - **(B)** $8,050
   - **(C)** $13,000
   - **(D)** $14,490

9. A machine is purchased for $20,000 and has an expected life of 7 years. At the end of 7 years, the salvage value is estimated to be $3,200. According to straight-line depreciation, what is the book value of the machine at the end of 4 years?
   - **(A)** $8,000
   - **(B)** $10,400
   - **(C)** $12,800
   - **(D)** $15,200

10. Using the relevant data in problem 10 above, calculate the book value at the end of 4 years using the MACRS method of depreciation.
    - **(A)** $5,900
    - **(B)** $5,240
    - **(C)** $4,920
    - **(D)** $7,740

189

11. The maintenance costs associated with an investment are $2,000 per year for the first 10 years and $1,000 per year thereafter. The investment has an infinite life. With interest at 10% per annum, the present worth of the annual disbursements is most nearly
(A) $14,000
(B) $16,000
(C) $18,000
(D) $22,000

12. A small South American country experiences inflation of 20% per month. The effective annual inflation rate is
(A) 20%
(B) 240%
(C) 790%
(D) 2,400%

13. With interest of 8% per year compounded semi-annually, the value of $1,000 after 5 years is most nearly
(A) $1,350
(B) $1,400
(C) $1,470
(D) $1,480

14. A personal computer system costs $18,000, and annual maintenance is $900. After 3 years the salvage value of the system is $3,000. If the interest rate is 8%, the equivalent uniform annual cost is
(A) $6,960
(B) $6,922
(C) $7,288
(D) $7,499

Answer Key

1. B	**8.** D
2. D	**9.** B
3. B	**10.** B
4. B	**11.** B
5. C	**12.** C
6. B	**13.** D
7. D	**14.** A

Answers Explained

1. B

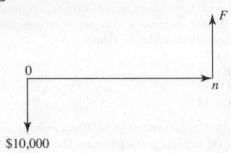

$F = \$10,000(F/P, 5\%, n)$
Answer is number where factor $(F/P, 6\%, n) = 2$.
Approx. 12 yr

2. D

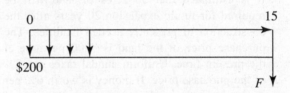

$F = \$200(F/A, 6\%, 15) + \$200(F/P, 6\%, 15)$
$= \$200(23.276) + \$200(2.3966)$
$= \$4,655.2 + \$479.3 = \$5,135$

3. B

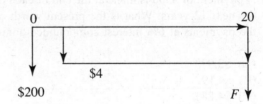

$F = \$200(F/P, 6\%, 20) + 4(F/A, 6\%, 20)$
$= \$200(3.207) + 4(36.786)$
$= \$788/\text{acre}$

4. B

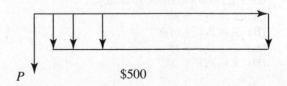

$P = \$500(P/A, 6\%, 12)$
$= \$500(8.3838) = \$4,192$

Practice Problems

5. C

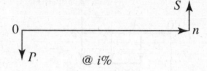

$$S = P(1 + i)^n$$

6. B

$$\$1,000 + \$1.50x = \$0.50 + \$15,000$$
$$x = \$14,000 \ @ \ 5,000/yr$$
$$\$14,000/5,000 = 2.8 \ yr$$

7. D

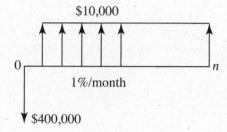

$$\$400,000 = \$10,000(P/A, \ 1\%, \ n)$$
$$(P/A, \ 1\%, \ n) = 40$$

Go to factor table showing P/A @ 1%.

$n = 52$

8. D

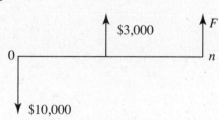

$$F = \$10,000(F/P, \ 15\%, \ 4)$$
$$= \$10,000(1.749) = 17,490$$
$$\underline{- \ 3,000}$$
$$\$14,490$$

9. B

Year	First Cost	Depreciation	Book Value
1	$20,000	$2,400	$17,600
2	17,600	2,400	15,200
3	15,200	2,400	12,800
4	12,800	2,400	10,400

10. B

Ignore salvage value in MACRS depreciation calculations.

Year	1st Cost	Factor	Depreciation	Book Value
1	$20,000	14.3	$2,860	$17,140
2		24.5	$4,900	$12,240
3		17.5	$3,500	$8,740
4		12.5	$2,500	$6,240

11. B

$$P = \$2,000(P/A, \ 10\%, \ 10) + \$1,000$$
$$(P/A, \ 10,\infty)(P/F, \ 10, \ 10)$$
$$= \$2,000(6.1446) + \$1,000(10)(.3855)$$
$$= \$12,289 + \$3,855$$
$$= \$16,144$$

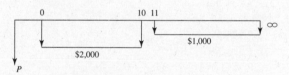

12. C

$$\text{Effective inflation rate} = (1 + r/m)^m - 1$$
$$= (1 + 0.2)^{12} - 1 = 7.916$$
$$= 790\%$$

13. D

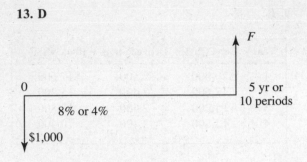

$F = \$1,000(F/P, 4\%, 10)$
   $= \$1,000(1.4802)$
   $= \$1,480.20$

14. A

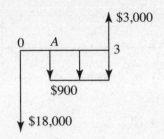

$A = \$18,000(A/P, 8\%, 3) + \$900 - \$3,000(A/F, 8\%, 3)$
   $= \$18,000(0.3880) + \$900 - 3,000(0.3080)$
   $= \$6,984 + \$900 - \$924 = \$6,960$

Engineering Mechanics: Statics

Masoud Olia, Ph.D., PE

Introduction

Statics is the study of bodies at rest. For a body to be at rest, the forces acting on the body have to balance each other. In this chapter, presentation of the concepts of static equilibrium of particles and rigid bodies is followed by discussion of the determination of external and internal forces for structures such as trusses and frames. Moments of inertia, centroids of plane areas, and basic concepts of friction are also discussed. The following is a complete list of this chapter's contents:

7.1 Forces

A **force** is considered a push or a pull exerted by one body on another. Force is a vector quantity, which has both magnitude and direction. Forces are classified as external, internal, concentrated, or distributed, as shown in Figure 7.1.

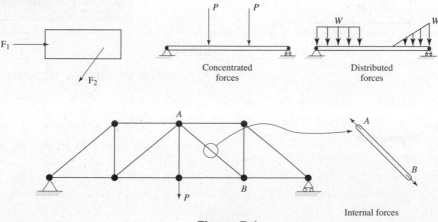

Concentrated forces

Distributed forces

Internal forces

Figure 7.1

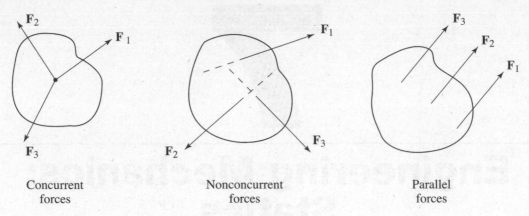

Concurrent forces

Nonconcurrent forces

Parallel forces

Figure 7.2

A force system can be classified as concurrent or nonconcurrent. If all the forces acting on a body pass through the same point, the force system is called concurrent. If all the forces acting on the body do not pass through a common point, the force system is nonconcurrent.

Resolution of a Force (2-D)

A force can be resolved along any two arbitrary axes. It is more convenient and efficient, however, to resolve a force along the *x*- and *y*-axes (Cartesian or rectangular coordinates system). The appropriate equations are as follows:

$$F_x = F \cos \theta, \qquad F_y = F \sin \theta \qquad [7.1]$$

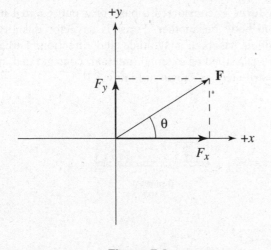

Figure 7.3

Example 7.1

Resolve the forces shown into their rectangular (*x* and *y*) components.

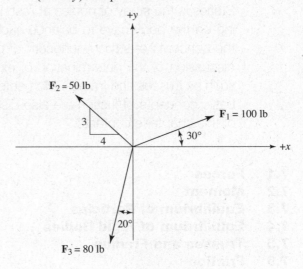

Figure 7.4

> Use "cosine" for the adjacent component and "sine" for the opposite component.

Solution:

$$F_{1x} = 100 \cos 30 = 86.6 \text{ lb}, \quad F_{1y} = 100 \sin 30 = 50 \text{ lb}$$

$$F_{2x} = -50\left(\frac{4}{5}\right) = -40 \text{ lb}, \quad F_{2y} = 50\left(\frac{3}{5}\right) = 30 \text{ lb}$$

$$F_{3x} = -80 \sin 20 = -27.36 \text{ lb}, \quad F_{3y} = -80 \cos 20$$
$$= -75.17 \text{ lb}$$

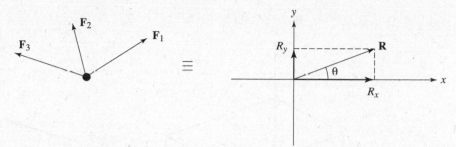

Figure 7.5

Resultant of a Concurrent Force System (2-D)

The resultant force R can be resolved into two rectangular components (R_x and R_y). Refer to Figure 7.5.

$$\mathbf{R} = \mathbf{F}_1 + \mathbf{F}_2 + \mathbf{F}_3 + \dots = \Sigma\mathbf{F}$$

$$R_x = \Sigma F_x, \qquad R_y = \Sigma F_y$$

$$R = \sqrt{R_x^2 + R_y^2}$$

$$\theta = \tan^{-1}\frac{R_y}{R_x} \qquad\qquad [7.2]$$

Example 7.2

Find the resultant of the forces shown.

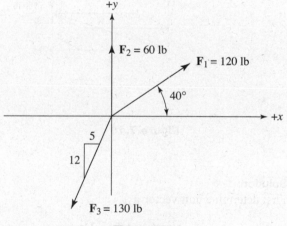

Figure 7.6

Solution:

$$R_x = F_{1x} + F_{2x} + F_{3x} = 120\cos 40 - 130\left(\frac{5}{13}\right) = 41.92\,\text{lb}$$

$$R_y = F_{1y} + F_{2y} + F_{3y} = 120\sin 40 + 60 - 130\left(\frac{12}{13}\right)$$

$$= 17.13\,\text{lb}$$

$$R = \sqrt{41.92^2 + 17.13^2} = 45.28\,\text{lb}$$

$$\theta = \tan^{-1}\frac{17.13}{41.92} = 22.22°$$

Resolution of a Force (3-D)

To resolve a force into its components (F_x, F_y, F_z), two methods are available.

Method 1. If the angles between the force (**F**) and the axes (x, y, z) are known (see Figure 7.7), then

$$F_x = F\cos\theta_x, \; F_y = F\cos\theta_y, \; F_z = F\cos\theta_z \quad [7.3]$$

Also, there is a relation between these angles such that

$$\cos^2\theta_x + \cos^2\theta_y + \cos^2\theta_z = 1 \qquad [7.4]$$

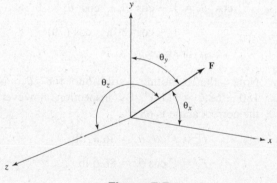

Figure 7.7

Example 7.3

Resolve the force shown into its components (x, y, z).

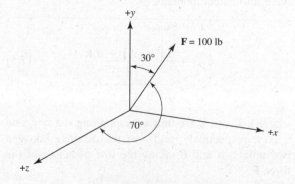

Figure 7.8

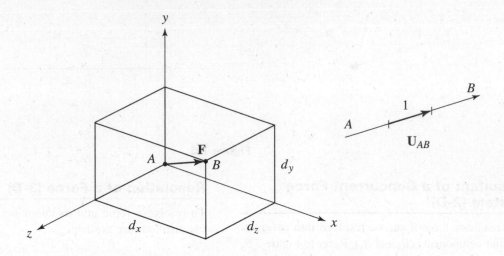

Figure 7.9

Solution:

In this problem, θ_y and θ_z are given, so use equation 7.4 to find θ_x.

$$\cos^2 \theta_x = 1 - \cos^2 \theta_y - \cos^2 \theta_z$$

$$= 1 - \cos^2(30) - \cos^2(70)$$

$$\theta_x = 68.6°$$

Note that another solution for θ_x is $180 - 68.6 = 111.4°$. By inspection, however, the correct angle is $68.6°$.

$$F_x = F \cos \theta_x = 36.47 \text{ lb}$$

$$F_y = F \cos \theta_y = 86.6 \text{ lb}$$

$$F_x = F \cos \theta_z = 34.2 \text{ lb}$$

Method 2. If the coordinates of two points along the line of action of the force are given (see Figure 7.9), use the unit vector $\hat{\mathbf{u}}_{AB}$ to express (resolve) the force into its components.

$$\mathbf{F} = F\hat{\mathbf{u}}_{AB}$$

$$\hat{\mathbf{u}}_{AB} = \frac{d_x\hat{\mathbf{i}} + d_y\hat{\mathbf{j}} + d_z\hat{\mathbf{k}}}{d} \qquad [7.5]$$

$$d = \sqrt{d_x^2 + d_y^2 + d_z^2}$$

where **i**, **j**, and **k** are unit vectors along the x, y, and z axes, respectively, and d is the distance between two points, A and B, along the line of action of the force **F**.

Example 7.4

Resolve the force shown into its components (x, y, z).

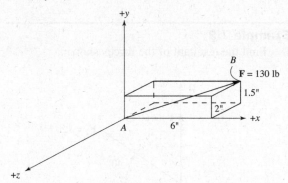

Figure 7.10

Solution:

First determine unit vector $\hat{\mathbf{u}}_{AB}$

$$\hat{\mathbf{u}}_{AB} = \frac{6\hat{\mathbf{i}} + 1.5\hat{\mathbf{j}} - 2\hat{\mathbf{k}}}{\sqrt{6^2 + (1.5)^2 + (-2)^2}}$$

$$\mathbf{F} = 130\,\hat{\mathbf{u}}_{AB}$$

$$= [120\hat{\mathbf{i}} + 30\hat{\mathbf{j}} - 40\hat{\mathbf{k}}]$$

where **i**, **j**, and **k** are unit vectors along the x, y, and z axes, respectively. Therefore, $F_x = 120$ lb, $F_y = 30$ lb, $F_z = -40$ lb.

Resultant of a Concurrent Force System (3-D)

This is similar to the two-dimensional case except that there is a third component (R_z). The equations are as follows:

$$R_x = \Sigma F_x, \quad R_y = \Sigma F_y, \quad R_z = \Sigma F_z$$

$$R = \sqrt{R_x^2 + R_y^2 + R_z^2} \qquad [7.6]$$

$$\theta_x = \cos^{-1}\frac{R_x}{R}, \; \theta_y = \cos^{-1}\frac{R_y}{R}, \; \theta_z = \cos^{-1}\frac{R_z}{R}$$

Example 7.5

Determine the resultant of the forces shown and find its direction ($\theta_x, \theta_y, \theta_z$)

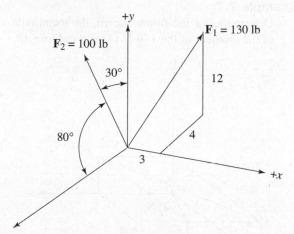

Figure 7.11

Solution:

First we determine θ_x for force $\mathbf{F}_2$.

$$\cos^2 \theta_x = 1 - \cos^2 \theta_y - \cos^2 \theta_z$$

$$= 1 - \cos^2(30) - \cos^2(80)$$

$\theta_x = 62°$ and also $\theta_x = 180° - 62° = 118°$

From the figure, the correct angle is 118°.

$F_{2x} = F_2 \cos \theta_x = -46.9 \text{ lb}$

$F_{2y} = F_2 \cos \theta_y = 86.6 \text{ lb}$

$F_{2z} = F_2 \cos \theta_z = 17.36 \text{ lb}$

$$\hat{\mathbf{u}} = \frac{3\,\hat{\mathbf{i}} + 12\,\hat{\mathbf{j}} - 4\,\hat{\mathbf{k}}}{13}$$

For force $\mathbf{F}_1$, we first determine the unit vector along its line of action.

$\mathbf{F}_1 = 130\,\hat{\mathbf{u}}$

$\quad = 30\,\hat{\mathbf{i}} + 120\,\hat{\mathbf{j}} - 40\,\hat{\mathbf{k}}$

then

$R_x = \Sigma F_x = 30 - 46.6 = -16.9 \text{ lb}$

$R_y = \Sigma F_y = 120 + 86.6 = 206.6 \text{ lb}$

$R_z = \Sigma F_z = -40 + 17.36 = -22.64 \text{ lb}$

$R = \sqrt{(-16.9)^2 + 206.6^2 + (-22.64)^2} = 208.82 \text{ lb.}$

To specify the direction of the resultant force, determine the angles that the resultant forms with the axes.

$$\cos \theta_x = \frac{R_x}{R} = \frac{-16.9}{208.82} \Rightarrow \theta_y = 94.64°$$

$$\cos \theta_y = \frac{R_y}{R} = \frac{206.9}{208.82} \Rightarrow \theta_x = 7.77°$$

$$\cos \theta_z = \frac{R_z}{R} = \frac{-22.64}{208.82} \Rightarrow \theta_z = 96.22°$$

7.2 Moment

The **moment** of a force is the measure of tendency of a rigid body to rotation about an axis. Moment is divided into two parts, moment of a single force and moment of a couple.

Moment of a Single Force

The moment of a single force can be about (1) a point or (2) an axis.

1. The moment of a force about a point is the product of the force and the perpendicular distance from the point to the line of action of the force. Therefore, the magnitude of the moment about point A is expressed by the equation

$$M_A = Fd \qquad [7.7]$$

where d is the perpendicular distance from point A to the line of action of force $\mathbf{F}$.

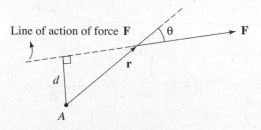

Figure 7.12

The direction of this moment is a clockwise rotation of force **F** about point A.

Moment can also be determined vectorially, and is given by the equation

$$\mathbf{M}_A = \mathbf{r} \times \mathbf{F} \qquad [7.8]$$

where **r** is the position vector from point A to anywhere along the line of action of the force.

Example 7.6

Find the moment of the 100-lb force shown about point A, using both the scalar and the vector approach.

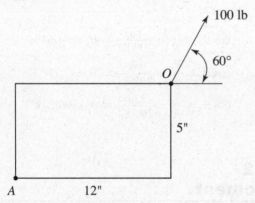

Figure 7.13

Solution:

Scalar approach:

Here we resolve the 100-lb force into its components and then determine the sum of the moment of the components about point A. This is known as Varignon's theorem.

CCW + :

$$M_A = -100 \cos 60(5) + 100 \sin 60(12) = 789.2 \text{ lb-in}$$

Therefore, the magnitude of the moment about A is 789.2 lb-in and its direction is counterclockwise.

Vector approach:

$$\mathbf{M}_A = \mathbf{r}_{AO} \times \mathbf{F} = (12\,\hat{i} + 5\,\hat{j}) \times (50\,\hat{i} + 86.6\,\hat{j})$$
$$\mathbf{M}_A = (12)(86.6)\hat{k} + (5)(50)(-\hat{k}) = 789.2\hat{k}$$

2. The moment of a force about an axis is the projection of the moment along that axis, shown in Figure 7.14. For projection along line AB the equations are as follows:

$$M_{AB} = \mathbf{M}_A \cdot \mathbf{u}_{AB}, \quad \mathbf{M}_{AB} = M_{AB}\,\mathbf{u}_{AB} \qquad [7.9]$$

Note that $\mathbf{M}_A$ can be determined about any point along line AB, $\mathbf{u}_{AB}$ is the unit vector along line AB, and M_{AB} is the magnitude of the moment about line AB. Also note that "·" is the symbol for the dot product.

Example 7.7

Determine, for the figure shown, the magnitude of the moment of the 140-N force about line AB.

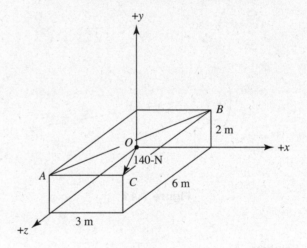

Figure 7.15

Solution:
We first describe **F** as a vector

$$\mathbf{F} = F\,\hat{u}_{oc} = 140 \frac{3\hat{i} + 2\hat{j} + 6\hat{k}}{7} = 60\hat{i} + 40\hat{j} + 120\hat{k}$$

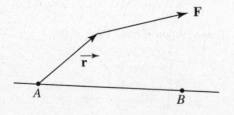

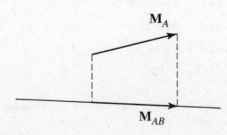

Figure 7.14

To be efficient, one should pick either $\mathbf{r}_{AC}$ or $\mathbf{r}_{BC}$ as position vector instead of $\mathbf{r}_{AO}$ or $\mathbf{r}_{BO}$.

$$\mathbf{M}_A = \mathbf{r}_{AC} \times \mathbf{F} = (3\,\widehat{\mathbf{i}}\,) \times (60\,\widehat{\mathbf{i}} + 40\,\widehat{\mathbf{j}} + 120\,\widehat{\mathbf{k}})$$

$$= -360\,\widehat{\mathbf{j}} + 120\,\widehat{\mathbf{k}}$$

$$M_{AB} = \mathbf{M}_A \cdot \mathbf{u}_{AB} = (-360\,\widehat{\mathbf{j}} + 120\,\widehat{\mathbf{k}}) \cdot \left(\frac{3\,\widehat{\mathbf{i}} - 6\,\widehat{\mathbf{k}}}{\sqrt{45}}\right)$$

$$= -107.33 \; N \cdot m$$

This number is the magnitude of the moment of force $\mathbf{F}$ about line AB.

Moment of a Couple

Two equal and parallel forces that have noncollinear lines of actions and are opposite in sense form a **couple** (Figure 7.16). The moment of a couple is the product of the force and the shortest distance between the two forces. Note in the equation below that the moment of a couple is the same with respect to any point in space.

$$M = Fd$$

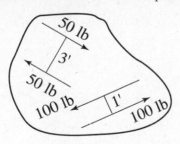

Figure 7.16

Example 7.8

Determine the resultant of the couples shown.

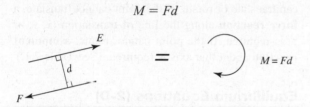

Figure 7.17

Solution:

CCW + :

$$M = (100)\,(1) - (50)(3) = -50 \; \text{lb-ft}$$

Therefore, the resultant moment (couple) is clockwise.

7.3 Equilibrium of Particles

Two-Dimensional Case

For a particle to be in equilibrium (at rest), the resultant of the forces acting on the particle must be equal to zero (i.e., there is no translation).

$$R_x = \Sigma F_x = 0, \quad R_y = \Sigma F_y = 0 \qquad [7.10]$$

Example 7.9

Two cables are tied together at C and loaded as shown. Determine the tension in each cable.

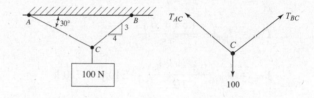

Figure 7.18

Solution:

First draw a free-body diagram of point C.

$$R_x = \Sigma F_x = \frac{4}{5}T_{BC} - T_{AC}\cos 30 = 0$$

$$R_y = \Sigma F_y = \frac{3}{5}T_{BC} + T_{AC}\sin 30 - 100 = 0$$

Solve simultaneously for the tensions:

$$T_{AC} = 87 \; \text{N} \quad \text{and} \quad T_{BC} = 94.17 \; \text{N}$$

Three-Dimensional Case

This is the same as the two-dimensional case except that there is a component in the z-direction.

$$R_x = \Sigma F_x = 0, \; R_y = \Sigma F_y = 0, \; R_z = \Sigma F_z = 0 \quad [7.11]$$

Example 7.10

A load W is supported by three cables as shown. Determine the magnitude of load W if the tension in cable CD is 900 N.

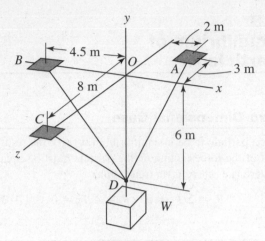

Figure 7.19

Solution:

It is given that $T_{DC} = 900$ N. Then:

$$\mathbf{T}_{DA} = T_{DA}\left(\frac{2\,\widehat{\mathbf{i}} + 6\,\widehat{\mathbf{j}} - 3\,\widehat{\mathbf{k}}}{7}\right),$$

$$\mathbf{T}_{DB} = T_{DB}\left(\frac{-4.5\,\widehat{\mathbf{i}} + 6\,\widehat{\mathbf{j}}}{7.5}\right),$$

$$\mathbf{T}_{DC} = 9_{OC}\left(\frac{6\,\widehat{\mathbf{j}} + 8\,\widehat{\mathbf{k}}}{10}\right) = 540\,\widehat{\mathbf{j}} + 720\,\widehat{\mathbf{k}}$$

$$R_x = \Sigma F_x = 0$$

$$\frac{2}{7}\,T_{DA} - \frac{4.5}{7.5}\,T_{DB} = 0$$

$$R_y = \Sigma F_y = 0$$

$$\frac{6}{7}T_{DA} + \frac{6}{7.5}\,T_{DB} + 540 - W = 0$$

$$R_z = \Sigma F_z = 0$$

$$-\frac{3}{7}T_{DA} + 720 = 0$$

Solve simultaneously for W:

$$W = 2{,}620 \text{ N}$$

7.4
Equilibrium of Rigid Bodies

Two-Force Members

If a member is under the action of only two forces (at different points on the body), the member is called a **two-force member.** For a straight two-force member, the force must act along the axis of the member (tension or compression).

If the force member is not straight, the force must act along the diagonal line connecting the two points where the two forces are acting. (See Figure 7.20.)

Connections (Supports) and Reactions

Rigid bodies either are connected to each other or rest on surfaces. In most cases it is necessary to determine the resulting reactions for different supports. To find the number and type of reaction for a given support, determine whether the point of connection can translate or rotate. If the point cannot translate, a **force reaction** along the line of translation (x, y, or z) is required. If the point cannot rotate, a **moment reaction** about that axis is required. (See Table 7.1.)

Equilibrium Equations (2-D)

For a rigid body to be in equilibrium, the following equations must be satisfied.

$$R_x = \Sigma F_x = 0, \quad R_y = \Sigma F_y = 0 \qquad [7.12]$$
$$\Sigma M = 0$$

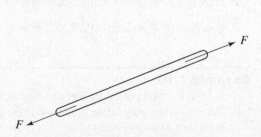

Straight two-force member (tension)

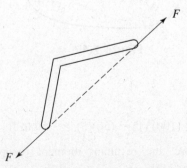

Nonstraight two-force member

Figure 7.20

This means that the body will neither translate nor rotate.

In equation 7.12 the moment can be taken about any point on the body. Note that for a two-dimensional case there are three equilibrium equations, so there cannot be more than three unknowns for a given free-body diagram (FBD).

Example 7.11

For the beam and loading shown, determine the reaction at each support.

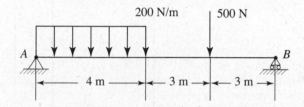

Figure 7.21

Solution:

Draw the free-body diagram for the beam and note that the uniform load can be replaced by an 800-N (area of the load) concentrated load at a distance 2 m from support A (centroid of the load).

$\Sigma M_A = 0 \quad CCW + :$

$B(10) - 500(7) - 800(2) = 0$; then $B = 510$ lb

$R_x = \Sigma F_x = 0, \quad A_x = 0$

$R_y = \Sigma F_y = 0,$

$A_y + 510 - 500 - 800 = 0 \Rightarrow A_y = 790$ N

Example 7.12

For the figure shown, determine the tension in cable BC and the reaction at A (neglect the weight of beam AB).

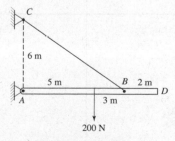

Figure 7.22

Solution:

$\Sigma M_A = 0$

$\dfrac{3}{5} T_{BC}(8) - 200(5) = 0$

$T_{BC} = 208.33$ N

$R_x = \Sigma F_x = 0$

$A_x - \dfrac{4}{5}(208.33) = 0$

$A_x = 166.66$ N

$R_y = \Sigma F_y = 0$

$A_y - 200 + \dfrac{3}{5}(208.33) = 0$

$A_y = 75$ N

Note that the horizontal component of tension BC passes through point A and has zero moment about that point.

Example 7.13

Determine the force in member BD of the structure shown. Also determine the components of reaction at point C.

You can also assume that F_{BD} is in tension, but you end up with a negative answer.

Solution:

Use the F_{BD} of member CDE and recognize that member BD is a two-force member in compression: Then,

$\Sigma M_C = 0 \quad CCW + :$

$-\dfrac{3}{5} F_{BD}(2) + \dfrac{4}{5} F_{BD}(3) - 300(5) = 0$

$F_{BD} = 1,250$ lb

$R_x = \Sigma F_x = 0$

$C_x + \dfrac{3}{5}(1,250) = 0$

$C_x = -750$ lb

$R_y = \Sigma F_y = 0$

$C_y - 300 + \dfrac{4}{5}(1,250) = 0$

$C_y = -700$ lb

Table 7.1. Supports for Rigid Bodies

Connections (Supports)	Reactions	Number of Unknowns
Hinge, pin	R_x R_y	2
Rough surface	R_x R_y	2
Roller, rocker smooth surface	Reaction perpendicular to surface R	1
Fixed	R_x M R_y	3
Ball and socket joint	R_x R_z R_y	3
Fixed	R_z R_y R_x M_z M_y M_x	6

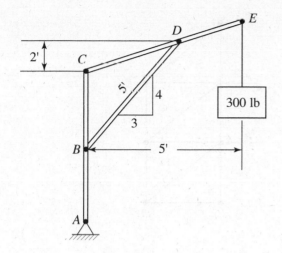

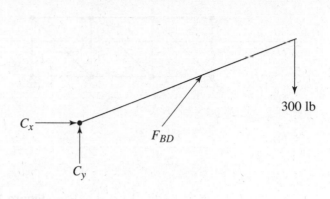

Figure 7.23

Equilibrium Equations (3-D)

In this case there are six equilibrium equations (three force equations and three moment equations).

$$R_x = \Sigma F_x = 0, \quad R_y = \Sigma F_y = 0, \quad R_z = \Sigma F_z = 0$$

$$\Sigma M_x = 0, \quad \Sigma M_y = 0, \quad \Sigma M_z = 0 \quad [7.13]$$

Note that for three-dimensional problems, we can't have more than six unknowns.

Example 7.14

In the figure, a plate weighing 135 N is supported by two cables (*CD* and *BE*) and a ball and socket joint at *A*. Determine the tension in each cable.

Solution:

$$\mathbf{T}_{CD} = T_{CD}\left(\frac{-8\,\hat{\mathbf{i}} + 4\,\hat{\mathbf{j}} - 8\,\hat{\mathbf{k}}}{12}\right)$$

$$\mathbf{T}_{BE} = T_{BE}\left(\frac{-6\,\hat{\mathbf{i}} + 3\,\hat{\mathbf{j}} + 2\,\hat{\mathbf{k}}}{7}\right)$$

$$\Sigma M_A = 0$$

$$8\,\hat{\mathbf{i}} \times T_{CD}\left(\frac{-8\,\hat{\mathbf{i}} + 4\,\hat{\mathbf{j}} - 8\,\hat{\mathbf{k}}}{12}\right) + 6\,\hat{\mathbf{i}}$$

$$\times T_{BE}\left(\frac{-6\,\hat{\mathbf{i}} + 3\,\hat{\mathbf{j}} + 2\,\hat{\mathbf{k}}}{7}\right) + (4\,\hat{\mathbf{i}} - 2.5\,\hat{\mathbf{j}})$$

$$\times (-135\,\hat{\mathbf{j}}) = 0$$

$$\left(\frac{16}{3}\,T_{CD} - \frac{12}{7}\,T_{BE}\right)\hat{\mathbf{j}} + \left(\frac{8}{3}\,T_{CD} + \frac{18}{7}\,T_{BE} - 540\right)\hat{\mathbf{k}} = 0$$

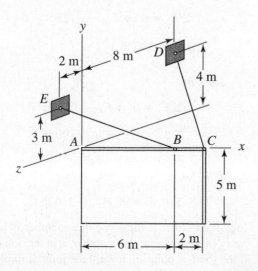

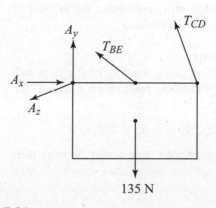

Figure 7.24

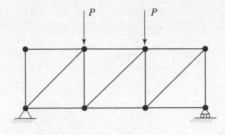

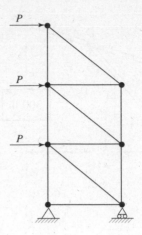

Figure 7.25

The first parenthesis represents $\sum M_y$ and the second one is $\sum M_z$.

Set each parenthesis equal to zero, and solve for the tensions:

$$T_{BE} = 157.5 \text{ N}, \qquad T_{CD} = 50.63 \text{ N}$$

Figure 7.26

7.5
Trusses and Frames

Trusses

A collection of rigid bodies that are pin connected to each other at their ends to form a new rigid body is known as a *truss*. Each member of a truss is a straight two-force body. The main objective is to determine the internal forces in some or all members of a truss. Each member can be in tension (T) or compression (C), depending on the loading or configuration of the truss.

Generally two methods are available to determine internal forces in a given truss, the method of joints and the method of sections.

Method of Joints

To use this method, draw the F_{BD} of the joint and apply the two-dimensional equilibrium equations for each joint. The resulting equations are

$$R_x = \sum F_x = 0, \quad R_y = \sum F_y = 0$$

Note that for a given joint there cannot be more than two unknowns.

Example 7.15

Using the method of joints, determine the force in each member of the truss shown.

Solution:

Using Figure 7.27:

Joint A:

$$\sum F_y = 0 \Rightarrow -150 + \frac{3}{5}F_{AC} = 0$$

$$F_{AC} = 250 \text{ lb C}$$

$$\sum F_x = 0 \Rightarrow F_{AB} - \frac{4}{5}(250) = 0$$

$$F_{AB} = 200 \text{ lb T}$$

Joint B:

$$\sum F_x = 0 \Rightarrow F_{BD} = 200 \text{ lb T}$$

$$\sum F_y = 0 \Rightarrow F_{BC} = 100 \text{ lb C}$$

Joint C:

$$\sum F_x = 0 \Rightarrow F_{CD} = 250 \text{ lb C}$$

Important Note: If the force in a member is pointing away from the joint, the member is in tension (*T*) and if the force is pointing toward the joint, it implies that the member is in compression (*C*).

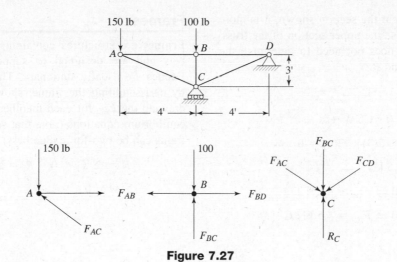

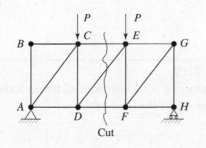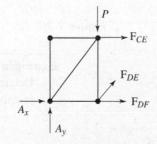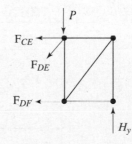

Figure 7.27

Method of Sections

This method is used when only the forces for a few members of a large truss are to be determined. Cut the truss in a section where the forces are to be determined (see Figure 7.28). Then consider the section to the left or right (up or down in other cases), draw the F_{BD}, and apply the two-dimensional equilibrium equations for rigid bodies. **Generally the section considered should not pass or cut more than three members at a time (since only three equilibrium equations are available).**

$$R_x = \Sigma F_x = 0, \quad R_y = \Sigma F_y = 0$$
$$\Sigma M = 0$$

Note that, for some trusses, it is necessary to determine the external forces (reactions) before solving for the internal forces.

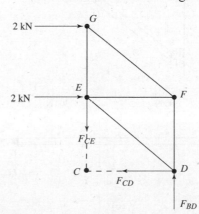

Figure 7.28

Example 7.16

Determine the forces in members *CE*, *CD*, and *BD* of the truss shown in Figure 7.29 (left).

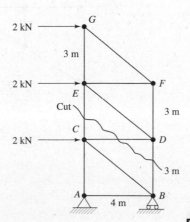

Figure 7.29

205

Here, after we cut the section shown, it is more convenient to use the upper section of the truss. Therefore, one does not need to determine the reactions at A and B.

Solution:

$$\Sigma M_D = 0 \quad \text{CCW} + :$$

$$F_{CE}(4) - 2(3) - 2(6) = 0$$

$$F_{CE} = 4.5 \text{ kN T}$$

$$\Sigma F_x = 0 \Rightarrow F_{CD} = 4 \text{ kN T}$$

$$\Sigma F_y = 0 \Rightarrow F_{BD} = 4.5 \text{ kN C}$$

Frames

Frames are structures containing multiforce members and are designed to support applied loads. Frames are usually stationary. The analysis is done by disassembling the frame, shown in Figure 7.30, drawing the F_{BD} for each member, and applying the equilibrium equations (note that some members of a frame can be two-force members).

$$R_x = \Sigma F_x = 0, \quad R_y = \Sigma F_y = 0$$

$$\Sigma M = 0$$

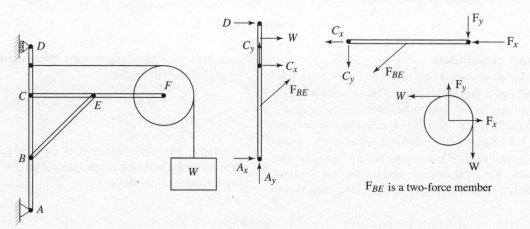

Figure 7.30

Example 7.17

Determine the components of all forces acting on member AC of the frame shown.

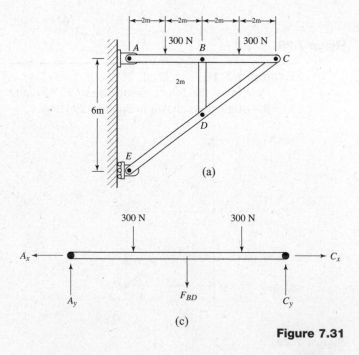

(a)

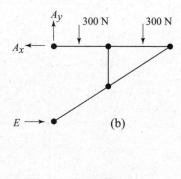

(b)

(c)

Figure 7.31

Solution:

First determine the reactions without disassembling the frame. (Figure 7.31b)

$$\Sigma M_A = 0$$

$$E(6) - 300(2) - 300(6) - 0$$

$$E = 400 \text{ N}$$

$$\Sigma F_x = 0 \Rightarrow A_x = 400 \text{ N} \leftarrow$$

$$\Sigma F_y = 0 \Rightarrow A_y = 600 \text{ N} \uparrow$$

Now use the F_{BD} of member AC (Figure 7.31c), noting that BD is a two-force member.

$$\Sigma M_C = 0$$

$$F_{BD}(4) + 300(2) + 300(6) - 600(8) = 0$$

$$F_{BD} = 600 \text{ N} \downarrow$$

$$\Sigma F_x = 0 \Rightarrow C_x = A_x = 400 \text{ N} \rightarrow$$

$$\Sigma F_y = 0 \Rightarrow C_y = 600 \text{ N} \uparrow$$

7.6
Friction

When two surfaces are in contact, the force tangent to the surface that opposes the direction of tendency to motion is called the **friction**. There are two types of friction: dry friction and fluid friction.

To understand the laws of dry friction, consider the figure shown below.

The maximum value of friction force (F_{max}) is related to the normal force (N) and the coefficient of static friction (μ_s):

$$F_{max} = \mu_s N \qquad\qquad [7.14]$$

Two Types of Friction Problems

There are two distinct types of friction problems:

1. The impending motion exists (the object is about to move). In this case friction has reached its maximum value, and $F = F_{max} = \mu_s N$. Apply the equilibrium equations for the particle or rigid body, and solve for the unknowns.
2. The impending motion is not known (you don't know or not whether the object is moving). In this case assume that the system is in equilibrium, and apply the appropriate equilibrium equations (particle or rigid body) to solve for the friction and normal forces. Two outcomes are possible.
   a. If $F < \mu_s N$, then the system is in equilibrium and the friction force calculated is correct.
   b. If $F > \mu_s N$, then the system is in motion and for ***particle problems only*** the correct value of friction force is $F = \mu_k N$, where μ_k is called the coefficient of kinetic friction. In general, μ_k is about 25% smaller than μ_s.

For rigid body problems, a dynamic analysis is used to determine the magnitude of the friction force.

Example 7.18

Determine the value of the friction force.

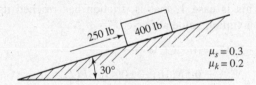

$$\mu_s = 0.3$$
$$\mu_k = 0.2$$

Figure 7.33

Solution:

This is case 2, that is, the impending motion is not known.

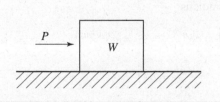 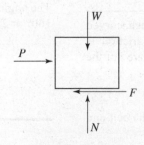 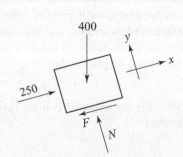

Figure 7.32

$\nwarrow \Sigma F_y = 0$

$N - 400 \cos 30 = 0 \Rightarrow N = 346.4$ lb

$\nearrow \Sigma F_x = 0$

$-F + 250 - 400 \sin 30 = 0 \Rightarrow F = 50$ lb

Now check: $\mu_s N = 0.3(364.4) = 103.92$ since $F < \mu_s N$, which implies that there is no motion and the friction force calculated is correct: $F = 50$ lb.

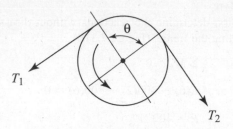

Figure 7.35

Example 7.19

If the bar weighs 50 N, determine the minimum value of load P that will initiate the motion (neglect friction at A and $\mu_s = 0.5$ at B)

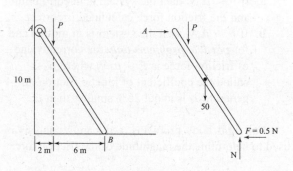

Figure 7.34

Solution:

This is case 1, that is, friction has reached its maximum value at B. Set $F = \mu_s N$.

$\Sigma M_A = 0 \quad$ CCW $+ :$

$N(8) - 0.5N(10) - 50(4) - P(2) = 0$

$3N - 2P = 200$

$\Sigma F_y = 0$

$N - 50 - P = 0$

Solve for P: $P = 50$ N

Belt Friction

Consider a belt that is pressed against a rough, curved surface and is pulled on both ends. If friction is considered, the tensions (T) on both ends are not the same. It can be shown that

$$T_2 = T_1 e^{\mu_s \theta} \qquad [7.15]$$

where θ is the angle of contact between the belt and the drum.

Example 7.20

A flat belt transmits a 30 lb-ft torque. Determine the minimum values of tension in both parts of the belt that will prevent slippage. The diameter of the drum is 6 in and $\mu_s = 0.3$.

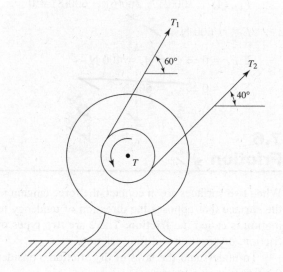

Figure 7.36

Solution:

Torque $= T = 30$ lb.ft $= 360$ lb.in

$$\Sigma M_{center} = 0$$

$$T_1(3) - T_2(3) - 360 = 0$$
$$\text{or } T_1 - T_2 = 120$$

The angle of contact between the belt and the drum is 160° or 2.793 radians:

$$\frac{T_1}{T_2} = e^{\mu_s \theta} \qquad \frac{T_1}{T_2} = e^{(2.793)(0.3)}$$

$$T_1 = 2.31\, T_2$$

Note that the angle ı must be in radians.

208

Solve simultaneously for T_1 and T_2:

$$T_1 = 211.6 \text{ lb}$$
$$T_2 = 91.6 \text{ lb}$$

Screw Friction

When a screw is subjected to axial loads, frictional forces are developed. To determine the moment needed to turn the screw, the following equation can be used:

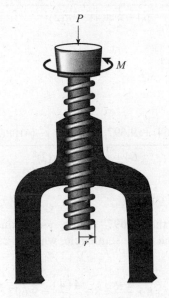

Figure 7.37

$$M = Pr \tan (\alpha \pm \phi) \qquad [7.16]$$

where

M is the external moment

P is the axial load

r is the mean radius of the thread

α is the lead angle and is equal to $\alpha = \tan^{-1} \dfrac{1}{2\pi r}$, where 1 is the lead of the screw

$\phi = \tan^{-1} \mu$

+ is for screw tightening

– is for screw loosening

Example 7.21

If the turnbuckle shown has a mean radius of 6 mm and lead of 3 mm, determine the moment, M, needed to draw the end screws closer together. Note that the coefficient of static friction $\mu_s = 0.2$.

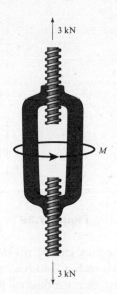

Figure 7.38

Solution:
We first determine the angle φ and α.

$$\phi = \tan^{-1} \mu = \tan^{-1}(0.2) = 11.31°$$

$$\alpha = \tan^{-1} \frac{1}{2\pi r} = \tan^{-1} \frac{3}{2\pi (6)} = 4.55°$$

Since the friction at two screws must be overcome, the moment required is twice the moment in equation 7.16.

$$M = 2P\,r \tan (\alpha + \phi) = 2(3000)(0.006) \tan (4.55 + 11.31) = 10.22 \text{ N} \cdot \text{m}$$

7.7 Center of Gravity

Center of gravity is defined as the point of a body where the weight of the body will pass through. For a homogeneous body the center of gravity is the same as the **centroid.**

In general, to determine the location of centroid of a plane area use the following equations:

$$\overline{X}_c = \frac{\int x_c dA}{A}, \quad \overline{Y}_c = \frac{\int y_c dA}{A} \qquad [7.17]$$

where $\int x_c\, dA$ and $\int y_c\, dA$ are called the first moments of the area with respect to the y-axis and x-axis, respectively.

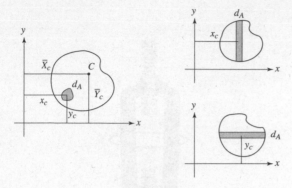

Figure 7.39

If the plane area can be divided into common shapes, as shown in Fig 7.40 (i.e., rectangles, semi-circles, triangles), then the discrete form of the equations can be used:

$$\overline{X}_c = \frac{\Sigma \, x_i A_i}{A}, \quad \overline{Y}_c = \frac{\Sigma \, y_i A_i}{A} \qquad [7.18]$$

where x_i and y_i are the coordinates of the centroid of common shapes relative to the reference selected.

Refer to Table 7.2 for the area and centroid of common shapes.

Example 7.22

Determine the center of gravity of the shaded plane area shown.

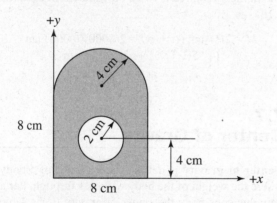

Figure 7.41

Solution:

Break the shaded area into:

Rectangle + semicircle − circle

$$\overline{X}_c = \frac{\Sigma \, x_i A_i}{A}$$

$$= \frac{(4)(64) + (4)\left[\frac{1}{2}\pi(4)^2\right] - (4)\left[\pi(2)^2\right]}{64 + \frac{1}{2}\pi(4)^2 - \pi(2)^2} = 4 \text{ cm}$$

$$\overline{Y}_c = \frac{\Sigma \, y_i A_i}{A}$$

$$= \frac{(4)(64) + (9.697)\left[\frac{1}{2}\pi(4)^2\right] - (4)\left[\pi(2)^2\right]}{64 + \frac{1}{2}\pi(4)^2 - \pi(2)^2}$$

$$= 5.87 \text{ cm}$$

Note that 9.697 is the y-coordinate of the centroid of the semicircle, which is calculated as follows:

$$y = 8 + \frac{4\,R}{3\pi} = 8 + \frac{4(4)}{3\pi} = 9.697 \text{ cm}$$

To find the centroid of lines (wires) and volumes, use the following similar equations.

For lines:

$$\overline{X}_c = \frac{\Sigma \, x_i L_i}{L}, \quad \overline{Y}_c = \frac{\Sigma \, y_i L_i}{L} \qquad [7.19]$$

For volumes:

$$\overline{X}_c = \frac{\Sigma \, x_i V_i}{V}, \quad \overline{Y}_c = \frac{\Sigma \, y_i V_i}{V} \qquad [7.20]$$

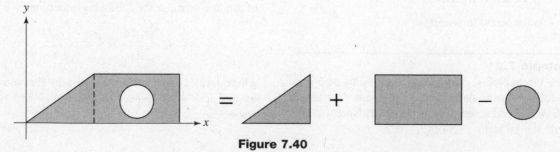

Figure 7.40

Example 7.23

Determine the centroid of the composite wire shown.

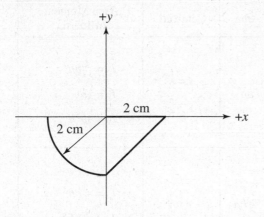

Figure 7.42

Solution:

We break the wire into two straight pieces and a quarter circular arc. Using Table 7.2 for the quarter circular arc

$$\bar{X}_c = \frac{\Sigma x_i L_i}{L} = \frac{(1)(2) + (1)(2.83) + (-1.273)(3.14)}{2 + 2.83 + 3.14}$$

$$= 0.1039 \text{ cm}$$

$$\bar{Y}_c = \frac{\Sigma y_i L_i}{L}$$

$$= \frac{(0)(2) + (-1)(2.83) + (-1.273)(3.14)}{2 + 2.83 + 3.14}$$

$$= -0.857 \text{ cm}$$

7.8 Moments of Inertia

Area Moment of Inertia

Area moment of inertia (I) is defined as the second moment of area and is determined as follows:

$$I_x = \int y^2 \, dA, \quad I_y = \int x^2 \, dA \qquad [7.21]$$

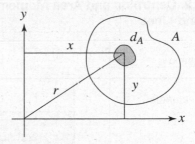

Figure 7.43

Area moment of inertia is a measure of resistance to bending and is needed to calculate stresses due to bending.

Refer to Table 7.2 for the equations of moments of inertia for common shapes.

Polar Moment of Inertia

Polar moment of inertia (J) is the second moment of area with respect to the z-axis. It represents a resistance to twisting and is needed to calculate shear stresses due to twisting.

$$J = I_z = \int r^2 \, dA = \int x^2 \, dA + \int y^2 \, dA$$

$$= I_x + I_y \qquad [7.22]$$

Note that in Figure 7.43, $r^2 = x^2 + y^2$

Mass Moment of Inertia

Mass moment of inertia (I) is similar to area moment of inertia, except that area (A) is replaced in equation 7.21 by mass (m).

$$I = \int r^2 \, dm \qquad [7.23]$$

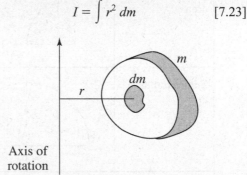

Figure 7.44

Mass moment of inertia is a measure of resistance of the body to rotational acceleration and is needed in dynamic problems to calculate forces and moments acting on a rigid body in rotation. Refer to Table 7.3 for the mass moment of inertia of common shapes.

Table 7.2. Centroids and Area Moments of Inertia of Common Shapes of Areas and Lines

Shape		Area	Centroid	Area Moment of Inertia
Rectangle		bh	$\bar{x} = \frac{b}{2}$ $\bar{y} = \frac{h}{2}$	$\bar{I}_c = \frac{1}{12}bh^3$ $I_x = \frac{1}{3}bh^3$ $I_y = \frac{1}{3}b^3h$
Triangle		$\frac{1}{2}bh$	$\bar{y} = \frac{h}{3}$	$\bar{I}_c = \frac{1}{36}bh^3$ $I_x = \frac{1}{12}bh^3$
Circular area		πR^2	$\bar{x} = 0$ $\bar{y} = 0$	$\bar{I}_x = I_y = \frac{\pi R^4}{4}$ $I_z = J = \frac{\pi}{2}R^4$
Semicircular area		$\frac{\pi R^2}{2}$	$\bar{y} = \frac{4R}{3\pi}$	$I_x = I_y = \frac{1}{8}\pi R^4$ $I_{x'} = \frac{(9\pi^2 - 64)R^4}{72\pi}$ $I_z = J = \frac{1}{4}\pi R^4$
Quarter circular area		$\frac{\pi R^2}{4}$	$\bar{x} = \frac{4R}{3\pi}$ $\bar{y} = \frac{4R}{3\pi}$	$I_x = I_y = \frac{1}{16}\pi R^4$ $I_z = J = \frac{1}{8}\pi R^4$
Semicircular arc			$\bar{x} = 0$ $\bar{y} = \frac{2R}{\pi}$	
Quarter-circular arc			$\bar{x} = \frac{2R}{\pi}$ $\bar{y} = \frac{2R}{\pi}$	

Table 7.3. Mass Moments of Inertia of Common Shapes

Shape		Mass Moment of Inertia
Slender rod		$\bar{I}_y = I_z = \frac{1}{12} m\ell^2$ $I_{y'} = \frac{1}{3} m\ell^2$ $I_x = 0$
Thin disk		$\bar{I}_x = \frac{1}{2} mr^2$ $\bar{I}_y = \frac{1}{4} mr^2$
Cylinder		$\bar{I}_x = \frac{1}{2} mr^2$ $\bar{I}_y = \frac{1}{12} m(\ell^2 + 3r^2)$
Rectangular prism		$\bar{I}_x = \frac{1}{12} m(a^2 + b^2)$ $\bar{I}_y = \frac{1}{12} m(\ell^2 + b^2)$ $\bar{I}_z = \frac{1}{12} m(\ell^2 + a^2)$ $\bar{I}_{y'} = \frac{1}{12} m(4\ell^2 + a^2)$
Solid sphere		$\bar{I}_x = \bar{I}_y = \bar{I}_z = \frac{2}{5} mr^2$

213

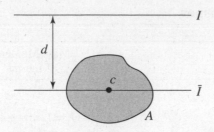

Figure 7.45

Parallel Axis Equation

To calculate moment of inertia with respect to an axis that is parallel to the centroidal axis (see Figure 7.45), the following equations can be used:

$$I = \bar{I} + Ad^2, \quad I = \bar{I} + md^2 \qquad [7.24]$$

where $\bar{I}$ is the either the centroidal mass or area moment of inertia.

In the parallel axis equation, $\bar{I}$ is the centroidal mass or area moment of inertia.

Example 7.24

For the figure shown, determine the area moment of inertia of the shaded area with respect to the x-axis.

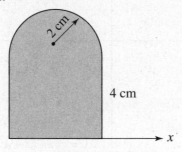

Figure 7.46

Solution:

$$I_{shape} = I_{rect} + I_{semicircle}$$

Using Table 7.2 for the rectangle:

$$I_x = \frac{1}{3}(4)(4)^3 = 85.33 \text{ cm}^4$$

For the semicircle, use the parallel axis equation:

$$I_x = \bar{I} + Ad^2$$

$$\frac{(9\pi^2 - 64)(2^4)}{72\pi} + \frac{\pi}{2}(2)^2\left[4 + \frac{4(2)}{3\pi}\right]^2 = 149.48 \text{ cm}^4$$

$$I_{x \text{ total}} = 85.33 + 149.48 = 234.81 \text{ cm}^4$$

Example 7.25

For the figure shown, determine the area moment of inertia with respect to the horizontal centroidal axis.

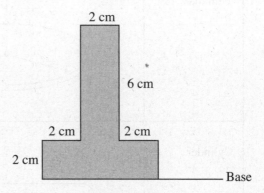

Figure 7.47

Solution:

Using the base as the reference first locate the centroid of the shape.

$$\bar{Y}_c = \frac{\Sigma y_i A_i}{A} = \frac{(6)(2)(1) + (6)(2)(5)}{(6)(2) + (6)(2)} = \frac{3 \text{ cm from}}{\text{the base}}$$

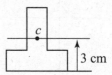

$$I = I_{flange} + I_{web}$$

Using parallel axis equation 7.24, the moment of inertia for the flange and the web can be determined as follows:

$$I_{flange} = \frac{1}{12}(6)(2)^3 + (6)(2)(2)^2 = 52 \text{ cm}^4$$

$$I_{web} = \frac{1}{12}(2)(6)^3 + (6)(2)(2)^2 = 84 \text{ cm}^4$$

$$I = 52 + 84 = 136 \text{ cm}^4$$

Example 7.25

For the figure shown, determine the mass moment of inertia with respect to the x-axis. Each bar has mass equal to m.

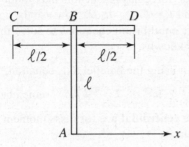

Figure 7.48

Solution:

$$I_{AB} = \frac{1}{12}m\ell^2 + m\left(\frac{\ell}{2}\right)^2 = \frac{1}{3}m\ell^2$$

$$I_{CD} = 0 + m\ell^2 = m\ell^2$$

$$I_{x\,\text{total}} = \frac{1}{3}m\ell^2 + m\ell^2 = \frac{4}{3}m\ell^2$$

Radius of Gyration

Radius of gyration (k) represents the distance from the axis at which the area or mass must be concentrated if the moment of inertia is to remain the same. The following equations are used to calculate the radius of gyration:

$$k = \sqrt{\frac{I}{A}}, \quad k = \sqrt{\frac{I}{m}} \qquad [7.25]$$

7.7 Summary

In this chapter the state of bodies at rest was presented. The important points discussed in this chapter are as follows:

1. For a particle to be in equilibrium, the resultant force acting on the particle must be equal to zero.
   - For a two-dimensional problem, this condition results in two equations, which are $\sum F_x = 0$ and $\sum F_y = 0$.

   - For a three-dimensional problem, this condition results in three equations, which are $\sum F_x = 0$, $\sum F_y = 0$, and $\sum F_z = 0$.

2. For a rigid body to be in equilibrium the resultant force and moment must be equal to zero.
   - For a two-dimensional problem, this condition results in three equations, which are $\sum F_x = 0$, $\sum F_y = 0$, and $\sum M = 0$.

   - For a three-dimensional problem, this condition results in six equations, which are $\sum F_x = 0$, $\sum F_y = 0$, $\sum F_z = 0$,

     $\sum M_x = 0$, $\sum M_y = 0$, and $\sum M_z = 0$.

3. When drawing a free-body diagram for a rigid body, remember to show all external reactions which are developed as a result of connections and supports. The type and number of these reactions can be determined as follows:
   - If the point on the rigid body cannot translate, a **force reaction** along the line of translation (x, y, or z) is required. If the point of connection can't rotate, a **moment reaction** about that axis is required.
   - Determine if a member is a two-force member. A two-force member is a member that is under the action of only two forces acting at different points on the member.

4. When analyzing a truss structure, the following important points should be considered:
   - If using the method of joints, remember that you can't have more than two unknown forces per joint, and use equations $\sum F_x = 0$ and $\sum F_y = 0$.

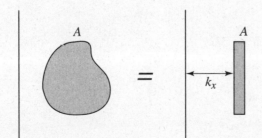

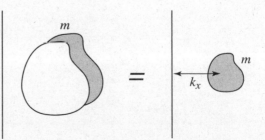

Figure 7.49

- If using the method of sections, remember not to cut more than three members at a time, and use equations $\sum F_x = 0$, $\sum F_y = 0$, and $\sum M = 0$.

5. For friction problems, remember the following points:
    - If the impending motion is not specified, assume the system is in equilibrium and then solve for friction and normal forces.
    - a) If $F < \mu_s N$, then the system is in equilibrium and the friction force calculated is correct.

    - b) If $F > \mu_s N$, then the system is in motion and for **particle problems only** the correct value of the friction force is $F = \mu_k N$.
- If the impending motion is specified, the friction force has reached its maximum value and $F = F_{max} = \mu_s N$. Now use the appropriate equilibrium equations to solve for the unknowns.

6. When using the parallel axis equation,

$$I = \overline{I} + Ad^2 \quad \text{or} \quad I = \overline{I} + md^2, \text{ remember that } \overline{I}$$

is the **centroidal** area or mass moment of inertia.

PRACTICE PROBLEMS

1. Determine the resultant of the forces.
   (A) 38 N
   (B) 121 N
   (C) 78 N
   (D) 525 N

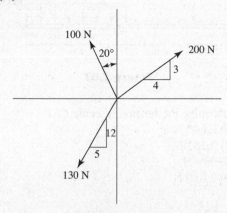

Figure 7.50

2. Determine the resultant of the forces exerted at point B if the tension in each cable is 450 N.
   (A) 735 N
   (B) 472 N
   (C) 375 N
   (D) 250 N

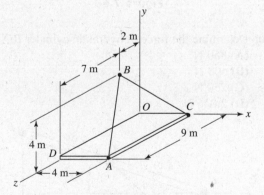

Figure 7.51

3. Determine the moment of force F about point A.
   (A) 30.4 N · m
   (B) 3.8 N · m
   (C) 16 N · m
   (D) 20 N · m

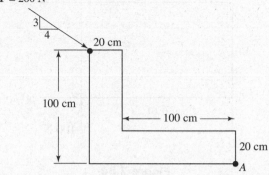

Figure 7.52

4. Determine the magnitude of moment of force $P = 50$ N about line OC.
   (A) 3,600 N · m
   (B) 424 N · m
   (C) 1,600 N · m
   (D) 0

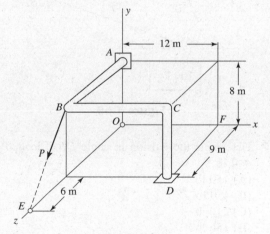

Figure 7.53

217

5. Determine the resultant of the couples.
   (A) 250 N · m
   (B) 550 N · m
   (C) 1,100 N · m
   (D) 500 N · m

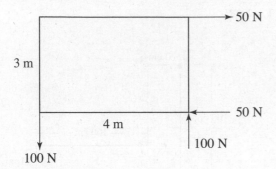

Figure 7.54

6. Determine the tension in cable *AC*.
   (A) 640 N
   (B) 863 N
   (C) 1,200 N
   (D) 550 N

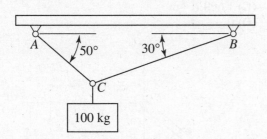

Figure 7.55

7. Determine the tension in cable *DB* if load *W* is 444 lb.
   (A) 150 lb
   (B) 450 lb
   (C) 210 lb
   (D) 156 lb

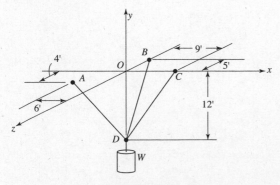

Figure 7.56

8. Determine the reaction at roller support *B*.
   (A) 365 N
   (B) 310 N
   (C) 435 N
   (D) 400 N

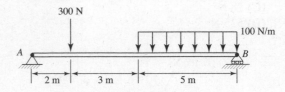

Figure 7.57

9. Determine the tension in cable *CE*.
   (A) 2,000 N
   (B) 2,253 N
   (C) 939 N
   (D) 2,080 N

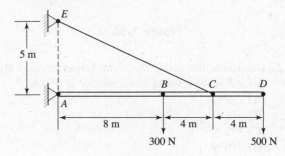

Figure 7.58

10. Determine the force in hydraulic cylinder *BD*.
    (A) 850 N
    (B) 608 N
    (C) 250 N
    (D) 2,026 N

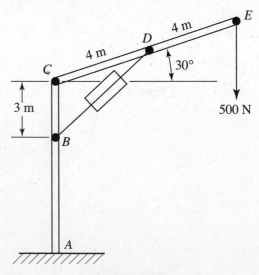

Figure 7.59

11. Determine the tension in cable *BE*.
   (A) 2,600 N
   (B) 1,310 N
   (C) 2,000 N
   (D) 917 N

13. Determine the force in member *CE* of the truss.
   (A) 30 kips
   (B) 24 kips
   (C) 48 kips
   (D) 18 kips

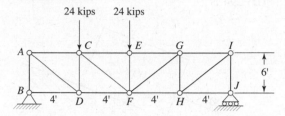

Figure 7.62

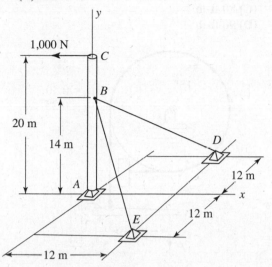

Figure 7.60

14. Determine the resultant reaction at *B* for the frame.
   (A) 300 N
   (B) 450 N
   (C) 200 N
   (D) 150 N

12. Determine the force in member *CE* of the truss.
   (A) 3 kN
   (B) 5 kN
   (C) 8 kN
   (D) 0

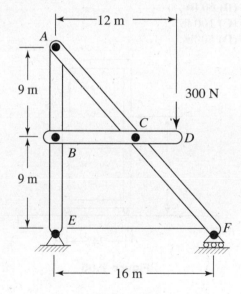

Figure 7.63

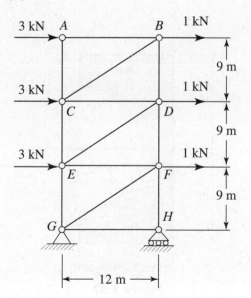

Figure 7.61

219

15. Determine the friction force if $\mu_s = 0.3$ and $\mu_k = 0.2$.
   (A) 170 N
   (B) 191 N
   (C) 850 N
   (D) 294 N

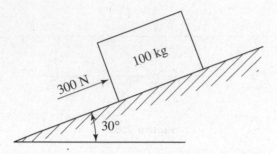

Figure 7.64

16. Determine the force required to move the cabinet. The casters at A and B are locked, the cabinet weighs 100 lb, $h = 30$ in, and $\mu_s = 0.3$.
   (A) 30 lb
   (B) 60 lb
   (C) 100 lb
   (D) 50 lb

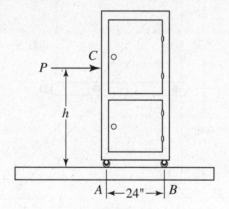

Figure 7.65

17. Determine the torque that must be applied to the flywheel in order to keep it rotating at a constant speed if $P = 8$ lb.
   (A) 45 ft-lb
   (B) 120 ft-lb
   (C) 80 ft-lb
   (D) 90 ft-lb

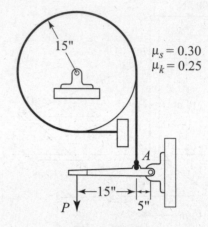

Figure 7.66

18. Determine the x-coordinate of the centroid of the plane area shown.
   (A) 2 cm
   (B) 6.25 cm
   (C) 2.3 cm
   (D) 0

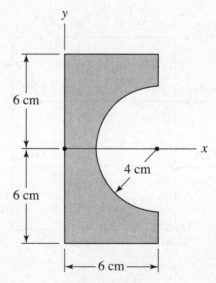

Figure 7.67

19. Determine the area moment of inertia with respect to the *x*-axis.
- **(A)** 435 cm⁴
- **(B)** 525 cm⁴
- **(C)** 382 cm⁴
- **(D)** 125 cm⁴

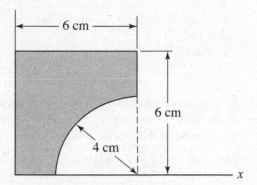

Figure 7.68

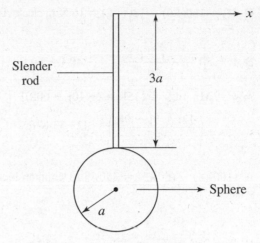

Figure 7.70

20. Determine the centroidal area moment of inertia with respect to the horizontal axis.
- **(A)** 36 cm⁴
- **(B)** 257 cm⁴
- **(C)** 293 cm⁴
- **(D)** 220 cm⁴

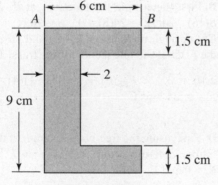

Figure 7.69

21. Determine the mass moment of inertia with respect to the *x*-axis. The sphere and the slender rod have the same mass, *m*.
- **(A)** 19.4 *ma²*
- **(B)** 24 *ma²*
- **(C)** .33 *ma²*
- **(D)** 16 *ma²*

Answer Key

1. B	**8.** C	**15.** B
2. A	**9.** B	**16.** A
3. C	**10.** D	**17.** D
4. B	**11.** B	**18.** C
5. A	**12.** A	**19.** C
6. B	**13.** B	**20.** C
7. C	**14.** D	**21.** A

Answers Explained

1. B $R_x = 200\left(\frac{4}{5}\right) - 100 \sin 20 - 130\left(\frac{5}{13}\right)$

$\qquad = 75.79 \text{ N}$

$\qquad R_y = 200\left(\frac{3}{5}\right) + 100 \cos 20 - 130\left(\frac{12}{13}\right)$

$\qquad = 93.97 \text{ N}$

$\qquad R = \sqrt{(75.79)^2 + (93.97)^2} = 120.7 \text{ N}$

2. A $T_{BA} = 450\dfrac{4\mathbf{i} - 4\mathbf{j} + 7\mathbf{k}}{9}$

$\qquad = 200\mathbf{i} - 200\mathbf{j} + 350\mathbf{k}$

$\qquad T_{BC} = 450\dfrac{4\mathbf{i} - 4\mathbf{j} - 2\mathbf{k}}{6}$

$\qquad = 300\mathbf{i} - 300\mathbf{j} - 150\mathbf{k}$

Also, $R_x = 500$ N, $R_y = -500$ N, $R_z = 200$ N.

$R = \sqrt{(500)^2 + (-500)^2 + (200)^2} = 734.8 \text{ N}$

3. C $M_A = 160(1.0) - 120(1.2) = 16$ N·m clockwise

4. B $\mathbf{P} = 50\,\dfrac{-8\,\mathbf{j} + 6\,\mathbf{k}}{10} = -40\,\mathbf{j} + 60\,\mathbf{k}$

$M_{OC} = \mathbf{M}_O \cdot \mathbf{u}_{OC} = [15\mathbf{k} \times (-40\mathbf{j} + 60\mathbf{k})]$

$\cdot \left(\dfrac{12\mathbf{i} + 8\mathbf{j} + 9\mathbf{k}}{17}\right) = 423.53$ N·m

5. A $(100)(4) - (50)(3) = 250$ N·m counterclockwise

6. B $R_x = \Sigma F_x = T_{BC} \cos 30 - T_{AC} \cos 50 = 0$
$R_y = \Sigma F_y = T_{BC} \sin 30 + T_{AC} \sin 50 - 981 = 0$

Solve simultaneously: $T_{AC} = 863$ N.

7. C $\mathbf{T}_{DA} = T_{DA}\left(\dfrac{-6\mathbf{i} + 12\mathbf{j} + 4\mathbf{k}}{14}\right)$

$\mathbf{T}_{DB} = T_{DB}\left(\dfrac{12\mathbf{j} - 5\mathbf{k}}{13}\right)$

$\mathbf{T}_{DC} = T_{DC}\left(\dfrac{9\mathbf{i} + 12\mathbf{j}}{15}\right)$

Then,

$R_x = 0 \Rightarrow -\dfrac{3}{7}T_{DA} + \dfrac{9}{15}T_{DC} + \dfrac{12}{15}T_{DC} = 0$

$R_y = 0 \Rightarrow \dfrac{6}{7}T_{DA} + \dfrac{12}{13}T_{DB} - 444 = 0$

$R_z = 0 \Rightarrow \dfrac{2}{7}T_{DA} - \dfrac{5}{13}T_{DB} = 0$

Solve simultaneously: $T_{DB} = 210$ lb.

8. C $\Sigma M_A = 0$,

$B(10) - 500(7.5) - 300(2) = 0$,

$B = 435$ N

9. B $\Sigma M_A = \mathbf{0}$,

$\left(\dfrac{5}{13}T_{CE}\right)(12) - (500)(16) - (300)(8) = 0$,

$T_{CE} = 2{,}253$ N

10. D Use the FBD of member *CDE*. Then:

Note that member *BD* is a straight two-force member and the angle between member *BD* and the horizontal axis can be determined by geometry and is equal to 55.3°.

$\Sigma M_c = 0$
$-(F_{BD} \cos 55.3)(2) + (F_{BD} \sin 55.3)(4 \cos 30) - 500(8 \cos 30) = 0$
$F_{BD} = 2{,}026$ N

11. B Because of symmetry the tension in cable *BE* is the same as tension in cable *BD*. Therefore, the moment at *A* about the *z*-axis should be used.

$\Sigma M_A = 0, \qquad 2\left(\dfrac{12}{22}T_{BE}\right)(14) - 1{,}000(20) = 0,$

$T_{BE} = 1{,}309.5$ N

12. A Cut through members *CE*, *DE*, and *DF*, and use the FBD of the upper section of the truss. Then:

$\Sigma M_D = 0, \qquad F_{CE}(12) - 3(9) - 1(9) = 0,$
$F_{CE} = 3$ kN T

13. B First, determine the reactions at *B*. $\Sigma M_J = 0$
$B(16) - 24(12) - 24(8) = 0 \quad B = 30$ N
Cut through members *CE*, *CF*, and *DF*, and use the FBD of the left section of the truss. Then:

$\Sigma M_F = 0, \qquad F_{CE}(6) + 30(8) - 24(4) = 0,$
$F_{CE} = 24$ kips C

14. D First, using the free-body diagram of the whole frame, determine the reactions at *E*. We get $E_x = 0$ and $E_y = 75$ N.

Draw the free-body diagram of member *BCD* and take the moment about point *C*:

$\Sigma M_c = 0 \quad B_y(8) - 300(4) = 0 \quad B_y = 150$ N

Draw the free-body diagram of *ABE* and take the moment about *A*:

$\Sigma M_A = 0 \quad B_x = 0$

Therefore, the resultant pin reaction at *B* is:

$B = \sqrt{(0)^2 + (150)^2} = 150$ N

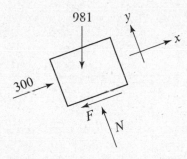

15. B Assume that the block is in equilibrium and draw the free-body diagram.

$\Sigma F_y = 0, \qquad N = 849.5 \text{ N}$
$\Sigma F_x = 0, \qquad F = 190.5 \text{ N}$

Since $F < \mu_s N$, the block is in equilibrium and the friction force is 190.5 N.

16. A Draw the free-body diagram of the cabinet.

$\Sigma F_x = 0 \quad P - F_A - F_B = 0 \quad P - \mu_s (N_A + N_B) = 0$
$\Sigma F_y = 0 \quad N_A + N_B = 0$
Solve for $P = 30$ lb

17. D Use the FBD of the lever, and take the moment about the pin:

$$\Sigma M_A = 0, \qquad T_1 = 32 \text{ lb}$$

Use equation 7.15 with $\mu_k = 0.25$ and $\beta = 270° = 3\pi/2$. Then $T_2 = 103.94$ lb.

Use the FBD of the drum, and take the moment about its center:

$$\Sigma M = 0, \qquad M = 89.925 \text{ ft-lb}$$

18. C Break the shaded area into:
Rectangle – semicircle

$$\overline{X}_c = \frac{\Sigma x_i A_i}{A}$$

$$= \frac{(3)(72) - \left[6 - \dfrac{4(4)}{3\pi}\right]\left[\dfrac{1}{2}\pi(4)^2\right]}{72 - \dfrac{1}{2}\pi(4)^2} = 2.3 \text{ cm}$$

19. C $I_x = \dfrac{6(6)^3}{3} - \dfrac{\pi(4)^4}{16} = 381.73 \text{ cm}^4$

20. C $I = 2I_{\text{flange}} + I_{\text{web}}$

$$I_{\text{flange}} = \frac{1}{12}(6)(1.5)^3 + 9(3.75)^2 = 128.25 \text{ cm}^4$$

$$I_{\text{web}} = \frac{1}{12}(2)(6)^3 = 36 \text{ cm}^4$$

$$I = 2(128.25) + 36 = 292.5 \text{ cm}^4$$

21. A $I_x = I_{\text{rod}} + I_{\text{sphere}}$

$$= \frac{1}{3}m(3a)^2 + \left[\frac{2}{5}ma^2 + m(4a)^2\right]$$

$$= 19.4 \ ma^2$$

Dynamics, Kinematics, and Vibrations

Mark Frisina, PE

Introduction

Dynamics, Kinematics, and Vibrations is the study of motion and the forces that govern the motion. The morning session of the FE examination typically contains 6 to 10 examination problems related to various dynamics topics, and the afternoon sessions for the Other Discipline and Mechanical Discipline exams have approximately 10% to 15% of the total questions in the area of dynamic systems. The following review modules are intended to assist the reader in reinforcing his or her knowledge of relevant dynamics principles and problem-solving techniques that are likely to appear on the FE exam.

8.1 **Kinematics of Particles**
8.2 **Kinetics of Particles**
8.3 **Work and Energy**
8.4 **Rigid Body Motion in a Plane**
8.5 **Vibrations**
8.6 **Summary**

8.1 Kinematics of Particles

Kinematics is the study of patterns of motion in space. This section focuses on describing two-dimensional plane motion and three-dimensional space motion by employing rectangular coordinates, tangential and normal coordinates, and radial and transverse components. The effect of forces on motion is outside the domain of kinematics. The most important parameters in particle kinematics are the vectors that represent position, velocity, and acceleration.

- $\mathbf{r}(t)$ is the *instantaneous* (time-dependent) *position vector* to a particle from a defined reference origin in space.
- *Instantaneous velocity* is defined as the time rate of change of the position vector and is represented by the relationship

$$\mathbf{v}(t) = \frac{d\mathbf{r}}{dt}, \qquad [8.1]$$

where $\mathbf{v}$ has typical units of m/s or ft/s.

- *Instantaneous acceleration* is represented by the symbol $\mathbf{a}(t)$ and is defined as the rate of change of the velocity vector. The governing equation for acceleration is

$$\mathbf{a}(t) = \frac{d\mathbf{v}(t)}{dt} = \frac{d^2r(t)}{dt^2} \qquad [8.2]$$

> Instantaneous **velocity and acceleration** are successive derivatives of the **position** equation.

225

Generally, the units for acceleration are m/s^2 or ft/sec^2.

$$\mathbf{v}(t) = \frac{dx(t)}{dt} i = \frac{dy(t)}{dt} j = \frac{dz(t)}{dt} k = \dot{x}\mathbf{i} + \dot{y}\mathbf{j} + \dot{z}\mathbf{k}$$

[8.3]

Acceleration vector:

$$\mathbf{a}(t) = \frac{d\mathbf{v}(t)}{dt} = \frac{d^2x(t)}{dt^2}\mathbf{i} + \frac{d^2y(t)}{dt^2}\mathbf{j} + \frac{d^2z(t)}{dt^2}$$

$$= \ddot{x}\mathbf{i} + \ddot{y}\mathbf{j} + \ddot{z}\mathbf{k}$$

[8.4]

Rectilinear Motion

Rectilinear motion is straight-line motion in space or in a plane. In short, this motion is restricted to a single line of action in space or in a plane. If a particle's motion is confined to an *x*-axis (horizontal direction) as an example of rectilinear motion, then, if the initial position and velocity are denoted by x_0 and v_0, respectively, general equations for conditions of variable velocity $v(t)$ and acceleration $a(t)$ are given for motion from time t_0 to t:

- *Instantaneous position:*

  $x(t) = x_0 + \int_0^t v(t)\, dt$ [8.5]

- *Instantaneous velocity:*

  $v(t) = v_0 + \int_0^t a(t)\, dt$ [8.6]

Uniform velocity occurs when acceleration is zero.

- For *uniform-velocity* (v = constant), straight-line motion, which requires $a = 0$:

  $x(t) = x_0 + v(t - t_0)$ [8.7]

- The particle's *displacement* Δx is defined simply as the change of position:

  $\Delta x = x(t) - x_0 = \int_0^t v(t)\, dt$ [8.8]

In FE exam problems in which a velocity versus time graph is provided, displacement can be readily obtained by calculating the *net area* of the graph during a specified time interval. This capability is especially useful when the velocity graph contains several different $v(t)$ functions.

For *uniform acceleration* (a = constant) in rectilinear motion, the velocity is a linear function; and

if the initial conditions x_0 and v_0 are given, the key equations are as follows:

$$x(t) = x_0 + v_0 t + \frac{at^2}{2} \quad \text{for } t_0 = 0 \qquad [8.9]$$

$$v(t) = v_0 + at \qquad [8.10]$$

$$v^2 = v_0{}^2 = 2a(x - x_0) \qquad [8.11]$$

Example 8.1

The rectilinear motion of a particle is described by the equation

$$x(t) = 6t^2 - 24t + 7$$

where *x* is in meters.

Solve for the displacement at the instant when the velocity is zero.

Solution:

First determine the particle's velocity equation.

$$v(t) = \frac{dx(t)}{dt} = 12t - 24$$

Velocity $v = 0$ at $t = 2.0$ s.

$\therefore$ Displacement $\Delta x = \int v(t)dt$

$\quad\quad = \int_0^2 (12t - 24)dt$

$\quad\quad = (6t^2 - 24t)\big|_0^2$

$\quad\quad = (24 - 48) - 0$

$\quad\quad = -24.0$ m

Plane Circular Motion

Plane circular motion requires that the particle traverse a path in a plane such that the *radius of curvature R is constant* with respect to a reference origin *C*, which is the center of curvature of the path. *Tangential* and *normal* coordinates are used to profile the circular-type motion; and with reference to Figure 8.1, it is important to recall that the tangential direction is along the line of action of the velocity and is typically positive when the motion results in an increase of angle ϕ in a counterclockwise direction. The normal direction is directed positively from the particle toward the center of curvature.

The **normal direction** always points inward toward the center of curvature.

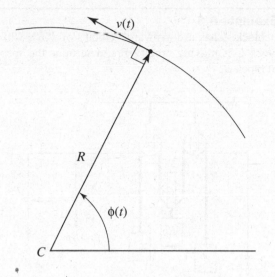

Figure 8.1

For circular motion, the governing equations are as follows:

- *Angular velocity:* $\omega = \dfrac{d\phi}{dt}$ (rad/s)   [8.12]

- *Angular acceleration:*

$$\propto = \dfrac{d\omega}{dt} \text{ (rad/s}^2) \qquad [8.13]$$

- *Tangential velocity:* $v(t) = R\omega$ [8.14]

- *Tangential acceleration:*

$$a_t = R\propto = \dfrac{R\,d\omega}{dt} \qquad [8.15]$$

- *Normal acceleration:*

$$a_n = \dfrac{v^2}{R} = R\omega^2 \qquad [8.16]$$

- *Total acceleration* (scalar):

$$a(t) = \sqrt{a_1^2 + a_n^2} \qquad [8.17]$$

- *Arc length* or distance along a curve:

$$s(t) = R\phi \qquad [8.18]$$

Example 8.2

A particle travels in a circular path along a plane surface in such a way that its radius of curvature is 60.0 m. If the angular measure is according to $\phi(t) = 4t^2 + 9$ rad, at what time will the particle's normal acceleration attain a value of 3,000 m/s^2?

Solution:

Since normal acceleration $a_v = v^2/R = 3,000$, the speed v of the particle is readily obtained with R specified.

$$v = \sqrt{3,000R} = \sqrt{180,000 \text{ m}^2/\text{s}^2} = 424.2 \text{ m/s}$$

$$= R\omega = (60.0 \text{ m})\left(\dfrac{d\phi}{dt}\right) = (60.0)(8.0t) = 424.2$$

$$\therefore t = \dfrac{424.2 \text{ m/s}}{480.0 \text{ m/s}^2} = 0.88 \text{ s}$$

Radial and Transverse Components of Plane Motion

Radial and transverse curvilinear motion occurs when the radius of curvature of a particle's path in a plane usually varies with time, except in the unique case of circular motion. In Figure 8.2 the *radial direction* is along the line of action from the reference origin to the particle, and the radial distance at any time t is denoted by $r(t)$. Note that the positive radial direction is directed away from the origin. The *transverse direction* is always perpendicular to the radial direction, and the positive transverse direction tends to cause an increase in the angle of the position vector in a counterclockwise sense as shown. Spiral motion and elliptical motion are classic examples of plane motion that are best profiled by radial and transverse components.

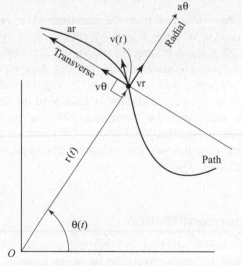

Figure 8.2

Radial and transverse components are effectively two-dimensional versions of cylindrical coordinates. FE exam problems related to radial and transverse components usually provide equations for radial distance $r(t)$ and angular measure $\theta(t)$, and typical problems often ask for computations of velocity and/or acceleration magnitudes using radial and transverse components. The governing equations of radial and transverse plane motion are as follows:

- *Radial velocity:* $\quad v_r = \dot{r} = \dfrac{dr(t)}{dt}$ [8.19]

- *Tranverse velocity:* $\quad v_\theta = r\dot{\theta}$ [8.20]

- *Total speed:* $\quad v(t) = \sqrt{v_r^2 + v_\theta^2}$ [8.21]

- *Radial acceleration:* $\quad a_r = \ddot{r} - r\dot{\theta}^2$ [8.22]

- *Transverse acceleration:*

$$a_\theta = r\ddot{\theta} + 2\dot{r}\dot{\theta} \qquad [8.23]$$

- *Total acceleration:*

$$a(t) = \sqrt{a_r^2 + a_\theta^2} \qquad [8.24]$$

Relative Plane Motion

Relative motion of one object with respect to another in a plane is determined by taking the vector difference of the individual absolute quantities of the two objects measured with respect to a common reference datum. Specifically, if the relative velocity of object T is desired with respect to (as viewed from) object S, and if $\mathbf{v}_s$ and $\mathbf{v}_t$ are the instantaneous velocity vectors for S and T, respectively, then note the specific order of the vector subtraction in the following equation:

Relative velocity of T with respect to S: $\mathbf{v}_{T/S} = \mathbf{v}_T - \mathbf{v}_S$

Bear in mind that the solution of a vector subtraction is often found by setting up a *vector triangle* and calculating the magnitude and direction of the desired relative velocity through the use of the law of cosines and law of sines. If relative position and/or relative acceleration is requested on an FE exam problem, first determine absolute position vectors or acceleration vectors at a specified time and then use the solution technique described above for relative velocity.

Dependent Motion

When the position of one particle depends on the position of another particle or other particles, the motion is said to be dependent. Dependent motion problems usually involve particles that are connected via some pulley arrangements. The main objective is to establish a relation between the velocities or acceleration of these connected particles. The following example will demonstrate how this is done:

Example 8.3

If block A has a downward velocity of 8 ft/s while block C is moving up at 2 ft/s, determine the speed of block B.

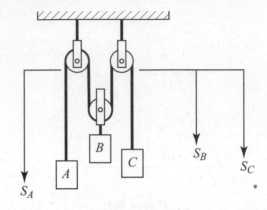

Figure 8.3

Solution:

We first write the equation relating the total length of the rope in terms of position vectors (s_A, s_B, s_C).

$$S_A + 2S_B + S_C + \text{constant} = L$$

To relate velocities, take the time derivative of the equation.

$$V_A + 2V_B + V_C = 0$$

Since $V_A = -8$ ft/s and $V_C = 2$ ft/s, then

$$-8 + 2\,V_B + 2 = 0 \quad V_B = 3 \text{ ft/s, so } B \text{ is moving}$$
upward.

Projectile Motion

Problems involving projectile motion are quite often represented on the dynamics sections of the FE exam. This type of curvilinear motion usually involves the following conditions:

- The particle's trajectory is confined to an *xy* plane.
- Only gravitational force acts immediately after the particle is initially projected; wind resistance and aerodynamic forces are neglected.
- The initial elevation y_0 of the object is given relative to a datum.
- Other initial conditions, such as velocity v_0 and the projection angle θ that the velocity makes with the horizontal, may be provided or need to be determined.

Figure 8.3 illustrates the initial conditions listed above. Note that the object may have two different times at the same elevation because of the quadratic equation for $y(t)$ given below. Be prepared to solve for the maximum horizontal distance (range) R, which occurs at $y = 0$, as well as for the time required to reach maximum height y_μ.

> The **maximum height** in projectile motion occurs when the vertical velocity v_y is zero.

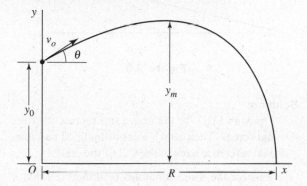

Figure 8.4

Projectile motion equations (assuming $t_0 = 0$):

$$x(t) = v_0 \cos \theta t$$

$$y(t) = y_0 + v_{v0}t - \frac{gt^2}{2} = y_0 + v_0 \sin \theta\, t - \frac{gt^2}{2}$$

$$v_x = v_0 \cos \theta = \text{constant}$$

$$v_y(t) = v_0 \sin \theta - gt$$

$$a_y = -g$$

Example 8.4

A football player claims that he can punt a ball over a 10.0-ft-high wall that is 35 yd away along a horizontal field. If the player kicks the ball from an elevation of 3.0 ft and the ball is projected at an initial speed of 60.0 ft/sec in a direction of 40° from the horizontal, will he successfully prove his claim?

Solution:

Assuming that wind resistance is negligible, first calculate the time when the ball reaches a horizontal position of 35 yd and then check the elevation at this position.

$$x(t) = v_x t = 35 \text{ yd} = 105.0 \text{ ft}$$

$$v_x = (60.0 \text{ ft/s})\cos 40° = 45.96 \text{ ft/s}$$

$$t = \frac{x(t)}{v_x} = \frac{105 \text{ ft}}{45.96 \text{ ft/s}} = 2.28 \text{ s}$$

Check whether elevation $y(t) \geq 10.0$ ft.

$$y(t) = y_0 + v_{y0}t - \frac{gt^2}{2}$$

$$= 3.0 + 60 \sin 40° t - \frac{(32.2)t^2}{2}$$

$$y(2.28) = 3.0 + 60(0.642)(2.28) - 16.1(2.28)^2$$
$$= 3.0 + 87.9 - 83.7$$

$$= 7.2 \text{ ft}$$

$\therefore$ The ball does not clear the wall.

8.2
Kinetics of Particles

Kinetics is the study of the effects of forces on motion. In particle dynamics, Newton's second law of motion is the governing principle.

Newton's Second Law of Motion

Newton's second law for a constant-mass object or particle: $\Sigma \mathbf{F} = m\mathbf{a}$

Here, $\Sigma \mathbf{F}$ is the *net external force* acting on the object and typically has units of pounds force or kilonewtons, and a is the *acceleration vector,* that must have the same instantaneous direction as the net force.

Since numerous FE problems in kinetics are centered on the second law principle, the following general solution procedure is strongly recommended:

1. Construct an effective *free-body diagram* (FBD) of the object being analyzed. This requires identifying and drawing all external forces (known and/or unknown) acting on the mass at their point of application.
2. Select an appropriate coordinate system and attach reference axes to the center of gravity (CG) of the body.
3. Resolve all known and/or unknown force components along each axis, paying strict attention to sign convention, and then write the second law equations along the established axial directions.

4. Solve the simultaneous component equations for the desired forces and/or accelerations, recognizing that only two unknowns can normally be solved in a two-dimensional problem.

Note also that conversion errors are often made in second-law computations involving mass or weight, and observe the following cautions:

- When using $a = \Sigma F/m$, be aware that 50 lbf/50 lbm $\neq$ 1.0 ft/sec^2. The universal conversion 1.0 lbf = 32.2 lbm ft/sec^2 should be carefully employed.
- Avoid miscalculations of weight using the simple formula $W = mg$.

Proper use of the conversion given above (also known as g_c) will prevent errors.

Rectilinear Motion

Straight-line motion problems involving a fixed-mass object subjected to a system of external forces are typically solved using scalar component versions of Newton's second law described above. For rectilinear motion in an *xyz* coordinate frame, where the object's instantaneous position is defined by the functions $x(t)$, $y(t)$, and $z(t)$, the governing component equations are as follows:

$$\Sigma F_x = ma_x = \frac{md^2 x(t)}{dt^2} = m\ddot{x}$$

$$\Sigma F_y = ma_y = m\ddot{y}$$

and

$$\Sigma F_z = m\ddot{z} \qquad [8.25]$$

Note that the magnitude of the total (net) force acting on the object is

$$\Sigma F = \sqrt{\Sigma F_x^2 + \Sigma F_y^2 + \Sigma F_z^2}$$

For motion along an *inclined plane*, a set of inclined and normal coordinates is employed, and the equations are

$$\Sigma F_I = ma_I$$

and

$$\Sigma F_n = ma_n \qquad [8.26]$$

Example 8.5

In Figure 8.5, a 150.0-kg crate is pulled up a frictionless inclined plane at 20° from the horizontal by a constant applied force F_A as shown. If the resulting acceleration is 7.0 m/s^2 along the incline, determine the magnitude of F_A.

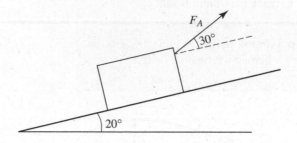

Figure 8.5

Solution:

Draw an FBD for the crate, and represent external forces. Then attach a set of inclined and normal reference axes to the CG of the crate.

Resolve the force components along the *I* and *n* directions as shown in Figure 8.6.

> Effective **free-body diagrams** are the key to problems involving **Newton's second law**.

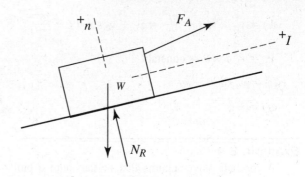

Figure 8.6

$$F_{AI} = F_A \cos 30° = 0.866 F_A$$

$$W_I = -W \sin 20° = -(150 \text{ kg})(9.81 \text{ m/s}^2)(0.342)$$
$$= -503.28 \text{ N}$$

$$F_{An} = F_A \sin 30° = 0.5 F_A$$

$$W_n = -W \cos 20° = -(1,471.5 \text{ N})(0.940)$$
$$= -1,383.2 \text{ N}$$

Write the second law equations:

$$\Sigma F_I = ma_I = (150 \text{ kg})(7.0 \text{ m/s}^2) = 1,050.0 \text{ N}$$

$$\begin{aligned} F_{AI} + W_I &= 0.866 F_A + (-503.28 \text{ N}) \\ &= 1,050.0 \text{ N} \end{aligned} \qquad (1)$$

$\Sigma F_\nu = ma_\nu = 0$ (assuming that crate remains in contact with incline)

$$\begin{aligned} 0.5F_A + (-1,383.2 \text{ N}) + N_P \\ = 0 \quad (N_P \geq 0 \text{ required for contact}) \end{aligned} \qquad (2)$$

Then, from equation (1)

$$F_A \frac{1,050 + 503.28}{0.866} = 1,793.6 \text{ N}$$

And, from equation (2)

$$N_P = -0.5(1,793.6) + 1,383.2 = 486.4 \text{ N} \geq 0$$

∴ Validates that $F_A = 1,793.6$ N

Plane Curvilinear Motion

For curvilinear motion in a plane with a fixed-mass object or particle, Newton's second law is employed to solve most FE problems. Although the majority of exam problems in this area relate to circular or arc-type motion, which is described by tangential and normal components given in the following section, occasionally a few problems involve variable radii of curvature, and these are best solved by employing radial and transverse coordinates. Central-force problems, such as gravity-driven motion, are in this category.

Noncircular motion:

$\Sigma F_r = ma_r$, valid for *radial forces* directed toward or away from the origin of the path.

$\Sigma F_\theta = ma_\theta$, valid for *transverse forces* perpendicular to the radial direction.

Use the expressions for a_ρ and a_θ given on page 228.

Tangential and Normal Forces

For circular-type motion in a plane involving a fixed-mass object, *tangential and normal forces* are involved in the solution of a variety of FE problems. Specific problem categories that often appear include

cable tension for an object traveling in a circular path, roller coasters requiring reaction forces, and centripetal forces/accelerations.

Tangential forces (circular or arc-type motion):
$\Sigma F_\tau = ma_\tau = mR\alpha$ where radius R is a constant.

Normal force(s): $\Sigma F_\nu = ma_\nu = \dfrac{mv^2}{R} = mR^2\omega$

Note that *centripetal force* F_ψ is equivalent to the net normal force, and thus

$$F_\psi = \frac{mv^2}{R} \qquad [8.27]$$

Example 8.6

A 1.8-kg ball attached to a 2.0-m inextensible cable swings in a vertical plane as illustrated in Figure 8.7. Determine the required speed of the ball (units of m/s) if the cable tension is 140.0 N when $\theta = 65°$.

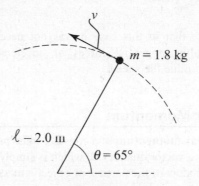

Figure 8.7

Solution:

Draw an FBD of the ball, as shown in Figure 8.8, and use the tangential and normal directions as references due to the circular motion. Since the cable tension is entirely along the normal direction, write the equation of motion in the *n*-direction, introducing the unknown speed *v*:

$$\Sigma F_\nu = \frac{mv^2}{R}$$

$$T + W_\nu = \frac{(1.8 \text{ kg})v^2}{2.0 \text{ m}}$$

$$140.0 \text{ N} + (1.8 \text{ kg})(9.81 \text{ m/s}^2) \cos 35° = (0.9 \text{ kg/m})v^2$$

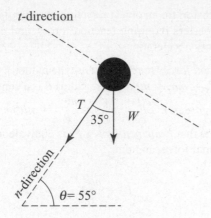

Figure 8.8

$$\frac{140.0 + 14.46}{0.9} \text{ N} \cdot \text{m/kg} = v^2$$

$$\therefore v = \sqrt{171.62 \text{ m}^2/\text{s}^2} = 13.1 \text{ m/s}$$

Note that in this case it was not necessary to involve tangential forces or accelerations to determine the speed.

Linear Momentum

The **linear momentum** of a constant-mass particle or object is a vector quantity **L,** which is simply defined as the product of the particle or object's mass and the instantaneous velocity vector:

$$\mathbf{L} = m\mathbf{v} \qquad [8.28]$$

Conventional units of **L** are lbf-sec or N · s.

Also, recall that for a fixed-mass object the net force acting equals the time rate of change of linear momentum. The equation

$$\Sigma \mathbf{F} = \frac{d\mathbf{L}}{dt} = \frac{md\mathbf{v}}{dt}$$

effectively restates Newton's second law.

An important ramification of the above equation is that, if the net force acting upon a fixed-mass object is zero, it is necessary that **L** = constant and the object's velocity remains constant. A clear example of this condition occurs when a puck travels across a horizontal "frictionless" air table at a constant speed.

For the dynamics section of the FE exam, an understanding of linear momentum concepts is essential in solving frequently represented *impact* and *impulse-momentum* problems (see pages 232–234).

Conservation of Linear Momentum

For a single particle having constant mass, it is required that the linear momentum remain constant (conserved) whenever the net force is zero over a specified period of time, as discussed above. Similarly, for a *system of constant-mass particles or rigid bodies,* if the net external force acting on the system is zero, then the total sum of the individual linear momentum vectors for all members within the system must remain constant or *conserved:*

$$\Sigma \mathbf{F}_{\text{sys}} = 0 \rightarrow \mathbf{L}_{\text{sys}} = \text{constant} \qquad [8.29]$$

For a system of masses $m_1, m_2, \ldots, m_n,$ *conservation of linear momentum* requires that

> **Linear momentum is conserved**
> when the net external force is zero.

$$m_1\mathbf{v}_1 + m_2\mathbf{v}_2 + \ldots + m_n\mathbf{v}_n$$
$$= m_1\mathbf{v}_1' + m_2\mathbf{v}_2' + \ldots + m_n\mathbf{v}_n'$$
$$= \text{constant} \qquad [8.30]$$

The prime notation for velocities in the above equation signifies a later time condition. Note that, when $\Sigma \mathbf{F}_{\text{sys}} = 0$, linear momentum must be conserved in any or all directions.

Scalar equations may be written in a particular direction (such as the y-direction) as follows:

$$m_1v_{1y} + m_2v_{2y} + \ldots + m_nv_{ny} = m_1v_{1y}'$$
$$+ m_2v_{2y}' + \ldots + m_nv_{ny}' = \text{constant}$$

Equations similar to the one given above will be used later in solving impact (collision) problems, in which the elasticity of the colliding objects must also be given.

In solving problems involving the conservation of linear momentum principle, it is imperative to pay close attention to the sign conventions of all scalar velocities.

Impulse-Momentum Principle

Impulse is a vector represented by the symbol **Imp** and defined by the equation

$$\mathbf{Imp}_{1\rightarrow 2} = \int_1^2 \mathbf{F(t)}dt \qquad [8.31]$$

where **F(t)** is the impulsive force acting upon a rigid body or particle during a time period from $t_1 \leq t \leq t_2$. For a constant-mass object, the *impulse equals the change of linear momentum* according to the equation

$$\mathbf{Imp}_{1\rightarrow 2} = m(\mathbf{v}_2 - \mathbf{v}_1) \qquad [8.32]$$

If the impulsive force acting upon the object is constant during a specified time period, then

$$\mathbf{F}(t_2 - t_1) = m(\mathbf{v}_2 - \mathbf{v}_1)$$

Scalar impulse/momentum equations may be written in any direction, such as the x-direction below, for a constant-mass and a time-invariant impulsive force:

$$\text{Imp}_x = F_x\,(t_2 - t_1) = m(\mathbf{v}_{2x} - \mathbf{v}_{1x}) \quad [8.33]$$

Units of impulse are lbf-sec or N · s, and caution must be exercised in converting from interim units of lbm-ft/sec or kg · m/s. Also, recall that positive impulsive forces result in an increase of momentum, whereas negative impulsive forces are directed opposite the object's initial velocity and cause a reduction in momentum.

Impulse-momentum problems on the FE exam often test for the impulse time period required for a given force to produce a known velocity change. Vector-triangle solution methods are useful in avoiding a set of tedious scalar equations.

Example 8.7

A 5.0-oz baseball has an initial horizontal velocity of 110.0 ft/sec at the instant that it is struck by a bat. If the ball is in contact with the bat for 0.15 sec, and the ball leaves the bat with a velocity of 140.0 ft/sec at 30° with respect to the horizontal, solve for the magnitude and direction of the average force applied to the ball.

Solution:

First compute the initial and final momentum vectors; then construct a vector triangle to solve for impulse **Imp**. Assume that the bat force is constant (average value), and consider the ball's weight to be negligible in comparison.

Initial momentum: $m\mathbf{v}_1$

$$= \left(\frac{5}{16}\ \text{lbm}\right)(110.0\ \text{ft/sec}) = 34.37\ \text{lbm-ft/sec}$$

Final momentum: $m\mathbf{v}_2$

$$= \left(\frac{5}{16}\ \text{lbm}\right)(140.0\ \text{ft/sec}) = 43.75\ \text{lbm-ft/sec}$$

In the vector triangle shown in Figure 8.9, impulse **Imp** is the resultant of the difference of the momentum vectors, and its magnitude is calculated from the law of cosines:

$$\text{Imp} =$$
$$\sqrt{(43.75)^2 (34.37)^2 - 2(43.75)(34.37)\cos 150°}$$
$$= \sqrt{5,699.76} = 75.49\ \text{lbm-ft/sec}$$
$$= \frac{75.49}{32.2} = 2.34\ \text{lbf-sec}$$

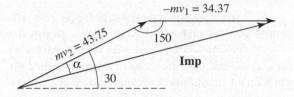

Figure 8.9

Direction of impulse:

$$\frac{\sin 150°}{75.49} = \frac{\sin \alpha}{34.37} \rightarrow \alpha = \sin^{-1}(0.227) = 13.16°$$

From the horizontal, the direction of **Imp** is $30.0 - 13.16 = 16.84°$, which must also be the direction of the bat (impulsive) force. Since

$$\mathbf{Imp} = \mathbf{F}(t_{2-}t_1) = \mathbf{F}(0.15\ \text{sec})$$
$$= 2.34\ \text{lbf-sec} \rightarrow \mathbf{F} = \frac{2.34}{0.15} = 15.63\ \text{lbf}$$

∴ $\mathbf{F} = 15.63$ lbf at $16.84°$ with respect to the horizontal

Impact

An **impact** is a collision of two or more rigid bodies in which the forces of collision are equal and opposite (Newton's action/reaction principle) and the net external force acting upon the system is taken to be zero. As discussed in a preceding section, the *total linear momentum of the system of colliding bodies must be conserved* in all directions. The elasticity of the materials that constitute the colliding objects has a dramatic effect on the velocity of each object immediately after the impact. A coefficient of restitution ε is an important parameter in the solution of impact problems and is defined as

$$0 \leq \varepsilon \leq 1.0$$

where $\varepsilon = 0$ represents a *perfectly plastic impact* (objects join together after collision), and $\varepsilon = 1.0$ represents a *perfectly elastic impact* (objects experience no permanent deformation and energy is conserved).

Typical impact problems on the FE exam involve one of the following situations:

- Direct central impact (head-on collision) in which the lines of action of the initial velocities before collision are collinear with the line joining the centers of mass.
- Oblique collision (off-center) in which the initial velocities have components that are both parallel and perpendicular to the line joining the mass centers.

FE impact problems usually ask for the postcollision velocity of each object based upon a given value of ε. If velocities before and after impact are known for each object, then the value of ε may be called for. For a direct central impact involving two objects having masses m_1 and m_2, the following key equations are applicable (primes denote the condition immediately after impact):

$$m_1v_1 + m_2v_2 = m_1v_1' + m_2v_2' \qquad [8.34]$$

$$v_2' - v_1' = \varepsilon(v_1 - v_2)$$

Oblique collisions employ a set of equations similar to those above, except that all velocities must be components along the line of impact (n-direction), which is perpendicular to the tangential plane of impact.

Example 8.8

A 15.0-kg sphere having an initial velocity of 48.0 m/s along a frictionless horizontal surface collides with a 28.0-kg rectangular block at the instant when the block has a velocity of 9.0 m/s in the same direction. If the block's velocity becomes 30.0 m/s immediately after the impact, solve for the coefficient of restitution ε.

Solution:

Let the sphere and the block be denoted as objects 1 and 2, respectively. The conservation of linear momentum of a system for any impact requires that

$$(15.0 \text{ kg})(48.0 \text{ m/s}) + (28.0 \text{ kg})(9.0 \text{ m/s})$$
$$= (15.0 \text{ kg})v_1' + (28.0 \text{ kg})(30.0 \text{ m/s})$$

$$(720 + 252)\text{kg} \cdot \text{m/s} = 15v_1' + 840 \text{ kg} \cdot \text{m/s}$$

$$v_1' = \frac{972 - 840}{15} = 8.8 \text{ m/s}$$

From the coefficient of restitution equation:

$$(30.0 - 8.8) \text{ m/s} = \varepsilon(48.0 - 9.0) \text{ m/s}$$

$$\therefore \varepsilon = \frac{21.2}{39.0} = 0.54$$

Angular Momentum

Angular momentum is a vector symbolized by $\mathbf{H}$ and defined as the "moment of momentum." For a constant-mass object that has position and velocity vectors $\mathbf{r}$ and $\mathbf{v}$, respectively, from a reference origin O, as shown in Figure 8.9, the angular momentum is given by the cross product:

$$\mathbf{H} = \mathbf{r} \times m\mathbf{v} \qquad [8.35]$$

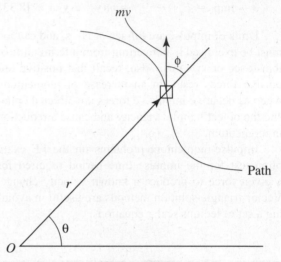

Figure 8.10

For most FE problems relating to angular momentum, both $\mathbf{r}$ and $m\mathbf{v}$ are in a two-dimensional plane and the direction of $\mathbf{H}$ is necessarily normal to this plane according to the right-hand rule for cross products. The magnitude of angular momentum H is calculated using the angle ϕ between $\mathbf{r}$ and $m\mathbf{v}$ or using the object's angular velocity ω ($\omega = d\theta/dt$):

$$H = mrv \sin \phi = mr^2\omega \qquad [8.36]$$

The units for H are lbm-ft^2/sec or kg $\cdot$ m^2/s.

It is important to recall that the net moment (torque) acting upon a fixed-mass object is related to the rate of change of angular momentum:

$$\Sigma\mathbf{M} = \dot{\mathbf{H}} \qquad [8.37]$$

For circular or arc-type motion ($r = R$), the magnitude of the applied moment is related to the object's angular acceleration α according to the equation

$$\Sigma M = mR^2\alpha \qquad [8.38]$$

Example 8.9

The cam follower mechanism illustrated in Figure 8.11 contains arm AC, which rotates at a constant angular velocity $\omega = 300$ rpm in a counterclockwise direction. If pin B is held against the cam lobes by a spring, and the distance of the pin from pivot point A is given by the equation

$$r = 100 \cos 3\theta + 250$$

where r is measured in millimeters, determine the angular momentum of the pin about the pivot after 4.0 s of motion if the initial angle θ is assumed zero and the pin has a mass of 0.3 kg.

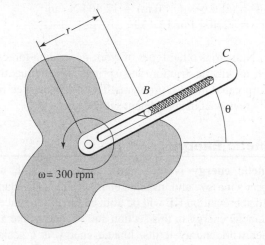

Figure 8.11

Solution:

Angular momentum for a fixed-mass object in plane motion is expressed as

$$H = mr^2\omega$$

In this case,

$$\omega = (300 \text{ rev/min})(2\pi \text{ rad/rev})(\text{min}/60 \text{ s})$$
$$= 31.41 \text{ rad/s}$$

since $\omega = \dfrac{d\theta}{dt} \rightarrow \theta = \int \omega\, dt = \omega t = 31.41t.$

Substitute for θ in the expression for r to obtain

$$r = 100 \cos 3(31.41t) + 250$$

At $t = 4.0$ s,

$$r = 100 \cos[94.23(4.0)] + 250$$
$$= 349.7 \text{ mm} \approx 0.350 \text{ m}$$

$$\therefore H = (0.3 \text{ kg})(0.350 \text{ m})^2(31.41 \text{ rad/s})$$
$$= 1.15 \text{ kg} \cdot \text{m}^2/\text{s}$$

Central Force Motion; Conservation of Angular Momentum

For motion of a fixed-mass object or particle in a plane as presented in the preceding section, a *central force* is defined as an external force whose line of action is always collinear with that of the position vector. Essentially, a central force is one that is continually directed toward or away from the object's reference origin and that produces *no net moment or zero torque*. Examples of central forces are cable tensions, gravitational forces, and spring/elastic forces. Referring to equation 8.37 (moment = angular momentum change), central force motion requires that *angular momentum remain constant* (be conserved).

The following equations summarize the conditions for conservation of angular momentum:

$$\Sigma\mathbf{M} = 0 \rightarrow \dot{\mathbf{H}} = 0$$

and

$$\mathbf{H} = \text{constant}$$

The magnitude of angular momentum is also constant.

$$H = mr^2\omega = \text{constant}$$
$$mr_1^2\omega_1 = mr_2^2\omega_2 \rightarrow \text{For } m = \text{constant, then}$$
$$r_1^2\omega_1 = r_2^2\omega_2 = \text{constant}$$

Clearly, an object's angular velocity increases substantially when the distance of the object from a reference origin or spin axis is reduced.

8.3
Work and Energy

Work and energy problems often constitute a significant percentage of the total problems in the dynamics section of the FE exam. Therefore, a careful review of concepts such as the principle of work and kinetic energy, potential energy, kinetic friction, and conservation of energy is strongly recommended.

Work of a Force

For particle and rigid-body dynamics, *work W* is a scalar quantity that is the dot product of an external force or net force with the displacement vector of the object or particle.

The governing equation for this type of mechanical work is

$$W = \int \mathbf{F} \, dr \qquad [8.39]$$

The units of work are ft-lbf or N · m (equivalent to joules). For *rectilinear plane motion,* represented in the majority of FE exam problems related to work and energy, the equations listed below are useful when a *single force or net force component along the line of action of the displacement is constant:*

$W = F_x \, \Delta x$, where Δx is the displacement along the x-direction

$W = F_y \, \Delta y$

$W = F_I \, \Delta I$ (along an inclined direction) $\qquad [8.40]$

Positive work occurs when the applied force or net force component is in the same direction as the displacement, whereas *negative work* is the result of a (net) force that acts in a direction opposite to the displacement direction. Additionally, bear in mind that forces or components of forces that are perpendicular to the displacement direction produce zero work.

For curvilinear motion, the work due to a force or net force is given by equation 8.39; however, numerical calculations are somewhat tedious and are not often required on the FE exam. Usually, it is advisable to employ the principle of work and kinetic energy (discussed next) to determine the work done along a curved path.

Example 8.10

A 600.0-kg milling machine that is being removed from a factory is lowered 14.0 m vertically by a rigging cable that has a constant tension of 5.0 kN. Determine the net work done on the machine during this process.

Solution:

An FBD of the machine is given in Figure 8.12, which shows the constant gravitational and cable tension forces **W** and **T**$_c$, respectively, and the positive y-direction is assigned downward.

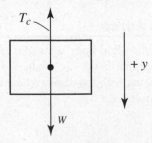

Figure 8.12

Since displacement Δy is given, the equation for net work is

$$W = F_y \Delta y$$

Here

$$F_y = W - T_c = (600.0 \text{ kg})(9.81 \text{ m/s}^2) - 5.0 \text{ kN}$$
$$= 5,886.0 - 5,000.0 = 886.0 \text{ N}$$

$\Delta y = 14.0$ m (positive displacement)

$$\therefore \; W = (886.0 \text{ N})(14.0 \text{ m}) = 12,404 \text{ N} \cdot \text{m}$$
$$= 12,404 \text{ J or } 12.4 \text{ kJ}$$

Numerous other types of work-generating forces, such as kinetic friction and springs, are frequently incorporated into FE problems. These will be specifically reviewed in subsequent sections.

Kinetic Energy

Kinetic energy is the energy associated with an object's mass *m* and instantaneous speed *v*. (The standard abbreviation KE will be utilized for all references to kinetic energy in this section and beyond.) Like all other work/energy terms, kinetic energy is a scalar and is defined by the following equation for constant-mass objects and/or rigid bodies:

$$\text{KE} = \frac{1}{2}mv^2 \qquad [8.41]$$

Note that, in the IP system of units, inputting values of mass in pounds mass and of speed in ft/sec into the KE equation *will not directly yield ft-lbf results.* Your ability to employ the correct conversion(s) will influence your degree of success in identifying the correct answers to KE-related problems. In the SI system, using kilograms and m/s for mass and speed terms in the KE equation will directly produce quantitative results in joules (or N · m).

The KE of a fixed-mass object can change only if a nonzero force or system of forces is applied that causes an acceleration or deceleration. This condition of zero net force also requires that the work done on an object having constant KE also be zero. In a later section, it will be shown that the amount of KE possessed by a rigid body is directly dependent on the object's potential energy for a conservative process involving only frictionless forces.

Potential Energy

Potential energy is categorized into two major areas. *Gravitational* potential energy is the energy associated with an object's elevation with respect to a reference datum. *Elastic* potential energy is the energy stored in a compressed or stretched spring. This section focuses on gravitational potential energy; linear springs will be reviewed in the next section.

With the standard symbol PE used for potential energy, gravitational PE due to an object's elevation h with respect to a datum is defined as follows:

$$PE = \mathcal{W}h = mgh \qquad [8.42]$$

where $\mathcal{W}$ is the weight or gravitational force on the object, and g is the acceleration of gravity in the environment for which weight is being measured.

Gravity work (W_g) is the work due to a gravitational force for a given vertical displacement or elevation change Δh.

Later, it will be shown that for a frictionless process in which only gravitational force acts, the change in PE for a rigid body is equal and opposite to the change in KE.

Linear Springs

From the equation

$$W_g = W \Delta h = W(h_1 - h_2) \qquad [8.43]$$

gravity work is related to potential energy change as

$$W_g = PE_1 - PE_2 = -\Delta PE.$$

An object that experiences an increase in elevation, and therefore a positive PE change (increase), is subjected to negative gravity work because the force and the displacement are in opposite directions. Conversely, when elevation (and PE) is required, gravity work is positive, as shown previously in Example 8.10.

A **linear spring** is a spring that has a *constant stiffness k*. Recall that the force due to a spring is given by the simple equation

$$F_s = -kx \qquad [8.44]$$

where x represents the spring *deformation,* which is the amount by which the spring is stretched or compressed from natural (undeformed) position. The direction of a spring force on an attached object is opposite to the direction of the deformation, and stiffness constant k is measured typically in N/m or lbf/

ft. A nonlinear spring exists when k varies with the spring length, as in an automobile coil spring.

The *potential energy* PE_s stored in a linear spring or an elastic material is due to the energy required to compress or stretch the spring and is computed from the equation

$$PE_s = \frac{1}{2}kx^2 \qquad [8.45]$$

Spring work W_s is related to the change in potential energy and is determined from the equation

$$W_s = PE_{s1} - PE_{s2} = \frac{1}{2}k(x_1{}^2 - x_2{}^2) \qquad [8.46]$$

where x_1 and x_2 signify the initial and final deformations, respectively.

Equation 8.46 is valid for both rectilinear and curvilinear motion and assumes no heat dissipation from the spring during compression or extension. Note that the work done by a spring is positive when the deformation decreases ($x_2 < x_1$) and the spring is unloading its potential energy.

Example 8.11

A linear spring having a stiffness of 750 N/m is initially compressed 0.8 m from its natural position. The spring is then stretched to a position such that it experiences a tension of 1.5 kN. Determine the spring work done during this process, assuming no heat dissipation.

Solution:

Since the spring work depends upon both the initial and final deformations, it is necessary first to obtain the final deformation x_2 from the equation for spring force and the given tension value.

$$x_2 = \frac{F_s}{k}$$

$$F_s = 1.5 \text{ kN} = 1,500.0 \text{ N}$$

$$x_2 = \frac{1,500.0 \text{ N}}{750.0 \text{ N/m}}$$

$$= 2.0 \text{ m stretched from natural position}$$

$$\therefore W_s = \frac{1}{2}k(x_1{}^2 - x_2{}^2)$$

$$= 375.0 \text{ N/m}[(0.8 \text{ m})^2 - (2.0 \text{ m})^2]$$

$$= -1,260 \text{ N} \cdot \text{m} = -1.26 \text{ kJ}$$

Note that the spring work is negative since the PE has increased because $x_2 > x_1$.

Kinetic Friction

Kinetic friction is the shearing force due to the sliding motion of a rigid body with respect to a surface of contact. The computation of kinetic friction simply involves determining the product of a *coefficient of kinetic friction μ_k* and the *normal reaction force N_R* acting upon the rigid body.

$$F_k = \mu_k N_R \qquad [8.47]$$

The value of μ_k is usually lower than the static friction coefficient μ_s for the same surfaces or objects in contact. It is also important to recognize that the direction of the kinetic friction force applied to a rigid body is opposite to the direction of the relative velocity of the body with respect to the contact surface or plane.

Most FE problems related to kinetic friction involve rectilinear motion, and a typical problem requires that the normal reaction be determined for an object on a horizontal or inclined surface in order to calculate F_k. Again, the use of effective free body diagrams is highly advantageous in solving problems incorporating kinetic friction.

The fact that kinetic friction forces oppose the velocity direction relative to the surface directly in contact with an object will cause a deceleration and consequently a decrease in the object's KE if the contacting surface is stationary. The *work done by a kinetic friction force is negative* relative to the surface of contact because this work opposes the displacement direction of an object, and kinetic friction work results in heat dissipation from a body. For this reason, kinetic friction forces are referred to as *nonconservative*.

Example 8.12

A mother pulls a sled containing her child across a horizontal snow-covered field as illustrated in Figure 8.13. The coefficient of kinetic friction μ_k between the base of the sled and the snow is estimated to be 0.15, and the pulling force is applied at a 30° angle with an assumed constant magnitude throughout the motion. The combined weight of child plus sled is known to be 65.0 lbf. If the sled travels 300.0 ft and the net work done on the child/sled system is 600.0 ft-lbf, determine:

A. the required pulling force
B. the friction force

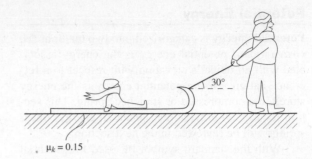

$\mu_k = 0.15$

Figure 8.13

Solution:
Draw an FBD of the child/sled system, as shown in Figure 8.13, that displays all of the external forces acting on the system and the attached *xy* coordinate reference frame.

A. Since net work and displacement are given, the net force in the displacement (*x*) direction can be readily obtained.

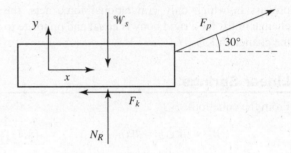

Figure 8.14

$$W_{net} = \Sigma F_x \Delta X$$

$$600.0 \text{ ft-lbf} = \Sigma F_x(300.0 \text{ ft}) \rightarrow \Sigma F_x = \frac{600.0 \text{ ft-lbf}}{300.0 \text{ ft}}$$
$$= 2.0 \text{ lbf}$$

Note that the work of pulling, F_p, is positive and the work of friction, F_k, is negative; therefore,

$$\Sigma F_x = F_{px} - F_k = F_p \cos 30° - \mu_k N_R$$
$$= 0.866F_p - 0.15 N_R = 2.0 \text{ lbf}$$

$\Sigma F_y = 0$ required for sled to maintain contact with surface.

$$\Sigma F_y = F_p \sin 30° - W_s + N_R = 0$$

$$0.5F_p - 65.0 \text{ lbf} + N_R = 0 \rightarrow N_R = 65.0 - 0.5F_p$$

Substitute for N_R in the equation for ΣF_x to obtain

$$0.866F_p - 0.15(65.0 - 0.5 F_p) = 2.0 \text{ lbf}$$

$$\therefore 0.941F_p = 11.75 \text{ lbf} \rightarrow F_p = 12.49 \text{ lbf}$$

B. Substitute for F_p above to obtain the normal reaction:

$$N_R = 65.0 - 0.5(12.49) = 58.75 \text{ lbf}$$

$$\therefore F_k = 0.15(58.75 \text{ lbf}) = 8.81 \text{ lbf}$$

Principle of Work and Kinetic Energy

One of the most useful problem-solving methodologies for the dynamics section of the FE exam is the **principle of work and kinetic energy**. This principle is derived from Newton's second law and effectively states that the *net work done on a rigid body is equal to the change in kinetic energy.* For any process involving conservative (frictionless) and/or nonconservative external forces acting upon a single object from state 1 to state 2, the governing equation is as follows:

$$W_{net} = \Delta KE = KE_2 - KE_1 \qquad [8.48]$$

Positive net work produces an **increase in kinetic energy**.

For a constant-mass object, the *work/kinetic energy equation* is written as

$$W_{net} = \frac{mv_2^2}{2} - \frac{mv_1^2}{2} = \frac{m}{2}(v_2^2 - v_1^2) \quad [8.49]$$

This equation is valid for both rectilinear and curvilinear motion; an increase in KE ($v_2 > v_1$) requires positive net work, whereas negative work results in a decrease in KE and velocity. A note of caution is needed with regard to converting KE values into work (ft-lbf) when using IP units. If initial and final velocities are known, net work can be computed directly without involving any forces or displacements. In most FE problems related to work/energy, however, it is necessary to determine forces and velocity (or KE) values at various stages of computation as illustrated in Example 8.13.

Example 8.13

A 0.25-kg cube initially at rest is projected from a compressed spring across a frictionless horizontal surface as shown in Figure 8.15. The cube then enters the 6.0-cm-radius circular loop and contains the minimum speed necessary at point C to allow it to barely maintain contact with the loop as it travels past the highest point, E. If the spring stiffness is 500.0 N/m, solve for the required initial deformation of the spring.

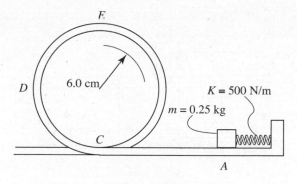

Figure 8.15

Solution:

The principle of work and kinetic energy is employed from the initial compressed position (point A) to the highest point E, where the object is assumed to have a negligible reaction force with the loop as a condition of minimum contact. The net work done on the cube is the difference in kinetic energy according to

$$W_{net} = KE_E - KE_A = \frac{1}{2}m(v_E^2 - v_A^2)$$

Since the cube is released from rest, $v_A = 0$ ($KE_A = 0$), and the net work done is simply the kinetic energy at E in this case:

$$W_{net} = \frac{1}{2}mv_E^2 = 0.125 \text{ kg}(v_E^2)$$

The net work done on the cube from A to E is the combination of the spring work and the gravity work:

$$W_{net} = W_s + W_g = \frac{1}{2}k(x_A^2 - x_B^2) + mg(h_A - h_E)$$

If the initial elevation is assumed to be zero, then $h_A = 0$ and $h_E = 2(6.0 \text{ cm}) = 0.12$ m. At point B, the spring is fully unloaded and undeformed $\rightarrow x_B = 0$. Substitute the above values in the work equation to obtain:

$$W_{net} = \frac{500.0 \text{ N/m}}{2}[x_A^2 - 0^2] + (0.25 \text{ kg})$$
$$(9.81 \text{ m/s}^2)(0 - 0.12 \text{ m})$$
$$= 250.0 \text{ N/m}(x_A^2) - 0.29 \text{ N} \cdot \text{m} = 0.125 \text{ kg}(v_E^2)$$

At point E, the cube is clearly in circular motion with radius $R = 0.06$ m and reaction force $N_R = 0$. The FBD of the cube at E in Figure 8.16 shows the forces acting in the normal direction:

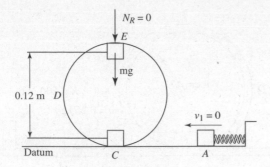

Figure 8.16

$$\Sigma F_n = \frac{mv_E^2}{R} = \frac{0.25\, v_E^2}{0.06\ \text{m}} = 4.17\, v_E^2 = mg + 0$$

$$4.17\, v_E^2 = (0.25\ \text{kg})(9.81\ \text{m/s}^2)$$

$$= 245\ \text{N} \rightarrow v_E = \sqrt{\frac{2.45}{4.17}} = 0.77\ \text{m/s}$$

Inserting the value of v_E into the work equation produces a solution for the spring deformation:

$$250.0\ \text{N/m}(x_A^2) = 0.125\ \text{kg}(0.77\ \text{m/s})^2 + 0.29\ \text{N}\cdot\text{m}$$

$$\rightarrow x_A^2 = \frac{(0.074 + 0.29)\ \text{N}\cdot\text{m}}{250.0\ \text{N/m}}$$

$$\therefore x_A = \sqrt{\frac{0.364}{250.0}} = 0.038\ \text{m}$$

$$= 3.8\ \text{cm}$$

Banking Angles

A frequently encountered FE exam topic for circular motion problems is the banking angles of roads or curved tracks. These angles are a very important consideration for vehicle safety in highway and/or public transportation design. Effectively, a **banking angle**, θ, with respect to a horizontal plane is the minimum angle required to keep a vehicle traveling at a *maximum rated speed* v_m from sliding outward, assuming that friction is negligible because of an icy or otherwise slippery road surface. If the radius of curvature R of a section of roadway is known, or if the path is circular, then Newton's second law analysis in the normal and vertical directions yields the *banking angle equation*:

$$v_m = \sqrt{gR \tan \theta}$$

or, equivalently,

$$\theta = \tan^{-1}\left(\frac{v_m^2}{gR}\right) \qquad [8.50]$$

where θ is measured in degrees.

Conservation of Energy

For motion of a rigid body in which only *conservative forces* (frictionless) are acting, the net work done on the body is independent of the path traveled and depends only on the initial and final positions. Gravity work and ideal spring work are examples of conservative processes for which there is no friction energy loss. From the principle of work and kinetic energy equation (equation 8.49), it can readily be shown that, if gravity is the only type of work done on the particle or rigid body, *the change in kinetic energy of the body or particle is equal and opposite to the change in the potential energy.*

In essence, this requires that, if an object subjected to gravity force experiences only a kinetic energy increase from state 1 to state 2, there must be a loss of potential energy in an amount equal to the kinetic energy rise between the same two states. Energy is simply converted from one form to another, and no energy is lost from the object to the surrounding environment. The *conservation of energy equation* given below (equation 8.51) indicates that the sum of the potential and kinetic energy possessed by an object is constant for all states if only conservative (frictionless) forces are applied:

> **Conservation of energy** requires a **frictionless** process.

$$\text{KE}_1 + \text{PE}_1 = \text{KE}_2 + \text{PE}_2 = \ldots = \text{KE}_n + \text{PE}_n$$
$$= \text{constant}$$

$$\Delta \text{KE} = -\Delta \text{PE}$$

$$\frac{1}{2}mv_1^2 + mgh_1 = \frac{1}{2}mv_2^2 + mgh_2$$
$$= \text{constant} \qquad [8.51]$$

For a rigid body or particle:

$$g(h_1 - h_2) = \frac{v_2^2 - v_1^2}{2}$$

With reference to equation 8.51, many FE dynamics problems involve a rigid body initially at rest ($v_1 = 0$) that experiences an elevation drop of ($h_1 - h_2$). The *final velocity* v_2 at the lower elevation h_2 is given (assuming no friction) by the equation

$$v_2 = \sqrt{2g\left(h_1 - h_2\right)} \qquad [8.52]$$

Note that, for free-fall (gravity) motion without air friction, equation 8.52 is often used to calculate terminal velocity.

Example 8.14

A 140.0-lbm ski jumper initially at rest begins to travel down the smooth ramp illustrated in Figure 8.17 from point A and attains a velocity of 65.0 mph at exit point B. Assuming that the skier experiences a normal reaction equal to 2.5 times his or her weight, determine:

A. the height h required to produce this stated speed
B. the radius of curvature ρ of the ramp at point B

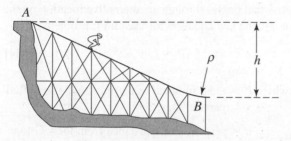

Figure 8.17

Solution:

A. Since gravity is the only work-producing force acting upon the skier if kinetic friction is considered negligible, the conservation of energy equation for terminal velocity (equation 8.52) may be employed, given that $v_B = 65.0$ mph $= 95.3$ ft/sec.

$$v_B = \sqrt{2g\left(h_A - h_B\right)} = \sqrt{2\left(32.2 \text{ ft/sec}^2\right)h}$$
$$= 95.3 \text{ ft/sec}$$

$$\therefore h = \frac{(95.3)^2}{64.4} = 141.0 \text{ ft}$$

B. For the normal reaction N_R at B, it was stated that

$$N_R = 2.5\,\mathcal{W}_s = 2.5(140.0 \text{ lbf}) = 350.0 \text{ lbf}$$

Since the radius ρ of the ramp is desired, apply Newton's second law equation to the skier (free body) for circular motion in the normal direction:

$$\Sigma F_n = \frac{mv_B^2}{\rho} = N_R - \mathcal{W}_s$$

$$\frac{(140.0 \text{ lbm})(95.3 \text{ ft/sec})^2}{\rho} = (350.0 - 140.0) \text{ lbf}$$

$$\therefore \rho = \frac{(140)(9{,}082.1)\left(\dfrac{1}{32.2}\right) \text{ ft-lbf}}{210.0 \text{ lbf}} = 188.0 \text{ ft}$$

8.4 Rigid Body Motion in a Plane

The motion of rigid bodies in a two-dimensional plane is a commonly tested topic on the FE dynamics section. Typical problems involve rolling cylinders and wheels, rods within perpendicular slots, gears, rotating 3-D bodies, and so on. The *center of gravity* of the rigid body being analyzed is an important parameter in the solution of numerous rigid-body dynamics problems, and terms evaluated at the CG will be denoted by the subscript c in this section and the following ones.

The fundamental modes of motion for a rigid body traveling in a plane are *translational* and *rotational* motion. Translational motion is effectively the rectilinear motion or displacement of the center of gravity and is governed by Newton's second law equation:

$$\Sigma \mathbf{F} = m\mathbf{a}_c$$

Rotational motion is the spin of the body about an axis that normally passes through the CG or a parallel axis (discussed on page 242), and the magnitude of rotation is measured using the familiar terms *angular velocity* (ω) and *angular acceleration* (α). The applied torque or net moment for plane rigid-body motion is also typically measured with respect to the CG and may be determined by using the following equation:

> **Net moment** applied is commonly referred to as **torque**.

$$\Sigma M_c = I_c\,\alpha \qquad [8.53]$$

where I_c is the mass moment of inertia, which is evaluated about a particular axis that passes through the CG of the object. The important concept of mass moment of inertia is reviewed in the next section.

Mass Moment of Inertia

The **mass moment of inertia** I for a three-dimensional rigid body is essentially a measure of the resistance of the body to angular acceleration. For any rigid body such as the one illustrated in Figure 8.18, mass moment of inertia is theoretically defined as the "second moment of mass" of all mass elements dm integrated throughout the entire mass m of the body:

$$I = \int_m r^2 \, dm$$

where r represents the moment-arm distance from any mass element to the axis about which I is being evaluated. Recall that the units for mass moment are conventionally $kg \cdot m^2$ or $lbm\text{-}ft^2$.

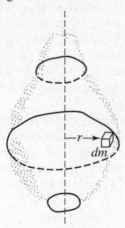

Figure 8.18

Clearly, performing the above integration for I is a tedious task, and therefore equations for the mass moments of inertia for various familiar shapes—cones, spheres, rectangular plates, disks, and so on—are provided in Table 8.1 for quick reference. The FE exam review pamphlet also contains this reference material for I. It is important to recognize that, for any given rigid-body geometry, the *formulas for I are directly dependent on a specific axis for evaluating I,* and this axis is usually chosen perpendicular to the plane in which the forces act on the body. For example, I_z is evaluated when forces are confined to an xy plane. Since the mass m of the object is required to compute numerical values of I, the product of an object's density and volume must be calculated to obtain mass if it is not directly given on an FE problem.

For *composite bodies* or shapes, such as a cylindrical shaft connected to a circular plate (disk) at each end, the composite mass moment of inertia about a given axis is the algebraic *sum of the individual moments of inertia* of the component shapes about the same axis. This computation will usually require the use of the parallel axis theorem, which is reviewed in the next section.

Parallel Axis Theorem

Whenever the mass moment of inertia must be evaluated about an axis that does not pass directly through the CG of a rigid body but is required for a *parallel axis* that passes through an alternative location on the body, the key equation for use on FE problems is

$$I' = I_c + md^2 \qquad [8.54]$$

where I' = mass moment of inertia about a specified parallel axis,

d = perpendicular distance from the CG to a point on the parallel axis,

I_c = mass moment of inertia about the CG.

Example 8.15

The T-shaped pendulum pictured in Figure 8.19 consists of two thin rods, each having a mass of 30 lbm, and is hinged at point O. Find the mass moment of inertia of the combined pendulum about an axis passing through the pin at O.

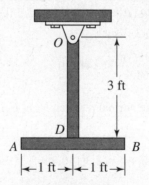

Figure 8.19

Solution:

The composite mass moment of inertia for the entire pendulum is the sum of the individual moments of inertia for rod OD and rod AB. Use Table 8.1 for mass moments of inertia for common geometries to obtain the equation for mass moment of inertia of a slender rod about an axis through one of its end points:

$$I_{\chi}' = \frac{ml^2}{3} = (I_O)_{OD} \quad \text{for rod OD}$$

$$(I_O)_{OD} = \frac{(30.0 \text{ lbm})(3.0 \text{ ft})^2}{3} = 90.0 \text{ lbm-ft}^2$$

For rod AB, compute the mass moment about point O by using the parallel axis theorem:

$$(I_O)_{AB} = (I_x)_{AB} + md^2$$

Since $d = 3.0$ ft from point D (the CG of AB) to the parallel axis through point O:

$$(I_O)_{AB} = \frac{ml^2}{12} + md^2$$

$$= \frac{(30.0 \text{ lbm})(2.0 \text{ ft})^2}{12} + (30.0)(3.0 \text{ ft})^2$$

$$= 280.0 \text{ lbm-ft}^2$$

$\therefore$ For the composite pendulum, the total moment of inertia I_O is summed as $I_O = 90.0 + 280.0 = 370.0$ lbm-ft^2.

Radius of Gyration

The **radius of gyration** of a rigid body of mass m is the *perpendicular distance k* from an axis passing through the CG to a point where the entire mass m can be concentrated to produce the same moment of inertia or gyration as the original body about this CG axis. Once I for a body of particular shape is obtained by using Table 8.1, the radius of gyration is readily computed from the relationship

$$I = mk^2 \quad [8.55]$$

Then

$$\text{Radius of gyration} \rightarrow k = \sqrt{\frac{I}{m}}$$

Rotation of a Rigid Body About a Fixed Axis

The rotation of a rigid body within a two-dimensional plane about a fixed axis through a point that is not necessarily the CG of the body is a topic frequently presented on the FE exam. Figure 8.20 shows a rigid-body rotation within an xy plane about an axis passing through point A, which is located at a distance r from the CG of the body.

If a constant net moment is applied, the body will experience a constant angular acceleration α according to the equation

$$\Sigma M_A = I_A \alpha \quad [8.56]$$

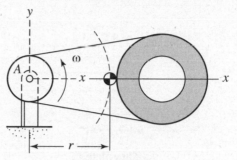

Figure 8.20

The above torque must be in a z-direction normal to the xy plane, and the moment of inertia about a z-axis (the axis of rotation) through point A is obtained by using the parallel axis theorem:

$$I_A = I_c + mr^2 \quad [8.57]$$

Since the motion of the CG is circular about point A, the acceleration of the CG with respect to A may be determined from tangential and normal acceleration equations. The net forces acting upon the rotating body in the x- and y-directions are then readily obtained:

$$a_{cx} = -r\omega^2 \rightarrow \Sigma F_x = -mr\omega^2$$

$$a_{cy} = r\alpha \rightarrow \Sigma F_y = mr\alpha \quad [8.58]$$

Since the net moment shown above is assumed constant, then necessarily the *angular acceleration α is constant* and the angular velocity ω changes at a constant rate. The value of angular velocity at any time t is therefore given by

$$\omega(t) = \omega_0 + \alpha t \quad [8.59]$$

where ω_0 is the initial angular velocity at $t = 0$ and ω_0 is expressed in rad/s.

The velocity of the CG around the axis is entirely tangential and is determined by using the equation

$$v_c = r\omega \quad [8.60]$$

Velocity and Acceleration: Absolute and Relative Motion

Rigid-body motion in a plane can be described by the simultaneous processes of *translation* and *rotation*. Many objects, such as automobile wheels, conveyor belts, and spinning baseballs, exhibit this type of

Table 8.1 Centroids and Moments of Inertia for Various Geometries

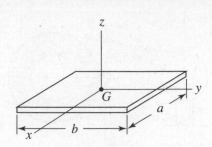

Thin plate

$$I_x = \tfrac{1}{12}mb^2 \qquad I_y = \tfrac{1}{12}ma^2 \qquad I_z = \tfrac{1}{12}m(a^2 + b^2)$$

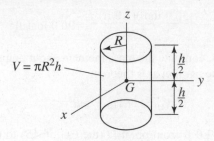

$V = \pi R^2 h$

Cylinder

$$I_x = I_y = \tfrac{1}{12}m(3R^2 + h^2) \qquad I_z = \tfrac{1}{2}mR^2$$

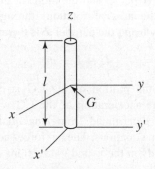

Slender rod

$$I_x = I_y = \tfrac{1}{12}m\ell^2 \qquad I_{x'} = I_{y'} = \tfrac{1}{3}m\ell^2 \qquad I_z = 0$$

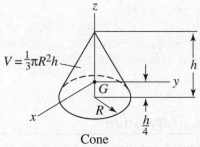

$V = \tfrac{1}{3}\pi R^2 h$

Cone

$$I_x = I_y = \tfrac{3}{80}m(4R^2 + h^2) \qquad I_z = \tfrac{3}{10}mR^2$$

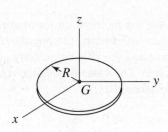

Thin circular disk

$$I_x = I_y = \tfrac{1}{4}mR^2 \qquad I_z = \tfrac{1}{2}mR^2$$

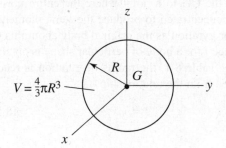

$V = \tfrac{4}{3}\pi R^3$

Sphere

$$I_x = I_y = I_z = \tfrac{2}{5}mR^2$$

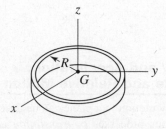

Thin ring

$$I_x = I_y = \tfrac{1}{2}mR^2 \qquad I_z = mR^2$$

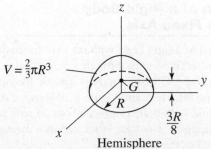

$V = \tfrac{2}{3}\pi R^3$

Hemisphere

$$I_x = I_y = 0.259mR^2 \qquad I_z = \tfrac{2}{3}mR^2$$

motion, and the FE exam usually features at least one problem in this category. An understanding of the *relative motion* of two or more points on a rotating/translating rigid body is advantageous in developing the proper solution of various problems.

In Figure 8.21, point A is translating in a plane with an absolute velocity of $\mathbf{v_A}$ as shown. If A is considered the instantaneous center of rotation of the body, with all other points on the body having a common angular velocity ω about a perpendicular axis through A, an arbitrary point B can be chosen on the body that necessarily has circular motion as viewed from A. This *relative velocity of B with respect to A* is denoted by the vector $\mathbf{v_{B/A}}$, and the magnitude of the relative velocity is calculated as the product of the straight distance $r_{B/A}$ and the angular velocity ω:

$$v_{B/A} = r_{B/A}\,\omega \qquad [8.61]$$

The relative velocity is then combined with the (absolute) translational velocity of A to produce the absolute velocity vector $\mathbf{v_B}$ at B:

$$\mathbf{v_B} = \mathbf{v_A} + \mathbf{v_{B/A}} \qquad [8.62]$$

The magnitude and direction of $\mathbf{v_B}$ are determined from the solution of the vector triangle illustrated in Figure 8.22 by using the law of cosines.

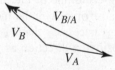

Figure 8.22

For *absolute and relative acceleration* of two points on a rigid body experiencing translational and rotational plane motion as described above, an equation similar to that used for relative velocity is employed for points A and B with the assumption that A is again the center of rotation.

$$\mathbf{a_B} = \mathbf{a_A} + \mathbf{a_{B/A}} \qquad [8.63]$$

If the absolute acceleration vector of A is known, the relative acceleration vector $\mathbf{a_{B/A}}$ is determined from the tangential and normal acceleration components $r_{B/A}\alpha$ and $r_{B/A}\omega^2$, respectively. Then a vector triangle is constructed to solve the absolute acceleration at B.

This methodology is useful in solving for the absolute acceleration of a point, B, on the circumference of a rolling wheel if the translational acceleration of the center of the wheel at point A is given, along with ω and α.

A commonly tested application of relative plane motion on the FE exam is **rolling motion of rigid bodies** in a plane. The rolling motion of a *circular wheel or disk* along a horizontal plane surface is quite commonly encountered on the FE, and Figure 8.23 illustrates a disk with *radius R* having a *center point C* and a *point P located at its outer edge.*

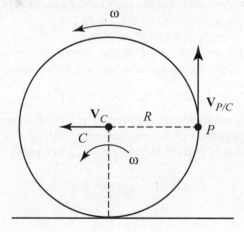

Figure 8.23

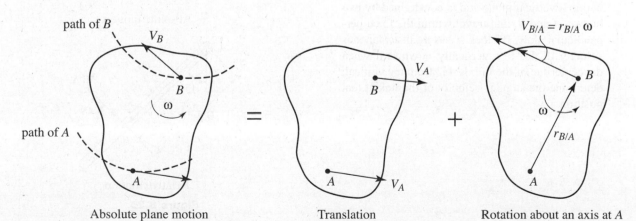

Absolute plane motion   Translation   Rotation about an axis at A

Figure 8.21

245

To analyze the absolute and relative motion of points C and P along the surface, we establish the tangential point of contact between the disk and the horizontal plane as the absolute reference or datum. At any time t, the *angular velocity* of the rolling disk is given by ω, and assuming a counterclockwise rotation of the disk, the angular velocity can be expressed as $\omega = \omega\mathbf{k}$, from the *right-hand rule* of directionality. Based upon the previously established equations 8.1 through 8.17 for plane circular motion, the following is a summary of the absolute and relative motions of the center C and tangential point P.

Absolute velocity of center C:

$$\mathbf{v}_c = -R\omega\mathbf{i}$$

Note: Center C moves with *tangential* velocity in the negative x-direction and has speed $R\omega$ at all times.

Relative velocity of point P with respect to C:

$$\mathbf{v}_{p/c} = \omega \times \mathbf{r}_{p/c}$$

Since P has *circular motion about center C* at all times and its velocity is vertical in the position illustrated, note that the relative velocity of P in the *specific position shown is* $\mathbf{v}_{p/c} = R\omega\mathbf{j}.$

Absolute velocity of point P:

For any position of the disk, the absolute velocity of P is the *vector sum* of $\mathbf{v}_c$ and $\mathbf{v}_{p/c}$. Specifically,

$$\mathbf{v}_p = \mathbf{v}_c + \mathbf{v}_{p/c} = \mathbf{v}_c + \omega \times \mathbf{r}_{p/c}$$

The *absolute speed v_p of tangential point P* is the scalar magnitude of the resultant of $\mathbf{v}_c$ and $\mathbf{v}_{p/c}$. For the position of P given in Figure 8.23 , we calculate

$$v_p = \sqrt{(R\omega)^2 + (R\omega)^2} = \sqrt{2}\ R\omega$$

Example 8.16

The rigid link illustrated in Figure 8.24 moves within a vertical plane and is constrained by two blocks, A and B, that travel within the fixed perpendicular slots. If block A has an instantaneous velocity of 4.0 m/s vertically downward when the centerline of the link is 45° from the vertical, determine the angular velocity of the link at that position.

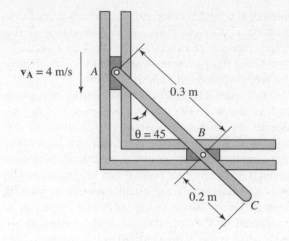

Figure 8.24

Solution:

Since the absolute velocity of block A is given, utilize the relative velocity equation for A and B on the assumption that A is the center of rotation (reference point) for the link:

$$\mathbf{v}_B = \mathbf{v}_A + \mathbf{v}_{B/A}$$

Clearly, the direction of $\mathbf{v}_B$ is horizontal (x-direction), and the direction of the relative velocity $\mathbf{v}_{B/A}$ is perpendicular to the centerline of the link, as shown in Figure 8.25, because of the circular motion of B about A. Therefore, it is necessary that $\mathbf{v}_{B/A}$ be directed at 45° from the positive x-direction as indicated in the figure.

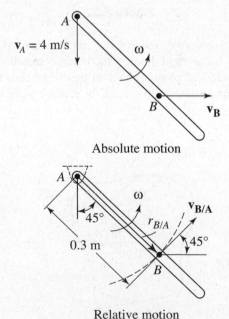

Absolute motion

Relative motion

Figure 8.25

The magnitude of this relative velocity is

$$v_{B/A} = r_{B/A}\omega = (\overline{AB})\omega = (0.3 \text{ m})\omega$$

Solve for the desired ω by writing the scalar components of the relative velocity equation in the x- and y-directions.

For the x-direction:

$$v_{BX} = v_{AX} + (v_{B/A})_X \rightarrow v_B = v_{BX}$$

because of the horizontal slot

$$v_B = 0 + (v_{B/A}) \cos 45°$$
$$= (0.3 \text{ m}) \, \omega \, (0.707) = (0.212 \text{ m}) \, \omega$$

For the y-direction:

$$v_{BY} = v_{AY} + (v_{B/A})_Y \rightarrow v_{AY} = v_A$$

because of the vertical slot

$$0 = -4.0 \text{ m/s} + (0.3 \text{ m})\omega \sin 45°$$
$$= -4.0 + (0.212 \text{ m})\omega$$
$$\therefore \omega = \frac{4.0 \text{ m/s}}{0.212 \text{ m}} = 18.86 \text{ rad/s}$$

Note that the magnitude of the absolute velocity of B must be the same as that of A because of the 45° angle. Calculate v_B from the x-component equation above to check the result for ω:

$$v_B = (0.212 \text{ m})(18.86 \text{ rad/s}) = 3.998 \approx 4.0 \text{ m/s}$$

which verifies ω.

8.5
Vibrations

Mechanical *vibration* problems involving linear spring/mass systems are frequently included on the FE exam. These problems typically involve single-degree-of-freedom systems that oscillate along a single direction (rectilinear motion). The following sections on free, damped, and torsional vibrations summarize the key concepts, equations, symbols, and problem applications for vibrations that are considered relevant to the FE exam.

The exhaustive mathematical derivations involving solutions of numerous ordinary differential equations are bypassed in order to focus on the interpretation of initial conditions and physical parameters involved in vibration problems. This concentration should enable you to identify readily the appropriate end-use equations that allow for efficient and accurate solution methodologies.

Free Vibrations

This type of motion is characterized by an oscillation that occurs with a linear spring and mass system without any external force or excitation. The system is initially stretched beyond a static equilibrium position and then released. Typical free-vibration problems provide the value of the initial displacement (deformation), denoted by x_0, and require solutions for the amplitude, velocity, period, and/or natural frequency at a given time or position.

The key parameters for free vibrations are represented by the symbols and units given below:

- Spring constant, k: lb/in or N/m.
- Velocity, $\dot{x}$: in/sec, ft/sec, mm/sec, etc.
- Mass of object, m: lb, kg, etc.
- Natural circular frequency,

$$\omega_n = \sqrt{\frac{k}{m}} : \text{rad/s} \qquad [8.64]$$

- Period of vibration, $\tau = 2\pi \sqrt{\dfrac{m}{k}} : \text{s} \qquad [8.65]$
- Natural frequency

$$f_n = \frac{1}{\tau} = \frac{1}{2\pi} \sqrt{\frac{k}{m}} : \text{rad/s} \qquad [8.66]$$

- Displacement from equilibrium, $x(t)$: in, mm, ft, etc.
- Amplitude of vibration,

$$X = \sqrt{x_0{}^2 + \left[\frac{\dot{x}_0}{\omega_n}\right]^2} : \text{in, mm, ft, etc.} \qquad [8.67]$$

Governing equation for *displacement of free vibrations* without the presence of damping:

$$x(t) = X \cos(\omega_n t - \phi) \qquad [8.68]$$

where ϕ is the phase angle calculated from

$$\phi = \tan^{-1}\left[\frac{\dot{x}_0}{x_0 \omega_n}\right].$$

The *instantaneous velocity* equation for an *undamped free vibration* is determined by

$$\dot{X}(t) = -X\omega_n \sin(\omega_n t - \phi) \qquad [8.69]$$

Damped Vibrations

The existence of a viscous force acting upon an oscillating system represents the condition of damped vibrations. For simplicity, the damping force F_d is regarded as the effect of damping resulting from the motion of a dashpot within a hydraulic cylinder. Key damping terms, symbols, and units are listed below:

- Damping constant, c: lbf-sec/ft, N · s/m, etc.
- Critical damping constant, $c_c = 2m\omega_n$ (this is the condition of most rapid damping)
- Damping factor, $\xi = \dfrac{c}{c_c}$
  ($\xi = 1$ represents critical damping)   [8.70]
- Frequency of damped vibration,

$$\omega_d = \omega_n \sqrt{1-\xi^2} : \text{rad/s} \qquad [8.71]$$

- Damping force $F_d = c\dot{x}$: lbf, N, etc.
- Natural frequency of a damped system,

$$f = \frac{\omega_d}{2\pi} = \frac{1}{\tau}.$$

  f is clearly the reciprocal of the damped period τ.
- Logarithmic decrement,

$$\delta = \xi\omega_n\tau = \ln\left(\frac{e^{-\xi\omega_n t}}{e^{-\xi\omega_n(t+\tau)}}\right) = \frac{2\pi\xi}{\sqrt{1-\xi^2}},$$

a dimensionless quantity. δ represents the natural logarithm of the ratio of two consecutive amplitudes which is the ratio of damped amplitude at time t to the amplitude at time $t + \tau$.

A **critically damped** system has the minimum amount of damping that results in a nonoscillating system.

Without derivation, the *displacement of a damped free vibration* is calculated from the following equation:

$$x(t) = Ce^{-\xi\omega_n\tau} \cos(\omega_d t - \phi) \qquad [8.72]$$

where $\phi = \tan^{-1}\left(\dfrac{\dot{x}_0 + x_0\xi\omega_n}{x_0\omega_d}\right)$, $C = \dfrac{x_0}{\cos\phi}$.

The *instantaneous velocity* equation for a *damped free vibration* is determined by

$$\dot{X}(t) = -C\omega_n e^{-\zeta\omega n t}\sin(\omega_n t - \phi) \qquad [8.73]$$

Torsional Vibrations

In a previous section, it was shown that, for rotational motion of a rigid body, the next torque applied is proportional to the angular acceleration:

$$\Sigma M = I\alpha$$

Torsional vibration problems use this basic principle, and this type of oscillatory motion typically involves a rectangular plate or disk that experiences an initial angular displacement θ while attached to a cylindrical rod. In Figure 8.26 the angular displacement of the rigid body generates a twisting moment or torque within the rod according to the equation

$$\Sigma M = -k_T\theta = I_0\ddot{\theta}, \qquad [8.74]$$

where k_T = *torsional stiffness* of the rod (N · m/rad or ft-lb/rad),

Also, $k_T = \sqrt{\dfrac{GJ}{L}}$

where G = shear modulus,

J = area polar moment of inertia,

L = length of the rod,

I_0 = mass moment of inertia of the oscillating body,

$\ddot{\theta}$ = the resulting angular acceleration (rad/s^2).

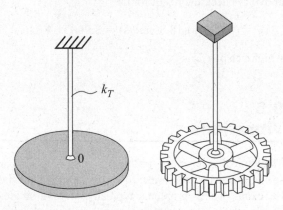

Figure 8.26

The *natural circular frequency* of a torsional vibration is given by

$$\omega_n = \sqrt{\frac{k_T}{I_0}}, \qquad [8.75]$$

and the *period* of the vibration is obtained from the equation

$$\tau = 2\pi\sqrt{\frac{I_0}{k}} \qquad [8.76]$$

8.6 Summary

Dynamics is the study of motion of particles and rigid bodies and the forces that govern their motion. The Dynamics section of the FE Exam tests students' problem-solving skills primarily in the topic areas of **Rectilinear Motion, Curvilinear Motion, Linear Momentum, Newton's Second Law Forces and Acceleration, Work and Energy, and Mechanical Vibrations.** For success in the Dynamics FE section, it is essential for students to have a deep understanding of **Kinematics of Particles** and to be very familiar with both straight line and curve motion in a two-dimensional plane. Various types of problems involving **position, velocity, and acceleration** in Cartesian coordinates (xy) and well as polar coordinates (r, θ, z) are to be expected. The ability to construct accurate free-body diagrams that represent all external forces and that include a proper coordinate system located at the center of gravity (or motion) is a highly advantageous skill for generating correct solutions.

Kinetics is another major topic area represented in the Dynamics section of the FE and knowledge of several Conservation Laws is highly beneficial. Specifically, the laws governing **Conservation of Linear Momentum** and **Conservation of Energy** form the basis for numerous types of questions relating to **Rigid Body Motion.** It is important to remember that collisions or **Impacts** between rigid bodies are a system that conserves linear momentum, and the elasticity of the colliding bodies influences the magnitude and direction of the motion immediately after the impact. Therefore, students' understanding of **Vectors** and their related scalar components will be an important skill in generating accurate results with problems in this area as well as in many other topics in Dynamics.

PRACTICE PROBLEMS

Questions 1–3
The rectilinear motion of a particle is described by the equation

$$X(t) = 2t^3 - 16t^2 + 7$$

where X is measured in meters.

1. The acceleration of the particle when the velocity is 280 m/s is
   (A) 88.0 m/s^2
   (B) 3.1 m/s^2
   (C) 1.6 m/s^2
   (D) 120.0 m/s^2

2. The displacement at the elapsed time of 20.0 s will be
   (A) 1,607.0 m
   (B) −7.0 m
   (C) 9,600.0 m
   (D) 40.0 m

3. The minimum velocity during the first 50.0 s of motion has a value of
   (A) 2.7 m/s
   (B) 0 m/s
   (C) −1,340.0 m/s
   (D) −42.7 m/s

Questions 4 and 5
A young boy shoots an arrow having a mass of 150 g at an angle of 60° with respect to (above) a horizontal. The arrow is initially at 1.0 m above a zero-elevation datum and leaves the bow with a velocity of 7.0 m/s. Air resistance is considered negligible.

4. The maximum height attained by the arrow will be
   (A) 2.43 m
   (B) 2.88 m
   (C) 4.61 m
   (D) 1.88 m

5. The total speed of the arrow after 0.75 s of motion is nearly
   (A) 1.3 m/s
   (B) 7.1 m/s
   (C) 3.7 m/s
   (D) 4.8 m/s

6. An object having a mass of 3.0 slugs is subjected to a net force of 80.0 lbf. The magnitude of the resulting acceleration is
   (A) 26.7 ft/sec^2
   (B) 0.8 ft/sec^2
   (C) 859.7 ft/sec^2
   (D) 2.7 ft/sec^2

7. A 24″ truck wheel is rolling at 600 rpm along a highway pavement. If the wheel lug nuts are located 6″ from the center of rotation and have a mass of 0.5 lbm, the angular momentum of each lug about the center must be
   (A) 2.30 ft-lbf-sec
   (B) 7.85 ft-lbf-sec
   (C) 73.80 ft-lbf-sec
   (D) 0.24 ft-lbf-sec

8. Two boats experiencing rectilinear motion are approaching each other in a harbor. Boat A is traveling due south at 6 mph, while boat B has a constant speed of 10 mph in a direction 20° west of north as shown in the figure.

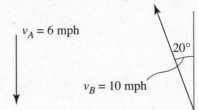

The magnitude and direction of the relative velocity of boat A as viewed from boat B should be

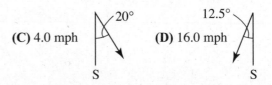

(A) 4.8 mph   (B) 15.8 mph

(C) 4.0 mph   (D) 16.0 mph

9. A bicycle racer enters a constant-radius curve on a flat track at a speed of 7.0 m/s. If the speed of the bicycle is increased at a constant rate of 1.5 m/s^2 along the curve and the total acceleration at the beginning of the curve is 1.7 m/s^2, the required radius of the curve is
   (A) 41.6 m
   (B) 60.0 m
   (C) 21.7 m
   (D) 76.5 m

10. A bowler rolls a 16.0-lbm ball at a single stationary pin having a mass of 3.0 lbm. If the ball strikes the pin with a speed of 9.0 ft/sec and the coefficient of restitution is 0.8, the speed of the pin immediately after the collision is necessarily
    (A) 7.2 ft/sec
    (B) 1.5 ft/sec
    (C) 13.6 ft/sec
    (D) 2.1 ft/sec

11. A child having a mass of 30.0 kg jumps downward onto a trampoline at a speed of 4.0 m/s. If the average impulsive force of the trampoline is 800.0 N and it acts upon the child for 0.5 s in a vertically upward direction, the child's velocity immediately after the trampoline force has been applied should be
    (A) 4.4 m/s
    (B) 17.2 m/s
    (C) 9.3 m/s
    (D) 5.8 m/s

Questions 12 and 13

As shown in the figure, a 40-kg crate is projected down a 35° inclined plane with an initial velocity of 3.0 m/s. The coefficient of kinetic friction between the incline and the crate is 0.15, and this condition applies also to the horizontal surface. For simplicity of calculations, the length of the crate is considered small.

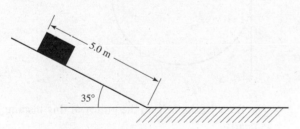

12. The net force acting on the crate along the inclined direction is
    (A) 18.1 N
    (B) 176.9 N
    (C) 273.3 N
    (D) 29.9 N

13. The kinetic energy of the crate at the instant when it has displaced 1.0 m across the horizontal surface will be approximately
    (A) 830 J
    (B) 1,125 J
    (C) 275 J
    (D) 1,005 J

14. A 750-g rectangular block moves across the top of a smooth cylindrical surface with an initial velocity of V_0 m/s. The radius of curvature of the cylinder is 1.6 m, as shown in the figure, and the block begins to lose contact with the surface at the position where $\theta = 30°$. The value of V_0 required to satisfy the stated conditions is

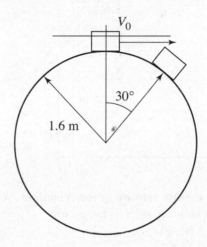

    (A) 3.1 m/s
    (B) 2.8 m/s
    (C) 5.6 m/s
    (D) 3.7 m/s

15. A cue ball strikes two billiard balls at rest simultaneously with a speed of 8.0 ft/s. Immediately after the collision, the cue ball stops in place, while one of the other balls moves with a velocity of 5.0 ft/s at 20° from the cue ball's line of action as diagrammed in the figure.

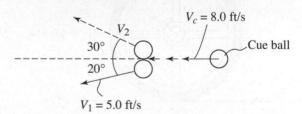

If the other ball that was struck moves at 30° in the direction illustrated, and all three balls have equal mass, the required post-collision speed V_2 of the second ball is
    (A) 3.0 ft/s
    (B) 6.6 ft/s
    (C) 7.2 ft/s
    (D) 3.8 ft/s

Dynamics, Kinematics, and Vibrations

16. An overhead garage door is guided by two rollers that travel within the perpendicular channels as shown in the figure. When $\theta = 35°$, the velocity of the wheel at Q is 0.5 m/s upward.

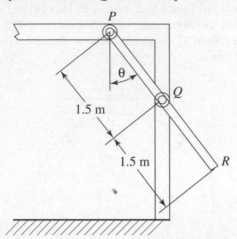

The angular velocity of the centerline PR of the door at this position is necessarily
(A) 3.60 rad/s
(B) 0.58 rad/s
(C) 0.33 rad/s
(D) 1.29 rad/s

17. A circular wheel rolls to the right along a horizontal surface and its center point C maintains a constant speed of 6.0 m/s as illustrated below. The motion of the wheel has no slipping at ground point G. Determine the absolute velocity of point P with respect to the ground reference.

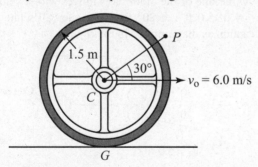

(A) 10.39 m/s
(B) 6.0 m/s
(C) 8.4 m/s
(D) 4.0 m/s

18. As shown in the figure, a 125-kg package is initially suspended above an undeformed ideal spring having a stiffness of 3 kN/m. The package is then released from rest and begins traveling downward.

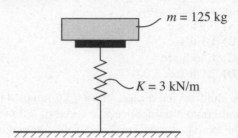

The speed of the package at the instant when the spring has been compressed 0.6 m will be
(A) 2.7 m/s
(B) 0 m/s
(C) 1.8 m/s
(D) 4.5 m/s

19. The pulley-belt drive system shown in the figure contains a pulley having a mass of 8.0 lbm and a radius of gyration of 6.0 in. At a given time, the angular acceleration of the pulley is known to be 30 rad/s^2.

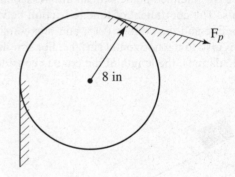

The force of the belt on the pulley at this instant is nearly
(A) 33.5 lbf
(B) 90.2 lbf
(C) 7.5 lbf
(D) 2.8 lbf

20. In the figure, the 16-kg cube is connected to a linear spring and moves without friction inside the chamber shown.

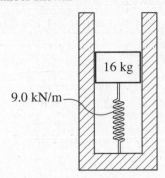

16 kg

9.0 kN/m

If the spring stiffness is 9.0 kN/m and the cube is given an initial downward velocity of 150 mm/s when the spring is compressed 4.0 mm, the amplitude of the resulting harmonic vibration should be

(A) 7.8 mm
(B) 14.1 mm
(C) 200.4 mm
(D) 55.2 mm

21. It is unlawful in most U.S. states to drop coins or other solid objects from skyscrapers or high outdoor platforms because of the dangerously high velocities attained. If air friction is considered negligible, the theoretical maximum speed of a coin thrown directly downward with an initial speed of 5.0 mph from an elevation of 800 ft above a reference datum will be

(A) 127.6 mph
(B) 165.4 mph
(C) 154.8 mph
(D) 143.9 mph

Answer Key

1. A
2. C
3. D
4. B
5. C
6. A
7. D
8. B
9. B
10. C
11. C
12. B
13. D
14. A
15. D
16. B
17. A
18. C
19. D
20. B
21. C

Answers Explained

1. **A** Given $X(t) = 2t^3 - 16t^2 + 7$

   Velocity $V(t) = \dfrac{dX}{dt} = 6t^2 - 32t$

   Acceleration $a(t) = \dfrac{dV}{dt} = 12t - 32$

   When $V = 280$ m/s, then $280 = 6t^2 - 32t$.

   Solve the quadratic equation for t: $t = 10$ s.

   Then:
   $$a(10) = 120 - 32 = 88.0 \text{ m/s}^2$$

2. **C** Displacement $\Delta X = X(t) - X_o$

   At $t = 20.0$ s:

   $\Delta X = X(20) - 7 = (16{,}000 - 6{,}400 + 7) - 7$
   $ = 9{,}600.0$ m

3. **D** Minimum velocity occurs when $a = 0$.

   $0 = 12t - 32$; then $t = 2.67$ s.

   $V(2.67) = 6(2.67)^2 - 32(2.67) = 42.77 - 85.44$
   $ = -42.7$ m/s.

4. B This problem involves projectile motion with no air friction.

Elevation $y(t) = y_o + V_{vo}t - \frac{1}{2}gt^2$

Velocity in y-direction $= V_v(t) = V_{vo} - gt$
$$= 7.0 \sin 60° - 9.81t$$

Maximum elevation occurs at $V_v = 0$.

$0 = 7.0(0.866) - 9.81t$; then $t = \frac{6.06}{9.81} = 0.62$ s.

$y(0.62) = 1.0 + 6.06(0.62) - 4.9(0.62)^2$
$= 1.0 + 3.76 - 1.88 = 2.88$ m

5. C Total speed of arrow is $V_T = \sqrt{V_\chi^2 + Vv^2}$

$V_\chi = 7.0 \cos 60° = 3.5$ m/s $=$ constant

$V_v(0.75) = 6.06 - 9.81(0.75) = -1.30$ m/s

$V_T = \sqrt{(3.5)^2 + (-1.3)^2} = 3.73$ m/s

6. A Here:

$a = \frac{\Sigma F}{m} = \frac{80.0 \text{ lbf}}{3.0 \text{ sl}} = \frac{80.0}{3.0}\left(\frac{1.0 \text{ sl-ft/s}^2}{\text{lbf}}\right)$

$= 26.67$ ft/s$^2 \approx 26.7$ ft/s^2

7. D Angular momentum $H = mr^2\dot{\theta}$

$$= 0.5\text{lbm}\left(\frac{6}{12}\text{ ft}\right)^2\dot{\theta}$$

$\dot{\theta} = 600$ rpm $\times \frac{2\pi}{60} = 62.8$ rad/s

$H = (0.5)(0.5)^2(62.8) = 7.85$ lbm-ft^2/s

$= \frac{7.85}{32.2} = 0.24$ ft-lbf-s

8. B Relative velocity of boat A with respect to boat B: $\mathbf{V}_{A/B} = \mathbf{V}_A + -\mathbf{V}_B$

Add velocity vectors tip to tail, and form a vector triangle.

Use the law of cosines to find the magnitude of relative velocity:

$\mathbf{V}_{A/B} = \sqrt{6^2 + 10^2 - 2(6)(10)\cos 160°}$
$= 15.8$ mph

Use the law of sines to find the direction of $\mathbf{V}_{A/B}$:

$\frac{\sin 160°}{15.8} = \frac{\sin a}{10.0}$; angle $a = 12.5°$ east of south.

9. B Use the tangential and normal (circular) coordinate system. Total acceleration magnitude is $a(t) = \sqrt{a_t^2 + a_n^2}$

Given $a_t = 1.5$, $a = 1.7$, and $V = 7.0$. Note that $a_n = \frac{V^2}{R}$. Then,

$$1.7 = \sqrt{(1.5)^2 + \left(\frac{V^2}{R}\right)^2}$$

$$\frac{V^2}{R} = \sqrt{(1.7)^2 - 2.25}$$

$$R = 60.0 \text{ m}$$

10. C Use conservation of linear momentum for impact:

$m_A V_A + m_B V_B = m_A V_A' + m_B V_B'$ (equation I)

Given $V_B = 0$, $V_A = +9.0$. Coefficient of restitution $e = 0.8$.

$V_B' - V_A' = e(V_A - V_B) = 0.8(9.0 - 0) = 7.2$ (equation II)

From equation I, $(16.0)(9.0) = 16V_A' + 3V_B'$ (equation III)

From equation II, $V_A' = V_B' - 7.2$. Substitute in equation II, and solve V_B' for the pin's postimpact velocity.

$144 = 16(V_B' - 7.2) + 3 V_B'$; $19V_B' = 259.2$. Then,

$$V_B' = 13.6 \text{ ft/s}$$

11. C Use the impulse/momentum equation in the vertical y-direction:

$\text{Imp}_y = mV_2 - mV_1$

$(-800.0 \text{ N})(0.5 \text{ s}) = 30.0 \text{ kg } (V_2 - 4.0)$. Note that the y-direction is positive downward.

$-400.0 \text{ N} \cdot \text{s} = 30.0V_2 - 120.0$; then:

$V_2 = \frac{-400.0 + 120.0}{30.0} = -9.33$ m/s upward

12. B Establish an FBD of the block; external forces are gravity, friction, and normal reaction. The kinetic friction force acts opposite the block's motion along the incline. Take the positive direction as downward along the inclined plane.

$\Sigma F_1 = W_1 - F_{F1} = (40.0)(9.81)\sin 35° - 0.15N_R$. Normal reaction N_R balances the weight component in the normal direction.

$N_R = W \cos \theta = (40.0)(9.81) \cos 35° = 321.4$ N; then:

$$\Sigma F_1 = 225.07 - 0.15(321.4) = 176.9 \text{ N}$$

13. D $KE_2 = KE_1 + W_{net\ 1\to2}$, and $W_{net\ 1\to2}$ is the work along the incline plus along the 1.0 m horizontal displacement.

$$KE_2 = \frac{1}{2}(40.0)(3.0)^2 + \left[\sum F_1(5.0 \text{ m})\right.$$

$$\left. + \sum F_X(1.0 \text{ m})\right]$$

$$= 180.0 \text{ N} \cdot \text{m} + (176.9 \text{ N})(5.0 \text{ m})$$
$$+ [-0.15(40)(9.81)(1.0 \text{ m})]$$

$$KE_2 = 180.0 + 884.5 - 58.8$$
$$= 1{,}005.6 \text{ N} \cdot \text{m} \approx 1{,}006 \text{ J}$$

14. A Use tangential and normal coordinates for this circular-type motion. Draw an FBD of the block, and note that the block has no normal reaction ($N_R = 0$) when $\theta = 30°$.

Since $\Sigma F_n = \dfrac{mV^2}{R}$, let V_2 represent the velocity

at the position when $\theta = 30°$. Use the FBD and substitute given values:

$$\sum F_n = \frac{(0.75 \text{ kg})V_2}{1.6 \text{ m}}$$

$W \cos 30° = 0.469 \dfrac{\text{kg}}{\text{m}} V_2^2\ (0.75 \text{ kg})(9.81 \text{ m/s}^2)$

$$\times\ (0.866) = 0.469 \frac{\text{kg}}{\text{m}} V_2^2$$

$$V_2 = \frac{(1.6 \text{ m})(6.37 \text{ N})}{0.75 \text{ kg}} = 3.69 \text{ m/s}$$

Use the conservation of energy equation from the top of the cylinder to point 2 to solve V_o:

$KE_1 + PE_1 = KE_2 + PE_2$. Set $PE_2 = 0$, and let $KE_1 = \dfrac{1}{2}mV_o^2$.

$PE_1 = mg(1.6 - 1.6 \cos 30°) = mg(0.214 \text{ m})$

Then $0.5mV_o^2 + mg(0.214) = 0.5m (3.69$ m/s$)^2 + 0$, and

$$V_o = \sqrt{9.43 \text{ m}^2/\text{s}^2} = 3.1 \text{ m/s}$$

15. D Use the conservation of linear momentum for the system of three balls in the x-direction:
$m_c V_c = m_1 V_{1x} + m_2 V_{2x} + m_c V_c' \to V_c' = 0$, and $m_c = m_1 = m_2$

$m_c(8.0 \text{ ft/sec}) = m_1(5.0 \cos 20°)$
$\qquad\qquad + m_2(V_2 \cos 30°)$
$8.0 = 4.7 + 0.866V_2$, and

$$V_2 = \frac{8.0 - 4.7}{0.866} = 3.8 \text{ ft/sec}$$

16. B Let a point C be located along a vertical line through point P that is the shortest perpendicular distance to point Q. Then, the given speed of point Q can be expressed as $V_Q = (CQ)\omega$, where distance $CQ = (1.5 \text{ m}) \sin 35° = 0.86 \text{ m}$, and ω is the desired angular speed of centerline PR.

Then 0.5 m/s $= (0.86 \text{ m})\omega$, and

$$\omega = 0.58 \text{ rad/s}.$$

17. A Using the principles and equations for rolling motion of a circular object, the angular velocity ω of the wheel is calculated from the speed of center C: $\mathbf{v_c} = -R\omega$. Therefore $\omega = \dfrac{v_c}{R} = 6.0$ m/s $= 4.0$ rad/s. The relative velocity of P with respect to the center is determined and 1.5 m then is added to the velocity of center C to obtain the resultant absolute velocity $\mathbf{v_p}$ of P.

$$\mathbf{v_{p/c}} = \omega \times \mathbf{r_{p/c}} = (4.0 \text{ rad/s}) (1.5 \text{ m})$$
$$= 6.0 \text{ m/s} \quad\searrow\quad 60°.$$

Solving the vector triangle shown using the Law of Cosines gives the absolute speed v_p.

$$v_p = \sqrt{6^2 + 6^2 - 2(6)(6)(\cos 120°)} = 10.39 \text{ m/s}.$$

The direction of velocity $\mathbf{v_p}$ is found from the Law of sines using the vector triangle above. $\theta = 30°$.

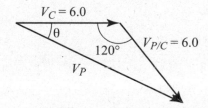

18. C Utilize the principle of work and kinetic energy to determine speed V_2 of the package. $W_{\text{net }1\rightarrow2} = KE_2 - KE_1$. Since $V_1 = 0$, $KE_1 = 0$.

The net work on the package is due to gravity and spring work.

$$W_{\text{net }1\rightarrow2} = (mg)(0.6 \text{ m}) + \frac{1}{2}k(X_1^2 - X_2^2)$$

$$= \frac{1}{2}mV_2^2$$

$(125.0 \text{ kg})(9.81)(0.6 \text{ m}) +$

$0.5 \ (3,000.0 \text{ N/m})[0 - (0.6 \text{ m})^2]$

$$= \frac{125.0 \text{ kg}}{2}V_2^2$$

Simplify:

$$V_2 = \sqrt{\frac{195.7}{62.5}} = 1.8 \text{ m/s}$$

19. D Rigid-body rotation requires $\Sigma M = I\alpha$, where $I = mk^2$.

$I = (8.0 \text{ lbm})(0.5 \text{ ft})^2 = 2.0 \text{ lbm-ft}^2$

$\sum M = (2.0 \text{ lbm-ft}^2)(30.0 \text{ rad/sec}^2)$

$$= \frac{60.0}{32.2} = 1.86 \text{ ft-lbf}$$

Since $\Sigma M = rF_p = 1.86$:

$$F_p = \frac{1.86}{0.67 \text{ ft}} = 2.8 \text{ lbf}$$

20. B Amplitude $X = \sqrt{x_o^2 + \left(\dfrac{\dot{x}_o}{\omega_n}\right)^2}$, and

$$\omega_n = \sqrt{\frac{k}{m}} = \sqrt{\frac{9,000 \text{ N/m}}{16.0 \text{ kg}}} = 23.7 \text{ rad/s. Then:}$$

$$X = \sqrt{(0.004)^2 + \left(\frac{0.15 \text{ m/s}}{23.7 \text{ rad/s}}\right)^2}$$

$$= \sqrt{0.00016 + 0.000040} = 0.0141 \text{ m} = 14.1 \text{ mm}$$

21. C $W_{\text{net}} = KE_2 - KE_1$

$$mg(\Delta h) = \frac{1}{2}m(v_2^2 - v_1^2)$$

$$2g(800.0 \text{ ft}) = v_2^2 - \left(5 \text{ mph} \times \frac{88 \text{ ft/s}}{60 \text{ mph}}\right)^2$$

$$51,520 \ \frac{\text{ft}^2}{\text{s}^2} = v_2^2 - 53.77$$

$$v_2 = \sqrt{51,573.8} = 227.1 \text{ ft/sec} = 154.8 \text{ mph}$$

9

Strength of Materials

Masoud Olia, Ph.D., PE

Introduction

Mechanics of materials is the branch of mechanics that relates external loads applied to a body in equilibrium to internal forces and the subsequent deformation of the body. In this chapter the concepts of stress and strain and their relationship are introduced. The deformation of axially loaded members is also presented. The torsion and the bending of structural members are analyzed. Structures under combined loading and stress transformation, as well as the design of columns, are discussed.

9.1 Normal Stress

Consider a bar with constant cross-sectional area (prismatic) under the action of a constant force that is applied at the centroid of the bar. The stress is defined as the load, P, divided by the area (load per unit area), A, and is called normal stress, denoted by σ and is "+". If the bar is in tension, the stress is called tensile stress; if the bar is compressed, the stress is called compressive stress and is "–". Stress has units of pounds per square inch (lb/in^2 or psi) in the English system and newtons per square meter (N/m^2) or pascals (Pa) in the metric system.

Also 1 kip = 1,000 lb
1 ksi = 1 kip/in^2 = 1,000 psi

$$\sigma = \frac{P}{A}$$

σ_t = tensile stress (+) [9.1]

σ_c = compressive stress (−)

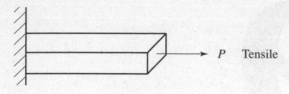

Figure 9.1

Example 9.1

For the composite bar shown below, determine the normal stress in each section.

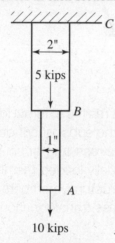

Figure 9.2

Solution:

First we have to determine the load in each section:

$P_{AB} = 10$ kips (tension)

$P_{BC} = 15$ kips (tension)

$$\sigma_{AB} = \frac{10}{\frac{\pi}{4}(1)^2} = 12.73 \text{ ksi (tensile)}$$

$$\sigma_{BC} = \frac{15}{\frac{\pi}{4}(2)^2} = 4.77 \text{ ksi (tensile)}$$

9.2
Shear Stress

If two plates are bolted or glued to each other, the stress parallel to the area of contact, A_s, is called the average shear stress and is denoted by τ_{ave}.

$$\tau_{ave} = \frac{V}{A_s} \qquad [9.2]$$

where V = shear load,

A_s = shear area.

For a bolted connection, A_S is the cross-sectional area of the bolt.

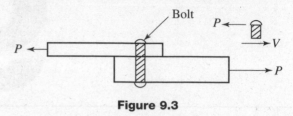

Figure 9.3

Example 9.2

If the allowable shearing stress in the bolt shown below is 350 MPa, determine the minimum diameter of the bolt.

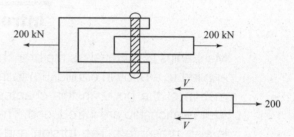

Figure 9.4

Solution:

This is a case of double shear.

$$V = \frac{P}{2} = \frac{200}{2} = 100 \text{ kN}$$

$$\tau = \frac{V}{A_s}, \quad A_s = \frac{V}{\tau_{all}} = \frac{100 \times 10^3}{350 \times 10^6} = 2.857 \times 10^{-4}$$

$$A_s = \frac{\pi}{4}d^2$$

$$d = \sqrt{\frac{4A_s}{\pi}} = \sqrt{\frac{4(2.857 \times 10^{-4})}{\pi}} = 0.01907 \text{ m}$$

9.3
Normal Strain

Consider again a prismatic bar under the action of a tensile load. The initial length of the bar is l, and after the load is applied the bar elongates by δ. Average normal strain is defined as the ratio of change in length to initial length and is denoted by ε_{ave}. If the bar is in tension, the strain is called tensile strain and is "+". If the bar is in compression, the strain is called compressive strain and is "–". Note that strain is dimensionless.

$$\varepsilon_{ave} = \frac{\delta}{\ell}$$

$$\varepsilon_t = \text{tensile strain} \qquad [9.3]$$

$$\varepsilon_c = \text{compressive strain}$$

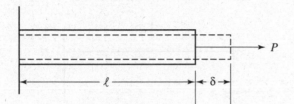

Figure 9.5

9.4
Shear Strain

Consider the stress element under the action of the shear stresses shown in the drawing below. The shear stress on each side (face) must be the same in order for the stress element to remain in equilibrium. Shear strain, denoted by γ, is defined as the change in the angle between two sides of the stress element from the 90° initial angle. Shear strain is a measure of distortion of the element under the action of shear stresses. Shear strain is dimensionless, and its unit must be radians.

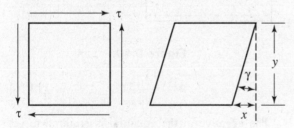

Figure 9.6

$$\tan \gamma \approx \gamma = \frac{x}{y} \qquad [9.5]$$

9.5
Factor of Safety

Factor of safety (F.S.) is defined as the ratio of the load or stress that causes failure to the applied load or stress. The applied load is usually referred to as the allowable load. The failure load can be taken as the yield or ultimate load.

$$F.S. = \frac{P_{fail}}{P_{all}} \text{ or } F.S. = \frac{\sigma_{fail}}{\sigma_{all}} \text{ or } F.S. = \frac{\tau_{fail}}{\tau_{all}} \qquad [9.6]$$

9.6
The Relationship
of Stress and Strain

The relationship of elastic stress to elastic strain is based on a law known as Hooke's law:

$$\sigma = E\varepsilon \qquad [9.7]$$

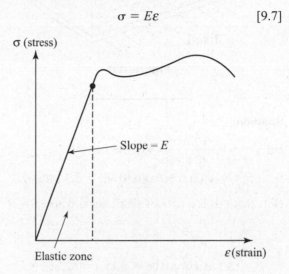

Figure 9.7

Here, E is a material property known as the modulus of elasticity or Young's modulus. For example, E for structural steel is about 30×10^6 psi or 200 GPa. If the relationship between stress and strain is represented graphically, then E is the slope of the straight line in the linear zone.

Also, Hooke's law can be applied to a member in pure shear (i.e., a shaft being twisted) and relates shear stress, τ, to shear strain, γ:

$$\tau = G\gamma \qquad [9.8]$$

where G is a material property known as modulus of rigidity and is the slope of the line in the linear zone of τ vs. γ plot.

Example 9.3

For the rectangular block shown below, $G = 500$ MPa. If the lower plate is fixed and the upper plate moves through 0.7 mm as a result of applied force P, determine the value of force P.

259

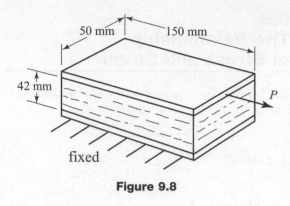

Figure 9.8

Solution:

$$\tan \gamma = \gamma = \frac{0.7 \text{ mm}}{42 \text{ mm}} = 0.0166 \text{ rad}$$

$$\tau = G\gamma = (500 \text{ MPa})(0.0166) = 8.3 \text{ MPa}$$

Note that this is a case of single shear where $V = P$

$$\tau = \frac{V}{A} = \frac{P}{A} \Rightarrow P = \tau A$$

$$P = (8.3 \times 10^6)(0.05 \times 0.15) = 62{,}250 \text{ N}$$

9.7
Plastic Behavior

In this chapter we will only consider loadings that will cause the material to behave elastically, and therefore Hooke's law is satisfied. If a material is loaded to its elastic limit and then the load is removed, the material will return to its initial state with no residual strains.

Some materials such as mild steel are designed to yield under a load and therefore are permanently deformed. The stress versus strain diagram for such materials can be modeled as elastic-perfectly plastic or elastoplastic. One should note that all subsequent equations due to axial loading, twisting, and bending of a member are valid only in the linear-elastic zone.

9.8
Thermal Strain

If a bar is constrained, as the temperature changes the bar will deform. The strain due to change in temperature is called thermal strain, and denoted by ε_T.

$$\varepsilon_T = \alpha \Delta T \qquad [9.9]$$

Here, α is the coefficient of thermal expansion, with units of 1/°F or 1/°C, and is defined as the strain due to a 1-degree change in temperature. Note that α, like E and G, is a material property. For example, for structural steel $\alpha = 12 \times 10^{-6} \frac{1}{°C}$

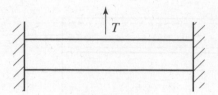

Figure 9.9

9.9
Poisson's Ratio

Consider a bar under the action of a load. The direction of the load is axial, and perpendicular to the axial direction is the lateral direction. When the load is applied, the bar will elongate in the axial direction, but at the same time it will contract in the lateral direction. Poisson's ratio, denoted by v, is defined as the negative of the ratio of the lateral strain to the axial strain.

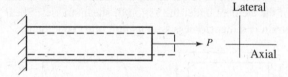

Figure 9.10

$$v = -\frac{\varepsilon_{\text{lateral}}}{\varepsilon_{\text{axial}}} \qquad [9.10]$$

The negative in the equation is required to get a "+" value for v. Note that v is a material property and, for example, $v = 0.3$ for structural steel. Also, Poisson's ratio v cannot be larger than 0.5.

The three properties E, G, and v are related to each other as follows:

$$G = \frac{E}{2(1 + v)} \qquad [9.11]$$

Example 9.4
The plastic rod shown in Figure 9.11 is 150 mm long and 12 mm in diameter. Determine the change in length and the change in diameter that occur when a 400-N load is applied if, for the plastic, $E = 2.5$ GPa and $v = 0.4$.

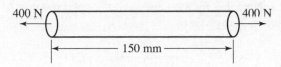

Figure 9.11

Solution:

$$\sigma = E\varepsilon_{\text{axial}}; \text{ where } \sigma = \frac{P}{A}$$

$$\varepsilon_{\text{axial}} = \frac{\sigma}{E} = \frac{P}{EA} = \frac{400}{\frac{\pi}{4}(0.012)^2(2.5\times10^9)} = 0.001415$$

$$\varepsilon_{\text{axial}} = \frac{\delta}{\ell}; \ \delta = \ell\varepsilon_{\text{axial}}$$

$$= 0.2122\text{-mm increase in length}$$

$$v = -\frac{\varepsilon_{\text{lateral}}}{\varepsilon_{\text{axial}}}$$

$$\varepsilon_{\text{lateral}} = -v\varepsilon_{\text{axial}} = -0.4(0.001415) = -0.000566$$

$$\varepsilon_{\text{lateral}} = \frac{\Delta d}{d}$$

$$\Delta d = d\varepsilon_{\text{lateral}} = (12 \text{ mm})(-0.000566)$$

$$\Delta d = -0.006792\text{-mm decrease in diameter}$$

9.10
Deformation in Axially Loaded Members

For an axially loaded member (refer to Figure 9.10) the axial deformation can be determined by using the following equation:

$$\sigma = E\varepsilon, \ \frac{P}{A} = E\frac{\delta}{\ell}; \ \delta = \frac{P\ell}{EA} \qquad [9.12]$$

Example 9. 5

The circular steel rod ($E = 30 \times 10^6$ psi) shown in Figure 9.12, which is 2 inches in diameter, is loaded axially. Determine the load P that causes a deformation of 0.1 in in the rod.

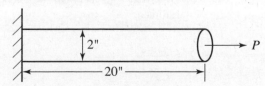

Figure 9.12

Solution:

$$\delta = \frac{P\ell}{EA} \Rightarrow P = \frac{\delta EA}{\ell}$$

$$P = \frac{(0.1)(30\times10^6)\left(\frac{\pi}{4}2^2\right)}{20} = 471.238 \text{ kips}$$

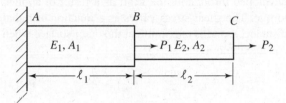

Figure 9.13

For a composite structure (Figure 9.13, stepped shaft), the summation form of the equation can be used to determine the total axial deformation:

$$\delta = \Sigma\frac{P\ell}{EA} \qquad [9.13]$$

Example 9.6

Using Figure 9.13, determine the total elongation of the stepped shaft for the following values:

$$P_1 = 10 \text{ kips}, P_2 = 5 \text{ kips}$$

$$E_1 = E_2 = 30 \times 10^6 \text{ psi}$$

$$d_1 = 2 \text{ in}, d_2 = 1 \text{ in}$$

$$\ell_1 = 20 \text{ in}, \ell_2 = 15 \text{ in}$$

Solution:
Using free-body diagram for each section AB and BC, the load in section AB, $P_{AB} = 15$ kips and the load in section BC, $P_{BC} = 5$ kips.

$$\delta = \Sigma\frac{P\ell}{EA} = \left(\frac{P\ell}{EA}\right)_{AB} + \left(\frac{P\ell}{EA}\right)_{BC}$$

$$\delta = \left[\frac{(15\times10^3)(20)}{(30\times10^6)\left(\frac{\pi}{4}2^2\right)}\right] + \left[\frac{(5\times10^3)(15)}{(30\times10^6)\left(\frac{\pi}{4}1^2\right)}\right]$$

$$= 0.006366 \text{ in}$$

9.11 Torsion

If a straight bar with constant circular cross section (i.e., shaft) is fixed at one end and is twisted by a torque at the other end, such a member is said to be in pure torsion.

When the shaft is twisted, shear stresses are developed throughout the shaft as a result of applied torque. The shear stress varies as a function of radial distance, r, and is maximum at the outer surface when r is maximum (i.e., $r = c$).

$$\tau = \frac{Tr}{J}, \qquad \tau_{max} = \frac{Tc}{J} \qquad [9.14]$$

Note that the shear stress is changing linearly as the function of radial position r.

Here, J is called the polar moment of inertia; for a circular cross section (shaft) $J = \pi d^4/32$ or $J =$

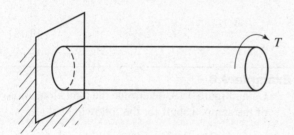

Figure 9.14

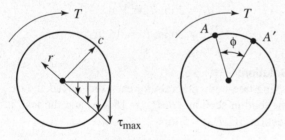

Figure 9.15

$\pi c^4/2$ and has units of inch^4 or meter^4. (Refer to Table 7.2 for formulas to determine J.)

Twisting will also cause the shaft to deform (distortion). The angle of twist, (ϕ), for a shaft of length L, under the action of torque T, is given by the equation

$$\phi = \frac{TL}{GJ} \qquad [9.15]$$

Here G is the shear modulus (material property), and ϕ is dimensionless and should be expressed in radians.

Note that for a hollow shaft the polar moment of inertia, J, is given by

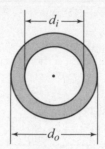

Figure 9.16

$$J = \frac{\pi(d_o^4 - d_i^4)}{32}, \quad J = \frac{\pi(c_o^4 - c_i^4)}{2} \qquad [9.16]$$

where d_o and c_o are the outside diameter and radius, respectively, and d_i and c_i are the inside diameter and radius, respectively.

Example 9.7

For the hollow cylinder shown in Figure 9.17, determine the torque, T, that causes a maximum shearing stress of 400 MPa.

Solution:

$$J = \frac{\pi(d_o^4 - d_i^4)}{32} = \frac{\pi(2^4 - 1.5^4)}{32}$$

$$J = 1.074 \text{ cm}^4 = 1.074 \times 10^{-8} \text{ m}^4$$

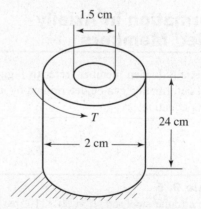

Figure 9.17

Note that τ_{max} occurs at the outside radius c_o

$$\tau_{max} = \frac{Tc_o}{J} \Rightarrow T = \frac{\tau_{max}J}{c_o}$$

$$T = \frac{(400 \times 10^6)(1.074 \times 10^{-8})}{0.01} = 429.6 \text{ N·m}$$

Example 9.8

Using Figure 9.17, determine the angle of twist for $T = 500\ N \cdot m$ and $G = 100\ GPa$.

Solution:

$$\phi = \frac{TL}{GJ} = \frac{(500)(0.24)}{(100 \times 10^9)(1.074 \times 10^{-8})} = 0.11\ rad$$

$$= 6.3°$$

For hollow, thin-walled shafts, shear stress can be determined by using the equation

$$\tau = \frac{T}{2\ A_m t} \qquad [9.17]$$

where t = thickness of the wall,

A_m = total mean area enclosed by the shaft measured to the midpoint of the wall.

Note that for noncircular sections, the shear stress distribution over the cross section is nonlinear and as a result the equation $\tau_{max} = \dfrac{T\ c}{J}$ cannot be used.

In rotating machines, the circular shafts are often used to transmit power developed by a machine. The power that is developed or transmitted depends on the torque and the angular speed of the shaft and is determined by the equation shown below:

$$P = T \omega \qquad [9.18]$$

In equation 9.18, P is the power, T is the torque, and ω is the angular speed. Power has units of watts or horsepower (hp). Note that ω should be converted to rad/sec and that 1 hp = 550 lb-ft/sec.

9.12
Bending

A beam is defined as a long, prismatic structure designed to support loads at various points along the member. In most cases, the loads are applied perpendicularly to the axis of the beam and cause only shear and bending in the beam.

Internal Forces and Moments

Consider the simply supported beam shown in Figure 9.18.

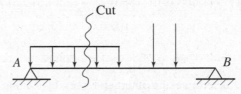

Figure 9.18

If the beam is cut at a certain section, it is observed that, for the left or right section to be in equilibrium, a shear load, V, and a bending moment, M, must be present at that section. The shear load and the bending moment are called internal forces. Note in Figure 9.19 that the directions of V and M for the left and right sections of the beam are indicated by the sign conventions used for positive shear and positive moment.

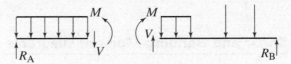

Figure 9.19

To determine the internal forces and moments, do the following:

1. *Determine the reactions at the supports.*
2. *Cut the section where the internal forces are to be calculated, and draw the free-body diagram.*
3. *Use equilibrium equations to determine the internal forces and moments.*

To find V, use $\sum F_y = 0$

To find M, use $\sum M = 0$

Note that the result is the same regardless of the section (right or left) used.

Example 9.9

For the beam shown below, determine the shear force and the bending moment at a point 2 m to the right of point A.

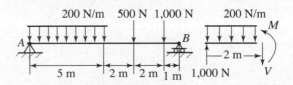

Figure 9.20

Solution:

First, calculate the reactions:

$$\Sigma M_A = 0$$

$$B(10) - (1,000)(9) - (500)(7) - (200)(5)(2.5) = 0$$

$$B = 1,500 \text{ N}$$

Then, $\Sigma F_y = 0$; $A = 1,000$ N

Now, determine the internal forces by cutting the beam at a point 2 m to the right of point A and applying the equilibrium equations.

Note that it is more efficient to use the left side of the beam.

$$\Sigma F_y = 0 \qquad -V + 1,000 - 200(2) = 0$$

$$V = 600 \text{ N}$$

$$\Sigma M_o = 0 \qquad M - 1,000(2) - 2(200)(1) = 0$$

$$M = 1,600 \text{ N} \cdot \text{m}$$

Shear and Bending-Moment Diagrams

Since determination of maximum shear and bending moment is important in design, it is more convenient to plot shear and bending moment along the beam than to cut many sections to calculate these values. The graphs obtained are called the shear diagram and the bending-moment diagram, respectively.

Drawing Shear and Bending-Moment Diagrams

The following relations hold:

$$\frac{dV}{dx} = -w, \qquad \int_{V_1}^{V_2} dV = -\int_{x_1}^{x_2} w\,dx \qquad [9.19]$$

where w is the distributed load acting on the beam. Net shear force $= -$(area under distributed load from x_1 to x_2)

$$\frac{dM}{dx} = V, \qquad \int_{M_1}^{M_2} dM = -\int_{x_1}^{x_2} V\,dx \qquad [9.20]$$

Net bending moment $=$ (area under shear load from x_1 to x_2)

Example 9.10

Draw the shear and bending-moment diagrams for the beam shown.

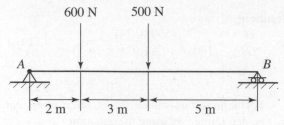

Figure 9.21

Solution:

First, determine the reactions.

$$\Sigma M_A = 0$$

$$B(10) - 500(5) - 600(2) = 0; B = 370 \text{ N}$$

$$\Sigma F_y = 0$$

$$A + B - 500 - 600 = 0$$

$$A = 730 \text{ N}$$

Then draw the diagrams.

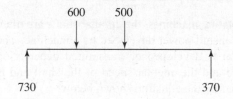

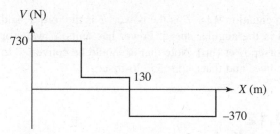

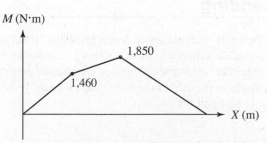

Figure 9.22

Example 9.11

Draw the shear and bending-moment diagrams for the beam shown.

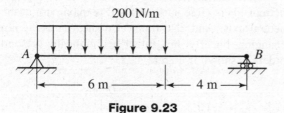

200 N/m

A B

← 6 m → ← 4 m →

Figure 9.23

Solution:

First, determine the reactions.

$$\Sigma M_A = 0$$

$$B(10) - (200)(6)(3) = 0 \Rightarrow B = 360 \text{ N}$$

$$\Sigma F_y = 0$$

$$A + B - 200(6) = 0$$

$$A = 840 \text{ N}$$

Then draw the diagrams.

Note that the slope of the decreasing line is equal to the negative of the distributed load (−200).

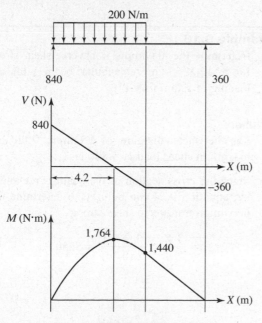

200 N/m

840   360

V (N)

840

← 4.2 → →X (m)

−360

M (N·m)

1,764

1,440

→ X (m)

Figure 9.24

Note that the position where shear is zero is determined by slope of the line in the shear diagram. Since the slope is −200, then $\frac{840}{x} = 200$ and $x = 2.4$ m .

Stresses Due to Bending

Consider a prismatic beam that is bent under the action of applied loads, as shown below.

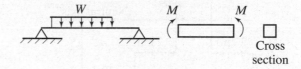

W M M

Cross
section

Figure 9.25

The bending stress, also known as the flexural stress, can be determined by using the equation

$$\sigma = \frac{My}{I} \qquad [9.21]$$

where σ = bending (flexural) stress,

M = bending moment at a given point,

y = distance from the centroid (neutral axis) to the point where the stress is to be calculated,

I = area moment of inertia with respect to the axis of bending through the centroid.

The maximum bending stress can be determined by using the equation

$$\sigma_{max} = \frac{Mc}{I} \qquad [9.22]$$

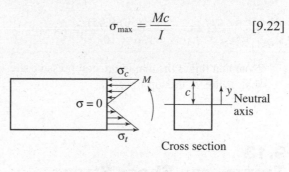

σ_c

M

$\sigma = 0$

c y Neutral axis

σ_t

Cross section

Figure 9.26

where c is the perpendicular distance from the neutral axis to a point farthest away from the neutral axis (top or bottom fiber).

Note that the stress distribution along the cross section is linear and the stress is zero at the centroid (neutral axis).

Example 9.12

Determine the maximum bending stress for the beam and loading of Example 9.10 with the cross section shown below.

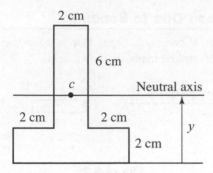

Figure 9.27

Solution:

For the moment diagram of Example 9.10,

$$M_{max} = 1,850 \text{ N·m}$$

To determine the maximum bending stress, first determine the area moment of inertia with respect to the neutral axis (refer to Example 7.25):

$$I = 136 \text{ cm}^4 = 136 \times 10^{-8} \text{ m}^4$$

and $C = 5$ cm $= 0.05$ m, which is the distance from the neutral axis (N.A.) to the top fiber.

Then calculate the maximum stress:

$$\sigma_{max} = \frac{M_{max}C}{I} = \frac{(1,850)(0.05)}{136 \times 10^{-8}} = 68 \text{ MPa}$$

Note that this is the maximum compressive stress that occurs at the top fiber.

9.13
Transverse Shear Stress

The stress developed in a structure as a result of a shear force, V, is called transverse shear stress and is denoted by τ.

$$\tau = \frac{VQ}{It}$$

[9.23]

where V = shear force at a given point,

Q = first moment of area with respect to the neutral axis $= \Sigma y A = \int y dA$,

I = area moment of inertia with respect to the neutral axis,

t = thickness of beam at position y for which the shear stress is calculated.

Note that the shear stress is zero at the top and bottom fibers (free surfaces) and is maximum at the neutral axis. And also transverse shear stress is not changing linearly but rather parabolicaly (second order) as shown in Figure 9.28.

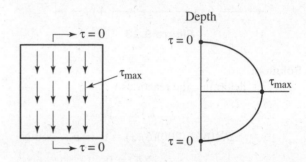

Figure 9.28

For the solid rectangular cross sections:

$$\tau_{max} = \frac{3}{2} \frac{V}{A}$$

[9.24]

where A = cross-sectional area.

For the solid circular cross sections:

$$\tau_{max} = \frac{4}{3} \frac{V}{A}$$

[9.25]

Example 9.13

Determine the maximum transverse shear stress for a 40×60-cm rectangular beam with the loading of Example 9.11.

Solution:

For the shear diagram of Example 9.11, the maximum shear load $V = 840$ N.

Since the cross section of the beam is rectangular, equation 9.24 can be used to determine the maximum transverse shear stress:

$$\tau_{max} = \frac{3}{2} \frac{V}{A} = \frac{3 (840)}{2 (0.24)} = 5250 \text{ Pa}$$

9.14
Combined Loading

Consider a structural member subjected to some combination of axial, bending, and torsional loads.

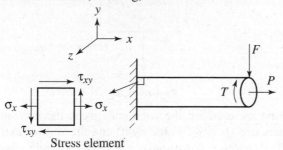

Figure 9.29

To determine the maximum stresses (normal and shear), first locate the point where the stresses are most severe. For a cantilever beam, this point is located at the fixed support. Once the stresses due to different loads are calculated, they can be represented on a very small element (stress element) and then maximum normal and shear stresses can be determined. The maximum normal stresses are called principal stresses, and the planes at which these stresses occur are called principal planes.

Example 9.14

Determine the normal and shearing stresses at points A and B in the figure below.

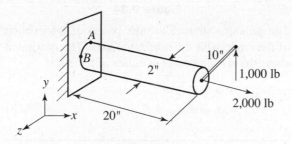

Figure 9.30

Solution:

$$T = M_x = 1,000(10) = 10,000 \text{ lb} \cdot \text{in}$$

$$M_z = 1,000 (20) = 20,000 \text{ lb} \cdot \text{in}$$

$$V_y = 1,000 \text{ lb}$$

$$P_x = 2,000 \text{ lb}$$

$$I_y = I_z = \frac{\pi}{4} (1)^4 = 0.7854 \text{ in}^4$$

$$I_x = J = 2I = 1.57 \text{ in}^4$$

$$A = \frac{\pi}{4} (2)^2 = 3.14 \text{ in}^2$$

At point A:

$$\sigma_A = -\frac{M_z c}{I_z} + \frac{P_x}{A} = -\frac{20,000(1)}{0.7854} + \frac{2000}{3.14}$$

$$= -24,828 \text{ psi (compressive)}$$

$$\tau_A = \frac{Tc}{J} = \frac{(10,000)(1)}{1.57} = 6,369 \text{ psi}$$

Note that since the stresses are larger at A, point A is the critical point on the shaft.

At point B:

$$\sigma_B = \frac{P_x}{A} = \frac{2000}{3.14} = 637 \text{ psi}$$

$$\tau_B = \frac{Tc}{J} - \frac{4}{3} \frac{V_y}{A} = \frac{(10,000)(1)}{1.57} - \frac{4(1,000)}{3(3.14)}$$

$$= 5,944 \text{ psi}$$

> Note that since point B is on the N.A., the stress due to bending is zero.

9.15
Principal and Maximum Shear Stresses

To determine the principal stresses and maximum shear stress and their planes:

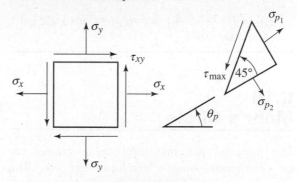

Figure 9.31

use the following equations:

$$\sigma_{p_1, p_2} = \sigma_{max,min} = \frac{\sigma_x + \sigma_y}{2} \pm \sqrt{\left(\frac{\sigma_x - \sigma_y}{2}\right)^2 + \tau_{xy}^2}$$

$$\tan 2\theta_p = \frac{2\tau_{xy}}{\sigma_x - \sigma_y} \qquad [9.26]$$

$$\tau_{max} = \frac{\sigma_{max} - \sigma_{min}}{2} = \sqrt{\left(\frac{\sigma_x - \sigma_y}{2}\right)^2 + \tau_{xy}^2}$$

$$\tan 2\theta_s = -\frac{\sigma_x - \sigma_y}{2\tau_{xy}}$$

Also note that

$$\theta_s = \theta_p \pm 45 \qquad [9.27]$$

where θ_p is the principal angle where maximum normal stress will occur and similarly θ_s is the angle where the maximum shear stress will act.

Example 9.15

Determine the principal and maximum shear stresses of Example 9.14 for point A.

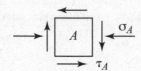

Figure 9.32

Solution:

$$\sigma_x = -24{,}828 \text{ psi}, \quad \sigma_y = 0$$

$$\tau_{xy} = -6{,}369 \text{ psi}$$

Using equation 9.25:

$$\sigma_{p_1, p_2} = \frac{-24{,}828}{2} \pm \sqrt{\left(\frac{-24{,}828}{2}\right)^2 + (-6{,}369)^2}$$

$$\sigma_1 = 1{,}538 \text{ psi}, \sigma_2 = -26{,}366 \text{ psi}$$

$$\tau_{max} = \sqrt{\left(\frac{-24{,}828}{2}\right)^2 + (-6{,}369)^2} = 13{,}952 \text{ psi}$$

9.16
Mohr's Circle

The principal and maximum shear stresses can be determined using a graphical approach. This approach involves drawing a circle that can be used to determine these stresses. Look at the given stress element shown below:

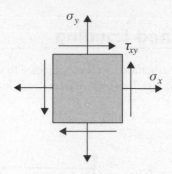

Figure 9.33

First we establish the coordinate axes: the horizon axis is σ (positive to the right) and the vertical axis is τ (positive downward). The Mohr's circle center has the coordinates of $C(\sigma_{ave}, 0)$, where $\sigma_{ave} = \frac{\sigma_x + \sigma_y}{2}$. The coordinates of point A along the right vertical side of the stress element is (σ_x, τ_{xy}). By using these two points, we can draw the circle as shown below:

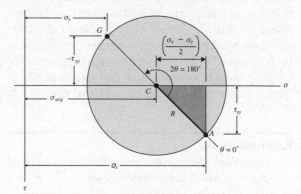

Figure 9.34

The principal stresses are the points of intersection of the circle with the horizontal axis. The maximum shear stress is equal to the radius of the circle.

$$\sigma_{p1} = \sigma_{ave} + R,$$
$$\sigma_{p2} = \sigma_{ave} - R, \text{ and } \tau_{max} = R \qquad [9.28]$$

Example 9.16

For the stress element shown below, draw Mohr's circle and determine the principal stresses and maximum shear stress.

Solution:

Point A (60, 30)

Center of the circle is at $C(\sigma_{ave}, 0)$

$\sigma_{ave} = (60 - 20)/2 = 20 \text{ MPa}$

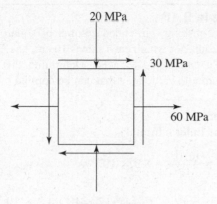

Figure 9.35

The Mohr's circle is shown below:

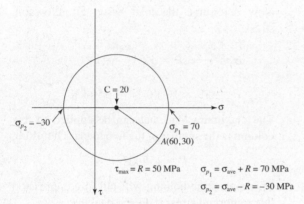

Figure 9.36

9.17
Thin-Walled Pressure Vessels

Cylindrical and spherical pressure vessels are used in industry for boilers or storage tanks. The term "thin-wall" refers to a tank for which the ratio of inner radius to thickness is equal to or larger than 10 ($r_i/t \geq 10$).

Cylindrical Vessels

For a cylinder of inner radius r and thickness t that is under pressure P, circumferential (hoop) and longitudinal stresses are developed in the vessel.

$$\sigma_{circ} = \sigma_{hoop} = \sigma_1 = \frac{Pr}{t} \qquad [9.29]$$

$$\sigma_{long} = \sigma_2 = \frac{Pr}{2t} \qquad [9.30]$$

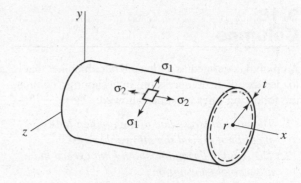

Figure 9.37

Example 9.16

A cylindrical pressure vessel has an inner radius of 2 m and a thickness of 3 cm. Determine the maximum internal pressure this vessel can sustain so that its circumferential stress component will not exceed 120 MPa.

Solution:

$$\sigma_1 = \frac{Pr}{t} \Rightarrow P = \frac{\sigma_1 l}{r}$$

$$P = \frac{(120 \times 10^6)(0.03)}{2} = 1.8 \text{ MPa}$$

Spherical Vessels

For a thin-walled spherical vessel subjected to inside pressure P:

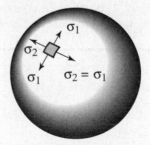

Figure 9.38

the stress at any location is given by the equation

$$\sigma_{long} = \sigma_1 = \sigma_2 = \frac{Pr}{2t} \qquad [9.31]$$

269

9.18 Columns

A column is defined as a long, slender member that is loaded axially in compression. In designing a column, the following rules must be followed:

1. *The normal stress due to the applied load should not exceed the allowable stress.*
2. *The axial deformation should not exceed the allowable deformation.*
3. *The load applied must be less than the critical load that causes buckling.*

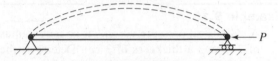

Figure 9.39

If a column is pinned at both ends, the smallest critical load P is given by the equation

> Note that I is the smallest moment of inertia about any axis.

$$P_{cr} = \frac{\pi^2 EI}{L^2} \qquad [9.32]$$

Equation 9.32 is known as *Euler's formula*.
For other end conditions, the equation

$$P_{cr} = \frac{\pi^2 EI}{L_e^2} \qquad [9.33]$$

can be used. Here, L_e is called the *effective length* and is shown in Figure 9.40 for four end conditions.

> The effective length is the distance between two consecutive inflection points on the elastic curve.

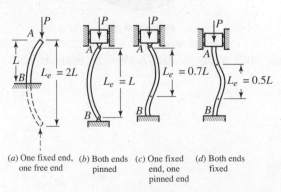

(a) One fixed end, one free end   (b) Both ends pinned   (c) One fixed end, one pinned end   (d) Both ends fixed

Figure 9.40

Example 9.18

A 2-m-long, pin-ended column of square cross section, for which each side is 10 cm, has $E = 12$ GPa and $\sigma_{all} = 15$ MPa. Determine the maximum allowable load that can be applied.

Solution:
Use Euler's formula.

$$I = \frac{(0.1)^4}{12} = 8.3 \times 10^{-6} \text{ m}^4$$

$$P_{cr} = \frac{\pi^2 EI}{L^2} = \frac{(3.14)^2(12 \times 10^9)(8.3 \times 10^{-6})}{2^2}$$

$$= 246.74 \text{ kN}$$

Now determine the load based on allowable stress.

$$\sigma = \frac{P}{A} \Rightarrow P = \sigma A$$

$$P = (15 \times 10^6)(0.1)^2 = 150 \text{ kN}$$

The maximum load that can be applied to this column is the smaller of the two loads. Therefore,

$$P = 150 \text{ kN}$$

This means the column will fail in compression due to normal stress before it buckles.

9.19 Deflection of Beams

There are several methods which can be used to determine the deflection of beams under transverse loads. Some of these analytical methods include the integration, superposition, moment-area, and energy methods. Here, we will only discuss the first two methods briefly.

1. Integration method.

For small deflections the differential equation of the elastic curve is given by

$$EI \frac{d^2 y}{dx^2} = M$$

or

$$EI \ y'' = M \qquad [9.34]$$

where y is the vertical displacement of the elastic curve (deflection) and M is the bending moment which in general varies as a function of position (x). Once the bending moment is determined, the differential equation can be solved for y by integrating twice. Using the proper boundary and matching conditions, the constants of integration can then be determined (Refer to Figure 9.41).

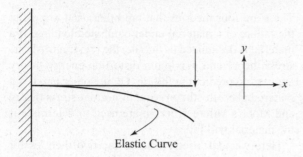

Figure 9.41

Example 9.19

For the cantilever beam, determine the maximum deflection and maximum slope.

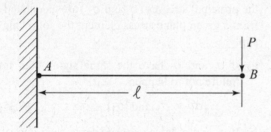

Figure 9.42

Solution:

First, determine the bending moment as a function of position (x). This could be done by considering either the left or right section of the beam after we cut a section. For convenience the right section is chosen.

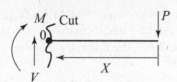

Figure 9.43

$$\sum M_o = 0 \quad M + Px = 0 \quad M = -Px$$

$$EI\ y'' = M = -Px$$

Integrate once:

$$EI\ y' = -P\frac{x^2}{2} + C_1$$

To find C_1, use the obvious boundary condition that at $x = 1$, $y' = 0$ (slope = 0)

Get:

$$C_1 = \frac{Pl^2}{2}$$

Integrate again:

$$EI\ y = -P\frac{x^3}{6} + P\frac{l^2}{2}x + C_2$$

To find C_2, use the other obvious boundary condition, which is at $x = 1$ $y = 0$ (deflection = 0). Get:

$$C_2 = -\frac{Pl^3}{3}$$

Therefore, the equation of the elastic curve is

$$y(x) = \frac{1}{EI}\left(-\frac{P}{6}x^3 + \frac{Pl^2}{2}x - \frac{Pl^3}{3}\right)$$

and the equation that describes the slope is

$$y'(x) = \frac{1}{EI}\left(-\frac{P}{2}x^2 + \frac{Pl^2}{2}\right)$$

To find the maximum deflection and maximum slope, which occurs at B, we set $x = 0$ which yields

$$y_{max} = y_B = -\frac{Pl^3}{3EI}$$

and

$$y'_{max} = y'_B = \frac{Pl^2}{2EI}$$

Note the (–) for the deflection implies downward displacement. The slope should be also negative (–), but since we used positive x direction to the left, we end up getting a positive (+), outcome.

2. Superposition method

Since the differential equation for the elastic curve is linear, the deflection for a series of loads acting on a beam can be superimposed. In this method, the given beam with its loading is divided into a series of beams with individual loads acting on them. Then the deflection at a certain point along the beam is determined by the algebraic sum of the deflections due to each load. Refer to the Beam Deflection Formulas table in the NCEES *FE Reference Handbook* for selected beams.

Example 9.20

For the beam shown, determine the maximum deflection at the center of the beam.

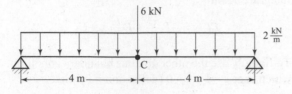

Figure 9.44

Use $E = 200$ GPa and $I = 50 \times 10^{-6}$ m^4

Solution:

This beam is divided into two beams: one with the only distributed load acting on it and the other with the concentrated load acting at its center. (See Figure 9.45.)

From the Beam Deflection Formulas table in the NCEES *FE Reference Handbook*, we have the following:

For beam 1:

$$y_{max} = y_c = -\frac{5wl^4}{384EI} = -\frac{5(2,000)(8)^4}{384(200 \times 10^9)(50 \times 10^{-6})}$$

$$= -0.0106 \text{ m}$$

For beam 2:

$$y_{max} = y_c = -\frac{pl^3}{48EI} = -\frac{(6,000)(8)^3}{348(200 \times 10^9)(50 \times 10^{-6})}$$

$$= -0.0064 \text{ m}$$

Then the actual deflection at point C is the algebraic sum of the two individual deflections.

$$Y_{max} = y_c = -(0.0106 + 0.0064) = -0.017 \text{ m}$$
$$\text{or } 17.0 \text{ mm}$$

9.20
Failure Theories

There are four theories that are often used to predict the failure of a material under combined loading. If a material is designated as ductile, the maximum-shear-stress theory and maximum-distortion-energy theory are used to predict the failure. On the other hand, if a material is brittle, the maximum-normal-stress theory and Mohr's failure criterion are used to determine if the material will fail.

Here we will briefly discuss the two theories for ductile materials and the maximum-normal-stress theory for brittle materials. Note that the equations to be presented are valid for a two-dimensional (plane stress) stress field.

1. Maximum-Shear-Stress Theory
   For ductile materials the failure is specified by the initiation of yielding. In this method, once the principal stresses (σ_1 and σ_2) are calculated from a given plane stress element the following apply:

   a) If σ_1 and σ_2 have the same signs, then for failure not to happen

   $$|\sigma_1| < \sigma_y \text{ and } |\sigma_2| < \sigma_y \qquad [9.35]$$

   b) If σ_1 and σ_2 have opposite signs, then for failure not to happen

   $$|\sigma_1 - \sigma_2| < \sigma_y \qquad [9.36]$$

   where σ_y is the yield stress and is a material property.

2. Maximum-Distortion Energy Theory (Von Mises)
   This is also known as Von Mises failure theory and is valid for ductile materials. For the material not to fail, the following equation has to be satisfied:

   $$\sigma_1^2 - \sigma_1\sigma_2 + \sigma_2^2 < \sigma_y^2 \qquad [9.37]$$

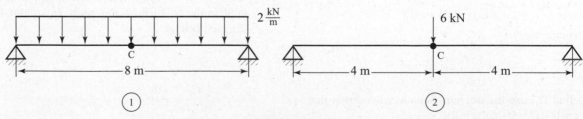

Figure 9.45

3. Maximum-Normal-Stress Theory

   For brittle materials fracture happens suddenly with not much yielding. In this theory the fracture occurs when the normal stress reaches the value of ultimate stress (σ_u). Once the principal stresses are calculated from a given plane stress element, the following applies:

   $$|\sigma_1| < \sigma_u \text{ and } |\sigma_2| < \sigma_u \qquad [9.38]$$

Example 9.20

The state of plane stress at a critical point in a machine bracket made of steel is shown. If the yield stress for steel is 36 ksi ($\sigma_y = 36$ ksi), determine if the failure occurs

a. Based on the maximum-shear-stress theory
b. Based on the maximum-distortion-energy theory

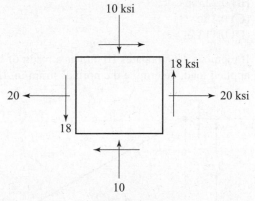

Figure 9.46

Solution:

First we determine the principal stresses by using equation 9.26:

$$\sigma_x = 20 \text{ ksi} \quad \sigma_y = -10 \text{ ksi} \quad \tau_{xy} = 18 \text{ ksi}$$

$$\sigma_{1,2} = \frac{\sigma_x + \sigma_y}{2} \pm \sqrt{\left(\frac{\sigma_x - \sigma_y}{2}\right)^2 + \tau_{xy}^2}$$

$$= \frac{20 + -10}{2} \pm \sqrt{\left(\frac{20 - -10}{2}\right)^2 + 18^2}$$

$$\sigma_1 = 28.43 \text{ ksi} \quad \text{and} \quad \sigma_2 = -18.43 \text{ ksi}$$

a. Based on the maximum-shear-stress theory, since the two principal stresses have opposite signs,

$$|\sigma_1 - \sigma_2| < \sigma_y \qquad |28.43 - -18.43| = 46.86 \text{ ksi}$$

which is larger than the yield stress of 36 ksi, so the bracket will fail.

b. Based on the maximum-distortion-energy theory,

$$\sigma_1^2 - \sigma_1\sigma_2 + \sigma_2^2 < \sigma_y^2$$

$$(28.43)^2 - (28.43)(-18.43) + (-18.43)^2 = 1671.9$$

which is greater than $(36)^2 = 1,296$. Therefore, the bracket will fail.

9.21
Summary

In this chapter the concepts of stress and strain and their relationship are introduced. Average normal stress is defined as the ratio of the centric load to the cross-sectional area. Average shear stress is defined as the ratio of the shear load (V) to the shear area (A_s). Deformation due to axial load as well as change in temperature is also presented in this chapter. Equations that calculate elastic stresses due to bending and twisting are discussed as well. Critical buckling load calculation for members in compression for different support conditions is presented. Deflection of beams using different methods, combined loading, stress transformation, and failure theories are also discussed and examples are shown.

PRACTICE PROBLEMS

1. Determine the normal stress in section *AB*.

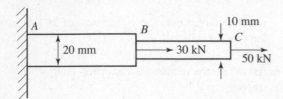

Figure 9.47

(A) 95 MPa
(B) 255 MPa
(C) 159 MPa
(D) 260 MPa

2. Determine the normal stress in link *AB* if *AB* has a cross-sectional area of 400 mm².

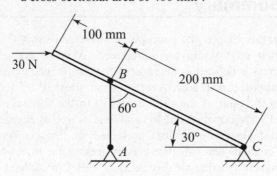

Figure 9.48

(A) 0.065 MPa
(B) 0.075 MPa
(C) 0.113 MPa
(D) 0.05 MPa

3. If the allowable shear stress for the bolts is 10 ksi, determine the minimum required diameter of each bolt.

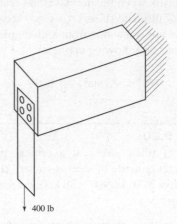

Figure 9.49

(A) 0.2 in
(B) 0.225 in
(C) 0.3 in
(D) 0.113 in

4. If member *AB* elongates 10 mm as a result of the applied load, determine the normal strain in *AB*.

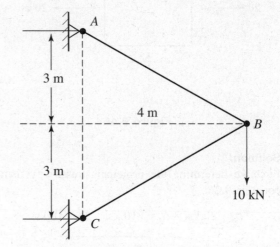

Figure 9.50

(A) 0.0033
(B) 0.002
(C) 0.0025
(D) 0.001

5. Determine the approximate vertical displacement of plate A. Each pad has cross-sectional dimensions of 30×30 mm and $G = 3$ MPa.

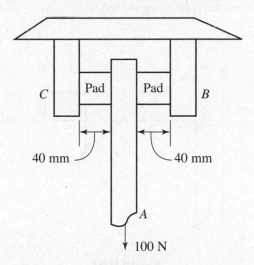

Figure 9.51

(A) 5 mm
(B) 0.74 mm
(C) 1.1 mm
(D) 0.2 mm

6. Determine the load required to decrease the diameter of the rod by 0.003 mm if $E = 4$ GPa and $\nu = 0.3$.

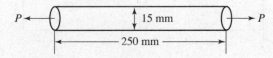

Figure 9.52

(A) 800 N
(B) 205 kN
(C) 253 N
(D) 471 N

7. Determine the change in temperature required to close the 2-mm gap between the sections.

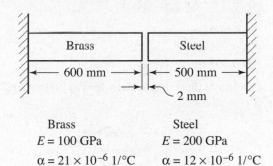

Brass	Steel
$E = 100$ GPa	$E = 200$ GPa
$\alpha = 21 \times 10^{-6}$ 1/°C	$\alpha = 12 \times 10^{-6}$ 1/°C

Figure 9.53

(A) 108 C°
(B) 55 C°
(C) 85 C°
(D) 20 C°

8. Determine diameter d if the deflection of point C is 0.04 in and $E = 30 \times 10^6$ psi for both sections.

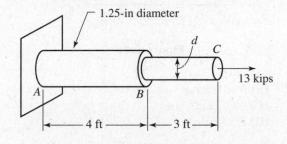

Figure 9.54

(A) 0.25 in
(B) 0.7 in
(C) 0.93 in
(D) 0.625 in

9. If the allowable shear stress is 840 KPa, determine the required diameter of a solid shaft under the action of a 1,500-N · m torque.
   (A) 0.209 m
   (B) 0.0091 m
   (C) 0.121 m
   (D) 0.263 m

10. Determine the angle of twist for the hollow shaft if $G = 80$ GPa.

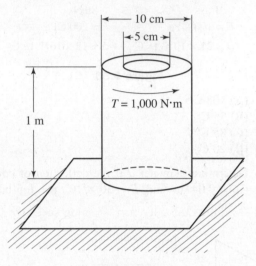

Figure 9.55

   (A) 0.0127 rad
   (B) 0.00136 rad
   (C) 0.001 rad
   (D) 0.52 rad

11. Determine the diameter of a solid shaft that is rotating at 1,500 RPM while transmitting 4 hp. The shaft has an allowable shear stress of 9.0 ksi.

   (A) 0.2 in
   (B) 0.456 in
   (C) 1.5 in
   (D) 0.78 in

12. Determine the bending moment at the center (point C) of the beam.

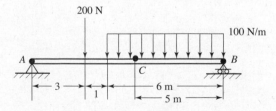

Figure 9.56

   (A) 600 N·m
   (B) 1,150 N·m
   (C) 1,600 N·m
   (D) 480 N·m

13. Determine the maximum magnitude of the bending moment in the beam. The bending moment at each end of the beam is zero, and there are no concentrated couples along the beam.

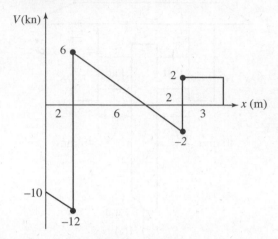

Figure 9.57

   (A) 18 kN · m
   (B) 6 kN · m
   (C) 22 kN · m
   (D) 4 kN · m

14. If the allowable bending stress is 5 ksi, determine h.

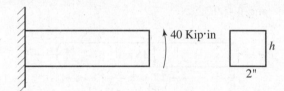

Figure 9.58

   (A) 4 in
   (B) 4.9 in
   (C) 3.2 in
   (D) 2 in

15. Determine the maximum bending stress for the beam.

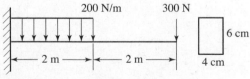

Figure 9.59

   (A) 66.7 MPa
   (B) 50 MPa
   (C) 25.2 MPa
   (D) 5.55 MPa

16. If the allowable bending stress is 10 ksi, determine the diameter of the solid shaft loaded as shown.

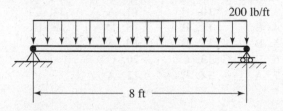

Figure 9.60

(A) 1.17 in
(B) 3.5 in
(C) 2.7 in
(D) 2.35 in

17. Determine the maximum transverse shear stress for the beam.

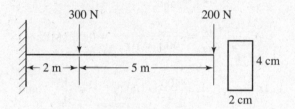

Figure 9.61

(A) 0.94 MPa
(B) 0.625 MPa
(C) 0.375 MPa
(D) 0.833 MPa

18. Determine the normal stress at point A for the shaft.

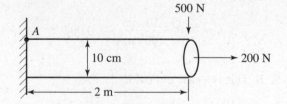

Figure 9.62

(A) 25.46 MPa
(B) 10.18 MPa
(C) 10.16 MPa
(D) 10.21 MPa

19. Determine the maximum normal stress for the stress element.

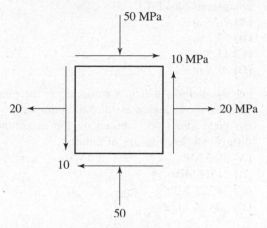

Figure 9.63

(A) 51.4 MPa
(B) 21.4 MPa
(C) 36.4 MPa
(D) 20 MPa

20. Determine the maximum shearing stress for the stress element.

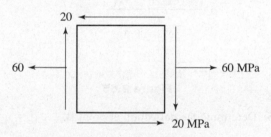

Figure 9.64

(A) 20 MPa
(B) 30 MPa
(C) 36 MPa
(D) 25 MPa

21. Determine the maximum pressure for a 60-cm-diameter cylinder, with thickness of 0.5 cm, if the allowable tensile stress is 150 MPa.
(A) 2.5 MPa
(B) 1.25 MPa
(C) 5 MPa
(D) 1.2 MPa

22. Determine the thickness of a 100-cm inner diameter sphere under a pressure of 600 KPa if the allowable stress is 150 MPa.
(A) 2 cm
(B) 1 cm
(C) 1.5 cm
(D) 0.5 cm

277

23. Determine the effective length of a pin-ended square (20×20 cm) column under a 150-kN compressive load if $E = 15$ GPa.
   (A) 7.25 m
   (B) 39.7 m
   (C) 11.5 m
   (D) 22.5 m

24. For the element which is subjected to the plane stresses shown, what is the minimum value of the yield stress (σ_y) based on the maximum-distortion-energy theory of failure?
   (A) 50.0 MPa
   (B) 215.4 MPa
   (C) 345.25 MPa
   (D) 298.66 MPa

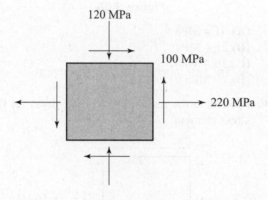

Figure 9.65

25. Determine the deflection at point B.

$E = 200$ GPa and $I = 20 \times 10^{-6}$ m^4, $\ell = 3$ m

   (A) 7.6 mm
   (B) 18.85 mm
   (C) 23.2 mm
   (D) 11.25 mm

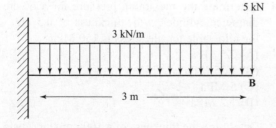

Figure 9.66

Answer Key

1. B	8. C	15. A	22. B
2. A	9. A	16. C	23. C
3. D	10. B	17. A	24. C
4. B	11. B	18. D	25. B
5. B	12. B	19. A	
6. D	13. C	20. C	
7. A	14. B	21. A	

Answers Explained

1. B $P_{AB} = 80$ kN in tension, and

$$\sigma_{AB} = \frac{P_{AB}}{A_{AB}} = \frac{80 \times 10^3}{\frac{\pi}{4}(0.02)^2} = 254.6 \text{ MPa}$$

2. A Draw an FBD of member *BC*, and take the moment about point *C*: $\Sigma M_c = 0$. Then $F_{AB} = 25.98$ kN, and

$$\sigma_{AB} = \frac{F_{AB}}{A_{AB}} = \frac{25.98}{400 \times 10^{-6}} = 0.065 \text{ MPa}$$

3. D Here, $V = \frac{400}{4} = 100$ lb $\tau = \frac{V}{A_s}$,   and

$$A_s = \frac{V}{\tau_{all}} = \frac{100}{10,000} = 0.01 \text{ in}^2. \text{ Then,}$$

$$d = \sqrt{\frac{4A_s}{\pi}} = \sqrt{\frac{4(.01)}{\pi}} = 0.1128 \text{ in}$$

4. B In this case: The length of *AB* is 5m = 5,000 mm

$$\epsilon = \frac{\delta}{l} = \frac{10 \text{ mm}}{5,000 \text{ mm}} = 0.002$$

5. B This is a case of double shear.

$$V = \frac{100}{2} = 50 \text{ N, and}$$

$$A = (30)(30) = 900 \text{ mm}^2 = 900 \times 10^{-6} \text{ m}^2$$

$$\tau = \frac{V}{A} = \frac{50}{900 \times 10^{-6}} = 0.055 \text{ MPa}$$

Also,

$$\tau = G\gamma$$

$$\gamma = \frac{\tau}{G} = \frac{0.055}{3} = 0.0185 \text{ rad}$$

Then,

$$y = (40 \text{ mm})(0.0185 \text{ rad}) = 0.74 \text{ mm}$$

6. D In this case:

$$\epsilon_{lateral} = \frac{-0.003}{15} = -0.0002 \quad \text{and}$$

$$\epsilon_{axial} = \frac{-\epsilon_{lateral}}{v} = \frac{-(0.0002)}{0.3} = 0.00066$$

Also,

$$\epsilon_{axial} = \frac{\sigma}{E} = \frac{P}{EA}$$

$$P = EA\epsilon_{axial} = 471 \text{ N}$$

7. A Here: $\delta = \Sigma \alpha l \, \Delta T$

$$\Delta T = \frac{\delta}{\Sigma \alpha l}$$

$$= \frac{2}{(21 \times 10^{-6})(600) + (12 \times 10^{-6})(500)}$$

$$= 108 \text{ C}°$$

8. C In this case:

$$\delta = \Sigma \frac{Pl}{EA} \quad \text{and}$$

$$0.04 = \frac{13,000}{30 \times 10^6} \left(\frac{48}{\frac{\pi}{4}(1.25)^2} + \frac{36}{\frac{\pi}{4}(d)^2} \right)$$

Then $d = 0.93$ in.

9. A For a solid shaft, $\tau = \frac{16\,T}{\pi\,d^3}$. Then,

$$d = \sqrt[3]{\frac{16T}{\pi d_\tau}} = \sqrt[3]{\frac{16(1,500)}{\pi(840,000)}} = 0.209 \text{ m}$$

10. B For the hollow shaft, $J = \frac{\pi}{32}(0.1^4 - 0.05^4) = 9.2 \times 10^{-6} \text{ m}^4$. Then,

$$\phi = \frac{TL}{GJ} = \frac{(1,000)(1)}{(80 \times 10^9)(9.2 \times 10^{-6})}$$

$$= 0.00136 \text{ rad}$$

11. B

$P = 4 \text{ hp} = 2,200 \text{ lb-ft/sec}$

$\omega = 1,500 \text{ RPM} = 157.08 \text{ rad/sec}$

$P = T\omega \quad T = P/\omega \quad T = \frac{2200}{157.08} = 14 \text{ lb-ft}$

$T = 168 \text{ lb-in}$

$\tau = \frac{Tc}{J} = \frac{16T}{\pi d^3} \quad d = \sqrt[3]{\frac{16T}{\pi\tau}} = \sqrt[3]{\frac{16(168)}{\pi(9,000)}} =$

0.456 in

12. B First, determine the reactions:

$$B = 480 \text{ N} \qquad A = 320 \text{ N}$$

Cut the beam at point C, and use either the left or the right side of the beam:

$$\Sigma M_C = 0 \quad \text{and} \quad M_C = 1,150 \text{ N·m}$$

13. C The area under the shear diagram is equal to the net bending moment, so $M_{max} = 22 \text{ kN·m}$

14. B In this case:

$$\sigma_{max} = \frac{M_{max}C}{I} \qquad 5,000 = \frac{40,000\left(\frac{h}{2}\right)}{\frac{1}{12}(2)\,h^3}$$

Then $h = 4.9$ in.

15. A The maximum moment will occur at the support and is equal to $1,600 \text{ N·m}$, so

$$\sigma_{max} = \frac{M_{max}C}{I} = \frac{(1,600)(0.03)}{\frac{1}{12}(0.04)(0.06)^3}$$

$$= 66.7 \text{ MPa}$$

16. C The maximum moment will occur at the center of the shaft and is equal to $1,600 \text{ lb-ft} = 19,200 \text{ lb-in}$. Then:

$$\sigma_{max} = \frac{M_{max}C}{I} = \frac{(19,200)\left(\frac{d}{2}\right)}{\frac{\pi}{64}d^4}$$

and $d = 2.7$ in.

17. A The maximum shear load $V = 500 \text{ N}$. For a rectangular cross section:

$$\tau_{max} = \frac{3V_{max}}{2A} = \frac{3(500)}{2(0.02)(0.04)} = 0.94 \text{ MPa}$$

18. D At point A:

$$\sigma_A = \frac{Mc}{I} + \frac{P}{A} = \frac{(1,000)(0.05)}{\frac{\pi}{64}(0.1)^4}$$

$$+ \frac{200}{\frac{\pi}{4}(0.1)^2} = 10.21 \text{ MPa}$$

19. A For the stress element:

$$\sigma_{p_1,p_2} = \frac{20-50}{2}$$

$$\pm\sqrt{\left[\frac{20-(-50)}{2}\right]^2 + (10)^2}$$

$$\sigma_{max} = 51.4 \text{ MPa}$$

20. C Here,

$$\tau_{max} = \sqrt{\left(\frac{60-0}{2}\right)^2 + (-20)^2} = 36 \text{ MPa}$$

21. A For the cylinder:

$$\sigma_1 = \frac{Pr}{t}$$

$$P = \frac{\sigma_1 t}{r} = \frac{(150 \times 10^6)(0.005)}{0.3} = 2.5 \text{ MPa}$$

22. B For the sphere:

$$\sigma = \frac{Pr}{2t}$$

$$t = \frac{Pr}{2\sigma} = \frac{600 \times 10^3 (0.5)}{2(15 \times 10^6)} = 0.01 \text{ m} = 1 \text{ cm}$$

23. C Here, $P_{cr} = \dfrac{\pi^2 EI}{L_e^2}$. Then,

$$L_e = \sqrt{\frac{\pi^2 EI}{P_{cr}}}$$

$$I = \frac{1}{12}(0.2)^4 = 1.3 \times 10^{-4} \text{ m}^4$$

$$L_e = \sqrt{\frac{\pi^2(15 \times 10^9)(1.3 \times 10^{-4})}{150 \times 10^3}} = 11.47 \text{ m}$$

24. C Solve for the principal stresses using the following equation [9.26]

$$\sigma_{1,2} = \frac{\sigma_x + \sigma_y}{2} \pm \sqrt{\frac{(\sigma_x - \sigma_y)^2}{2} + \tau_{xy}^2}$$

$$= \frac{220 + -120}{2} \pm \sqrt{\frac{(220 - -120)^2}{2} + 100^2}$$

$$\sigma_1 = 247.23 \text{ MPa} \quad \text{and} \quad \sigma_2 = -147.23 \text{ MPa}$$

Then use the equation for the maximum-distortion-energy theory [9.37]

$$\sigma_1^2 - \sigma_1\sigma_2 + \sigma_2^2 = \sigma_y^2$$

Solve for $\sigma_y = 345.25$ MPa

25. B Divide the loading into two parts: one with only the distributed load and the other with just the concentrated load acting at point B.

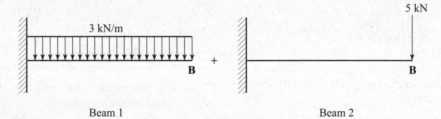

Figure 9.67

Using the Beam Deflection Formulas table in the NCEES *FE Reference Handbook*, we determine the deflection for each beam at point B:

$$y_1 = -\frac{wl^4}{8EI} = -\frac{(3,000)(3)^4}{8(200 \times 10^9)(20 \times 10^{-6})} = -0.0076 \text{ m}$$

$$y_2 = -\frac{pl^3}{3EI} = -\frac{(5,000)(3)^3}{3(200 \times 10^9)(20 \times 10^{-6})} = -0.01125$$

The total deflection is the sum of the two individual deflections due to each loading.

$$y_B = y_1 + y_2 = -0.0076 + (-0.01125) = -0.01885 \text{ m or } 18.85 \text{ mm downward}$$

10

Material Properties

Richard J. Murphy, Ph.D.

Introduction

This chapter is intended as a review of the important basic concepts in materials science. The presentation of the various content areas assumes that most of these topics have been covered in a full course in materials science. It may be useful to use the text from that course as a companion to the review presented here.

The chapter follows a conventional order of topics. It begins with atomic structure, followed by bonding, crystallography, and phase equilibrium. After sections on ferrous metallurgy, solid solubility, diffusion, and corrosion, the various behaviors are covered: mechanical, magnetic, and electrical. Finally, there are brief reviews of polymers and composites.

10.1 Atomic Structure

The atomic structure of an atom involves the spatial and energetic arrangement of the electrons in the atom. The nucleus is not included in the description of

> Atomic structure is arrangement of electrons, *not* arrangement of atoms.

atomic structure. (It is important to remember that the term *atomic structure* refers to the arrangement of electrons in an atom, *not* the arrangement of atoms. Atom arrangement is what crystal structure is all about.)

For each atom, the number of electrons and the number of protons match the atomic number. The electrons are particles in constant motion around the nucleus. By virtue of their dual nature, they are also waves. Electrons possess energy that is due to their motion (mass and velocity) and their wavelengths. Because the wavelength of each electron must be an integral fraction of the path length of that electron around the nucleus, the energies of the electrons in an atom are not spread over a range of values; rather, they are limited to specific values with no possibility of intermediate energies. This characteristic is what is meant by quantization.

Later refinements of atomic theory expand many of the early concepts. The electrons occupy energy levels in such a way that no more than two electrons can have the same energy (occupy the same orbital), and those two must have different spins. This rule is called the **Pauli exclusion principle.** The gross quantization is designated by the principal quantum number n ($n = 1, \ldots, n$). The energy of each level (according to the Bohr atomic model) is given by equation 10.1:

$$E = -\frac{13.6Z^2}{n^2} \text{ eV} \qquad [10.1]$$

where Z is the atomic number and n is the principal quantum number of the energy level.

When electrons are excited from one energy level to another, energy is absorbed. The magnitude of the energy absorbed, that is, the difference between the two energy levels, is calculated by using equation 10.2:

$$\Delta E = E_2 - E_1 = -13.6Z^2\left(\frac{1}{2^2} - \frac{1}{1^2}\right)$$

$$= \frac{13.6Z^2}{4} = -3.4Z^2 \text{ eV} \qquad [10.2]$$

The wavelength of radiation absorbed in this electron excitation can then be determined using equation 10.3:

$$\lambda = \frac{hc}{\Delta E} \qquad [10.3]$$

where h is Planck's constant and c is the velocity of light.

Within the energy level designated by each value of n, there is further quantization of energy, indicated by the letters s, p, d, and f in the order of increasing energy. These subquantizations are called orbitals (*not* orbits!). It is at this level that a connection

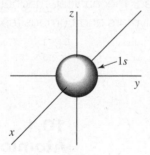

(a) The s-orbitals are spherically symmetrical.

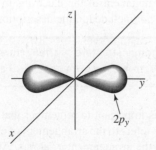 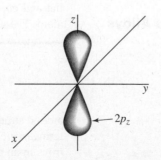 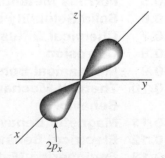

(b) The p-orbitals have a dumbbell shape

Figure 10.1 The shapes of (a) the s-orbital and (b) the three p-orbitals. (From Schaeffer et al. *The Science and Design of Engineering Materials*, 1994, McGraw-Hill. Reproduced with permission of the McGraw-Hill Companies.)

is made between electron energy and spatial distribution. Energy levels with the designation s represent a spherically symmetric distribution of the probability of location of an electron, as shown in Figure 10.1a. The p levels are actually comprised of three separate, mutually perpendicular dumbbell-shaped probability distributions, each parallel to the x-, y-, or z-axis, as shown in Figure 10.1b. Each orbital can contain up to two and only two electrons. This is a restatement of the Pauli exclusion principle: each orbital can contain no more than two electrons, and those two electrons must have different spins.

$4s$ orbitals fill before the $3d$ orbitals.

Table 10.1. The Electron Configurations of the First Thirty Elements

Element	Atomic Number	Electron Configuration
Hydrogen	1	$1s^1$
Helium	2	$1s^2$
Lithium	3	$1s^2 2s^1$
Beryllium	4	$1s^2 2s^2$
Boron	5	$1s^2 2s^2 2p^1$
Carbon	6	$1s^2 2s^2 2p^2$
Nitrogen	7	$1s^2 2s^2 2p^3$
Oxygen	8	$1s^2 2s^2 2p^4$
Fluorine	9	$1s^2 2s^2 2p^5$
Neon	10	$1s^2 2s^2 2p^6$
Sodium	11	$1s^2 2s^2 2p^6 3s^1$
Magnesium	12	$1s^2 2s^2 2p^6 3s^2$
Aluminum	13	$1s^2 2s^2 2p^6 3s^2 3p^1$
Silicon	14	$1s^2 2s^2 2p^6 3s^2 3p^2$
Phosphorus	15	$1s^2 2s^2 2p^6 3s^2 3p^3$
Sulfur	16	$1s^2 2s^2 2p^6 3s^2 3p^4$
Chlorine	17	$1s^2 2s^2 2p^6 3s^2 3p^5$
Argon	18	$1s^2 2s^2 2p^6 3s^2 3p^6$
Potassium	19	$1s^2 2s^2 2p^6 3s^2 3p^6 4s^1$
Calcium	20	$1s^2 2s^2 2p^6 3s^2 3p^6 4s^2$
Scandium	21	$1s^2 2s^2 2p^6 3s^2 3p^6 4s^2 3d^1$
Titanium	22	$1s^2 2s^2 2p^6 3s^2 3p^6 4s^2 3d^2$
Vanadium	23	$1s^2 2s^2 2p^6 3s^2 3p^6 4s^2 3d^3$
Chromium	24	$1s^2 2s^2 2p^6 3s^2 3p^6 4s^1 3d^5$
Manganese	25	$1s^2 2s^2 2p^6 3s^2 3p^6 4s^2 3d^5$
Iron	26	$1s^2 2s^2 2p^6 3s^2 3p^6 4s^2 3d^6$
Cobalt	27	$1s^2 2s^2 2p^6 3s^2 3p^6 4s^2 3d^7$
Nickel	28	$1s^2 2s^2 2p^6 3s^2 3p^6 4s^2 3d^8$
Copper	29	$1s^2 2s^2 2p^6 3s^2 3p^6 4s^1 3d^{10}$
Zinc	30	$1s^2 2s^2 2p^6 3s^2 3p^6 4s^2 3d^{10}$

This orbital notation can be used to identify the energy levels of each electron in any atom, as seen in Table 10.1. In each electron configuration, the magnitude of n is given first, followed by the designation of s and, if appropriate, of p and d. The number of electrons with each orbital designation (i.e., with this energy) is shown as a superscript. The sum of the superscripts is, therefore, the number of electrons in the atom.

10.2 Bonding

Material behavior is largely determined by the types of bonding that occur among the various atoms in the material. In some cases where molecules exist in the material, the more important bonds may be those among the molecules. Usually, the stronger bonding, called interatomic bonding, occurs between atoms. There are three types of interatomic bonding: covalent, ionic, and metallic. Each takes place because of special aspects of the structure of the atom.

Covalent Bonding

Consider the hydrogen atom. This atom has one electron of type s, designated as $1s^1$. Remember that the s orbital is a spherical cloud of probability of location; therefore, if that atom is in the vicinity of another hydrogen atom, the electron of the second atom can occupy the s orbital of the first atom because the two orbitals overlap to some extent. It can be shown by advanced calculations that the total energy of the pair of atoms, with both s orbitals sharing two electrons, is lower than the sum of the energies of the two atoms taken separately. This fact indicates that the two atoms would rather be together, that is, bonded, than apart. But the two atoms cannot overlap completely because of the strong repulsive force of the two nuclei. Therefore, the two atoms assume an equilibrium separation that balances the attractive and repulsive forces. The total energy curve representing this bonding event for two atoms is shown in Figure 10.2.

In this case, if a third hydrogen atom approached close enough to overlap the s orbitals of the first two atoms, it would provide a third electron to the s orbital. The Pauli exclusion principle states that an orbital can have no more than two electrons. Therefore, the third atom cannot enter into a bond relation with this pair and must move on. That is the reason why hydrogen gas exists as a diatomic molecule.

Covalent bonding can be defined as bonding that occurs between atoms with half-filled orbitals. For instance, chlorine has an atomic number of 17 and the atomic structure $1s^2 2s^2 2p^6 3s^2 3p^5$. Notice that each of the orbitals is filled with two electrons except one of the three $3p$ orbitals, which is half-filled. This is a covalent bonding orbital. It can bond with a half-filled orbital of any other atom, for example, another chlorine atom. This is the reason why chlorine gas exists as a diatomic molecule.

Carbon has only two half-filled orbitals, a configuration unfavorable for forming a three-dimensional network of bonded carbon atoms. In an excited state, however, carbon atoms can possess four half-filled bonding orbitals, and thereby form covalent bonds with four neighbors that also have bonding orbitals. This is how diamond, silicon, methane, and carbon tetrachloride form, as well as the carbon-carbon bonds in polymers. This type of bond is very strong.

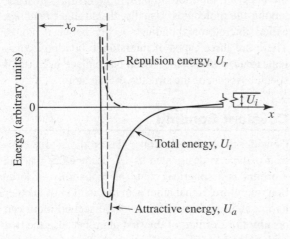

Figure 10.2 The energy versus atom separation curve for the bonding of two atoms or ions. (From Schaeffer et al. *The Science and Design of Engineering Materials*, 1994, McGraw-Hill. Reproduced with permission of the McGraw-Hill Companies.)

Ionic Bonding

Ionic bonding consists of electrostatic attraction between oppositely charged ions. Ions are atoms that have lost or gained electrons, giving them a positive or negative charge. Ions are formed most easily in atoms that have s electrons as the highest energy level or that are one or two electrons short of filling all the p orbitals. In the former case, the s electrons can be excited away from the atom, leaving a positively charged ion (cation), as in Na^+ and Ca^{2+}. In the latter case, the atom can attract and capture extra free electrons to completely fill the p orbitals, giving it a negative charge (anion), as in F^- and O^{2-}.

A measure of the ease with which atoms attract electrons to make negative ions is called **electronegativity.** The atoms on the far left of the Periodic Table, such as sodium and calcium, have low electronegativities and are considered electropositive; the atoms on the far right, such as fluorine and oxygen, have high electronegativities. When ions of opposite charge are present together, they can experience strong electrostatic attractive forces, which, in balance with the repulsive forces from nucleus-nucleus interaction, result in large net bonding energy. The larger the difference in electronegativity, the stronger the bond. This equilibrium of bond energies is shown in Figure 10.2.

Metallic Bonding

Metallic bonding occurs in materials whose atoms have s electrons as their highest energy electrons. When many atoms of this type come together, the energy of the system is lowered by a process in which the s orbitals overlap and the s electrons become free from their atoms to move throughout the material, while the electrons in the ion cores remain strongly bound. This type of bonding gives rise to common metallic properties such as high thermal and electrical conductivities.

It is not easy to compare the relative strengths of the three types of bonding. In general, however, average bond strengths increase in the order metallic, ionic, covalent. All three types are strong in comparison to the much weaker intermolecular bonds.

10.3 Crystallography

The arrangements of atoms in a solid material fall into two general classes:

1. Entirely periodic, predictable, and reproducible
2. Quite random.

The first arrangement is characteristic of a crystal, while the second is found in amorphous solids, usually referred to as glasses. The arrangement of atoms in crystals is the subject of crystallography.

Bravais Lattices

The subject is simplified if one considers arrangements of points in space, called **lattices,** rather than of atoms. It has been shown that there are fourteen ways of arranging points in space so as to satisfy the symmetry and reproducible periodicity that a

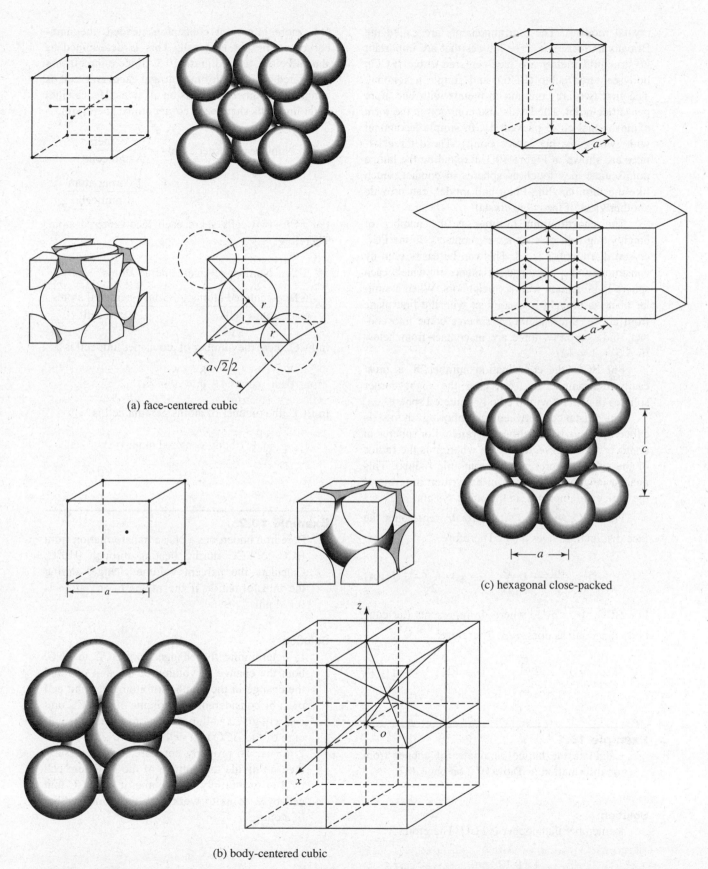

(a) face-centered cubic

(c) hexagonal close-packed

(b) body-centered cubic

Figure 10.3 Various representations of the metallic crystal structures. (From Schaeffer et al. *The Science and Design of Engineering Materials*, 1994, McGraw-Hill. Reproduced with permission of the McGraw-Hill Companies.)

crystal requires. These arrangements are called the Bravais lattices. The three lattices that are important in structural metals are face-centered cubic (FCC), body-centered cubic (BCC), and simple hexagonal. The first two are common in metals with one atom per lattice point. The last is also common in the form of hexagonal close packed (HCP, simple hexagonal with two atoms per lattice point). These three lattices are shown in Figure 10.3. Expanding the lattice points until they touch as spheres in contact, much like the familiar Ping-Pong ball model, can provide another view of the cubic crystal.

The **coordination number** is the number of neighboring spheres each sphere contacts. In the FCC crystal that number is 12. This can be easily seen by constructing the close-packed plane, in which each sphere is in contact with six neighbors. When a similar plane is brought into contract with the first plane from above, three additional spheres come into contact, and, similarly, three are in contact from below (6 + 3 + 3 = 12).

For BCC, the coordination number, 8, is most easily determined by observing the eight corner spheres in contact with the body-centered sphere.

The distance between neighboring atoms is called the **nearest-neighbor distance.** For spheres in contact, it is represented as $2r$, where r is the radius of the sphere (representing the atomic radius). This nearest-neighbor distance can be written in terms of the lattice parameter, a, in the cubic system.

For FCC, $4r = a\sqrt{2}$, where $4r$ represents the face diagonal, as does $a\sqrt{2}$. Therefore,

$$2r = \frac{a\sqrt{2}}{2} = \frac{a}{\sqrt{2}}; \qquad r = \frac{a}{2\sqrt{2}} \quad [10.4]$$

For BCC, $4r = a\sqrt{3}$, where $4r$ represents the cube body diagonal, as does $a\sqrt{3}$. Therefore:

$$2r = \frac{a\sqrt{3}}{2}; \qquad r = \frac{a\sqrt{3}}{4} \quad [10.5]$$

Example 10.1

Calculate the lattice parameter of copper from the information in Table 10.2 on page 287.

Solution:

Remember that copper is FCC. Therefore:

$$a = \frac{4r}{\sqrt{2}} = \frac{4 \times 0.128 \text{ nm}}{1.414} = 0.362 \text{ nm}$$

One more important concept is needed, the number of atoms per unit cell. This is determined as shown below, using Figure 10.3. Since unit cells are abstracted from an infinite array, each unit cell is connected to other unit cells on all sides. Hence eight other unit cells share each corner atom. Therefore:

$$\text{Number of corner atoms} = \frac{8 \text{ corners}}{8 \text{ unit cells}}$$

$$= \frac{1 \text{ corner atom}}{1 \text{ unit cell}}$$

For FCC, two cells share each face-centered atom. Therefore,

$$\text{Number of face-centered atoms}$$

$$= \frac{6 \text{ face-centered atoms}}{2 \text{ unit cells}} = \frac{3 \text{ face-centered atoms}}{1 \text{ unit cell}}$$

In FCC, then, the number of atoms per unit cell is

$$\frac{8}{8} + \frac{6}{2} = 4.$$

In BCC, the number of atoms per unit cell is

$$\frac{8}{8} + 1 \text{ body-centered atom} = 2.$$

Example 10.2

Pure iron undergoes a phase transformation from BCC to FCC during heating through 910°C. Calculate the percent volume changes during the transformation if the radius of the atom is 0.124 nm.

Solution:

In determining the change from BCC to FCC, both the change in volume of the unit cell and the change in the number of atoms per unit cell must be considered. The volume of the FCC unit cell will give a volume of 4 atoms, while the volume of the BCC unit cell will give a volume of 2 atoms. To properly compare, it is necessary, then, to divide the volume of the FCC unit cell by 2 or to multiply the volume of the BCC unit cell by 2 so as to work with the same number of atoms.

286

$$V_{BCC} = a^3_{BCC}; \qquad V_{FCC} = a^3_{FCC};$$

$$\frac{\Delta V}{V} = \frac{a^3_{BCC} - \left(\dfrac{a^3_{FCC}}{2}\right)}{a^3_{BCC}} = 1 - \frac{1}{2}\left(\frac{a^3_{FCC}}{a^3_{BCC}}\right)$$

$$a^3_{BCC} = \left(\frac{4r}{\sqrt{3}}\right)^3; \qquad a^3_{FCC} = \left(\frac{4r}{\sqrt{2}}\right)^3;$$

$$\frac{\Delta V}{V} = 1 - \frac{1}{2}\left(\frac{\dfrac{4r}{\sqrt{2}}}{\dfrac{4r}{\sqrt{3}}}\right)^3 = 1 - \frac{3\sqrt{3}}{4\sqrt{2}}$$

$$\frac{\Delta V}{V} = 1 - 0.919 = 0.081 = 8.1\%$$

Note: The percent volume change is independent of atom size.

Density Calculation

It is now a simple process to calculate the density of a crystalline element. The equation for density is simply

$$\rho = \frac{m}{V}\ \text{g/cm}^3 \qquad\qquad [10.6]$$

The only information needed is the type of structure, the mass of the atoms in the unit cell, and the volume of the unit cell.

The mass of the atoms in the unit cell is the product of the mass of each atom and the number, n, of atoms in the unit cell. The mass of each atom is the gram-atomic weight (GAW—the weight, in grams, of 1 mole of atoms) divided by Avogadro's number:

$$m = \text{mass of atoms in unit cell} = \frac{n \times \text{GAW}}{6.023 \times 10^{23}}\ \text{g}$$

$$[10.7]$$

The volume of the unit cell is, for cubic structures, the cube of the lattice parameter, a^3. The volume can also be obtained from the atom radius using equation 10.4 or 10.5:

$$\text{For BCC, } V = \left(\frac{4r}{\sqrt{3}}\right)^3; \quad \text{for FCC, } V = \left(\frac{4r}{\sqrt{2}}\right)^3$$

$$[10.8]$$

The crystal structures, atomic masses, atomic radii, and densities of four cubic crystals are given in Table 10.2.

Table 10.2 Data for Selected Crystals

Crystal	Structure	Atomic Mass (amu)	Atomic Radius (nm)	Density (g/cm³)
Cu	FCC	63.55	0.128	8.93
α-Fe	BCC	55.85	0.124	7.87
Mo	BCC	95.94	0.136	10.22
Al	FCC	26.98	0.143	2.7

Example 10.3

Calculate the density of aluminum from the data in Table 10.2.

Solution:

Aluminum is FCC. (You must know this. So also are copper, nickel, silver, gold, and austenitic stainless steel.) Therefore, there are 4 atoms per unit cell.

$$\text{Density} = \frac{\text{mass of atoms in unit cell}}{\text{volume of unit cell}}$$

$$= \frac{4\ \text{atoms/unit cell} \times 27\ \text{g/mol} / 6.024 \times 10^{23}\ \text{atoms/mol}}{a^3\ \text{cm}^3/\text{unit cell}}$$

$$= 2.7\ \text{g/cm}^3$$

Atomic Packing Factor

The atomic packing factor (apf) is the volume of the atoms in a unit cell divided by the volume of the unit cell.

For FCC:

$$\text{apf} = \frac{4 \times \dfrac{4}{3}\pi r^3}{a^3} = \frac{4 \times \dfrac{4}{3}\pi r^3}{\left(\dfrac{4r}{\sqrt{2}}\right)^3}$$

$$= \frac{\sqrt{2}\pi}{6} = 0.74 \qquad\qquad [10.9]$$

For BCC:

$$\text{apf} = \frac{2 \times \frac{4}{3}\pi r^3}{a^3} = \frac{4 \times \frac{4}{3}\pi r^3}{\left(\frac{4r}{\sqrt{3}}\right)^3} = \frac{\sqrt{3}\pi}{8} = 0.68$$

[10.10]

Miller Indices: The Language of Crystals

To describe the planes and directions in a cubic crystal, a special notation called the Miller indices has been developed. The method for determining the indices of a plane is as follows.

> Draw a plane so that it does *not* pass through the origin of the unit cell.

1. Construct the cubic unit cell using the rectangular coordinate system (x, y, z). Draw the plane of interest within the unit cell in such a way that it does not pass through the origin.
2. Note the intercepts with the x-, y-, and z-axes as integer or fractional multiples of the cell edge, and designate these numerical values as a, b, c, respectively.
3. Take the reciprocals of a, b, c, and call these values h, k, and l, respectively.
4. Enclose the reciprocals hkl in parentheses as a name for the plane.

For example, if a plane is parallel to the z-axis, the z-intercept, c, is ∞. Then the reciprocal, l, is 0. (See Example 10.4.)

The method for determining the indices for crystallographic directions is somewhat different.

> Draw a direction so that it passes through the origin of the unit cell.

1. Construct the unit cell in an x-, y-, z-coordinate system. Draw the direction of interest within the unit cell in such a way that it passes through the origin.
2. Determine the x-, y-, z-coordinates of the point at which the direction intersects the cell away from the origin, and designate them as integer or fractional multiples of the length of the cube edge.
3. Enclose coordinates x, y, z in brackets as a name for the direction. For example, a cube edge is [100], [010], or [001].

Example 10.4

Determine the Miller indices of the plane and the direction shown in Figure 10.4.

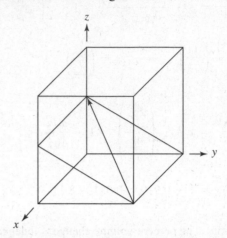

Figure 10.4 A sketch of a plane and a direction in a cubic unit cell

Solution:

For the plane:

1. Construct a right-handed coordinate system using the edges of the unit cell. In this case the x-axis is pointing toward the viewer.
2. Select an origin that does not lie in the plane. In this case the origin can be the lower, back, left corner of the cell.
3. Note that the x-intercept is ∞; the y-intercept is 1; the z-intercept is $\frac{1}{2}$.
4. Take the reciprocals of these intercepts in order; the results are 0, 1, 2.
5. Since this is a plane, collect the reciprocals in parentheses without commas: (012).

For the direction:

1. Use the same construction as in step 1 for the plane.
2. Select the origin at the tail of the direction. In this case it is the lower, front, right corner.
3. Locate the point at which the head of the direction line intersects the unit cell, and write the coordinates of this point. In this case:

$$x = -1, y = -1, z = \frac{1}{2}.$$

4. Collect the coordinates in square brackets:

$$[\bar{1}\,\bar{1}\,\tfrac{1}{2}].$$

5. Clear fractions: $[\bar{2}\,\bar{2}\,1]$.

10.4
Phase Equilibrium

The relationship among phases in a material is strictly governed by the thermodynamic variables temperature, T, pressure, P, and phase composition (the relative amounts of components in solution within the phase). This relationship is summarized by the **equilibrium phase diagram**, which shows, at 1 atmosphere pressure, the conditions of temperature and alloy composition at which the various phases are stable or in equilibrium. For example, Figure 10.5 is the equilibrium phase diagram for the copper-nickel system. If you trace across the diagram at 1,000°C, you find only one phase, alpha, for all compositions. This is a solid region. At 1,200°C a traverse of the diagram begins with the liquid phase (designated as L), then enters the region containing liquid and alpha, and concludes with the alpha phase. This means that, in the copper-nickel system at 1,200°C, the liquid phase is stable over a range of compositions up to the boundary of the two-phase region called the liquidus. Also, it means that the alpha phase is stable over a range of compositions from pure nickel back to the boundary of the two-phase region, called the solidus; and that, within the two boundaries, the two phases, liquid and alpha, are in equilibrium. It is in this two-phase region that the **inverse lever rule** must be applied to gain information inside the region.

Now consider Figure 10.6. To analyze the copper-nickel system at 1,200°C, you draw a horizontal line, called a tie-line, through the two-phase liquid-plus-solid region from one boundary to the next. The endpoints of the tie-lines then indicate the compositions of the two phases that are in equilibrium. These compositions, called the phase compositions, are read by tracing down to the horizontal axis of the phase diagram.

If an alloy composition is chosen that falls within the endpoints of this tie-line at 1,200°C, you know from the phase diagram that the alloy consists of two phases. You can determine the amounts of the phases in the alloy by applying the inverse lever rule. Label the alloy composition C_0, the composition of the solid C_S, and the composition of the liquid C_L; then the relationship for the fraction of liquid is given by equation 10.11:

$$f_L = \frac{C_S - C_0}{C_S - C_L} \qquad [10.11]$$

Notice that the equation simply relates a segment of the tie-line to the entire length of the tie-line. The relationship is inverse in the sense that, if you are interested in the fraction of the phase on the left, you take the ratio of the right-hand segment of the tie-line to the entire length of the tie-line. If, however, you are seeking a fraction of the phase on the right, you take the left-hand end of the tie-line and compare that to the entire length of the tie-line.

This inverse lever rule can be used in any two-phase region in any phase diagram. It does not matter how complex the phase diagram; $f_L + f_S = 1$.

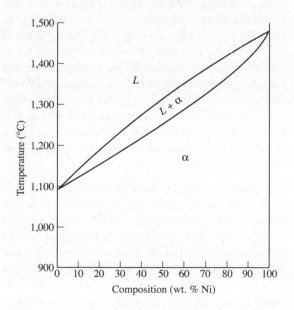

Figure 10.5 The copper-nickel equilibrium phase diagram

The inverse lever rule is used to determine the relative amounts of phases in a two-phase region.

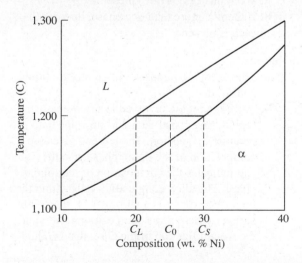

Figure 10.6 A portion of the copper-nickel equilibrium phase diagram

289

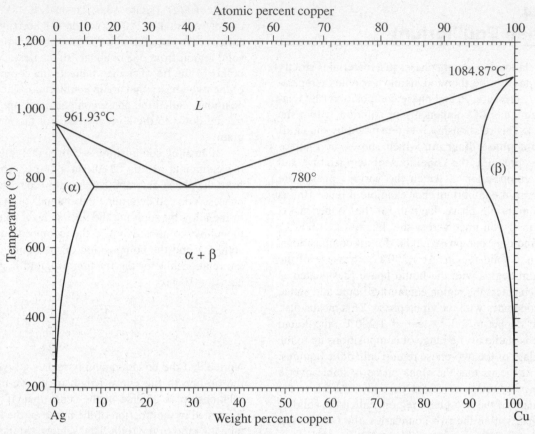

Figure 10.7 The copper-silver equilibrium phase diagram. (Used with permission of ASM International.)

Example 10.5

Consider an alloy containing 24 wt. % Ni that was heated to the liquid region and then slowly cooled to 1,200°C.

A. What phase(s) is (are) present?
B. If there is more than one phase, how much of each is present?

Solution:

A. This alloy is composed of two phases, liquid and solid.
B. Since the region involved is a two-phase region, first construct a tie-line at the temperature in question. The left end gives the composition of the liquid phase, and the right end indicates the composition of the solid phase. The alloy composition falls within the line. According to equation 10.11, the combination of these three compositions will yield the weight fraction of the alloy that is liquid, f_L.

$$f_L = \frac{30 - 24}{30 - 20} = \frac{6}{10} = 60\%$$

Then the amount of solid is 40%.

Some materials exhibit phase diagrams that are somewhat more complicated than the one in Figure 10.5. In some binary alloy systems, the addition of the second component lowers the melting point of the solvent at both ends of the phase diagram, so that the range of the liquid phase is extended downward in the middle, resulting in a phase diagram like that in Figure 10.7. This is called a **eutectic phase diagram** and is characterized by the feature that liquid is stable at temperatures below the melting points of both pure elements.

The important feature of the eutectic phase diagram is the horizontal line at which the liquid region becomes ever more narrow until it finally reaches a point. This line of constant temperature, called the eutectic temperature, connects three important points: the composition of the alpha phase, which is stable at this temperature, and the composition of the liquid phase. All three of these phases are in equilibrium. When the alloy passes through the eutectic temperature, the liquid phase of the eutectic composition changes into a two-phase solid mixture of alpha and beta, called the eutectic mixture or, simply, the eutectic. All the liquid is gone and has been replaced by the eutectic.

The transformation of phases that occurs during the eutectic transformation is represented by equation 10.12.

$$L - \alpha + \beta \qquad [10.12]$$

Example 10.6

Using the phase diagram in Figure 10.7, consider an alloy of 20 wt. % Cu that has been slowly cooled from the liquid to 779°C. Calculate the fraction of the alloy that is eutectic.

Solution:

The amount of eutectic in the alloy at this temperature is equal to the amount of liquid in the alloy at 781°C, that is, just before the transformation to complete solid. To determine this amount of liquid, draw a tie line at 781°C in the liquid + alpha region. Sketch the vertical line representing the alloy composition of 20 wt. % Cu. Then use equation 10.11 to calculate the weight fraction of liquid. The amount of eutectic after solidification is the amount of liquid before solidification:

$$f_L = \frac{C_0 - C_\alpha}{C_L - C_\alpha} = \frac{20 - 9}{29 - 9} = \frac{11}{20} = 0.55 = 55\%$$

10.5
Ferrous Metallurgy

The iron-iron carbide phase diagram shown in Figure 10.8 is the basis for all steel and cast-iron metallurgy.

Consider the pure iron axis over the entire temperature region shown. Pure iron exists in two different structures, FCC and BCC, depending on the temperature. It is very important to remember that iron is BCC at all temperatures up to 912°C, at which point it changes to FCC. It remains FCC until 1,430°C, at which point it returns to BCC until it melts at 1,536°C. The lower temperature BCC phase is called alpha, or ferrite; the FCC phase is called gamma, or austenite; and the high-temperature BCC phase is called delta.

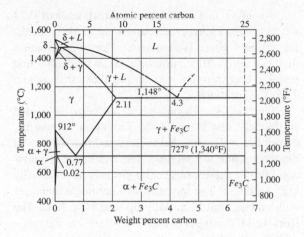

Figure 10.8 The iron-iron carbide equilibrium phase diagram. (Used with permission of ASM International.)

The phase diagram shows the degree of solubility of carbon in each of these phases. Notice that the ferrite region is very narrow, indicating low solubility of carbon, and that the austenite region is quite wide (up to 2 wt. % C max), indicating quite high solubility. The phase that occurs at 6.67 wt. % C is called Fe_3C, or cementite, and has virtually no solubility width; it exists only at 6.67 wt. % C.

To review the simple heat treatments available in the Fe-Fe_3C system, consider first an alloy containing 0.77 wt. % C. When heated above the eutectoid temperature, 727°C, the alloy is single-phase austenite. This part of the process is called austenitization. When the alloy is slowly cooled to the eutectoid temperature, the austenite undergoes a phase transformation to form a two-phase mixture of ferrite and cementite, called pearlite. When the transformation is complete, so that the microstructure is completely pearlitic, the alloy may continue to cool below 727°C. This heat treatment, called normalization, involves austenitization followed by air-cooling.

An alloy that contains less carbon than 0.77 wt. % is called a hypoeutectoid steel. These alloys are austenitized by heating above the A_3 temperature. The A_3 line is the boundary separating the austenite region from the ferrite plus austenite region. For any alloy, this boundary is the A_3 temperature. When they are slowly cooled below the A_3 temperature, ferrite called primary ferrite forms in the austenite. When the alloy crosses the A_1 temperature, the eutectoid temperature, the remaining austenite transforms to pearlite. The final microstructure, then, is ferrite and pearlite. The inverse lever rule determines the relative amounts of the two constituents. If the alloy is low carbon, say 0.2 wt. %, the final microstructure is nearly all ferrite with very little pearlite.

An alloy that contains more carbon than 0.77 wt. % is called a hypereutectoid steel. In these steels, the primary phase is cementite. The microstructure of these alloys after normalization is cementite and pearlite. Cast irons are iron-carbon alloys that contain 2 wt. % C or more.

On the other hand, when austenite is cooled sufficiently fast to avoid the formation of pearlite, martensite will form below a threshold called the martensite start temperature. The martensite is a non-equilibrium phase whose chief virtue is its potentially high hardness. The hardness of martensite increases with carbon content, usually up to 0.8 wt. % C. The Rockwell C hardness of hardened steel is in the range of $R_c50 - R_c60$.

Stainless steel is an iron-chromium alloy that contains more than 12 percent chromium. The three classes of stainless steel are ferritic, martensitic, and austenitic.

Chromium stablilizes the ferrite phase in these alloys, so ferritic stainless steel is an iron-chromium alloy with high chromium content. Therefore, ferritic stainless steel is very corrosion resistant. When the carbon content is elevated, the iron-chromium-carbon alloy can be heat treated to form a hard, corrosion-resistant martensite. This alloy is martensitic stainless steel. Although not as hard as iron-carbon martensite, its corrosion resistance makes martensitic stainless steel useful for cutlery.

When nickel is alloyed with iron and chromium, it stabilizes austenite to room temperature. Such an alloy, called austenitic stainless steal, is both corrosion resistant and face centered cubic in structure. As such, it is paramagnetic rather than ferromagnetic, as are all other iron alloys at room temperature.

10.6
Solid Solubility in Metal Alloys

Solute atoms can reside in the solvent lattice, to some extent, in either of two ways: substitutionally or interstitially.

Substitutional solute atoms replace solvent atoms on their lattice sites. The extent varies according to thermodynamic and other factors. For instance, copper and nickel atoms can completely exchange with each other in the FCC lattice and are thus considered completely soluble. Copper and silver, on the other hand, can tolerate only limited exchange, and are thus considered only slightly soluble.

Interstitial solute atoms occupy positions between the solvent atoms in the solvent lattice. For this reason, interstitial solutes are limited to small atoms such as carbon, hydrogen, and nitrogen. The extent of solubility is also limited because only a small fraction of the interstitial sites can be occupied.

A vacancy is a vacant lattice site, that is, no atom is occupying the site. Every metal has an equilibrium concentration of vacancies that increases exponentially with increasing temperature up to the melting point.

10.7
Chemical Diffusion

At any temperature above absolute zero, atoms are vibrating in their lattice positions. An atom can change position by moving to a neighboring vacancy or, in the case of an interstitial solute atom, a neighboring interstitial site. The process is a simple jump from one site to the next. By this chemical diffusion process, atoms can move through the material.

Fick's first law gives the rate of atomic motion in terms of the rate of mass transfer:

$$J = -D\frac{dc}{dx} \qquad [10.13]$$

where $\frac{dc}{dx}$ is the concentration gradient of the solution (the variation of the amount of solute, c, with distance), $J = \frac{1}{A}\frac{dm}{dt}$ is the mass flux in the x-direction (the mass of atoms moving through a section of area A), and D is the diffusion constant of the diffusing solute/solvent system.

The magnitude of D indicates a material system's rate of diffusion and is an exponential function of temperature, since the rate of atom jumping increases greatly with increasing temperature as the amplitude of atom vibration increases. The variation with temperature is given by equation 10.14:

$$D = D_0 \exp \frac{-Q}{RT} \qquad [10.14]$$

Here, D_0, the diffusion constant, and Q, the activation energy, are material constants for any given solute/solvent combination. R is the universal gas constant, expressed in the units of Q; and T is absolute temperature. The higher the activation energy, the lower the diffusion coefficient, D, and the greater its dependence on temperature.

Since interstitial atoms, such as hydrogen and carbon, are small and always have unoccupied interstitial sites as neighbors, the activation energy is lower and the diffusion rate is higher by several orders of magnitude than for substitutional solutes, principally because the chance of a neighboring site being vacant in the substitutional case is quite low.

Table 10.3 lists diffusion constants and activation energies for three systems.

Table 10.3. Diffusion Constants and Activation Energies for Some Systems

Diffusion System	D_0 (m^2/sec)	Q (kJ/mol)
C in α-Fe (BCC)	3.9×10^{-7}	80
C in γ-Fe (FCC)	7×10^{-6}	131
Cu in Al	6.5×10^{-5}	136

Since Q is given here in kilojoules per mole, the value of R should be 8.31 J/mol · K.

10.8 Corrosion

In this section the basic elements of galvanic corrosion of metals are discussed. The process by which metals lose mass is called **oxidation**. In this process, metal atoms lose one or more electrons and become metal ions as expressed in equation 10.15:

$$M \rightarrow M^{n+} + n\bar{e} \qquad [10.15]$$

where M refers to any metal, and $\bar{e}$ is an electron. This oxidation process occurs at the anode. Thus, the anode corrodes.

Simultaneously with the anode reaction, a reduction process must be operating so as to conserve electrons. In this reaction, equation 10.16, electrons are consumed:

$$M^{n+} + n \rightarrow M \qquad [10.16]$$

Since the electrons pass from the anode to the cathode by electrical conduction, the anode and the cathode must be electrically connected.

Typical anode reactions are as follows:

> Anode reactions produce metal ions.

$$Cu \rightarrow Cu^{2+} + 2e^-$$
$$Ni \rightarrow N^+ + e^-$$
$$Al \rightarrow Al^{3+} + 3e^- \qquad [10.17]$$
$$H_2 \rightarrow 2H^+ + 2e^-$$

Typical cathode reactions are these:

$$Cu^{2+} + 2e^- \rightarrow Cu$$
$$Ni^+ + e^- \rightarrow Ni$$
$$Al^{3+} + 3e^- \rightarrow Al \qquad [10.18]$$
$$2H^+ + 2e^- \rightarrow H_2$$

Corrosion can be viewed, then, as an electrochemical reaction consisting of the combination of an oxidation and a reduction process. Two half-cell reactions, like those shown above, can be combined to form a complete reaction. For example,

$$Zn^{2+} + 2e^- \rightarrow Zn \qquad \text{Anode}$$
$$2H^+ + 2e^- \rightarrow H_2 \qquad \text{Cathode}$$

By reversing the direction of the anode reaction to represent oxidation, and then combining the two half-cell reactions, a complete reaction can be obtained. For example:

$$Zn + 2HCl \rightarrow ZnCl_2 + H_2$$
$$\text{or} \qquad [10.19]$$
$$Zn + 2H^+ \rightarrow Zn^{2+} + H_2$$

In the first reaction zinc is the anode and is oxidized, and hydrogen in hydrogen chloride is reduced to be liberated as hydrogen gas at the surface of the metal conductor, which acts as the cathode. In some corrosion cells, the cathode half-cell reaction can be the plating out of metal ions onto the metal cathode. In most cases, however, the cathode half-cell reaction involves water and hydrogen. Some typical cathode reactions are as follows:

$$O_2 + 4H^+ + 4e^- \rightarrow 2H_2O \quad pH < 7$$
$$O_2 + 2H_2O + 4e^- \rightarrow 4(OH^-) \quad pH \geq 7 \quad [10.20]$$
$$2H^+ + 2e^- \rightarrow H_2 \quad pH < 7$$

To determine the relative tendency of an anode-cathode pair to corrode, a concept of electrode potential is used. In the standard Emf Series shown in Table 10.4, the electrode potentials of the reduction half-cell reactions of various metals are listed with the hydrogen gas electrode serving as a reference of 0 volt. When two dissimilar metals are placed in contact, the one that is lower in the table is the anode and corrodes. Care should always be taken to avoid using in service dissimilar metals that will be in electrical contact.

When iron is exposed to oxygen and water, an oxidation reaction occurs: $Fe \rightarrow Fe^{2+} + 2e^-$. In the presence of oxygen and water, Fe^{2+} is further oxidized to Fe^{3+}, resulting in $Fe(OH)_3$. This dehydrates to $Fe_2O_3 \cdot nH_2O$, which is rust.

Corrosion Protection

Some common methods of corrosion control are cathodic protection, anodizing, and coating. In cathodic protection, a more corrosive metal is put in contact with the piece to be protected, rendering the piece cathodic. The sacrificial anode, if properly sized, will preferentially corrode over an acceptable period before being replaced. Zinc coating (galvanizing) is a form of cathodic protection. The zinc is anodic to steel and thereby protects the steel while slowly corroding away.

Another form of cathodic protection is called *impressed current*. In large structures such as pipelines, where galvanic sacrificial anodes are impractical and costly, anodes are connected to a DC power source. The negative terminal is connected to the structure being protected. The positive terminal to the distributed electrodes is imbedded in proximity to the structure.

Anodizing is another method of corrosion protection. It is an electrolytic process that increases the thickness of the surface oxide layer of metal parts. In this process, a direct current is passed through an electrolyte with the work piece as the positive electrode (anode). Anodizing produces a thick, hard, protective corrosion barrier.

Coatings of various types are also used for corrosion protection, especially those containing chromates and phosphates. A wide range of polymer-based coatings are also widely used.

Use of corrosion-resistant alloys is clearly a common means of avoiding or reducing corrosion. Using stainless steel is the most effective method because of its high chromium content. Stainless steel offers a wide range of properties, as discussed in Section 10.5.

10.9
Mechanical Behavior

The Stress-Strain Curve

The response of a material to applied stress or applied strain is called mechanical behavior. For crystalline structural materials, that is, all metals and many ceramics, this behavior is of two types: linear elastic and plastic. Each type is characterized by the properties that relate the applied stress to the response variable, whatever it may be.

The most common of these relations is between stress and strain, and the relevant properties are measured on the uniaxial stress-strain curve. Keep in mind that stress, σ, in this case is defined as follows:

$$\sigma = \frac{\text{load}}{\text{area}} \qquad [10.21]$$

where load is the applied load in the axial direction, and, for engineering stress, area is the original area; for true stress, area is the instantaneous area.

Strain, ε, is defined here as:

Engineering strain: $\qquad \varepsilon = \frac{\Delta L}{L_0} \qquad [10.22]$

where ΔL is the change in length, and L_0 is the original or gauge length of the sample. Also,

True strain: $\qquad \varepsilon = \ln\frac{L}{L_0} \qquad [10.23]$

where L_0 is the original length, and L is the instantaneous length.

An example of the engineering stress-strain curve for a typical engineering alloy is shown in Figure 10.9. From it some very important properties can be determined. The elastic modulus, the yield strength, the ultimate tensile strength, and the fracture strain are all clearly exhibited in an accurately constructed stress-strain curve. The elastic modulus, E (Young's modulus), is the slope of the elastic portion of the curve (the steep, linear region) because E is the proportionality constant relating stress and strain during elastic deformation: $\sigma = E\varepsilon$. The 0.2% offset yield strength is the stress value, $\sigma_{0.2\%ys}$ of the intersection of a line (called the offset) constructed parallel to the elastic portion of the curve but offset to the right by a strain of 0.002. It represents the onset of plastic deformation. The ultimate tensile strength is the engineering stress value, or σ_{uts}, at the maximum of the engineering stress-strain curve. It represents the maximum load, for that original area, that the sample can sustain without undergoing the instability of necking, which will lead inexorably to fracture. The fracture strain is the engineering strain value at which fracture occurred. However, a lot more information is contained in a stress-strain curve.

Table 10.4 Standard EMF Series

	Electrode Reaction	Standard Electrode Potential, V^0 (V)
	$Au^3 + 3e^- \rightarrow Au$	+1.420
	$O_2 + 4H^+ + 4e^- \rightarrow 2H_2O$	+1.229
	$Pt^{2+} + 2e^- \rightarrow Pt$	~ +1.2
	$Ag^+ + e^- \rightarrow Ag$	+0.800
Increasingly inert (cathodic)	$Fe^{3+} + e^- \rightarrow Fe^{2+}$	+0.771
	$O_2 + 2H_2O + 4e^- \rightarrow 4(OH^-)$	+0.401
	$Cu^{2+} + 2e^- \rightarrow Cu$	+0.340
	$2H^+ + 2e^- \rightarrow H_2$	0.000
	$Pb^{2+} + 2e^- \rightarrow Pb$	−0.126
	$Sn^{2+} + 2e^- \rightarrow Sn$	−0.136
	$Ni^{2+} + 2e^- \rightarrow Ni$	−0.250
	$Co^{2+} + 2e^- \rightarrow Co$	−0.277
	$Cd^{2+} + 2e^- \rightarrow Cd$	−0.403
	$Fe^{2+} + 2e^- \rightarrow Fe$	−0.440
	$Cr^{3+} + 3e^- \rightarrow Cr$	−0.744
Increasingly active (anodic)	$Zn^{2+} + 2e^- \rightarrow Zn$	−0.763
	$Al^{3+} + 2e^- \rightarrow Al$	−1.662
	$Mg^{2+} + 2e^- \rightarrow Mg$	−2.363
	$Na^+ + e^- \rightarrow Na$	−2.714
	$K^+ + e^- \rightarrow K$	−2.924

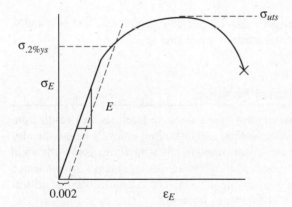

Figure 10.9 An example of the engineering stress-strain curve for a typical engineering alloy

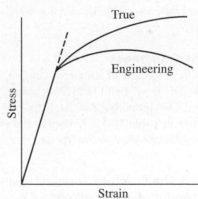

Figure 10.10 The true stress-strain curve and the engineering stress-strain curve for the same material

At the outset, though, a clear distinction must be made between a true stress-true strain curve and an engineering stress-engineering strain curve. The difference is shown in Figure 10.10, in which are plotted, on the same axes, the true stress-strain curve and the engineering stress-strain curve for the same material. The difference is also evident in the definitions of true stress-true strain and engineering stress-engineering strain. The engineering stress is the load applied to the sample divided by a constant, the original area. The true stress is the load applied to the sample divided by a variable, the instantaneous area. Note that the true stress always rises in the plastic region, whereas the engineering stress rises and then falls after going through a maximum.

The maximum represents a significant difference between the engineering stress-strain curve and the true stress-strain curve. In the engineering stress-strain curve, this point indicates the beginning of necking.

Hardness

Since hardness is a measure of a material's resistance to permanent indentation, load in relation to indentation depth is a measure of strength. Strong materials have high hardnesses and vice versa.

As an approximation, the relationship between hardness and strength is as follows:

$$UTS(MPa) = 3.5 \times HB$$
$$UTS(psi) = 500 \times HB$$

where HB = Brinell Hardness Number (without units)

UTS(MPa) = ultimate tensile strength in MPa
UTS(psi) = ultimate tensile strength in psi

Plastic Deformation of Crystalline Materials

The mechanism for plastic deformation in a crystalline material is the motion of dislocations through the crystal in response to shear stress of an adequate level. The magnitude of the stress required for this process is a measure of the ease of dislocation motion in that material. The yield strength of a crystalline material, then, can be viewed as the stress required to begin dislocation motion.

If there is an ease to dislocation motion, significant amounts of plastic deformation can be achieved and the material is considered **ductile.** If, however, there is so much resistance to dislocation motion that plastic deformation cannot occur without the stress rising to the fracture level, then the material is considered **brittle.** Being strong or hard is usually associated with being brittle; ductility is associated with being less strong.

Fracture Toughness

The fracture toughness of a material is a property that represents a material's ability to absorb energy before fracturing. The property, K_{IC}, is listed for several metal alloys in Table 10.5. The higher the value, the more energy is absorbed in the fracture process.

Table 10.5 The Fracture Toughness and Yield Strength of Some Alloys

Material	K_{IC}(MPA$\sqrt{m}$) / (ksi$\sqrt{in}$)	Yield Strength (MPa)/(ksi)
4340 Steel (hardened)	77/70	1552/225
2024 Aluminum alloy T3	45/41	338/49
Ti-6Al-4V	60/55	814/118

The fracture stress for a piece of material depends on the fracture toughness value for that material *and* the size of the largest crack in that piece. The geometry and crack size notation are shown in Figure 10.11. The relation is as follows:

$$K_{IC} = \sigma_f \sqrt{\pi a}$$

where σ_f = tensil fracture stress

K_{IC} = fracture toughness

a = crack depth of a surface crack

$2a$ = total width of an internal crack

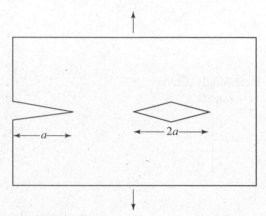

Figure 10.11 Sketch of crack size and orientation with vertical applied stress

Metal Fatigue

Metal alloys are subject to fractures that result from cyclic loading and unloading at levels below the ultimate tensile strength and sometimes below the yield strength. During cyclic loading, cracks are nucleated and propagate until the crack lengths become critical for spontaneous fracture.

A common way to represent fatigue behavior is by the S-N curve, shown in Figure 10.12. This curve shows the variation of the cyclic stress level at failure with the log of the number of cycles at failure. Since this data shows significant scatter, these curves are useful in showing the effects of environmental, alloying, and temperature on fatigue behavior. In many ferrous alloys, the curve flattens at low stress levels, indicating that there is a stress level below which fatigue failure will not occur. This stress level is called the *endurance limit*. Most nonferrous alloys do not exhibit an endurance limit.

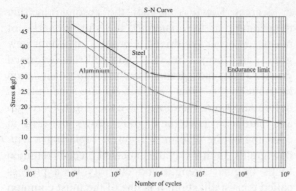

Figure 10.12 Typical S-N curves for ferrous and nonferrous alloys (File: S-N curves.PNG from Wikimedia Commons, the free media repository)

Strengthening Mechanisms

Since the yield strength of a material has been defined as the stress at which plastic deformation begins, it follows that any means by which dislocation motion can be impeded will raise the stress required to initiate plastic deformation, and thereby raise the strength of the material. The various processes that accomplish this purpose are called strengthening mechanisms.

Strain Hardening

One of the most common mechanisms for strengthening a crystalline material is strain hardening. In this case, dislocation motion itself becomes responsible for impeding dislocation motion. This mechanism involves two concurrent actions: the intersection of dislocations during motion impedes further motion, and the multiplication of dislocations during motion increases the rate of intersection. Therefore, in strain hardening, as plastic strain (dislocation motion) increases, the stress required to continue the motion must also increase. This result can be observed in the positive slope of the plastic region of the true stress-true strain curve shown in Figure 10.10.

Solid Solution Strengthening

Dislocations are more difficult to move through the lattice of a crystalline solid solution than through the lattice of the pure crystal. The resistance to motion increases with solute content. Although there is a significant variation in the degree to which various solutes contribute to the strengthening effect, most solutes serve to increase this effect. Therefore, in general, the purest crystals have lowest strength (neglecting strain hardening).

Second-Phase Strengthening

Dislocation motion is further impeded by small particles of one or more additional phases. The dislo-

cations have difficulty passing through the particles and therefore must bow out between them in order to continue to move. Thus, the distance between these particles determines strength; as interparticle spacing decreases, strength increases.

A very common use of second-phase strengthening is the precipitation hardening of a group of aluminum alloys known as heat-treatable alloys. The phase diagram of an aluminum-copper alloy of this type is shown in Figure 10.13. The dashed line, C_0, indicates the particular alloy that is suited to precipitation hardening treatment. When this alloy is heated into the single-phase alpha region, the alloy becomes a solid solution, single phase. When cooled into the alpha plus theta region, the theta phase precipitates out of solution as a dispersion throughout the matrix. When the alpha solution is cooled slowly, large particles of theta form; when the alpha is cooled rapidly to room temperature, formation of theta is suppressed. Finally, if the supercooled alpha is reheated to a relatively low temperature, the theta will nucleate and grow as a very fine dispersion within the alpha matrix. This fine dispersion with small interparticle spacing will contribute significantly to the strength of the aluminum alloy.

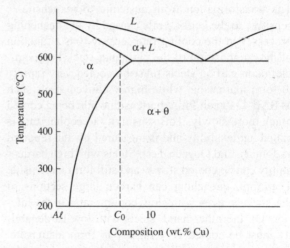

Figure 10.13 The aluminum-rich portion of the aluminum-copper equilibrium phase diagram

The designation of this second-phase strengthening heat treatment is T6.

Grain Refinement

Fine-grained polycrystals are stronger than coarse-grained polycrystals because the limited volume of crystal in a small grain impedes dislocation motion. Higher stress is required in the small (fine) grain for the same amount of dislocation motion as in a large (coarse) grain.

An ASTM numbering scheme quantifies grain size. A grain size is designated by a number from 1 to 10; the larger the number, the smaller the size. The quantitative relationship is given by equation 10.24:

$$N = 2^{n-1} \qquad [10.24]$$

where N is the average number of grains per square inch at $100\times$ magnification and n is the ASTM grain-size number.

Quenching and Tempering of Martensite

Tempered martensite in steel is one of the most effectively strengthened phases found in materials. The crystal structure of martensite is more complicated than the structures of most metals and the microstructure is very fine because of the manner in which the martensite forms. During tempering, the fine precipitates of carbides also contribute to strength and hardness. In addition, there is a large amount of carbon in solution, even after tempering, so that the phase is strengthened by solid solution. It is clear that many different mechanisms contribute to the strengthening of martensite.

Hardenability

The transformation from austenite to martensite is sensitive to the cooling rate caused by the quenching process. But the cooling rate sensitivity is a function of the composition of the particular steel. For example, plain carbon steels must be cooled very rapidly to form martensite, while highly alloyed steels such as SAE 4340 can form martensite while being cooled much more slowly. This sensitivity to cooling rate is called hardenability and is measured by and reported as Jominy End Quench Test. Steels with high hardenability can be cooled slowly and still form martensite. Therefore, quenching can harden large sections of these steels, even with slow cooling rates in the interior. On the other hand, steels with low hardenability must be cooled very rapidly to form martensite. Therefore, only small sections of these steels can be hardened in the interior because the cooling rate during quenching must be kept high.

Annealing

When metal alloys have undergone plastic deformation, the length of dislocations in the crystals increases in proportion to the amount of plastic strain. Since the dislocations also possess elastic strain energy in proportion to their length, as dislocation length increases, so does the total energy of the metal. This increase in elastic strain energy is called stored energy of plastic deformation.

When a plastically deformed piece of metal is heated to a temperature above half its melting point, in degrees Kelvin (one-half the absolute melting temperature of the material), the stored energy of plastic deformation drives a process in which new grains are formed in the midst of the deformed grains. These new grains have the same dislocation distribution as was present before the plastic deformation and, therefore, have the original strength and hardness. Each new grain grows outward to consume all the deformed material, replacing it with new grains. In this way the annealed piece is returned to its original strength and ductility by a process called recrystallization. It is one of the most common annealing processes.

Cold Working Versus Hot Working

When a metal alloy is plastically deformed at a temperature above half its absolute melting point, recrystallization occurs simultaneously with plastic deformation. In this situation, referred to as hot working, strain hardening cannot occur during the deformation. On the other hand, plastic deformation carried out at a temperature below between one-third to one-half the absolute melting point, referred to as cold working, is accompanied with the usual strain hardening.

Surface Condition

The surface condition of a material can have a significant effect on a material's behavior. Several of the categories of surface effects will be discussed here.

The degree of smoothness of the surface of a material can affect the fatigue life of a metal. Fatigue cracks have been shown to nucleate at sites of surface roughness. Polishing is often used to resist fatigue fracture at critical locations.

Surface hardening is an important process, especially in steels. There are two types of processes in this category. The first includes carburization and nitriding. The carbon and/or nitrogen are diffused into the surface of the work piece to produce a concentration gradient to a desired depth. When the piece is heat treated, the higher surface concentration leads to a harder surface layer.

The second version of surface hardening of steel involves the local heat treatment of only the surface. Induction and flame hardening are examples of this type.

Another class of surface hardening involves the use of residual stress distributions that put the surface layer into compression. In this way, surface cracks experience reduced tension stress in service, allowing for higher service loads than without residual surface compression. This is accomplished by producing temperature gradients using surface heating. Thermal expansion gradients produce thermal stresses that, if high enough, cause plastic flow and result in surface compression.

10.10
Thermal Mechanical Behavior

Thermal stresses arise in solid materials because of thermal expansion in conjunction with temperature gradients. When a heat flux is imposed on a solid, a temperature gradient occurs because of the finite thermal conductivity. For a given heat flux, the lower the thermal conductivity is, the higher the temperature gradient. Therefore, the temperature gradient in conjunction with the thermal expansion coefficient produces a volume gradient, which is a thermal stress gradient. With low thermal conductivity and a high thermal expansion coefficient, thermal stresses can become sufficient to cause brittle fracture in the case of low ductility or cause plastic flow in the case of a ductile material. The latter results in residual stresses. Thermal cycling can exacerbate the problem. In the early stage, cracks can nucleate because of thermal stresses and then propagate to fracture, much the same as if fatigue occurs.

10.11
Magnetic Behavior

The magnetic behavior of materials is divided generally into three groups: diamagnetic, paramagnetic, and ferromagnetic. When an uninformed person refers to a material being nonmagnetic, he or she is referring, without distinction, to the first two types of behavior. When a material is said to be magnetic, its behavior is ferromagnetic.

In all materials, the electrons in the atoms have magnetic moments because of their motions about the atoms and about their own axes of spin. Each of these motions produces a small magnetic moment that, with no applied magnetic field, is canceled. When a magnetic field is applied, however, one type of material experiences alignment of electron magnetic moments to oppose the applied field. This behavior is **diamagnetism.** In another type of material, the weak permanent magnetic moments of the atoms align with the applied magnetic field. This behavior is **paramagnetism.** Both of these magnetic responses to a magnetic field are *nonpermanent* and *weak.* When the applied field falls to zero, the net magnetic moment of the material also becomes zero.

Copper and silver are diamagnetic, while aluminum and titanium are paramagnetic.

In certain metals (iron, cobalt, and nickel) and their alloys, a quite different behavior, **ferromagnetism,** is observed. In these metals, the magnetic moments from electron spin align themselves within the atom and from atom to atom, resulting in a *strong* magnetic moment in the absence of an applied magnetic field. If all the moments of all the atoms aligned themselves across a bar of such a material, the result would be a permanent magnet, with a north and a south pole.

In some of these materials, however, the bar divides itself into domains. Each domain would contain aligned atoms and possess north and south poles, but these would exactly cancel the north and south poles of neighboring domains. The *bar* would thus have no *net* moment, even though each domain was permanently aligned. When a magnetic field is applied, however, the domain boundaries reposition themselves to create a net magnetic moment to align with the applied field. Now, if the material is a hard magnet, the domains do not return to their original sizes and shapes. This bar will then be a strong permanent magnet. If the material is a soft magnet, the domains will reposition themselves when the applied field is removed so that again there is no *net* magnetic moment.

In most materials, the difference between hard and soft magnets is microstructural. The microstructural effect is in the ease of motion of the domain wall boundaries, called Bloch walls.

Thus, an electromagnet is a soft magnet, meaning that the applied field can control the net magnetic moment. A permanent magnet is a hard magnet in that the net magnetic moment is always present because the domain boundaries are prevented by the micro-structure from moving.

To test a material for magnetic behavior, a permanent magnet is used as a test magnet. If the test magnet is attracted to the material, the material is ferromagnetic. If no attraction occurs, the material is either paramagnetic or diamagnetic. Those two types cannot be distinguished by feel.

A ferromagnetic material loses its ferromagnetic behavior and becomes paramagnetic when heated above the Curie temperature. For iron this is 768°C, just below the BCC (α) − FCC (γ) transformation temperature, 912°C. In steel, the FCC (γ) phase, austenite, can be stable below 768°C. In such cases, the paramagnetic behavior is related to the structure, not to the temperature; stable austenite, even at low temperature, will be paramagnetic rather than ferromagnetic.

10.12
Electrical Behavior

The electrical behavior of materials is divided among four types: metals (conductors), insulators, semiconductors, and superconductors.

Metals

Metal atoms are held together by metallic bonding, a very complicated process that has an important consequence—high electrical conductivity. The bonding mechanism produces free electrons, which are not bound to any particular atom in the solid—hence the term *free*. It is these free electrons that give rise to high electrical conductivity. The governing law of conductivity is Ohm's law:

$$V = IR \qquad [10.25]$$

where V is voltage in volts, I is current in amperes, and R is resistance in ohms. Since R is dependent on the cross-sectional area and length of the conductor, it is not a material property. This is shown in the relation

$$R = \frac{\rho L}{A} \qquad [10.26]$$

where ρ is the resistivity, in ohm-meters, of the material and is thus a material property. The conductivity of a material, σ, is the reciprocal of ρ:

$$\sigma = \frac{1}{\rho} \, (\text{ohm-meter})^{-1} \qquad [10.27]$$

If conductivity is considered as the motion of charged carriers, electrons for example, the conductivity of a conductor is given by equation 10.28:

$$\sigma = \eta \, |e| \, \mu \qquad [10.28]$$

where η is the number of charge carriers per unit volume (free electrons in metals), $|e|$ is the electron charge, and μ is the charge mobility.

In the case of metals, the carrier concentration, η, remains constant, while the mobility, μ, decreases, as the temperature increases because of the increase in the frequency of collisions of the electrons with the increasing amplitude of atom vibration. This relationship results in the rule that, for metals, *resistivity increases as temperature increases*. In fact, that rule becomes the basis for the definition of a metal.

Insulators

In some materials, such as diamond, usually involving covalent bonding, the electrons are bound to their atoms. Hence, there are no free electrons and so η in equation 10.28 is nearly zero. The result is negligible conductivity or very high resistivity. These materials will act as insulators.

Semiconductors

In some materials, called semiconductors, the electrons are bound; but with increases in temperature and/or applied voltage, the electrons can be excited into allowed but unoccupied energy states. In such states, the electrons can act as free conduction electrons. For every excitation event, there is produced a vacant bound state that is abandoned when the electron is excited to the conduction state. Other bound electrons can then migrate into the vacant bound state (called a hole), giving rise to electron mobility or conduction. However, in this case, convention assigns positive current to hole motion, even though electrons are moving. The situation can be compared to a moving line of chairs with one chair vacant. If a person moves into the vacant chair when it arrives next to him, the person (electron) will be moving, say, from left to right, while the vacant chair (hole) will be moving from right to left.

In an intrinsic semiconductor such as pure silicon, the number of free electrons and the number of holes are always equal. This is a very important concept.

With these semiconductors, in equation 10.28 the term η, the carrier concentration, increases with increasing temperature because of the increasing number of excited electrons. The increase is very much larger than the decrease in mobility, so the net result is an increase in conductivity. Therefore, *conductivity increases* and *resistivity decreases* with increasing temperature. This is the basis for the definition of an intrinsic semiconductor; this definition is the converse of the metal definition.

In microelectronics, extrinsic semiconductors are the common ones. In these, impurity atoms, called dopants, control the carrier concentration. Selected solute atoms in silicon, for example, can be fully positively ionized at room temperature to provide one free conduction electron per dopant atom. In this case the free electron concentration remains roughly constant and equal to the dopant concentration. Since, in this case, the dopant atoms contribute *electrons* to the conduction process, the semiconductor is called an n-type (n for negative or electron conduction). If, however, another type of dopant that can be fully negatively ionized at room temperature is dissolved in silicon, it will capture one bound electron, thus creating one hole per dopant atom. In this case the doped silicon is p type (p for positive or hole conduction). In the former case the dopant atoms are called donors, and in the latter case they are called acceptors. Remember:

Dopant Type	Extrinsic Behavior
Donors	n-type
Acceptors	p-type

When a p-type semiconductor is electronically joined to an n-type semiconductor, a p-n junction is formed. This junction, called a diode, passes current proportional to the voltage when in forward bias; but when the voltage is reversed, in reverse bias, the diode passes only a small constant current. This behavior is the basis for the operation of a solid-state rectifier.

Superconductors

A limited group of materials, composed of metals and ceramics, displays a very complicated behavior called superconductivity. When these materials are cooled below a critical temperature, T_c, resistivity drops nearly to zero for all temperatures below T_c, as shown in Figure 10.14. The range of T_c extends from a low near 1 K for some niobium alloys to a high approaching 200 K for some of the newer ceramics.

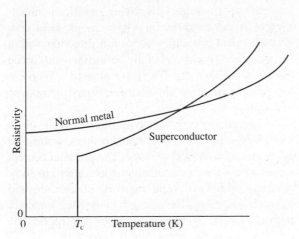

Figure 10.14 Resistivity versus temperature for a superconducting alloy

10.13
Engineering Materials

In addition to metals and alloys covered in previous sections on behavior, this section will cover aspects of glasses, polymers, and composites.

Glasses

When atoms or molecules cannot arrange themselves in an orderly array they are considered amorphous. Such is the case in most liquids, but there are many solids that display this amorphous structural characteristic. Such solids are called glasses. The most familiar of these is soda-lime glass. Here the SiO_2 atom groups, along with sodium and calcium ions, bond in such a variety of ways that a regular array is rarely attained.

Unlike crystals, glasses do not exhibit melting points. Rather, they undergo a gradual transition from solid to viscous behavior, that is, time-dependent flow. The temperature at which the transition begins during heating is called the glass transition temperature, T_g. The glass is brittle at temperatures below T_g, since dislocations are extremely difficult to move. Above T_g, the material gradually becomes viscous, and the viscosity decreases as the temperature increases until, in the upper limit, the viscosity is that of common liquids. Just above T_g, the viscosity is extremely high and displays a leathery behavior, the type used in the well-known glass blowing process. There is no sharp transition in viscosity in this region.

Polymers

Polymers are materials comprised of long-chain molecules. The chains are principally bonded together by means of weak intermolecular bonds. The resulting ease with which the chains slide relative to one another is the origin of the low strength of polymers. The chain nature of the molecules also makes it difficult for polymers to crystallize, so they are amorphous or partially crystalline.

The structures of the polymer chains vary greatly among polymers and the strength of the intermolecular bonds is determined by the degree of irregularity of the chain structure, called polarity. For example, the most regular chain structure, and the lowest polarity, is found in polyethylene, and these chains are the most weakly bonded. As the irregularity of the side groups increases, the polarity increases and the bond strength increases. For example, in polymethylmethacrylate (Plexiglass) and polystyrene the polarities are high, as are the bond strengths. The glass transition temperature can be interpreted as the temperature at which the interchain bonds begin to break during heating. It follows, then, that the weaker the bonds the lower the T_g.

In general polymers fall into either of two classes, thermoplastic and thermosetting. Thermoplastic polymers are those whose chains are interconnected only by weak intermolecular bonds. When heated above the glass transition temperature, they become viscous and can be easily shaped. When cooled to near or below T_g, they behave as nonviscous solids and retain their shape. Examples of this class are polyethylene, polypropylene, polystyrene, and nylon. The glass transition temperatures for several polymers are shown in Table 10.6. Remember that polymers below T_g are brittle.

Table 10.6 Glass Transition Temperatures for Several Thermoplastic Polymers

Material	T_g (°C) (approx. values)
Polyethylene	−90
Polypropylene	−18
Polystyrene	75–100
Polyvinylchloride	82

In thermosetting polymer the chains are additionally interconnected by strong covalent interatomic bonds, called cross-links. These cross-links can be direct interatomic bonds between chains or short chain segments connecting the longer chains. These cross-links are difficult to break, so the ability of the chains to slide past each other permanently is greatly reduced, if not completely removed. Once the cross-links are in place (the polymer is cured), the polymer cannot undergo shape change, even at increased temperature. This also means they are brittle. Examples of this class are two-part polymers such as epoxies and polyester resins. Rubber is also a thermosetting polymer; vulcanization is one of the cross-linking processes. But in rubber, the long chain molecules are tightly coiled, allowing large elastic deflection with no permanent shape change.

Composites

Composite materials are mixtures of two or more mutually insoluble phases that are produced by methods other than phase transformation heat treatments.

The promising benefit of these materials lies in the variety of combinations of properties and shapes that cannot otherwise be obtained in conventional materials. Some involve the use of fibers embedded in a binding matrix, such as fiber-reinforced polymers and metals. Others use particulates mixed with a minority binding phase, such as in cemented carbides. Composite structures are usually designed to optimize the relation between material properties and service requirements.

The physical properties of composites usually follow the rule of mixtures, wherein the total property of the composite is the sum of the products of the volume fraction of each component and that component's property. For example, the density and specific heat capacity of a composite may be determined as follows:

$$\rho_c = \Sigma f_i \rho_i$$

$$C_c = \Sigma f_i C_i$$

ρ_c = density of composite

ρ_i = density of component i

C_c = specific heat capacity of composite

f_i = volume fraction of component i

C_i = specific heat capacity of component i

Although the rule of mixtures is often used as an approximation, the mechanical properties of composites are more difficult to predict from the mechanical properties of the components because of the complexities of particle shape and interaction. These materials are capable of displaying significant anisotropy, e.g., the condition in which the physical and/or mechanical properties vary with direction within the material. This is especially so in fiber-reinforced materials wherein the fibers are aligned. Wood is an example of a natural fiber-reinforced composite material.

Concrete is a common composite, which involves a predesigned mixture of portland cement, water, and aggregate such as sand or stone. The portland cement reacts with water, i.e., undergoes hydration, to form a three-dimensional solid network of calcium and aluminum silicates. The strength of the "set" concrete depends on the water/cement ratio; the higher the ratio, the lower the strength of the concrete.

Manufacturing Processes

Manufacturing processes are the steps that take the materials, selected by design, through to the finished product. They are primarily directed to final shape but can also involve property development.

The scope of the field of manufacturing processes is far too broad to be covered in a chapter of this nature. Instead, a list of the general categories is presented with comments.

Casting. Used for complex shapes; favors low-melting alloys; many different techniques

Molding. Used primarily for polymeric, ceramic, and glassy materials

Forming. Shape change using ductility; hot and cold; can also be used to affect final properties

Machining. Shape change by material removal; necessary, at least, for finishing to final dimensions

Joining. Usually necessitated by final shape or assembly; welding and brazing; heat-affected zone

Powder Metallurgy. Green strength low; must be sintered; not fully dense

Rapid Manufacturing. Produces finished solid shapes directly from a digital model by adding, layer by layer, a metal or polymer using successive 2D digital profiles to guide an energy source such as a high-energy laser or other curative agent; also called 3D printing and additive manufacturing

10.14 Summary

This chapter has covered, in summary form, the important topics in Materials Science and Engineering. The material presented is intended to be a review of important material and will not be effective as an introduction to any topic. Atomic structure, bonding, crystal structure, corrosion, and phase equilibrium, including ferrous metallurgy, are the foundational topics in the first part of the chapter. The review moves to mechanical, electrical, and magnetic behavior, followed by brief reviews of polymers and composite materials.

PRACTICE PROBLEMS

1. In regard to corrosion, which of the following is NOT true?
   (A) The cathode loses mass.
   (B) Oxidation occurs at the anode.
   (C) The anode and cathode must be electrically connected.
   (D) Water usually plays an important role.

2. During the tension deformation of a metal test bar, necking occurs when
   (A) plastic deformation begins.
   (B) the ultimate tensile strength is reached.
   (C) strain hardening ceases to operate.
   (D) the heat of deformation causes recrystallization.

3. Which of the following is NOT related to recrystallization?
   (A) hot working
   (B) cold working
   (C) elastic modulus
   (D) melting temperature

4. The temperature above which a ferromagnetic material must be heated to lose its ferromagnetism is called the
   (A) Curie temperature.
   (B) saturation point.
   (C) transition temperature.
   (D) critical temperature.

5. Which of the following does NOT involve dislocations?
   (A) fracture strain
   (B) Young's modulus
   (C) 0.2% offset yield strength
   (D) ductile-brittle transition temperature

6. What is the diffusion coefficient of carbon in iron at 700°C?
   (A) 4.5×10^{-13} m^2/s
   (B) 1.97×10^{-11} m^2/s
   (C) 3.8×10^{-10} m^2/s
   (D) 6.44×10^{-13} m^2/s

7. Consider the phase diagram shown in Figure 10.7. If the alloy containing 50 wt. % Cu is slowly cooled to just below the eutectic temperature, what fraction is eutectic?
   (A) 34.4%
   (B) 65.6%
   (C) 50.6%
   (D) 77.1%

8. Consider the phase diagram shown in Figure 10.5. At 1,300°C an alloy contains 50% liquid in equilibrium with 50% solid. What is the composition of the alloy?
   (A) 45% Ni
   (B) 58% Ni
   (C) 51% Ni
   (D) 25% Ni

9. The volume of the unit cell of molybdenum, calculated by using the information in Table 10.2, is
   (A) 0.057 nm^3.
   (B) 0.011 nm^3.
   (C) 0.317 nm^3.
   (D) 0.031 nm^3.

10. A ferromagnetic material must
    (A) be a permanent magnet.
    (B) contain iron.
    (C) have ionic bonding.
    (D) have a large saturation magnetization.

11. Given the following data regarding nickel: structure is FCC; atomic mass is 58.71 amu; atomic radius is 0.125 nm, what is the theoretical density of this metal?
    (A) 8.9 g/cm^3
    (B) 6.2 g/cm^3
    (C) 2.7 g/cm^3
    (D) 5.1 g/cm^3

12. What are the indices of the direction along the face diagonal of a cubic crystal?
    (A) $\langle 111 \rangle$
    (B) $\langle 110 \rangle$
    (C) $\{111\}$
    (D) $\{110\}$

13. The diffusion rate of carbon in iron at 1,000°C is
    (A) 2.6×10^{-9} m^2/s
    (B) 1.8×10^{-12} m^2/s
    (C) 2.93×10^{-11} m^2/s
    (D) 7.3×10^{-11} m^2/s

14. As the temperature increases,
    (A) the resistivity of a metal decreases.
    (B) the diffusion coefficient of a solute decreases.
    (C) the conductivity of an intrinsic semiconductor increases.
    (D) the resistivity of a superconductor approaches 0.

15. A semiconductor with donor impurities
    (A) conducts by hole migration.
    (B) is an n-type semiconductor.
    (C) is an intrinsic semiconductor.
    (D) acts as a rectifier.

16. Heat treating of steel most frequently involves which of the following?
    (A) martensite
    (B) pearlite
    (C) ferrite
    (D) spherodite

17. At room temperature low-carbon steel is almost completely
    (A) martensite.
    (B) pearlite.
    (C) ferrite.
    (D) spherodite.

18. If the ASTM grain-size number increases from 3 to 9, by what factor does the number of grains in a $100\times$ field of view change?
    (A) 64
    (B) -49
    (C) 27
    (D) -9

19. The strengthening process involving the formation of fine distribution of a second phase is called
    (A) quench and temper.
    (B) solid solution strengthening.
    (C) precipitation hardening.
    (D) case hardening.

20. In a ferromagnetic alloy, easy migration of domain boundaries usually results in
    (A) permanent magnetism.
    (B) good conductivity.
    (C) a very weak net magnetic moment.
    (D) superconductivity at low temperatures.

Answer Key

1. A	6. B	11. A	16. A
2. B	7. B	12. B	17. C
3. C	8. B	13. C	18. A
4. A	9. D	14. C	19. C
5. B	10. D	15. B	20. C

Answers Explained

1. **A** Reduction occurs at the cathode. The half-cell reaction is $M^{n+} + ne \rightarrow M$. This is characterized by the plating of metal ions from the electrolyte or the generation of hydrogen gas from dissolved hydrogen ions in the electrolyte.

2. **B** Necking is associated with the maximum value in the engineering stress-strain curve. The engineering stress coordinate of this maximum is defined as the ultimate tensile stress.

3. **C** Hot working is associated with plastic deformation carried out at temperatures above approximately one-half of the melting point in kelvin units. In this temperature range, recrystallization occurs whether the metal was worked below and then heated to this temperature or was worked above this temperature. Recrystallization has nothing to do with elastic deformation or the attendant property elastic modulus.

4. **A** This is the definition of the Curie temperature.

5. **B** Dislocation motion is the cause of plastic deformation. Fracture strain and ductile-brittle transition both involve dislocations because the type of fracture, ductile or brittle, depends on whether or not dislocations are able to move in crystalline materials. Young's modulus, on the other hand, pertains to elastic deformation and hence does not involve dislocations. It involves only the force-separation interaction from the interatomic or intermolecular bonding.

6. **B** Use the equation $D = D_0 \, e^{-\frac{Q}{RT}}$, and choose the values for D_0 and Q. At 700°C, iron is BCC. From Table 10.3, $D_0 = 3.9 \times 10^{-7}$ m²/s and $Q = 80$ kJ/mol. Then:

$$D_{700C} = 3.9 \times 10^{-7} \, e^{-\frac{80,000}{8.31 \cdot 973}} \text{ m}^2/\text{s}$$
$$= 1.97 \times 10^{-11} \text{ m}^2/\text{s}$$

7. **B** In Figure 10.7, the eutectic composition is 28 wt. % Cu, and the composition of the copper-rich phase is 92 wt. % Cu. Use the inverse lever rule given in equation 10.11; then the fraction of the alloy that is eutectic is

$$f_{\text{eut}} = \frac{92 - 50}{92 - 28} = \frac{42}{64} = 65.6\%$$

8. **B** In Figure 10.5, the composition of liquid and solid at 1,300°C are 47 wt. % Ni and 64 wt. % Ni respectively. Use the lever rule to give equation 10.11:

$$0.5 = \frac{64 - x}{64 - 47}$$

$$x = 55.5 \text{ wt. % Ni}$$

9. D Molybdenum is BCC. For this crystal structure, the lattice parameter, a, is related to the body diagonal by the equation

$$4r = a\sqrt{3}$$

or

$$a = \frac{4r}{\sqrt{3}}$$

Also, volume = a^3. Then

$$V = \left(\frac{4 \cdot 0.136}{\sqrt{3}}\right)^3 = 0.031 \text{ nm}^3$$

10. D A ferromagnetic material exhibits a large magnetic response to an applied magnetic field. The response need not be permanent; it may last only as long as the field is applied, or it may last long after the applied field is removed. Cobalt and nickel are ferromagnetic.

11. A The designation FCC indicates 4 atoms per unit cell, so the mass of the atoms in the nickel unit cell is given by the equation

$$\frac{4 \cdot 58.71}{6.023 \times 10^{23}} = 3.90 \times 10^{-22} \text{ g}$$

The volume of the unit cell is a^3. In terms of atom radius, $a = \frac{4r}{\sqrt{2}}$. Therefore, the volume is

$$V = \left(\frac{4r}{\sqrt{2}}\right)^3 = 0.044 \text{ nm}^3$$

Then, since density is mass divided by volume:

$$\rho = \frac{3.90 \times 10^{-22} \text{ g}}{0.044 \text{ nm}^3} = 8.9 \text{ g/cm}^3$$

12. B The arrow in the figure below represents the face diagonal of a cubic unit cell. To obtain the indices, determine the coordinates of the point at which one end of the arrow intersects the unit cell while the other end is at the origin. In this case, the coordinates are $1a$, $1a$, $0a$. Now, collect the coefficients of a in angle brackets: in this case, $<110>$, since it represents any or all of the face diagonals.

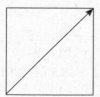

13. C Apply the same method as in Problem 6, but use the FCC values in Table 10.3, since iron is FCC at temperatures above 912°C. The equation is

$$D = D_0 e^{-\frac{Q}{RT}}$$

Then, since $D_0 = 7 \times 10^{-6}$ m^2/s and $Q = 131$ kJ/mol:

$$D_{1,000°C} = 7 \times 10^{-6} e^{-\frac{131,000}{8.31 \cdot 1273}} \text{ m}^2/\text{s}$$
$$= 2.93 \times 10^{-11} \text{ m}^2/\text{s}$$

14. C As the temperature increases, the number of conductive electrons in the conduction bond increases. Therefore, the conductivity increases by equation 10.28.

15. B An extrinsic semiconductor with donor impurities is an n-type semiconductor, since conduction is by free electrons excited into the conduction band.

16. A The most common purpose for heat-treating steel is the formation of martensite, the very hard and strong phase in iron-carbon alloys.

17. C At room temperature low-carbon steel is ferrite and cementite. However, the lever law, applied at this temperature range for low-carbon steel, would yield ferrite as the overwhelming majority phase.

18. A The equation is $N = 2^{n-1}$; therefore, for $n = 3$, $N = 4$, and for $n = 9$, $N = 256$. The number of grains per square inch in a 100× field of view increases from 4 to 256, an increase by a factor of 64.

19. C Age hardening, also known as precipitation hardening, involves the formation of a second phase by precipitation during an aging heat treatment for the purpose of strengthening. The other processes listed involve single phases only.

20. C When ferromagnetic domain boundaries migrate easily, the result is the partitioning of the piece into regions with a cancelling or almost cancelling of magnetic moments. The net magnetic moment in this case is therefore very weak.

11

Fluid Mechanics

Hameed Metghalchi, Sc.D.

Farzan Parsinejad, Ph.D.

Introduction

Fluid mechanics is the study of the changes that occur in a fluid that is subjected to forces. There are two major divisions within the field of fluid mechanics: fluid statics and fluid dynamics. **Fluid statics** is the study of the state of a fluid subjected to balanced forces such that the fluid remains in an equilibrium condition. **Fluid dynamics** is the study of fluid in motion and the interactions of the fluid with the objects it may encounter.

11.1
Definitions

As with any other field of study, fluid mechanics has a vocabulary of terms that must be understood before the basic principles can be introduced.

What Is a Fluid?

A **fluid** is a substance that will deform *continuously* when a shear stress is applied to it, no matter how small the shear stress is. A substance in a liquid form or vapor phase is a fluid.

Control Volume

All of the principles of fluid mechanics are defined in relation to a control volume. The **control volume** is an imaginary region in space defined by its boundaries, which may be stationary or moving. In general, material may cross a boundary of the control volume. Though the boundaries of a control volume are completely arbitrary, when possible they are chosen in a way that simplifies analysis.

Stress

The forces that act upon a fluid may be categorized as either body forces or surface forces. *Body forces* are forces that act on all fluid particles within the fluid and occur without physical contact. Examples include electromagnetic and gravitational forces. *Surface forces* are forces that are applied directly to the surface of the fluid through physical contact.

The total force per unit area acting on the fluid at any point within its volume is an important parameter, and has been given the name stress. The **stress** at any point P is defined as follows:

$$\sigma(P) = \lim_{\delta A \to 0} \frac{\delta \mathbf{F}}{\delta \mathbf{A}} \qquad [11.1]$$

where $\sigma(P)$ = stress at point P,
$\delta \mathbf{A}$ = an infinitesimal area at point P,
$\delta \mathbf{F}$ = the force acting on infinitesimal area $\delta \mathbf{A}$.

Both the area and the force in equation 11.1 are vector quantities; that is, they have magnitude and direction. These quantities are usually expressed by three components with respect to some coordinate system.

In general, to describe the stress of a fluid at any point with respect to a coordinate system, nine components are required. For convenience, stress is often expressed in terms of two vector components, in relation to the differential area ($\delta \mathbf{A}$). The component normal (or perpendicular) to $\delta \mathbf{A}$ is the normal stress (σ_n), and the component along $\delta \mathbf{A}$ is the shear stress (τ).

In a stationary fluid, the normal stress is the fluid pressure, whose magnitude is independent of the direction of the differential area used. The shear stress is related to fluid motion.

11.2
Fluid Properties

The **fluid properties** are the fluid characteristics whose values are used to predict the reaction of a fluid that is subjected to applied forces. The list that follows does not include all the possible fluid properties, but it defines the terms used throughout this chapter.

Velocity

Velocity is the time rate of change of the position of a fluid particle. The velocity of a fluid is a vector field quantity. For each point within the volume that makes up the fluid, a value can be assigned to the magnitude and direction of the fluid velocity. The velocity of a fluid will appear in three forms throughout this chapter. These forms are as follows:

- Velocity ($\mathbf{V}$) = the vector quantity of fluid velocity at a given point in the fluid. Both magnitude and direction are associated with this quantity.
- Speed (V) = the magnitude of the vector velocity at a given point in the fluid.
- Average speed ($\overline{V}$) = the average fluid speed through a surface of a control volume.

Mach Number

The **Mach number** (M_a) is a dimensionless parameter used to describe fluid speed. It is defined as the ratio of fluid speed to the speed of sound:

$$M_a = \frac{V}{c} \qquad [11.2]$$

where M_a = Mach number (dimensionless),
V = fluid speed (ft/sec or m/s),
c = speed of sound (ft/sec or m/s).

For a perfect gas, the speed of sound is defined as

$$c = \sqrt{kRT} \qquad [11.3]$$

where c = speed of sound (ft/sec or m/s),
k = specific heat ratio (c_p/c_v),
R = gas constant (ft-lbf/lb mole-°R or kJ/kg · K),
T = absolute temperature of the gas (K or °R).

In equation 11.3:

$$R = \frac{\overline{R}}{M} \qquad [11.4]$$

where $\overline{R}$ = the universal gas constant:
= 1,545 ft-lbf/lb mole-°R,
= 8.314 kJ/(mol · K),
M = molecular weight of the gas.

Example 11.1

What is the Mach number of a jet of oxygen gas at standard conditions (1 atm and 25°C) with a speed of 450 m/s? (Assume that for oxygen $k = 1.40$ and $R = 0.260$ kJ/kg · K.)

Solution:

Oxygen at standard condition can be modeled as a perfect gas; therefore, from equation 11.3, the speed of sound is as follows:

$$c = \sqrt{kRT}$$

$$= \sqrt{1.40\left(0.260 \, \frac{kj}{kg \cdot K}\right)[25 + 273K]\left(\frac{1,000N \cdot M}{kJ}\right)\left(\frac{\frac{kg \cdot m}{s^2}}{N}\right)}$$

$$= 329.35 \text{ m/s}$$

Now determine the Mach number directly from equation 11.2:

$$M_a = \frac{V}{c}$$

$$= \frac{450 \text{ m/s}}{329.35 \text{ m/s}}$$

$$= 1.37$$

Acceleration

Acceleration is the time rate of change in the velocity of a fluid particle. Acceleration is also a vector field quantity, and can act to change either the magnitude or the direction of the velocity.

Pressure

There are two reference values for pressure to which all other values of pressure are related. **Gauge pressure** is defined in reference to the ambient atmospheric pressure, and **absolute pressure** is defined in reference to true zero pressure. The two values for pressure are related by the equation:

$$p_a = p_g + p_{atm} \qquad [11.5]$$

where p_a = absolute pressure (lbf/in^2 = psia or N/m^2 = kPa),

p_g = gauge pressure (lbf/in^2 = psig or N/m^2 = kPa),

p_{atm} = ambient atmospheric pressure:
= 14.696 psia or 101.3 kPa at standard sea-level conditions.

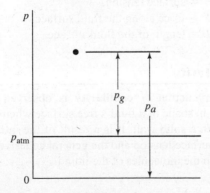

Figure 11.1 Definition of gauge pressure

Example 11.2

What will a pressure gauge read if the fluid in a vessel is at 200 psia? (Assume that the vessel is in an environment at standard sea-level pressure.)

> Most of the pressure measurement units measure the gauge pressure.

Solution:

From equation 11.5:

$$p_g = p_a - p_{atm}$$

$$= 200 \text{ psi} - 14.7 \text{ psi}$$

$$= 185.3 \text{ psig}$$

Density

The **density** of a fluid (ρ) is the mass contained in one unit of volume. The density of a fluid at a given location is expressed as

$$\rho = \lim_{\Delta V \to 0}\left(\frac{\Delta m}{\Delta V}\right) \qquad [11.6]$$

where ρ = fluid density (lbm/ft^3 or kg/m^3),

Δm = mass contained in infinitesimal volume ΔV (lbm or kg),

ΔV = infinitesimal volume (ft^3 or m^3).

Analysis of fluid systems is often simplified by assuming that the density is the same for every point within the volume of the fluid, as in incompressible flow analysis.

Specific Volume

The **specific volume** of a fluid (v) is the volume of space occupied by a unit mass of the fluid. It is the reciprocal of the density and is expressed as:

$$v = \frac{1}{\rho} \qquad [11.7]$$

where v = specific volume (ft^3/lbm or m^3/kg).

Specific Weight

The **specific weight** (γ) of a fluid is defined as the weight of a unit volume of fluid. It is related to the fluid density as follows:

$$\gamma = \rho g \qquad [11.8]$$

where γ = specific weight (lbf/ft^3 or N/m^3),
g = acceleration due to gravity:
= 32.17 ft/sec^2 or 9.81 m/s^2 at sea level.

Specific Gravity

Specific gravity is a dimensionless expression for the density of a fluid. Its value is the ratio of the fluid density to the density of liquid water at 4°C:

$$SG = \frac{\rho}{\rho_{water}} = \frac{\gamma}{\gamma_{water}} \qquad [11.9]$$

where ρ_{water} = 1,000 kg/m^3,
= 62.4 lbm/ft^3,
γ_{water} = 9,810 N/m^3,
= 62.4 lbf/ft^3.

Example 11.3

What is the density of an oil if its specific gravity is 0.86?

Solution:

From equation 11.9:

$$\rho = (SG)(\rho_{water})$$
$$= (0.86)(62.4 \text{ lbm/ft}^3)$$
$$= 53.7 \text{ lbm/ft}^3$$

Viscosity

When a shear force is applied to a fluid, fluid motion will occur. The magnitude of the fluid velocity is related to the magnitude of the shear stress. For most common fluids the shear stress within the fluid is directly proportional to the velocity gradient. **Viscosity** (μ) is the name given to the proportionality constant:

> Viscosity of a fluid shows how resistant the fluid is to motion.

$$\tau_n(P) = \mu \frac{\partial V}{\partial x_n} \qquad [11.10]$$

where $\tau_n(P)$ = shear stress on a differential area normal to direction n,
μ = fluid dynamic viscosity (described below),
$\dfrac{\partial V}{\partial x_n}$ = velocity gradient in direction n.

Fluids of this type are called *Newtonian fluids*. It should be recognized from equation 11.10 that, for a given shear stress, as the viscosity of a fluid increases, the magnitude of the velocity gradient decreases.

There are many fluids, such as toothpaste, blood, and many paints, that cannot be classified as Newtonian fluids. For many non-Newtonian fluids the shear stress can be related to the velocity gradient of the fluid through a power law relationship. Two examples of such non-Newtonian fluids are *psuedoplastic* fluids, for which the viscosity decreases as the velocity gradient increases, and *dilatant* fluids, for which the viscosity increases as the velocity gradient increases.

For a power law (non-Newtonian) fluid,

$$\tau_t(P) = K \left(\frac{\partial V}{\partial X_i} \right)^n \qquad [11.11]$$

where K = consistency index,
n = power law index: $n < 1$ for a psuedoplastic fluid, $n > 1$ for a dilatant fluid.

Surface Tension

At the surface of any fluid (or at the boundary between two fluids), there is a force that tends to hold the surface together. This force is called the **surface tension** of the fluid, which is defined as the tensile force between two points on the surface of the fluid, per unit length of distance between the two points:

$$\sigma = \frac{F}{L} \qquad [11.12]$$

where σ = surface tension,
F = force along the fluid surface,
L = length of the fluid surface.

Capillarity

Capillary action or **capillarity** is observed as the change in shape of a fluid's free surface when in contact with a solid wall. It is a result of the interaction of the surface tension and the general cohesive forces between the molecules of the liquid.

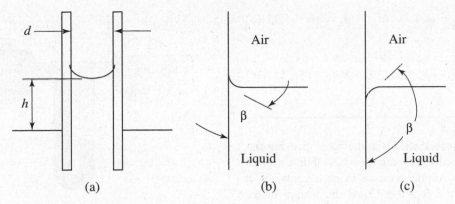

Figure 11.2 Illustration of a meniscus. (a) Meniscus in a small tube.
(b) Surface tension dominates cohesive forces. (c) Cohesive forces dominate surface tension.

The curved fluid surface that generally forms as a result of capillary action is called a *meniscus,* shown in Figure 11.2(a). If the area of a free surface is large, then, rather than forming a complete meniscus, the edges of the surface will be curved as shown in Figures 11.2(b) and (c). Whether or not the surface forms a complete meniscus, the slope of the curve at the wall of the tube is called the *angle of contact, (β)*. The angle of contact is an indication of the relationship between the surface tension and the cohesive forces of the liquid. If the angle is less than 90°, as in Figure 11.2(b), the surface tension is stronger than the cohesive forces; if the angle is greater than 90°, as in Figure 11.2(c), the cohesive forces are stronger than the surface tension.

When a small-diameter tube is placed in a fluid for which the surface tension is stronger than the molecular cohesive forces, the fluid level inside the tube will rise a certain distance, as shown in Figure 11.2(a). The height of this column is given by these equations:

$$h = \frac{4\sigma \cos \beta}{\rho d g} \qquad [11.13]$$

and

$$h = \frac{4\sigma \cos \beta}{\gamma d} \qquad [11.14]$$

where h and d are as shown in Figure 11.2(a).

11.3
Fluid Statics

Fluid statics is the study of fluid systems with no fluid motion (no shear stresses). The primary chal-

lenge of fluid statics is to determine the pressure distribution within a fluid.

Hydrostatic Pressure

When a fluid is not moving, the stress acting on each fluid particle becomes independent of the direction from which it is evaluated. In other words, for stress as defined in equation 11.1, the stress in a stationary fluid at some point has some value that is the same no matter in what direction the differential area (δA) is taken. This value is the **hydrostatic pressure** of the fluid at that point.

The hydrostatic pressure of a fluid is a function of height within the volume, regardless of the shape of the container. This is due to the fact that, at any point in a fluid, the fluid must support the weight of the fluid above it. For any two locations at the same height, the hydrostatic pressure will be the same. This is the principle behind the operation of two devices used to measure hydrostatic pressure, manometers and barometers.

For the control volume shown in Figure 11.3, the hydrostatic pressure at point B in the fluid is given by the equations:

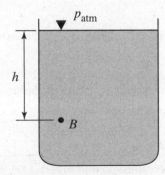

Figure 11.3 Hydrostatic pressure at point B.

$$p_g = \rho g h \quad \text{or} \quad p_g = \gamma h \quad [11.15]$$

and

$$p_a = p_{atm} + \rho g h \quad [11.16]$$

Example 11.4

The two vessels shown in the accompanying figure are filled with water ($\rho = 62.4$ lbm/ft^3). What is the hydrostatic pressure in each container at points A and B 2 feet below the surface of the water?

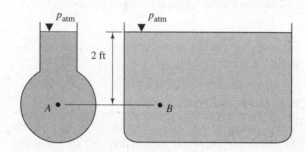

Solution:

Since the hydrostatic pressure in a fluid is a function of its depth only and is independent of the shape of the container, the pressure at point A will equal the pressure at point B.

From equation 11.15:

$$p_g = \rho g h$$

$$= \left(62.4\frac{\text{lbm}}{\text{ft}^3}\right)\left(32.17\frac{\text{ft}}{\text{s}^2}\right)$$

$$\times (2 \text{ ft})\left(\frac{\text{ft}^2}{144 \text{ in}^2}\right)\left(\frac{\text{lbf}}{32.17\frac{\text{lbm ft}}{\text{s}^2}}\right)$$

$$= 0.866 \text{ lbf/in}^2 = 0.866 \text{ psig}$$

The term $\left(\dfrac{\text{lbf}}{32.17\frac{\text{lbm ft}}{\text{s}^2}}\right)$ in the above solution

represents a unit conversion between pounds mass and pounds feet. This is the inverse of the quantity g_c.

Manometry

A **manometer**, illustrated in Figure 11.4, is a device used to measure a pressure difference. It consists of a U-shaped tube with each end open to one of the two environments whose pressure we wish to compare.

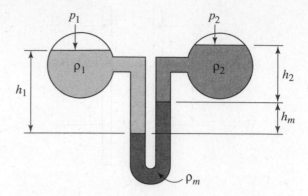

Figure 11.4 Diagram of a manometer, used to measure the difference in hydrostatic pressure between vessel 1 and vessel 2.

The difference in pressure between these two environments is related to the difference in height between the two legs of the manometer as follows:

$$p_1 - p_2 = \rho_m g h_m + (\rho_2 g h_2 - \rho_1 g h_1) \quad [11.17]$$

or

$$p_1 - p_2 = \gamma_m h_m + (\gamma_2 h_2 - \gamma_1 h_1) \quad [11.18]$$

If the density or specific weight of either fluid 1 or fluid 2 in Figure 11.4 is small compared to the density or specific weight of the manometer fluid, its contribution to equations 11.17 and 11.18 can be neglected. For example, if fluid 1 is air (density = 1.23 kg/m^3), fluid 2 is water (density = 1,000 kg/m^3), and the manometer fluid is oil (density = approx. 830 kg/m^3), then the effect of the air can be neglected, and equations 11.17 and 11.18 can be reduced to

$$p_1 - p_2 = \rho_m g h_m + \rho_2 g h_2 \quad [11.19]$$

and

$$p_1 - p_2 = \gamma_m h_m + \gamma_2 h_2 \quad [11.20]$$

If, however, the densities or specific weights of *both* fluids are small compared to the density or specific weight of the manometer fluid, equations 11.17 and 11.18 reduce to

$$p_1 - p_2 = \rho_m g h_m \quad [11.21]$$

and

$$p_1 - p_2 = \gamma_m h_m \quad [11.22]$$

When one end of the manometer is exposed directly to the atmosphere, the device is called an open manometer, illustrated in Figure 11.5. Equations 11.17 and 11.18 apply to an open manometer if p_2 is replaced with atmospheric pressure p_{atm}.

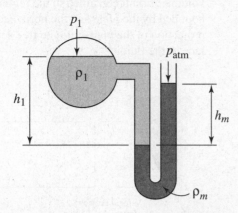

Figure 11.5 Diagram of an open manometer.

For the manometer shown in Figure 11.5, the pressure at the surface of fluid 1(p_1) is given by

$$p_1 = p_{atm} + \rho_m g h_m - \rho_1 g h_1 \qquad [11.23]$$

and

$$p_1 = p_{atm} + \gamma_m h_m - \gamma_1 h_1 \qquad [11.24]$$

Example 11.5

The open manometer shown in the accompanying figure is used to measure the pressure (p_1) in an air tank. The manometer fluid is mercury, which has a density of 13,550 kg/m³. What is the pressure in the vessel?

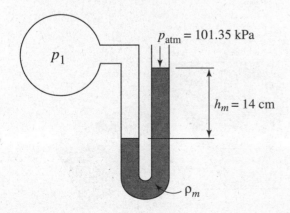

Solution:

The density of air is much smaller than the density of mercury, so the contribution of air can be neglected. From equation 11.23:

> Density of air at standard pressure and temperature (1 atm, 0°C) is 1.168 kg/m³. Density of mercury at the same conditions is 13,534 kg/m³.

$$p_1 = p_{atm} + \rho_m g h_m$$

$$= 101.35 \text{ kPa} + \left(13{,}550 \ \frac{\text{kg}}{\text{m}^3}\right)\left(9.81 \ \frac{\text{m}}{\text{s}^2}\right)$$

$$\times (14 \text{ cm})\left(\frac{\text{m}}{100 \text{ cm}}\right)\left(\frac{\text{N}}{\frac{\text{kg} \cdot \text{m}}{\text{s}^2}}\right)\left(\frac{\text{kPa}}{1{,}000 \ \frac{\text{N}}{\text{m}^2}}\right)$$

$$= 119.96 \text{ kPa}$$

Barometers

A **barometer,** illustrated in Figure 11.6, is a device used to measure atmospheric pressure. Atmospheric pressure acts on the exposed surface of the barometer fluid to maintain the height of a column of the fluid in a tube. The pressure in the empty section at the top of the tube is equal to the vapor pressure of the barometer fluid at the ambient temperature.

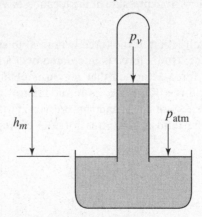

Figure 11.6 Diagram of a barometer.

The height of the fluid column is related to the atmospheric pressure as follows:

$$p_{atm} - p_v = \rho g h_m \qquad [11.25]$$

or

$$p_{atm} - p_v = \gamma h_m \qquad [11.26]$$

where p_v = vapor pressure of the manometer fluid at the ambient temperature.

The barometer fluid is usually chosen so that $p_v \approx 0$.

Forces on Submerged Surfaces

At any location on the surface of a submerged object, the pressure at that point is proportional to the point's distance below the surface. This pressure is determined as follows:

$$p = p_0 + \rho g z \qquad [11.27]$$

or

$$p = p_0 + \gamma z \qquad [11.28]$$

where p = pressure at the given location,
p_0 = pressure at the surface of the fluid,
z = vertical distance between the point and the surface of the fluid.

For an arbitrary region on the surface of a submerged object, each differential area, $d\mathbf{A}$, is acted upon by a force expressed as

$$d\mathbf{F} = p\, d\mathbf{A} \qquad [11.29]$$

where $d\mathbf{A}$ = differential area on the surface of a submerged object,
$d\mathbf{F}$ = force acting on $d\mathbf{A}$,
p = fluid pressure at the location of $d\mathbf{A}$.

The direction of the force is normal to and into the surface. If this force is integrated over a region, the total force acting on that region is obtained. If this region is a flat plate, as in the control volume shown in Figure 11.7, and the pressure at the surface of the fluid is zero, this force is expressed as follows:

$$\mathbf{F} = \mathbf{F}_x + \mathbf{F}_y \qquad [11.30]$$

$$|\mathbf{F}_x| = \frac{p_1 + p_2}{2} A_v \qquad [11.31]$$

$$|\mathbf{F}_y| = \rho g V_f \qquad [11.32]$$

where $\mathbf{F}$ = total force exerted on the plate by the fluid, directed normal into the surface of the plate,
$|\mathbf{F}_x|$ = magnitude of the horizontal component of $\mathbf{F}$,
$|\mathbf{F}_y|$ = magnitude of the vertical component of $\mathbf{F}$,

p_1 = gauge pressure at the top edge of the plate,
p_2 = gauge pressure at the bottom edge of the plate,
A_v = vertical projection of the plate area,
V_f = volume of fluid contained in the region bounded by the plate and the horizontal projection of the plate onto the free surface of the fluid.

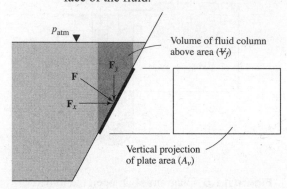

Figure 11.7 A flat-plate region of a submerged surface.

Equations 11.30 through 11.32 apply when the contribution of the atmospheric pressure to the force on a submerged surface is negligible. When this effect is not negligible, the gauge pressures in equation 11.31 should be replaced with absolute pressures, and a surface-force term ($P_{atm}A_h$) should be added to equation 11.32. In this term, A_h is the horizontal projection of the submerged plate onto the free surface of the fluid.

Example 11.6

Find the total force acting on the submerged plate shown in the accompanying figure. Assume that the back side of the plate is also exposed to p_{atm}.

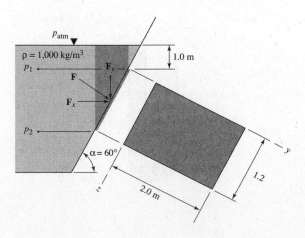

Solution:

The total force acting on the plate is directed normal to the plate, and into the plate, as shown in the figure. The magnitude of this force can be determined from equation 11.30:

$$\mathbf{F} = \mathbf{F}_x + \mathbf{F}_y$$

From equation 11.31:

Note that pressure increases linearly with depth.

$$|\mathbf{F}_x| = \frac{p_1 + p_2}{2} A_v$$

$$p_1 = \rho g h_1$$

$$= \left(1{,}000 \ \frac{kg}{m^3}\right)\left(9.81 \ \frac{m}{s^2}\right)(1.0 \ m)$$

$$\times \left(\frac{N}{\frac{kb \cdot m}{s^2}}\right)\left(\frac{kPa}{1{,}000 \ \frac{N}{m^2}}\right)$$

$$= 9.81 \ kPa$$

$$p_2 = \rho g h_2$$

$$h_2 = 1.0 \ m + (1.2 \ m) \sin 60$$

$$= 2.04 \ m$$

$$p_2 = \left(1{,}000 \ \frac{kg}{m^3}\right)\left(9.81 \ \frac{m}{s^2}\right)(2.04 \ m)$$

$$\times \left(\frac{N}{\frac{kg \cdot m}{s^2}}\right)\left(\frac{kPa}{1{,}000 \ \frac{N}{m^2}}\right)$$

$$= 20.01 \ kPa$$

$$A_v = (2.0 \ m)[(1.2 \ m) \sin 60]$$

$$= 2.08 \ m^2$$

$$|\mathbf{F}_x| = \frac{9.81 \ kPa + 20.01 \ kPa}{2}$$

$$\times (2.08 \ m^2)\left(\frac{1{,}000 \ \frac{N}{m^2}}{kPa}\right)$$

$$= 31{,}010 \ N$$

From equation 11.33:

$$|\mathbf{F}_y| = \rho g V_f$$

$$V_f = \left\{ [(1.0 \ m)(1.2 \ m)\cos 60] \right.$$

$$\left. + \frac{[(1.2 \ m)\cos 60][(1.2 \ m)\sin 60]}{2} \right\}(2.0 \ m)$$

$$= 1.82 \ m^3$$

$$|\mathbf{F}_y| = \left(1{,}000 \ \frac{kg}{m^3}\right)\left(9.81 \ \frac{m}{s^2}\right)(1.82 \ m^3)\left(\frac{N}{\frac{kg \cdot m}{s^2}}\right)$$

$$= 17{,}850 \ N$$

The total force can be expressed as

$$|\mathbf{F}| = \sqrt{|\mathbf{F}_x|^2 + |\mathbf{F}_y|^2}$$

$$= \sqrt{(31{,}010 \ N)^2 + (17{,}850 \ N)^2}$$

$$= 35{,}780 \ N$$

The Center of Pressure

For a region on the surface of a submerged rigid body, the **center of pressure** is the location in this region through which a concentrated point force (**F** from the preceding section) will have the same effect on the submerged body as the distributed pressure force over the entire region. This definition is illustrated in Figure 11.8.

The coordinates' center of pressure relative to the centroid of the submerged area (y^* and z^*) is given by the following equations:

$$y^* = \frac{\rho g I_{y_c z_c} \sin \alpha}{F}, \qquad z^* = \frac{\rho g I_{y_c} \sin \alpha}{F} \quad [11.33]$$

or

$$y^* = \frac{\gamma I_{y_c z_c} \sin \alpha}{F}, \qquad z^* = \frac{\gamma I_{y_c} \sin \alpha}{F} \quad [11.34]$$

where I_{y_c} − area moment of inertia about an axis parallel to the y-axis and passing through the centroid of the area. The value of I_{y_c} is given by the equation

$$I_{y_c} = I_y - A y_c^2 \quad [11.35]$$

from the parallel axis theorem, where

$$I_y = \int_A y^2 \, dA \qquad [11.36]$$

In equations 11.33 and 11.34, $I_{y_c z_c}$ is the area product of inertia about an axis parallel to the y- and z-axes and passing through the centroid of the area. The value of $I_{y_c z_c}$ is given by the equation

$$I_{y_c z_c} = I_{yz} - A y_c z_c \qquad [11.37]$$

from the parallel axis theorem, where

$$I_{yz} = \int_A xy \, dA \qquad [11.38]$$

Values for the moment of inertia and product of inertia for several common shapes can be found in the dynamics chapter (Chapter 8) of this book.

Example 11.7

Determine the center of pressure for the submerged surface of Example 11.6, as described in the figure for that example.

Solution:

From equation 11.33:

$$y^* = \frac{\rho g I_{y_c z_c} \sin \alpha}{F}, \qquad z^* = \frac{\rho g I_{y_c} \sin \alpha}{F}$$

$$I_{y_c z_c} = I_{yz} - A y_c z_c$$

$$= \frac{b^2 h^2}{4} - (bh)\left(\frac{b}{2}\right)\left(\frac{h}{2}\right)$$

$$= \frac{b^2 h^2}{4} - \frac{b^2 h^2}{4}$$

$$= 0$$

---

y^* and z^* can be used in calculating moment of pressure force about a point or line, i.e., a hinge on a surface.

$$y^* = 0 \text{ m}$$

$$I_{y_c} = I_y - A y_c^2$$

$$= \frac{b^3 h}{3} - (bh)\left(\frac{b}{2}\right)^2$$

$$= \frac{b^3 h}{3} - \frac{b^3 h}{4}$$

$$= \frac{b^3 h}{12}$$

$$= \frac{(0.20 \text{ m})^3 (0.12 \text{ m})}{12}$$

$$= 8 \times 10^{-5} \text{ m}^4$$

$$z^* =$$

$$\frac{\left(1,000 \, \frac{\text{kg}}{\text{m}^3}\right)\left(9.81 \, \frac{\text{m}}{\text{s}^2}\right)\left(8 \times 10^{-5} \text{ m}^4\right) \sin 60}{36.06 \text{ N}}$$

$$\times \left(\frac{\frac{\text{N}}{\text{kg} \cdot \text{m}}}{\text{s}^2}\right)$$

$$= 0.0188 \text{ m}$$

$$= 1.88 \text{ cm}$$

Archimedes' Principle and Buoyancy

Any object that is completely or partially submerged will experience a buoyant force. This is a force directed in a direction opposite to the weight of the object, which is the result of the total pressure force of the fluid acting on the submerged body.

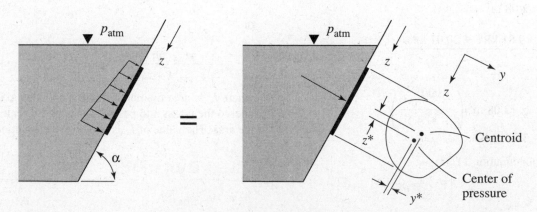

Figure 11.8 Distributed pressure force and equivalent point force on a submerged surface.

According to **Archimedes' principle,** the magnitude of the buoyant force is equal to the weight of the fluid displaced by the object. If the object is floating on the surface of a fluid, it will displace a volume of the fluid that has the same weight as the object.

Example 11.8

A 25-kg cube is submerged in water ($\rho = 1,000$ kg/m^3), as shown in part (a) of the accompanying figure. What is the magnitude of the buoyant force acting on the cube?

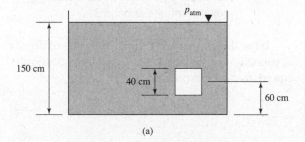

(a)

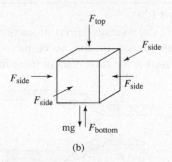

(b)

Solution:

From the free-body diagram in part (b) of the figure, it can be seen that the force acting on each side of the cube is counteracted by the force on the opposite side; therefore the total force acting on the cube must be directed vertically. Take forces directed up as positive; then the total resultant force is given by the equation

$$F_{total} = F_{bottom} - F_{top} - mg$$

Calculate the force on each face of the cube by multiplying the hydrostatic pressure on that face by the area of the face:

$$F_{bottom} = A_{bottom}(\rho g h)_{water}$$

$$= (40 \text{ cm})^2\left(\frac{m}{100 \text{ cm}}\right)^2\left[\left(1,000 \frac{kg}{m^3}\right)\left(9.81 \frac{m}{s^2}\right)\right.$$

$$\left.\times (110 \text{ cm})\left(\frac{m}{100 \text{ cm}}\right)\left(\frac{N}{\frac{kg \cdot m}{s^2}}\right)\right]$$

$$= 1,727 \text{ N}$$

$$F_{top} = A_{top}(\rho g h)_{water}$$

$$= (40 \text{ cm})^2\left(\frac{m}{100 \text{ cm}}\right)^2\left[\left(1,000 \frac{kg}{m^3}\right)\left(9.81 \frac{m}{s^2}\right)\right.$$

$$\left.\times (70 \text{ cm})\left(\frac{m}{100 \text{ cm}}\right)\left(\frac{N}{\frac{kg \cdot m}{s^2}}\right)\right]$$

$$= 1,099 \text{ N}$$

The weight of the cube is

$$mg = (25 \text{ kg})\left(9.81 \frac{m}{s^2}\right)\left(\frac{N}{\frac{kg \cdot m}{s^2}}\right)$$

$$= 245 \text{ N}$$

The total force is

$$F_{total} = F_{bottom} - F_{top} - mg$$

$$= 1,727 \text{ N} - 1,099 \text{ N} - 245 \text{ N}$$

$$= 383 \text{ N}$$

As an alternative to the approach shown above, Archimedes' principle may be used to solve the problem. The cube will experience a buoyant force equal to the weight of the water it displaces:

> In general, V_{cube} is replaced with total wetted volume of an object in fluid.

$$F_{buoyant} = \rho_{water} V_{cube} g$$

$$= \left(1,000 \frac{kg}{m^3}\right)(0.40 \text{ m})^3\left(9.81 \frac{m}{s^2}\right)\left(\frac{N}{\frac{kg \cdot m}{s^2}}\right)$$

$$= 628 \text{ N}$$

The weight of the cube will counteract the buoyant force, so the total resultant force acting on the cube is

$$F_{total} = F_{buoyant} - mg$$

$$= 628 \text{ N} - 245 \text{ N}$$

$$= 383 \text{ N}$$

11.4
Fluid Dynamics

Fluid dynamics is the study of fluid in motion and the interactions of the fluid with the objects it may encounter. The basic principles of mass and energy conservation and Newton's second law of motion are applied to fluid systems to allow prediction of the flow characteristics.

Conservation of Mass (the Continuity Equation)

In every fluid system, mass is conserved. This conservation can be stated in terms of a mass balance equation:

$$\left(\frac{dm}{dt}\right)_{cv} = \dot{m}_{in} - \dot{m}_{out} \qquad [11.39]$$

A common class of fluid problem is the steady-state problem, in which the values of all of the fluid properties within the control volume do not change with time. For a steady-state control volume:

$$\left(\frac{dm}{dt}\right)_{cv} = 0 \qquad [11.40]$$

so that

$$\dot{m}_{in} = \dot{m}_{out} \qquad [11.41]$$

For the arbitrary control volume shown in Figure 11.9, conservation of mass is expressed as

$$\dot{m}_1 = \dot{m}_2 \qquad [11.42]$$

where $\dot{m}_1$ = mass flow rate through cross-sectional area of control volume at point 1,
$\dot{m}_2$ = mass flow rate through cross-sectional area of control volume at point 2.

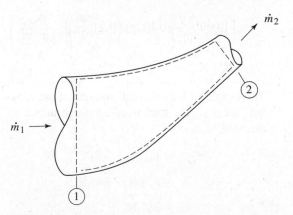

Figure 11.9 Conservation of mass in an arbitrary control volume.

Equivalent expressions of equation 11.42 are as follows:

$$\rho_1 A_1 \overline{V}_1 = \rho_2 A_2 \overline{V}_2 \qquad [11.43]$$

and

$$\rho_1 Q_1 = \rho_2 Q_2 \qquad [11.44]$$

where A_1 = cross-sectional area of control volume at point 1,
A_2 = cross-sectional area of control volume at point 2,
$\overline{V}_1$ = average fluid speed at point 1,
$\overline{V}_2$ = average fluid speed at point 2,
Q_1 = volumetric flow rate of fluid entering control volume at point 1,
Q_2 = volumetric flow rate of fluid entering control volume at point 2.

If the flow is incompressible, the value of density is constant, and conservation of mass, or **continuity equation,** can be expressed as:

$$Q = A_1 \overline{V}_1 = A_2 \overline{V}_2 \qquad [11.45]$$

Example 11.9
What is the average speed at section 2 for the diffuser shown in the accompanying figure? (Assume that the flow is incompressible.)

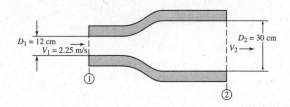

Solution:
From equation 11.40:

$$A_1 \overline{V}_1 = A_2 \overline{V}_2$$

$$\overline{V}_2 = \frac{A_1}{A_2}\overline{V}_1 = \frac{D_1^2}{D_2^2}\overline{V}_1$$

$$= \frac{(0.12\ \text{m})^2}{(0.30\ \text{m})^2}\left(2.25\ \frac{\text{m}}{\text{s}}\right)$$

$$= 0.36\ \text{m/s}$$

Impulse-Momentum Principle

Newton's second law of motion ($\mathbf{F} = m\mathbf{a}$) can be applied to fluid systems through a principle known as the **impulse-momentum principle.** This principle is expressed mathematically for a control volume as follows:

$$\Sigma\left(\frac{d\mathbf{P}}{dt}\right)_{cv} = \Sigma\mathbf{F} \qquad [11.46]$$

where $\dfrac{d\mathbf{P}}{dt}$ = rate of momentum entering or leaving the control volume = $\dot{m}\mathbf{V}$,

$\Sigma\mathbf{F}$ = total external force acting on the control volume.

For a steady-state system, the change in momentum of a control volume can be expressed as

$$\left(\frac{d\mathbf{P}}{dt}\right)_{cv} = \left(\frac{d\mathbf{P}}{dt}\right)_{in} - \left(\frac{d\mathbf{P}}{dt}\right)_{out}$$
$$= (\dot{m}\mathbf{V})_{in} - (\dot{m}\mathbf{V})_{out} \qquad [11.47]$$

so that

$$\Sigma\mathbf{F} = (\dot{m}\mathbf{V})_{in} - (\dot{m}\mathbf{V})_{out} \qquad [11.48]$$

The impulse-momentum principle can be applied to the case of a pipe bend, enlargement, or contraction to find the total force exerted on a fluid by such a device.

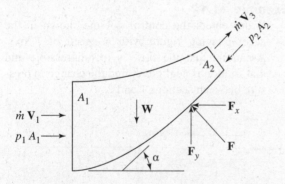

Figure 11.10 Free-body diagram of fluid in a general pipe bend, enlargement, or contraction.

Equation 11.46 can be applied to the control volume described in Figure 11.10 to produce these equations:

These relationships are valid for both compressible and incompressible fluids/ flows.

$$|\mathbf{F}_x| = (\dot{m}\overline{V}_1 + p_1A_1) + (\dot{m}\overline{V}_2 - p_2A_2)\cos\alpha$$
$$[11.49]$$
$$|\mathbf{F}_y| = \mathbf{W} - (\dot{m}\overline{V}_2 - p_2A_2)\sin\alpha \qquad [11.50]$$

Example 11.10

For the control volume shown in Figure 11.10 with the parameters listed below, find the *x*- and *y*-components of the resultant force on the control volume.

$\rho = 1{,}000 \text{ kg/m}^3 \qquad\qquad \overline{V}_1 = 4 \text{ m/s}$

$V_{cv} = 1.25 \text{ m}^3 \qquad p_1 - 12 \text{ kPa} \qquad p_2 = 10 \text{ kPa}$

$\alpha = 40° \qquad\qquad A_1 = 0.50 \text{ m}^2 \qquad A_2 = 0.30 \text{ m}^2$

Solution:

From equation 11.49:

$$|\mathbf{F}_x| = (\dot{m}\overline{V}_1 + p_1A_1) + (\dot{m}\overline{V}_2 - p_2A_2)\cos\alpha$$
$$\dot{m} = \rho\overline{V}_1A_1$$
$$= \left(1{,}000\ \frac{\text{kg}}{\text{m}^3}\right)\left(4\ \frac{\text{m}}{\text{s}}\right)(0.50\ \text{m}^2)$$
$$= 2{,}000\ \text{kg/s}$$
$$\overline{V}_2 = \frac{\overline{V}_1A_1}{A_2}$$
$$= \frac{\left(4\ \dfrac{\text{m}}{\text{s}}\right)(0.50\ \text{m}^2)}{0.30\ \text{m}^2}$$
$$= 6.67\ \text{m/s}$$
$$|\mathbf{F}_x| = \left[\left(2{,}000\ \frac{\text{kg}}{\text{s}}\right)\left(4\ \frac{\text{m}}{\text{s}}\right)\left(\frac{\text{N}}{\frac{\text{kg}\cdot\text{m}}{\text{s}^2}}\right)\right.$$
$$\left. + (12{,}000\ \text{Pa})(0.50\ \text{m}^2)\left(\frac{\text{N}}{\frac{\text{m}^2}{\text{Pa}}}\right)\right]$$

$$+ \left[\left(2{,}000\ \frac{\text{kg}}{\text{s}}\right)\left(6.67\ \frac{\text{m}}{\text{s}}\right)\left(\frac{\text{N}}{\frac{\text{kg}\cdot\text{m}}{\text{s}^2}}\right)\right.$$
$$\left. - (10{,}000\ \text{Pa})(0.30\ \text{m}^2)\left(\frac{\text{N}}{\frac{\text{m}^2}{\text{Pa}}}\right)\right]\cos 40$$
$$= (8{,}000\ \text{N} + 6{,}000\ \text{N})$$
$$\qquad + (13{,}333\ \text{N} - 3{,}000\ \text{N})\cos 40$$
$$= 21{,}915\ \text{N}$$

From equation 11.50:

$$|\mathbf{F}_y| = |\mathbf{W}| - (\dot{m}\overline{V}_2 - p_2 A_2)\sin \alpha$$

$$|\mathbf{W}| = \rho V_{cv} g$$

$$= \left(1{,}000 \ \frac{kg}{m^3}\right)(1.25 \ m^3)\left(9.81 \ \frac{m}{s^2}\right)\left(\frac{N}{\frac{kg \cdot m}{s^2}}\right)$$

$$= 12{,}263 \ N$$

$$|\mathbf{F}_y| = 12{,}263 \ N - (13{,}333 \ N - 3{,}000 \ N)\sin 40$$

$$= 5{,}621 \ N$$

Conservation of Energy (the Bernoulli or Field Equation)

Where there is continuous (steady) flow, the idea of conservation of energy can be expressed using the Bernoulli equation. The **Bernoulli equation,** known also as the **field equation,** is an expression of the conservation of energy, neglecting frictional losses and is applied between two sections on a streamline. The Bernoulli equation is expressed as:

$$\frac{p_1}{\gamma} + \frac{\overline{V}_1^{\,2}}{2g} + z_1 = \frac{p_2}{\gamma} + \frac{\overline{V}_2^{\,2}}{2g} + z_2 \quad [11.51]$$

Any real flow will experience frictional losses, which are not accounted for in the Bernoulli equation. Equation 11.51 can be modified to include the effect of frictional losses as follows:

$$\frac{p_1}{\gamma} + \frac{\overline{V}_1^{\,2}}{2g} + z_1 = \frac{p_2}{\gamma} + \frac{\overline{V}_2^{\,2}}{2g} + z_2 + h_f$$
$$[11.52]$$

Here, h_f is the major head loss, and its value is determined using empirical relationships. For flow in a horizontal pipe with constant cross-sectional area, equation 11.52 reduces to

$$h_f = \frac{p_1 - p_2}{\gamma} \quad [11.53]$$

Example 11.11

For the system described in Example 11.10, find the change in pressure between sections 1 and 2 using the Bernoulli equation. (Assume that frictional losses are negligible.)

Solution:
From equation 11.51:

> Note that points 1 and 2 are both on the same level, i.e., $Z_1 = Z_2$.

$$\frac{p_1}{\rho} + \frac{\overline{V}_1^{\,2}}{2} = \frac{p_2}{\rho} + \frac{\overline{V}_2^{\,2}}{2}$$

$$p_2 - p_1 = \frac{(\overline{V}_1^{\,2} - \overline{V}_2^{\,2})\rho}{2}$$

$$= \frac{\left[\left(2.25 \ \frac{m}{s}\right)^2 - \left(0.36 \ \frac{m}{s}\right)^2\right]\left(1{,}000 \ \frac{kg}{m^3}\right)}{2}$$

$$\times \left(\frac{N}{\frac{kg \cdot m}{s^2}}\right)\left(\frac{kPa}{1{,}000 \ \frac{N}{m^2}}\right)$$

$$= 2.47 \ kPa$$

Example 11.12

Water enters the control volume shown in the accompanying figure with a speed of 5 m/s. Assuming that the flow is incompressible and that friction is negligible, find the change in pressure between sections 1 and 2.

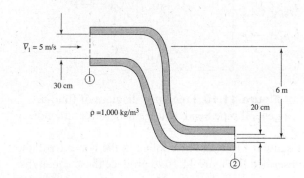

Solution:
From equation 11.51:

$$\frac{p_1}{\rho g} + \frac{\overline{V}_1^{\,2}}{2g} + z_1 = \frac{p_2}{\rho g} + \frac{\overline{V}_2^{\,2}}{2g} + z_2$$

$$p_2 - p_1 = \rho g\left[\left(\frac{\overline{V}_1^{\,2} - \overline{V}_2^{\,2}}{2g}\right) + (z_1 - z_2)\right]$$

$$= \frac{\rho(\overline{V}_1^{\,2} - \overline{V}_2^{\,2})}{2} + \rho g(z_1 - z_2)$$

Calculate $\overline{V}_2$ using the continuity equation (as expressed in equation 11.45):

$$A_1\overline{V}_1 = A_2\overline{V}_2$$

$$\overline{V}_2 = \frac{A_1}{A_2}\overline{V}_1 = \frac{D_1^2}{D_2^2}\overline{V}_1$$

$$= \frac{(0.30\text{ m})^2}{(0.20\text{ m})^2}\left(5\,\frac{\text{m}}{\text{s}}\right)$$

$$= 11.25\text{ m/s}$$

The change in pressure is

$$p_2 - p_1 = \frac{\rho(\overline{V}_1^2 - \overline{V}_2^2)}{2} + \rho g(z_1 - z_2)$$

$$= \frac{\left(1,000\,\frac{\text{kg}}{\text{m}^3}\right)\left[\left(5\,\frac{\text{m}}{\text{s}}\right)^2 - \left(11.25\,\frac{\text{m}}{\text{s}}\right)^2\right]}{2}$$

$$+ \left(1,000\,\frac{\text{kg}}{\text{m}^3}\right)\left(9.81\,\frac{\text{m}}{\text{s}^2}\right)(6\text{ m})$$

$$= \left(-50,781\,\frac{\text{kg}}{\text{m}\cdot\text{s}^2} + 41,880\,\frac{\text{kg}}{\text{m}\cdot\text{s}^2}\right)$$

$$\times \left(\frac{\text{N}}{\frac{\text{kg}\cdot\text{m}}{\text{s}^2}}\right)\left(\frac{\text{kPa}}{1,000\,\frac{\text{N}}{\text{m}^2}}\right)$$

$$= -8.90\text{ kPa}$$

Static Versus Stagnation Pressure

The pressure used in the Bernoulli equation is the **static pressure** of a fluid. It is the pressure that would be measured by a device that moves with the fluid. A stationary device that measures pressure in a moving fluid would read a different pressure, known as the **stagnation pressure**. A relationship between the static pressure and the stagnation pressure of a fluid can be derived from the Bernoulli equation:

$$\frac{p_0}{\rho} = \frac{p_s}{\rho} + \frac{V^2}{2} \qquad [11.54]$$

or

$$p_0 = p_s + \frac{1}{2}\rho V^2 \qquad [11.55]$$

where p_s = static pressure of the fluid,
$\quad p_0$ = stagnation pressure of the fluid,
$\quad V$ = speed of the fluid at the point where the measurement is taken.

Grade Line (Hydraulic Gradient) and Energy Line

If a pipe is fitted with a series of open manometers along its length, the static fluid pressure at each location can be visualized. With flowing fluid, the pressure at each successive location will be lower, as shown in Figure 11.11. If an imaginary line is drawn through the meniscus of each manometer, that line is the **grade line** or **hydraulic gradient**.

Alternatively, the hydraulic gradient can be thought of as a plot of static pressure as a function of distance along the pipe. If the stagnation pressure rather than the static pressure of a fluid flowing through a pipe is plotted as a function of distance along the pipe, the line that results is the **energy line**.

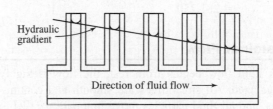

Figure 11.11 Description of the hydraulic gradient.

Reynolds Number

The **Reynolds number** (Re) is a nondimensional number used to characterize fluid flow. It is a ratio of the inertial forces to viscous forces acting on the fluid and is defined as follows:

$$\text{Re} = \frac{\rho\overline{V}D}{\mu} \qquad [11.56]$$

or

$$\text{Re} = \frac{\overline{V}D}{\nu} \qquad [11.57]$$

where ν = kinematic viscosity, defined as μ/ρ.

The parameter D used in equations 11.56 and 11.57 is a characteristic length. For flow in a tube the characteristic length is the pipe diameter. For flow over a flat plate, the characteristic length is the length of the plate in the direction of flow.

For a power-law non-Newtonian fluid, the Reynolds number is defined as

$$\text{Re}' = \frac{\rho\overline{V}^{2-n}D^n}{K\left(\frac{3n+1}{4n}\right)^n 8^{n-1}} \qquad [11.58]$$

where K = consistency index,
$\quad n$ = power law index: $n < 1$ for a pseudoplastic fluid, $n > 1$ for a dilatant fluid.

Example 11.13

Find the Reynolds number for the water exiting the control volume of Example 11.12. Assume that the viscosity of water is 8×10^{-3} N · s/m^2.

Solution:

From equation 11.56:

$$\text{Re} = \frac{\rho \overline{V} D}{\mu}$$

$$= \frac{\left(1,000 \, \frac{\text{kg}}{\text{m}^3}\right)\left(11.25 \, \frac{\text{m}}{\text{s}}\right)(0.20 \, \text{m})}{8 \times 10^{-3} \, \frac{\text{N} \cdot \text{s}}{\text{m}^2}}$$

$$= 281{,}250$$

Laminar Flow

As fluid flow begins from rest, the flow is highly organized. All streamlines are smooth and continuous, and all fluid particles move in a path parallel to the mean bulk flow. This type of flow, called **laminar flow,** generally occurs in fluid flows with low Reynolds numbers, or in fluids with high viscosity and/or low density flowing at low speed. In laminar flow, the fluid flow field can be thought of as a collection of fluid layers. In each layer, all fluid particles are moving with the same velocity, and there is no macroscopic mixing of fluid layers.

Turbulent Flow

As the flow rate is increased, the general character of the flow begins to change significantly. Random, three-dimensional fluctuations in flow velocity will begin to appear, superimposed over the local mean flow velocity. This type of flow, called **turbulent flow,** results in significant fluid mixing.

The lowest Reynolds number for which turbulent flow can be expected is called the *critical Reynolds number* (Re$_c$). For flow in a pipe, the value of Re$_c$ is approximately 2,300. For pipe flow at Reynolds numbers greater than approximately 4,000, the flow is considered fully turbulent. The range of Reynolds numbers between these limits is the transitional region, where it is difficult to predict which type of flow will occur.

11.5 Internal Flow

In this section we consider fluid flowing through a closed pipe or duct. Experimental observation has shown that fluid particles adjacent to the walls of the pipe have zero velocity. This is called the *no-slip condition* (no fluid is slipping against the wall of the pipe). The maximum fluid velocity occurs along the centerline of the pipe.

The variation in fluid velocity is caused by the viscous forces within the fluid, or the shear force between fluid particles. The shear forces perform work against the moving fluid, reducing the fluid energy as it travels through the pipe. This loss of energy due to friction, often called *head loss* (h_f), is observed as a drop in fluid pressure in the pipe.

As fluid enters or leaves a pipe or other internal passage, the velocity profile in the passage, is a function of the fluid's location along the length of the passage and the geometry of the opening. Far from the entrance and exit, however, the velocity profile becomes uniform for any position along the length of the pipe, and independent of the entrance and exit geometry. Flow in this region is fully developed.

For fully developed laminar flow, illustrated in Figure 11.12, the velocity of a fluid particle in a pipe is a function of the radial location of that particle in the pipe and of the pressure drop through the tube ($\Delta p/L$). Some relationships for flow in a pipe are as follows:

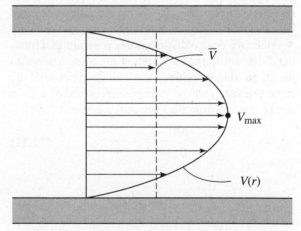

Figure 11.12 Fully developed laminar-flow velocity profile for flow in a pipe.

- Speed distribution:

$$V(r) = -\frac{R^2}{4\mu}\frac{\Delta p}{L}\left[1 - \left(\frac{r}{R}\right)^2\right] \quad [11.59]$$

- Average speed:

$$\overline{V} = \frac{Q}{A} = -\frac{R^2}{8\mu}\frac{\Delta p}{L} \quad [11.60]$$

- Maximum speed:

$$V_{max} = -\frac{R^2}{4\mu}\frac{\Delta p}{L} = 2\overline{V} \quad [11.61]$$

- Volumetric flow rate:

$$Q = -\frac{\pi R^4}{8\mu}\frac{\Delta p}{L} \quad [11.62]$$

- Shear stress distribution:

$$\tau(r) = \frac{r}{2}\frac{\Delta p}{L} \quad [11.63]$$

and

$$\frac{\tau(r)}{\tau_w} = \frac{r}{R} \quad [11.64]$$

where r = radial coordinate of a coordinate system with its origin on the centerline of the pipe,

R = inside radius of the pipe.

Darcy Equation

To estimate the frictional losses for flow in a horizontal pipe with a constant cross-sectional area, the Bernoulli equation (equation 11.51) reduces to

$$p_1 - p_2 = \rho g h_f = \gamma h_f \quad [11.65]$$

The frictional head loss is estimated using the **Darcy equation:**

$$h_f = f\frac{L}{D}\frac{\overline{V}^2}{2g} \quad [11.66]$$

For laminar flow:

$$f = \frac{64}{Re}$$

The friction factor (f) for transitional or turbulent flow can be found in the Moody or Stanton diagram, a copy of which is included at the end of this chapter. Another parameter required for use of the Moody diagram is the relative roughness (e/D). The roughness (e) of a pipe is a function of the material and the process by which the pipe is manufactured. Some examples of roughness values are given in Table 11.1.

Table 11.1 Roughness Values for Selected Materials

Material	e (ft)	e (mm)
Riveted steel	0.003–0.03	0.9–9.0
Concrete	0.001–0.01	0.3–3.0
Cast iron	0.00085	0.25
Galvanized iron	0.0005	0.15
Galvanized steel or wrought iron	0.000015	0.046
Drawn tubing	0.000005	0.0015

Example 11.14

Using the Darcy equation, calculate the head loss in a horizontal pipe of constant cross-sectional area if the inside diameter of the pipe is 6 in and its length is 200 ft. Water with a density of 62.4 lbm/ft^3 and a viscosity of 1.67×10^{-5} lbf-s/ft^2 flows into the pipe at a speed of 2 ft/s. The roughness of the pipe is 0.01 in. (Assume that the flow is incompressible.)

Solution:

To use the Darcy equation, first calculate the Reynolds number and relative roughness.

$$Re = \frac{\rho \overline{V} D}{\mu}$$

$$= \frac{\left(62.4\,\frac{lbm}{ft^3}\right)\left(2\,\frac{ft}{s}\right)(6\,in)\left(\frac{ft}{12\,in}\right)}{1.67 \times 10^{-5}\,\frac{lbf\text{-}s}{ft^2}}$$

$$\times \left(\frac{lbf}{32.17\,\frac{lbm\text{-}ft}{s^2}}\right)$$

$$= 9{,}670$$

$$\frac{e}{D} = \frac{0.01\,in}{6\,in}$$

$$= 0.00167$$

Use the Moody diagram to determine that the friction factor is 0.024. Now use the Darcy equation to find the head loss.

Reynolds number tells us whether we should look up the friction factor in the laminar or turbulent section on the Moody diagram. In this example, the flow is laminar.

$$h_f = f \cdot \frac{L}{D} \cdot \frac{\overline{V}^2}{2g}$$

$$= 0.024 \left[\frac{200 \text{ ft}}{6 \text{ in} \left(\frac{\text{ft}}{12 \text{ in}} \right)} \right] \left[\frac{\left(2 \frac{\text{ft}}{\text{s}} \right)^2}{2 \left(32.17 \frac{\text{ft}}{\text{s}^2} \right)} \right]$$

$$= 0.596 \text{ ft}$$

Hagen-Poiseuille Equation

Pressure loss for laminar flow in a circular pipe can be related to volumetric flow rate by using the **Hagen-Poiseuille equation:**

$$Q = \frac{\pi R^4 \, \Delta p_f}{8 \mu L} \qquad [11.67]$$

or

$$Q = \frac{\pi D^4 \, \Delta p_f}{128 \mu L} \qquad [11.68]$$

where Q = volumetric flow rate (ft³/sec or m³/s),
R = inside radius of the pipe,
D = inside diameter of the pipe,
Δp_f = pressure drop in the pipe due to frictional losses.

Head Losses in Pipe Bends, Enlargements, Contractions, Entrances, and Exits

Minor head losses (h_{fm}) occur any time that the size or direction of a pipe changes. The value of the minor head loss can be found by using the following equation:

$$h_{fm} = C \frac{\overline{V}^2}{2g} \qquad [11.69]$$

The loss coefficient (C) is determined by the geometry of the change and can vary greatly. A table is commonly used to determine the loss coefficient. Figure 11.13 gives the loss coefficients for some types of entrances and exits.

The Bernoulli equation is modified as follows to determine the change in energy of a fluid for flow through a pipe, including both major and minor head losses:

$$\frac{p_1}{\gamma} + \frac{\overline{V}_1^2}{2g} + z_1 = \frac{p_2}{\gamma} + \frac{\overline{V}_2^2}{2g} + z_2 + h_f + h_{fm}$$

$$[11.70]$$

Example 11.15

If an elbow with loss coefficient $C = 0.75$ is placed at the end of the pipe of Example 11.14, what is the minor head loss through the elbow?

Solution:
From equation 11.69:

$$h_{fm} = C \frac{\overline{V}^2}{2g}$$

$$= 0.75 \left[\frac{\left(2 \frac{\text{ft}}{\text{s}} \right)^2}{2 \left(32.17 \frac{\text{ft}}{\text{s}^2} \right)} \right]$$

$$= 0.047 \text{ ft}$$

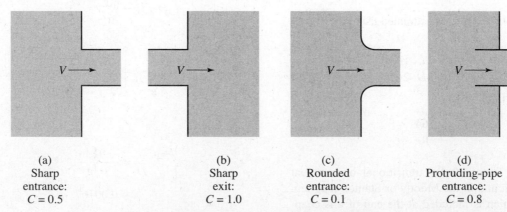

(a)	(b)	(c)	(d)
Sharp entrance:	Sharp exit:	Rounded entrance:	Protruding-pipe entrance:
$C = 0.5$	$C = 1.0$	$C = 0.1$	$C = 0.8$

Figure 11.13 Loss coefficients for several types of pipe entrances and exits.

Multiple-Path Pipeline

A common technique used to increase the capacity of a pipe is to add another pipe in parallel.

It is possible to predict how much fluid will flow through each path. The fluid pressures before and after the branch must be equal. Therefore, the flow will divide itself in such a way that the head losses through each path will be equal.

For the multiple-path pipeline shown in Figure 11.14:

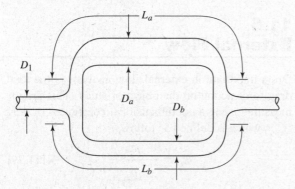

Figure 11.14 Diagram of a multiple-path pipeline.

$$h_{fa} = h_{fb} \qquad [11.71]$$

$$f_a \frac{L_a}{D_a} \frac{\overline{V}_a^2}{2g} = f_b \frac{L_b}{D_b} \frac{\overline{V}_b^2}{2g} \qquad [11.72]$$

From the continuity equation (equation 11.45):

$$\frac{\pi}{4} D_1^2 \overline{V}_1 = \frac{\pi}{4} D_a^2 \overline{V}_a + \frac{\pi}{4} D_b^2 \overline{V}_b \qquad [11.73]$$

When the flow rate of the fluid entering the control volume and the diameters and lengths of the pipes are known, equations 11.72 and 11.73 can be solved simultaneously to determine $\overline{V}_a$ and $\overline{V}_b$.

Example 11.16

For the multiple-path pipeline shown in the accompanying figure, what is the average speed through pipe A ($\overline{V}_A$)? (Assume that the friction factor in pipe A is equal to the friction factor in pipe B.)

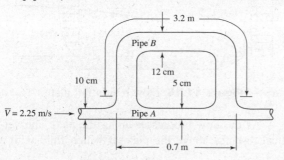

Solution:

From equation 11.72:

$$f_A \frac{L_A}{D_A} \frac{\overline{V}_A^2}{2g} = f_B \frac{L_B}{D_B} \frac{\overline{V}_B^2}{2g}$$

$$\frac{\overline{V}_A}{\overline{V}_B} = \sqrt{\frac{D_A}{D_B} \frac{L_B}{L_A}}$$

$$\sqrt{\frac{5 \text{ cm}}{12 \text{ cm}} \frac{3.2 \text{ m}}{0.7 \text{ m}}}$$

$$= 1.38$$

From equation 11.73:

$$\overline{V}A = \overline{V}_A A_A + \overline{V}_B A_B$$

$$= \overline{V}_A A_A + \frac{\overline{V}_A}{1.38} A_B$$

$$= \overline{V}_A \left(A_A + \frac{A_B}{1.38} \right)$$

$$\overline{V}_A = \overline{V} \left(\frac{\frac{\pi D^2}{4}}{\frac{\pi D_A^2}{4} + \frac{\pi D_B^2}{4(1.38)}} \right)$$

$$= \overline{V} \left(\frac{D^2}{D_A^2 + \frac{D_B^2}{1.38}} \right)$$

$$= 2.25 \frac{\text{m}}{\text{s}} \left[\frac{(10 \text{ cm})^2}{(5 \text{ cm})^2 + \frac{(12 \text{ cm})^2}{1.38}} \right]$$

$$= 1.74 \text{ m/s}$$

Flow in Noncircular Ducts

Flow in noncircular ducts can be analyzed using the same techniques used for round ducts. All that is needed is to replace the tube diameter with a new parameter called the *hydraulic radius* (R_h) or *hydraulic diameter* (D_h). The equations for this parameter are as follows:

$$R_h = \frac{\text{cross-sectional area}}{\text{wetted perimeter}} \qquad [11.74]$$

$$D_h = 4\left(\frac{\text{cross-sectional area}}{\text{wetted perimeter}}\right) = 4R_h \quad [11.75]$$

For instance, the Reynolds number for flow in a noncircular duct is expressed as:

$$Re_{D_h} = \frac{\rho \overline{V} D_h}{\mu} \qquad [11.76]$$

Example 11.17

Find the hydraulic diameter of a rectangular duct 24 in wide and 15 in high.

Solution:

$$D_h = 4\left(\frac{\text{cross-sectional area}}{\text{wetted perimeter}}\right)$$

$$= 4\left[\frac{wh}{2(w + h)}\right]$$

$$= 4\left[\frac{(24 \text{ in})(15 \text{ in})}{2(24 \text{ in} + 15 \text{ in})}\right]$$

$$= 18.46 \text{ in}$$

Open-Channel Flow

The relationships developed until now have been for flow inside a closed pipe. Most of these, such as the Darcy friction factor, have been determined through experimentation. As the basic geometry of the fluid system to be analyzed changes, we can expect to need new empirical relationships. Two such relationships for open-channel (there is free surface) flow are Manning's equation and the Hazen-Williams equation.

Manning's equation:

$$\overline{V} = \frac{1}{n}R^{2/3}S^{1/2} \qquad [11.77]$$

Hazen-Williams equation:

$$\overline{V} = 0.849CR^{0.63}S^{0.54} \qquad [11.78]$$

where n = Manning roughness coefficient,
   C = Hazen-Williams roughness coefficent,
   R = hydraulic radius of the channel (ft or m),
   S = slope of the energy grade line (ft/ft or m/m).

11.6
External Flow

Often important in external flow analysis is the total drag force acting on the object of study. This drag is measured using a nondimensional coefficient of drag (C_D), which is defined as follows:

$$C_D = \frac{F_D}{\frac{1}{2}\rho V_\infty^2 A} \qquad [11.79]$$

where C_D = nondimensional coefficient of drag,
   F_D = total drag force exerted on the object by the fluid,
   ρ = density of the fluid,
   V_∞ = speed of the fluid far from the object,
   A = area over which the drag force is acting.

If the coefficient of drag is given, the drag force can be calculated using this equation:

$$F_D = \frac{C_D\rho V_\infty^2 A}{2} \qquad [11.80]$$

Uniform Flow over a Flat Plate

Flow over a flat plate demonstrates a concept important to the study of fluid dynamics, the boundary layer. As the fluid flow encounters the first edge of the plate, the vertical flow distribution is uniform, as shown in Figure 11.15.

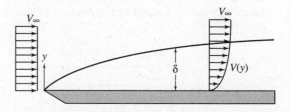

Figure 11.15 Flow over a flat plate.

As the flow progresses along the plate, the vertical distance above the plate at which fluid with a speed of V_∞ is first encountered increases. This distance is called the boundary layer thickness (δ). In the boundary layer, the fluid velocity and the shear stress in the fluid vary with distance above the surface of the plate.

The characteristic length used for uniform flow over a flat plate is the length of the plate. The Reynolds number for flow over a flat plate is expressed as follows:

$$Re_L = \frac{\rho V_\infty L}{\mu} \qquad [11.81]$$

The coefficient of drag for flow over a flat plate is given by these equations:

$$C_D = \frac{1.33}{Re_L^{0.5}} \quad (10^4 < Re_L < 5 \times 10^5)$$

and $\qquad\qquad\qquad\qquad\qquad\qquad$ [11.82]

$$C_D = \frac{0.031}{Re_L^{1/7}} \quad (10^6 < Re_L < 10^9)$$

Example 11.18

A square plate with 50 cm on a side is placed in an airstream with a speed of 12 m/s.

A. What is the coefficient of drag for the plate?

B. What is the total drag force on the plate? (Assume that the density of air is 1.225 kg/m³, and that the viscosity is 2×10^{-5} N · s/m².)

Solution:

A. From equation 11.81:

$$Re_L = \frac{\rho V_\infty L}{\mu}$$

$$= \frac{\left(1.225 \frac{kg}{m^3}\right)\left(12 \frac{m}{s}\right)(0.50\,m)}{2 \times 10^{-5} \frac{N \cdot s}{m^2}}\left(\frac{N}{\frac{kg \cdot m}{s^2}}\right)$$

$$= 3.67 \times 10^5$$

From equation 11.82:

$$C_D = \frac{1.33}{Re_L^{0.5}}$$

$$= \frac{1.33}{(3.67 \times 10^5)^{0.5}}$$

$$= 2.19 \times 10^{-3}$$

B. The total force is given by equation 11.80:

$$F_D = \frac{C_D \rho V_\infty^2 A}{2}$$

$$= \frac{(2.19 \times 10^{-3})\left(1.225 \frac{kg}{m^2}\right)\left(12 \frac{m}{s}\right)^2 (0.50\,m)^2}{2}$$

$$\times \left(\frac{N}{\frac{kg \cdot m}{s^2}}\right)$$

$$= 4.84 \times 10^{-2}\,N$$

11.7 Devices

The principles of fluid mechanics discussed so far can be used to describe the operation of many industrial devices. A few examples follow.

Deflector and Blades

Many devices use the kinetic energy of a high-velocity fluid to produce power, which is commonly transmitted to a series of curved surfaces or blades. The impulse-momentum principle is used to describe their operation. We will neglect the viscous drag of the fluid at the blade surface; therefore, the fluid leaving the trailing edge of the blade will have the same speed as the fluid approaching the leading edge of the blade.

Fixed Blade

Figure 11.16 shows a fluid jet acting on a stationary blade. Since the blade does not move, the direction of the fluid stream that leaves the blade is determined by the angle of the trailing edge of the blade. The blade's reaction force is related to the change in momentum of the fluid. It is assumed that the speed of the fluid stream leaving the blade is equal to the speed of the fluid stream entering the blade.

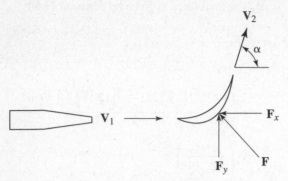

Figure 11.16 Fluid jet flowing over a stationary blade.

$$|V_2| = |V_1| \quad [11.83]$$

$$|F_x| = \dot{m}(|V_1| - |V_2| \cdot \cos \alpha) \quad [11.84]$$

$$|F_y| = \dot{m}|V_2| \cdot \sin \alpha \quad [11.85]$$

where $|V_1|$ = magnitude of the fluid velocity approaching the blade,

$|V_2|$ = magnitude of the fluid velocity leaving the blade.

Example 11.19

Find the total reaction force on a fixed blade if 5 kg/s of water with a speed of 6 m/s enters the blade, and the blade angle α is 65°.

Solution:

From equation 11.84:

$$|F_x| = \dot{m}(|V_1| - |V_2| \cdot \cos \alpha)$$

$$= \left(5 \frac{kg}{s}\right)\left[6 \frac{m}{s} - 6 \frac{m}{s}(\cos 65)\right]\left(\frac{N}{\frac{kg \cdot m}{s^2}}\right)$$

$$= 17.32 \text{ N}$$

From equation 11.85:

$$|F_y| = \dot{m}|V_2| \cdot \sin \alpha$$

$$= \left(5 \frac{kg}{s}\right)\left(6 \frac{m}{s}\right)\sin 65 \left(\frac{N}{\frac{kg \cdot m}{s^2}}\right)$$

$$= 27.19 \text{ N}$$

The magnitude of the reaction force is

$$|F| = \sqrt{|F_x|^2 + |F_y|^2}$$

$$= \sqrt{(17.32 \text{ N})^2 + (27.19 \text{ N})^2}$$

$$= 32.23 \text{ N}$$

Moving Blade

When the blade is moving in the same direction as the fluid jet, as is the case in the rotating portions of turbomachinery, the velocity of the blade affects the change in fluid momentum that occurs. This situation is illustrated in Figure 11.17.

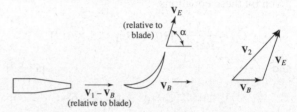

Figure 11.17 Fluid jet flowing over a moving blade.

The relationships between the velocities and the forces on the blade are given by the following equations:

$$V_E = V_1 - V_B \quad [11.86]$$

$$V_2 = V_E + V_B \quad [11.87]$$

$$|F_x| = -\dot{m}(|V_1| - |V_B|)(1 - \cos \alpha) \quad [11.88]$$

$$|F_y| = \dot{m}(|V_1| - |V_B|) \cdot \sin \alpha \quad [11.89]$$

where V_1 = velocity of the fluid jet approaching the blade,

V_B = velocity of blade,

V_E = velocity of fluid leaving the blade, relative to the blade,

V_2 = true velocity of fluid leaving the blade.

Example 11.20

Find the total reaction force on a moving blade if 5 kg/s of water with a speed of 6 m/s enters the blade, the blade is moving at a speed of 1.4 m/s, and the blade angle α is 65°.

Solution:

From equation 11.88:

$$|F_x| = \dot{m}(|V_1| - |V_B|)(1 - \cos \alpha)$$

$$= \left(5 \frac{kg}{s}\right)\left(6 \frac{m}{s} - 1.4 \frac{m}{s}\right)(1 - \cos 65)\left(\frac{N}{\frac{kg \cdot m}{s^2}}\right)$$

$$= 13.28 \text{ N}$$

From equation 11.89:

$$|\mathbf{F}_y| = \dot{m}(|\mathbf{V}_1| - |\mathbf{V}_B|)(\sin\alpha)$$

$$= \left(5\,\frac{\text{kg}}{\text{s}}\right)\left(6\,\frac{\text{m}}{\text{s}} - 1.4\,\frac{\text{m}}{\text{s}}\right)(\sin 65)\left(\frac{\text{N}}{\frac{\text{kg}\cdot\text{m}}{\text{s}^2}}\right)$$

$$= 20.85\,\text{N}$$

The magnitude of the reaction force is:

$$|\mathbf{F}| = \sqrt{|\mathbf{F}_x|^2 + |\mathbf{F}_y|^2}$$

$$= \sqrt{(13.28\,\text{N})^2 + (20.85\,\text{N})^2}$$

$$= 24.72\,\text{N}$$

Impulse Turbine

An **impulse turbine,** shown schematically in Figure 11.18, is a disk with blades located radially around its edge. Fluid strikes the blades in a stream tangent to the turbine. The reaction force of the fluid against the blades begins to turn the turbine around its axis, and this motion can be used to generate power.

The amount of power generated is related to the angle of blade deflection (α). Maximum power is generated at $\alpha = 180°$. The turbine power is related to the mass flow rate of the fluid and to its velocity by the equation

$$\dot{W} = Q\rho(|\mathbf{V}_1 - \mathbf{V}_B|)(1 - \cos\alpha)|\mathbf{V}_B| \qquad [11.90]$$

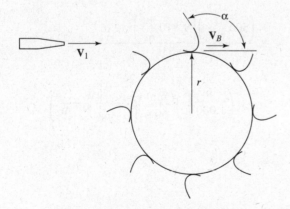

Figure 11.18 An impulse turbine.

Maximum power is expressed as

$$\dot{W}_{\max} = Q\rho\left(\frac{|\mathbf{V}_1|^2}{4}\right)(1 - \cos\alpha) \qquad [11.91]$$

> These relationships are valid for both compressible and incompressible fluids/flows.

or, when $\alpha - 180°$:

$$\dot{W}_{\max} = \frac{Q\rho|\mathbf{V}_1|^2}{2} \qquad [11.92]$$

The speed of the blade is related to the rotational speed of the turbine by the equation

$$|\mathbf{V}_B| = r\omega \qquad [11.93]$$

where r = radius of the turbine (ft or m),
  ω = rotational speed of the turbine (rad/s).

Example 11.21

Find the power of an impulse turbine if 25 kg/s of water leaves the nozzle with a speed of 65 m/s, and the turbine is spinning at 500 rpm. The radius of the turbine is 50 cm, and the blade angle α is 120°.

Solution:

The rotational speed of the turbine is

$$\omega = \left(500\,\frac{\text{rev}}{\text{min}}\right)\left(\frac{2\pi\,\text{rad}}{\text{rev}}\right)\left(\frac{\text{min}}{60\,\text{s}}\right)$$

$$= 52.36\,\text{rad/s}$$

From equation 11.93, the blade velocity is

$$|\mathbf{V}_B| = r\omega$$

$$= (0.50\,\text{m})\left(52.36\,\frac{\text{rad}}{\text{s}}\right)$$

$$= 26.18\,\text{m/s}$$

From equation 11.90:

$$\dot{W} = \dot{m}(|\mathbf{V}_1 - \mathbf{V}_B|)(1 - \cos\alpha)|\mathbf{V}_B|$$

$$= \left(25\,\frac{\text{kg}}{\text{s}}\right)\left(65\,\frac{\text{m}}{\text{s}} - 26.18\,\frac{\text{m}}{\text{s}}\right)(1 - \cos 120)$$

$$\times \left(26.18\,\frac{\text{m}}{\text{s}}\right)\left(\frac{\text{N}}{\frac{\text{kg}\cdot\text{m}}{\text{s}^2}}\right)\left(\frac{\text{kW}}{1{,}000\,\frac{\text{N}\cdot\text{m}}{\text{s}}}\right)$$

$$= 38.11\,\text{kW}$$

Jet Propulsion

A fluid jet used for propulsion is an example of the practical use of the impulse-momentum principle. The momentum of the fluid jet leaving the control volume results in a force that acts on the control volume in a direction opposite to that of the fluid jet. In the system described in Figure 11.19, the potential energy of the fluid is converted to kinetic energy as the fluid flows out of the orifice. As the fluid is expelled, it leaves the system with a certain amount of momentum. As a result, a force is exerted on the control volume in the direction opposite to the flow out of the orifice. This force is given by the equation

$$F = Q\rho(\overline{V} - 0) \quad [11.94]$$

or

$$F = \rho\overline{V}^2 A \quad [11.95]$$

where F = force on the control volume,
$\overline{V}$ = average speed of the fluid through the nozzle,
A = cross-sectional area of the nozzle exit.

By applying the Bernoulli equation between the free surface and the outlet of this system, we calculate the force as follows:

$$\frac{p_1}{\gamma} + (z_2 + h) + \frac{0^2}{2g} = \frac{p_1}{\gamma} + z_2 + \frac{\overline{V}^2}{2g} \quad [11.96]$$

$$\overline{V}^2 = 2gh \quad [11.97]$$

$$F = 2\rho ghA \quad [11.98]$$

Pumps

Pumps use mechanical energy to increase the energy of a fluid. The amount of energy that is added to the fluid per unit time is expressed as

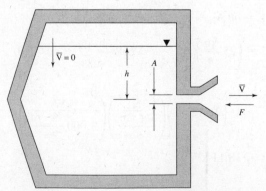

Figure 11.19 Jet of fluid exiting a container.

$$P = Q\gamma h \quad [11.99]$$

where P = pump power (ft-lbf/sec or W),
Q = volumetric flow rate of fluid that is pumped (ft³/sec or m³/s),
h = change in fluid head (ft or m).

Because of inefficiencies in the pump, the amount of energy that must be supplied to the pump is somewhat greater than the amount of energy that is transferred to the fluid. The ratio of power transferred to the fluid to power supplied to the pump is the pump's efficiency (η). The amount of power that must be supplied to a pump can be found by using this equation:

$$P_{\text{req}} = \frac{Q\gamma h}{\eta} \quad [11.100]$$

Example 11.22

A pump that is 63% efficient is required to lift 500 L/min of water from a stream into a tank that is 12 m above the surface of the stream. What is the minimum power input required for this pump?

Solution:
From equation 11.100:

$$P_{\text{req}} = \frac{Q\gamma h}{\eta}$$

$$= \frac{\left(500 \,\frac{\text{L}}{\text{min}}\right)\left(9,810 \,\frac{\text{N}}{\text{m}^3}\right)(12 \,\text{m})}{0.63}$$

$$\times \left(\frac{\text{m}^3}{1,000 \,\text{L}}\right)\left(\frac{\text{min}}{60 \,\text{s}}\right)\left(\frac{\text{kW}}{1,000 \,\frac{\text{N} \cdot \text{m}}{\text{s}}}\right)$$

$$= 1.56 \,\text{kW}$$

11.8
Fluid Measurements

A number of devices have been developed to allow measurement of fluid velocity and flow rate (Q). A few of the most common types are described below.

Pitot Tube

A **Pitot tube** is a device used to measure the stagnation pressure of a moving fluid. It is a tube with a small-diameter hole at the end that is pointed directly into a flowing fluid, as shown in Figure 11.20.

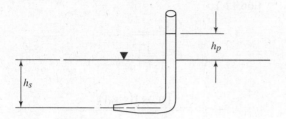

Figure 11.20 Illustration of a Pitot-tube flow-measurement device.

The speed of the fluid is computed from the stagnation pressure using one of these equations:

$$V = \sqrt{\frac{2\left(p_0 - p_s\right)}{p}} \qquad [11.101]$$

or

$$V = \sqrt{\frac{2g\left(p_0 - p_s\right)}{\gamma}} \qquad [11.102]$$

For the system shown in Figure 11.20, the static pressure and the stagnation pressure at the opening of the Pitot tube are given by the following relationships:

$$p_s = \rho g h_s + p_{atm} = \gamma h_s + p_{atm} \qquad [11.103]$$

and

$$p_0 = p_s + \rho g h_p \qquad [11.104]$$

By combining equations 11.102 and 11.104, the fluid speed can be related directly to h_p:

$$h_p = \frac{V^2}{2g} \qquad [11.105]$$

Although equation 11.101 was developed for an incompressible fluid, it may also be used with reasonable accuracy for a compressible fluid if the Mach number is less than approximately 0.3.

Example 11.23

A. For the Pitot tube shown in Figure 11.20, what is the stagnation pressure of the fluid if $h_p = 13.6$ cm and $h_s = 8$ cm? [Assume that the fluid is water ($\rho = 1,000$ kg/m³), and that $p_{atm} = 101.3$ kPa.]

B. What is the fluid speed?

Solution:

A. From equation 11.103:

$$p_s = \rho g h_s + p_{atm}$$

$$= \left(1,000 \ \frac{kg}{m^3}\right)\left(9.81 \ \frac{m}{s^2}\right)(8 \ cm)$$

$$\times \left(\frac{m}{100 \ cm}\right)\left(\frac{N}{\frac{kg \cdot m}{s^2}}\right)\left(\frac{kPa}{1,000 \ \frac{N}{m^2}}\right) + 101.3 \ kPa$$

$$= 102.1 \ Pa$$

From equation 11.104:

$$p_0 = p_s + \rho g h_p$$

$$= 102.1 \ kPa + \left(1,000 \ \frac{kg}{m^3}\right)\left(9.81 \ \frac{m}{s^2}\right)(13.6 \ cm)$$

$$\times \left(\frac{m}{100 \ cm}\right)\left(\frac{N}{\frac{kg \cdot m}{s^2}}\right)\left(\frac{kPa}{1,000 \ \frac{N}{m^2}}\right)$$

$$= 103.4 \ kPa$$

B. From equation 11.101:

$$V = \sqrt{\frac{2\left(p_0 - p_s\right)}{p}}$$

$$= \sqrt{\frac{2\left(103.4 \ kPa - 102.1 \ kPa\right)\left(1,000 \ \frac{N}{m^2}}{kPa}\right)\left(\frac{kg \cdot m}{\frac{s^2}{N}}\right)}{1,000 \ \frac{kg}{m^3}}}$$

$$= 1.61 \ m/s$$

Orifice Flow Meter

The **orifice flow meter,** illustrated in Figure 11.21, is a flat plate, with a small hole in the center, placed in a pipe. Pressure measurements are taken at fixed locations upstream and downstream of the orifice. The flow rate of the fluid is related to the difference in value between the upstream and downstream pressure measurements.

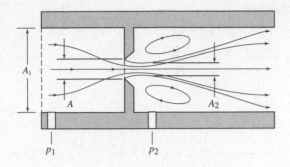

Figure 11.21 Diagram of an orifice flow meter.

The volumetric flow rate through an orifice flow meter is given by the following equation:

$$Q = CA \sqrt{2g\left[\left(\frac{P_1}{\gamma} + z_1\right) - \left(\frac{P_2}{\gamma} + z_2\right)\right]} \qquad [11.106]$$

The flow meter coefficient (C) is given by the equation

$$C = \frac{C_v C_c}{\sqrt{1 - C_c^2 \left(\frac{A}{A_1}\right)^2}} \qquad [11.107]$$

The coefficients C_v and C_c are given in Table 11.2.

Table 11.2 Various Types of Orifices and Their Nominal Coefficient Values

	Sharp Edge	Rounded	Short Tube	Borda
$V \to$		$V \to$	$V \to$	$V \to$
Minimum value of C (meter coefficient)	0.61	0.98	0.80	0.51
C_c (coefficient of contraction)	0.62	1.00	1.00	0.52
C_v (coefficient of velocity)	0.98	0.98	0.80	0.98

An orifice flow meter provides a simple and efficient means to measure the flow rate. Unfortunately, however, this simplicity comes at a price; the sudden contraction of the flow produces large head losses.

Example 11.24

Water flows through a rounded orifice flow meter in a pipe having a 20-cm diameter. The orifice is in a horizontal section of the pipe and has an area of 0.03 m². The pressure just upstream of the orifice is 75 kPa, and the pressure just downstream is 15 kPa. What is the volumetric flow rate of the water?

Solution:

First determine the meter coefficient. From equation 11.107:

$$C = \frac{C_v C_c}{\sqrt{1 - C_c^2 \left(\frac{A}{A_1}\right)^2}}$$

$$= \frac{(0.98)(1.00)}{\sqrt{1 - (1.00)^2 \left(\frac{0.03\,\text{m}^2}{\pi (0.25\,\text{m})^2}\right)}}$$

$$= 1.065$$

Then use equation 11.106:

$$Q = CA \sqrt{2\left(\frac{p_1 - p_2}{\rho}\right)}$$

$$= (1.065)(0.03\,\text{m}^2) \times$$

$$\sqrt{2\left[\left(\frac{75\,\text{kPa} - 15\,\text{kPa}}{1{,}000\,\frac{\text{kg}}{\text{m}^3}}\right)\right]\left(1{,}000\,\frac{\text{N}}{\text{m}^2}\right)\left(\frac{\text{kg} \cdot \text{m}}{\text{s}^2}\right)}$$

$$= 0.35\,\text{m}^3/\text{s}$$

Venturi Flow Meter

Like the orifice flow meter, the **venturi flow meter**, illustrated in Figure 11.22, uses a pressure drop across a flow restriction of known geometry to measure the flow rate.

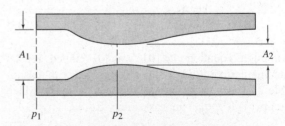

Figure 11.22 Diagram of a venturi flow meter.

The volumetric flow rate through a venturi flow meter is given by the following equation:

$$Q = \frac{C_v A_2}{\sqrt{1 - \left(A_2/A_1\right)^2}} \sqrt{2g\left[\left(\frac{p_1}{\gamma} + z_1\right) - \left(\frac{p_2}{\gamma} + z_2\right)\right]}$$

$$[11.108]$$

The coefficient of velocity (C_v) is obtained through experimentation.

The gradual contraction of the venturi flow meter largely reduces the head loss to the fluid stream. The shape of this device, however, is much more complicated to manufacture than an orifice.

Submerged Orifice

The rate of flow through a submerged orifice, illustrated in Figure 11.23, is related to the hydrostatic pressure at each side of the orifice as shown in equation 11.106 or 11.107.

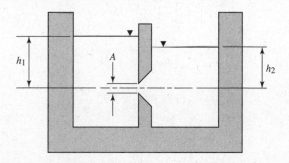

Figure 11.23 An orifice submerged below the surface of a fluid.

$$Q = C_c C_v A \sqrt{2g(h_1 - h_2)} \qquad [11.109]$$

or

$$Q = CA \sqrt{2g(h_1 - h_2)} \qquad [11.110]$$

Orifice Discharging into the Atmosphere

If the fluid on the downstream side of an orifice discharges into the ambient atmosphere, as illustrated in Figure 11.24, the atmosphere's contribution to equations 11.108 and 11.109 can be neglected, and the flow rate can be expressed as follows:

$$Q = CA\sqrt{2gh} \qquad [11.111]$$

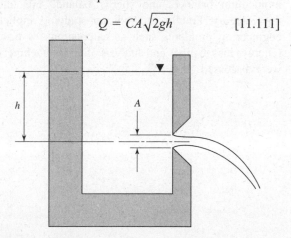

Figure 11.24 Diagram of an orifice discharging into the atmosphere.

11.9
Dimensional Analysis — Dimensional Homogeneity

A dimensionally homogeneous relationship is one whose value is independent of the fundamental dimensions used (the units used to measure mass, length, or time). Dimensional homogeneity is assured for relationships that are placed in a dimensionless form. The *Buckingham PI* theorem tells how many dimensionless groups are required to describe a relationship. It states that, if the relationship has n independent variables, involving r different fundamental dimensions (mass, length, or time), then $n - r$ dimensionless groups are required.

11.10
Similitude

Model testing plays a large role in fluid dynamic analysis. The reason is that the equations of fluid flow are sufficiently complex that it is impracticable to solve them for all but the simplest flow conditions. Furthermore, testing of a full-size prototype is often prohibitively expensive.

We can be assured that the model we use for testing predicts accurately the fluid interaction of the full-scale prototype if we can prove that there is similarity between the two. Three kinds of similarity are required:

- *Geometric similarity:* The model must have the same shape as the full-scale prototype. Each of the geometric parameters that describe the shape and size of the model differs from the corresponding geometric parameter of the full-scale prototype by a constant scale factor.
- *Kinematic similarity:* The character of the fluid flow must be the same for the model and the full-scale prototype. In other words, the direction of flow of each fluid particle near the model must be the same as that of the fluid particle at the corresponding location near the full-scale prototype. The velocities of the two corresponding fluid particles must be related by a constant scale factor.
- *Dynamic similarity:* The forces that interact with the model must be similar to the forces that act on the full-scale prototype. For each force that acts on the full-scale prototype,

there must be a corresponding force acting on the model. Each corresponding force must act in the same direction, through the corresponding location, and with a magnitude that differs by a constant scale factor. The model and the prototype must have geometric and kinematic similarity if dynamic similarity is to be achieved.

Dynamic similarity is achieved when the following conditions are met:

$$\left[\frac{F_I}{F_p}\right]_p = \left[\frac{F_I}{F_p}\right]_m \quad \text{or} \quad \left[\frac{\rho \overline{V}^2}{p}\right]_p = \left[\frac{\rho \overline{V}^2}{p}\right]_m \quad [11.112]$$

$$\left[\frac{F_I}{F_V}\right]_p = \left[\frac{F_I}{F_V}\right]_m \quad \text{or} \quad \left[\frac{\rho \overline{V} l}{\mu}\right]_p$$

$$= \left[\frac{\rho \overline{V} l}{\mu}\right]_m \quad \text{or} \quad [\text{Re}]_p = [\text{Re}]_m \quad [11.113]$$

$$\left[\frac{F_I}{F_G}\right]_p = \left[\frac{F_I}{F_G}\right]_m \quad \text{or} \quad \left[\frac{\overline{V}^2}{lg}\right]_p$$

$$= \left[\frac{\overline{V}^2}{lg}\right]_m \quad \text{or} \quad [\text{Fr}]_p = [\text{Fr}]_m \quad [11.114]$$

$$\left[\frac{F_I}{F_E}\right]_p = \left[\frac{F_I}{F_E}\right]_m \quad \text{or} \quad \left[\frac{\rho \overline{V}^2}{E}\right]_p$$

$$= \left[\frac{\rho \overline{V}^2}{E}\right]_m \quad \text{or} \quad [\text{Ca}]_p = [\text{Ca}]_m \quad [11.115]$$

$$\left[\frac{F_I}{F_T}\right]_p = \left[\frac{F_I}{F_T}\right]_m \quad \text{or} \quad \left[\frac{\rho l \overline{V}^2}{\sigma}\right]_p$$

$$= \left[\frac{\rho l \overline{V}^2}{\sigma}\right]_m \quad \text{or} \quad [\text{We}]_p = [\text{We}]_m \quad [11.116]$$

where F_I = inertia/force,
F_p = pressure force,
F_V = viscous force,
F_G = gravity force,
F_E = elastic force,
F_T = surface-tension force,
l = characteristic length,
Re = Reynolds number,
Fr = Froude number,
Ca = Cauchy number,
E = modulus of elasticity,
σ = surface tension,
We = Weber number.

The subscript p refers to the full-scale prototype, and the subscript m refers to the model.

Example 11.25

If a one-fifth-scale boat model is to be tested in a tow tank, at what speed should the model be towed if a speed of 15 knots (7.72 m/s) is to be simulated?

Solution:

From equation 11.113:

$$\left[\frac{\rho \overline{V} l}{\mu}\right]_p = \left[\frac{\rho \overline{V} l}{\mu}\right]_m$$

$$\overline{V}_m = \left(\frac{\rho_p}{\rho_m}\right)\left(\frac{\mu_m}{\mu_p}\right)\left(\frac{l_p}{l_m}\right)\overline{V}_p$$

$$= (1)(1)\left(\frac{1}{5}\right)\left(7.72 \frac{\text{m}}{\text{s}}\right)$$

$$= 1.54 \text{ m/s}$$

11.11
Summary

In this chapter, principles for analyzing problems in fluid mechanics were discussed. The chapter started with definitions of fundamental constituents of fluid mechanics and then presented two categories of problems: hydrostatics (study of fluids at rest) and dynamics (study of fluids at motion).

In the discussion of hydrostatics, the concepts of fluid pressure, buoyancy, and barometry were presented. This was followed by fluid dynamics problems. Conservation laws such as mass balance, momentum balance, and energy balance fall into this category. Finally, methods for studying applied engineering problems such as calculations of head (energy) loss in pipes and analysis of turbo machinery were discussed.

PRACTICE PROBLEMS

1. What is the Mach number of a fluid stream flowing at a velocity of 370 ft/sec? ($c = 1{,}080$ ft/sec)
   (A) 0.66 (B) 2.92
   (C) 0.34 (D) 3.4

2. What is the gauge pressure of air in a vessel if the absolute pressure is 340 kPa and the ambient atmospheric pressure is 101 kPa?
   (A) 441 kPa (B) 105 kPa
   (C) 340 kPa (D) 239 kPa

3. The specific weight of a fluid is
   (A) the ratio of the density of the fluid to the density of water
   (B) the mass of the fluid per unit volume
   (C) the weight of the fluid per unit volume
   (D) the ratio of the weight of the fluid to the weight of water

4. The viscosity of a fluid
   (A) relates an applied normal stress to the velocity gradient in the fluid
   (B) relates an applied shear stress to the velocity gradient in the fluid
   (C) appears constant for all values of the velocity gradient in a dilatant fluid
   (D) appears constant for all values of the velocity gradient in a pseudoplastic fluid

5. What is the pressure at a point 18,000 ft below the surface of the ocean?

$$(\gamma = 64.0 \text{ lb/ft}^3)$$

   (A) 1,150,000 psig
   (B) 250 psig
   (C) 8,000 psig
   (D) 258,000 psig

6. In the figure, what is the absolute pressure (p_A) in vessel A?

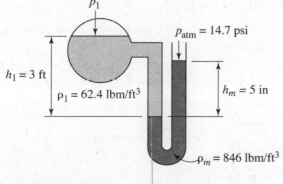

   (A) 15.8 psia (B) 18.4 psia
   (C) 13.6 psia (D) 11.0 psia

7. For the barometer shown in the figure, what is the atmospheric pressure if $h_m = 20$ cm, $\rho_m = 13{,}550$ kg/m^3, and $p_v = 0$ kPa?

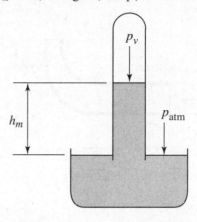

   (A) 128 kPa
   (B) 26.6 kPa
   (C) 4.36 kPa
   (D) 82.4 kPa

8. A cylindrical bar with a density of 41.6 lbm/ft^3 and a length of 3 ft floats in water ($\rho = 62.4$ lbm/ft^3) in such a way that the end surfaces of the bar are parallel to the surface of the water. How high will the top of the bar extend above the surface of the water?
   (A) 1 ft
   (B) 2 ft
   (C) 1.5 ft
   (D) 0.5 ft

9. For steady flow of a compressible fluid in the expander shown in the figure, what is the density at the exit section (ρ_2)?

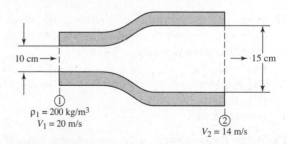

   (A) 190 kg/m^3
   (B) 62 kg/m^3
   (C) 642 kg/m^3
   (D) 127 kg/m^3

335

10. For the steady-state system shown in the figure, what is the force in the spring?

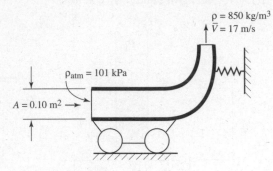

$\rho = 850 \text{ kg/m}^3$
$\bar{V} = 17 \text{ m/s}$

$\rho_{atm} = 101 \text{ kPa}$

$A = 0.10 \text{ m}^2$

(A) 43,495 N
(B) 24,575 N
(C) 16,895 N
(D) 24,565 N

11. For the control volume shown in the figure, what is the change in pressure $(p_1 - p_2)$ if the specific weight of the fluid is 9,810 N/m^3? (Assume that frictional losses are negligible.)

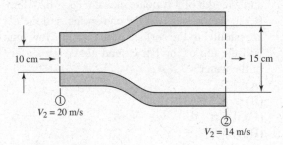

10 cm

15 cm

① $V_2 = 20 \text{ m/s}$

② $V_2 = 14 \text{ m/s}$

(A) −102 kPa
(B) −204 kPa
(C) 102 kPa
(D) 204 kPa

12. What is the stagnation pressure of water $(\rho = 1,000 \text{ kg/m}^3)$ flowing with a velocity of 28 m/s if the static pressure is 300 kPa?
(A) −92 kPa
(B) 692 kPa
(C) 92 kPa
(D) 1,084 kPa

13. Water flows through a 20-cm-diameter pipe. If the critical Reynolds number is 2,300, what is the maximum value of the average velocity for which laminar flow can be assumed?
(A) 0.047 m/s
(B) 1.2 m/s
(C) 0.0092 m/s
(D) 0.75 m/s

14. Water flows through a 17-cm-diameter horizontal pipe $(f = 0.017)$ at a rate of 0.34 m^3/s. What is the frictional head loss per meter of pipe length?
(A) 1.31 m
(B) 1.14 m
(C) 0.02 m
(D) 0.78 m

15. What is the minor head loss for the pipe entrance shown in the figure?

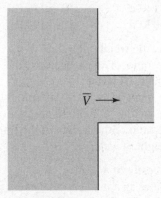

$\bar{V} \rightarrow$

$\bar{V} = 165 \text{ m/s}$
$C = 0.5$

(A) 1.28 m
(B) 16.81 m
(C) 2,775 m
(D) 693.81 m

16. What is the hydraulic diameter of a square duct that measures 80 cm on a side?
(A) 2.4 m
(B) 1.2 m
(C) 0.80 m
(D) 0.40 m

17. What is the coefficient of drag for a flow of air $(\rho = 1.225 \text{ kg/m}^3, \mu = 1.8 \times 10^{-5} \text{ N} \cdot \text{s/m}^2)$ over a flat plate that is 93 cm long, if $V_\infty = 86$ m/s?
(A) 3.38×10^{-3}
(B) 5.70×10^{-3}
(C) 2.19×10^{-3}
(D) 8.27×10^{-3}

18. What is angle θ for the moving blade shown in the figure?

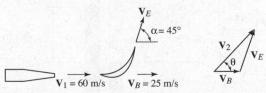

V_E
$\alpha = 45°$

V_2
θ
V_E
V_B

$V_1 = 60 \text{ m/s}$ $V_B = 25 \text{ m/s}$

(A) 57.82°
(B) 63.55°
(C) 32.18°
(D) 26.45°

19. If 16 kg/s of water, traveling at 67 m/s, enters the impulse turbine shown in the figure, which has a 26-cm diameter and a blade angle α of 120°, what is the maximum power that can be generated?

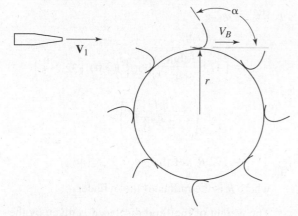

(A) 134 W
(B) 402 W
(C) 26.93 kW
(D) 8.98 kW

20. What is the pumping power required for the system shown in the figure? (Neglect friction in the pipes.)

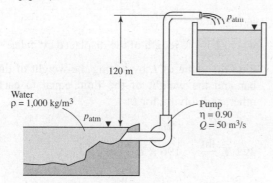

(A) 58,900 kW   **(B)** 65,400 kW
(C) 82,300 kW   **(D)** 53,000 kW

21. What is the volumetric flow rate (Q) of water ($\gamma = 9{,}810$ N/m³) for the venturi flow meter shown in the figure?

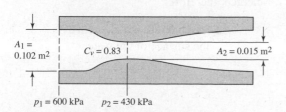

(A) 0.232 m³/s   **(B)** 0.248 m³/s
(C) 0.733 m³/s   **(D)** 0.467 m³/s

Answer Key

1. C	**8.** A	**15.** D
2. D	**9.** D	**16.** C
3. C	**10.** D	**17.** A
4. B	**11.** A	**18.** C
5. C	**12.** B	**19.** C
6. A	**13.** C	**20.** B
7. B	**14.** B	**21.** A

Answers Explained

1. C From equation 11.2:

$$M = \frac{\overline{V}}{c}$$

$$= \frac{370 \, \frac{\text{ft}}{\text{s}}}{1{,}080 \, \frac{\text{ft}}{\text{s}}}$$

$$= 0.343$$

2. D From equation 11.5:

$$p_a = p_g + p_{atm}$$

or

$$p_g = p_a - p_{atm}$$
$$= (340 \text{ kPa}) - (101 \text{ kPa})$$
$$= 239 \text{ kPa}$$

3. C The specific weight of a fluid is defined as the weight of a unit volume of fluid. It is related to the fluid density by the equation

$$\gamma = \rho g$$

where γ = specific weight (lbf/ft³ or N/m³),
       g = acceleration due to gravity (32.17 ft/s² or 9.81 m/s² at sea level).
       ρ = density

4. B It can be shown that the shear force in most common fluids can be expressed using equation 11.10:

$$\tau_n(P) = \mu \frac{\partial V}{\partial x_n}$$

Therefore choice **B** is correct. In a dilatant fluid, the viscosity appears to increase as the velocity gradient increases. In a pseudoplastic fluid, the viscosity appears to decrease as the velocity gradient increases.

5. C From equation 11.15: $p_g = \gamma h$

$$= \left(64.0 \, \frac{\text{lbf}}{\text{ft}^3}\right)(18,000 \text{ ft})\left(\frac{\text{ft}^2}{144 \text{ in}^2}\right)$$

$$= 8,000 \text{ lbf/in}^2 \text{ or psig}$$

6. A From equation 11.23:

$$p_1 = p_{\text{atm}} + \rho_m g h_m - \rho_1 g h_1$$

Then

$$\rho_m g h_m = \left(846 \, \frac{\text{lbm}}{\text{ft}^3}\right)\left(32.2 \, \frac{\text{ft}}{\text{s}^2}\right)(5 \text{ in})$$

$$\times \left(\frac{\text{lbf}}{32.2 \, \frac{\text{lbm-ft}}{\text{s}^2}}\right)\left(\frac{\text{ft}^3}{1,728 \text{ in}^3}\right)$$

$$= 2.45 \text{ lbf/in}^2$$

Also,

$$\rho_1 g h_1 = \left(62.4 \, \frac{\text{lbm}}{\text{ft}^3}\right)\left(32.2 \, \frac{\text{ft}}{\text{s}^2}\right)(3 \text{ ft})$$

$$\times \left(\frac{\text{lbf}}{32.2 \, \frac{\text{lbm-ft}}{\text{s}^2}}\right)\left(\frac{\text{ft}^2}{144 \text{ in}^2}\right)$$

$$= 1.30 \text{ lbf/in}^2$$

so

$$p_1 = 14.7 \, \frac{\text{lbf}}{\text{in}^2} + 2.45 \, \frac{\text{lbf}}{\text{in}^2} - 1.30 \, \frac{\text{lbf}}{\text{in}^2}$$

$$= 15.85 \text{ lbf/in}^2 \text{ or psia}$$

7. B From equation 11.25:

$$p_{\text{atm}} - p_v = \rho_m g h_m$$

Then

$$p_{\text{atm}} = \rho_m g h_m$$

$$= \left(13,550 \, \frac{\text{kg}}{\text{m}^3}\right)\left(9.81 \, \frac{\text{m}}{\text{s}^2}\right)(0.20 \text{ m})$$

$$\times \left(\frac{\text{N}}{\frac{\text{kg}\cdot\text{m}}{\text{s}^2}}\right)\left(\frac{\text{kPa}}{1,000 \, \frac{\text{N}}{\text{m}^2}}\right)$$

$$= 26.6 \text{ kPa}$$

8. A According to Archimedes' principle, an object floating in a fluid will displace an amount of fluid whose weight is equal to the weight of the object. The weight of the cylindrical bar is given by the equation

$$W_{\text{bar}} = m_{\text{bar}} g$$

$$= \rho_{\text{bar}} V_{\text{bar}} g$$

$$= \left(41.6 \, \frac{\text{lbm}}{\text{ft}^3}\right)[(\pi R^2)(3 \text{ ft})]\left(32.2 \, \frac{\text{ft}}{\text{s}^2}\right)$$

$$\times \left(\frac{\text{lbf}}{32.2 \, \frac{\text{lbm-ft}}{\text{s}^2}}\right)$$

$$= 392 \, R^2 \text{ lbf/ft}^2$$

where R is the radius of the cylinder.

The weight of the fluid displaced is given by the equation

$$W_{\text{disp}} = m_{\text{disp}} g$$

$$= \rho_{\text{water}} V_{\text{disp}} g$$

$$= \left(62.4 \, \frac{\text{lbm}}{\text{ft}^3}\right)[(\pi R^2)(L)]\left(32.2 \, \frac{\text{ft}}{\text{s}^2}\right)\left(\frac{\text{lbf}}{32.2 \, \frac{1 \text{bm-ft}}{\text{s}^2}}\right)$$

$$= 196 \, R^2 L \text{ lbf/ft}^3$$

where L is the length of the displaced cylinder.

Find the value of L by setting the weight of the bar and the weight of the fluid equal to each other, and solving for L:

$$W_{\text{bar}} = W_{\text{disp}}$$

$$392 \, R^2 \, \frac{\text{lbf}}{\text{ft}^2} = 196 \, R^2 L \, \frac{\text{lbf}}{\text{ft}^3}$$

$$L = \frac{392}{196} \, \frac{R^2}{R^2} \, \frac{\frac{\text{lbf}}{\text{ft}^2}}{\frac{\text{lbf}}{\text{ft}^3}}$$

$$= 2 \text{ ft}$$

This result is the amount of the bar that is submerged; therefore:

1 ft of bar will extend above the surface of the water.

9. **D** From the principle of mass conservation, equation 11.43 gives:

$$\rho_1 A_1 \overline{V}_1 = \rho_2 A_2 \overline{V}_2$$

or

$$\rho_2 = \frac{\rho_1 A_1 \overline{V}_1}{A_2 \overline{V}_2}$$

$$= \left(200 \ \frac{kg}{m^3}\right)\left(\frac{\frac{\pi(0.10 \ m)^2}{4}}{\frac{\pi(0.15 \ m)^2}{4}}\right)\left(\frac{20 \ \frac{m}{s}}{14 \ \frac{m}{s}}\right)$$

$$= 127 \ kg/m^3$$

10. **D** From application of the impulse-momentum principle and a summation of forces in the horizontal direction, the force in the spring is given by the equation

$$F_{spring} = \dot{m} \overline{V}$$

Also,

$$\dot{m} = \rho \overline{V} A$$

$$= \left(850 \ \frac{kg}{m^3}\right)\left(17 \ \frac{m}{s}\right)(0.10 \ m^2)$$

$$- 1{,}445 \ kg/s$$

so

$$F_{spring} = \left(1{,}445 \ \frac{kg}{s}\right)\left(17 \ \frac{m}{s}\right)\left(\frac{N}{\frac{kg \cdot m}{s^2}}\right)$$

$$= 24{,}565 \ N$$

11. **A** Use the Bernoulli equation (equation 11.51):

$$\frac{p_1}{\gamma} + \frac{\overline{V}_1^2}{2g} + z_1 = \frac{p_2}{\gamma} + \frac{\overline{V}_2^2}{2g} + z_2$$

or

$$p_1 - p_2 = \gamma \left[\frac{(\overline{V}_2^2 - \overline{V}_1^2)}{2g} + (z_2 - z_1)\right]$$

$$= \left(9{,}810 \ \frac{N}{m^3}\right)\left[\frac{\left(14 \ \frac{m}{s}\right)^2 - \left(20 \ \frac{m}{s}\right)^2}{2\left(9.81 \ \frac{m}{s^2}\right)} + 0 \ m\right]$$

$$\times \left(\frac{kPa}{1{,}000 \ \frac{N}{m^2}}\right)$$

$$= -102 \ kPa$$

12. **B** From equation 11.55:

$$p_o = p_s + \frac{1}{2}\rho V^2$$

$$= (300 \ kPa) + \frac{1}{2}\left(1{,}000 \ \frac{kg}{m^3}\right)\left(28 \ \frac{m}{s}\right)^2\left(\frac{N}{\frac{kg \cdot m}{s^2}}\right)$$

$$\times \left(\frac{kPa}{1{,}000 \ \frac{N}{m^2}}\right)$$

$$= 692 \ kPa$$

13. **C** From equation 11.56:

$$Re = \frac{\rho \overline{V} D}{\mu}$$

or

$$\overline{V} = \frac{\mu Re}{\rho D}$$

$$= \frac{\left(8 \times 10^{-4} \ \frac{N \cdot s}{m^2}\right)(2{,}300)}{\left(1{,}000 \ \frac{kg}{m^3}\right)(0.20 \ m)}\left(\frac{\frac{kg \cdot m}{s^2}}{N}\right)$$

$$= 0.0092 \ m/s$$

14. **B** From equation 11.66: $h_f = f\left(\frac{L}{D}\right)\left(\frac{\overline{V}^2}{2g}\right)$

Find the average velocity through the pipe as follows:

$$\overline{V} = \frac{Q}{A}$$

$$= \frac{0.34 \ \frac{m^3}{s}}{\frac{\pi(0.17 \ m)^2}{4}}$$

$$= 14.98 \ m/s$$

Then,

$$h_f = (0.017)\left[\frac{1 \ m}{0.17 \ m}\right]\left[\frac{\left(14.98 \ \frac{m}{s}\right)^2}{2\left(9.81 \ \frac{m}{s^2}\right)}\right]$$

$$= 1.14 \ m$$

15. D From equation 11.69: $h_{fm} = C\dfrac{\overline{V}^2}{2g}$

$$= (0.5)\frac{\left(165\,\dfrac{\text{m}}{\text{s}}\right)^2}{2\left(9.81\,\dfrac{\text{m}}{\text{s}^2}\right)}$$

$$= 693.81 \text{ m}$$

16. C From equation 11.75:

$$D_h = 4\left(\frac{\text{cross-sectional area}}{\text{wetted perimeter}}\right)$$

$$= 4\left(\frac{(0.80\text{ m})^2}{4(0.80\text{ m})}\right)$$

$$= 0.80 \text{ m}$$

17. A From equation 11.82:

$$C_D = \frac{1.33}{\text{Re}_L^{0.5}}\ (10^4 < \text{Re}_L < 5 \times 10^5)$$

or

$$C_D = \frac{0.031}{\text{Re}_L^{1/7}}\ (10^6 < \text{Re}_L < 10^9)$$

The Reynolds number for this flow is given by equation 11.81:

$$\text{Re}_L = \frac{\rho V_\infty L}{\mu}$$

$$= \frac{\left(1.225\,\dfrac{\text{kg}}{\text{m}^3}\right)\left(86\,\dfrac{\text{m}}{\text{s}}\right)(0.93\text{ m})}{\left(1.8 \times 10^{-5}\,\dfrac{\text{N}\cdot\text{s}}{\text{m}^2}\right)}$$

$$= 5.443 \times 10^6$$

Therefore the second equation for the coefficient of drag is used:

$$C_D = \frac{0.031}{\text{Re}_L^{1/7}}$$

$$= \frac{0.031}{(5.443 \times 10^6)^{1/7}}$$

$$= 3.381 \times 10^{-3}$$

18. C If friction along the face of the blade is neglected, the velocity leaving the blade has the same magnitude as the velocity entering the blade, directed as shown in the diagram. Solve for angle θ as follows:

$$\mathbf{V}_{2x} = \mathbf{V}_B + \mathbf{V}_E \sin \alpha$$

$$= 25\frac{\text{m}}{\text{s}} + \left(60\,\frac{\text{m}}{\text{s}}\right)\sin(45°)$$

$$= 67.43 \text{ m/s}$$

and

$$\mathbf{V}_{2y} = \mathbf{V}_E \cos \alpha$$

$$= \left(60\,\frac{\text{m}}{\text{s}}\right)\cos(45°)$$

$$= 42.43 \text{ m/s}$$

Then,

$$\theta = \tan^{-1}\left(\frac{V_{2y}}{V_{2x}}\right)$$

$$= \tan^{-1}\left(\frac{42.43\,\dfrac{\text{m}}{\text{s}}}{67.43\,\dfrac{\text{m}}{\text{s}}}\right)$$

$$= 32.18°$$

19. C The maximum power that can be generated by the impulse turbine shown in the figure is given by equation 11.91:

$$\dot{W}_{\text{max}} = Q\rho\left(\frac{|V_1^2|}{4}\right)(1 - \cos \alpha)$$

$$= \dot{m}\left(\frac{|V_1^2|}{4}\right)(1 - \cos \alpha)$$

$$= \left(16\,\frac{\text{kg}}{\text{s}}\right)\left[\frac{\left(67\,\dfrac{\text{m}}{\text{s}}\right)^2}{4}\right][1 - \cos(120°)]$$

$$\times \left(\frac{\text{N}}{\dfrac{\text{kg}\cdot\text{m}}{\text{s}^2}}\right)\left(\frac{\text{kW}}{1{,}000\,\dfrac{\text{N}\cdot\text{m}}{\text{s}}}\right)$$

$$= 26.93 \text{ kW}$$

20. B Use the Bernoulli equation:

$$\frac{p_1}{\gamma} + \frac{\overline{V}_1^{\,2}}{2g} + z_1 + h_m = \frac{p_2}{\gamma} + \frac{\overline{V}_2^{\,2}}{2g} + z_2$$

In this case, $p_1 = p_2$ and $\overline{V}_1 = \overline{V}_2$, so

$$h_m = z_2 - z_1 \\ = 120 \text{ m}$$

The pumping power is then given by the equation

$$\dot{W} = \rho g h Q$$

$$= \left(1{,}000 \ \frac{\text{kg}}{\text{m}^3}\right)\left(9.81 \ \frac{\text{m}}{\text{s}^2}\right)(120 \text{ m})\left(50 \ \frac{\text{m}^3}{\text{s}}\right)$$

$$\times \left(\frac{\text{N}}{\frac{\text{kg·m}}{\text{s}^2}}\right)\left(\frac{\text{kW}}{1{,}000 \ \frac{\text{N·m}}{\text{s}}}\right)$$

$$= 58{,}860 \text{ kW}$$

But this is the power required to pump the water. The pump will need somewhat more power to account for its irreversibilities:

$$\eta = \frac{\dot{W}}{\dot{W}_{\text{supplied}}}$$

$$\dot{W}_{\text{supplied}} = \frac{\dot{W}}{\eta}$$

$$= \frac{58{,}860 \text{ kW}}{0.90}$$

$$= 65{,}400 \text{ kW}$$

21. A The volumetric flow rate through a venturi flow meter is given by equation 11.108:

$$Q = \frac{C_v A_2}{\sqrt{1-(A_2/A_1)}} \sqrt{2g\left[\left(\frac{p_1}{\gamma}+z_1\right) - \left(\frac{p_2}{\gamma}+z_2\right)\right]}$$

$$= \frac{(0.83)(0.015 \text{ m}^2)}{\sqrt{1-\left(\frac{0.015 \text{ m}^2}{0.012 \text{ m}^2}\right)^2}}$$

$$= \sqrt{2\left(9.81 \ \frac{\text{m}}{\text{s}^2}\right)\left[\left(\frac{(600 \text{ kPa})}{9{,}810 \ \frac{\text{N}}{\text{m}^3}}\right)\left(\frac{1{,}000 \ \frac{\text{N}}{\text{m}^2}}{\text{kPa}}\right) - \left(\frac{(430 \text{ kPa})}{9{,}810 \ \frac{\text{N}}{\text{m}^3}}\right)\left(\frac{1{,}000 \ \frac{\text{N}}{\text{m}^2}}{\text{kPa}}\right)\right]}$$

$$= 0.232 \text{ m}^3/\text{s}$$

MOODY (STANTON) DIAGRAM

	e, (ft)	e, (mm)
Riveted steel	0.003-0.03	0.9-9.0
Concrete	0.001-0.01	0.3-3.0
Cast iron	0.00085	0.25
Galvanized iron	0.0005	0.15
Commercial steel or wroght iron	0.00015	0.046
Drawn tubing	0.000005	0.0015

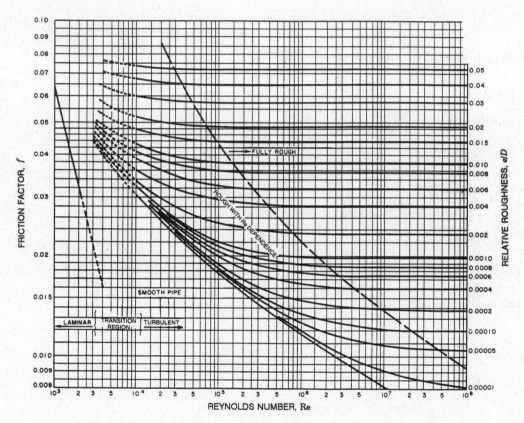

Reproduced from "Friction Factors for Pipe Flow," *MOODY DIAGRAM*, in *Trans. ASME 66*, 1944. Used with permission of ASME.

12

Electricity and Magnetism

Frederick F. Driscoll

Introduction

This chapter reviews the basic concepts of electrical engineering. It is divided into the ten sections listed below.

- Current, Voltage, and Power
- Resistance
- Direct-Current (DC) Circuit Analysis
- Inductance
- Capacitance
- Resistor-Capacitor (R-C) and Resistor-Inductor (R-L) Circuits
- Alternating-Current (AC) Circuit Analysis
- Operational Amplifiers
- Electric and Magnetic Fields
- DC Motors

12.1 Current, Voltage, and Power

Our physical world may be interpreted in terms of matter and energy, both of which exist in a variety of forms. Matter undergoes change; energy either causes or is a result of the change. *Matter* has been defined as anything that occupies space and possesses mass. *Energy,* on the other hand, is the ability to do work. Although energy itself cannot be measured, its results can.

Matter exists either in a solid, a liquid, or a gaseous state. Since most electric components are made of solids, circuit analysis usually begins with an introduction to atomic structure or at least the properties of electric charge.

Electric Charge

One of the unique properties of an atom's electron is its negative **electric charge.** The charge of a single electron is the smallest electric charge that exists. For purposes of analysis it is more practical to group a large number of electrons together; hence the basic quantity of electric charge is the coulomb (C). One coulomb equals the accumulated charge of 6.24×10^{18} electrons and is represented by q or Q.

Note: It is conventional to use lower-case letters to represent values that vary with time and to use upper-case letters to represent values that do not vary with time.

> Charge flows easily through conductors.

Materials that are good conductors of electric current are composed of atoms with only one valence electron. Valence electrons are the electrons in an atom's outermost energy level. In a metal, valence electrons are not tightly bound to the nucleus of the atom and are said to be free or mobile. Good conductors, such as silver, copper, and aluminum, have high concentrations of free electrons. Materials such as air, glass, mica, and rubber have considerably fewer free electrons and are referred to as insulators.

Current

To understand **electric current,** consider a portion of a conductor as shown in Figure 12.1a. First consider the case where no outside energy except thermal energy is applied to the conductor. Thermal energy is always present because a temperature of absolute

zero cannot be achieved. Under normal conditions, free electrons are said to be *uniformly distributed* and to have *random motion.*

> Free electrons are uniformly distributed and have random motion within a material.

Uniform distribution means that free electrons will not be concentrated in one particular location but will permeate throughout the material as a gas diffuses through a closed container. Thus at any instant any incremental volume of the same size will contain the same number of free electrons.

Random motion means that no specific direction can be assigned to the free electrons. Therefore the probability of finding a particular number of free electrons moving in one direction is the same as finding an equal number of free electrons moving in the opposite direction. Thus the net motion in any one direction is zero. For free electrons to have direction, energy other than thermal energy must be applied to one end of the conductor. When a positive charge is connected to one end of the conductor and a negative charge to the other, a potential difference is created between the ends of the conductor, as shown in Figure 12.1b. Since like charges repel and unlike charges attract, electrons will travel toward the positive end of the conductor and away from the negative end.

Even when the application of different charges to a conductor has caused the free electrons to move toward the positive charge, they do not move in a straight path. Instead, they follow a zigzag pattern similar to random motion. Unlike true random motion, however, there is now a definite drift toward the positive charge. This net motion of free electrons in one direction constitutes an electric current known as a *drift current.* This current depends on the rate

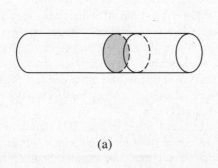

(a)

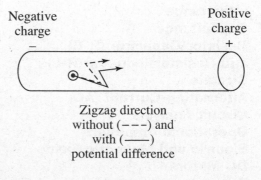

Negative charge
−

Positive charge
+

Zigzag direction
without (−−−) and
with (——)
potential difference

(b)

Figure 12.1

at which the free electrons move. The term "rate" implies a quantity divided by time. It should be noted that "current" is a rate-of-flow term and not a velocity term. Thus, current depends, not on the speed of individual electrons, but rather on the net motion of a large number of free electrons.

> The unit for current is the ampere (A) which is equivalent to 1 coulomb per second.

The basic unit of current is the ampere (A). A definition of ampere is as follows: If 1 coulomb of charge passes through a cross-sectional area of a conductor in 1 second, there is 1 ampere of current.

$$1 \text{ ampere} = \frac{1 \text{ coulomb}}{1 \text{ second}} \qquad [12.1a]$$

This relationship may be expressed mathematically as

> Current is the time rate of change of charge. Its units are amperes (A).

$$i = \frac{dq}{dt} \qquad [12.1b]$$

where i is the instantaneous current, and dq/dt is the rate of change of charge with respect to time (coulombs per second).

An electron has a negative charge; it drifts from a negative terminal to a positive one. This principle is known as *electron flow*. The analysis of electric circuits is usually based, however, on the direction of current from an energy source's positive terminal through the circuit to the source's negative terminal. This is known as the **conventional current direction.**

Voltage

For an electric current to exist in a conductor, the ends of the conductor must be connected to different charges. Only then is it possible to have a movement of free electrons in a specific direction. A potential difference exists between two points in a circuit whenever one point has a more positive charge than the other point. It is this potential difference that causes current.

Voltage potential is referred to as either a potential rise or a potential fall. A voltage rise or potential rise occurs when work is done on the charge by an element; an example is an energy source. A voltage

drop or potential fall exists when work is done by the charge on an element; an example is a load. The distinction between potential rise and potential fall is established by polarity markings across an element as shown in Figure 12.2.

> Electric circuits consist of sources and loads connected by conductors.

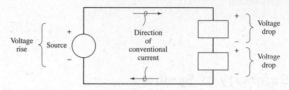

Figure 12.2

The basic unit of voltage, either a potential rise or a potential fall, is the volt (V). A definition of voltage is as follows: If 1 joule of energy is expended in moving 1 coulomb of charge from one point to another in an electric circuit, the voltage is 1 volt.

> Voltage is a measure of the energy transferred per unit of charge when the charge is moved from one point to a second point in the circuit.

$$1 \text{ volt} = \frac{1 \text{ joule}}{1 \text{ coulomb}} \qquad [12.2a]$$

The mathematical expression for voltage is

$$v = \frac{dw}{dq} \qquad [12.2b]$$

where v is the instantaneous voltage and dw/dq is the change in energy with respect to change in charge.

Power

The goal of an electric circuit is to transfer energy from one point to another in order to accomplish useful work. Hence electric circuits are composed of (1) sources and (2) loads. The *source* supplies the electrical energy, while the *load* converts the energy into work to accomplish what the circuit or system has been designed for. The rate at which energy is transferred from the source to the load is called power. **Power** is defined as energy per unit time, and the basic unit of power is the watt (W).

> Power is the rate at which energy is transferred.

$$1 \text{ watt} = \frac{1 \text{ joule}}{1 \text{ second}} \qquad [12.3a]$$

Although power is defined as the transfer of 1 joule of electrical energy in 1 second, it may be determined from measured values of voltage and current.

$$\text{Power} = \text{voltage} \times \text{current} \qquad [12.3b]$$

Using the symbols for voltage and current gives the expression

$$p = vi \qquad [12.3c]$$

Summary of Symbols

The expressions of charge, current, voltage, energy, and power are summarized in Table 12.1.

Summary of basic electrical quantities.

Table 12.1 Summary of Expressions

Quantity	Symbol	Unit	Equation
Charge	q	coulomb	$q(t) = \int i(t)\, dt$
Current	i	ampere	$i(t) = dq/dt$
Voltage	v	volt	$v(t) = dw/dq$
Energy	w	joule	$w(t) = \int p(t)\, dt$
Power	p	watt	$p(t) = dw/dt$
			$p(t) = v(t) \cdot i(t)$

Electric Circuits

This section has summarized many concepts in the analysis of electric circuits that consist of energy sources and loads. The energy sources will be primarily voltage sources, and the loads will contain passive components or active devices along with passive components.

Passive components, which are capable only of storing or dissipating energy, consist of resis-

tance, inductance (self and mutual), and capacitance. Resistance converts electrical energy into other energy forms. Self-inductance stores energy in a magnetic field, and mutual inductance transfers energy from one part of a circuit to another through a magnetic field. Capacitance stores energy in an electric field.

Operational amplifiers, introduced in Section 12.8, are one type of active device used in the design of electric circuits. This review concentrates primarily, however, on passive circuits.

12.2 Resistance

As previously mentioned, the application of potential energy to a conductor causes a net movement of free electrons in one direction. Even under the influence of potential energy, the path of a single electron is not straight but is a zigzag pattern similar to random motion. One cause of the zigzag pattern is collisions of the electron with other atoms. Since energy cannot be destroyed, the externally applied potential energy is transformed into kinetic energy of the electron.

When an electron collides with an atom, heat is produced in a thermodynamic process called *Joule heating*. The collisions, which simultaneously produce heat, also constitute an opposition to the flow of current. Opposition to current flow is **resistance,** as shown in Figure 12.3a. For circuit analysis, the primary characteristic of resistance is energy conversion (dissipation of heat or other forms of energy).

Every material used as a conductor has some degree of resistance. Some materials conduct electric charge more efficiently, that is, with less resistance and production of heat, than others. Copper, for example, is an excellent conductor, while steel is a good conductor but not an excellent one.

Resistance is a characteristic of that part of a device that transforms electrical energy into heat. Resistance is designated by the letter R and the symbol shown in Figures 12.3b and 12.3c. The basic unit of resistance is the ohm (Ω). A resistor (same letter and symbol designation) is the actual physical device that is inserted

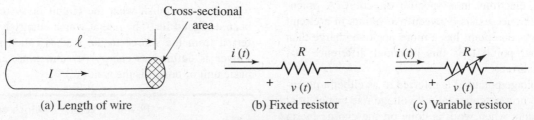

(a) Length of wire   (b) Fixed resistor   (c) Variable resistor

Figure 12.3

into a circuit or system to limit the flow of current. It produces a voltage drop and dissipates energy. A resistor may be classified as either fixed or variable. A *fixed* resistor is one whose value is constant, while a *variable* resistor is one whose value can be changed. The circuit symbols for a fixed and a variable resistor are shown in Figures 12.3b and 12.3c, respectively.

The resistance of a material with a uniform cross section is given by the equation

> The resistance of a material is directly proportional to its resistivity and length and inversely proportional to its cross-sectional area.

$$R = \frac{\rho l}{A} \qquad [12.4]$$

where R = resistance (Ω),
ρ = resistivity ($\Omega \cdot m$),
l = length of the material (m),
A = cross-sectional area (perpendicular to current flow) (m^2)

The units of measurement are indicated in parentheses.

This equation is illustrated in Figure 12.3a.

Typical values of resistivity for various materials are listed in Table 12.2.

Table 12.2 Resistivities of Four Materials

Material	Resistivity, ρ ($\Omega \cdot m$)
Silver	1.468×10^{-8}
Copper	1.72×10^{-8}
Aluminum	2.83×10^{-8}
Insulators	10^6 to 10^{16}

The resistivity of a material depends on the temperature. For most metals, resistivity increases with an increase in temperature. In Equations 12.5 and 12.6, ρ_0 and R_0 are the resistivity and resistance values, respectively, at temperature T_0, and α is the temperature coefficient of resistance. The units of α are $1/°C$.

$$\rho = \rho_0[1 + \alpha(T - T_0)] \qquad [12.5]$$

$$R = R_0[1 + \alpha(T - T_0)] \qquad [12.6]$$

Ohm's Law

For a linear resistor, the voltage drop across a resistor is directly proportional to the current and is given by Ohm's law:

> The voltage across a resistor is directly proportional to the current through the resistor.

$$v(t) = i(t)R \qquad [12.7]$$

If the current and voltage do not vary with time, Ohm's law is usually written as follows:

$$V = IR \qquad [12.8]$$

Power Dissipated by a Resistor

The power dissipated by a resistor is determined by using the equation

$$P = VI = \frac{V^2}{R} = I^2R \qquad [12.9]$$

The unit of power is the watt (joules per second).

Resistors in Series

Resistors connected in series have the same current through each element, as shown in Figure 12.4. The sum of the voltages across the resistors equals the source voltage. This relationship is known as *Kirchhoff's voltage law:*

$$V_T = V_1 + V_2 + V_3 \qquad [12.10]$$

Applying Ohm's law to each resistor yields

$$V_T = IR_1 + IR_2 + IR_3 \qquad [12.11a]$$

or

$$V_T = I(R_1 + R_2 + R_3) \qquad [12.11b]$$

The total resistance "seen" by the source is

$$R_T = R_1 + R_2 + R_3 \qquad [12.12]$$

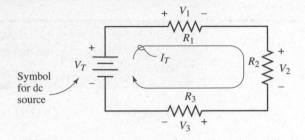

Figure 12.4

If the circuit contains n resistors in series, the total resistance is expressed as

$$R_T = R_1 + R_2 + \ldots + R_n \qquad [12.13]$$

Resistors in Parallel

Two or more resistors are in parallel if they experience the same voltage drop. Figure 12.5 shows three resistors in parallel. Kirchhoff's current law states that the sum of the currents entering a node equals the sum of the currents leaving that node. Therefore, for the circuit of Figure 12.5:

$$I_T = I_1 + I_2 + I_3 \qquad [12.14]$$

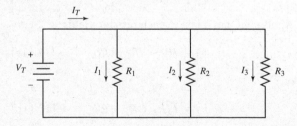

Figure 12.5

Since the voltage across each resistor is the same, equation 12.14 can be divided by V_T to obtain

$$I_T = \frac{V_T}{R_1} + \frac{V_T}{R_2} + \frac{V_T}{R_3} \qquad [12.15a]$$

or

$$I_T = V_T \left(\frac{1}{R_1} + \frac{1}{R_2} + \frac{1}{R_3} \right) \qquad [12.15b]$$

where each term is the reciprocal of Ohm's law and:

$$\frac{1}{R_T} = \frac{1}{R_1} + \frac{1}{R_2} + \frac{1}{R_3} \qquad [12.16a]$$

Rearranging yields the equivalent resistance:

$$R_T = \frac{1}{\dfrac{1}{R_1} + \dfrac{1}{R_2} + \dfrac{1}{R_3}} \qquad [12.16b]$$

Equations 12.16a and 12.16b may be expanded as follows to n resistors in parallel:

$$\frac{1}{R_T} = \frac{1}{R_1} + \frac{1}{R_2} + \ldots + \frac{1}{R_n} \qquad [12.17a]$$

and

$$R_T = \frac{1}{\dfrac{1}{R_1} + \dfrac{1}{R_2} + \ldots + \dfrac{1}{R_n}} \qquad [12.17b]$$

If there are only two resistors in parallel, the equivalent resistance can be determined by a *product-over-the-sum rule:*

$$R_T = \frac{R_1 R_2}{R_1 + R_2} \qquad [12.18]$$

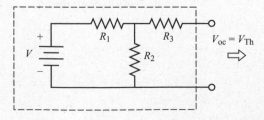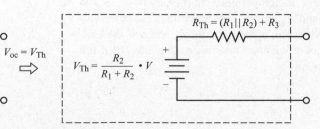

Figure 12.6

348

The short-circuit current is I_N.

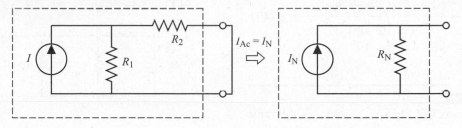

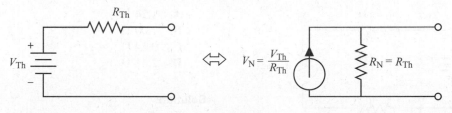

(a) Norton equivalent circuit

(b) Equivalent circuits

Figure 12.7

Note: The product-over-the-sum rule is for only two resistors in parallel, but it can be used more than once. For example, if three resistors are connected in parallel, the product over the sum rule can be applied to find the equivalent resistance for the first two resistors and that answer used with the third resistor and the product-over-the-sum rule applied again.

Thévenin and Norton Equivalent Circuits

Thévenin Circuit

Any two terminals of a circuit may be represented by a series combination of a voltage source and a resistor. The value of the equivalent voltage source is the voltage appearing at the open-circuit terminals. The value of the equivalent resistance is equal to the resistance "seen" looking back into the network with all energy sources replaced by their internal resistances (ideal voltage sources have zero internal resistance, and ideal current sources have infinite internal resistance). This series combination of a voltage source, V_{Th}, and a resistor, R_{Th}, is known as a *Thévenin equivalent circuit* and is shown in Figure 12.6. Note that the equivalence exists only at the terminals. Example 12.7 is an application.

Alternatively, the Thévenin equivalent resistance is equal to the open-circuit voltage, V_{oc}, divided by the short-circuit current, I_{sc}.

Norton Circuit

Any two terminals of a circuit may be represented by a parallel combination of a current source and a resistor. The value of the equivalent current source is the short-circuit current appearing at the terminals. The value of the equivalent resistance is equal to the resistance "seen" looking back into the network with all energy sources replaced by their internal resistances (ideal voltage sources have zero internal resistance, and ideal current sources have infinite internal resistance). This parallel combination of a current source, I_N, and a resistor, R_N, is known as a *Norton equivalent circuit* and is shown in Figure 12.7a. Figure 12.7b shows equivalent circuits.

Example 12.1

What is the resistance of 2,000 m of copper wire that has a resistivity of $1.72 \times 10^{-8} \ \Omega \cdot$ m and a diameter of 4 mm?

A. 0.086 Ω
B. 2.74 Ω
C. 1.37 Ω
D. 3.24 Ω

Solution:

Apply equation 12.4 to obtain

$$R = \frac{\rho l}{A} = \frac{(1.72 \times 10^{-8} \ \Omega \cdot m)(2,000 \ m)}{\frac{\pi}{4}(4 \times 10^{-3} \ m)^2} = 2.74 \ \Omega$$

Answer is B.

Example 12.2

For the circuit of Figure 12.8, what is the value of current I_T?

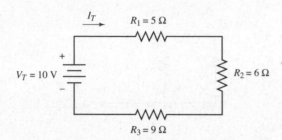

Figure 12.8

A. 1.5 A
B. 4.0 A
C. 2.0 A
D. 0.5 A

Solution:

Use equation 12.12 to solve for the total resistance "seen" by the voltage source:

$$R_T = R_1 + R_2 + R_3 = 20 \ \Omega$$

Apply Ohm's law to obtain

$$I_T = \frac{10 \ V}{20 \ \Omega} = 0.5 \ A$$

Answer is D.

Example 12.3

For the circuit of Figure 12.8, what is the value of the power dissipated by R_2?
A. 1.5 W
B. 4.0 W
C. 1.25 W
D. 3.0 W

Solution:

From the solution of Example 12.2, $I_T = 0.5$ A. Apply equation 12.9 to obtain

$$P = I_T^2 R_2 = (0.5 \ A)^2 \times 6 \ \Omega = 1.5 \ W$$

Answer is A.

Example 12.4

What is the equivalent resistance for the circuit of Figure 12.9?

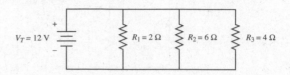

Figure 12.9

A. 1.24 Ω
B. 1.09 Ω
C. 0.24 Ω
D. 2.41 Ω

Solution:

For three resistors in parallel, use equation 12.16b. Then,

> Try using the product-over-the-sum rule twice—let $R_1 \parallel R_2 = R_a$, then $R_a \parallel R_3 = R_T$.

$$R_T = \frac{1}{\frac{1}{2} + \frac{1}{6} + \frac{1}{4}} = 1.09 \ \Omega$$

Answer is B.

Example 12.5

If a 9-V battery produces 10 A of current when a short circuit is placed across its terminals, what is the battery's internal resistance?
A. 1.11 Ω
B. 0.45 Ω
C. 0.90 Ω
D. 1.45 Ω

Solution:

Apply Ohm's law (equation 12.8):

$$R = \frac{V_T}{I} = \frac{9 \ V}{10 \ A} = 0.9 \ \Omega$$

Answer is C.

Example 12.6

What is the value of I_T in the circuit of Figure 12.10?

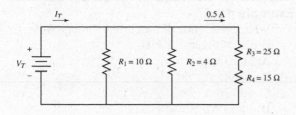

Figure 12.10

A. 0.5 A
B. 3.0 A
C. 2.5 A
D. 7.5 A

Solution:

Since R_3 and R_4 are in series, $R_{eq} = 25\ \Omega + 15\ \Omega = 40\ \Omega$. The voltage across this combination is $(0.5\ \text{A})(40\ \Omega) = 20\ \text{V}$. Since R_1 and R_2 are in parallel with this combination, 20 V appears across R_1 and R_2. Apply Ohm's law to obtain

$$I_1 = \frac{20\ \text{V}}{10\ \Omega} = 2\ \text{A} \quad \text{and} \quad I_2 = \frac{20\ \text{V}}{4\ \Omega} = 5\ \text{A}$$

Then apply Kirchhoff's current law, equation 12.14 yielding:

$$I_T = I_1 + I_2 + I_3 = 2\ \text{A} + 5\ \text{A} + 0.5\ \text{A} = 7.5\ \text{A}$$

Answer is D.

Example 12.7

What are the Thévenin equivalent voltage and resistance between terminals A and B of Figure 12.11?

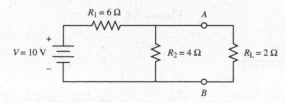

Figure 12.11

A. 10 V, 4.0 Ω
B. 4.0 V, 2.0 Ω
C. 6.0 V, 4.2 Ω
D. 4.0 V, 2.4 Ω

Solution:

Remove the 2-Ω-load resistor, and "look back" into the circuit toward the left. The voltage drop across the 4-Ω resistor is the Thévenin equivalent voltage:

$$V_{\text{Th}} = \frac{4\ \Omega}{6\ \Omega + 4\ \Omega} \times 10\ \text{V} = 4\ \text{V}$$

This equation is known as the *voltage division law,* but the same answer can be determined by calculating the total current and using Ohm's law.

To determine the Thévenin equivalent resistance, replace the voltage source, V_T, by its internal resistance, 0 Ω. This puts the 6-Ω resistor in parallel with the 4-Ω resistor. Apply equation 12.18 to obtain

$$R_{\text{Th}} = \frac{6\ \Omega \times 4\ \Omega}{6\ \Omega + 4\ \Omega} = 2.4\ \Omega$$

Answer is D.

Example 12.8

What is the value of the current through the 6-Ω resistor in the circuit of Figure 12.12?

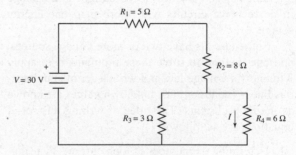

Figure 12.12

A. 2.0 A
B. $\frac{2}{3}$ A
C. 5.0 A
D. $\frac{4}{3}$ A

Solution:

To determine the total current drawn from the voltage source, determine the total resistance, R_T:

$$R_T = R_1 + R_2 + \frac{R_3 \times R_4}{R_3 + R_4}$$

$$= 5\ \Omega + 8\ \Omega + \frac{3\ \Omega \times 6\ \Omega}{3\ \Omega + 6\ \Omega} = 15\ \Omega$$

and the total current:

$$I_T = \frac{30 \text{ V}}{15 \text{ }\Omega} = 2 \text{ A}$$

The total current flows through the 5-Ω and 8-Ω resistors and then divides between the 3-Ω resistor and the 6-Ω resistor:

$$I_{6\Omega} = \frac{3 \text{ }\Omega}{3 \text{ }\Omega + 6 \text{ }\Omega} \times 2 \text{ A} = \frac{2}{3} \text{ A}$$

The last equation is known as the *current division law*. The same result can be obtained by using Ohm's law and the equivalent resistance of the parallel combination of the 3-Ω and 6-Ω resistors.

Answer is B.

12.3
Direct-Current (DC) Circuit Analysis

This section applies Ohm's law and Kirchhoff's voltage and current laws to solve current and voltage in dc resistive circuits with more than one energy source.

Some circuits have two or more voltage sources, and one method to solve these problems is to apply Kirchhoff's voltage law and write a set of simultaneous linear equations. The unknown values are known as the loop currents. The rules of writing this set of equations are as follows:

1. Determine the number of loop currents by using the expression $(T - N) + 1$, where T is the total number of elements in the circuit and N is the number of nodes in the circuit.
2. Assign a loop current for each independent closed path.
3. Apply Kirchhoff's voltage law to each loop. If you are applying Kirchhoff's voltage law to one loop, and an element has two or more loop currents flowing through it, the total voltage drop across the element equals the sum (or difference) of the voltage drops due to the individual loop currents.

With reference to the circuit shown in Figure 12.13a, the three steps are as follows:

1. The circuit has six elements and four nodes; therefore it can be analyzed using three loop currents (I_A, I_B, and I_C).
2. A set of loop currents is shown in the figure.

3. Kirchhoff's voltage law is applied to each loop.

Now you have three simultaneous equations with three unknowns.

Example 12.9
What is the value of the current through R_1 in the circuit of Figure 12.13b?
A. 0.833 A
B. 0.345 A
C. 0.776 A
D. 1.33 A

> The direction of a loop current may be either clockwise or counterclockwise but once the choice is made do not change it while you are solving the problem.

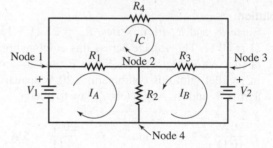

$$(R_1 + R_2) I_A + R_2 I_B - R_1 I_C = V_1$$

$$+ R_2 I_A + (R_2 + R_3) I_B + R_3 I_C = V_2$$

$$- R_1 I_A + R_3 I_B + (R_1 + R_3 + R_4) I_C = 0$$

(a)

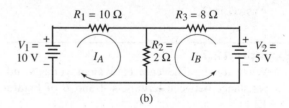

(b)

Figure 12.13

Solution:
The set of equations is as follows:

Loop A: $(10 \text{ }\Omega + 2 \text{ }\Omega)I_A + (2 \text{ }\Omega)I_B = 10 \text{ V}$

Loop B: $(2 \text{ }\Omega)I_A + (8 \text{ }\Omega + 2 \text{ }\Omega)I_B = 5 \text{ V}$

Since R_1 has only one loop current flowing through it, only loop current I_A need be calculated.

$$I_A = \frac{\begin{vmatrix} 10 & 2 \\ 5 & 10 \end{vmatrix}}{\begin{vmatrix} 12 & 2 \\ 2 & 10 \end{vmatrix}} = \frac{100 - 10}{120 - 4} = \frac{90}{116} = 0.776 \text{ A}$$

Answer is C.

Example 12.10

What is the value of the current through the 4-Ω resistor in the circuit of Figure 12.14a?

A. 0.2 A
B. 4.0 A
C. 0.85 A
D. 1.5 A

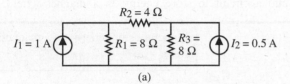

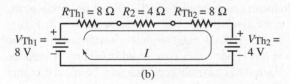

(b)

Figure 12.14

Solution:

This circuit can be solved by converting the Norton equivalent circuits into a Thévenin equivalent circuit. Thus Figure 12.14a may be redrawn as Figure 12.14b.

Now apply Kirchhoff's voltage law to Figure 12.14b:

> **Note:** The current through R_1 and R_2 in the original circuit is not 0.2A.

$$I_{4\,\Omega} = \frac{V_{\text{Th}_1} - V_{\text{Th}_2}}{R_1 + R_2 + R_3} = \frac{8 \text{ V} - 4 \text{ V}}{8\,\Omega + 4\,\Omega + 8\,\Omega} = 0.2 \text{ A}$$

Answer is A.

12.4
Inductance

The second passive element mentioned in Section 12.1 is **inductance.** The unique property of this element is that it stores energy in a *magnetic field.* Let us, then, examine some fundamental properties of a magnetic field.

Before 1819, electricity and magnetism were considered independent natural phenomena. However, in that year Hans Christian Oersted (1777–1851) observed an interaction between electricity and magnetism. Thus the new science of electromagnetism was born. Oersted's experiment showed that an electric current caused a magnetic compass needle to move. Because the wire through which the electric current flowed did not come into direct contact with the compass needle, Oersted concluded that the interaction must involve the surrounding space.

Whenever there is an interaction between two points in space with no visible contact between them, the surrounding space is called a field. The two important properties of a field are intensity (magnitude) and direction. In Oersted's experiment the current had an effect on the magnetic compass because of a field.

We now know that, whenever a current flows in a conductor, a magnetic field is created. The flow of current dictates both the intensity and the direction of the magnetic field. The larger the current value, the

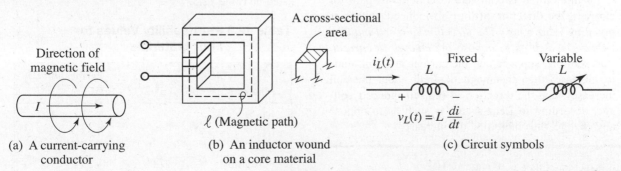

(a) A current-carrying conductor (b) An inductor wound on a core material (c) Circuit symbols

Figure 12.15

greater the intensity, **H,** of the magnetic field, which is a vector quantity. The direction of the magnetic field is determined by the *right-hand rule:* If the thumb of the right hand points in the direction of conventional current, the fingers of the right hand will curl around the conductor in the direction of the magnetic field. (See Figure 12.15a and b.)

Because it is a human trait to want to visualize every concept, it is convenient to have some way of depicting a field. Michael Faraday (1791–1867), a pioneer in the science of electromagnetism, suggested adopting "lines of force" to represent a magnetic field. Although these lines are imaginary, they are the best way to visualize a magnetic field.

For a current-carrying conductor, Figure 12.15a shows magnetic lines of force and their direction. The direction of the magnetic field is always perpendicular to the direction of current, and the field must always form a complete loop.

If a current-carrying conductor is wound in the form of a coil, not only will the magnetic lines of force encircle each infinitesimal portion of the conductor, but also they will add to produce a magnetic field surrounding the entire coil.

Up to this point, the discussion of the interaction between electricity and magnetism has dealt with the fact that the flow of current in a conductor produces a magnetic field about the conductor. Now consider another interaction between electricity and magnetism; that is, a changing magnetic field surrounding a conductor induces a voltage in the conductor. This induced voltage is called a *self-induced voltage* because it is induced in the same conductor that is carrying the current.

Note that a voltage, not a current, is induced in the conductor. The existence of current depends on whether or not there is a complete circuit. (If the induced voltage occurs in another conductor, the principle is known as *mutual inductance.* Applications of mutual inductance are given in section 12.7.)

Ideal Transformers

If the circuit is complete (a continuous path for current), the direction of the self-induced voltage is given by Lenz's law: *The direction of a self-induced voltage is such as to oppose the change in current.* For example, suppose that the current in a coil tends to increase; then the magnetic field about the coil increases, and the direction of the self-induced voltage, according to Lenz's law, is such as to oppose any change, thus inhibiting the increase.

The unit of inductance is the henry (H).

On the other hand, when the current in a circuit tends to decrease, the magnetic field also decreases and the direction of the self-induced voltage tends to keep the current from decreasing. Thus, when the magnetic field is collapsing, it gives energy back to the circuit, temporarily acting as an energy source. This opposition to change in current is an effect of electromagnetic induction.

The circuit element that has the property of opposing any change in current in a circuit by storing energy in a magnetic field is called *self-inductance.* The basic unit for inductance, mutual or self, is the henry (H), named in honor of Joseph Henry. Inductance, the electrical property of storing energy in a magnetic field, is represented by the symbol L.

An *inductor* is the physical device connected into a circuit to store energy in a magnetic field.

Inductors store energy in a magnetic field.

Inductors may be classified as fixed or variable; the circuit symbols for both types are shown in Figure 12.15c. A *fixed inductor* has a nonadjustable value and has either an air core or an iron core. A variable inductor has an iron core that can be moved within the coil to vary the magnetic field and thereby vary its inductance.

An inductor stores energy in a magnetic field. In terms of physical parameters, inductance is calculated by using the equation

$$L = \frac{\mu A N^2}{l} \qquad [12.19]$$

where μ = permeability of the magnetic material (H/m),
A = cross-sectional area of the path (m²),
N = number of turns of the coil,
l = length of the path (m).

Values of permeability for various materials are listed in Table 12.3.

Table 12.3 Permeability Values for Four Materials

Material	Permeability, μ (H/m)
Air	1.257×10^{-6}
Iron	6.53×10^{-3}
Steel	8.80×10^{-3}
Permalloy	1.257×10^{-1}

Voltage/Current Relationship for an Inductor

The voltage across an inductor depends on the value of the inductor times the rate of change of current through the inductor. The equation is as follows:

$$v_L(t) = L\frac{di_L}{dt} \qquad [12.20]$$

Energy Stored in an Inductor

The energy stored by an inductor is given by the expression

$$Energy = \frac{1}{2}LI^2 \qquad [12.21]$$

Inductors in Series

Inductors connected in series combine in the same way as resistors in series; that is, for n inductors in series the equivalent inductance is as follows:

$$L_T = L_1 + L_2 + \dots + L_n \qquad [12.22]$$

> Inductors in series and parallel are combined using the same rules as for resistors.

Inductors in Parallel

Inductors connected in parallel combine in the same way as resistors in parallel.

$$L_{eq} = \cfrac{1}{\dfrac{1}{L_1} + \dfrac{1}{L_2} + \dots + \dfrac{1}{L_n}} \qquad [12.23]$$

If only two inductors are connected in parallel, the product over the sum rule can be used:

$$L_{eq} = \frac{L_1 L_2}{L_1 + L_2} \qquad [12.24]$$

Example 12.11

What is the equivalent inductance of the circuit shown in Figure 12.16?

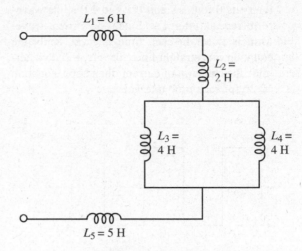

Figure 12.16

A. 13 H
B. 21 H
C. 12.5 H
D. 15 H

Solution:

Apply equations 12.22 and 12.24 to obtain

$$L_T = L_1 + L_2 + \frac{L_3 \times L_4}{L_3 + L_4} + L_5$$

$$= 6\,H + 2\,H + \frac{4\,H \times 4\,H}{4\,H + 4\,H} + 5\,H = 15\,H$$

Answer is D.

Example 12.12

The current waveform through a 1-H coil is shown in Figure 12.17. What is the voltage waveform? (Choose one of the waveforms given in Figure 12.18 as the answer.)

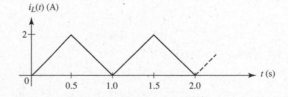

Figure 12.17

A. a square wave
B. a triangular wave
C. a pulse wave
D. a triangular wave with dc offset

Solution:

The waveform is periodic; that is, it repeats after 1 s. Determine the voltage waveform for the time intervals 0 to 0.5 s and 0.5 s to 1.0 s; the waveform repeats after 1 s. Since the current waveform is linear in each time interval, apply the equation of a straight line ($y = mx + b$) to determine the equation of current; then apply equation 12.20 for each time interval.

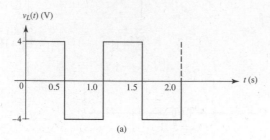

$v_L(t)$ (V)

(a)

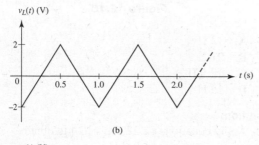

$v_L(t)$ (V)

(b)

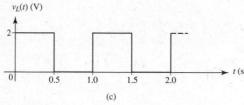

$v_L(t)$ (V)

(c)

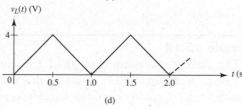

$v_L(t)$ (V)

(d)

Figure 12.18

For the time interval 0 to 0.5 s:

$$i(t)_{0-0.5\,s} = \frac{(2-0)A}{(0.5-0)s}t + 0 = 4\frac{A}{s} \times t$$

Apply equation 12.20 to obtain

$$v_L(t) = 1\,H \times \frac{d}{dt}(4\,\frac{A}{s}\,t) = 4\,V$$

For the time interval 0.5 s to 1.0 s:

$$i(t)_{0.5\,s-1.0\,s} = \frac{(0-2)\,A}{(1-0.5)\,s} \times t + 4\,A$$

$$= -4\,\frac{A}{s} \times t + 4\,A$$

Apply equation 12.20 to obtain

$$v_L(t) = 1\,H \times \frac{d}{dt}\left(-4\,\frac{A}{s}\,t + 4\right) = -4\,V$$

Plot the voltage values for each time interval; hence, the voltage waveform is a square wave.

Answer is A.

12.5
Capacitance

The third passive element mentioned in Section 12.1 is **capacitance.** Its properties are as unique as those of resistance and inductance. Whenever there is a difference of charge in a circuit, a voltage (or potential difference) exists. A *capacitor* stores energy in an *electric field,* which is the space between a positive and a negative charge. As in the case of a magnetic field, the "lines of force" concept may be used to

> A positive charge on one plate is balanced by a negative charge on the other plate.

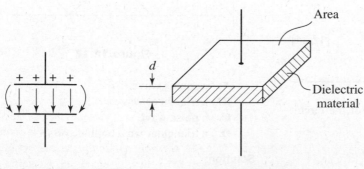

(a) Direction of electric field (b) Physical device (c) Circuit symbols for a capacitor

Figure 12.19

visualize an electric field. Unlike the lines of force of a magnetic field, which form closed loops, the lines of force of an electric field always *begin and end* on charges.

Like the lines of force of a magnetic field, the lines of force of an electric field indicate both strength and direction, and electric field intensity, **E,** is a vector quantity. The greater the number of lines, the more intense the electric field. The strength of an electric field, measured from a point charge, is directly proportional to the charge and indirectly proportional to the distance squared. Thus the strength of an electric field is greatest closest to the charge and decreases as one moves away from it.

The direction of the electric field is the path that a positive charge would travel if placed in the field. Since a basic phenomenon of charge is that like charges repel and unlike charges attract, the direction of the lines of force is always *from the positive charge to the negative charge,* as shown in Figure 12.19a.

An electric field can be created in a circuit by placing two conducting plates in parallel and having one plate more positive than the other, as in Figure 12.19a. The material between the two plates is nonconducting, or insulating. This insulating material is called a *dielectric.* Some dielectrics in common use are air, Bakelite™, ceramic, Formica™, glass, kraft, polyethylene, and Teflon™.

Thus, capacitance is the property of storing energy in an electric field. The energy is stored between the parallel plates of a capacitor, which is a physical device having the property of capacitance. The unit for capacitance is the farad (F).

The capacitance for a pair of parallel plates such as those shown in Figure 12.19b is given by the equation

$$C = \frac{\varepsilon A}{d} \qquad [12.25]$$

where C = capacitance (F),
ε = dielectric constant (F/m),
A = area of each plate (m^2),
d = distance between plates (m).

The dielectric constants of several materials are listed in Table 12.4.

Table 12.4 Dielectric Constants of Six Materials

Material	Dielectric Constant ϵ (10^{-11} F/m)
Air	0.885
Pyrex	3.54
Mica	7.85
Mylar	4.16
Quartz	3.54
Tantalum dioxide	150

To examine how one plate of a capacitor can be made positive with respect to the other plate, consider a capacitor connected to a battery through a switch. When the switch is closed, electrons from the negative terminal of the battery collect on the bottom plate. To maintain equilibrium, an equal number of electrons leave the top plate and drift toward the positive terminal of the battery. While this is happening, the capacitor is said to be *charging.* This flow of charges will continue until the voltage across the capacitor equals the source voltage. When the capacitor voltage and applied voltage are equal, the capacitor is said to be *charged.*

The capacitor will remain in this charged condition even after the source is removed. For the capacitor to return to a neutral or uncharged state, an external path must be provided for the flow of charge. This external path may range from a short circuit (zero resistance) to a complicated network. While the capacitor is returning to a neutral (or uncharged) state, it is said to be *discharging.* Although a capacitor supplies energy to a circuit while it is discharging, it is not classified as an energy source because a capacitor cannot *transform* some other form of energy into electrical energy; it can only *store* energy.

Since the movement of charges in one direction constitutes an electric current, the current associated with a capacitor is present only when the capacitor is either charging or discharging, for only at these times is there a movement of electrons. It should be emphasized that charge flows, not *between* the plates of a capacitor, but only in the external circuit, although it is common to refer to the current in the capacitor or the capacitor current.

Capacitors may be classified as either fixed or variable. The circuit and letter symbols are shown in Figure 12.19c. A *fixed capacitor* has a nonadjustable value. A *variable capacitor* has one set of plates that are movable, thereby changing the value of the capacitance.

Both fixed and variable capacitors have maximum voltage ratings specified by the manufacturer. If the voltage across the plates of a capacitor exceeds the maximum rating, the dielectric will break down and permit a charge to flow between the plates.

In terms of charge and voltage, capacitance is the ratio of the charge stored on the plates of a capacitor to the voltage measured across the capacitor:

$$C = \frac{q_C(t)}{v_C(t)} \qquad [12.26]$$

Voltage/Current Relationship for a Capacitor

Capacitor current is the value of the capacitor times the time rate of change in voltage across the capacitor:

$$i_C(t) = C\frac{dv_C}{dt} \qquad [12.27]$$

Energy Stored in a Capacitor

The energy stored in a capacitor is given by the equation

$$\text{Energy} = \frac{1}{2}CV^2 \qquad [12.28]$$

Capacitors in Series

Capacitors connected in series combine in the same way as resistors in parallel:

Capacitors in series combine like resistors in parallel.

$$C_{eq} = \frac{1}{\dfrac{1}{C_1} + \dfrac{1}{C_2} + \ldots + \dfrac{1}{C_n}} \qquad [12.29]$$

For two capacitors in series the product over the sum rule may be used:

$$C_{eq} = \frac{C_1 C_2}{C_1 + C_2} \qquad [12.30]$$

Capacitors in Parallel

Capacitors connected in parallel combine in the same way as resistors in series:

Capacitors in parallel combine like resistors in series.

$$C_{eq} = C_1 + C_2 + \ldots + C_n \qquad [12.31]$$

Example 12.13

What is the value of capacitance for the capacitor shown in Figure 12.20 if the dielectric constant is 4.16×10^{-11} F/m?

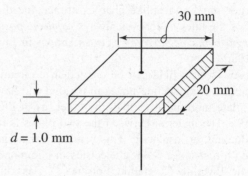

Figure 12.20

A. 0.25 μF
B. 25 pF
C. 2.5 μF
D. 4.1 nF

Solution:
Apply equation 12.25 to obtain

$$C = \frac{(4.16 \times 10^{-11}\text{F/m})(0.02 \text{ m} \times 0.03 \text{ m})}{0.001 \text{ m}} = 25 \text{ pF}$$

Answer is B.

Example 12.14

The charge on the plates of a 1.0-μF capacitor is 50 μC. What is the energy stored by the capacitor?

A. 1.25 mJ
B. 25 mJ
C. 5.5 mJ
D. 10.5 mJ

Solution:

Calculate the voltage across the capacitor by rearranging equation 12.26:

$$v_C(t) = \frac{q_C(t)}{C} = \frac{50\ \mu C}{1\ \mu F} = 50\ V$$

Now apply equation 12.28:

$$\text{Energy} = \frac{1}{2}\ CV^2 = \frac{1}{2} \times 1\ \mu F \times (50\ V)^2 = 1.25\ mJ$$

Answer is A.

Example 12.15

What is the total capacitance between terminals A and B for the circuit of Figure 12.21?

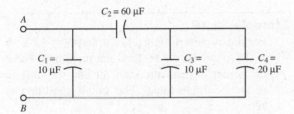

Figure 12.21

A. 10 μF
B. 5.0 μF
C. 30 μF
D. 60 μF

Solution:

For the parallel combination of C_3 and C_4, the equivalent capacitance is 30 μF. This capacitance value is in series with C_2; use equation 12.30:

$$C_{eq} = \frac{C_2 C_{3,4}}{C_2 + C_{3,4}} = \frac{60\ \mu F \times 30\ \mu F}{60\ \mu F + 30\ \mu F} = 20\ \mu F$$

Combine this equivalent capacitance value with C_1 to obtain 30 μF.

Answer is C.

12.6
Resistor-Capacitor (R-C) and Resistor-Inductor (R-L) Circuits

R-C Circuits

Figure 12.22 shows a resistor and a capacitor connected in series through a switch to a dc voltage

source. When the switch is closed, the voltage across the capacitor increases exponentially. The rate of rise depends on the circuit's time constant, τ. The time constant $\tau = RC$ and is expressed in seconds. In one time constant, the voltage across the capacitor reaches 63.3% of its steady-state value. For all practical purposes the steady-state value is reached in 5τ. The value of resistance as "seen" by the capacitor is symbolized as R.

> In one time constant, the voltage across the capacitor has charged to 63.3% of its final value.

The equation for voltage across the capacitor, where the voltage increases or decreases exponentially is

$$v_C(t) = V_f + (V_i - V_f)e^{-t/\tau} \qquad [12.32]$$

where V_f is the maximum value that C will charge (or discharge) to, and V_i is the initial voltage across the capacitor.

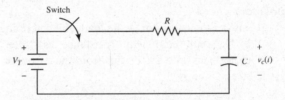

Figure 12.22

If the initial voltage across the capacitor is zero ($V_i = 0$), then $v_C(t)$ is given by the equation

$$v_C(t) = V_f(1 - e^{-t/\tau}) \qquad [12.33]$$

Example 12.16

Consider the circuit of Figure 12.22. The capacitor is initially uncharged, and the circuit values are $V_T = 10$ V, $R = 2\ \Omega$, and $C = 1/8$ F. What is the voltage across the capacitor 0.5 s after the switch is closed?

A. 5.52 V
B. 6.67 V
C. 7.54 V
D. 8.65 V

Solution:

The time constant for the circuit is

$$\tau = RC = 2 \times \frac{1}{8} = \frac{1}{4}\ s$$

359

Apply equation 12.33 to obtain

$$v_C(t) = 10 \text{ V} \times (1 - e^{-0.5/0.25}) = 8.65 \text{ V}$$

Answer is D.

Example 12.17

What is the time constant for the circuit of Figure 12.23?

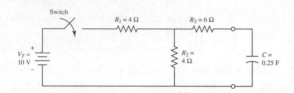

Figure 12.23

A. 2.0 s
B. 1.5 s
C. 4.5 s
D. 5.0 s

Solution:

Consider the capacitor as the load, and determine the Thévenin equivalent resistance as "seen" looking back into the circuit from the capacitor. The equation is

$$R_{Th} = \frac{R_1 \times R_2}{R_1 + R_2} + R_3 \frac{4 \, \Omega \times 4 \, \Omega}{4 \, \Omega + 4 \, \Omega} + 6 \, \Omega = 8 \, \Omega$$

Then the time constant is

$$R_{Th} \times C = 8 \, \Omega \times 0.25 \text{ F} = 2 \text{ s}$$

Answer is A.

R-L Circuits

Figure 12.24 shows a resistor and an inductor connected in series through a switch to a dc voltage source. When the switch is closed, the inductor's magnetic field opposes any change in current; therefore current flow increases exponentially. The rate of increase in current depends on the circuit's time constant, τ. The time constant $\tau = L/R$ and is expressed in seconds. In one time constant, the current reaches 63.3% of its steady-state value. For all practical purposes the steady-state value is reached in 5τ.

The general equation for current through the inductor where the current increases or decreases exponentially is

$$i_L(t) = I_f + (I_i - I_f)e^{-t/\tau} \qquad [12.34]$$

where I_f is the maximum value of current that can exist in the inductor, $I_f = V_T/R$ for the circuit shown in Figure 12.24, and I_i is the initial condition that is the value of current through the inductor before the switch is closed. Note that the initial-condition value may be zero.

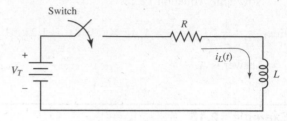

Figure 12.24

Example 12.18

For the circuit of Figure 12.24, let $V_T = 15$ V, $R = 3 \, \Omega$, and $L = 3$ H. Find the maximum value of current through the inductor and the time to reach the final value. The initial condition is zero.

A. 5 A, 2 s
B. 5 A, 5 s
C. 3 A, 1 s
D. 2.67 A, 3 s

Solution:

The final or maximum value of current that can flow in the circuit of Figure 12.24 is given by the equation

$$I_{max} = \frac{V_T}{R} = \frac{15 \text{ V}}{3 \, \Omega} = 5 \text{ A}$$

For all practical purposes, the time to reach I_{max} is

$$5\tau = 5\left(\frac{L}{R}\right) = 5\left(\frac{3 \text{ H}}{3 \, \Omega}\right) = 5 \text{ s}.$$

Answer is B.

12.7 Alternating-Current (AC) Circuit Analysis

This section introduces ac voltage and current waveforms and ac circuit analysis. The analysis of ac circuits requires the use of complex numbers and complex algebra to solve circuit problems. Therefore you may need to refer to Chapter 1, Mathematics, for a review of complex numbers.

Sinusoidal Waveform

In the United States commercially generated electrical energy is produced in the form of a periodic alternating voltage waveform, called a *sinusoidal waveform,* as shown in Figure 12.25, where T represents the period of the wave or the time for one complete cycle.

The period, T, of a sinusoidal waveform is the time for one complete cycle.

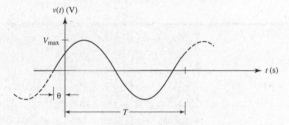

Figure 12.25

The reciprocal of the period is the frequency, f, of the wave, which is given by the equation

Frequency, f, has the unit of hertz (Hz) and ω is angular frequency with units of radians per second (rad/s).

$$f = \frac{1}{T} \qquad [12.35a]$$

The unit of frequency is the hertz (Hz). This wave is periodic because it repeats every T seconds. The wave of Figure 12.25 can be expressed mathematically as

Note: electrical engineers often express the argument of a sinusoid in mixed units, ωt is in radians per second and the phase angle in degrees. However, to add ωt and θ, they must be in the same units.

$$v(t) = V_{max} \sin(\omega t + \theta) \qquad [12.35b]$$

where $\omega = 2\pi f$ and has units of radians per second, and θ is the phase angle, expressed in radians.

Example 12.19

Write the equation for the sinusoidal voltage waveform shown in Figure 12.26.
- **A.** $4(\sin 4t + 45°)$ V
- **B.** $2(\sin 25.12t + \frac{\pi}{4})$ V
- **C.** $8(\sin 25.25t + 22°)$ V
- **D.** $4(\sin 12t + \frac{\pi}{8})$ V

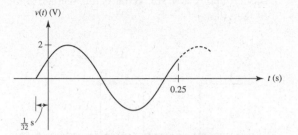

Figure 12.26

Solution:

Since the period, T, is 0.25 s, then the frequency, f, is 4 Hz ($f = 1/0.25$ s $= 4$ Hz), and $\omega = 2\pi(4) = 25.13$ rad/s. The phase angle is determined by converting 1/32 s to radians by setting up a ratio of $t/T = \theta/2\pi$:

$$\frac{\frac{1}{32}}{0.25} = \frac{\theta}{2\pi}$$

Solve for θ to obtain $\pi/4$ rad.

Hence the equation is

$$v(t) = 2\left(\sin 25.13t + \frac{\pi}{4}\right) \text{ volts}$$

Answer is B.

Average Values

The average value of a periodic waveform (either voltage or current) is given by equation 12.36a:

$$X_{eq} = \frac{1}{T}\int_0^T x(t)dt \qquad [12.36a]$$

where $x(t)$ is the waveform expression in the time domain, and T is the period of the waveform.

If the waveform is symmetrical about the time axis (x-axis), the average value is zero. Equation 12.36a states that the average value equals the area under the waveform divided by the period, or

$$X_{eq} = \frac{\text{area}}{\text{period}} \qquad [12.36b]$$

Equation 12.36b is useful when determining the average value for pulse, triangular, and other similarly shaped waveforms that are formed by straight lines only.

Example 12.20

. For Figure 12.27, the maximum value is 2 V and the frequency is 4 Hz. What is the average value of the voltage?

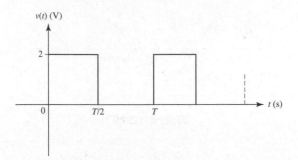

Figure 12.27

A. 1.0 V
B. 1.25 V
C. 2.15 V
D. 0.5 V

Solution:

Rearrange equation 12.35a to find the period of the wave:

$$T = \frac{1}{4} = 0.25 \text{ s} \quad \text{or} \quad \frac{T}{2} = 0.125 \text{ s}$$

Apply equation 12.36b to obtain

$$V_{avg} = \frac{2 \text{ V} \times 0.125 \text{ s}}{0.25 \text{ s}} = 1.0 \text{ V}$$

Answer is A.

Effective or RMS Values

An alternating voltage or current usually requires knowledge of its root mean square (rms) value. The rms value of voltage results in the same amount of heating effect to a load as a dc voltage of the same value would produce. Hence the effect is the same. The general equation for calculating rms values is as follows:

$$X_{\text{rms}} = \sqrt{\frac{1}{T} \int_0^T x^2(t)\, dt} \qquad [12.37]$$

Example 12.21

What is the rms value for the waveform shown in Figure 12.27? The amplitude is 2 V, and the period is 0.25 s.

A. 2 V
B. 2.5 V
C. $\sqrt{2}$ V
D. $2\sqrt{2}$ V

Solution:

Refer to Figure 12.27 and Example 12.20. The appropriate expression is

$$v(t) = \begin{cases} 2 \text{ V} & \text{for } 0 < t < 0.125 \text{ s} \\ 0 \text{ V} & \text{for } 0.125 \text{ s} < t < 0.25 \text{ s} \end{cases}$$

> The rms value of a periodic sinusoid is the peak value (maximum value) divided by the square root of two. This rule applies only to sinusoid waveforms and not to other periodic waveforms such as the pulse wave of Figure 12.27.

Apply equation 12.37 to obtain

$$v_{\text{rms}}(t) = \sqrt{\frac{1}{25} \left[\int_0^{0.125} 4\, dt + \int_{0.125}^{0.25} 0\, dt \right]} = \sqrt{2} \text{ V}$$

Answer is C.

Impedance

Impedance, Z, is defined as the opposition to ac current and is expressed in ohms (Ω). It is the ratio of total ac voltage to total ac current when both are expressed as phasors. Impedance is a complex quantity. Since it has both magnitude and phase, it can be expressed in either rectangular or polar form (magnitude and angle). In rectangular form, the real part is resistive and the imaginary part is reactive. In circuit analysis, the components are considered ideal. Therefore, a resistive element, R, has neither inductance nor capacitance associated with it. Similarly, an inductive element has neither resistance nor capacitance, and a capacitive element has neither resistance nor inductance. In rectangular form, impedance is expressed as

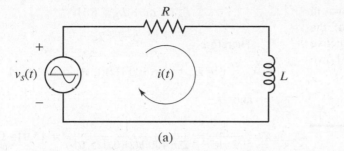

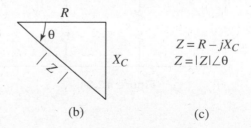

Figure 12.28

To determine the impedance of a circuit, the inductance and capacitance values and their complex impedance values must be determined:

$Z_2 = j\omega L$ and $Z_c = 1/j\ \omega C = -j/\omega C$

$$Z = R \pm jX \qquad [12.38]$$

where R is the resistance value, and X is the reactive value. Reactance may be either inductive or capacitive.

Inductive Reactance

The magnitude of inductive reactance is given by the expression

$$X_L = \omega L = 2\pi f L \qquad [12.39]$$

Thus inductive reactance depends on the value of L, the frequency, and the constant 2π.

Figure 12.28a shows a series R-L circuit driven by an ac voltage source. The phasor diagram and impedance expressions for this circuit are given in Figures 12.28b and 12.28c, respectively.

For an inductor, voltage leads current by 90°. Mathematically, this lead is represented by a $+j$ term. The impedance of an inductor is expressed as

$$Z_L = jX_L = j\omega L = j2\pi f L \qquad [12.40]$$

To find the equivalent impedance of a circuit, use the same rule as for resistors; that is, add impedances in series. For impedances in parallel, use the reciprocal rule.

Capacitive Reactance

The magnitude of capacitive reactance is given by the equation

$$X_C = \frac{1}{\omega C} = \frac{1}{2\pi f C} \qquad [12.41]$$

Thus capacitive reactance depends on the reciprocal of C, the frequency, and the constant 2π.

Figures 12.29a–c show a series R-C circuit driven by an ac voltage source, the phasor diagram, and the impedance expressions for this circuit.

For a capacitor, voltage lags current by 90°. A lag of 90° is represented by a $-j$ or $1/j$ term in an impedance formula. The impedance of a capacitor is expressed as

$$Z_C = -jX_C = \frac{-j}{\omega C} = \frac{-j}{2\pi f C} \qquad [12.42]$$

Series R-L-C Circuit

Figure 12.30a shows a series R-L-C circuit. The total impedance of this circuit is given by the equation

$$Z = R + jX_L - jX_C \qquad [12.43]$$

If the magnitude of the inductive reactance, X_L, is greater than the capacitive reactance, X_C, the pha-

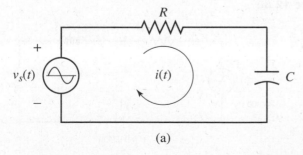

Figure 12.29

sor diagram will have a positive angle, as shown in Figure 12.30b. If, however, the magnitude of the capacitive reactance is greater than the inductive reactance, the phase angle will be negative, as shown in Figure 12.30c.

Parallel R-L-C Circuits

The reciprocal of impedance is known as admittance and is represented by the symbol, Y.

Figure 12.31 shows a parallel R-L-C circuit. The total impedance of this circuit is given by the equation

$$Z = \frac{1}{\frac{1}{R} + \frac{1}{jX_L} + jX_c} \qquad [12.44]$$

Example 12.22

What is the total impedance of the circuit shown in Figure 12.32?

A. $2.2 \angle 48.7° \text{ k}\Omega$
B. $6.4 \angle -36.6° \text{ k}\Omega$
C. $5.19 \angle 54.1° \text{ k}\Omega$
D. $4.73 \angle 62.3° \text{ k}\Omega$

Solution:

First determine the equivalent inductance value for L_1 and L_2.

$$L_{eq} = L_1 + L_2 = 8 \text{ H}$$

Therefore

$$X_{eq} = 2\pi f L_{eq} = 2\pi(400 \text{ Hz})(8 \text{ H}) = 20,106 \ \Omega$$

and

$$X_C = \frac{1}{2\pi f C} = \frac{1}{2\pi(400 \text{ Hz})(0.025 \ \mu\text{F})} = 15,915 \ \Omega$$

The total reactance is $X_T = X_{eq} - X_C = 4.19 \text{ k}\Omega$, and the impedance is

$$Z = (2.2 + j\,4.19) \text{ k}\Omega = 4.73 \angle 62.3° \text{ k}\Omega$$

Answer is D.

Example 12.23

What is the ac current in a circuit if the impedance is $(5 + j4) \ \Omega$ and the voltage source, V_{rms}, is 10 V?

A. $2.04 \angle 38.7° \text{ A}$
B. $1.56 \angle -38.7° \text{ A}$
C. $1.25 \angle 54.3° \text{ A}$
D. $1.11 \angle -42.6° \text{ A}$

Solution:

Divide the voltage by the total impedance to determine the ac current:

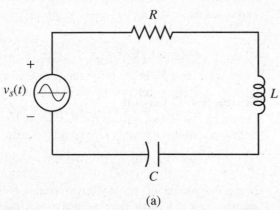

(a)

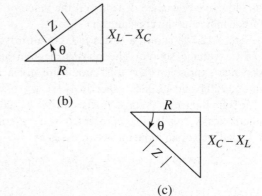

(b)

(c)

Figure 12.30

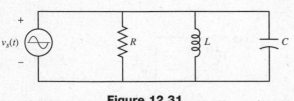

Figure 12.31

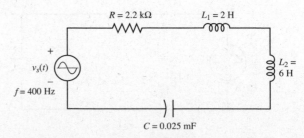

Figure 12.32

364

$$I_{rms} = \frac{10 \text{ V}}{(5 + j4) \; \Omega} = (1.22 - j0.976) \text{ A}$$

$$= 1.56 \angle -38.7° \text{ A}$$

Answer is B.

Example 12.24

What is the total impedance, Z_T, of the circuit shown in Figure 12.33?

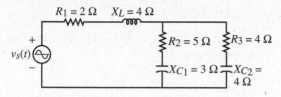

Figure 12.33

A. $(4.28 + j\,2.22) \; \Omega$
B. $(11 - j\,3) \; \Omega$
C. $(2.54 - j\,3.87) \; \Omega$
D. $(6.28 + j\,9.31) \; \Omega$

Solution:
Determine the impedance of each branch.

$$Z_1 = R_1 + j\,X_L = (2 + j4) \; \Omega$$
$$Z_2 = R_2 - j\,X_{C_1} = (5 - j3) \; \Omega$$
$$Z_3 = R_3 - j\,X_{C_2} = (4 - j4) \; \Omega$$

Then

$$Z_T = Z_1 + \frac{Z_2 Z_3}{Z_2 + Z_3}$$

$$= (2 + j4) + \frac{(5 - j3)(4 - j4)}{(5 - j3) + (4 - j4)}$$

$$= (4.28 + j2.22) \; \Omega$$

Answer is A.

Example 12.25

What is the total current, I_T, of the circuit shown in Figure 12.34?

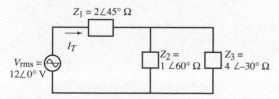

Figure 12.34

A. $(1 + j3) \text{ A}$
B. $(4 - j2) \text{ A}$
C. $4.04 \angle -45.3° \text{ A}$
D. $5.13 \angle 22.7° \text{ A}$

Solution:
The total impedance for the circuit is

$$Z_T = Z_1 + \frac{Z_2 Z_3}{Z_2 + Z_3}$$

$$= 2\angle 45° + \frac{(1\angle 60°)(4\angle -30°)}{1\angle 60° + 4\angle -30°} = 2.97\angle 45.3° \; \Omega$$

Then

$$I_T = \frac{12.0\angle 0° \text{ V}}{2.97\angle 45.3° \; \Omega} = 4.04\angle -45.3° \text{ A}$$

Answer is C.

Complex Power

The topic of this section is the product of voltage times current for ac circuits. For dc circuits, this product is the power dissipated by the resistance. For ac circuits that contain resistors as well as inductors and/or capacitors, the power dissipated by the resistance is known as the **real power;** but the product of voltage and current for the reactive elements, as well as the product of the total voltage and current, must be considered. The unit for power is the watt (W), and power is symbolized by the letter P.

The real power dissipated by an ac circuit is due to the resistance in the circuit and is given by the equation

$$P = V_{rms}I_{rms} \cos \theta \qquad [12.45]$$

where θ is the phase angle measured from V to I. The term $\cos \theta$ is referred to as the power factor, pf. Since $\cos \theta$ is positive for both positive and negative angles, the power factor is referred to as either leading (when the current leads the voltage) or lagging (when the current lags the voltage). For this analysis it is customary to use voltage as the reference. Therefore, for an inductive circuit the power factor is lagging, and for a capacitive circuit the power factor is leading.

The volt-ampere product for the reactive element(s) in the circuit is known as reactive power, is given the symbol Q, and has units of volt-amperes reactive (vars). The equation for reactive power is as follows:

$$Q = V_{rms}I_{rms} \sin \theta \qquad [12.46]$$

The total volt-ampere value of a circuit is the vector addition of the real power and the reactive power.

The complex power, S, for an ac circuit with both resistive and reactive elements is given by the expression

$$S = P \pm j Q \qquad [12.47]$$

and the units are volt-amperes (va). Volt-amperes represents a complex number, and the phase angle is either positive or negative depending on whether the circuit is capacitive or inductive, respectively, as shown in Figure 12.35.

For the power triangle, reactive power, Q, is positive for capacitive loads and negative for inductive loads.

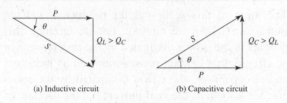

(a) Inductive circuit   (b) Capacitive circuit

Figure 12.35

Example 12.26

What is the volt-ampere value for the circuit of Figure 12.36a?
A. $(31.9 - j\,24)$ va
B. $(16.4 + j\,21.1)$ va
C. $(7.8 - j\,16.5)$ va
D. $(21.1 + j\,16.4)$ va

Solution:

The current, I_{rms}, is given by the equation

$$I_{rms} = \frac{20\angle 60° \text{ V}}{(8 + j6)\ \Omega} = 2\angle 23.1° \text{ A}$$

The angle from V to I is $60° - 23.1° = 36.9°$. See Figure 12.36b. Apply equation 12.45 to obtain the real power:

$$P = (20)(2)\cos 36.9° = 31.9 \text{ W}$$

Then apply equation 12.46 to determine the reactive volt-amperes:

$$Q = (20)(2)\sin 36.9° = 24 \text{ vars}$$

Since the circuit is inductive, the power factor is lagging. Apply equation 12.47 to obtain

$$S = (31.9 - j24) \text{ va}$$

The triangle shown in Figure 12.36c is often referred to as a power triangle.

Answer is A.

Example 12.27

What is the power factor for the circuit of Figure 12.36?
A. 0.60
B. 0.34
C. 0.82
D. 0.79

Solution:

From the solution of Example 12.26, $\theta = 36.9°$. Then:

$$\text{Power factor} = \cos \theta = \cos 36.9° = 0.79$$

Answer is D.

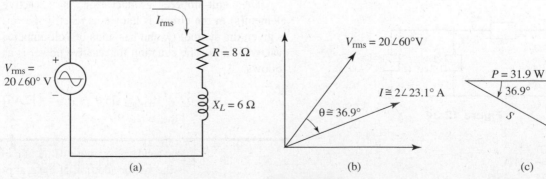

(a)   (b)   (c)

Figure 12.36

Resonance

When a resistor, an inductor, and a capacitor are connected in series, there is a frequency at which $X_L = X_C$. At this frequency the circuit is said to be in resonance, and the frequency is called the series resonant frequency, f_0. At f_0, the impedance of the circuit is at a minimum and is equal to R. At resonance, the current is at a maximum; also, the voltage drops across, L and C are at a maximum. At frequencies below resonance, X_C is greater than X_L. At frequencies above resonance, X_L is greater than X_C. The resonant frequency, f_0, depends on the value of L and C and is given by the equation

> Resonance occurs when the inductive reactance equals the capacitive reactance.

$$\omega_0 = 2\pi f_0 = \frac{1}{\sqrt{LC}} \qquad [12.48]$$

where ω_0 is in radians per second and f_0 is in hertz. Since at resonance $X_L = X_C$, then

$$\omega_0 L = \frac{1}{\omega_0 C} \qquad [12.49]$$

and the impedance of the series circuit at resonance is given by the equation

$$Z = R \qquad [12.50]$$

A figure of merit to compare resonant circuits is known as the quality factor, Q. For a series resonant (tuned) circuit, the quality factor is expressed as

$$Q = \frac{\omega_0 L}{R} = \frac{1}{\omega_0 CR} \qquad [12.51]$$

Note: When analyzing a resonant circuit, the symbol Q represents the quality factor, not charge or reactive power.

Figure 12.37 shows how the current in a series circuit varies as the frequency is changed. The ratio of the resonant frequency to the quality factor is known as the bandwidth, BW. The equation for bandwidth is

$$BW = \frac{\omega_0}{Q} = \frac{R}{L} \text{ rad/s} \qquad [12.52a]$$

or

$$BW = \frac{R}{2\pi L} \text{ Hz} \qquad [12.52b]$$

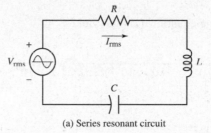

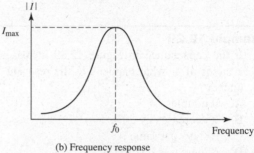

(a) Series resonant circuit

(b) Frequency response

Figure 12.37

Figure 12.38 is a parallel *R-L-C* circuit. The frequency at which $X_L = X_C$ is also known as the resonant frequency and is given by equation 12.48. For this circuit, the voltage is a maximum at resonance, and the quality factor is expressed as

$$Q = \omega_0 RC = \frac{R}{\omega_0 L} \qquad [12.53]$$

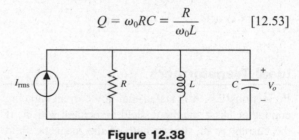

Figure 12.38

Example 12.28

What value of capacitance will cause the circuit of Figure 12.39 to have a resonant frequency of 1.2 MHz?

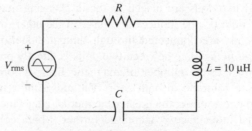

Figure 12.39

A. 0.054 μF
B. 0.00176 μF
C. 0.0034 μF
D. 0.012 μF

Solution:

Rearrange equation 12.48 to yield

$$C = \frac{1}{L(2\pi f_0)^2} = \frac{1}{(10 \times 10^{-6})(2\pi \times 1.2 \times 10^6)^2}$$

$$= 0.00176 \ \mu F$$

Answer is B.

Example 12.29

If the capacitance in Figure 12.39 increases by a factor of 4, what happens to the resonant frequency?

A. It doubles.
B. It decreases by 50%.
C. It stays the same.
D. It increases by 50%.

Solution:

From equation 12.48:

$$f_0 = \frac{1}{2\pi\sqrt{L \times 4C}} = \frac{1}{4\pi\sqrt{LC}} = \frac{\frac{1}{2}}{2\pi\sqrt{LC}}$$

The resonant frequency is halved.

Answer is B.

Ideal Transformers

In Section 12.4, it is stated that any current-carrying conductor has a magnetic field associated with it. If the current is ac, the strength of the magnetic field is constantly increasing and decreasing. Likewise, if a changing magnetic field exists about the conductor, an ac voltage is induced in it. (A current exists in the conductor provided that a closed path exists.) The magnetic field is increased by winding the conductor in the form of a coil, thus creating the passive element of self-inductance.

If two coils are placed so that a changing magnetic field of one induces a voltage in the other, the two coils are connected through mutual induction. *Mutual inductance* between two coils is directly proportional to the number of magnetic lines of force that are common to both coils. The closer the coils are to one another, the greater the mutual inductance. Mutual inductance is increased further if both coils can be wound on a common iron core, as shown in Figure 12.40. If the coils are separated, mutual induc-

tance is reduced. The circuit element that uses the principle of mutual inductance is the transformer.

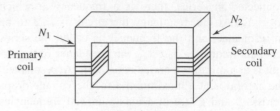

Figure 12.40

This review section deals with **ideal transformers** (those that have no losses). The iron-core transformers used in power and audio systems closely approximate the ideal. For the ideal transformer, reflected impedance, current ratio, and voltage ratio are quickly and easily calculated in terms of the number of turns on the primary and secondary coils, N_1 and N_2, respectively.

Figure 12.41 shows the schematic for an ideal transformer, and the equations for this device are as follows:

$$a = \frac{N_1}{N_2} \qquad [12.54]$$

where a represents the ratio of N_1 to N_2.

Transformers are used to step voltages up or down at various points in a system.

$$a = \left|\frac{V_P}{V_S}\right| = \left|\frac{I_S}{I_P}\right| \qquad [12.55]$$

$$Z_P = a^2 Z_S \qquad [12.56]$$

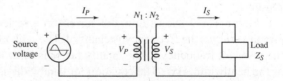

Figure 12.41

Note: Voltage is directly proportional to the turns ratio, and the current is inversely proportional to the turns ratio. Hence, for an ideal transformer the power on the primary side equals the power on the secondary side.

Example 12.30

For the circuit of Figure 12.42, what is the primary current?

A. 2.4 A
B. 1.6 A
C. 2.0 A
D. 0.1 A

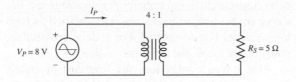

Figure 12.42

Solution:

Find the reflected resistance on the primary side of the transformer by applying equation 12.56:

$$R_P = a^2 R_S = \left(\frac{4}{1}\right)^2 \times 5\ \Omega = 80\ \Omega$$

Apply Ohm's law:

> If the voltage is stepped down, the current is stepped up. Similarly, if the voltage is stepped up, the current is stepped down.

$$I_P = \frac{V_P}{R_P} = \frac{8\ \text{V}}{80\ \Omega} = 0.1\ \text{A}$$

Answer is D.

Example 12.31

What is the primary voltage in the circuit of Figure 12.43?

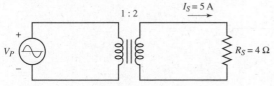

Figure 12.43

A. 10 V
B. 80 V
C. 5 V
D. 20 V

Solution:

Apply Ohm's law to R_S:

$$V_S = I_S R_S = (5\ \text{A})(4\ \Omega) = 20\ \text{V}$$

Rearrange equation 12.55 to obtain

$$V_P = a \times V_S = \left(\frac{1}{2}\right) \times 20\ \text{V} = 10\ \text{V}$$

Answer is A.

Example 12.32

What is the value of C_P in the circuit of Figure 12.44?

A. 0.001 μF
B. 0.01 μF
C. 1.0 μF
D. 10.0 μF

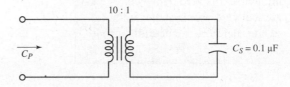

Figure 12.44

Solution:

Refer to equation 12.56, $Z_P = a^2 Z_S$. For a capacitor, this equation is as follows:

$$C_P = \frac{1}{a^2} \times C_S = \frac{1}{(10)^2} \times 0.1\ \mu\text{F} = 0.001\ \mu\text{F}$$

Remember that, for a capacitor, capacitive reactance is $X_C = 1/\omega C$.

Answer is A.

Polyphase Systems

Polyphase systems are used for the generation and distribution of large quantities of power. Some advantages of these systems are that induction motors have an inherent starting torque because the voltages are out of phase, total power is constant, the systems are more efficient, and for a fixed distance the line loss is less. Although polyphase systems may have any number of voltages not in phase, most polyphase systems are three phase. Three-phase (3-ϕ) motors are smaller than single-phase motors of the same horsepower, operate with less vibration, and have constant torque.

Figure 12.45 shows the waveforms produced from a three-phase generator. The waveforms are 120° out of phase with one another.

> Three-phase distribution systems are often used by businesses and industry.

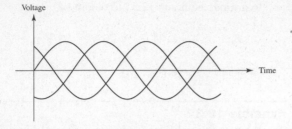

Figure 12.45

In a balanced wye-wye connected system, the neutral current is zero. However, most practical wye-wye systems are not balanced and therefore the neutral current is not zero.

Figure 12.46a shows three single windings. These windings may be connected either in a wye (Y) connection (also called a star connection), as shown in Figure 12.46b, or in a delta (Δ) connection, as shown in Figure 12.46c. In the wye configuration, if the common connection is brought out (used), it is referred to as the neutral. In the delta configuration, there is no neutral. Generators are not connected in a delta configuration, but transformer windings are. For a wye connection, voltages may be specified as line-to-line or as line-to-neutral. For a delta connection, voltages are measured only line-to-line.

For a wye connection, the line current equals the corresponding phase current and is given by the equation

$$I_L = I_{Ph} = \frac{V_{Ph}}{Z_{Ph}} \qquad [12.57]$$

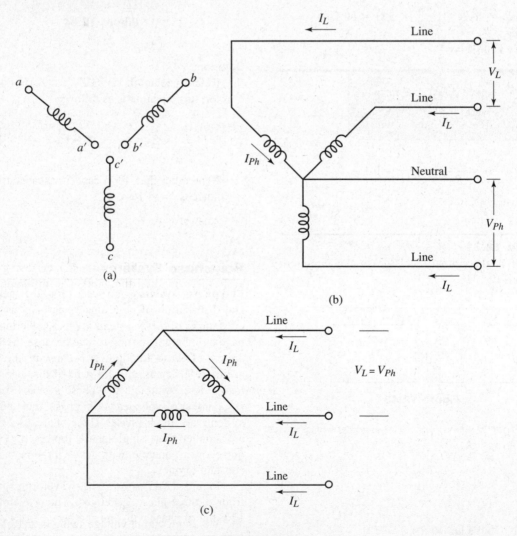

Figure 12.46

370

The line-to-line and phase voltages for a wye connection are 30° out of phase; the relationship between their magnitudes is expressed as

$$|V_L| = \sqrt{3}\,|V_{Ph}| \qquad [12.58]$$

The analysis of a delta configuration is similar to that for the wye connection. Any equation involving voltage or current for the wye has the same mathematical form as its counterpart for current or voltage for the delta. For a delta connection, the line-to-line and phase voltages are equal, and the equation is

$$V_L = V_{Ph} = I_{Ph}Z_{Ph} \qquad [12.59]$$

The line-to-line and phase currents for a delta connection are 30° out of phase; the relationship between their magnitudes is expressed as

$$|I_L| = \sqrt{3}\,|I_{Ph}| \qquad [12.60]$$

Example 12.33

A balanced delta-connected load is wired to a 208-V line-to-line system as shown in Figure 12.47. What is the magnitude of the line current?
A. 20.8 A
B. 18.0 A
C. 36.0 A
D. 12.0 A

Solution:
Rearrange equation 12.59 to obtain

$$I_{Ph} = \frac{V_{Ph}}{Z_{Ph}} = \frac{208\ \text{V}}{(6 - j8)\ \Omega} = 20.8\angle 53.1°\ \text{A}$$

Apply equation 12.60:

$$|I_L| = \sqrt{3}\,|I_{Ph}| = \sqrt{3} \times 20.8\ \text{A} = 36.0\ \text{A}$$

Answer is C.

Example 12.34

If the load impedances in Figure 12.47 are connected in a wye configuration, what is the magnitude of the line current?

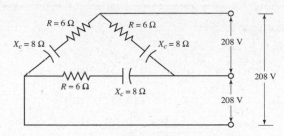

Figure 12.47

A. 20.8 A
B. 18.0 A
C. 36.0 A
D. 12.0 A

Solution:
Rearrange equation 12.58:

$$|V_{Ph}| = \frac{|V_L|}{\sqrt{3}} = \frac{208\ \text{V}}{\sqrt{3}} = 120\ \text{V}$$

Apply equation 12.57:

$$|I_L| = \frac{120\ \text{V}}{(6 - j8)\ \Omega} = 12\ \text{A}$$

Answer is D.

12.8 Operational Amplifiers

The operational amplifier (op amp), whose circuit symbol is shown in Figure 12.48, is a high-gain, direct-coupled amplifier with an open-loop gain, A_{OL}, of several hundred thousand. The differential input voltage, $E_d = v_1 - v_2$, is a small voltage, in the order of microvolts, that drives the amplifier. The output voltage, V_0, is the differential input voltage, E_d, times the open-loop gain:

$$V_0 = A_{OL} \times E_d = A_{OL} \times (v_1 - v_2) \qquad [12.61]$$

where v_1 is the voltage measured at the noninverting input (+) terminal of the op amp with respect to ground, and v_2 is the voltage measured at the inverting input (−) terminal with respect to ground. For most op amps, the maximum output voltage swing is usually 1 or 2 volts less than the power supply's voltages, which are typically set at ± 15 volts. Therefore the maximum output voltage swing is ± 13 V.

The difference between the input voltages is also referred to as the differential signal.

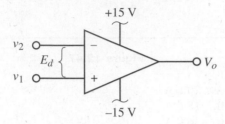

+15 V

v_2

$-$

E_d

$+$

v_1

V_o

−15 V

Figure 12.48

Example 12.35

What is a typical maximum input differential range for E_d if an op amp's open-loop gain is 250,000?

A. ± 60 μV
B. ± 52 μV
C. ± 30 μV
D. ± 45 μV

Solution:

Rearrange equation 12.61 to obtain

$$E_d = \frac{V_{OL}}{A_{OL}} = \frac{\pm 13 \text{ V}}{250 \times 10^3} = \pm 52 \ \mu\text{V}$$

Answer is B.

For most applications of op amps, an external resistor is connected between the output terminal and the op amp's (−) terminal. This type of circuit connection is called negative feedback. The advantage of negative feedback is that circuit performance can be expressed in terms of the op amp's closed-loop gain, A_{CL}, which depends only on the external resistors. By adding a negative-feedback circuit, designers can ignore the changes in A_{OL} from op amp to op amp as long as A_{OL} is at least 10 times larger than A_{CL}. This review considers circuits using the characteristics of an ideal op amp, which are summarized below, and negative feedback.

Ideal Op Amp Characteristics

1. The open-loop gain, A_{OL}, is infinite.
2. The differential input voltage, E_d, is zero.
3. The current drawn by either the (+) or the (−) input terminal is negligible.
4. The output impedance of the op amp is negligible.

Inverting Amplifiers

Figure 12.49 shows one of the most commonly used op amp circuits. The amplifier's closed-loop gain, A_{CL}, is set by the values of R_f and R_i according to the equation

$$A_{CL} = -\frac{R_f}{R_i} \qquad [12.62]$$

where the minus sign indicates that the output and input voltage polarities are 180° out of phase. Therefore, the circuit is said to have a negative gain. The op amp circuit of Figure 12.49 can amplify both ac and dc signals.

Operational amplifiers are most often used with negative feedback where part of the output signal is returned to the input.

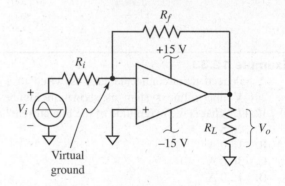

R_f

R_i

+15 V

$+$

V_i

$-$

$+$

R_L

V_o

−15 V

Virtual ground

Figure 12.49

Characteristics of an Inverting Amplifier

1. The input resistance of the circuit equals resistance R_i.
2. The magnitude of the closed-loop gain is the ratio of R_f/R_i.
3. The output voltage will be the negative of the input signal.
4. The voltage at the inverting input (−) is at virtual ground because E_d is of the order of microvolts.

Example 12.36

What must be the value of R_f in the circuit of Figure 12.50 for the magnitude of the closed-loop gain to be 15?

A. 7.5 kΩ
B. 1.5 kΩ
C. 150 kΩ
D. 75 kΩ

For the ideal inverting op-amp amplifier, the magnitude of the voltage gain is determined by the ratio of the feedback resistor, R_f, to the input resistor, R_i.

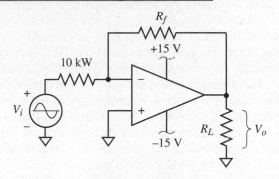

Figure 12.50

Solution:

Rearrange equation 12.62 to obtain

$$R_f = |A_{CL}| \times R_i = 15 \times 10 \text{ k}\Omega = 150 \text{ k}\Omega$$

Answer is C.

Noninverting Amplifier

The op amp circuit of Figure 12.51 has the input voltage applied to the noninverting input (+) terminal. Hence this circuit is referred to as a noninverting amplifier. The gain of this circuit is given by the equation

$$A_{CL} = 1 + \frac{R_f}{R_i} = \frac{R_f + R_i}{R_i} \qquad [12.63]$$

The output signal of a noninverting amplifier is in phase with the input signal.

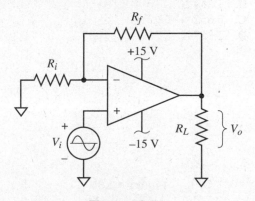

Figure 12.51

Characteristics of a Noninverting Amplifier

1. The input resistance of the circuit (as "seen" by the input voltage source) is infinite.
2. The magnitude of the closed-loop gain is $1 + R_f/R_i$.
3. The output voltage will be in phase with the input signal.

Example 12.37

What is the gain of the op amp circuit of Figure 12.51 if $R_i = 10 \text{ k}\Omega$ and $R_f = 22 \text{ k}\Omega$?

A. -2.2
B. 2.2
C. -3.2
D. 3.2

Solution:

Apply equation 12.63 to obtain

$$A_{CL} = 1 + \frac{22 \text{ k}\Omega}{10 \text{ k}\Omega} = 3.2$$

Answer is D.

Inverting Summing Amplifier

The op amp circuit of Figure 12.52 has three input voltages applied to the inverting input (−) terminal. This circuit is referred to as an inverting summing amplifier. The output voltage of this circuit is given by the equation

$$V_0 = -\left(\frac{R_f}{R_1} \times V_1 + \frac{R_f}{R_2} \times V_2 + \frac{R_f}{R_3} \times V_3 \right)$$

$$[12.64]$$

The circuit of Figure 12.52 allows adding different input signals with gain.

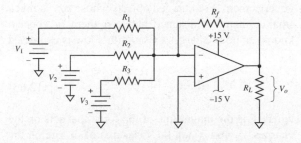

Figure 12.52

Example 12.38

What is the value of R_f in the circuit of Figure 12.53?

A. 16.66 kΩ
B. 45 kΩ
C. 33.33 kΩ
D. 4.62 kΩ

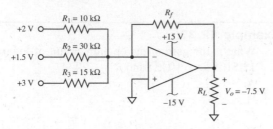

Figure 12.53

Solution:

Factor R_f from equation 12.64, and rearrange terms:

$$R_f = \frac{V_0}{-\left(\dfrac{V_1}{R_1} + \dfrac{V_2}{R_2} + \dfrac{V_3}{R_3}\right)}$$

$$= \frac{-7.5\ \text{V}}{-(0.2 + 0.05 + 0.2) \times 10^{-3}} = 16.66\ \text{k}\Omega$$

Answer is A.

12.9
Electric and
Magnetic Fields

This section reviews some of the fundamental principles of electric and magnetic fields.

Electric Fields: Section 12.5 introduced the concept that like charges repel and unlike charges attract. Charles de Coulomb (1736–1806) first measured these attractions and repulsions and deduced the law that governs them. For charged objects whose size is much smaller than the distance between them (a concept known as point charges), Coulomb's law is expressed as

$$\mathbf{F} = \frac{1}{4\pi\ \varepsilon_0} \times \frac{q_1 q_2}{r^2} \qquad [12.65]$$

where **F** is the magnitude of the force that acts on the charges, q_1 and q_2 are the measures of charge on the objects, r is the distance between the charged objects, and ε_0 is known as the permittivity constant (for air $\varepsilon_0 = 8.854 \times 10^{-12}$ coulomb^2/newton · meter^2 or

farad per meter). Note that force, **F**, is a vector quantity. The direction of force can be determined by the direction that a positive point charge would move when released. If there are multiple point charges bound to an object, vector addition must be used to find the resultant force on the object.

In the vicinity of any point charge, there is an electric field, **E**. The units for **E** are newtons per coulomb (N/C) or volts per meter (V/m). The lines of force that constitute the electric field are lines of flux, **Ψ**. The direction of electric flux, by convention, is the direction in which a positive charge would move if released in the electric field. Figure 12.54 shows how to visualize the electric field between a positive and negative charge. The magnitude of the electric field intensity is given by the equation

$$\mathbf{E} = \frac{q_1}{4\pi\ \varepsilon_0\ r^2} \qquad [12.66]$$

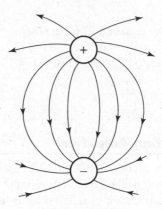

Figure 12.54

Example 12.39

Figure 12.55 shows point charges q_1 and q_2. What is the magnitude of the force acting on q_1? (Assume that $q_1 = -2.0 \times 10^{-6}$ C and $q_2 = +3.0 \times 10^{-6}$ C.)

A. 0.416 N
B. 1.254 N
C. 0.852 N
D. 2.357 N

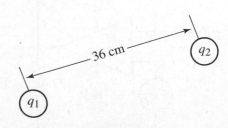

Figure 12.55

Solution:

The distance between the charges is $r = 36$ cm $= 3.6 \times 10^{-1}$ m. Apply equation 12.65 to obtain

$$|F| = \frac{(2.0 \times 10^{-6} \text{ C})(3.0 \times 10^{-6} \text{ C})}{4\pi(8.85 \times 10^{-12} \text{ C}^2/\text{N} \cdot \text{m}^2)(3.6 \times 10^{-1} \text{ m})^2}$$

$= 0.416$ N

Answer is A.

Magnetic Fields: A magnetic field exists in the space around a magnet. The lines of magnetic flux, external to the magnet, are directed from the north to the south pole as shown in Figure 12.56. The total amount of magnetic flux is measured in webers (Wb) and is represented by the symbol ϕ. The magnitude of the flux density, **B** (weber/meter^2, or tesla (T)), is calculated by dividing the magnetic flux by an area perpendicular to it and is given by the equation

$$\mathbf{B} = \frac{\phi}{A} \qquad [12.67]$$

The magnitude or strength of the magnetic field, **H,** is measured in amperes per meter and is expressed as

$$\mathbf{H} = \frac{\mathbf{B}}{\mu_0} \qquad [12.68]$$

where μ_0 is the permeability of air ($\mu_0 = 4\pi \times 10^{-7}$ H/m).

Magnetic flux lines form closed paths. The lines help us to visualize the strength of the magnetic field. If they are drawn close together, the field is considered strong, and if they are drawn farther apart, the field is considered weak.

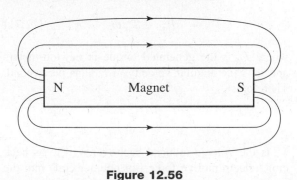

Figure 12.56

Example 12.40

Figure 12.57 shows an inductor wound on an iron-core ring whose cross-sectional diameter is 4 cm. If the total magnetic flux has been measured to be 5×10^{-5} Wb, what is the flux density?

A. 1.59×10^{-4} Wb/m^2
B. 6.53×10^{-4} Wb/m^2
C. 3.98×10^{-4} Wb/m^2
D. 2.75×10^{-4} Wb/m^2

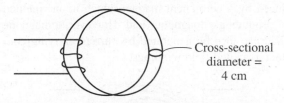

Cross-sectional diameter = 4 cm

Figure 12.57

Solution:

Apply equation 12.67 to obtain

$$B = \frac{5 \times 10^{-5} \text{ Wb}}{\frac{1}{4}\pi(0.04 \text{ m})^2} = 3.98 \times 10^{-4} \text{ Wb/m}^2$$

Answer is C.

12.10 DC Motors

A motor is a machine that converts electric energy into mechanical energy. There are dc and ac electric motors. All motors use the interaction of electricity and magnetism. The three basic processes that occur are (1) a current flowing in a conductor creates a magnetic field around the conductor; (2) a current-carrying conductor in an external magnetic field forces the conductor to move (this generates motor torque), and (3) a conductor moving in a magnetic field has a voltage induced into it. In a motor, all these interactions occur.

Figure 12.58 illustrates the first effect of a current-carrying conductor in a permanent magnetic field. The current-carrying conductor is moving at right angles to the permanent magnetic field. In this figure, the current flows in the conductor into the page and its magnetic field is in a clockwise direction around the conductor as given by the right-hand rule (the thumb of the right hand points in the direction of current flow and the fingers curl around the conductor in the direction of the magnetic field). The magnetic field of the current-carrying conductor in Figure 12.58 interacts with the permanent magnetic field as shown. Due to this interaction, the magnetic field above and below the conductor increases and decreases, respectively, and the strength of the magnetic field above the conductor forces it in a downward direction. This force generates motor torque. As the conductor moves in the magnetic field the third process applies, that is, a voltage will be induced into the conductor and will oppose the input voltage to the motor. In this example, the magnetic field was produced by a permanent magnet which is one method of producing a magnetic field. However, a magnetic field can also be generated by current-carrying conductors called field windings.

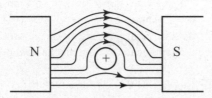

Figure 12.58 Effect of a current-carrying conductor in a magnetic field.

The basic structure of an electric motor includes two major components, a stator and a rotor, separated by an air gap. The stator does not move; it is the stationary part and is normally the motor's outer frame. The conductors that produce the magnetic field are referred to as the field windings and are usually wired into the stator. The rotor is that part of the motor that is free to move; it is the rotating part. The rotor is normally the motor's inner part and the conductors wired in this part are known as the armature. See Figure 12.59.

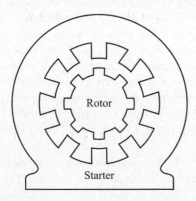

Figure 12.59 Basic structure of an electric motor.

An advantage of a dc motor is the ability to control shaft speed by controlling the input voltage. Dc motors are identified by the way the field winding is connected. A dc motor that has the field winding

connected in parallel with the armature is identified as a shunt motor. If the field winding is connected in series with the armature, the motor is identified as a series motor. A third type of dc motor is a compound motor; this motor employs both shunt and series field windings. A fourth type of dc motor has a separately excited field. See Figure 12.60.

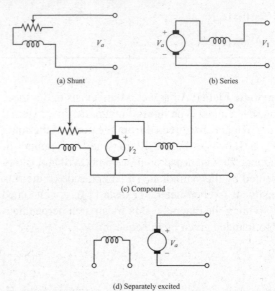

Figure 12.60 Field winding connections for dc motors.

For a lossless motor, the input electrical power, P_{in}, to the armature equals the mechanical output power, P_{out}.

$$P_{in} = P_{out} \qquad [12.69]$$

where

$$P_{in} = V_a I_a \qquad [12.70]$$

V_a is the voltage induced into the armature and I_a is the armature current. The mechanical output power is given by

$$P_{out} = T_m \, \omega \qquad [12.71]$$

where T_m is the generated torque in newton-meters and ω is the angular velocity in radians per second. Therefore,

$$V_a I_a = T_m \, \omega \qquad [12.72]$$

V_a is the voltage induced back (referred to as back emf—electromotive force—or counter emf) into the conductor moving in a magnetic field. In terms of motor parameters, it is expressed as

$$V_a = K_a \Phi \omega \text{ volts} \qquad [12.73]$$

where K_a = a constant depending on the motor design
Φ = magnetic flux per pole
ω = the speed of the armature

The motor constant K_a is given by

$$K_a = \frac{P Z_a}{2\pi a} \qquad [12.74]$$

where P is the number of poles, Z_a is the total number of conductors in the armature winding, and a is the number of parallel paths through the armature winding. For a separately excited dc motor, the electrical power into the motor is

$$P_t = V_t I_a \text{ watts} \qquad [12.75]$$

where V_t = the input dc voltage to the motor (also referred to as the terminal voltage).
I_a = armature current

An equivalent circuit for a dc motor with a separately excited field winding and a movable current-carrying conductor called the armature is shown in Figure 12.61. R_a is the equivalent armature resistance, L_a is the equivalent armature inductance, and I_a is the armature current. The armature current can be expressed as

$$I_a = \frac{V_t - V_a}{R_a} \qquad [12.76]$$

where the armature inductance is considered negligible.

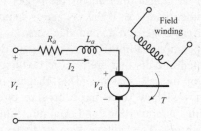

Figure 12.61 Equivalent circuit of a dc motor and a separately excited field winding.

Example 12.41

Determine the back emf of the dc motor shown in Figure 12.61 if the terminal voltage is 24 V and the armature current is 2 A. The armature resistance is 1.2 Ω and the armature inductance is negligible.
A. 26.4 V
B. 21.6 V
C. 24.2 V
D. 25.2 V

Solution:
Applying Kirchhoff's voltage law to the armature circuit of Figure 12.62, yields

$$V_a = V_t - I_a R_a = 24 \text{ V} - (2\text{A})(1.2\ \Omega) = 21.6 \text{ V}$$

Answer is B.

Multiplying each term of the fundamental motor equation, given in Example 12.41, by I_a, yields

$$V_a I_a = V_t I_a - I^2_a R_a \qquad [12.77]$$

The term $V_t I_a$ is the electrical input power and the $I^2_a R_a$ term is the power lost as heat in the armature circuit. The term $V_a I_a$ represents the power delivered to the armature to develop mechanical power although some of the power will be lost due to mechanical or rotational losses of the motor. The torque generated because of the mechanical power is

$$T_m = K_a \Phi I_a \text{ newton-meters} \qquad [12.78]$$

Example 12.42

What is the electrical power developed by the motor given in Example 12.41?
A. 43.2 W
B. 48.4 W
C. 44.6 W
D. 42.8 W

Solution:
From the solution of Example 12.41, V_a = 21.6 V and the electrical power developed is

$$V_a I_a = (21.6\text{V})(2\text{A}) = 43.2 \text{ W}$$

Answer is A.

Example 12.43

What is the immediate effect on the torque developed by a separately excited dc motor described in Example 12.41 if the field flux is reduced by 5%?
A. Decrease of 5 %
B. Increase by 33 %
C. Decrease by 50 %
D. Increase by 10 %

Solution:

The torque developed when the armature current is 2A is

$$T_m = K_a \, \Phi \, I_a = K_a \, \Phi \, (2 \text{ A})$$

and the armature voltage is

$$V_a = V_t - I_a R_a = 24 \text{ V} - (2 \text{ A})(1.2 \ \Omega) = 21.6 \text{ V}$$

If Φ is reduced by 5%, then the armature voltage is also reduced by 5% because the rotor speed cannot change instantly. Therefore, the new armature voltage, V_a, is

$$V_a = 0.95(21.6 \text{ V}) = 20.52 \text{ V}$$

and the armature current, I_a, is now

$$I_a = \frac{V_t - V_a}{R_a} = \frac{24\text{V} - 20.52\text{V}}{1.2} = 2.9\text{A}$$

The new value of torque is

$$T_m = K_a(0.95) \, \Phi \, (2.9 \text{ A}) = K_a \, \Phi \, (2.755)$$
$$\text{newton-meters}$$

and the change in torque is

$$\frac{T_{new}}{T_{original}} = \frac{K_a \Phi \ (2.755)}{K_a \Phi \ (2.0)} = 1.33$$

Therefore, the torque increases by 33%.

Answer is B.

12.11
Summary

Electrical engineering, similar to other engineering fields, covers a vast amount of material. The objective of this chapter was to introduce the reader to some of the underlying principles that are helpful in solving circuit problems. Circuits are composed of sources and loads. Most of this chapter analyzed loads composed of passive elements—resistors, inductors, and capacitors—but the last section did introduce circuits that also included an integrated circuit—the operational amplifier. Resistors dissipate energy while inductors and capacitors can store energy in a magnetic or electric field, respectively. The concepts of charge, current, voltage, power, and energy were introduced and then shown how they are related. Current is the time rate of change of charge, voltage is a measure of the energy per unit of charge, and power is the rate at which energy is transferred. Ohm's law and Kirchhoff's voltage and current laws were then used to analyze circuits as well as to show how the same elements in series and parallel are combined to determine an equivalent value.

Some circuits are best analyzed using either a Thévenin or Norton equivalent circuit, but circuits with multiple sources may need to be solved differently, such as by using loop analysis. These circuit analysis techniques were applied to resistive circuits. Circuits that include resistors, inductors, and/or capacitors and have an ac sinusoidal input were analyzed by first determining the inductive and capacitive reactance for the inductor and capacitor elements and then applying the previously mentioned circuit analysis techniques. Resistor and inductor or resistor and capacitor circuits driven by a dc input voltage source were also analyzed, but these circuits have a transient output response and this response is either an exponential rise or exponential decay.

Businesses and industry often receive their power as three-phase and the power source and loads are connected either as a wye or as a delta. Transformers are usually used to step up or step down the voltage. For ideal transformers the power on the secondary side of the transformer equals the power on the primary side; therefore, if the voltage is stepped down, the current is stepped up and vice versa.

Operational amplifiers are high-gain dc coupled integrated circuits that are a basic building block in many of today's systems. This chapter introduced and showed how three basic op-amp circuits—the inverting, noninverting, and summing amplifiers—are analyzed. The voltage gain of these circuits is set by resistors external to the op amp. This chapter ended with a short introduction to electric and magnetic fields.

PRACTICE PROBLEMS

1. What is the resistance of 1,000 m of copper wire that has a resistivity of 1.72×10^{-8} $\Omega \cdot$ m and a diameter of 2 mm?
   (A) 6.97 Ω
   (B) 3.41 Ω
   (C) 5.47 Ω
   (D) 1.23 Ω

2. If the values given in Problem 1 are for a temperature at 25°C, what is the resistance of the wire at 75°C? The resistance temperature coefficient at 25°C is 0.0039/°C.
   (A) 6.54 Ω
   (B) 8.32 Ω
   (C) 4.07 Ω
   (D) 1.47 Ω

3. If the diameter of the wire in Problem 1 is increased by a factor of 10, what is the resistance of the 1,000 m of wire?
   (A) 69.7 mΩ
   (B) 54.7 mΩ
   (C) 92.1 mΩ
   (D) 0.033 Ω

Problems 4–7 are based on the following figure:

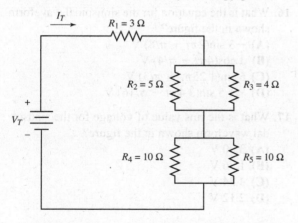

4. What is the total resistance for the circuit shown in the figure?
   (A) 15.3 Ω
   (B) 8.70 Ω
   (C) 12.8 Ω
   (D) 10.2 Ω

5. If current I_T in the figure is 1.5 A, what is V_T?
   (A) 15.3 V
   (B) 4.50 V
   (C) 10.6 V
   (D) 21.4 V

6. If current I_T in the figure is 1.5 A, what is the value of the current through R_3?
   (A) 0.66 A
   (B) 0.75 A
   (C) 0.83 A
   (D) 1.11 A

7. Consider that the parallel combination of R_4 and R_5 in the figure is the load. What are the Thévenin equivalent voltage and the resistance as "seen" by R_4 and R_5? (Assume $V_T = 10$ V.)
   (A) 5 V, 12 Ω
   (B) 10 V, 5.2 Ω
   (C) 5 V, 1.27 Ω
   (D) 10 V, 12 Ω

Problems 8 and 9 are based on the following figure:

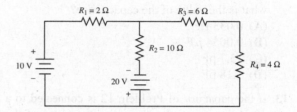

8. What is the value of current through R_2 in the circuit shown in the figure?
   (A) 2.0 A
   (B) 1.67 A
   (C) 2.43 A
   (D) 3.0 A

9. What is the value of power dissipated by R_4 in the figure?
   (A) 2.77 W
   (B) 1.83 W
   (C) 2.16 W
   (D) 0.73 W

10. The current through a 10-mH coil changes at the rate of 2 A/s. What voltage appears across the inductor?
    (A) 0.5 V
    (B) 20 mV
    (C) 50 mV
    (D) 0.2 V

379

11. What is the total energy stored in the circuit shown in the figure below?

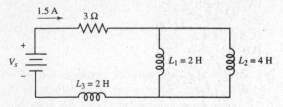

(A) 4.5 J
(B) 3.75 J
(C) 12.0 J
(D) 5.5 J

12. Two parallel circular disks, each with a diameter of 12 mm, are separated by 4 mm. If the dielectric constant between them is 4.16×10^{-11} F/m, what is the value of the capacitance?
(A) 0.033 μF
(B) 0.0054 μF
(C) 15.7 pF
(D) 1.18 pF

13. If the capacitor of Problem 12 is connected to a 12-V dc source, the charge on the plates of the capacitor is
(A) 14.1×10^{-12} C
(B) 64.8×10^{-8} C
(C) 20.3×10^{-12} C
(D) 128.6×10^{-8} C

14. If the initial current through the inductor in the figure below was 0 A, how long has the switch been closed?

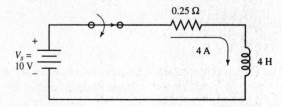

(A) 120 s
(B) 3.64 s
(C) 1.68 s
(D) 8.33 min

15. What is the capacitor voltage 5 s after the switch in the figure below is closed? The initial capacitor voltage is 2 V.

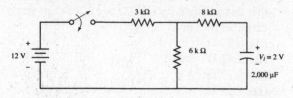

(A) 8.17 V
(B) 4.56 V
(C) 3.33 V
(D) 2.96 V

Problems 16 and 17 are based on the following figure:

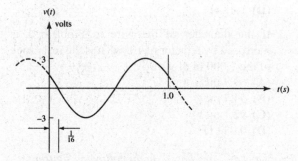

16. What is the equation for the sinusoidal waveform shown in the figure?
(A) $-3 \sin(2\pi t - \pi/8)$ V
(B) $3 \cos(4\pi t - \pi/4)$ V
(C) $6 \sin(1.25\pi t + \pi/3)$ V
(D) $-1.5 \sin(3.4\pi t - \pi/16)$ V

17. What is the rms value of voltage for the sinusoidal waveform shown in the figure?
(A) 3.00 V
(B) 1.76 V
(C) 4.24 V
(D) 2.12 V

18. What is the average value of voltage for the circuit shown in the figure below?

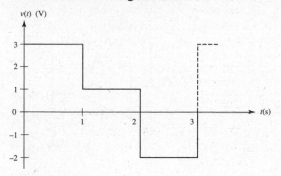

v(t) (V)

(A) 1.33 V
(B) 0.67 V
(C) 0.33 V
(D) 2.17 V

19. What is the total impedance, Z_T, of the circuit shown in the figure below?

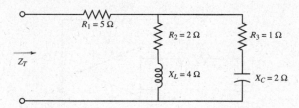

(A) $8.16 \angle 15.1° \ \Omega$
(B) $7.47 \angle -11.9° \ \Omega$
(C) $(6.13 - j1.2) \ \Omega$
(D) $(5.67 + j1.33) \ \Omega$

20. What is the value of the current if the impedance of a circuit is $(4 - j2) \ \Omega$ and the voltage source is $18 \angle +30° \ V$?
(A) $3.0 \angle 30° \ A$
(B) $(4.32 - j2.08) \ A$
(C) $5.12 \angle -46° \ A$
(D) $(2.22 + j3.36) \ A$

21. What is the value of current I_2 in the circuit shown in the figure below?

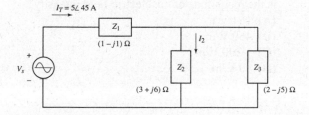

(A) $(2.42 + j3.61) \ A$
(B) $6.38 \angle -23.5° \ A$
(C) $(4.35 - j2.99) \ A$
(D) $2.67 \angle +17.4° \ A$

Problems 22–25 are based on the following figure:

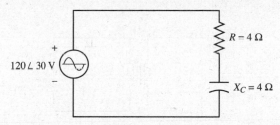

22. What is the power dissipated (the real power) by the circuit in the figure?
(A) 1.8 kW
(B) 3.2 kW
(C) 0.9 kW
(D) 2.7 kW

23. What is the power factor for the circuit in the figure?
(A) 0.866
(B) 0.258
(C) 0.707
(D) 0.512

24. If the frequency of the voltage source in the figure is 60 Hz, what value must an inductor connected in series have to make the circuit resonant?
(A) 0.088 H
(B) 402 mH
(C) 0.12 H
(D) 10.6 mH

25. A series resonant circuit has a Q value of 20 and a bandwidth of 15 kHz. What is the resonant frequency?
(A) 30.0 kHz
(B) 47.7 kHz
(C) 25.1 kHz
(D) 53.6 kHz

Problems 26 and 27 are based on the following figure:

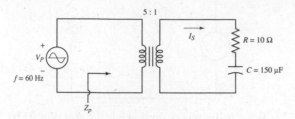

26. What is the impedance measured on the primary side of the transformer in the circuit shown in the figure?
(A) $508 \angle -60.5° \ \Omega$
(B) $(110 - j165) \ \Omega$
(C) $(250 - j302) \ \Omega$
(D) $324 \angle -42.5° \ \Omega$

27. If $V_{prms} = 120 \angle 0°$ V in the circuit shown in the figure, what is the value of the secondary current?
(A) $2.18 \angle 24.6°$ A
(B) $(2.4 - j1.34)$ A
(C) $(0.582 + j1.03)$ A
(D) $3.82 \angle -60.5°$ A

28. What is the magnitude of the line current for the balanced delta connection in the circuit shown in the figure below?

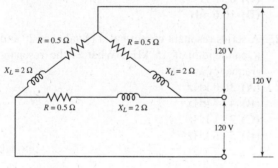

(A) 101 A
(B) 25.5 A
(C) 45.0 A
(D) 85.5 A

29. What is the output voltage, V_o, for the circuit shown in the figure below?

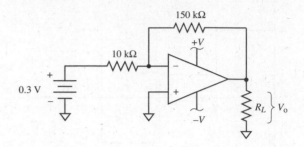

(A) 4.5 V
(B) -3.0 V
(C) 3.0 V
(D) -4.5 V

30. What is the output voltage, V_o, for the circuit shown in the figure below?

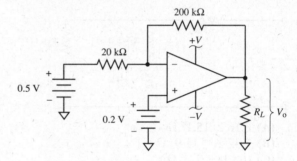

(A) -2.8 V
(B) -5.2 V
(C) 5.2 V
(D) 3.8 V

31. If two point charges, q_1 and q_2, are 2.0 m apart, what is the magnitude of the force acting on q_1?
(A) 0.84×10^{-6} N
(B) 6.52×10^{-6} N
(C) 3.6×10^{-6} N
(D) 1.2×10^{-6} N

32. For the point charges given in Problem 31, what is the magnitude of the electric field intensity?
(A) 164 V/m
(B) 460 V/m
(C) 1,200 V/m
(D) 90 V/m

Answer Key

1. C	9. D	17. D	25. B
2. A	10. B	18. B	26. A
3. B	11. B	19. B	27. C
4. D	12. D	20. D	28. A
5. A	13. A	21. C	29. D
6. C	14. C	22. A	30. A
7. B	15. C	23. C	31. C
8. C	16. A	24. D	32. D

Answers Explained

1. C Apply equation 12.4:

$$R = \frac{\rho l}{A} = \frac{(1.72 \times 10^{-8}\ \Omega \cdot m)(1,000\ m)}{\frac{\pi}{4}(2 \times 10^{-3}m)^2}$$

$$= 5.47\ \Omega$$

2. A The resistance at 25°C is 5.47 Ω. Apply equation 12.6:

$$R = 5.47\ \Omega \times [1 + 0.0039/°C \\ \times (75°C - 25°C)] = 6.54\ \Omega$$

3. B Apply equation 12.4:

$$R = \frac{\rho l}{A} = \frac{(2.72 \times 10^{-8}\Omega \cdot m)(1,000\ m)}{\frac{\pi}{4}(20 \times 10^{-3}\ m)^2}$$

$$= 54.7\ m\Omega$$

4. D Apply equation 12.18 to R_2 and R_3 and then again to R_4 and R_5. Then apply equation 12.12:

$$R_T = 3 + \frac{5 \times 4}{5 + 4} + \frac{10 \times 10}{10 + 10} = 10.2\ \Omega$$

5. A Apply Ohm's law; equation 12.8 yields

$$V_T = I_T \times R_T = (1.5\ A)(10.2\ \Omega) = 15.3\ V$$

6. C Use the current division law:

$$I_3 = \frac{5\ \Omega}{5\ \Omega + 4\ \Omega} \times 1.5\ A = 0.83\ A$$

7. B When R_4 and R_5 are removed as the load, $I_T = 0$ A and

$$V_{Th} = V_T = 10\ V.$$

To solve for R_{Th}, replace V_T by 0 Ω; then

$$R_{Th} = 3\ \Omega + \frac{5\ \Omega \times 4\ \Omega}{5\ \Omega + 4\ \Omega} = 5.2\ \Omega$$

8. C Assume that both loop currents are chosen in a clockwise direction.

Loop *A*: Apply the Kirchhoff voltage law to the loop containing the two voltage sources and R_1 and R_2.

Loop *B*: Apply the Kirchhoff voltage law to the outside loop. (*Note:* Through R_1, loop currents I_A and I_B flow.) The set of simultaneous equations is:

$$12I_A + 2I_B = 30\ V$$
$$2I_A + 12I_B = 10\ V$$

Only loop current I_A flows through R_2; therefore only I_A has to be determined:

$$I_A = \frac{\begin{vmatrix} 30 & 2 \\ 10 & 12 \end{vmatrix}}{\begin{vmatrix} 12 & 2 \\ 2 & 12 \end{vmatrix}} = \frac{360 - 20}{144 - 4} = 2.43\ A$$

9. D Solve for loop current I_B:

$$I_B = \frac{\begin{vmatrix} 12 & 30 \\ 2 & 10 \end{vmatrix}}{\begin{vmatrix} 12 & 2 \\ 2 & 12 \end{vmatrix}} = \frac{120 - 60}{144 - 4} = 0.42\ A$$

Now apply equation 12.9:

$$P = I^2 R_4 = (0.42\ A)^2 \times 4\ \Omega = 0.73\ W$$

10. B Apply equation 12.20:

$$v_L = L\frac{di}{dt} = (10 \times 10^{-3}\ H)(2\ A/s) = 20\ mV$$

11. B Calculate the total inductance by applying equations 12.24 and 12.22:

$$L_{eq} = \frac{L_1 L_2}{L_1 + L_2} + L_3 = \frac{2\ H \times 4\ H}{2\ H + 4\ H} + 2\ H$$

$$= 3\frac{1}{3}\ H$$

Apply equation 12.21:
Total energy stored $= \frac{1}{2} \times \frac{10}{3}\ H$

$$\times (1.5\ A)^2 = 3.75\ J$$

12. D Apply equation 12.25:

$$C = \frac{(4.16 \times 10^{-11}\ F/m) \times \frac{1}{4} \times \pi(0.012\ m)^2}{0.004\ m}$$

$$= 1.18\ pF$$

13. A Rearrange equation 12.26:

$$Q = C \times V = (1.18 \times 10^{-12}\text{ F})(12\text{ V})$$
$$= 14.1 \times 10^{-12}\text{ C}$$

14. C First determine the maximum value of current and the time constant:

$$I_{max} = \frac{V}{R} = \frac{10\text{ V}}{0.25\ \Omega} = 40\text{ A},$$

$$\tau = \frac{L}{R} = \frac{4\text{ H}}{0.25\ \Omega} = 16\text{ s}$$

Substitute in equation 12.34 with $I_i = 0$ A:

$$4 = 40(1 - e^{-t/16})$$

Solve for t:

$$t = 1.68\text{ s}$$

15. C Solve for the Thévenin equivalent circuit values, using C as the load:

$$V_{Th} = \frac{6\ \Omega}{3\ \Omega + 6\ \Omega} \times 12\text{ V} = 8\text{ V},$$

$$R_{Th} = \frac{3\text{ k}\Omega \times 6\text{ k}\Omega}{3\text{ k}\Omega + 6\text{ k}\Omega} + 8\text{ k}\Omega = 10\text{ k}\Omega$$

Since V_{Th} is V_f and the resistance "seen" by C is $R_{Th} + 8$ k$\Omega = 10$ kΩ:

$$\tau = R_T \times C = 10 \times 10^3\ \Omega \times 2{,}000$$
$$\times 10^{-6}\text{ F} = 20\text{ s}$$

Apply equation 12.32 with $V_i = 2$ V:

$$v_c(t) = 8 + (2 - 8)e^{-5/20} = 3.33\text{ V}$$

16. A The waveform in the figure can be expressed as a negative sine wave with a phase shift.

Solve for frequency, f: $f = \frac{1}{T} = \frac{1}{1\text{ s}} = 1$ Hz; then

$$\omega = 2\pi f = 2\pi\text{ rad/s}.$$

Set up a ratio to solve for the phase angle in radians per second:

$$\frac{\theta}{2\pi} = \frac{\frac{1}{16}}{1}$$

$$\theta = \frac{1}{16} \times 2\pi = \frac{\pi}{8}\text{ rad/s}$$

Substitute in equation 12.35(b):

$$v(t) = -3\sin\left(2\pi t - \frac{\pi}{8}\right)\text{ V}$$

17. D For sinusoidal waveforms the rms value equals 0.707 of the peak voltage. Hence,

$$V_{rms} = 0.707 V_{max} = 0.707(3\text{ V}) = 2.121\text{ V}$$

This value can also be determined by applying equation 12.37.

18. B Apply equation 12.36(b):

$$V_{avg} = \frac{(3\text{ V} \times 1\text{ s}) + (1\text{ V} \times 1\text{ s}) + (-2\text{ V} \times 1\text{ s})}{3\text{ s}}$$

$$= \frac{2}{3}\text{ V or }0.67\text{ V}$$

19. B Apply this equation:

$$Z_T = Z_1 + \frac{Z_2 \times Z_3}{Z_2 + Z_3}$$

Substitute the circuit values:

$$Z_T = 5\ \Omega + \frac{(2 + j4)\ \Omega \times (1 - j2)\ \Omega}{(2 + j4)\ \Omega + (1 - j2)\ \Omega}$$
$$= (7.31 - j1.54)\ \Omega = 7.47\angle{-11.9°}\ \Omega$$

20. D Apply Ohm's law for ac circuits:

$$I_{rms} = \frac{V_s}{Z_T} = \frac{18\angle 30°\text{ V}}{(4 - j2)\ \Omega} = (2.22 + j3.36)\text{ A}$$

21. C Apply the current division law:

$$I_2 = \frac{Z_3}{Z_2 + Z_3} \times I_T$$

$$= \frac{(2 - j5)\ \Omega}{(3 + j6)\ \Omega + (2 - j5)\ \Omega} \times 5\angle 45°\text{ A}$$

$$= 5.28\angle{-34.5°}\text{ A} = (4.35 - j2.99)\text{ A}$$

22. A Apply Ohm's law for ac circuits:

$$I_{rms} = \frac{120\angle 30°\text{ V}}{(4 - j4)\ \Omega} = (5.49 + j20.5)\text{ A}$$

$$= 21.2\angle 75°\text{ A}$$

The phase angle from the voltage phasor to the current phasor is 30° to 75°, or 45°.

Apply equation 12.45:

$$P = V_{rms} \times I_{rms} \times \cos\theta = 120\text{ V} \times 21.2\text{ A}$$
$$\times \cos(45°) = 1.8\text{ kW}$$

23. C For the circuit shown:

$$pf = \cos\theta = \cos(45°) = 0.707$$

24. D At resonance, $X_L = X_C$. Since $X_C = 4\ \Omega$ for this problem and $X_L = 2\pi f L$, then

$$L = \frac{4\ \Omega}{2\pi \times 60\ \text{Hz}} = 10.6\ \text{mH}$$

25. B Rearrange equation 12.52(a) and solve for f_0:

$$f_0 = \frac{BW \times Q}{2\pi} = \frac{(15 \times 10^3\ \text{Hz})(20)}{2\pi}$$

$$= 47.7\ \text{kHz}$$

26. A From equation 12.56, $Z_p = a^2 Z_s$. For this problem $Z_s = R - jX_C$, where

$$X_C = \frac{1}{2\pi \times 60\ \text{Hz} \times 150 \times 10^{-6}\ \text{F}}$$

$$= 17.68\ \Omega$$

Therefore, $Z_s = R - jX_C = (10 - j17.68)\ \Omega$, and

$$Z_p = \left(\frac{5}{1}\right)^2 (10 - j17.68)\ \Omega = (250 - j442)\ \Omega$$

$$= 508\angle{-60.5°}\ \Omega$$

27. C Rearrange equation 12.55 to determine the secondary voltage:

$$V_s = \frac{V_p}{a} = \frac{120\angle 0°\ \text{V}}{5} = 24\angle 0°\ \text{V}$$

Apply Ohm's law:

$$I_s = \frac{V_s}{Z_s} = \frac{24\angle 0°\ \text{V}}{(10 - j17.68)\ \Omega}$$

$$= (0.582 + j1.03)\ \text{A}$$

28. A Rearrange equation 12.59:

$$I_{Ph} = \frac{V_{Ph}}{Z_{Ph}} = \frac{120\angle 0°\ \text{V}}{(0.5 + j2)\ \Omega} = 58.2\angle{-76°}\ \text{A}$$

Apply equation 12.60:

$$|I_L| = \sqrt{3} \times I_{Ph} = \sqrt{3} \times 58.2 = 101\ \text{A}$$

29. D The voltage gain for an inverting amplifier is given by equation 12.62:

$$A_{CL} = -\frac{150\text{k}\Omega}{10\text{k}\Omega} = -15$$

The output voltage is determined as follows:

$$V_o = A_{CL} \times \text{source voltage} = -15 \times 0.3\ \text{V}$$
$$= -4.5\ \text{V}$$

30. A Determine the output voltage due to each voltage source alone, and then add the results.

For the inverting amplifier:

$$V_{o1} = 0.5\ \text{V} \times \frac{-200\ \text{k}\Omega}{20\ \text{k}\Omega} = -5.0\ \text{V}$$

For the noninverting amplifier:

$$V_{o2} = 0.2\ \text{V} \times \frac{200\ \text{k}\Omega + 20\ \text{k}\Omega}{20\ \text{k}\Omega} = 2.2\ \text{V}$$

Total output voltage =
$$V_o = V_{o1} + V_{o2} = -5.0\ \text{V} + 2.2\ \text{V} = -2.8\ \text{V}$$

31. C Apply equation 12.65; the magnitude of the force is

$$F = \frac{(4 \times 10^{-8}\ \text{C})(4 \times 10^{-8}\ \text{C})}{4\pi(8.85 \times 10^{-12}\ \text{C/N·m})(2\ \text{m})^2}$$

$$= 3.6 \times 10^{-6}\ \text{N}$$

32. D Apply equation 12.66:

$$E = \frac{4 \times 10^{-8}\ \text{C}}{4\pi(8.85 \times 10^{-12}\ \text{C/N·m})(2\ \text{m})^2}$$

$$= 90\ \text{V/m}$$

13

Thermodynamics

Hameed Metghalchi, Sc.D.

Farzan Parsinejad, Ph.D.

Introduction

Thermodynamics is the study of the properties of matter—in particular, the relationships among these properties and the changes in their values that either occur spontaneously or as a result of interaction with other objects. The ultimate goal of thermodynamics is to understand the operation of thermal systems, to be able to design these systems, and to recognize opportunities for improving the performance of existing systems.

This presentation is divided into five major sections:

- Definitions of terms: The vocabulary of thermodynamics is developed to provide a framework for the explanation of thermodynamic principles.
- Laws and Principles: The basic laws and principles that describe the relationships among the properties of matter are presented.
- Material Systems: Several types of materials that can form the contents of thermodynamic systems are introduced. For certain special cases, the general expressions for some of the basic laws and principles are simplified, and some new relationships are introduced.
- Applications: The basic principles are applied to systems frequently encountered, and specific methods for analyzing these systems are presented.
- Heat Transfer: The three modes of heat transfer are presented.

13.1
Definitions of Terms

System: A system is specified by a description of the material it contains and the boundaries that contain it. There are two main classifications of systems.

Open System: Material may enter into or exit from the system. Many sources call this type of system a *control volume.*

Closed System: Material may not enter into or exit from the system. Many sources call this type of system a *control mass.*

Property: A property is an attribute of a system to which a value can be assigned to describe the system. A necessary feature of a property is that the value can be stated at a given time independently of its value at any other time, and independently of the process by which the system arrived at its state.

Intensive properties are independent of the mass of the system; examples are temperature and pressure.

Extensive properties are proportional to the mass of the system; an example is volume.

Specific properties are determined by dividing the value of an extensive property by the mass of the system. A specific property is frequently denoted by the lower case version of the letter used to represent the extensive property, for example,

$$\nu = \frac{V}{m} \qquad [13.1]$$

where V = volume of system (ft or m^3),

m = mass of system (lbm or kg),

ν = specific volume of system (ft^3/lbm or m^3/kg).

State: The state of a system is its condition at an instant of time, determined by the values of a set of independent properties. For a system composed of a single, macroscopically homogeneous substance, the equilibrium state is specified by the values of any two independent properties. The values of all other properties can be determined from these.

Interaction: Changes in the state of a system may either take place spontaneously, without a corresponding change in the environment, or occur through interaction with the environment. All interactions take place through the boundaries of the system. The types of interactions studied here are work interactions, heat interactions, and bulk-flow interactions.

Process: A process is a description of the evolution of a system from one state to another. Some special types of processes are:

Reversible: A process for which there exists another process that can return the system to its original state, leaving no net effect on the system's environment. A true reversible process cannot be achieved in practice. All real

processes will experience losses, or transfers of energy that cannot be recovered without changing the state of the environment. The utility of investigating reversible processes is that they provide a limit to the change in state that can occur in any process.

Irreversible: A process for which there does not exist another process that can return the system to its original state, without changing the state of the environment. All real processes are irreversible.

Adiabatic: A process for which all transfers of energy occur through work interactions.

Constant-property: A process during which the value of one property remains constant. When such a process occurs, the analysis of the process is usually simplified.

Isometric: Constant-volume process (energy transfer to a closed rigid container).

Isothermal: Constant-temperature process (condensation or evaporation in a heat exchanger).

Isobaric: Constant-pressure process (weighted piston and cylinder).

Isentropic: Constant-entropy process (ideal turbine).

Isenthalpic: Constant-enthalpy process (flow through a valve).

13.2
Laws and Principles

A few general ideas comprise the science of thermodynamics. Two of these ideas, the first and second laws of thermodynamics, are introduced, along with the concept of the balance equation, in this section.

Mass Balance Equation

A thermodynamic system can frequently be analyzed solely through an accounting of all possible changes in the values of additive properties. In general, the time rate of change of the value of an additive property is equal to the algebraic sum of the rate at which an amount of the property crosses the system boundary and the rate at which an amount of the property is produced or destroyed. For example, in the case of the additive property mass, the balance equation is:

$$\frac{dm}{dt} = \dot{m}_{\text{transferred}} + \dot{m}_{\text{produced or destroyed}} \qquad [13.2]$$

where $\dot{m}_{\text{transferred}}$ is the algebraic sum of the mass-flow rates of all material streams into or out of the system.

The sign convention for mass-flow rate is positive if the flow of material is into the system, and negative if the flow of material is out of the system. By definition, $\dot{m}_{\text{transferred}} = 0$ for a closed system.

When nuclear reactions can be neglected, mass is neither produced nor destroyed. In terms of equation 13.2, $\dot{m}_{\text{produced or destroyed}} = 0$.

Energy and the Energy Balance Equation

In mechanics, the concept of energy is well defined. There is kinetic energy, related to the motion of a rigid body, and there is potential energy, related to the position of a rigid body in a gravity field.

In thermodynamics, the concept of energy is a little more complicated. Each individual molecule of the substance that makes up a system can possess a finite amount of energy, stored in a number of different modes. For example, each molecule can have kinetic energy due to its translational motion, kinetic energy due to its rotational motion, and potential energy due to vibration of the molecule's atoms with respect to each other. The total energy for a substance is the summation of all types of energy for each molecule that makes up the substance. Because the number of energy modes associated with each molecule can become large, and the number of molecules associated with any real system is extremely large, it is not possible to perform this summation directly.

For a macroscopic system, the easiest types of energy to consider separately are kinetic and potential energy. The energies associated with all other modes of energy storage are grouped together in the property named internal energy and denoted by U. The internal energy of a substance is related to its total energy by the following equation:

$$E = U + \frac{m\overline{V}^2}{2} + mgz \qquad [13.3]$$

where
E = total energy of system (Btu or kJ),
U = internal energy of system (Btu or kJ),
m = mass of system (lbm or kg),
$\overline{V}$ = velocity of material in system (ft/s or m/s),
g = acceleration due to gravity:
$\quad$ = 32.17 ft/s^2 at sea level,
$\quad$ = 9.81 m/s^2 at sea level,
z = height in gravity field (ft or m);

or, in terms of mass-specific properties (properties per unit mass):

$$e = u + \frac{\overline{V}^2}{2} + gz \qquad [13.4]$$

where
e = specific energy of the system (Btu/lbm or kJ/kg),
u = specific internal energy of the system (Btu/lbm or kJ/kg).

The balance equation for energy contains the essence of the principle known as the *first law of thermodynamics* and is given as:

$$\frac{dE}{dt} = \dot{E}_{\text{transferred}} + \dot{E}_{\text{produced or destroyed}} \qquad [13.5]$$

Energy is commonly transferred through work (W), heat (Q), and bulk-flow interactions. All three types of transfer are possible for an open system; bulk flow is not allowed in a closed system.

Work Interaction

Work (W) is energy transfer due to a mechanical interaction between a system and its surroundings. The value of work is considered positive if the system performs work on the surroundings (energy is removed from the system). The types of work commonly encountered are shaft work, boundary work, and flow work, given by the following equations:

$$\dot{E}_{t_{\text{work}}} = -\dot{W}$$

Shaft work: $\quad \dot{W}_s = \tau\omega \qquad [13.6]$

Boundary work: $\quad \dot{W}b = p\dfrac{dV}{dt}$

Flow work: $\quad \dot{W}_f = \dot{m}pv$

where
$\dot{W}$ = rate of energy transfer from a work interaction,
τ = torque applied to shaft,
ω = rotational velocity of shaft (rad/sec),
$\dot{m}$ = mass flow rate entering system,
t = time,
p = pressure in system,
V = volume of system,
v = specific volume of material in system.

A pressure-volume ($p-v$) diagram (Figure 13.1) can be used to graphically compute the boundary work done on or by a closed system during a reversible process, as shown below:

$$W_{b_{rev}} = \int_1^2 p\,dV \qquad [13.7a]$$

or

$$w_{b_{rev}} = \int_1^2 p\,dv \qquad [13.7b]$$

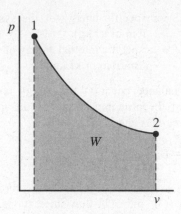

Figure 13.1 Graphical computation of energy transferred in a work interaction

The work done on or by an open system during a reversible process is

$$w_{rev} = -\int_1^2 v\, dp \qquad [13.8]$$

Heat Interaction

A heat interaction is a transfer of energy that occurs when two systems at different temperatures are placed in contact with each other. In all cases, the transfer of energy is from the high-temperature system to the low-temperature system. We choose to use the sign convention that the energy transfer from a heat interaction is positive if the energy is entering the system.

$$\dot{E}_{t\,\text{Heat}} = \dot{Q} \qquad [13.9]$$

where $\dot{Q}$ = rate of energy transfer from a heat interaction.

Bulk-Flow Interaction

For each material stream that enters the system, two modes of energy transfer occur. The energy of the material that enters the system is added to the system, and the material performs flow work on the system. The total energy transferred in a bulk-flow interaction is given by:

$$\dot{E}_{t\,\text{bulk flow}} = \dot{m}\left(u + \frac{\overline{V}^2}{2} + gz\right) + \dot{m}\,pv \qquad [13.10]$$

If enthalpy ($h = u + pv$) is introduced, this equation becomes

$$\dot{E}_{t\,\text{bulk flow}} = \dot{m}\left(h + \frac{\overline{V}^2}{2} + gz\right) \qquad [13.11]$$

When there are multiple inlet and outlet streams, each of these streams contributes to the transfer of energy. In this case, the total energy transferred because of bulk flow is the sum of the amounts of energy transferred for all fluid streams, or

$$\dot{E}_{t\,\text{bulk flow}} = \Sigma\, \dot{m}_i\left(h_i + \frac{\overline{V}^2_i}{2} + gz_i\right)$$
$$- \Sigma\, \dot{m}_e\left(h_e + \frac{\overline{V}^2_i}{2} + gz_e\right) \qquad [13.12]$$

where the subscript i indicates flow into the system, and the subscript e indicates flow out of the system. Again, when nuclear reactions are negligible:

$$\dot{E}_{\text{produced or destroyed}} = 0 \qquad [13.13]$$

Note that, if nuclear reactions cannot be neglected, energy can be produced, but only at the cost of some mass. Conversely, energy can be destroyed, but only if there is a net gain in mass. When nuclear reactions are considered, there must be a conservation of mass and energy together. In other words, the balance equations for mass and energy may only be considered together, and the term for mass produced or destroyed must be related to the term for energy produced or destroyed.

All Interactions Combined (The First Law of Thermodynamics)

The balance equations for all types of interaction may be combined to produce an overall balance equation for energy:

$$\frac{dE}{dt} = \dot{Q} - \dot{W} + \Sigma\, \dot{m}_i\left(h_i + \frac{\overline{V}_i^2}{2} + gz_i\right)$$
$$- \Sigma\, \dot{m}_e\left(h_e + \frac{\overline{V}_e^2}{2} + gz_e\right) \qquad [13.14]$$

This relationship is commonly known as the **first law of thermodynamics** and is valid when nuclear reactions are negligible.

Example 13.1

A container is filled with water at standard conditions (101 kPa, 25°C) at a rate of 3 kg/min. The velocity of the water as it enters the container through an opening 0.50 m above the average water level is 12 m/s. What is the rate of energy transfer to the container?

Solution:
From equation 13.11:

$$\dot{E}_{t\,\text{bulk flow}} = \dot{m}\left(h + \frac{\overline{V}^2}{2} + gz\right)$$

Then:

$$\dot{E}_{t \text{ bulk flow}} = \left(3 \, \frac{\text{kg}}{\text{min}}\right)\Bigg[419 \, \frac{\text{kJ}}{\text{kg}} + \frac{(12 \text{ m/s})^2}{2}$$

$$\times \left(\frac{\text{kj}}{1{,}000 \, \frac{\text{kg} \cdot \text{m}^2}{\text{s}^2}}\right)$$

$$+ \, (9.81 \text{ m/s}^2)(0.50 \text{ m})\left(\frac{\text{kj}}{1{,}000 \, \frac{\text{kg} \cdot \text{m}^2}{\text{s}^2}}\right)\Bigg]\left(\frac{1 \text{ min}}{60 \text{ s}}\right)$$

$$= 21 \, \frac{\text{kJ}}{\text{s}}$$

$$= 21 \text{ kW}$$

Note: In this case, the contributions of kinetic energy transfer (0.072 kJ/kg) and potential energy transfer (0.005 kJ/kg) are negligible compared to the contribution of enthalpy transfer (419 kJ/kg).

Example 13.2

A container that holds 3 kg of water is stirred with a mixer that is powered by the lowering of a 50-kg mass for a distance of 8 m. The mixing raises the internal energy of the water, a change that is accompanied by an increase in temperature. If the change in temperature is related to the change in internal energy of liquid water according to this equation:

$$\Delta U = mc_v \, \Delta T$$

what is the change in temperature?

Solution:

$$\Delta U = E_{t \text{ work}} = -W$$

and

$$W = mg(z_2 - z_1)$$

$$= (50 \text{ kg})\left(9.81 \, \frac{\text{m}}{\text{s}^2}\right)(-8 \text{ m})$$

$$= -3{,}924 \text{ N} \cdot \text{m}$$

$$= -3.92 \text{ kJ}$$

$$\Delta U = 3.92 \text{ kJ}$$

Liquid water has very high specific heat, and its temperature does not change much.

$$\Delta T = \frac{\Delta U}{m_{H_2O} \, c_{v_{H_2O}}}$$

$$= \frac{3.92 \text{ kJ}}{(3 \text{ kg})\left(4.18 \, \frac{\text{kJ}}{\text{kg} \cdot \text{K}}\right)}$$

$$= 0.31 \text{ K}$$

Cyclic Machines

A cyclic machine is a device, or a collection of devices, that allows a continuous exchange of energy between systems, often changing energy from one form into another. The machine must operate on a thermodynamic cycle, that is, a series of processes in which the ending state of the last process is equivalent to the beginning state of the first process.

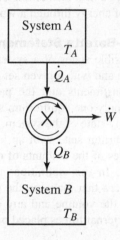

Figure 13.2 Thermodynamic cyclic of machine X, interacting with two thermal reservoirs, system A and system B

System A and system B in Figure 13.2 are thermal reservoirs. These are idealized systems that will reject or receive any amount of energy through reversible heat interactions, yet their temperatures will remain the same. A common example of a thermal reservoir is the ocean. Although there are many machines that utilize heat interactions with the ocean (e.g., a water-cooled boat engine or a power-generation plant), the overall temperature of the ocean remains unchanged by interactions with these machines.

Second Law of Thermodynamics

There have been many statements of the **second law of thermodynamics**. All have been descriptions of observations on all types of real thermodynamic systems that have never been violated. These observations are that (1) the transfer of energy occurs only in a preferred direction, namely, from higher temperature to lower temperature in a heat interaction, or from higher pressure to lower pressure in a boundary work interaction, and (2) that no real process is 100% efficient. Three statements of the second law, modified to agree with the language of this discussion, are given below.

Clausius Statement

It is impossible to devise a cyclic machine that produces, as its only effect, the transfer of energy through a heat interaction from a low-temperature body to a high-temperature body.

Kelvin-Planck Statement

It is impossible to devise a cyclic machine that will have no other effect than to extract energy from a reservoir through a heat interaction and to produce an equal amount of energy through a work interaction.

Gyftopolous-Baretta Statement

Among all possible states of a system with a given value of energy and with a given set of values of the amounts of constituents and the parameters, there exists one and only one equilibrium state. Moreover, starting from any state of the system, it is possible to reach an equilibrium state with an arbitrarily specified set of values of the amounts of constituents and the parameters. In this definition, the constituents are the substances that make up the system, and the parameters are the volume and any other properties that describe external forces placed on the system.

Availability

Availability (Φ), also called *exergy* or *available energy*, is the theoretical maximum amount of energy that can be extracted from a system in the form of useful work. The availability of the system cannot be specified by itself, but must be specified with reference to an environment large enough that interactions with the system do not measurably change the state of the environment.

The concept of the cyclic machine discussed earlier can serve also to help demonstrate the concept of availability. Assume that a thermodynamic system is connected to the environment through a *reversible* (ideal) cyclic machine, as shown in Figure 13.3.

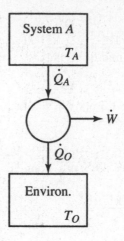

Figure 13.3 Cyclic machine that demonstrates the concept of availability

For the purposes of this description, it is assumed that the temperature of the system (T_A) is greater than the temperature of the environment (T_O), and that the system and the environment interact with the cyclic machine through heat interactions. As this cyclic machine begins to operate, energy is transferred from the thermodynamic system and into the environment. In addition, energy is transferred from the cyclic machine via the work interaction. Because the system has a finite extent, its temperature will begin to drop as energy is removed. The cyclic machine will continue to operate until the temperature of the system equals the temperature of the environment.

The total amount of energy transferred through the work interaction, from the time the cyclic machine begins to operate until it stops, is equal to the system's availability (exergy) in its initial state. Because energy is transferred from the cyclic machine both through the work interaction and the heat interaction into the environment, it is clear that the availability of system A is less than the amount of energy transferred from the system.

The state of the system at which the value of availability is zero is the state at which the cyclic machine stops operating. This state, known as the *dead state* of the system, occurs when the values of certain key properties of the system (e.g., pressure and temperature) are equal to those of the environment.

The balance equation for availability is as follows:

$$\frac{d\Phi}{dt} = \dot{\Phi}_{transferred} - \dot{\Phi}_{destroyed} \qquad [13.15]$$

Entropy and the Entropy Balance Equation

Entropy is related to the difference between the amount of system energy and the amount of system availability, or the amount of energy that cannot be used to produce work. This definition can be represented as follows:

$$S = S_{ref} + \frac{1}{T_o}[(E - E_{ref}) + p_o(V - V_{ref})$$
$$- (\Phi - \Phi_{ref})] \qquad [13.16]$$

where T_o is the absolute temperature of the environment, and the subscript ref indicates that the values are taken at an arbitrary reference state of the system.

The balance equation for entropy is

$$\frac{dS}{dt} = \dot{S}_{transferred} + \dot{S}_{produced} \qquad [13.17]$$

Other references may use $\dot{S}_{generated}$ or $\dot{\sigma}$ instead of $\dot{S}_{produced}$. This term is related to the irreversibilities of the process as follows:

$$\dot{I} = T_o\dot{S}_p = T_o\dot{S}_g = T_o\dot{\sigma} = \dot{\Phi}_{destroyed} \qquad [13.18]$$

The balance equation for entropy for all three types of interactions considered here (work, heat, and bulk flow) is

$$\frac{dS}{dt} = \Sigma \frac{\dot{Q}}{T_b} + \Sigma \dot{m}_i s_i - \Sigma \dot{m}_e s_e + \dot{S}_{produced} \qquad [13.19]$$

where $\Sigma\dfrac{\dot{Q}}{T_b}$ = total rate of entropy transfer through heat interactions,

$\Sigma \dot{m}_i s_i - \Sigma \dot{m}_e s_e$ = total rate of entropy transfer through bulk-flow interactions.

In equation 13.19, T_b is the temperature of the system boundary at the point where the heat interaction occurs. This temperature may or may not be constant with respect to time and position on the system boundary. There is no transfer of entropy in a work interaction. The entropy-produced term is related to the irreversibilities that occur in any real process.

Clausius Inequality

A historical statement, closely related to the second law of thermodynamics, can be derived from the entropy balance equation for a closed system operating in a cycle. This statement is known as the **Clausius inequality,** and is given as follows:

$$\oint \frac{\delta Q}{T} \leq 0 \qquad [13.20]$$

Example 13.3

What is the rate of entropy production for the cyclic machine diagramed in Figure 13.4?

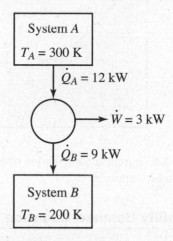

Figure 13.4 Entropy production for a cyclic machine

Solution: From equation 13.19:

$$\frac{dS}{dt} = \Sigma \frac{\dot{Q}}{T_b} + \dot{S}_{produced}$$

$$0 = \Sigma \frac{\dot{Q}}{T_b} + \dot{S}_{produced}$$

or

$$\dot{S}_{produced} = -\Sigma \frac{\dot{Q}}{T_b}$$

$$= -\left(\frac{\dot{Q}_A}{T_A} - \frac{\dot{Q}_B}{T_B}\right)$$

$$= -\left(\frac{12,000 \text{ W}}{300 \text{ K}} - \frac{9,000 \text{ W}}{200 \text{ K}}\right)$$

$$= 5 \text{ W/K}$$

> The increase of entropy always reduces the work output.

Temperature-Entropy (T-S) Diagram

A temperature-entropy (*T-S*) diagram (Figure 13.5) can be used to compute graphically the energy transfer from a reversible heat interaction, as shown below:

$$Q_{rev} = \int_1^2 T \, dS \qquad [13.21]$$

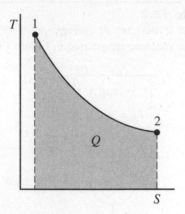

Figure 13.5 Graphical computation of energy transferred in a heat interaction

Availability Balance Equation

The balance equation for availability is

$$\frac{d\Phi}{dt} = \dot{\Phi}_{transferred} - \dot{\Phi}_{destroyed} \qquad [13.22]$$

Availability is never produced, but irreversibilities in the system destroy availability. By the same process used to determine the balance equations for the other additive properties discussed, the balance equation for availability can be derived. This equation is

$$\frac{d\Phi}{dt} = -\dot{W} + \sum\left(1 - \frac{T_o}{T_b}\right)\dot{Q} + \sum \dot{m}_i\,\Psi_i - \sum \dot{m}_e\,\Psi_e$$
$$+ p_o\frac{dV}{dt} - T_o\,\dot{S}_{produced} \qquad [13.23]$$

where Ψ is the *flow availability,* or the amount of availability transferred during a bulk-flow interaction, and is given by:

$$\Psi = (h - h_o) + \frac{\overline{V}^2}{2} + gz - T_o(s - s_o) \quad [13.24]$$

Flow availability is also the maximum amount of energy that the flowing material can transfer through a work interaction.

Example 13.4

If the heat and work interactions are assumed to be reversible, what is the rate of change of availability for the system shown in Figure 13.6?

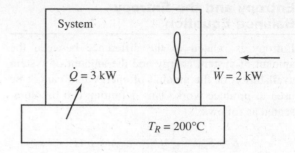

Figure 13.6 A system that undergoes a work interaction and a heat interaction

Solution:

$$T_o = 25°C$$

$$P_o = 101 \text{ kPa}$$

$$\frac{d\Phi}{dt} = -\dot{W} + \left(1 - \frac{T_o}{T}\right)\dot{Q}$$

From equation 13.23:

$$= 2 \text{ kW} + \left[1 - \frac{(25 + 273)\text{ K}}{(200 + 273)\text{ K}}\right](3 \text{ kW})$$

$$= 3.11 \text{ kW}$$

Properties

We now switch our focus from definitions of the basic laws of thermodynamics to descriptions of the substances that make up a thermodynamic system. The first type of substance we will consider is the simple substance. A different type of substance, the ideal gas, is discussed in Section 13.3.

We define a simple substance as a substance that is macroscopically homogeneous and is isotropic. "Homogeneous" means that there is no way to distinguish material in any location of the system from material in any other location. "Isotropic" means that there is no preferred spatial direction in which any interaction will behave differently than if applied from any other spatial direction. This definition can apply to any single substance, such as water, or to a homogeneous mixture of substances, such as air, which is a mixture of oxygen, nitrogen, and other gases.

Enthalpy

Often, when certain types of systems are investigated, a group of properties will arise naturally. One such property is enthalpy, which appears naturally when bulk-flow interactions are considered. By analyzing the energy balance equation for a bulk-flow interaction, we find that the energy transferred is related to the change in value of the following property grouping: $U + pV$. This grouping of properties occurs frequently enough that it has been given its own name, *enthalpy*, and it has become traditional to use this property directly in the analysis of thermodynamic systems.

The expresion for enthalpy is:

$$H = U + pV \qquad [13.25]$$

or, in terms of specific properties:

$$h = u + pv \qquad [13.26]$$

Enthalpy has units of energy (Btu or kJ). Specific enthalpy has units of energy per unit mass (Btu/lbm or kJ/kg).

Gibbs and Hemholtz Free Energy

Just as the idea of enthalpy naturally arises when analyzing bulk flow interactions, two other properties naturally appear when the thermodynamics of chemical reactions are examined.

Gibbs free energy, defined by equation 13.27, appears when chemical reactions at constant pressure and temperature are examined.

$$G = H - TS \qquad [13.27]$$

or, in terms of specific properties:

$$g = h - Ts \qquad [13.28]$$

Hemholtz free energy, defined by equation 13.29, appears when chemical reactions occurring at constant volume and temperature are examined.

$$A = U - TS \qquad [13.29]$$

or, in terms of specific properties:

$$a = u - Ts \qquad [13.30]$$

Both Gibbs free energy and Hemholtz free energy have units of energy (Btu or kJ). In their mass-specific forms, the units are energy per unit mass (Btu/lbm or kJ/kg).

Specific Heat

With current experimental techniques and devices, the thermodynamic properties that are easiest to measure are pressure, volume, and temperature. Given that the equilibrium state of any system comprised of a fixed amount of a simple substance is determined by the values of two independent properties, we know that the internal energy of a system can be expressed as a function of volume and temperature $[(u = u(v, T))]$, and that enthalpy can be expressed

as a function of pressure and temperature $[h = h(p, T)]$. These relationships are used to define the thermodynamic property of specific heat.

There are two versions of the property called *specific heat*, the constant-volume and the constant-pressure specific heat, defined as follows:

Constant-volume specific heat: $c_V = \dfrac{\partial u}{\partial T}\bigg)_V \quad [13.31]$

Constant-pressure specific heat: $c_p = \dfrac{\partial h}{\partial T}\bigg)_p \quad [13.32]$

The values of both of these properties vary greatly with temperature for most systems: when small changes in temperature are considered, however, a constant specific heat may be assumed. For solids and incompressible liquids, the difference between the values of c_p and c_V is negligible.

13.3
Material Systems

While the first and second laws of thermodynamics do describe the thermodynamic behavior of all substances, further information is gained by studying how different substances react to thermodynamic interactions.

Ideal Gases

Any equation that states the relationships among the values of the properties of a substance at a given state is called an *equation of state*. It has been observed that relationships among the values of the properties of gases in an equilibrium state, at low pressure, can be expressed in a very simple equation.

Although this equation represents an idealized gas, for most real gases at high temperatures and low pressures the error is negligible. Some gases, such as helium, hydrogen, and nitrogen, behave like an ideal gas for most equilibrium states, while others, such as steam, behave like an ideal gas only at low pressures. A fair indication of the likelihood that a gas will behave like an ideal gas can be obtained by looking at its molecular structure. Monatomic and diatomic gases will almost always behave like an ideal gas; but as the complexity of the molecule increases, the values of the properties of the gas will begin to deviate from those predicted by the ideal gas equation.

The ideal gas equation of state is

$$pv = RT \qquad [13.33a]$$

or

$$pV = mRT \qquad [13.33b]$$

or

$$pV = N\overline{R}T \qquad [13.33c]$$

where p = pressure (lbf/in^2 = psi or N/m^2 = Pa),
v = specific volume (ft^3/lbm or m^3/kg),
V = volume of the system (ft^3 or m^3),
m = mass of gas in the system (lbm or kg),
N = number of moles of gas in the system,
R = gas constant (kJ/kg · K or Btu/lbm − °R),
$\overline{R}$ = universal gas constant:
= 1,545 ft-lbf/(lbmol-°R),
= 8.314 J/(mol · K),
T = temperature.

While the universal gas constant, $\overline{R}$ is, as its name implies, the same for all gases, the gas constant R is specific to each gas and can be found from the equation

$$R = \frac{\overline{R}}{M} \qquad [13.34]$$

where M = molecular weight.
For an ideal gas:

$$R = c_p - c_v \qquad [13.35]$$

Example 13.5

If 4 kg of carbon monoxide gas is introduced into a 5.3-m^3 vessel at 25°C, what is the pressure of the gas?

Solution:
From equations 13.33b and 13.34:

$$p = \frac{m\overline{R}T}{MV}$$

$$= \left[\frac{(4 \text{ kg})\left(8.314 \dfrac{\text{J}}{\text{mol} \cdot \text{K}}\right)(25 + 273) \text{ K}}{\left(28 \dfrac{\text{g}}{\text{mol}}\right)(5.3 \text{ m}^3)\left(\dfrac{1 \text{ kg}}{1,000 \text{ g}}\right)}\right]\left(\frac{\text{N} \cdot \text{m}}{\text{J}}\right)$$

$$\times \left(\frac{\text{kPa}}{1,000 \dfrac{\text{N}}{\text{m}^2}}\right)$$

$$= 66.78 \text{ kPa}$$

Example 13.6

What is the gas constant for argon gas?

Solution:
From equation 13.34:

$$R = \frac{\overline{R}}{M}$$

$$= \frac{8.314 \dfrac{\text{J}}{\text{mol} \cdot \text{K}}}{40 \dfrac{\text{g}}{\text{mol}}}\left(\frac{1,000 \text{ g}}{1 \text{ kg}}\right)$$

$$= 208\frac{\text{J}}{\text{kg} \cdot \text{K}}$$

$$= 0.208 \frac{\text{kJ}}{\text{kg} \cdot \text{K}}$$

Any two states of an ideal gas can be related as follows:

$$\frac{p_1 v_1}{T_1} = \frac{p_2 v_2}{T_2} \qquad [13.36]$$

The balance equations discussed earlier can be applied to ideal gas systems. When these balance equations are combined with the equation of state, many simplifications and interesting results are found.

For an ideal gas system:

$$\left(\frac{\partial h}{\partial p}\right)_T = 0$$

and

$$\left(\frac{\partial u}{\partial v}\right)_T = 0 \qquad [13.37]$$

or

$$h = h(T) \text{ only}$$

and

$$u = u(T) \text{ only}$$

which leads to

$$c_p = c_p(T) \text{ only}$$

and

$$c_v = c_v(T) \text{ only}.$$

The case where the specific heat can be considered constant (i.e., when there are relatively small changes in temperature) leads to the following equations:

$$\Delta u = c_v \Delta T$$

and

$$\Delta h = c_p \Delta T \qquad [13.38]$$

$$\Delta s = c_p \ln\left(\frac{T_2}{T_1}\right) - R \ln\left(\frac{p_2}{p_1}\right) \qquad [13.39]$$

and

$$\Delta s = c_v \ln\left(\frac{T_2}{T_1}\right) + R \ln\left(\frac{v_2}{v_1}\right) \qquad [13.40]$$

Example 13.7

A sample of nitrogen undergoes an isobaric process in which the temperature is raised from 10°C to 45°C. What is the change in entropy per kilogram of nitrogen for this process?

Solution:

From equation 13.39:

$$\Delta s = c_p \ln\left(\frac{T_2}{T_1}\right) - R \ln\left(\frac{p_2}{p_1}\right)$$

Since

$$\ln\left(\frac{p_2}{p_1}\right) = 0$$

Then

$$\Delta s = c_p \ln\left(\frac{T_2}{T_1}\right)$$

$$= \left(1.04\frac{kJ}{kg \cdot K}\right) \ln\left[\frac{(45 + 273)K}{(10 + 273)K}\right]$$

$$= 0.121 \frac{kJ}{kg \ K}$$

Isentropic Process

For any isentropic process of an ideal gas system with constant specific heat:

$$p_1 v_1{}^k = p_2 v_2{}^k \qquad [13.41]$$

$$T_1 p_1{}^{(1-k)/k} = T_2 p_2{}^{(1-k)/k} \qquad [13.42]$$

$$T_1 v_1{}^{(k-1)} = T_2 v_2{}^{(k-1)} \qquad [13.43]$$

where $k = \frac{c_p}{c_v} = $ specific heat ratio.

If the isentropic process occurs in a closed system, then,

$$w = \frac{p_2 v_2 - p_1 v_1}{1 - k}$$

$$= \frac{R(T_2 - T_1)}{(1 - k)} \qquad [13.44]$$

$$= \frac{RT_1}{k - 1}\left[1 - \left(\frac{p_2}{p_1}\right)^{(k-1)/k}\right]$$

Example 13.8

If an ideal gas sample ($k = 1.40$) experiences an isentropic process starting from standard conditions (14.69 psia, 77°F) where the temperature drops to 30°F, what is the final pressure?

Solution:

From equation 13.42:

$$T_1 p_1{}^{(1-k)/k} = T_2 p_2{}^{(1-k)/k}$$

Then,

$$\frac{T_1}{T_2} = \left(\frac{p_2}{p_1}\right)^{(1-k)/k}$$

or

$$\left(\frac{T_1}{T_2}\right)^{k/(1-k)} = \frac{p_2}{p_1}$$

$$p_2 = p_1\left(\frac{T_1}{T_2}\right)^{k/(1-k)}$$

$$= (14.69 \ psia)\left[\frac{(77 + 460)°R}{(30 + 460)°R}\right]^{1.40/(1-1.40)}$$

$$= 10.66 \ psia$$

Example 13.9

If, starting at 200°F, an ideal gas undergoes an isentropic process in which the specific volume quadruples, what is the final temperature? ($k = 1.40$)

Solution:

From equation 13.43:

$$\frac{T_2}{T_1} = \left(\frac{v_1}{v_2}\right)^{k-1}$$

Then, $\quad T_2 = T_1\left(\frac{v_1}{v_2}\right)^{k-1}$

$$= [(200 + 460)°R]\left(\frac{1}{4}\right)^{1.40-1}$$

$$= 379°R$$

$$= -81°F$$

Isobaric Process in a Closed System

For an isobaric (constant-pressure) process:

$$w_b = \int_1^2 p \ dv$$

From equation 13.7a:

$$w_b = p \int_1^2 dv \qquad [13.45]$$

$$= p\Delta v$$

and (**Charles's law**):

$$\frac{T_1}{v_1} = \frac{T_2}{v_2} = \text{constant} \qquad [13.46]$$

Isometric Process in a Closed System
For an isometric (constant-volume) process:
From equation 13.7a:

$$w_b = \int_1^2 p\, dv = 0$$

and

$$\frac{T_1}{p_1} = \frac{T_2}{p_2} = \text{constant} \qquad [13.47]$$

Isothermal Process in a Closed System
For an isothermal (constant-temperature) process:

$$w_b = \int_1^2 p\, dv$$

$$= \int_1^2 \frac{RT}{v}\, dv$$

$$= RT \ln\left(\frac{v_2}{v_1}\right)$$

Then:

$$w_b = RT \ln\left(\frac{v_2}{v_1}\right) = RT \ln\left(\frac{p_1}{p_2}\right) \qquad [13.48]$$

and (**Boyle's law**):

$$p_1 v_1 = p_2 v_2 = \text{constant} \qquad [13.49]$$

Example 13.10
A piston and cylinder system contains 3 kg of air at 15°C. In an isothermal compression in which the final volume is one-third of the initial volume, how much boundary work has been performed?

Solution:
From equation 13.48:

$$w_b = RT \ln\left(\frac{v_2}{v_1}\right)$$

Then,

$$w_b = \left(\frac{8.314 \frac{J}{\text{mol} \cdot K}}{29 \frac{g}{\text{mol}}}\right)[(15 + 273)\, K] \ln\left(\frac{v_2}{3v_2}\right)$$

$$\times \left(\frac{1,000\, g}{kg}\right)\left(\frac{kJ}{1,000\, J}\right)$$

$$= -90.7 \frac{kJ}{kg} \quad \text{(work performed on the gas)}$$

> The negative sign means work is done on the system.

and

$$W_b = mw_b$$

$$= (3\, kg)\left(-90.7 \frac{kJ}{kg}\right)$$

$$= -272\, kJ$$

Polytropic Process in a Closed System
A polytropic process is one for which:

$$p_1 v_1^n = p_2 v_2^n = \text{constant} \qquad [13.50]$$

where n is a constant ($n \neq 1$). Note that, when the process is isentropic, $n = k$.

In this case, the work performed during the process is calculated as follows:

$$w = \frac{p_2 v_2 - p_1 v_1}{1 - n} \qquad [13.51]$$

Mach Number
The Mach number is a dimensionless parameter used to describe the velocity of an ideal gas. It is defined as the ratio of fluid velocity to the speed of sound, or

$$M_a = \frac{\overline{V}}{c} \qquad [13.52]$$

where M_a = Mach number (dimensionless),
$\overline{V}$ = fluid velocity (ft/s or m/s),
c = speed of sound (ft/s or m/s).

For an ideal gas, the speed of sound is defined as

$$c = \sqrt{kRT} \qquad [13.53]$$

where c = speed of sound (ft/s or m/s),
  k = specific heat ratio (c_p/c_v),
  R = gas constant (ft-lbf/lbm · °R or kJ/kg · K),
  T = absolute temperature of the gas (K or °R).

Ideal Gas Mixtures

When a substance is composed of a mixture of ideal gases, determining the values of its properties and the relationships among its equilibrium states becomes a bit more complicated. To define the mixture, it is necessary to specify the amount of each constituent in the mixture. The common way to do this is to calculate the *mole fraction,* that is, the ratio of the number of moles of each component to the total number of moles in the mixture. The mole fraction is given by:

$$x_i = \frac{N_i}{\sum_i N_i} \qquad [13.54]$$

so that

$$\sum_i x_i = 1$$

An equivalent method of defining the composition of the mixture is to determine the *mass fraction,* given by:

$$y_i = \frac{m_i}{\sum_i m^i} \qquad [13.55]$$

so that

$$\sum_i y_i = 1$$

It is possible to convert one of these two measures of composition into the other by use of the following formulas:

Mass fraction from mole fraction:

$$y_i = \frac{x_i M_i}{\sum_i x_i M_i} \qquad [13.56]$$

Mole fraction from mass fraction:

$$x_i = \frac{y_i/M_i}{\sum_i y_i/M_i} \qquad [13.57]$$

Once the composition is defined, it is possible to define the parameters required to apply the ideal gas equation to the mixture, molar mass, and gas constant:

Molar mass of mixture: $M_{\text{mix}} = \sum_i x_i M_i$ [13.58]

Gas constant of mixture: $R_{\text{mix}} = \dfrac{\overline{R}}{M_{\text{mix}}}$ [13.59]

Example 13.11

An ideal gas mixture has the following composition, in terms of mole fractions:

Butane	0.39
Methane	0.42
Propane	0.19

What is the mass fraction of the butane?

Solution:

From equation 13.56:

$$y_i = \frac{x_i M_i}{\Sigma x_i M_i}$$

$$= \frac{(0.39)(58)}{(0.39)(58) + (0.42)(16) + (0.19)(44)}$$

$$= 0.60$$

Analysis of the ideal gas mixture can now be accomplished as if the mixture were a pure ideal gas:

$$pv = R_{\text{mix}} T \qquad [13.60a]$$

or

$$pV = \sum_i m_i R_{\text{mix}} T \qquad [13.60b]$$

or

$$pV = \sum_i N_i \overline{R} T \qquad [13.60c]$$

It is sometimes helpful to determine the contribution of each gas to the total pressure or volume of the system. These contributions, called the partial pressure and partial volume of the constituent, are given by:

$$p_i = x_i p = \frac{m_i R_i T}{V} \qquad [13.61]$$

so that

$$\sum_i p_i = p$$

and

$$V_i = x_i V = \frac{m_i R_i T}{p} \qquad [13.62]$$

so that

$$\sum_i V_i = V$$

Equation 13.61 leads to **Dalton's law:**

$$pV = (p_1 + p_2 + p_3 + \dots + p_n)V \qquad [13.63]$$

Example 13.12

What is the partial pressure of nitrogen in the following mixture (composition given in terms of mass fraction) in a 3-m^3 vessel at 40°C, if the total mass of the mixture is 3 kg?

Argon	0.44
Hydrogen	0.29
Nitrogen	0.27

Solution:

From equation 13.61: $p_i = \dfrac{m_i R_i T}{V}$

Then

$$m_i = y_i m$$
$$= (0.27)(3 \text{ kg})$$
$$= 0.81 \text{ kg}$$

Also

$$R_i = \frac{\overline{R}}{M_i}$$

$$= \frac{8.314 \, \dfrac{\text{J}}{\text{mol K}}}{28 \, \dfrac{\text{g}}{\text{mol}}} \left(\frac{\text{kJ}}{1{,}000 \text{ J}}\right)\left(\frac{1{,}000 \text{ g}}{\text{kg}}\right)$$

$$= 0.297 \, \frac{\text{kJ}}{\text{kg K}}$$

$$p_i = \frac{(0.81 \text{ kg})\left(0.297 \, \dfrac{\text{kJ}}{\text{kg K}}\right)[(40 + 273) \text{ K}]}{3 \text{ m}^3}$$

$$= 25.1 \, \frac{\text{kJ}}{\text{m}^3}\left(\frac{1{,}000 \text{ N} \cdot \text{m}}{\text{kJ}}\right)\left(\frac{\text{kPa}}{1{,}000 \, \dfrac{\text{N}}{\text{m}^2}}\right)$$

$$= 25.1 \text{ kPa}$$

The specific properties of the mixture are obtained from a mass-weighted average of the specific properties of each constituent:

$$u_{\text{mix}} = \sum_i (y_i u_i)$$

and

$$h_{\text{mix}} = \sum_i (y_i h_i) \qquad [13.64]$$

Simple Compressible Substances

While the ideal gas assumption works for most simple substances at high energy levels, many substances we commonly encounter, such as water, behave in significantly different ways. Water from the tap is a liquid. If, however, we place the water in the freezer, it becomes ice, a solid. If we put the water in a pot and boil it, it becomes steam, a gas. These are the three principal phases in which any substance can exist: solid, liquid, and gas.

An explanation for these phases can be found if we examine the microscopic structure of a substance. At the lowest energy levels of the substance, its molecules arrange themselves into a close-packed, organized structure, called a lattice. Because the energy level is low and the molecules are close together, the intermolecular forces dominate, and the lattice structure is maintained. Because the molecules cannot be easily separated or displaced, the substance appears macroscopically as a solid.

Add energy to the substance through an interaction with its environment, and a change begins to take place. The energy in some molecules will become great enough that the lattice structure will begin to break down. At this stage, the substance still has some structure, but its molecules are easily displaced. The substance now appears macroscopically as a liquid.

Add even more energy to the substance, and eventually all of its molecules will overcome the intermolecular forces. When this happens, the substance will lose all signs of structure, and its molecules will be separated by far greater distances. At this point the molecules can move about randomly, and the substance appears macroscopically as a gas.

Because of this phase-change behavior, we can expect that the equation of state that describes the equilibrium states of most substances will be a great deal more complicated than that of an ideal gas. For the most part we rely on the vast quantity of experimental data that exists to predict the equilibrium states of substances, using either tabulated data, empirical correlations, or graphical thermodynamic surfaces.

The equilibrium state of a simple compressible substance is a function of at most two independent properties. Mathematically, such a relationship can be represented as a surface in three-dimensional space with each axis representing the value of a thermodynamic property. This representation has become a tradition in thermodynamics, with the pressure expressed as a function of the specific volume and the temperature [$p = p(v, T)$]. The projections of such a surface on the p–v and p–T planes are shown in Figure 13.7.

The equilibrium states of most substances can be described with such a diagram. As well as showing the equilibrium states, these diagrams indicate the relationships among phases. The dome-shaped region of the p v diagram is the vapor dome. States that fall on the left edge of the dome are completely liquid but are on the verge of boiling. For these states, the substance is called a saturated liquid. States that fall on the right edge of the dome are completely gaseous but are on the verge of condensing. For these states, the substance is called a saturated vapor. The two regions are separated by a point at the top of the dome called the critical point. Under the dome, the substance exists in two different homogeneous states; a portion of the substance is a saturated liquid, and the other portion is a saturated vapor. Both portions exist at the same pressure and temperature. There is only one point on the thermodynamic surface at which the substance can exist in all three phases, the triple point.

If lines of constant temperature are drawn on the p–v diagram, an important feature of compressible substances becomes apparent. The constant-temperature lines under the vapor dome are exactly horizontal, indicating that, as long as two phases (liquid and vapor) are present, changes in state that occur at constant pressure will also occur at constant temperature. Any change in the amount of energy will serve to move the overall composition of the substance either closer to the saturated liquid phase or closer to the saturated vapor phase. The same type of behavior is observed as the substance changes between solid and liquid states.

Thermodynamic Tables

Because use of these graphical surfaces for analysis can be cumbersome, the equilibrium-state data for many common substances have been compiled in tabular form. The most common of these tables are the steam tables for water; examples are given at the end of this chapter. Usually there are two tables, one for the saturated liquid and saturated vapor equilibrium states (page 417), and one for the superheated vapor (steam) equilibrium states (page 419) of water. These tables give the values of specific volume, specific internal energy, specific enthalpy, and specific entropy as functions of temperature and pressure. The saturated (water) table lists values as functions of either pressure or temperature.

The information available in these tables is limited because the values of only a few states are given. What happens when information is needed about states not listed in the table? For analysis, it is usually sufficient to linearly interpolate the values of the properties for a state that falls between two states given in the tables.

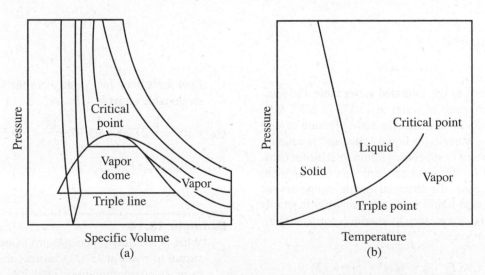

Figure 13.7 The p–v (a) and p–T (b) projections of a thermodynamic surface

Example 13.13

What is the specific volume of steam at a pressure of 100 kPa and a temperature of 430°C?

Solution:

The superheated steam table (page 419) gives information about properties at a pressure of 100 kPa and a temperature of 400°C or 500°C, as follows:

T (°C)	v (m³/kg)	u (kJ/kg)	h (kJ/kg)	s kJ/(kg · K)
400	3.103	2967.9	3278.2	8.5435
430	v_{430}	u_{430}	h_{430}	s_{430}
500	3.565	3131.6	3488.1	8.8342

By linear interpolation,

$$\frac{v_{430} - 3.103}{3.565 - 3.103} = \frac{430 - 400}{500 - 400}$$

$$v_{430} = 3.242 \text{ m}^3/\text{kg}$$

Similar calculations can be made for each of the other properties:

$$u_{430} = 3,017.0 \text{ kJ/kg}$$

$$h_{430} = 3,341.2 \text{ kJ/kg}$$

$$s_{430} = 8.6307 \text{ kJ/(kg} \cdot \text{K)}$$

Example 13.14

What volume does 50 kg of water at 130°C and 400 kPa occupy?

Solution:

From equation 13.1:

$$V = mv$$

According to the saturated water table, the saturation pressure of water at 130°C is 270.1 kPa. Since this is less than the state pressure of 400 kPa, the water is in liquid state, and is called a compressed or subcooled liquid. A table for compressed liquid water properties is not supplied; but because the thermodynamic properties of compressed liquid water are primarily functions of temperature, the following approximations can be made:

$$v = v_f(T), \quad u = u_f(T), \quad s = s_f(T)$$

$$h = h_f(T) + v_f(T)[p - p_{sat}(T)]$$

where $v_f(T)$ is the saturated liquid-specific volume at temperature T,

$u_f(T)$ is the saturated liquid-specific internal energy at temperature T,

$s_f(T)$ is the saturated liquid-specific entropy at temperature T,

$h_f(T)$ is the saturated liquid-specific enthalpy at temperature T,

$p_{sat}(T)$ is the saturation pressure of water at temperature T.

From the saturated water table:

$$v_f(130°C) = 0.001070 \frac{\text{m}^3}{\text{kg}}$$

so

$$V = (50 \text{ kg})\left(0.001070 \frac{\text{m}^3}{\text{kg}}\right) = 0.0535 \text{ m}^3$$

Example 13.15

What is the energy transferred to a container of water vapor during a reversible heat interaction at a constant temperature of 700°C in which the pressure is increased from 100 kPa to 600 kPa?

Solution:

From equation 13.21:

$$Q_{rev} = \int_{s_1}^{s_2} T \, ds$$

Then,

$$Q_{rev} = T \int_{s_1}^{s_2} ds$$

$$= T(s_2 - s_1)$$

Find the values for s_2 and s_1 in the superheated steam table on page 419:

$$Q_{rev} = [(700 + 273) \text{ K}]\left(8.5107 \frac{\text{kJ}}{\text{kg} \cdot \text{K}} - 9.3398 \frac{\text{kJ}}{\text{kg} \cdot \text{K}}\right)$$

$$= -806.7 \frac{\text{kJ}}{\text{kg}}$$

Example 13.16

What is the flow (open system) availability of a stream of water at 75°C? Consider the dead state to be standard conditions (101 kPa and 25°C), and neglect contributions of potential and kinetic energy.

Solution:

From equation 13.24:

$$\Psi = (h - h_o) - T_o(s - s_o)$$

Find the values for h and s in the saturated steam table (h_f and s_f at $T = 75°C$):

$$\Psi = \left(313.93\ \frac{kJ}{kg} - 104.89\ \frac{kJ}{kg}\right) - [(25 + 273)\ K]$$

$$\times \left[\left(1.0155\ \frac{kJ}{kg \cdot K}\right) - \left(0.3674\ \frac{kJ}{kg \cdot K}\right)\right]$$

$$= 15.91\ \frac{kJ}{kg}$$

Vapor-Liquid Mixtures

The saturated vapor-liquid (water) table provides data for the region of the thermodynamic surface under the vapor dome. Property data are listed either as functions of temperature or as functions of pressure. Since pressure and temperature are related for a saturated substance, only one value is needed to determine the state of a saturated substance. Usually the value of the other property will be given as a reference.

Each of the dependent properties in the table (specific volume, specific internal energy, specific enthalpy, and specific entropy) is listed for two states, the saturated liquid state with no vapor present, denoted with an f subscript, and the saturated vapor state with no liquid present, denoted with a g subscript. Some values have an fg subscript, which indicates the difference between the values of the property at the saturated liquid and at the saturated vapor state (i.e., $h_{fg} = h_g - h_f$).

As a substance evaporates from a saturated liquid to a saturated vapor, or condenses from a saturated vapor to a saturated liquid, it can be considered as a mixture of two homogeneous substances, a saturated liquid and a saturated vapor. All of the energy added to or removed from the system goes into changing the phase of the fluid. For analysis it is often necessary to determine average values of the properties for the vapor-liquid mixture. Since each of the properties listed in the table is additive in its extensive form, and is reported in its mass specific intensive form, the total value for the substance can be calculated by using a mass-weighted average of the value for each state. This gives rise to a parameter called *quality*, defined as:

$$x = \frac{\text{mass of substance in vapor phase}}{\text{total mass of liquid and vapor combined}}$$

or

$$x = \frac{m_g}{m_f + m_g} \qquad [13.65]$$

The equations that give the mass weighted average of the values of the mass specific properties for a liquid-vapor mixture are as follows:

$$v_{mix} = v_f + x(v_g - v_f)$$
$$= v_f + xv_{fg} \qquad [13.66a]$$

$$u_{mix} = u_f + x(u_g - u_f)$$
$$= u_f + xu_{fg} \qquad [13.66b]$$

$$h_{mix} = h_f + x(h_g - h_f)$$
$$= h_f + xh_{fg} \qquad [13.66c]$$

$$s_{mix} = s_f + x(s_g - s_f)$$
$$= s_f + xs_{fg} \qquad [13.66d]$$

Example 13.17

A vapor-liquid mixture of water at 85°C has a specific volume of 0.75 m³/kg.
A. What is the quality of the mixture?
B. What is the average specific internal energy of the mixture?

Solutions:

A. From equation 13.66a:

$$x = \frac{v_{mix} - v_f}{v_g - v_f}$$

so that

$$x = \frac{0.75\ \dfrac{m^3}{kg} - 0.001033\ \dfrac{m^3}{kg}}{2.828\ \dfrac{m^3}{kg} - 0.001033\ \dfrac{m^3}{kg}}$$

$$= 0.265$$

B. From equation 13.66b:

$$u_{mix} = u_f + xu_{fg}$$

so that

$$u_{mix} = 355.84\ \frac{kJ}{kg} + (0.265)\left(2132.6\ \frac{kJ}{kg}\right)$$

$$= 921\ \frac{kJ}{kg}$$

Psychrometrics

Often, when an environmental system is analyzed, the substance that makes up the system is found to be air. Normally air is assumed to be a homogeneous mixture of gases, and the values of its thermodynamic properties can be found in tabulated form. However, the values in these tables are for a standard composition of gases, and the composition of the air in an environment can change greatly over time. The component most likely to affect the thermodynamic properties of air is the amount of water vapor the air contains, or its humidity.

To deal with this variation, air is often considered to be a mixture of two gases: dry air, whose thermodynamic properties can be found in standard air tables, and water vapor, whose thermodynamic properties are given in standard steam tables. As with ideal gas mixtures, the pressure of the dry air-water vapor mixture is the sum of the partial pressures of its components, or:

$$p = p_a + p_v \qquad [13.67]$$

where p = total pressure of the mixture,
  p_a = partial pressure of the dry air,
  p_v = partial pressure of the water vapor.

The composition of an air-water vapor mixture can be expressed in terms of the mass ratio or the mole ratio of the two components, just as with an ideal gas mixture. The mass ratio measure is called the *specific humidity* and is given by:

$$\omega = \frac{m_v}{m_a} = 0.622\frac{p_v}{p - p_v} \qquad [13.68]$$

where ω = specific humidity or humidity ratio,
  m_v = mass of the water vapor,
  m_a = mass of the dry air.

The coefficient 0.622 is determined from the ratio of molar masses of water to dry air.

Relative humidity is defined as:

$$\phi = \frac{p_v}{p_g} \qquad [13.69]$$

where ϕ = relative humidity
  p_g = saturation pressure of water at the temperature of the air-water vapor mixture.

Dew-Point Temperature

The composition of the air-water vapor mixture is limited to some degree by the equilibrium states of water; that is, there is a maximum amount of water vapor that the mixture can hold. Any water added to the system in excess of this maximum will exist as a liquid. Figure 13.8 illustrates the relationship of the dew-point temperature to the temperature of the air-water mixture.

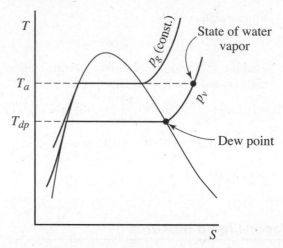

Figure 13.8 *T–s* diagram for water that illustrates the principles of psychrometry

To understand this idea, consider a system that consists of water only at a given temperature. If the pressure of the water is below the saturation pressure for that temperature, the water will exist as a vapor. If the pressure is greater than the saturation pressure, the water will exist as a liquid. For an air-water vapor mixture, the pressure of the water vapor is its partial pressure. As water is added to the system, the partial pressure will increase until it is equal to the saturation pressure of water at the given temperature. At this point, the mixture cannot accept any more water in the vapor phase. Any additional water added to the system will exist in a liquid phase.

The saturation pressure of water is a function of temperature. Therefore, for any given pressure of a water system below the critical pressure, there exists a temperature for which that pressure is the saturation pressure. For an air-water vapor mixture, there exists a temperature for which the partial pressure of the water vapor is equal to the saturation pressure of water. This temperature is called the *dew-point temperature*. If the temperature of the mixture is reduced below its dew point, some of the water vapor will begin to condense into a liquid; this condensation will continue until the partial pressure of water vapor is equal to the saturation pressure of water at the lower temperature. When this happens in our environment, the condensed water collects on solid particles in the air and becomes fog.

Example 13.18

An air-water vapor mixture at a temperature of 80°C has a relative humidity of 27%. What is the dew-point temperature of the mixture?

Solution:

From equation 13.69:

$$p_v = \phi p_g$$

where p_g is the saturation pressure of water at the temperature given, or 46.39 kPa. Then,

$$p_v = 0.27 \, (46.39 \text{ kPa})$$

$$= 12.53 \text{ kPa}$$

$$T_{dp} = T_{sat}(12.53 \text{ kPa}) \approx 50°C$$

Wet-Bulb Temperature

A common device used to experimentally determine the amount of water vapor in the air, or the humidity, is a sling psychrometer. This device consists of two glass-tube thermometers, mounted side by side, with a handle attached to the top of the thermometers. The bulb of one thermometer is covered with a fabric wick, which is saturated with water. This is the wet bulb thermometer; the other is the dry-bulb thermometer. The device is operated by swinging it by its handle.

Because there will be some energy transfer due to evaporation of the water in the wick of the wet bulb thermometer, the two thermometers will display different temperatures. This difference is used to determine the humidity of the air. The temperature read from the wet-bulb thermometer is called the *wet-bulb temperature* of the air-water vapor mixture; the temperature read from the dry-bulb thermometer is the dry-bulb temperature.

Psychrometric Chart

A psychrometric chart relates the values of specific volume, specific enthalpy, relative humidity, wet- and dry-bulb temperatures, and humidity ratio for equilibrium states of a mixture at a given pressure. Given the values of any two of these properties, we can determine the values of all others. An example of a psychometric chart is found on page 420.

13.4 Applications

All of the definitions, laws, and concepts given above are combined to analyze and predict the performance of real mechanical devices. Some examples of these devices are discussed below.

Energy-Conversion Devices

There are many energy-conversion devices to which the above laws and relationships can be applied. Each of these systems is a steady-state system, meaning that the rate of change of every property at each location in the device will remain zero. This steady-state assumption greatly simplifies analysis.

The balance equations applied to a steady-state system are as follows:

Mass balance:

$$\frac{dm}{dt} = \dot{m}_{transferred} = 0 \quad \text{or} \quad \sum \dot{m}_i = \sum \dot{m}_e \quad [13.70]$$

Energy balance:

$$\frac{dE}{dt} = \dot{Q} - \dot{W} + \sum \dot{m}_i \left(h_i + \frac{\overline{V}_i^2}{2} + gz_i \right)$$

$$- \sum \dot{m}_e \left(h_e + \frac{\overline{V}_e^2}{2} + gz_e \right) = 0 \quad [13.71]$$

Entropy balance:

$$\frac{dS}{dt} = \sum \frac{\dot{Q}}{T_b} + \sum \dot{m}_i s_i - \sum \dot{m}_e s_e + \dot{S}_{produced} = 0 \quad [13.72]$$

Nozzles and Diffusers

Nozzles and diffusers, diagrammed in Figure 13.9, are devices used to change the velocity and pressure of a fluid stream; therefore we can expect kinetic energy terms to be significant. Analysis is often simplified by the assumption that effects due to changes in potential energy are negligible.

From mass balance:

$$\dot{m}_i = \dot{m}_e = \dot{m} \quad [13.73]$$

From energy balance:

$$h_e - h_i = \frac{\overline{V}_i^2 - \overline{V}_e^2}{2} \quad [13.74]$$

From entropy balance:

$$\dot{S}_{produced} = \dot{m}(s_e - s_i) \quad [13.75]$$

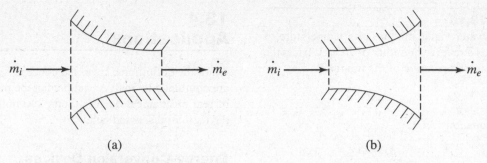

Figure 13.9 A subsonic nozzle (a) and a subsonic diffuser (b)

The efficiency of the nozzle or diffuser is given by:

$$\eta = \frac{\text{change in kinetic energy of the actual device}}{\text{change in kinetic energy for an ideal device}}$$

$$= \frac{\overline{V}_e^2 - \overline{V}_i^2}{2(h_i - h_{es})} \qquad [13.76]$$

where η = nozzle or diffuser efficiency,
          h_{es} = enthalpy at isentropic exit state.

In a perfectly efficient nozzle, no entropy will be produced, and the value of entropy will be the same at the outlet as at the inlet. In a real nozzle, however, some entropy will be produced, and the value of enthalpy at the exit state will be somewhat higher than that of a perfectly efficient nozzle.

Turbines, Pumps, and Compressors

Turbines, pumps, and compressors, shown in Figure 13.10, are used either to produce mechanical work from some of the internal energy of a substance or to use mechanical work to increase the internal energy of a substance. In a turbine, energy is removed from the substance through a mechanical work interaction; in a pump or compressor, energy is transferrred to the substance through a mechanical work interaction. Whatever device is considered, the conversion of energy is often assumed to take place adiabati-

cally, that is, with no heat interaction, and changes in kinetic and potential energy are often negligible.

From mass balance:

$$\dot{m}_i = \dot{m}_e = \dot{m} \qquad [13.73]$$

From energy balance:

$$\dot{W} = \dot{m}(h_i - h_e) \qquad [13.77]$$

From entropy balance:

$$\dot{S}_{\text{produced}} = \dot{m}(s_e - s_i) \qquad [13.75]$$

The efficiency of turbines, pumps, and compressors can be found by using the following equation:

$$\eta_{\text{turbine}} = \frac{w_{\text{actual}}}{w_{\text{isentropic}}} = \frac{h_i - h_e}{h_i - h_{es}} \qquad [13.78]$$

and

$$\eta_{\text{comp, pump}} = \frac{w_{\text{isentropic}}}{w_{\text{actual}}} = \frac{h_{es} - h_i}{h_e - h_i} \qquad [13.79]$$

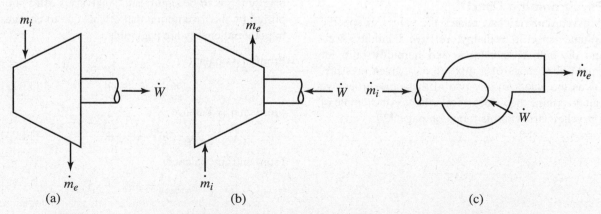

Figure 13.10 Schematic representation of (a) a turbine, (b) a compressor, and (c) a pump

Example 13.19

A 30-kg/s stream of steam enters a turbine at 800 kPa and 600°C, and exits the turbine at 50 kPa and 300°C

A. What is the rate of energy transferred through turbine work?

B. What is the rate of entropy production?

Solutions:

A. From equation 13.77:

$$\dot{W} = \dot{m}(h_i - h_e)$$

$$= \left(30\frac{\text{kg}}{\text{s}}\right)\left(3{,}699.4\frac{\text{kJ}}{\text{kg}} - 3{,}075.5\frac{\text{kJ}}{\text{kg}}\right)\left(\frac{\text{kW}}{\text{kJ/s}}\right)$$

$$= 19 \text{ kW}$$

B. From equation 13.75:

$$\dot{S}_{\text{produced}} = \dot{m}(s_e - s_i)$$

$$= \left(30\frac{\text{kg}}{\text{s}}\right)\left(8.5373\frac{\text{kJ}}{\text{kg}\cdot\text{K}}\right.$$

$$\left. - 8.1333\frac{\text{kJ}}{\text{kg}\cdot\text{K}}\right)\left(\frac{\text{kW}}{\text{kJ/s}}\right)$$

$$= 12\frac{\text{kW}}{\text{K}}$$

Example 13.20

How much power is required for an ideal pump ($\eta = 1$) to increase the pressure of a 7-kg/s stream of water at 25°C from 50 kPa to 600 kPa?

Solution:

For a reversible process, equation 13.8 gives:

$$w_{\text{rev}} = -\int_1^2 v\, dp$$

Assume that liquid water is incompressible; then:

$$\dot{W} = -\dot{m}v\int_{p_1}^{p_2} dp$$

$$= -\dot{m}v(p_e - p_i)$$

$$= -\left(7\frac{\text{kg}}{\text{s}}\right)\left(0.001003\frac{\text{m}^3}{\text{kg}}\right)(600 \text{ kPa} - 50 \text{ kPa})$$

$$\times \left(\frac{1{,}000 \text{ N/m}^2}{\text{kPa}}\right)\left(\frac{\text{J}}{\text{N}\cdot\text{m}}\right)\left(\frac{\text{kW}}{1{,}000 \text{ J/s}}\right)$$

$$= -3.86 \text{ kW}$$

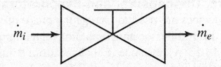

It does not take a lot of power to pressurize liquid.

The pump must supply 3.86 kW of power to the system.

Valves

A valve, represented in Figure 13.11, is a device that can be used to lower the pressure in a fluid system. In most cases the amount of energy transferred through heat interactions is negligible, as are changes in kinetic and potential energy.

$$\dot{m}_i \longrightarrow \quad\quad\quad \longrightarrow \dot{m}_e$$

Figure 13.11 Schematic representation of a valve

From mass balance:

$$\dot{m}_i = \dot{m}_e = \dot{m} \qquad\qquad [13.73]$$

From energy balance:

$$h_i = h_e \qquad\qquad [13.80]$$

From entropy balance:

$$\dot{S}_{\text{produced}} = \dot{m}(s_e - s_i) \qquad\qquad [13.75]$$

Example 13.21

Steam flows through a valve. At the inlet of the valve, the steam pressure is 600 kPa, the temperature is 700°C, and the mass flow rate is 4 kg/s. At the outlet of the valve, the steam pressure is 100 kPa. At what rate is entropy produced?

Solution:

From equation 13.75:

$$\dot{S}_{\text{produced}} = \dot{m}(s_e - s_i)$$

Find the values of entropy and enthalpy at the inlet state ($p_i = 600$ kPa and $T_i = 700$°C) in the superheated steam table. Then:

$$s_i = 8.51\frac{\text{kJ}}{\text{kg}\cdot\text{K}}$$

$$h_i = 3925.3\frac{\text{kJ}}{\text{kg}}$$

Because flow through a valve is isenthalpic (h = constant), the state of the fluid at the valve outlet can be determined from p_e and $h_e = h_i$. Using the superheated steam table, find the value of entropy at the outlet:

$$s_e = 9.34 \frac{\text{kJ}}{\text{kg} \cdot \text{K}}$$

Then:

$$\dot{S}_{\text{produced}} = \left(4 \frac{\text{kg}}{\text{s}}\right)\left(9.34 \frac{\text{kJ}}{\text{kg} \cdot \text{K}} - 8.51 \frac{\text{kJ}}{\text{kg} \cdot \text{K}}\right)$$

$$= 3.32 \frac{\text{kJ}}{\text{s K}}$$

Boilers, Condensers, and Evaporators

These devices are used to change the phase of a substance. In a boiler or an evaporator, diagrammed in Figure 13.12, a substance is changed from a liquid to a vapor. In a condenser, a vapor is changed to a liquid. These devices are often used to take advantage of the relatively large amounts of energy transfer required to change the phase of many common fluids. For the most part, boilers, condensers, and evaporators transfer their energy through heat interactions. Changes in potential and kinetic energy are often negligible.

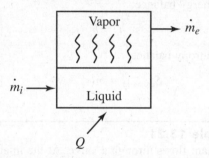

Figure 13.12 Diagram of a boiler or an evaporator

From mass balance:

$$\dot{m}_i = \dot{m}_e = \dot{m} \qquad [13.73]$$

From energy balance:

$$\dot{Q} = \dot{m}(h_e - h_i) \qquad [13.81]$$

From entropy balance:

$$\dot{S}_{\text{produced}} = \dot{m}(s_e - s_i) - \frac{\dot{Q}}{T} \qquad [13.82]$$

where T is the temperature at which the heat interaction occurs.

Example 13.22

A steady-state boiler operates with 5000 kW of energy supplied through heat interaction. If 1.5 kg/s of water at 800 kPa and 25°C enters the boiler, what is the temperature of the steam leaving the boiler?

Solution:

From equation 13.81:

$$h_e = h_i + \frac{\dot{Q}}{\dot{m}}$$

As in Example 13.15, h_i is approximated by $h_f(25°C)$ or $h_i = 104.89$ kJ/kg. Then:

$$h_e = 104.89 \frac{\text{kJ}}{\text{kg}} + \frac{5{,}000 \text{ kW}\left(\frac{\text{kJ/s}}{\text{kW}}\right)}{1.5 \frac{\text{kg}}{\text{s}}}$$

$$= 3{,}438 \frac{\text{kJ}}{\text{kg}}$$

The superheated steam table shows that at 800 kPa, $T_e \approx 480°C$.

Heat Exchangers

A heat exchanger, diagrammed in Figure 13.13, is a device that allows energy to be exchanged between two substances through a heat interaction. Analysis is simplified by assuming that no other interaction takes place; in particular, that there is no heat interaction between the heat exchanger and its environment.

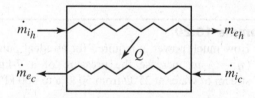

Figure 13.13 Diagram of a heat exchanger

From mass balance:

$$\dot{m}_{ih} = \dot{m}_{eh} = \dot{m}_h \quad \text{and} \quad \dot{m}_{ic} = \dot{m}_{ec} = \dot{m}_c \quad [13.83]$$

where h = hot stream, and c = cold stream.
From energy balance:

$$\dot{m}_h(h_i - h_e)_h = \dot{m}_c(h_e - h_i)_c \qquad [13.84]$$

In many cases the change in temperature through a heat exchanger is small enough that the constant-pressure specific heat may be considered constant, and the following equation may be used in place of equation 13.84:

$$\dot{m}_h c_p(T_i - T_e)_h = \dot{m}_c c_p(T_e - T_i)_c \qquad [13.85]$$

From entropy balance:

$$\dot{S}_{\text{produced}} = \dot{m}_c(s_e - s_i)_c - \dot{m}_h(s_i - s_e)_h \qquad [13.86]$$

Heat transfer processes are always irreversible unless the temperature difference between two streams is close to zero.

Example 13.23

In a heat exchanger, 10 kg/s of water enters the hot side at 80°C and 150 kPa, and leaves at 40°C; 12 kg/s of water enters the cold side at 15°C. If there is no pressure drop in the heat exchanger [$(p_e = p_i)_c$ and $(p_e = p_i)_h$], at what temperature does the cold water leave the heat exchanger?

Solution:

From equation 13.85:

$$\dot{m}_h c_p (T_i - T_e)_h = \dot{m}_c c_p (T_e - T_i)_c$$

so

$$T_{ec} = \frac{\dot{m}_h c_p (T_i - T_e)_h + \dot{m}_c c_p T_{ic}}{\dot{m}_c c_p}$$

or

$$T_{ec} = T_{ic} + \frac{\dot{m}_h c_p}{\dot{m}_c c_p}(T_i - T_e)_h$$

$$= 15°C + \frac{\left(10\ \dfrac{kg}{s}\right)\left(4.18\ \dfrac{kJ}{kg\ K}\right)}{\left(12\ \dfrac{kg}{s}\right)\left(4.18\ \dfrac{kJ}{kg\ K}\right)}(80°C - 40°C)$$

$$= 48°C$$

Mixers and Separators

Mixers (see Figure 13.14) and separators are devices that combine or separate material streams. For the purposes of this discussion, the mixer or separator will experience no work or heat interactions with its surroundings, and changes in kinetic and potential energy are negligible.

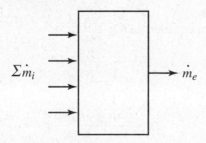

Figure 13.14 Schematic diagram of a mixer

From mass balance:

$$\sum \dot{m}_i = \dot{m}_e \qquad [13.87]$$

From energy balance:

$$\sum \dot{m}_i h_i = \dot{m}_e h_e \qquad [13.88]$$

From entropy balance:

$$\dot{m}_e s_e - \sum \dot{m}_i s_i = \dot{S}_{produced} \qquad [13.89]$$

Note that the above equations are valid for a mixer only. For a separator, the summations will be over the outlet terms instead of the inlet terms.

Thermodynamic Cycles

The cyclic machine discussed earlier was a generalization of the many cyclic devices used in industry. All of the devices discussed here operate on a thermodynamic cycle, that is, a series of processes for which the end state of the last process is equivalent to the start state of the first process.

As with energy-conversion devices, it is often desirable to determine the figure of merit for the thermodynamic cycles. The figure of merit for a power-producing thermodynamic cycle is the *thermal efficiency,* while the figure of merit used for refrigeration and heat pump cycles is the *coefficient of performance* (COP).

Carnot Cycle

A Carnot-cycle machine undergoes four processes, which may occur with a cycle of four steady-state components or in a piston and cylinder system. The two types of systems are shown in Figures 13.15 and 13.16.

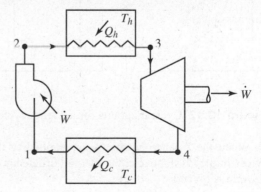

Figure 13.15 Carnot cycle composed of steady-state devices

The processes of an ideal Carnot cycle are as follows:

Process 1–2: Energy is added to the system in a reversible, adiabatic work interaction (isentropic compression).

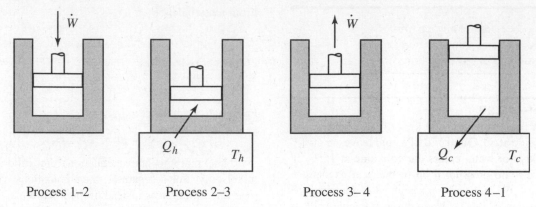

Process 1–2   Process 2–3   Process 3–4   Process 4–1

Figure 13.16 Piston and cylinder Carnot cycle

Process 2–3: Energy is added to the system in a reversible heat interaction with the high-temperature reservoir.

Process 3–4: Energy is removed from the system in a reversible, adiabatic work interaction (isentropic expansion).

Process 4–1: Energy is removed from the system in a reversible heat interaction with the low-temperature reservoir.

All of the devices considered above are idealized concepts of real devices, which can never be achieved. Irreversibilities in all real devices cause a deviation in performance from that of a Carnot cycle. Figure 13.17 shows plots of a typical Carnot cycle on (a) a p–v and (b) a T–s diagram.

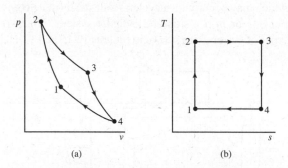

(a)   (b)

Figure 13.17 Cycle diagrams for a Carnot cycle

When the balance equations are applied to the devices in the Carnot cycle, the thermal efficiency of the cycle is given by

$$\eta_{\text{Carnot}} = \frac{\dot{W}_{\text{net}}}{\dot{Q}_i} = \frac{\sum \dot{W}}{\dot{Q}_{\text{in}}} = \frac{\sum \dot{Q}}{\dot{Q}_{\text{in}}}$$

$$= \frac{\dot{Q}_{23} + \dot{Q}_{41}}{\dot{Q}_{23}} \qquad [13.90]$$

(Remember that $\dot{Q}_{41}$ is negative because energy is leaving the system.)

Using equation 13.21, we obtain:

$$Q_{\text{rev}} = \int_1^2 T \, dS$$

Equation 13.90 can be rewritten as

$$\eta_{\text{Carnot}} = \frac{T_H(s_3 - s_2) + T_L(s_1 - s_4)}{T_H(s_3 - s_2)} \qquad [13.91]$$

Since the T–s diagram in Figure 13.17 (b) shows that $s_2 = s_1$ and $s_3 = s_4$, equation 13.91 can be simplified to

$$\eta_{\text{Carnot}} = 1 - \frac{T_L}{T_H} \qquad [13.92]$$

Since the Carnot cycle represents the idealized cyclic machine, no other cycle can operate at the same temperatures with a higher thermal efficiency.

Example 13.24

A machine that operates using a Carnot cycle with a thermal efficiency of 34% is in contact with a low-temperature reservoir at 20°F. What is the temperature of the high-temperature reservoir?

Solution:

From equation 13.92:

$$T_H = \frac{T_L}{1 - \eta_{\text{Carnot}}}$$

Then:

$$T_H = \frac{(20 + 460)°\text{R}}{1 - 0.34}$$

$$= 727°\text{R}$$

$$= 267°\text{F}$$

Example 13.25

For a Carnot cycle, if energy is added at the high-temperature heat interaction at a rate of 4 kW, and energy is removed at the low-temperature heat interaction at a rate of 1.3 kW, what is the efficiency of the cycle?

Solution:

From equation 13.90:

$$\eta_{Carnot} = \frac{\dot{Q}_{23} + \dot{Q}_{41}}{\dot{Q}_{23}}$$

so

$$\eta_{Carnot} = \frac{4 \text{ kW} + (-1.3 \text{ kW})}{4 \text{ kW}}$$

$$= 0.67 \text{ or } 67\%$$

Rankine Cycle (Heat Engine)

The most common example of a Rankine cycle is a simple steam power plant. The overall effect is to use energy supplied from the high-temperature reservoir to provide useable energy through a heat interaction.

The processes of an ideal Rankine cycle are as follows:

Process 1–2: The pressure of the working fluid, in liquid phase, is raised with an ideal or isentropic pump (energy added to the system through a work interaction).

Process 2–3: The liquid working fluid is boiled in a heat exchanger connected to the high-temperature reservoir (energy added to the system through a heat interaction).

Process 3–4: The working-fluid vapor is expanded in an ideal or isentropic turbine (energy removed from the system in a work interaction).

Process 4–1: The working-fluid vapor is condensed into a liquid in a heat exchanger connected to the low-temperature reservoir (energy removed from the system in a heat interaction).

The thermal efficiency of a Rankine cycle must be lower than that of a Carnot cycle operating between the same high and low temperatures. By applying the balance equations to each of the devices in the Rankine cycle, the efficiency can be found. The equation for this efficiency is:

$$\eta_{Rankine} = \frac{(h_3 - h_4) - (h_2 - h_1)}{h_3 - h_2} \qquad [13.93]$$

Vapor-Compression Refrigeration/Heat Pump Cycle

This cycle uses energy provided through a work interaction to transfer energy between two systems at different temperatures. If the purpose of the system is to keep an environment at low temperature, the cycle is called a *refrigeration cycle;* if the purpose is to keep an environment at high temperature, the cycle is called a *heat pump.*

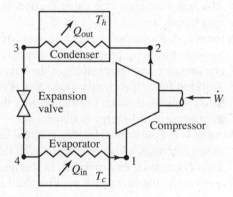

Figure 13.19 Schematic representation of an ideal vapor-compression refrigeration/heat pump system

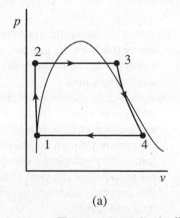

(a)

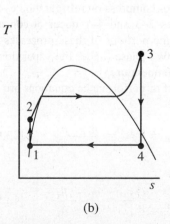

(b)

Figure 13.18 Cycle diagrams for an ideal Rankine cycle

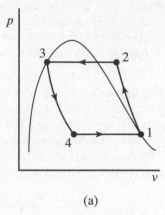

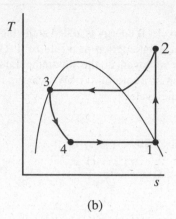

(a)   (b)

Figure 13.20 Cycle diagrams for a vapor-compression refrigeration cycle

The components of an ideal vapor-compression refrigeration system are as follows:

  Process 1–2, Compressor: Energy is added to the working fluid, which is in its vapor phase, through an ideal work interaction.

  Process 2–3, Condenser: Energy is removed from the system through a heat interaction with the high temperature environment. In this heat exchanger, the vapor is condensed to a liquid.

  Process 3–4, Expansion valve: The very irreversible throttling process is used to reduce the pressure and temperature of the liquid working fluid. If the expansion valve is insulated to prevent a heat interaction with the environment, no energy is transferred.

  Process 4–1, Evaporator: Energy is added to the system through a heat interaction with the low-temperature environment. In this heat exchanger, the two-phase working fluid is boiled to a vapor.

Figure 13.20 shows the typical (a) p–v and (b) T–s diagrams for a vapor compression refrigeration cycle. Note that processes 2–3 and 4–1 occur at constant pressure, and that the portions of these processes for which the fluid is two-phase (under the vapor dome) occur at constant temperature.

The coefficient of performance equations are as follows:

$$\text{COP}_{\text{ref}} = \frac{\dot{Q}_{4-1}}{\dot{W}_{1-2}} = \frac{h_1 - h_4}{h_2 - h_1} \qquad [13.94]$$

and

$$\text{COP}_{\text{hp}} = \frac{\dot{Q}_{2-3}}{\dot{W}_{1-2}} = \frac{h_2 - h_3}{h_2 - h_1} \qquad [13.95]$$

The values for entropy in equations 13.94 and 13.95 are determined using a P–h diagram for the refrigerant being used. An example of such a diagram for Refrigerant HFC-134a is provided at the end of this chapter.

Otto Cycle

The Otto cycle is an air-standard cycle that approximates the performance of the typical spark-ignition internal-combustion engine used in an automobile. The main differences between the two are (1) that the compression and the expansion of the piston in the Otto cycle are considered isentropic, unlike the actual process in a spark-ignition engine, and that (2) the working fluid in the Otto cycle is air rather than an air-fuel mixture and products of combustion. Although in an actual spark-ignition engine the values of the working fluid's thermodynamic properties at times vary greatly from those of air, qualitative information can be obtained using a fluid for which large quantities of thermodynamic state data exist.

The processes of an ideal Otto cycle, which is diagrammed in Figure 13.21, are as follows:

  Process 1–2: Isentropic (constant-entropy) compression of the gas in the cylinder. Energy is added to the system through a work interaction.

  Process 2–3: Isometric (constant-volume) addition of energy. In the internal-combustion engine, energy is added to the system through combustion, which is initiated by energizing the spark plug. For the purposes of the Otto cycle, it is assumed that energy is added to the system through a heat interaction.

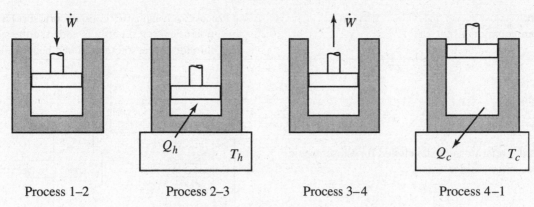

Process 1–2 Process 2–3 Process 3–4 Process 4–1

Figure 13.21 Operation of an Otto cycle

Process 3–4: Isentropic expansion of the gas in the cylinder. Energy is removed from the system through a work interaction.

Process 4–1: Isometric rejection of energy. In the internal-combustion engine, the spent air and combustion products are replaced by fresh air and fuel. For the purposes of the Otto cycle, it is assumed that energy is removed from the system through a heat interaction.

Figure 13.22 shows typical (a) p–v and (b) T–s diagrams for an Otto cycle. Note that processes 2–3 and 4–1 occur at constant specific volume, while processes 1–2 and 3–4 occur at constant specific entropy.

An important parameter of the Otto cycle is the compression ratio, defined as

$$r = \frac{v_1}{v_2} \qquad [13.96]$$

Using the ideal gas relationships for an isentropic process, we find that:

$$\frac{T_2}{T_1} = \left(\frac{v_1}{v_2}\right)^{k-1} = r^{k-1} \text{ (constant } k\text{)} \qquad [13.97]$$

$$\frac{T_4}{T_3} = \left(\frac{v_3}{v_4}\right)^{k-1} = \frac{1}{r^{k-1}} \text{ (constant } k\text{)} \qquad [13.98]$$

The thermal efficiency of the Otto cycle is given by:

$$\eta = 1 - \frac{1}{r^{k-1}} \text{ (constant } k\text{)} \qquad [13.99]$$

where k is the specific heat ratio, $k = \frac{c_p}{c_v}$.

Example 13.26

What is the thermal efficiency of an Otto cycle in which air is the working fluid, if the compression process raises the pressure from 100 kPa to 1.8 MPa?

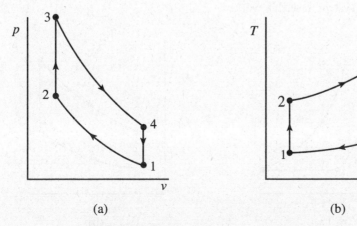

(a) (b)

Figure 13.22 Cycle diagrams for an Otto cycle

Solution:

From equation 13.99:

$$\eta = 1 - \frac{1-1}{r^{k-1}}$$

where

$$r = \frac{v_1}{v_2}.$$

Use the relationships developed for an isentropic process for an ideal gas:

$$r^{k-1} = \left(\frac{v_1}{v_2}\right)^{k-1} = \left(\frac{p_2}{p_1}\right)^{\frac{k-1}{k}}$$

$$\eta = 1 - \frac{1}{\left(\frac{p_2}{p_1}\right)^{\frac{k-1}{k}}}$$

$$= 1 - \frac{1}{\left(\frac{1,800 \text{ kPa}}{100 \text{ kPa}}\right)^{\frac{1.4-1}{1.4}}}$$

$$= 0.56 \text{ or } 56\%$$

Otto cycle is an ideal cycle that is used to simulate the process in a spark-ignition engine. The real engine has an efficiency in the range of 20–30%.

Brayton Cycle

The Brayton cycle, diagrammed in Figure 13.23, is used to approximate the operation of a gas turbine cycle. Like the Otto cycle, this is an air-standard cycle, with isentropic work interactions.

The processes of an ideal Brayton cycle are as follows:

Process 1–2: Isentropic (constant-entropy) compression, which occurs in a compressor. Energy is added to the system through a work interaction.

Process 2–3: Isobaric (constant-pressure) addition of energy. Energy is added to the system through a heat interaction in a heat exchanger.

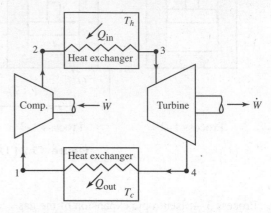

Figure 13.23 Diagram of a system that operates on a Brayton cycle

Process 3–4: Isentropic expansion in the turbine. Energy is removed from the system through a work interaction.

Process 4–1: Isobaric rejection of energy. Energy is removed from the system through a heat interaction in a heat exchanger.

Figure 13.24 shows typical (a) p–v and (b) T–s diagrams for a Brayton cycle. Note that processes 2–3 and 4–1 occur at constant pressure, while processes 1–2 and 3–4 occur at constant specific entropy.

The thermal efficiency of the Brayton cycle is given by this equation:

$$\eta = 1 - \frac{1}{\left(\frac{p_2}{p_1}\right)^{\frac{k-1}{k}}} \quad \text{(constant } k\text{)} \quad [13.100]$$

where k is the specific heat ratio, $k = \frac{c_p}{c_v}$.

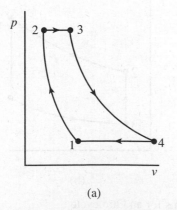

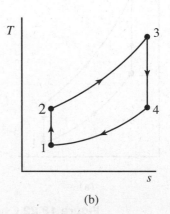

(a)   (b)

Figure 13.24 Cycle diagrams for a Brayton cycle (a) p–v and (b) T–s

COMBUSTION PROCESSES

Combustion is a complex sequence of chemical reactions between a fuel and an oxidant accompanied by the production of heat or both heat and light in the form of either a glow or flames. In a complete combustion reaction, a compound reacts with an oxidizing element, and the products are compounds of each element in the fuel with the oxidizing element.

A combustion equation is written and balanced for reacting compounds (i.e., fuel molecule and oxidizer). If in an ideal situation the amount of oxidizer (i.e., oxygen, air) is sufficient to react with all the fuel and in the end no unburned fuel remains, then the combustion is *stoichiometric*. For example, the stoichiometric combustion of methane in oxygen is:

$$CH_4 + 2O_2 \rightarrow CO_2 + 2\,H_2O + HEAT$$

Combustion in Air

In the large majority of the real world uses of combustion, the oxygen (O_2) oxidant is obtained from the ambient air and the resultant flue gas from the combustion contains nitrogen. For each mole of oxygen there will be 3.76 moles of nitrogen. Therefore, the stoichiometric combustion of methane in air is

$$CH_4 + 2O_2 + 2(3.76N_2) \rightarrow CO_2 + 2\,H_2O + 7.52N_2$$

In reality, combustion processes are never perfect or complete. In flue gases from combustion of carbon (as in coal combustion) or carbon compounds (as in combustion of hydrocarbons, wood, etc.), both unburned carbon (as soot) and carbon compounds (CO and others) will be present. Also, when air is the oxidant, some nitrogen will be oxidized to various, mostly harmful, nitrogen oxides (NO_x). Also, the amount of air that supplies oxygen can be more or less than the stoichiometric combustion of fuel and air. The air–fuel ratio is then defined to show the disproportionate amount of oxidant as follows:

$$A/F = \frac{\text{mass of air}}{\text{mass of fuel}}$$

$$= \frac{\text{number of moles of air}}{(\text{number of moles of fuel})} \frac{28.96}{(\text{Molecular weight of fuel})}$$

$$\text{Percent Theoretical Air} = \frac{(A/F)_{\text{actual}}}{(A/F)_{\text{stoichiometric}}} \times 100$$

where $(A/F)_{\text{stoichiometric}}$ is the air-fuel ratio calculated from the stoichiometric combustion equation.

Example 13.27

Calculate the air–fuel ratio of complete combustion of isooctane (C_8H_{18}) with 200% theoretical air.

Solution:

The stoichiometric reaction of isooctane and air can be written as

$$C_8H_{18} + 12.5O_2 + 12.5(3.76N_2) \rightarrow$$
$$8CO_2 + 9H_2O + 47N_2$$

$$(A/F)_{\text{stoichiometric}} = \frac{12.5(1 + 3.76)}{1}$$

$$\frac{28.96}{(8(12) + 18(1))} = 15.12$$

$$(A/F)_{\text{actual}} = (15.12)\,(2) = 30.24$$

13.5 Summary

Fundamentals of general thermodynamics have been presented. System (closed and open), property (intensive and extensive), state, interaction, and process (reversible and irreversible) have been defined. Energy, exergy (available energy), and entropy have been defined. Work and heat interactions have been discussed thoroughly. Balance equations for mass, energy, entropy, and exergy have been developed. First and second laws of thermodynamics have been presented. Properties of ideal and non-ideal gases have been calculated. Energy conversion devices such as nozzle, diffuser, turbine, pump, compressor, boiler, condenser, evaporator, and heat exchanger have been analyzed.

Thermodynamic cycles such as Carnot cycle, Rankine cycle, vapor-compression refrigeration/heat pump cycle, Otto cycle, and Brayton cycle have been discussed in detail. The combustion process has been reviewed.

Specific Heat Data for Gases

Substance	Molecular Weight	c_p		c_v		k
		Btu/ lbm · °R	kJ/kg · K	Btu/ lbm · °R	kJ/kg · K	
Air	29	0.240	1.00	0.171	0.718	1.40
Argon	40	0.125	0.520	0.0756	0.312	1.67
Butane	58	0.415	1.72	0.381	1.57	1.09
Carbon dioxide	44	0.203	0.846	0.158	0.657	1.29
Carbon monoxide	28	0.249	1.04	0.178	0.744	1.40
Ethane	30	0.427	1.77	0.361	1.49	1.18
Helium	4	1.25	5.19	0.753	3.12	1.67
Hydrogen	2	3.43	14.3	2.44	10.2	1.40
Methane	16	0.532	2.25	0.403	1.74	1.30
Neon	20	0.246	1.03	0.148	0.618	1.67
Nitrogen	28	0.248	1.04	0.177	0.743	1.40
Octane vapor	114	0.409	1.71	0.392	1.64	1.04
Oxygen	32	0.219	0.918	0.157	0.658	1.40
Propane	44	0.407	1.68	0.362	1.49	1.12
Steam	18	0.455	1.87	0.335	1.41	1.33

Specific Heat Data for Liquids

Substance	c_p		Density	
	Btu/lbm · °R	kJ/kg · K	lbm/ft³	kg/m³
Ammonia	1.146	4.80	38	602
Mercury	0.033	0.139	847	13,560
Water	1.000	4.18	62.4	997

Specific Heat Data for Solids

Substance	c_p		Density	
	Btu/lbm · °R	kJ/kg · K	lbm/ft³	kg/m³
Aluminum	0.215	0.900	170	2,700
Copper	0.092	0.386	555	8,900
Ice*	0.502	2.11	57.2	917
Iron	0.107	0.450	490	7,840
Lead	0.030	0.128	705	11,310

*Properties for ice taken at 0°C (32°F).

Saturated Water – Temperature Table

Temp.	Saturation Pressure	Specific Volume (m³/kg)		Internal Energy (kJ/kg)			Enthalpy (kJ/kg)			Entropy (kJ/kg · K)		
(°C)	(kPa)	Saturated Liquid	Saturated Vapor	Saturated Liquid	Evap.	Saturated Vapor	Saturated Liquid	Evap.	Saturated Vapor	Saturated Liquid	Evap.	Saturated Vapor
T	P_{sat}	v_f	v_g	u_f	u_{fg}	u_g	h_f	h_{fg}	h_g	s_f	s_{fg}	s_g
0.01	0.6133	0.001000	206.14	0	2375.3	2375.3	0.01	2501.3	2501.4	0	9.1562	9.1562
5	0.8721	0.001000	147.12	20.97	2361.3	2382.3	20.98	2489.6	2510.6	0.0761	8.9496	9.0257
10	0.12276	0.001000	106.38	42.00	2347.2	2389.2	42.01	2477.7	2519.8	0.1510	8.7498	8.9008
15	1.7051	0.001001	77.93	62.99	2333.1	2396.1	62.99	2465.9	2528.9	0.2245	8.5569	8.7814
20	2.339	0.001002	57.79	83.95	2319	2402.9	83.96	2454.1	2538.1	0.2966	8.3706	8.6672
25	3.169	0.001003	43.36	104.88	2304.9	2409.8	104.89	2442.3	2547.2	0.3674	8.1905	8.558
30	4.246	0.001004	32.89	125.78	2290.8	2416.6	125.79	2430.5	2556.3	0.4369	8.0164	8.4533
35	5.628	0.001006	25.22	146.67	2276.7	2423.4	146.68	2418.6	2565.3	0.5053	7.8478	8.3531
40	7.384	0.001008	19.52	167.56	2262.6	2430.1	167.57	2406.7	2574.3	0.5725	7.6845	8.257
45	9.593	0.001010	15.26	188.44	2248.4	2436.8	188.45	2394.8	2583.2	0.6387	7.5261	8.1648
50	12.349	0.001012	12.03	209.32	2234.2	2443.5	209.33	2382.7	2592.1	0.7038	7.3725	8.0763
55	15.754	0.001015	9.568	230.21	2219.9	2450.1	230.23	2370.7	2600.9	0.7679	7.2234	7.9913
60	19.94	0.001017	7.671	251.11	2205.5	2456.6	251.13	2358.5	2609.6	0.8312	7.0784	7.9096
65	25.03	0.001020	6.197	272.02	2191.1	2463.1	272.06	2346.2	2618.3	0.8935	6.9375	7.831
70	31.19	0.001023	5.042	292.95	2176.6	2569.6	292.98	2333.8	2626.8	0.9549	6.8004	7.7553
75	38.58	0.001026	4.131	313.90	2162.0	2475.9	313.93	2321.4	2635.3	1.0155	6.6669	7.6824
80	47.39	0.001029	3.407	334.86	2147.4	2482.2	334.91	2308.8	2643.7	1.0753	6.5369	7.6122
85	57.83	0.001033	2.828	355.84	2132.6	2488.4	355.90	2296.0	2651.9	1.1343	6.4102	7.5445
90	70.14	0.001036	2.361	376.85	2117.7	2494.5	376.92	2283.2	2660.1	1.1925	6.2866	7.4791
95	84.55	0.001040	1.9820	397.88	2102.7	2500.6	697.96	2270.2	2668.1	1.2500	6.1659	7.4159
100	101.35	0.001044	1.6729	418.94	2087.6	2506.5	419.04	2257.0	2676.1	1.3069	6.0480	7.3549
105	120.82	0.001048	1.4194	440.02	2072.3	2512.4	440.15	2243.7	2683.8	1.3630	5.9328	7.2958
110	143.27	0.001052	1.2102	461.14	2057.0	2518.1	461.3	2230.2	2691.5	1.4185	5.8202	7.2387
115	169.06	0.001056	1.0366	482.30	2041.4	2523.7	482.48	2216.5	2699	1.4734	5.7100	7.1833
120	198.53	0.001060	0.8919	503.50	2025.8	2529.3	503.71	2202.6	2706.3	1.5276	5.6020	7.1296
125	232.1	0.001065	0.7706	524.74	2009.9	2534.6	524.99	2188.5	2713.5	1.5813	5.4962	7.0775
130	270.1	0.001070	0.6685	546.02	1993.9	2539.9	546.31	2174.2	2720.5	1.6344	5.3925	7.0269
135	313.0	0.001075	0.5822	567.35	1977.7	2545.0	567.69	2159.6	2727.3	1.6870	5.2907	6.9777
140	361.3	0.001080	0.5089	588.74	1961.3	2550.0	589.13	2144.7	2733.9	1.7391	5.1908	6.9299
145	415.4	0.001085	0.4463	610.18	1944.7	2554.9	610.63	2129.6	2740.3	1.7907	5.0926	6.8833
150	475.8	0.001091	0.3928	631.68	1927.9	2559.5	632.20	2114.3	2746.5	1.8418	4.9960	6.8379
155	543.1	0.001096	0.3468	653.24	1910.8	2564.1	653.84	2098.6	2752.4	1.8925	4.9010	6.7935
160	617.8	0.001102	0.3071	674.87	1893.5	2568.4	675.55	2082.6	2758.1	1.9427	4.8075	6.7502
165	700.5	0.001108	0.2727	696.56	1876.0	2572.5	697.34	2066.2	2763.5	1.9925	4.7153	6.7078
170	791.7	0.001114	0.2428	718.33	1858.1	2576.5	719.21	2049.5	2768.7	2.0419	4.6244	6.6663
175	892.0	0.001121	0.2168	740.17	1840.0	2580.2	741.17	2032.4	2773.6	2.0909	4.5347	6.6256
	(MPa)											
180	1.0021	0.001127	0.19405	762.09	1821.6	2583.7	763.22	2015.0	2778.2	2.1396	4.4461	6.5857
185	1.1227	0.001134	0.17409	784.10	1802.9	2587.0	785.39	1997.1	2782.4	2.1879	4.3586	6.5465
190	1.2544	0.001141	0.15654	806.19	1783.8	2590.0	807.62	1978.8	2786.4	2.2359	4.2720	6.5079
195	1.3978	0.001149	0.14105	828.37	1764.4	2592.8	829.98	1960.0	2790.0	2.2835	4.1863	6.4698
200	1.5538	0.001157	0.12736	850.65	1744.7	2595.3	852.45	1940.7	2793.2	2.3309	4.1014	6.4323
205	1.7230	0.001164	0.11521	873.04	1724.5	2597.5	875.04	1921.0	2796.0	2.3780	4.0172	6.3952
210	1.9062	0.001173	0.10441	895.53	1703.9	2599.5	897.76	1900.7	2798.5	2.4248	3.9337	6.3585
215	2.1040	0.00118	0.09479	918.14	1682.9	2601.1	920.62	1879.9	2800.5	2.4714	3.8507	6.3221
220	2.318	0.001190	0.08619	940.87	1661.5	2602.4	943.62	1858.5	2802.1	2.5178	3.7683	6.2861
225	2.548	0.001199	0.07849	963.73	1639.6	2603.3	966.78	1836.5	2803.3	2.5639	3.6863	6.2503
230	2.795	0.001209	0.07158	986.74	1617.2	2603.9	990.12	1813.8	2804.0	2.6099	3.6047	6.2146
235	3.06	0.001219	0.06537	1009.89	1594.2	2604.1	1013.62	1790.5	2804.2	2.6558	3.5233	6.1791
240	3.344	0.001229	0.05976	1033.21	1570.8	2604.0	1037.32	1766.5	2803.8	2.7015	3.4422	6.1437
245	3.648	0.001124	0.05471	1056.71	1546.7	2603.4	1061.23	1741.7	2803.0	2.7472	3.3612	6.1083

Saturated Water—Temperature Table (Continued)

Temp.	Saturation Pressure	Specific Volume (m³/kg)		Internal Energy (kJ/kg)			Enthalpy (kJ/kg)			Entropy (kJ/kg · K)		
		Saturated Liquid	Saturated Vapor	Saturated Liquid	Evap.	Saturated Vapor	Saturated Liquid	Evap.	Saturated Vapor	Saturated Liquid	Evap.	Saturated Vapor
(°C)	(kPa)											
T	P_{sat}	v_f	v_g	u_f	u_{fg}	u_g	h_f	h_{fg}	h_g	s_f	s_{fg}	s_g
250	3.973	0.001125	0.05013	1080.39	1522.0	2602.4	1085.36	1716.2	2801.5	2.7927	3.2802	6.0730
255	4.319	0.001263	0.04598	1104.28	1596.7	2600.9	1109.73	1689.8	2799.5	2.8383	3.1992	6.0375
260	4.688	0.001276	0.042211	1128.39	1470.6	2599.0	1134.37	1662.5	2796.9	2.8838	3.1181	6.0019
265	5.081	0.001289	0.03877	1152.74	1443.9	2596.6	1159.28	1634.4	2793.6	2.9294	3.0368	5.9662
270	5.499	0.001302	0.03564	1177.36	1416.3	2593.7	1184.51	1605.2	2789.7	2.9751	2.9551	5.9301
275	5.942	0.001317	0.03279	1202.25	1387.9	2590.2	1210.07	1574.9	2785	3.0208	2.873	5.8938
280	6.412	0.001332	0.03017	1227.46	1358.7	2586.1	1235.99	1543.6	2779.6	3.0668	2.7903	5.8571
285	6.909	0.001348	0.02777	1253.00	1328.4	2581.4	1262.31	1511.0	2773.3	3.113	2.707	5.8199
290	7.436	0.001366	0.02557	1278.92	1297.1	2576.0	1289.07	1477.1	2766.2	3.1594	2.6227	5.7821
295	7.993	0.001384	0.02354	1305.2	1264.7	2569.9	1316.3	1441.8	2758.1	3.2062	2.5375	5.7437
300	8.581	0.001404	0.02167	1332.0	1231.0	2563.0	1344.0	1404.9	2749.0	3.2534	2.4511	5.7045
305	9.202	0.001425	0.019948	1359.3	1195.9	2555.2	1372.4	1366.4	2738.7	3.301	2.3633	5.6643
310	9.856	0.001447	0.018350	1387.1	1159.4	2546.4	1401.3	1326.0	2727.3	3.3493	2.2737	5.623
315	10.547	0.001472	0.016867	1415.5	1121.1	2536.6	1431.0	1283.5	2714.5	3.3982	2.1821	5.5804
320	11.274	0.001499	0.015488	1444.6	1080.9	2525.5	1461.5	1238.6	2700.1	3.448	2.0882	5.5362
330	12.845	0.001561	0.012996	1505.3	993.7	2498.9	1525.3	1140.6	2665.9	3.5507	1.8909	5.4417
340	14.586	0.001638	0.010797	1570.3	894.3	2464.6	1594.2	1027.9	2622.0	3.6594	1.6763	5.3357
350	16.513	0.001740	0.008813	1641.9	776.6	2418.4	1670.6	893.4	2563.9	3.7777	1.4335	5.2112
360	18.651	0.001893	0.006945	1725.2	626.3	2351.5	1760.5	720.3	2481.0	3.9147	1.1379	5.0526
370	21.03	0.002213	0.004925	1844.0	384.5	2228.5	1890.5	441.6	2332.1	4.1106	0.6865	4.7971
374.14	22.09	0.003155	0.003155	2029.6	0	2029.6	2099.3	0	2099.3	4.4298	0	4.4298

Superheated Steam Table

Temp. T (°C)	Specific Volume v (m³/kg)	Internal Energy u (kJ/kg)	Enthalpy h (kJ/kg)	Entropy s (kJ/kg · K)	Specific Volume v (m³/kg)	Internal Energy u (kJ/kg)	Enthalpy h (kJ/kg)	Entropy s (kJ/kg · K)
	$P = 0.01$ MPa (45.81°C)				$P = 0.05$ MPa (45.81°C)			
sat	14.674	2437.9	2584.7	8.1502	3.240	2483.9	2645.9	7.5939
50	14.869	2443.9	2592.6	8.1749				
100	17.196	2515.5	2687.5	8.4479	3.418	2511.6	2682.5	7.6947
150	19.512	2587.9	2783.0	8.6882	3.889	2585.6	2780.1	7.9401
200	21.825	2661.3	2879.5	8.9038	4.356	2659.9	2877.7	8.1580
250	24.136	2736.0	2977.3	9.1002	4.820	2735.0	2976.0	8.3556
300	26.445	2812.1	3076.5	9.2813	5.284	2811.3	3075.5	8.5373
400	31.063	2968.9	3279.6	9.6077	6.209	2968.5	3278.9	8.8642
500	35.679	3132.3	3489.1	9.8978	7.134	3132.0	3488.7	9.1546
600	40.295	3302.5	3705.4	10.1608	8.057	3302.2	3705.1	9.4178
700	44.911	3479.6	3982.7	10.4028	8.981	3479.4	3928.5	9.6599
800	49.526	3663.8	4159.0	10.6281	9.904	3663.6	4158.9	9.8852
900	54.141	3855.0	4396.4	10.8396	10.828	3854.9	4396.3	10.0967
1000	58.757	4053.0	4640.6	11.0393	11.751	4052.9	4640.5	10.2964
1100	63.372	4257.5	4891.2	11.2287	12.674	4257.4	4891.1	10.4859
1200	67.987	4467.9	5147.8	11.4091	13.597	446708	5147.7	10.6662
1300	72.602	4683.7	5409.7	11.5811	14.521	4683.6	5409.6	10.8382
	$P = 0.10$ MPa (99.63°C)				$P = 0.20$ MPa (120.23°C)			
sat	1.6940	2506.1	2675.5	7.3594	0.8857	2529.5	2706.7	7.1272
100	1.6958	2506.7	2676.2	7.3614				
150	1.9364	2582.8	2776.4	7.6134	0.9596	2576.9	2768.8	7.2795
200	2.172	2658.1	2875.3	7.8343	1.0803	2654.4	2870.5	7.5066
250	2.406	2733.7	2974.3	8.0333	1.1988	2731.2	2971.0	7.7086
300	2.639	2810.4	3074.3	8.2158	1.3162	2808.6	3071.8	7.8926
400	3.103	2967.9	3278.2	8.5435	1.5493	2966.7	3276.6	8.2218
500	3.565	3131.6	3488.1	8.8342	1.7814	3130.8	3487.1	8.5133
600	4.028	3301.9	3704.4	9.0976	2.013	3301.4	3704.0	8.7770
700	4.490	3479.2	3928.2	9.3398	2.244	3478.8	3927.6	9.0194
800	4.952	3663.5	4158.6	9.5652	2..475	3663.1	4158.2	9.2449
900	5.414	3854.8	4396.1	9.7767	2.705	3854.5	4395.8	9.4566
1000	5.875	4052.8	4640.3	9.9764	2.937	4052.5	4640.0	9.6563
1100	6.337	4257.3	4891.0	10.4659	3.168	4257.0	4890.7	9.8458
1200	6.799	4467.7	5147.6	10.3463	3.399	4467.5	5147.5	10.0262
1300	7.260	4683.5	5409.5	10.5183	3.630	4683.2	5409.3	10.1982
	$P = 0.40$ MPa (143.63°C)				$P = 0.60$ MPa (158.85°C)			
sat	0.4625	2553.6	2738.9	6.8959	0.3157	2567.4	2756.8	6.7600
150	0.4708	2564.5	2752.8	6.9299				
200	0.5342	2646.8	2860.5	7.1706	0.3520	2638.9	2850.1	6.9665
250	0.5951	2726.1	2964.2	7.3789	0.3938	2720.9	2957.2	7.1816
300	0.6548	2804.8	3066.8	7.5662	0.4344	2801.0	3061.6	7.3724
350					0.4742	2881.2	3156.7	7.5464
400	0.7726	2964.4	3273.4	7.8985	0.5137	2962.1	3270.3	7.7079
500	0.8893	3129.2	3484.9	8.1913	0.5920	3127.6	3482.8	8.0021
600	1.0055	3300.2	3702.4	8.4558	0.6697	3299.1	3700.9	8.2674
700	1.1215	3477.9	3926.5	8.6987	0.7472	3477.0	3925.3	8.5107
800	1.2372	3662.4	4157.3	8.9244	0.8245	3661.8	4156.5	8.7367
900	1.3529	3853.9	4395.1	9.1362	0.9017	3853.4	494.4	8.9486
1000	1.4685	4052.0	4639.4	9.3360	0.9788	4051.5	4638.8	9.1485
1100	1.5840	4256.5	4890.2	9.5256	1.0559	4256.1	4889.6	9.3381
1200	1.6996	4467.0	5146.8	9.7060	1.1330	4466.5	5146.3	9.5185
1300	1.8151	4682.8	5408.8	9.8780	1.2102	4682.3	5408.3	9.6906
	$P = 0.80$ MPa (170.43°C)				$P = 1.00$ MPa (179.91°C)			
sat	0.2404	2576.8	2769.1	6.6628	0.1944	2583.6	2778.1	6.5865
200	0.2608	2630.6	2839.3	6.8158	0.2060	2621.9	2827.9	6.6940
250	0.2931	2715.5	2950.0	7.0384	0.2327	2709.9	2942.6	6.9247
300	0.3241	2797.2	3056.5	7.2328	0.2579	2793.2	3051.2	7.1229
350	0.3544	2878.2	3161.7	7.4089	0.2825	2875.2	3157.7	7.3011
400	0.3843	2959.7	3267.1	7.5716	0.3066	2957.3	3263.9	7.4651
500	0.4433	3126.0	3480.6	7.8673	0.3541	3124.4	3478.5	7.7622
600	0.5018	3297.9	3699.4	8.1333	0.4011	3296.8	3697.9	8.0290
700	0.5601	3476.2	3924.2	8.3770	0.4478	3475.3	3923.1	8.2731
800	0.6181	3661.1	4155.6	8.6033	0.4943	3660.4	4154.7	8.4996
900	0.6761	3852.8	4393.7	8.8153	0.5407	3852.2	4392.9	8.7118
1000	0.7340	4051.0	4638.2	9.0153	0.5871	4050.5	4637.6	8.9119
1100	0.7919	4255.6	4889.1	9.2050	0.6335	4255.1	4888.6	9.1017
1200	0.8497	4466.1	5145.9	9.3855	0.6798	4465.6	5145.4	9.2822
1300	0.9076	4681.8	5407.9	9.5575	0.7261	4681.3	5407.4	9.4543

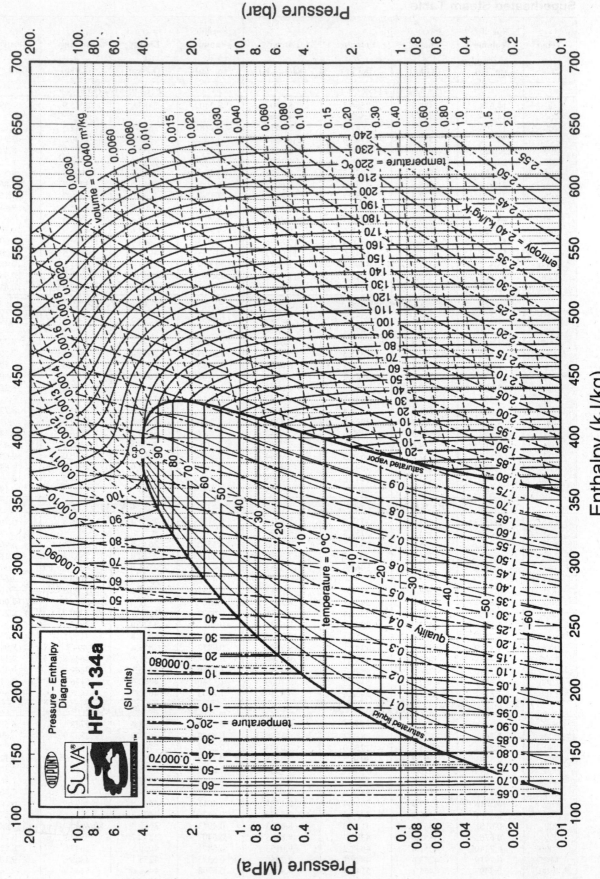

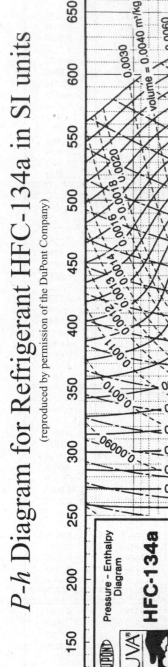

P-h Diagram for Refrigerant HFC-134a in SI units

(reproduced by permission of the DuPont Company)

Pressure (bar)

Pressure (MPa)

Enthalpy (kJ/kg)

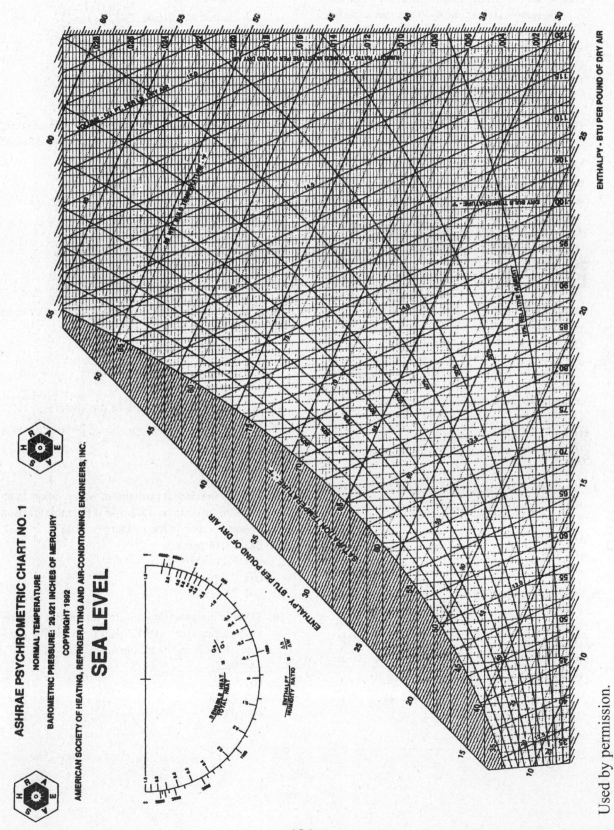

ASHRAE PSYCHROMETRIC CHART NO. 1

NORMAL TEMPERATURE

BAROMETRIC PRESSURE: 29.921 INCHES OF MERCURY

COPYRIGHT 1992

AMERICAN SOCIETY OF HEATING, REFRIGERATING AND AIR-CONDITIONING ENGINEERS, INC.

SEA LEVEL

PRACTICE PROBLEMS

1. The availability of a system
   (A) is the maximum amount of energy that may be removed from the system in an ideal heat interaction
   (B) is the maximum amount of energy that may be added to a system in an ideal work interaction.
   (C) is the maximum amount of energy that may be removed from a system and its environment in an ideal work interaction.
   (D) remains unchanged in an ideal work interaction.

2. What is the change in availability for a system in a closed, rigid container in which 37kJ of energy is transferred to the system through a reversible work interaction, and 42kJ of energy is transferred from the system through a reversible heat interaction with a thermal reservoir at 30°C? Take the dead state of the system to be standard conditions (101 kPa and 25°C).
   (A) 5 kJ
   (B) 36.3 kJ
   (C) 42 kJ
   (D) 37.7 kJ

3. What is the mass of oxygen gas in a container with a volume of 25 ft³, at a pressure of 14.7 psia and a temperature of 75°F?
   (A) 2 lbm
   (B) 0.0004 lbm
   (C) 0.06 lbm
   (D) 5 lbm

4. What is the Mach number of air at 25°C flowing with a velocity of 75 m/s?
   (A) 0.75
   (B) 0.04
   (C) 1.27
   (D) 0.22

5. What is the mass of water at 400 kPa and 500°C contained in a 6-m³ vessel?
   (A) 6 kg
   (B) 6.75 kg
   (C) 6000 kg
   (D) 60 kg

6. An ideal gas mixture contains the following components:

   Oxygen 1 kg

   Argon   3 kg

   Carbon monoxide 2 kg

   What is the mass fraction of argon?
   (A) 20%
   (B) 30%
   (C) 40%
   (D) 50%

7. What is the molar mass of the following mixture, where the composition is defined in terms of mole fractions?

   Argon   0.28

   Helium 0.34

   Neon   0.38

   (A) 4
   (B) 22
   (C) 20
   (D) 40

8. A 6-m³ vessel contains the following mixture of ideal gases at 20°C:

   Nitrogen 0.1 kmol

   Oxygen   0.3 kmol

   Helium   0.05 kmol

   What is the total pressure of the vessel?
   (A) 100 kPa
   (B) 183 kPa
   (C) 137 kPa
   (D) 12.5 kPa

9. What is the partial pressure of water vapor in an air-water mixture at 15.2 psia if the mole fraction of water vapor in the mixture is 1.6%?
   (A) 0.024 psia
   (B) 0.24 psia
   (C) 2.4 psia
   (D) 24 psia

10. What is the specific humidity of an atmospheric air-water mixture at 15.2 psia if the partial pressure of water in the mixture is 0.22 psia?
    (A) 0.22
    (B) 0.05
    (C) 0.022
    (D) 0.009

11. What is the velocity of a stream of nitrogen gas at the exit of a diffuser if the inlet velocity is 120 m/s, the inlet temperature is 30°C, and the outlet temperature is 36°C?
    (A) 44 m/s
    (B) 119.9 m/s
    (C) 121.1.m/s
    (D) 85. m/s

12. What is the power output of a turbine if 0.5 kg/s of air enters at 750 kPa and 80°C and exits at 370 kPa and 37°C?
    (A) 43 kW
    (B) 380 kW
    (C) 21 kW
    (D) 37 kW

13. An ideal air compressor pressurizes air from standard conditions (14.7 psia, 77°F) to 80 psia. What is the air temperature at the exit?
    (A) 865°F
    (B) 411°F
    (C) 125°F
    (D) 419°F

14. In a heat exchanger, 8 lbm/s of water enters the cold side at 35°F and exits at 63°F. Hot water enters the other side of the heat exchanger at 90°F and exits at 46°F. What is the flow rate of the hot water?
    (A) 5 lbm/sec
    (B) 8 lbm/sec
    (C) 10 lbm/sec
    (D) 15 lbm/sec

15. A Carnot engine operates between 400°C and 30°C. What is the efficiency of the engine?
    (A) 92.5%
    (B) 60%
    (C) 55%
    (D) 43%

16. What is the efficiency of an air standard Brayton cycle that operates with a 4:1 pressure ratio?
    (A) 48%
    (B) 43%
    (C) 38%
    (D) 33%

17. What type of cycle is represented by the following process?
    1-Energy added to the system through an isentropic work interaction.
    2-Energy removed from the system through an isobaric heat interaction.
    3-Isenthalpic expansion of the working fluid.
    4-Energy added to the system through an isobaric heat interaction.
    (A) ideal Rankine cycle
    (B) ideal Otto cycle
    (C) ideal vapor-compression refrigeration cycle
    (D) Carnot cycle

18. The average specific volume of water at 80°C is 1 m^3/kg. What is the quality of the water?
    (A) 0
    (B) 0.3
    (C) 0.6
    (D) 1

19. The change in entropy of air from 25°C and 1 atm pressure to 200°C and 4 atm pressure is
    (A) 0 kJ/kg · K
    (B) 0.07 kJ/kg · K
    (C) 0.69 kJ/kg · K
    (D) 1.69 kJ/kg · K

20. Air expands isentropically in a piston cylinder device from 300°C and 800 kPa to 100 kPa. The work done by the air is
    (A) 133 kJ/kg
    (B) 181 kJ/kg
    (C) 253 kJ/kg
    (D) 316 kJ/kg

21. Air expands isentropically in a turbine from 300°C and 800 kPa to 100 kPa. The work output of the turbine is
    (A) 133 kJ/kg
    (B) 181 kJ/kg
    (C) 253 kJ/kg
    (D) 316 kJ/kg

22. Steam expands isentropically from 1 MPa and 350°C to atmospheric pressure in a steam turbine. What is the work output of the turbine?
    (A) 482 kJ/kg
    (B) 502 kJ/kg
    (C) 1,500 kJ/kg
    (D) 2,739 kJ/kg

23. Refrigerant 134a is isentropically compressed in a compressor from the saturated vapor state at 0.1 MPa pressure to 1 MPa pressure. The work required to run the compressor is
(A) 10 kJ/kg
(B) 30 kJ/kg
(C) 50 kJ/kg
(D) 70 kJ/kg

24. What is the coefficient of performance of a Carnot refrigerator operating between 20°F and 75°F?
(A) 0.36
(B) 1.2
(C) 3.5
(D) 8.7

25. What is the coefficient of a Carnot heat pump operating between 1°C and 20°C?
(A) 0.5
(B) 1.05
(C) 7.8
(D) 15.4

26. What is the closed-system availability of steam at 300°C and 0.1 MPa pressure if the environment is at 25°C and 1 atm pressure?
(A) 362 kJ/kg
(B) 514 kJ/kg
(C) 634 kJ/kg
(D) 631 kJ/kg

27. Energy is added to air in a closed system in a reversible and isothermal heat interaction. The pressure of the air decreases from 300 kPa to 100 kPa while the temperature is kept constant at 100°C. The amount of energy transferred in this process is
(A) 31 kJ/kg
(B) 80 kJ/kg
(C) 116 kJ/kg
(D) 410 kJ/kg

Answer Key

1. C	9. B	17. C	25. D
2. B	10. D	18. B	26. C
3. A	11. A	19. B	27. C
4. D	12. C	20. B	
5. B	13. B	21. C	
6. D	14. A	22. B	
7. C	15. C	23. C	
8. B	16. D	24. D	

Answers Explained

1. C Availability (Φ), also called exergy or available energy, is defined as the theoretical maximum amount of energy that can be extracted from a system and its environment in the form of useful work.

2. B Use equation 13.23; the change in availability is given by the expression

$$\Delta\phi = \left(1 - \frac{T_o}{T_b}\right)Q - W$$

$$= \left[1 - \frac{(25 + 273)\text{K}}{(30 + 273)\text{K}}\right](-42 \text{ kJ}) + 37 \text{ kJ}$$

$$= 36.3 \text{ kJ}$$

3. A From equation 13.33b: $pV = mRT$

or

$$m = \frac{pV}{RT}$$

The only property not given in the problem statement is the gas constant R, which is determined using the property data for oxygen at the end of the chapter and equation 13.35:

$$R = c_p - c_v$$

$$= 0.219\frac{\text{Btu}}{\text{lbm-R}} - 0.157\frac{\text{Btu}}{\text{lbm-R}}$$

$$= 0.062\frac{\text{Btu}}{\text{lbm-R}}$$

Then,

$$m = \frac{pV}{RT}$$

$$= \frac{\left(14.7\frac{\text{lbf}}{\text{in}^2}\right)(25 \text{ ft}^3)}{\left(0.062\frac{\text{Btu}}{\text{lbm-R}}\right)(75 + 460)\text{R}}\left(\frac{144 \text{ in}^2}{\text{ft}^2}\right)$$

$$-\left(\frac{\text{Btu}}{778 \text{ lbf-ft}}\right)$$

$$= 2.05 \text{ lbm}$$

4. D From equation 13.53, the speed of sound in this air stream is determined by

$$c = \sqrt{kRT}$$

Use the property data table at the end of the chapter to determine k and R: $k = 1.40$, and

$$R = c_p - c_v$$
$$= 1.00\frac{kJ}{kg \cdot K} - 0.718\frac{kJ}{kg \cdot K}$$
$$= 0.282\frac{kJ}{kg \cdot K}$$

Then,

$$c = \sqrt{(1.4)\left(0.282\ \frac{kj}{kg \cdot K}\right)(25+273)\,K\left[\left(\frac{1,000\,N \cdot m}{kj}\right)\left(\frac{kg \cdot m}{\frac{s^2}{N}}\right)\right]}$$

$$= 343\ m/s$$

From equation 13.52:

$$M_a = \frac{\overline{V}}{c}$$
$$= \frac{75\frac{m}{s}}{343\frac{m}{s}}$$
$$= 0.22$$

5. B At these conditions, the water is in the form of superheated steam. To determine the mass of water in the vessel, use the equation

$$m = \rho V$$

or

$$m = \frac{V}{v}$$

From the superheated steam table, $v = 0.8893$ m^3/kg.

Then,

$$m = \frac{6\ m^3}{0.8893\ \frac{m^3}{kg}}$$
$$= 6.75\ kg$$

6. D The mass fraction is given by equation 13.55:

$$y_i = \frac{m_i}{\Sigma m_i}$$
$$= \frac{3\ kg}{1\ kg + 3\ kg + 2\ kg}$$
$$= 0.50 = 50\%$$

7. C The gas constant for a mixture of ideal gases is given by equation 13.59:

$$R_{mix} = \frac{\overline{R}}{M_{mix}}$$

where, from equation 13.58:

$$M_{mix} = \Sigma x_i M_i$$

From the property data at the end of the chapter, the molar mass for each constituent is as follows:

Argon: 40 g/mol
Helium: 4 g/mol
Neon: 20 g/mol

Then,

$$M_{mix} = (0.28)\left(40\ \frac{g}{mol}\right) + (0.34)\left(4\ \frac{g}{mol}\right)$$
$$+ (0.38)\left(20\ \frac{g}{mol}\right)$$
$$= 20.16\ g/mol$$

8. B Determine the partial pressure for each constituent from the ideal gas law:

$$p_i = \frac{n_i \overline{R}\,T}{V}$$

For nitrogen:

$$p_{N_2} = \frac{(0.1\ kmol)\left(8.314\ \frac{kJ}{kmol \cdot K}\right)(20 + 273)K}{6\ m^3}$$

$$\times \left(\frac{kPa}{1,000\ \frac{N}{m^2}}\right)\left(\frac{1,000\ N \cdot m}{kJ}\right)$$

$$= 40.6\ kPa$$

For oxygen:

$$p_{O_2} = \frac{(0.3 \text{ kmol})\left(8.314 \dfrac{\text{kJ}}{\text{kmol·K}}\right)(20 + 273)\text{K}}{6 \text{ m}^3}$$

$$\times \left(\frac{\text{kPa}}{1{,}000 \dfrac{\text{N}}{\text{m}^2}}\right)\left(\frac{1{,}000 \text{ N·m}}{\text{kJ}}\right)$$

$$= 121.8 \text{ kPa}$$

For helium:

$$p_{He} = \frac{(0.05 \text{ kmol})\left(8.314 \dfrac{\text{kJ}}{\text{kmol·K}}\right)(20 + 273)\text{K}}{6 \text{ m}^3}$$

$$\times \left(\frac{\text{kPa}}{1{,}000 \dfrac{\text{N}}{\text{m}^2}}\right)\left(\frac{1{,}000 \text{ N·m}}{\text{kJ}}\right)$$

$$= 20.3 \text{ kPa}$$

The total pressure is the sum of the partial pressures for the three constituents:

$$p = p_{N_2} + p_{O_2} + p_{He}$$
$$= 40.6 \text{ kPa} + 121.8 \text{ kPa} + 20.3 \text{ kPa}$$
$$= 182.7 \text{ kPa}$$

9. B The partial pressure is given by the equation

$$p_v = x_v p$$
$$= (0.016)(15.2 \text{ psia})$$
$$= 0.243 \text{ psia}$$

10. D The specific humidity for the air-water mixture is given by equation 13.68:

$$\omega = 0.622\frac{p_v}{p - p_v}$$
$$= 0.622\frac{0.22 \text{ psia}}{14.7 \text{ psia} - 0.22 \text{ psia}}$$
$$= 0.009$$

11. A From equation 13.74:

$$h_e - h_i = \frac{\overline{V}_i{}^2 - \overline{V}_e{}^2}{2}$$

The change in enthalpy is related to the change in temperature by equation 13.38:

$$h_e - h_i = c_p(T_e - T_i)$$

Find the value of specific heat in the property data tables at the end of the chapter. Then com-

bine and rearrange the equations to determine the outlet velocity:

$$\overline{V}_e =$$

$$\sqrt{\left(120 \frac{\text{m}}{\text{s}}\right)^2 - 2\left(1.04 \frac{\text{kj}}{\text{kg·K}}\right)(6\text{·K})\left(\frac{10^3 \text{ J}}{\text{kJ}}\right)\left(\frac{\text{N·m}}{\text{J}}\right)\left(\frac{\text{kJ·m}}{\text{N}-\text{s}^2}\right)}$$

$$= 43.82 \text{ m/s}$$

12. C The power output in a turbine, or the work performed, is given by equation 13.77:

$$\dot{W} = \dot{m}(h_i - h_e)$$

The change in enthalpy is related to the change in temperature by equation 13.38:

$$h_i - h_e = c_p(T_i - T_e)$$

Combine these two equations to find the power output:

$$\dot{W} = \dot{m}c_p(T_i - T_e)$$
$$= \left(0.5\frac{\text{kg}}{\text{s}}\right)\left(1.00 \frac{\text{kJ}}{\text{kg·K}}\right)(80° \text{ C} - 37° \text{ C})$$
$$= 21.5 \frac{\text{kJ}}{\text{s}} = 21.5 \text{ kW}$$

13. B An ideal compressor operates on an isentropic process. Use equation 13.42 to relate the pressures and temperatures of such a process:

$$T_1 p_1{}^{1-k/k} = T_2 p_2{}^{1-k/k}$$

Rearrange this equation as follows:

$$V = T_1\left(\frac{p_2}{p_1}\right)^{k-1/k}$$
$$= (77 + 460)R\left(\frac{80 \text{ psia}}{14.7 \text{ psia}}\right)^{\frac{1.40-1}{1.40}}$$
$$= 871 \, R = 411°\text{F}$$

14. A From applying an energy balance to the heat exchanger, equation 13.84:

$$\dot{m}_h(h_i - h_e)_h = \dot{m}_c(h_e - h_i)_c$$
$$\dot{m}_h \, c_p(T_i - T_e)_h = \dot{m}_c \, c_p(T_e - T_i)_c$$

Rearrange the second equation:

$$\dot{m}_h = \dot{m}_c \frac{c_p(T_e - T_i)_c}{c_p(T_i - T_e)_h}$$

$$= \left(8\frac{\text{lbm}}{\text{s}}\right)\frac{\left(1.000\dfrac{\text{Btu}}{\text{lbm-R}}\right)(63°\text{F} - 35°\text{F})}{\left(1.000\dfrac{\text{Btu}}{\text{lbm-R}}\right)(90°\text{F} - 46°\text{F})}$$

$$= 5 \text{ lbm/s}$$

15. C The efficiency of a Carnot engine is given by equation 13.92:

$$\eta = 1 - \frac{T_L}{T_h}$$

$$= 1 - \frac{(30 + 273)K}{(400 + 273)K}$$

$$= 0.55 = 55\%$$

16. D The efficiency of a Brayton cycle is given by equation 13.100:

$$\eta = 1 = \frac{1}{\left(\frac{p_2}{p_1}\right)^{\frac{k-1}{k}}}$$

For air, $k = 1.40$, so the efficiency is:

$$\eta = 1 - \frac{1}{(4)^{\frac{1.4-1}{1.4}}}$$

$$= 0.327 = 33\% \, .$$

17. C The key to this answer is step 3. The only system discussed that includes an isenthalpic expansion of the working fluid is the vapor compression refrigeration cycle.

18. B The average specific volume of a two-phase water mixture is given by equation 13.66a:

$$v_{mix} = v_f + x(v_g - v_f)$$

The quality is then given by the equation

$$x = \frac{v_{mix} - v_f}{v_g - v_f}$$

From the saturated water table at the end of the chapter:

$v_f = 0.001029 \text{ m}^3/\text{kg}$ and $v_g = 3.407 \text{ m}^3/\text{kg}$

Then,

$$x = \frac{1.00\,\frac{m^3}{kg} - 0.001029\,\frac{m^3}{kg}}{3.407\,\frac{m^3}{kg} - 0.001029\,\frac{m^3}{kg}}$$

$$= 0.293$$

19. B The change in entropy for a process of an ideal gas is given by equation 13.39:

$$\Delta s = c_p \ln\left(\frac{T_2}{T_1}\right) - R \ln\left(\frac{p_2}{p_1}\right)$$

Determine the specific heat and the gas constant from the property data tables at the end of the chapter:

$$c_p = 1.00\frac{kJ}{kg\cdot K}$$

and

$$R = c_p - c_v$$

$$= 1.00\frac{kJ}{kg\cdot K} - 0.718\frac{kJ}{kg\cdot K}$$

$$= 0.282 \text{ kJ/kg·K}$$

The change in entropy is then:

$$\Delta s = c_p \ln\left(\frac{T_2}{T_1}\right) - R \ln\left(\frac{p_2}{p_1}\right)$$

$$= \left(1.00\frac{kJ}{kg\cdot K}\right)\ln\left(\frac{(200+273)K}{(25+273)K}\right)$$

$$- \left(0.282\frac{kJ}{kg\cdot K}\right)\ln\left(\frac{4 \text{ atm}}{1 \text{ atm}}\right)$$

$$= 0.071 \text{ kJ/kg·K}$$

20. B Calculate the boundary work performed in an isentropic process in a closed system by using equation 13.44:

$$w = \frac{RT_1}{k-1}\left[1 - \left(\frac{p_2}{p_1}\right)^{\frac{k-1}{k}}\right]$$

where $k = 1.40$, and

$$R = c_p - c_v$$

$$= 1.00\frac{kJ}{kg\cdot K} - 0.718\frac{kJ}{kg\cdot K}$$

$$= 0.282 \frac{kJ}{kg\cdot K}$$

Then,

$$w = \frac{\left(0.282\frac{kJ}{kg\cdot K}\right)(300 + 273)K}{1.40 - 1}$$

$$\times \left[1 - \left(\frac{100 \text{ kPa}}{800 \text{ kPa}}\right)^{\frac{1.40-1}{1.40}}\right]$$

$$= 181 \text{ kJ/kg}$$

21. C Calculate the boundary work performed in an isentropic process in an open system by using the equation

$$w = \left(\frac{k}{k-1}\right) RT_1 \left[1 - \left(\frac{p_2}{p_1}\right)^{\frac{k-1}{k}} \right]$$

where $k = 1.40$, and

$$R = c_p - c_v$$

$$= 1.00 \frac{kJ}{kg \cdot K} - 0.718 \frac{kJ}{kg \cdot K}$$

$$= 0.282 \frac{kJ}{kg \cdot K}$$

Then:

$$w = \frac{1.40}{1.40 - 1} \left[\left(0.282 \frac{kJ}{kg \cdot K} \right) (300 + 273) K \right]$$

$$\times \left[1 - \left(\frac{100 \text{ kPa}}{800 \text{ kPa}} \right)^{\frac{1.40-1}{1.40}} \right]$$

$$= 253 \text{ kJ/kg}$$

22. B From equation 13.77:

$$w = h_i - h_e$$

From the superheated steam tables:

$$h_i = 3{,}157.7 \text{ kJ/kg}$$

and

$$s_i = 7.3011 \text{ kJ/kg·K}$$

Since the expansion is isentropic, and the entrance value of entropy is lower than the saturation value of entropy at atmospheric pressure, some condensation will occur. To find the value of enthalpy at the exit state, the quality of the exit steam must be estimated. From equation 13.66d:

$$x = \frac{s_i - s_f}{s_g - s_f}$$

From the saturated water table:

$$s_f = 1.3069 \text{ kJ/kg·K}$$
$$\text{and}$$
$$s_g = 7.3549 \text{ kJ/kg·K}$$

Then:

$$x = \frac{7.3011 \frac{kJ}{kg \cdot K} - 1.3069 \frac{kJ}{kg \cdot K}}{7.3549 \frac{kJ}{kg \cdot K} - 1.3069 \frac{kJ}{kg \cdot K}}$$

$$= 0.991$$

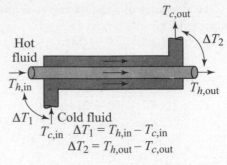

$T_{c,\text{in}}$ $\Delta T_1 = T_{h,\text{in}} - T_{c,\text{in}}$
$\Delta T_2 = T_{h,\text{out}} - T_{c,\text{out}}$

(*a*) Parallel-flow heat exchangers

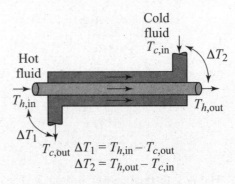

$T_{c,\text{out}}$ $\Delta T_1 = T_{h,\text{in}} - T_{c,\text{out}}$
$\Delta T_2 = T_{h,\text{out}} - T_{c,\text{in}}$

(*b*) Counter-flow heat exchangers

Figure 13.34 Concentric tube heat exchangers

Also, from equation 13.66c:

$$h_e = h_f + x(h_g - h_f)$$

and from the saturated water table:

$$h_f = 419.04 \text{ kJ/kg} \quad \text{and} \quad h_g = 2676.1 \text{ kJ/kg}$$

Then:

$$h_e = 419.04 \frac{kJ}{kg} + 0.991 \left(2{,}676.1 \frac{kJ}{kg} - 419.04 \frac{kJ}{kg} \right)$$

$$= 2{,}655.8 \text{ kJ/kg}$$

and

$$w = h_i - h_e$$

$$= 3{,}157.7 \frac{kJ}{kg} - 2{,}655.8 \frac{kJ}{kg}$$

$$= 501.9 \text{ kJ/kg}$$

23. C From equation 13.77:

$$w = h_i - h_e$$

Determine the enthalpy values using the *p-h* diagram for HFC-134a, provided at the end of the chapter. Locate the entrance state at the intersection of the $p = 0.1$ MPa line and the right side of the vapor dome. Locate the exit state by following the closest constant entropy line until it intersects with the $p = 1.0$ MPa line. The values of entropy are as follows:

$$h_i = 380 \text{ kJ/kg} \quad \text{and} \quad h_g = 430 \text{ kJ/kg}$$

Then,

$$w = h_i - h_e$$
$$= 380 \frac{\text{kJ}}{\text{kg}} - 430 \frac{\text{kJ}}{\text{kg}}$$
$$= -50 \text{ kJ/kg}$$

(Note the negative value, indicating that the compressor is adding energy to the system through a work interaction.)

24. D The coefficient of performance for a Carnot refrigeration system is given by the equation

$$\text{COP} = \frac{T_C}{T_H - T_C}$$
$$= \frac{(20 + 460)\text{R}}{(75 + 460)\text{R} - (20 + 460)\text{R}}$$
$$= 8.73$$

25. D The coefficient of performance for a Carnot heat pump system is given by the equation

$$\text{COP} = \frac{T_H}{T_H - T_C}$$
$$= \frac{(20 + 273)\text{R}}{(20 + 273)\text{R} - (1 + 273)\text{R}}$$
$$= 15.42$$

26. C The closed-system availability is given by the equation

$$A = (u - u_0) + p_0(v - v_0) - T_0(s - s_0)$$

From the steam tables at the end of the chapter:

$$u = 2{,}810.4 \text{ kJ/kg}, \quad v = 2.639 \text{ m}^3\text{/kg},$$
$$s = 8.2158 \text{ kJ/kg·K}$$

From the saturated water table at the end of the chapter:

$$u_0 = 104.88 \text{ kJ/kg}, \quad v_0 = 0.001003 \text{ m}^3\text{/kg},$$
$$s_0 = 0.3674 \text{ kJ/kg·K}$$

The dead-state pressure is $p_0 = 101.35$ kPa, and $T_0 = (25 + 273)$K.

Then,

$$A = \left(2{,}810.4 \frac{\text{kJ}}{\text{kg}} - 104.88 \frac{\text{kJ}}{\text{kg}}\right)$$
$$+ (101.35 \text{ kPa})\left(\frac{1{,}000 \frac{\text{N}}{\text{m}^2}}{\text{kPa}}\right)\left(\frac{\text{kJ}}{1{,}000 \text{ N·m}}\right)$$
$$\times \left[2.639 \frac{\text{m}^3}{\text{kg}} - 0.001003 \frac{\text{m}^3}{\text{kg}}\right]$$
$$- (298 \text{ K})\left[8.2158 \frac{\text{kJ}}{\text{kg·K}} - 0.3674 \frac{\text{kJ}}{\text{kg·K}}\right]$$
$$= 2{,}705.52 \frac{\text{kJ}}{\text{kg}} + 267.36 \frac{\text{kJ}}{\text{kg}} - 2{,}338.82 \frac{\text{kJ}}{\text{kg}}$$
$$= 634 \text{ kJ/kg}$$

27. C The first law of thermodynamics is often stated as follows:

$$\Delta u = q - w$$

For the given process:

$$\Delta u = c_v \Delta T$$
$$= 0$$

which means that

$$q = w = \int p\, dv$$

or, from equation 13.48:

$$q = w = RT \ln\left(\frac{p_1}{p_2}\right)$$

where

$$R = c_p - c_v$$
$$= 1.00 \frac{\text{kJ}}{\text{kg·K}} - 0.718 \frac{\text{kJ}}{\text{kg·K}}$$
$$= 0.282 \text{ kJ/kg·K}$$

Then,

$$q = w = \left(0.282 \frac{\text{kJ}}{\text{kg·K}}\right)[(100 + 273)\text{K}]$$
$$\times \ln\left(\frac{300 \text{ kPa}}{100 \text{ kPa}}\right)$$
$$= 115.6 \text{ kJ/kg}$$

14

Heat Transfer

Hameed Metghalchi, Sc.D.

Farzan Parsinejad, Ph.D.

Introduction

Heat transfer is energy in transit, which occurs as a result of a temperature gradient or difference. This temperature difference is thought of as a driving force that causes heat energy to flow. Heat transfer occurs by three basic mechanisms or modes: conduction, convection, and radiation.

14.1 Conduction

14.2 Convection

14.3 Radiation

14.4 Summary

14.1 Conduction

Conduction heat transfer is defined as heat transfer in solids and fluids without bulk motion as shown in Figure 14.1. Heat conduction generally takes place in solids, though it may occur in fluids without bulk motion or with rigid body motion. In fluids, conduction is due to the collusions of the molecules during their random motion. In solids, it is due to the combination of vibrations of molecules in a lattice and the energy transport by free electrons.

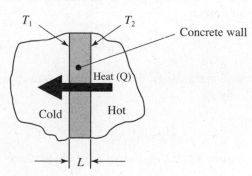

Figure 14.1 Heat conduction through a concrete wall

It is observed that the rate of heat conduction through a wall ($\dot{Q}$) with constant thickness is proportional to the temperature difference ($T_2 - T_1$) between the surfaces and the area normal to the heat flow direction (A) and is inversely proportional to the thickness of the wall (L). In other words:

$$\dot{Q} = -kA\left(\frac{dT}{dx}\right) \qquad [14.1]$$

Therefore, the rate of conduction through a plane wall is

$$\dot{Q} = -kA\left(\frac{T_2 - T_1}{L}\right) \qquad [14.2]$$

where

k is thermal conductivity of the wall's material (W/m·K),

A is cross-sectional area of the wall (m^2),

L is the thickness of the wall (m),

T_1, T_2 are the surface temperatures of the wall (K).

Example 14.1

The front of a slab of lead ($k = 35$ W/m·K) is kept at 110°C and the back is kept at 50°C. If the area of the slab is 0.4 m^2 and it is 0.03 m thick, calculate the heat transfer rate.

Solution:

Using equation 14.2,

$$\dot{Q} = -(35)(0.4)\left(\frac{50-110}{0.03}\right) = 28 \text{ kW}$$

Thermal Resistance

Equation 14.2 is analogous to the relation for electric current flow:

$$I_e = \left(\frac{V_2 - V_1}{R}\right) \qquad [14.3]$$

Based on this analogy, the system can be drawn schematically as:

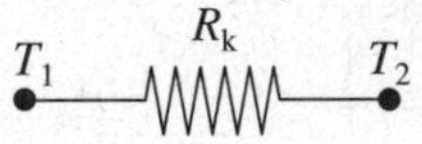

$$T_1 \overset{R_k}{\wedge\!\wedge\!\wedge} T_2$$

Figure 14.2: Conductive thermal resistance

$$\dot{Q} = \frac{\Delta T}{R_{total}} \qquad [14.4]$$

where $R_{total} = \sum R$ and

$R = \dfrac{L}{KA}$ for plane wall conduction (K/W),

L = wall thickness

$R = \dfrac{\ln(R_2/R_1)}{2\pi KL}$ for cylindrical wall conduction (K/W),

L = cylinder length

$R = \dfrac{1}{hA}$ for convection (K/W)

Composite Wall (Materials in Series)

Figure 14.3 shows a composite wall with three different materials and different thicknesses.

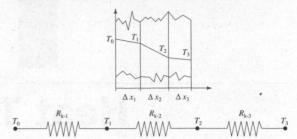

Figure 14.3 Composite wall with material in series

Heat transfer rate through material 1:

$$\dot{Q}_1 = \frac{k_1 A}{\Delta x_1}(T_0 - T_1) = \frac{(T_0 - T_1)}{R_1} \qquad [14.5]$$

Heat transfer rate through material 2:

$$\dot{Q}_2 = \frac{k_2 A}{\Delta x_2}(T_1 - T_2) = \frac{(T_1 - T_2)}{R_2} \qquad [14.6]$$

Heat transfer rate through material 3:

$$\dot{Q}_3 = \frac{k_3 A}{\Delta x_3}(T_2 - T_3) = \frac{(T_2 - T_3)}{R_3} \qquad [14.7]$$

As the system is steady state and no internal heat generated, the heat flows entering and exiting each layer are equal. Therefore,

$$\dot{Q}_1 = \dot{Q}_2 = \dot{Q}_3 \qquad [14.8]$$

By combining equations 14.5 to 14.8:

$$\dot{Q} = \frac{(T_0 - T_3)}{R_1 + R_2 + R_3} = \frac{(T_0 - T_3)}{\sum R_i} \qquad [14.9]$$

where $R_1 = \dfrac{\Delta x_1}{k_1 A}$, $R_2 = \dfrac{\Delta x_2}{k_2 A}$, and $R_3 = \dfrac{\Delta x_3}{k_3 A}$.

Example 14.2

A wall of a house measures 3 m by 6 m, has no windows, and consists of 0.6 cm thick oak paneling (0.147 W/m·K) and 5 cm of white pine (0.11 W/m·K). The inside temperature of the wall is 25°C, and the outside temperature is –10°C. Determine the heat loss through the wall.

Solution:

We can use equation [14.9] for two resistances:

$$\dot{Q} = \frac{(T_0 - T_3)}{R_1 + R_2} = \frac{25 - (-10)}{\dfrac{0.006}{0.17(3)(6)} + \dfrac{0.05}{0.11(3)(6)}} = 1298 \, W$$

Heat Transfer To/From a Circular Duct

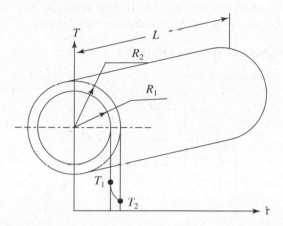

Figure 14.4 Heat conduction through a cylindrical wall

Figure 14.4 shows a circular duct. Assuming that the system is at steady state, there is no internal heat generation, and temperature varies only with r, Fourier's law of conduction is expressed in the following form:

$$\dot{Q} = \frac{2\pi k L}{\ln(R_2 / R_1)} (T_1 - T_2) \qquad [14.10]$$

Therefore, the thermal resistance is shown as:

$$R = \frac{\ln(R_2 / R_1)}{2\pi k L} \qquad [14.11]$$

Critical Insulation Radius

The radius of insulation when the heat loss from the pipe or wire is minimum can be shown by:

$$r_{cr} = \frac{k_{insulation}}{h_{\infty}} \qquad [14.12]$$

here h_{∞} is the convection heat transfer coefficient.

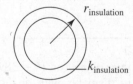

Figure 14.5 Heat conduction through a cylindrical wall

General Conduction Equation Based on Cartesian Coordinates

If we apply the first law of thermodynamics (energy balance) to a control volume, we get:

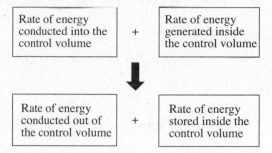

By combining equations relevant to each rate of energy change:

$$q''' = -k \left[\frac{\partial^2 T}{\partial x^2} + \frac{\partial^2 T}{\partial y^2} + \frac{\partial^2 T}{\partial z^2} \right] + \frac{1}{\alpha} \frac{\partial T}{\partial t} \qquad [14.13]$$

where q''' is the heat generation rate per unit volume and $\alpha = k/\rho C$ (m²/s) is called the thermal diffusivity of the material. Note that the thermal conductivity (k) represents how well a material conducts heat and the heat capacity (ρC) represents how much energy a material stores per unit volume. The larger the diffusivity is, the faster the propagation of heat into the medium. A small value of thermal diffusivity means that heat is mostly absorbed by the material and a small amount of heat will be conducted further.

Plane Wall with Internal Heat Generated

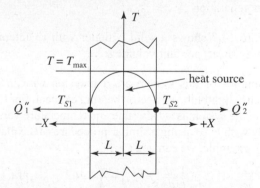

Figure 14.6 Plane wall with internal heat generation

Figure 14.6 shows a plane wall with an internal energy (heat) per unit volume q'''. The heat source is at the center plane. Thus, we can expect a temperature profile that is symmetric about the center. For this case, the process is steady state and has one-dimensional heat flow. Therefore, equation 14.3 can be simplified as:

$$\frac{d^2T}{dx^2} + \frac{q'''}{k} = 0 \qquad [14.14]$$

This can be solved, giving the general solution:

$$T(x) = \frac{q'''L^2}{2k}\left(1 - \frac{x^2}{L^2}\right) + \left(\frac{T_{S2} - T_{S1}}{2}\right)\left(\frac{x}{L}\right) + \left(\frac{T_{S2} + T_{S1}}{2}\right)$$
$$[14.15]$$

$$\dot{Q}_1'' + \dot{Q}_2'' = 2q'''L \qquad [14.16]$$

where

$$\dot{Q}_1'' = k(dT / dx)_{-L} \qquad [14.17]$$

$$\dot{Q}_2'' = k(dT / dx)_L \qquad [14.18]$$

Cylinder with Internal Heat Generation

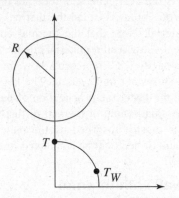

Figure 14.7 Solid cylinder with uniform internal heat generation

Figure 14.7 shows a solid cylinder with an internal energy (heat) per unit volume q'''.

Examples of this system are a tungsten wire in an electric lightbulb, an electric heating element. For this case, the system is steady state and temperature varies only with r. By using a similar procedure as the plane wall example, we can get:

$$\frac{1}{r}\frac{d}{dr}\left(r\frac{dT}{dr}\right) + \frac{q'''}{k} = 0 \qquad [14.19]$$

The temperature profile can be achieved solving the above equation subject to boundary condition $T(R) = T_W$.

$$T(r) = \frac{q'''R^2}{4k}\left(1 - \frac{r^2}{R^2}\right) + T_W \qquad [14.20]$$

Heat Transfer From Extended Surface

The purpose of adding an extended surface is to help dissipate heat. Fins are usually added to a heat transfer device to increase the rate of heat removal. This is because of the increase of the heat transfer area.

Figure 14.8 shows a straight fin.

Rectangular Fin

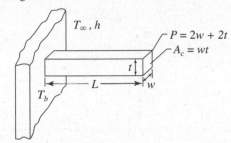

Pin Fin

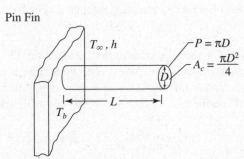

Figure 14.8 A uniform fin

The total dissipated heat can be calculated with the following relation:

$$\dot{Q} = \sqrt{hpkA_C}\,(T_b - T_\infty)\tanh mL_C \qquad [14.21]$$

where h is the convective heat transfer coefficient, p is the exposed perimeter, k is the thermal conductivity, A_C is the cross-sectional area, T_b is the temperature at the base of the fin, T_∞ is the fluid temperature, $m = \sqrt{hp/kA_C}$, and $L_C = L + A_C/p$ is the corrected length.

Example 14.3

A 15 cm long copper rod ($k = 389.15$ W/m·K), 0.625 cm in diameter with an insulated tip is connected to a wall. The wall temperature is 204.4°C, while the ambient (fluid) temperature is 21°C. The h value is 28.4 W/m²·K. Determine the heat loss from the rod.

Solution:

$$m = \sqrt{hp / kA_C}$$

$$= \sqrt{28.4(\pi)(0.00625) / 389.15(\pi(0.00625)^2 / 4)}$$

$$= 6.83$$

Then from equation 14.21:

$$\dot{Q} = \sqrt{hpkA_C}\,(T_W - T_\infty)\tanh mL$$

$$\dot{Q} = \sqrt{28.4(\pi)(0.00625)(389.15)(\pi(0.00625)^2 / 4)}$$
$$(204.4 - 21)\tanh(6.83(0.15)) = 11.55 \text{ W}$$

Unsteady-State Heat Conduction

Here we consider transient conduction problems in which no internal heat is generated. Temperature will therefore vary with location within the system and with time. Temperature and heat transfer variation of the system depend on the system's *internal resistance* and *surface resistance*.

Assume we have a slab with an initial temperature of T_i and it is left in fluid stream at T_∞. Heat is transferred by convection at the surface. As the surface temperature decreases, heat is transferred from the center of the slab to the surface and then to the fluid. If the system itself is copper or the volume is small, the temperature response within the slab will be considerably different than if the system is glass or the volume is large. The response has to do with the *internal resistance of the material*. Furthermore, if the convection coefficient is very high, the surface temperature becomes almost identical to the fluid temperature quickly. Alternatively for a low convection coefficient, a large temperature difference exists between the surface and the fluid. The value of the convection coefficient controls what is known as the *surface resistance* to heat transfer.

Thus, the temperature variation within the system is dependent on the internal and surface resistances. The larger the internal resistance is or the smaller the surface resistance is, the larger the temperature variation within the system and vice versa. A *Biot number* is defined as:

$$\text{Bi} = \frac{\text{Convection at the surface within the body}}{\text{Conduction within the body}}$$

$$\text{Bi} = \frac{hL_C}{k}, \quad L_C = \frac{V\,(\text{volume})}{A_s\,(\text{surface area})} \quad [14.22]$$

System with Negligible Internal Resistance (Lumped Capacitance Method)

For this case $\text{Bi} < 0.1$ and the temperature profile within the body is quite uniform. Therefore, the rate of change in internal energy of the body is equal to the rate of heat taken away from the surface by convection:

$$\rho VC\,(dT / dt) = hA_S(T - T_\infty) \quad [14.23]$$

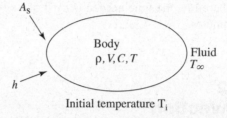

Figure 14.9. A system with negligible internal resistance

By solving equation 14.23, the temperature variation with time is

$$T - T_\infty = (T_i - T_\infty)\,e^{-(hA_S / \rho CV)t} = (T_i - T_\infty)\,e^{-\frac{t}{\tau_c}}$$
$$[14.24]$$

where τ_c = time constant = $\rho c V / h A_s$

Total heat transferred up to time t is

$$Q_{\text{total}} = \rho CV\,(T_i - T) \quad [14.25]$$

Example 14.4

An aluminum ($\rho = 2{,}707$ kg/m^3), $c = 900$ J/kg·K, $k = 204$ W/m·K) sphere weighing 7 kg and initially at $T_i = 260°$C is suddenly immersed in a fluid at 10°C. If $h = 50$ W/m^2·K, determine the time required to cool the aluminum to 90°C. L_c for a sphere is $r_0/3$.

Solution:

First we check to determine if the lumped capacitance assumption is valid. In other words, we calculate the Bi number.

$$V_{\text{sphere}} = \frac{4}{3}\pi r_0^3 \quad \text{and}$$

$$V = \frac{\text{mass}}{\text{density}} = \frac{7}{2{,}707} = 0.00258 \text{ m}^3$$

$$r_0 = \left[\frac{3(0.00258)}{4\pi}\right]^{1/3} = 0.085 \text{ m}$$

So that $L_c = r_0/3 = 0.028$ m

$$\text{Bi} = \frac{hL_c}{k} = \frac{50(0.028)}{204} = 0.007 < 0.1$$

So we can use equation 14.24:

$$\frac{T - T_\infty}{T_i - T_\infty} = e^{-(hA/\rho cV)t}$$

$$\frac{90 - 10}{260 - 10} = e^{-\left(\frac{50(4(0.085)^2)}{2,707(900)(0.00258)}\right)t}$$

$$\Rightarrow 3.125 = e^{(-0.00074)t}$$

Therefore, the time needed is $t = 1,580$ s $= 26.3$ minutes.

14.2
Convection

Convection is the mode of heat transfer between a solid surface and the adjacent fluid that is in motion. Convection involves the combined effects of conduction and fluid motion. The faster the fluid motion is, the greater the convective heat transfer. In the absence of any bulk fluid motion, heat transfer between a solid surface and the adjacent fluid is by pure conduction. The presence of bulk motion of the fluid enhances the heat transfer between the solid surface and the fluid.

Convection is called *forced convection* if the fluid is forced to flow over the surface by external means such as a fan, a pump, or the wind. In contrast, convection is called *natural or free convection* if the fluid motion is caused by buoyancy force that is induced by density difference due to the variation of temperature in the fluid. Heat transfer from a solid surface can be obtained from:

$$\dot{Q} = hA(T_W - T_\infty) \qquad [14.26]$$

where h is convective heat transfer coefficient (W/m²·K)

A is heat transfer area (m²)

T_w is surface temperature of the wall (K)

T_∞ is the bulk fluid temperature (K)

Therefore thermal resistance due to convection is

$$R = \frac{1}{hA} \qquad [14.27]$$

Example 14.5

Hot water flows through a steel pipe ($k = 50$ W/m·K) 5 cm i.d. and a 6.25 cm o.d. The average temperature of the water is 111°C, and the ambient air temperature is 33°C. The convective heat transfer coefficient between the water and the inside pipe surface is 1,135 W/m²·K and that between the ambient air and outside pipe surface is 28.5 W/m²·K.

a. Determine the rate of heat loss per linear meter of pipe.

b. Calculate the minimum thickness of insulation ($k = 0.09$ W/m·K) needed so that the outer surface of the insulation is not more than 55°C and the heat loss from hot water is not more than 44 W per linear meter of the pipe.

Solution:

a. Equation 14.9 can be used, in which convective heat resistances replace conductive ones:

$$\frac{\dot{Q}}{L} = L\frac{(T_{in} - T_{out})}{R_1 + R_2 + R_3} = \frac{(T_{in} - T_{out})}{\frac{1}{h_i\pi d_i} + \frac{\ln(r_o/r_i)}{2\pi k} + \frac{1}{h_o\pi d_o}}$$

$$= \frac{(111 - 33)}{\frac{1}{1135\pi(0.05)} + \frac{\ln(0.03125/0.025)}{2\pi(50)} + \frac{1}{28.5\pi(0.0625)}}$$

$$\frac{\dot{Q}}{L} = 421.56 \text{ W/m}$$

b. This time we write the heat flow from the interior temperature to the insulation outer temperature:

$$\frac{\dot{Q}}{L} = 44 = \frac{(T_{in} - T_{ins})}{\frac{1}{h_i\pi d_i} + \frac{\ln(r_o/r_i)}{2\pi k_{pipe}} + \frac{\ln(r_{ins}/r_o)}{2\pi k_{ins}}}$$

$$= \frac{(111 - 55)}{\frac{1}{1,135\pi(0.05)} + \frac{\ln(0.03125/0.025)}{2\pi(50)} + \frac{\ln(r_{ins}/0.03125)}{2\pi(0.09)}}$$

By solving for r_{ins}, we find $r_{ins} = 6.40$ cm, which means the thickness is $6.40 - 6.25 = 0.15$ cm.

Table 14.1: Typical values of convective heat transfer coefficients

Type of Convection	h (W/m²K)
Natural convection (air)	5–15
Natural convection (water)	500–1,000
Force convection (air)	10–200
Force convection (oil)	20–2000
Force convection (water)	300–20,000
Water boiling	3000–100,000
Water condensing	500–10,000

External Flow

All properties are calculated based on the average temperature of the fluid and the body.

In a flat plate in parallel flow:

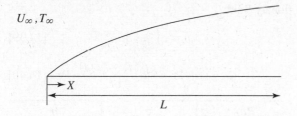

Figure 14.10 Parallel flow over a flat plate

$$\mathrm{Re}_L = \frac{\rho U_\infty L}{\mu} \qquad [14.28]$$

$$\overline{Nu}_L = \frac{\bar{h}L}{k} = 0.6640\,\mathrm{Re}_L^{1/2}\,\mathrm{Pr}^{1/3} \qquad \left(\mathrm{Re}_L < 10^5\right) \qquad [14.29]$$

$$\overline{Nu}_L = \frac{\bar{h}L}{k} = 0.0366\,\mathrm{Re}_L^{0.8}\,\mathrm{Pr}^{1/3} \qquad \left(\mathrm{Re}_L > 10^5\right) \qquad [14.30]$$

In a cylinder cross flow:

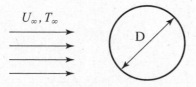

Figure 14.11 Cross flow over a cylinder

$$\mathrm{Re}_D = \frac{\rho U_\infty D}{\mu} \qquad [14.31]$$

$$\overline{Nu}_D = \frac{\bar{h}D}{k} = C\,\mathrm{Re}_D^n\,\mathrm{Pr}^{1/3} \qquad [14.32]$$

where

Re_D	C	n
1–4	0.989	0.330
4–40	0.911	0.385
40–4,000	0.683	0.466
4,000–40,000	0.193	0.618
40,000–400,000	0.0266	0.805

Flow over a sphere of diameter D:

$$\overline{Nu}_D = 2 + \left(0.4\,\mathrm{Re}_D^{1/2} + 0.06\,\mathrm{Re}_D^{2/3}\right)\mathrm{Pr}^{0.4}\left(\frac{\mu}{\mu_s}\right)^{1/4}$$

$$\text{for} \qquad \begin{bmatrix} 0.71 \le \mathrm{Pr} \le 380 \\ 3.5 \le \mathrm{Re}_D \le 7.6 \times 10^4 \\ 1.0 \le (\mu/\mu_s) \le 3.2 \end{bmatrix}$$

$$[14.33]$$

In the above equations, all properties except μ_s are evaluated at T_∞.

For a freely falling droplet, the correlation becomes simpler:

$$\overline{Nu}_D = 2 + 0.6\,\mathrm{Re}_D^{1/2}\,\mathrm{Pr}^{1/3} \qquad [14.34]$$

Internal Flow

Laminar flow in a circular tube:

$$\mathrm{Re}_D = \frac{\rho U_\infty D}{\mu} < 2,300$$

a. $Nu_D = 4.36$ in a fully developed laminar flow with uniform heat flux

b. $Nu_D = 3.66$ in a fully developed laminar flow with uniform surface temperature

c. $Nu_D = 1.86\left(\dfrac{\mathrm{Re}_D\,\mathrm{Pr}}{\frac{L}{D}}\right)^{1/3}\left(\dfrac{\mu_b}{\mu_s}\right)^{14}$ for not fully

developed flow and uniform surface temperature where

L = length of tube

μ_b = dynamic viscosity of fluid at bulk temperature of fluid, T_b

μ_s = dynamic viscosity of fluid at inside surface temperature of the tube, T_s

In turbulent flow in a circular pipe:

$$Nu_D = 0.023\, Re_D^{0.8}\, Pr^{1/3}\left(\frac{\mu_b}{\mu_s}\right)^{0.14} \qquad [14.35]$$

For $Re_D > 10^4$, $Pr > 0.7$, and either uniform surface temperature or uniform heat flux:

With liquid metals ($0.003 < Pr < 0.05$):

$$Nu_D = 6.3 + 0.0167\, Re_D^{0.85}\, Pr^{0.93}$$
for uniform heat flux

$$Nu_D = 7.0 + 0.0250\, Re_D^{0.80}\, Pr^{0.80}$$
for constant wall temperature $\qquad$ [14.36]

In non-circular ducts, equivalent (hydraulic) diameter (D_H) should be used in place of diameter, D.

$$D_H = \frac{4 \times (\text{cross-sectional area})}{\text{wetted perimeter}} \qquad [14.37]$$

For example,

$$\text{Circular duct: } D_H = \frac{4 \times \left(\frac{\pi}{4}D^2\right)}{\pi D} = D$$

$$\text{Square duct with side L: } D_H = \frac{4 \times \left(L^2\right)}{4L} = L$$

$D_H = D_o - D_i$ for circular annulus where D_o is outside diameter and D_i is inside diameter.

Natural (Free) Convection

Natural or free convection occurs when there is no forced velocity. Body force such as buoyancy force due to density difference (in turn due to temperature gradient) causes free convection current.

In a flat plate in a large body of stationary fluid, convective heat transfer coefficient can be calculated using equation 14.38. This equation can also apply to a vertical cylinder of a sufficiently large diameter in a large body of stationary fluid.

$$h = c\left(\frac{k}{L}\right)Ra^n \qquad [14.38]$$

where

L = length of the plate (cylinder) in the vertical direction

Ra_L = Rayleigh number = Gr_L (Grashof number =

$\frac{g\beta(T_s - T_\infty)L^3}{\upsilon^2}$) $\times Pr$ (Prandtl number = $\frac{\upsilon}{\alpha}$)

T_s = Surface temperature

T_∞ = Fluid temperature

β = Volumetric thermal expansion coefficient

$\left(\beta = -\frac{1}{\rho}\left(\frac{\partial \rho}{\partial T}\right)_p\right)$, which relates density change to temperature change

Ideal gas ($p = \rho RT$), $\beta = \frac{1}{T}$

In this correlation, $\beta = \frac{2}{T_s + T_\infty}$

Kinematic viscosity ($\upsilon = \frac{\mu}{\rho}$): $\frac{m^2}{s}$

Range of Ra_L	c	n
10^4–10^9	0.59	1/4
10^9–10^{13}	0.10	1/3

In a long horizontal cylinder in a large body of stationary fluid:

$$\bar{h} = c\left(\frac{k}{D}\right)Ra_D^{\,n} \qquad [14.39]$$

$$Ra = Gr_D\, Pr = \frac{g\beta(T_s - T_\infty)D^3}{\upsilon^2}Pr \quad [14.40]$$

Ra_D	c	n
10^{-3}–10^2	1.02	0.148
10^2–10^4	0.850	0.188
10^4–10^7	0.480	0.250
10^7–10^{12}	0.125	0.333

Boiling

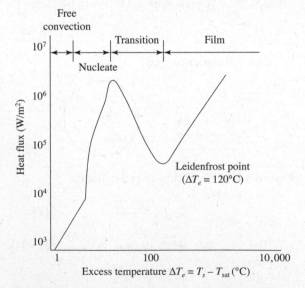

Figure 14.12 Boiling curve for water at 1 atm. Surface heat flux as a function of excess temperature.

In film boiling for excess temperature beyond the Leidenfrost point:

$$\overline{Nu}_D = \frac{\overline{h}_{conv}D}{k_v} = c\left[\frac{g\left(\rho_l - \rho_v\right)h'_{fg}D^3}{\upsilon_v k_v\left(T_s - T_{sat}\right)}\right]^{1/4}$$

[14.41]

$C = 0.62$ for horizontal cylinder

$C = 0.67$ for sphere

h'_{fg} = convected latent heat, which includes sensible energy required to maintain vapor temperature = $h_{fg} + 0.8c_{p,v}\left(T_s - T_{sat}\right)$

Properties k_v, ρ_v, υ_v are those of vapor, and ρ_l is density of liquid.

Condensation of a Pure Vapor

On a vertical surface:

$$\overline{Nu}_L = \frac{\overline{h}L}{k_l} = 0.943\left[\frac{\rho_l^2 g h_{fg}L^3}{\mu_l k_l\left(T_{sat} - T_s\right)}\right]^{0.25}$$

[14.42]

where

ρ_l = density of liquid phase of fluid (kg/m^3)

g = gravitational acceleration (9.81 m/s^2)

h_{fg} = latent heat of vaporization (J/kg)

L = length of surface (m)

μ_l = dynamic viscosity of liquid phase of fluid (kg/(s-m))

K_l = thermal conductivity of liquid phase of fluid (W/m·K)

T_{sat} = saturation temperature of fluid (K)

T_s = temperature of vertical surface (K)

Note: Evaluate all liquid properties at the average temperature between the saturated temperature, T_{sat}, and the surface temperature, T_s.

With outside horizontal tubes:

$$\overline{Nu}_D = \frac{\overline{h}D}{k} = 0.729\left[\frac{\rho_l^2 g h_{fg}D^3}{\mu_l k_l\left(T_{sat} - T_s\right)}\right]^{0.25}$$

[14.43]

where

D = tube outside diameter (m)

Note: Evaluate all liquid properties at the average temperature between the saturated temperature, T_{sat}, and the surface temperature, T_s.

14.3 Radiation

Radiation is the energy emitted by matter in the form of electromagnetic waves as a result of the changes in the electron configurations of the atoms or molecules. Unlike conduction and convection, the transfer of energy by radiation does not require the presence of an intervening medium. In fact, energy transfer by radiation is the fastest (at the speed of light) and suffers no attenuation in a vacuum. This is exactly how the energy of the sun reaches Earth.

The maximum rate of radiation that can be emitted from a surface at an absolute temperature (T_s) is given by the *Stefan-Boltzmann law* as:

$$\dot{Q} = \varepsilon\sigma\,AT_s^4$$

[14.44]

where $\sigma = 5.67 \times 10^{-8}$ W/(m^{2}K^4) is the *Stefan-Boltzmann constant* and ε is the emissivity of the surface. An idealized surface, which emits radiation at a maximum rate, has $\varepsilon = 1$ and is known as a *blackbody*. The radiation emitted by actual surfaces is less than that emitted by the blackbody. The value of ε is in the range $0 < \varepsilon < 1$ and is a measure of how closely a surface approximates a blackbody.

For a small body (1) that is small compared to its surroundings (2), the net rate of radiative heat transfer from the body ($\dot{Q}_{12}$) is shown by:

$$\dot{Q}_{12} = \varepsilon\sigma\,A\left(T_1^4 - T_2^4\right)$$

[14.45]

Table 14.2
Emissivity of some materials at 300 K

Material	ε
Aluminum foil	0.07
Anodized aluminum	0.82
Polished copper	0.03
Polished gold	0.03
Polished silver	0.02
Polished stainless steel	0.17
Water	0.96
Black paint	0.98
White paint	0.90
White paper	0.92–0.97
Asphalt pavement	0.85–0.93
Human skin	0.95
Wood	0.82–0.92
Soil	0.93–0.96

Another important radiation property of a surface is its *absorptivity* (α), which is the fraction of the radiation energy incident on a surface absorbed by the surface. Like emissivity, its value is in the range $0 < \alpha < 1$. A blackbody absorbs the entire radiation incident on it. That is, a blackbody is a perfect absorber ($\alpha = 1$) as well as a perfect emitter. In practice, α and ε are assumed to be independent from the temperature and wavelength of the radiation. The average absorptivity of a surface is taken to be equal to its average emissivity.

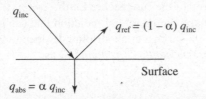

Figure 14.13 The absorption of radiation incident on an opaque surface of absorptivity

Figure 14.13 shows incident, reflected, and absorbed energy on a surface. The rate at which a surface absorbs radiation is determined from:

$$q_{abs} = \alpha q_{inc} \qquad [14.46]$$

where q_{inc} is the rate at which radiation is incident on the surface. For nontransparent surfaces, the portion of incident radiation not absorbed by the surface is reflected to the surroundings.

The difference between the rate of radiation emitted by the surface and the rate of radiation absorbed is the net radiation heat transfer. If the rate of radiation absorption is greater than the rate of radiation emission, the surface is said to be gaining energy by radiation. Otherwise, the surface is said to be losing energy by radiation.

Real bodies are frequently approximated as a "gray body," which is one for which $\alpha = \varepsilon$.

Radiation View Factors

The above equations for blackbodies and gray bodies (equations 14.45 and 14.46) assume that the small body could see only the large enclosing body and nothing else. Hence, all radiation leaving the small body would reach the large body. For the case where two objects can see more than just each other, then one must introduce a view factor *F*. The heat transfer calculations become significantly more involved.

The view factor F_{12} is used to parameterize the fraction of thermal power leaving object 1 and reaching object 2. Specifically, this quantity is equal to:

$$\dot{Q}_{1 \to 2} = A_1 F_{12} \varepsilon_1 \sigma T_1^4 \qquad [14.47]$$

Likewise, the fraction of thermal power leaving object 2 and reaching object 1 is given by:

$$Q_{2 \to 1} = A_2 F_{21} \varepsilon_2 \sigma T_2^4 \qquad [14.48]$$

The case of two blackbodies in thermal equilibrium can be used to derive the following *reciprocity relationship* for view factors:

$$A_1 F_{12} = A_2 F_{21} \qquad [14.49]$$

Thus, once one knows F_{12}, F_{21} can be calculated immediately. Radiation view factors can be analytically derived for simple geometries

Heat Transfer Between Two Finite Gray Bodies

The heat flow transferred from Object 1 to Object 2 where the two objects see only a fraction of each other and nothing else is given by:

$$\dot{Q} = \left(\frac{1-\varepsilon_1}{\varepsilon_1} + \frac{1}{F_{12}} + \left(\frac{1-\varepsilon_2}{\varepsilon_2} \right) \frac{A_1}{A_2} \right)^{-1} A_1 \sigma \left(T_1^4 - T_2^4 \right)$$

$$[14.50]$$

This equation demonstrates the usage of F_{12}. However, it represents a nonphysical case since positioning two finite objects such that they can see only a portion of each other and "nothing" else is impossible. On the contrary, the complementary view factor $(1 - F_{12})$ cannot be neglected as radiation energy sent in those directions must be accounted for in the thermal bottom line

Heat Exchangers

The purpose of heat exchange between two fluids that are different temperatures and separated by a solid wall occurs in many engineering applications. The device used to implement this exchange is termed a *heat exchanger*. Specific application is found in space heating and air-conditioning.

Heat exchangers are typically classified according to *flow arrangement* and type of construction. Figures 14.14 shows two simple parallel-flow and counterflow heat exchangers.

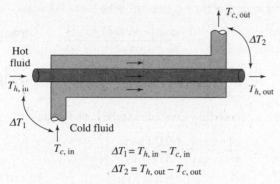

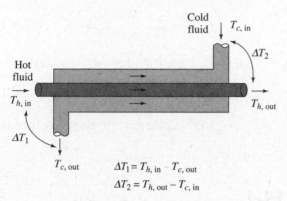

(a) Parallel-flow heat exchangers

$$\Delta T_1 = T_{h,\,in} - T_{c,\,in}$$
$$\Delta T_2 = T_{h,\,out} - T_{c,\,out}$$

(b) Counterflow heat exchangers

$$\Delta T_1 = T_{h,\,in} - T_{c,\,out}$$
$$\Delta T_2 = T_{h,\,out} - T_{c,\,in}$$

Figure 14.14 Concentric tube heat exchangers

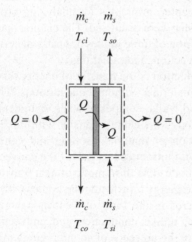

Figure 14.15 A typical counterflow heat exchanger

The *rate of heat transfer* between the two fluid streams in the heat exchanger is shown as Q in Figure 14.15 and in equation 14.51.

$$Q = \left(\dot{m}c_p\right)_s \left(T_{so} - T_{si}\right) = \left(\dot{m}c_p\right)_c \left(T_{ci} - T_{co}\right)$$
[14.51]

UA-Value and LMTD

The unit's *overall conductance* or UA value is defined as the product of the overall heat transfer coefficient and the heat transfer area. For counterflow applications, the heat transfer rate is defined as the product of overall conductance and the log-mean temperature difference, LMTD:

$$Q = UA\cdot\Delta T_{lm}$$
[14.52]

where the *log-mean temperature difference* is equal to:

$$\Delta T_{lm} = \frac{\Delta T_{out} - \Delta T_{in}}{\ln\left(\dfrac{\Delta T_{out}}{\Delta T_{in}}\right)}$$
[14.53]

For counterflow in tubular heat exchangers:

$$\Delta T_{lm} = \frac{\left(T_{Ho} - T_{Ci}\right) - \left(T_{Hi} - T_{Co}\right)}{\ln\left(\dfrac{T_{Ho} - T_{Ci}}{T_{Hi} - T_{Co}}\right)}$$
[14.54]

T_{Hi} = inlet temperature of the hot fluid (K)

T_{Ho} = outlet temperature of the hot fluid (K)

T_{Ci} = inlet temperature of the cold fluid (K)

T_{Co} = outlet temperature of the cold fluid (K)

For parallel flow in tubular heat exchangers:

$$\Delta T_{lm} = \frac{\left(T_{Ho} - T_{Co}\right) - \left(T_{Hi} - T_{Ci}\right)}{\ln\left(\dfrac{T_{Ho} - T_{Co}}{T_{Hi} - T_{Ci}}\right)}$$
[14.55]

Example 14.6

Hot oil that has a specific heat of 2,093.4 J/kg·K flows through a conterflow heat exchanger at rate of 6.3 kg/s with an inlet temperature of 211°C and an outlet temperature of 83°C. Cold oil that has a specific heat of 1,674.7 J/kg·K flows in at a rate of 10.1 kg/s and leaves at 166.7°C. Determine the area of the heat exchanger necessary to handle the load if the overall heat transfer coefficient based on the inside area is 766.6 W/m²·K.

Solution:

From equation 14.51 we can find the unknown temperature T_{ci}:

$$\left(\dot{m}c_p\right)_s \left(T_{so} - T_{si}\right) = \left(\dot{m}c_p\right)_c \left(T_{ci} - T_{co}\right) \text{ or}$$
$$\left(6.3(2,093.4)\right)_s \left(83 - 211\right)$$
$$= \left(10.1(1,674.7)\right)_c \left(T_{ci} - 166.7\right)$$

From this we can find:

$T_{ci} = 48.9°$ C

To find LMTD, use equation 14.54:

$$\Delta T_{lm} = \frac{(83 - 48.9) - (211 - 166.7)}{\ln\left(\frac{83 - 48.9}{211 - 166.7}\right)} = 10.56°C$$

The heat transfer rate can be calculated from equations 14.51 and 14.52.

$$Q = \left(\dot{m}c_p\right)_s (T_{so} - T_{si}) = UA \cdot T_{lm}$$

$$Q = 6.3(2,093.4)_s(211 - 83) = 766.6(A)(10.56)$$

Therefore, the total area required is

$A = 77.57$ m^2

Effectiveness

The *heat exchanger effectiveness*, ε, is defined as the ratio of the rate of heat transfer in the exchanger, Q, to the maximum theoretical rate of heat transfer:

$$\varepsilon = \frac{Q}{Q_{max}} \qquad [14.56]$$

The maximum theoretical rate of heat transfer is limited by the fluid stream with the smallest heat capacity rate:

$$\varepsilon = \frac{\left(\dot{m}c_p\right)_s (T_{so} - T_{si})}{\left(\dot{m}c_p\right)_{min} (T_{ci} - T_{si})} \qquad [14.57]$$

where $(\dot{m}c_p)_{min}$ is the smaller of $(\dot{m}c_p)_s$ or $(\dot{m}c_p)_c$

NTU

The *number of transfer units* (NTU) is an indicator of the actual heat transfer area or physical size of the heat exchanger. The larger the value of NTU is, the closer the unit is to its thermodynamic limit. NTU is defined as:

$$NTU = \frac{UA}{\left(\dot{m}c_p\right)_{min}} \qquad [14.58]$$

Effectiveness and NTU relations can be shown by the following equations:

$$c_r = \frac{c_{min}}{c_{max}} = \text{heat capacity ratio}$$

$$\dot{c} = \dot{m}c_p = \text{heat capacity rate (W/K)}$$

For a parallel flow concentric tube heat exchanger:

$$\varepsilon = \frac{1 - \exp\left[-NTU\,(1 + c_r)\right]}{1 + c_r} \qquad [14.59]$$

$$NTU = -\frac{\ln\left[1 - \varepsilon\,(1 + c_r)\right]}{1 + c_r} \qquad [14.60]$$

For a counterflow concentric tube heat exchanger:

$$\varepsilon = \frac{1 - \exp\left[-NTU\,(1 - c_r)\right]}{1 - c_r \exp\left[-NTU\,(1 - c_r)\right]} \qquad (c_r < 1)$$

$$[14.61]$$

$$\varepsilon = \frac{NTU}{1 + NTU} \qquad (c_r = 1) \qquad [14.62]$$

$$NTU = \frac{1}{c_r - 1} \ln\left(\frac{\varepsilon - 1}{\varepsilon c_r - 1}\right) \qquad (c_r < 1)$$

$$[14.63]$$

$$NTU = \frac{\varepsilon}{1 - \varepsilon} \qquad (c_r = 1) \qquad [14.64]$$

14.4
Summary

In this chapter, principles for analyzing problems in heat transfer were discussed. The chapter started with the definition of three modes of heat transfer: conduction, convection, and radiation.

Conduction is the transfer of energy flux in solid or fluid materials without bulk motions. This could be in solid walls, composite walls, cylinders, or fins. Convection is the energy transfer with forced or free bulk motion of material. Boiling and condensation were briefly introduced as part of the convective heat transfer mode. The third mode of heat transfer, radiation, is the energy transfer among bodies with different temperatures in the form of electromagnetic waves. Radiation, unlike conduction and convection, does not require the presence of an intervening medium.

Finally, some applications of various heat transfer equations in analyzing heat exchangers were introduced.

PRACTICE PROBLEMS

1. What is the thermal resistance of the composite wall shown in the figure if the exposed surface area is 20m²?

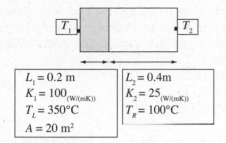

$L_1 = 0.2$ m
$K_1 = 100_{(W/(mK))}$
$T_L = 350°C$
$A = 20$ m²

$L_2 = 0.4$m
$K_2 = 25_{(W/(mK))}$
$T_R = 100°C$

(A) 0.009 K/W
(B) 0.0009 K/W
(C) 0.012 K/W
(D) 0.0012 K/W

2. What is the convective heat transfer coefficient, h_o, of the system in the figure?

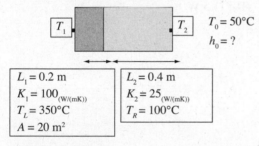

$T_0 = 50°C$
$h_0 = ?$

$L_1 = 0.2$ m
$K_1 = 100_{(W/(mK))}$
$T_L = 350°C$
$A = 20$ m²

$L_2 = 0.4$ m
$K_2 = 25_{(W/(mK))}$
$T_R = 100°C$

(A) 100 W/(m²K)
(B) 200 W/(m²K)
(C) 280 W/(m²K)
(D) 320 W/(m²K)

3. A stone with emissivity of 0.9 in Arizona reaches 52°C in the summer sun. What is the radiant heat energy transfer from the stone if its surface area is 200 cm²?
(A) 6 W
(B) 8 W
(C) 10 W
(D) 12 W

4. What is the radius of the insulation surrounding a copper wire ($k = 0.08$ W/(mK)) on a calm day with $h = 2$ W/(m²k)?
(A) 2 cm
(B) 4 cm
(C) 6 cm
(D) 8 cm

5. Wind chill is related to heat transfer on a windy day. Consider a 4 mm thick tissue ($k = 0.2$ W/m·K) with an interior temperature of 37°C. The convection heat transfer coefficient is 20 W/m²·K for a calm day and 70 W/m²·K for a windy day. In both cases, ambient air temperature is –10°C. What is the heat flux on a windy day?
(A) 1,240 W/m²
(B) 1,260 W/m²
(C) 1,280 W/m²
(D) 1,300 W/m²

6. Wind chill is related to heat transfer on a windy day. Consider a 4 mm thick tissue ($k = 0.2$ W/m·K) with an interior temperature of 37°C. The convection heat transfer coefficient is 20 W/m²·K for a calm day and 70 W/m²·K for a windy day. In both cases, ambient air temperature is –10°C. What is the skin outer surface temperature on a windy day?
(A) 10°C
(B) 12°C
(C) 14°C
(D) 16°C

7. What is the average convective heat transfer coefficient of air ($U_\infty = 10$ m/s, $k = 26.3 \times 10^{-3}$ W/m·K, $p = 1.1614$ kg/m³, $\mu = 184.6 \times 10^{-7}$ N·s/m², Pr = 0.707) for a flow over a 10-meter plate?
(A) 100 W/m²-K
(B) 24 W/m²-K
(C) 50 W/m²-K
(D) 4 W/m²-K

8. Engine oil at 350 K ($k = 0.138$ W/m·K, $\rho = 853.9$ kg/m³, $\mu_{350} = 0.0356$ N·s/m², $\mu_{300} = 0.486$ N·s/m², Pr = 546) flows as fully developed with a speed of 1 m/s in a 0.5 m diameter pipe with a uniform temperature of 300 K. Calculate Nu based on diameter.
(A) 4.36
(B) 3.66
(C) 239
(D) 600

9. For a circular annulus with an inside radius of 5 cm and an outside radius of 9 cm, determine the hydraulic (equivalent) diameter.
(A) 10 cm
(B) 19 cm
(C) 4 cm
(D) 8 cm

10. For a sphere of diameter D inside a cubical box of length $L = D$, determine the view factor F_{21}.

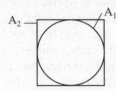

(A) 1
(B) π
(C) $\pi/2$
(D) $\pi/6$

11. Determine the log-mean temperature difference for a counterflow heat exchanger when hot fluid comes in at 200°C and leaves at 150°C while cold fluid comes in at 20°C and leaves at 50°C.

(A) 50°C
(B) 90°C
(C) 140°C
(D) 180°C

12. A schematic of two coaxial cylinders is shown. Assume that both cylinders behave as black bodies and neglect conduction and convection, at the ends of the cylinders. Determine the net radiation per unit length of the two cylinders.

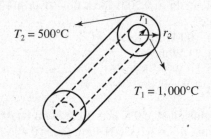

(A) 33.4 kW/m
(B) 80.8 kW/m
(C) 50.1 kW/m
(D) 101.9 kW/m

13. Consider an oil cooler for an internal combustion engine to cool SAE-30 lubricating oil from 70°C to 55°C using water at an inlet temperature of 27°C with a temperature rise of 12°C. The design heat load is 200 kW. Assuming an average overall heat transfer coefficient $U = 700$ W/m^{2}C based on the outer surface area of the tubes, determine the heat transfer surface area required for a single pass, parallel-flow heat exchanger.

(A) 18.5 m^2
(B) 10.5 m^2
(C) 5.5 m^2
(D) 9.5 m^2

14. A cylindrical fin, with a cross-section diameter of 1 cm and a length of 5 cm ($k = 43$ W/mC) transfers energy from a wall at 250°C to a fluid at 20°C ($h = 140$ W/m^{2}C). Determine the rate of heat transfer from the fin.

(A) 25.6 W
(B) 15.1 W
(C) 26.5 W
(D) 11.5 W

Answer Key

1. B	5. C	9. D	13. B
2. C	6. A	10. D	14. C
3. D	7. B	11. C	
4. B	8. C	12. C	

Answers Explained

1. B

$$R_{tot} = \sum R = \frac{L_1}{K_1 A} + \frac{L_2}{K_2 A} = \frac{1}{A}\left(\frac{L_1}{k_1} + \frac{L_2}{k_2}\right) = 0.0009 \ \frac{\text{K}}{\text{W}}$$

2. C

$$\dot{Q} = \frac{T_1 - T_2}{R_1 + R_2} = \frac{(T_2 - T_o)}{R_o} : \frac{T_1 - T_2}{\dfrac{L_1}{K_1 A} + \dfrac{L_2}{K_2 A}} = \frac{T_2 - T_o}{\dfrac{1}{h_o A}}$$

$$\rightarrow h_o = 280 \ \frac{\text{W}}{\text{m}^2\text{K}}$$

3. D $\dot{Q} = \varepsilon \sigma A T^4 = 12 \text{W}$

4. B $r_{cr} = \dfrac{k_{ins}}{h_\infty} = 0.04 \ \text{m} = 4 \ \text{cm}$

5. C $q'' = \dfrac{37 - (-10)}{\dfrac{0.004}{0.2} + \dfrac{1}{70}} = 1282 \ \text{W/m}^2$

6. A $T_s = 37 - \dfrac{\dfrac{0.004}{0.2}(37-(-10))}{\dfrac{0.004}{0.2} + \dfrac{1}{70}} = 9.6\,°\text{C}$

7. B

$$\text{Re} = \frac{\rho U_\infty L}{\mu} = \frac{1.1614 \ kg}{\text{m}^3}\left|\frac{10 \ \text{m}}{\text{s}}\right|\frac{10 \ \text{m}}{}\left|\frac{\text{m}^2}{184.6\times10^{-7} \ \text{N}-\text{s}}\right|\frac{\text{N}-\text{s}^2}{\text{kg}-\text{m}}$$

$$= 6.29\times10^6$$

$$\bar{h} = \frac{k}{L}\left(0.0366 \ \text{Re}_L^{0.8} \ \text{Pr}^{1/3}\right) = \frac{26.3\times10^{-3} \ \text{W}}{10 \ \text{m}^2\text{K}}\left(0.0366\left(6.29\times10^6\right)^{0.8}(0.707)^{1/3}\right)$$

$$= 0.09626\times10^{-3}\left(6.29\times10^6\right)^{0.8}(0.707)^{1/3} = 24$$

8. C

$$\text{Re}_D = \frac{\rho U D}{\mu} = \frac{853.9 \ \text{kg}}{\text{m}^3}\left|\frac{1 \ \text{m}}{\text{s}}\right|\frac{0.5 \ \text{m}}{}\left|\frac{\text{m}^2}{0.0356 \ \text{N}-\text{s}}\right|\frac{\text{N}-\text{s}^2}{\text{kg}-\text{m}}$$

$$= 11,993$$

$$Nu_D = 0.023 \ \text{Re}_D^{0.8} \ \text{Pr}^{1/3}\left(\frac{\mu_b}{\mu_s}\right)^{0.14}$$

$$= 0.023(11,993)^{0.8}(546)^{1/3}\left(\frac{0.0356}{0.486}\right)^{0.14}$$

$$= 239$$

9. D $D_H = D_o - D_i = 2(R_o - R_i) = 2(9 - 5) = 8$

10. D By inspection, $F_{12} = 1$. All radiation from the sphere will reach inside the surface of the cube.

$$F_{12}A_1 = F_{21}A_2$$

$$F_{21} = F_{12}\frac{A_1}{A_2} = 1\times\frac{\pi D^2}{6D^2} = \frac{\pi}{6}$$

11. C

$$\Delta T_{lm} = \frac{(T_{Ho} - T_{Ci}) - (T_{Hi} - T_{Co})}{\ln\left(\dfrac{T_{Ho} - T_{Ci}}{T_{Hi} - T_{Co}}\right)}$$

$T_{Hi} = 200\,°\text{C}$

$T_{Ho} = 150\,°\text{C}$

$T_{Ci} = 20\,°\text{C}$

$T_{Co} = 50\,°\text{C}$

$$\Delta T_{lm} = \frac{(150 - 20) - (200 - 50)}{\ln\left(\dfrac{150 - 20}{200 - 50}\right)} = \frac{-20}{\ln\left(\dfrac{130}{150}\right)} = 139.76$$

12. C $\dot{Q} = \sigma A_1 F_{12}\left(T_1^4 - T_2^4\right)$

$$= \sigma(2\pi r_1 L) F_{12}\left(T_1^4 - T_2^4\right)$$

The shape factor, F_{12}, for a blackbody completely within another blackbody is 1.

$$\frac{\dot{Q}}{L} = \sigma(2\pi r_1) F_{12}\left(T_1^4 - T_2^4\right)$$

$$= \left(5.67\times10^{-8} \ \frac{\text{W}}{\text{m}^2\text{K}^4}\right)(2\pi)(0.1 \ \text{m})(1)\left[1273^4 - 773^4\right]$$

$$= 80.79 \ \text{kW/m}$$

13. B $\dot{Q} = UA\Delta T_{lm}$

For parallel-flow heat exchanger

$$\Delta T_{lm} = \frac{(T_{Ho} - T_{Co}) - (T_{Hi} - T_{Ci})}{\ln\left(\dfrac{T_{Ho} - T_{Co}}{T_{Hi} - T_{Ci}}\right)}$$

$T_{Hi} = 70\,°\text{C}$

$T_{Ho} = 55\,°\text{C}$

$T_{Ci} = 27\,°\text{C}$

$T_{Co} = 39\,°\text{C}$

$$\Delta T_{lm} = \frac{(55 - 39) - (70 - 27)}{\ln\left(\dfrac{55 - 39}{70 - 27}\right)} = \frac{-27}{\ln\left(\dfrac{16}{43}\right)} = 27.31$$

$$A = \frac{\dot{Q}}{U\Delta T_{lm}} = \frac{200,000}{700(27.31)} = 10.462 \ \text{m}^2$$

14. C

$$\dot{Q} = \sqrt{hpkA_c}\ (T_b - T_\infty)\tanh{(mL_c)}$$

$$A_c = \frac{\pi}{4}d^2 = \frac{\pi}{4}(1\ \text{cm})^2 = \frac{\pi}{4}\ \text{cm}^2, \qquad p = \pi d = \pi\ \text{cm}$$

$$L_c = L + \frac{A_c}{p} = L + \frac{\frac{\pi}{4}d^2}{\pi d} = L + \frac{d}{4} = 5 + \frac{1}{4} = 5.25\ \text{cm}$$

$$m = \sqrt{\frac{hp}{kA_c}} = \sqrt{\frac{140\left(\pi \times 10^{-2}\right)}{43\left(\frac{\pi}{4}\times 10^{-4}\right)}} = 36.088\ \frac{1}{\text{m}}$$

$$\dot{Q} = \sqrt{140\left(\pi \times 10^{-2}\right)43\left(\frac{\pi}{4}\times 10^{-4}\right)}\,(250 - 20)$$
$$\times \tanh\left(36.088 \times 5 \times 10^{-2}\right)$$

$$\dot{Q} = 26.539\ \text{W}$$

15

Dynamic Systems and Controls

Nader Jalili, Ph.D.

Introduction

Dynamic systems are referred to as physical systems that are **time-varying** in nature or possess a quantity that undergoes motion. Hence, they are mathematically governed by differential equations. Controls refer to the notion of incorporating external stimuli (force, torque, motion, etc.) to improve the system's response to a desired level. Obviously, such control action requires measuring the system's states and output so a set of desired characteristics (e.g., steady-state error, settling time, overshoot) can be achieved. This chapter provides a brief overview of these three topics along with related example case studies. The chapter is designed to prepare the reader by reinforcing the fundamental concepts while also emphasizing practical aspects and problem-solving techniques.

15.1 Dynamic Systems Overview

Mathematical modeling of a dynamic system refers to describing the system in terms of governing (differential) equations. These equations are typically obtained from either a direct approach or numerical methods (e.g., finite element method). Concerning the direct approaches, there are two different modeling strategies: the Newtonian approach and the analytical method. The former method is based on deriving the equations of motion using the free-body diagram (FBD) of the system and taking into account the effects of external forces applied on the boundary of the system. This typically requires a system decomposition exercise. The dynamic system is considered to be built up based on its components. The second modeling approach, the analytical approach, is an energy-based modeling framework. Interactions among different fields (e.g., electrical, mechanical, magnetic) can be conveniently established and presented.

Linear vs. Nonlinear Models

Most natural and practical systems are nonlinear in nature. Examples include gearboxes with inherent backlash, machine components with dry frictions, and linear systems possessing dead zones (due to manufacturing deficiencies) or undergoing large-amplitude vibrations. The linearized models developed for these naturally nonlinear systems are our own idealization, which may not be justifiable. However, for small-amplitude motions considered in this chapter, linear assumptions are made.

Lumped-parameters vs. Distributed-parameters Models

Similar to linear and nonlinear modeling viewpoints, physical systems can be mathematically modeled as either discrete or continuous systems. All real systems are made of physical parameters that cannot be assumed isolated and, hence, are continuous by their nature. An idealization of these naturally continuous systems is the discretization of these systems into many isolated components that can be described by independent degrees of freedom (DOF). Figure 15.1 demonstrates this idealization process on a flexible beam where only one mode (fundamental) of vibration is considered when discretizing this continuous system.

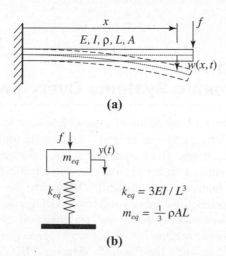

(a)

(b)

Figure 15.1 Schematic representation of idealistic discretization (b) of a naturally continuous system (a)

Discrete or lumped-parameters systems are governed by ordinary differential equations (ODE) since the modeling can be performed for each of the inde-

pendent DOF and isolated parameters. Without going into the details and referring interested readers to [Jalili 2010]*, the motions of a *continuous or distributed-parameters system* can be conveniently described by a finite number of *displacement* variables, governed by partial differential equations (PDE). In order not to disturb the focus of this chapter, we consider only a lumped-parameters modeling approach where we deal with only ODE.

Dynamic systems, in general, could include mechanical, electrical, electromechanical, thermodynamics, and fluidic systems. Since the focus of this chapter is more on dynamic response characteristics rather than modeling these systems itself, only a brief overview of dynamic models of mechanical systems is presented. Other systems can be modeled using appropriate first principles. Once these models are obtained, a generalization technique (e.g., state-space representation) can be used to describe these apparently different systems in a standard form to be solved using any applicable technique. This process will be clarified in the following subsections.

Dynamic Models of Mechanical Systems

Mechanical elements are typically classified into three groups: inertia and mass elements, damping and frictional elements, and stiffness elements.

In *inertia and mass elements,* inertia is represented by quantities such as mass and mass moment of inertia. These quantities resist the motion (acceleration to be more specific). Newton's second law in translation or rotation can be utilized to realize this relationship. In other words, the more mass or inertia present, the less acceleration for a given input force or torque. For *linear translation*:

$$\sum (\text{forces}) = \frac{d(\text{linear momentum})}{dt}$$

$$\Rightarrow F = \frac{d}{dt}(mv) = m\frac{d}{dt}v = ma = m\ddot{x}$$

or $\sum (\text{forces}) = (\text{mass})(\text{linear acceleration})$ for constant mass.

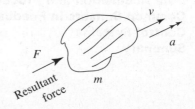

*Jalili, N., *Piezoelectric-based Vibration Control: From Macro to Micro/Nano Scale Systems* (New York: Springer, 2010).

For *linear rotation*:

$$\sum (\text{torque}) = \frac{d(\text{angular momentum})}{dt}$$

$$\Rightarrow T = \frac{d}{dt}(J\omega) = J\frac{d}{dt}\omega = J\alpha = J\ddot{\theta}$$

or $\sum (\text{torque}) = (\text{mass moment of inertia})(\text{angular acceleration})$ for constant inertia.

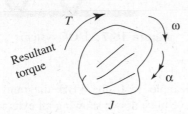

In *damping and frictional elements*: damping or frictional elements are classified as energy dissipative elements associated with some form of friction. There are many types of damping. However, the most common types of damping are viscous or air damping, Coulomb or dry friction, and structural damping. Due to the nonlinear nature of dry friction and complexity of structural damping, we consider only the viscous or air damping type here. This is also in line with our linear assumptions where only linear cause-and-effect elements are considered here.

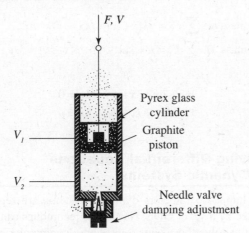

Figure 15.2 A commercial linear air damper

In *viscous damping,* the generated force in this type of damping is proportional to the relative velocities of its two ends and always resistive to the motion.

- Linear translation:

$$F = cV = c(V_2 - V_1) = c(\dot{x}_2 - \dot{x}_1)$$

where c is the linear damping constant and measured in N/m/s or kg/s, and $V = V_2 - V_1$ is the relative velocity.

- Linear rotation:

$$T = c_t\omega = c_t(\omega_2 - \omega_1) = c_t(\dot{\theta}_2 - \dot{\theta}_1)$$

where c_t is the torsional damping constant and measured in N·s.

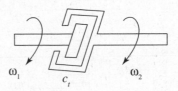

Figure 15.3 Schematic of a rotational (torsional) damper

Notice the direction of damping force, $V_2 - V_1$ or $V_1 - V_2$ ($\omega_2 - \omega_1$ or $\omega_1 - \omega_2$ for the case of rotational elements) is selected such that the damping force is always "resistive" to motion.

In *stiffness elements*, these elements are represented by stress/strain-induced elements when subjected to force and act as a compliance effect.

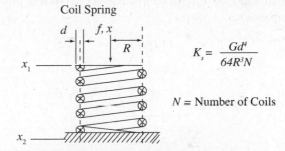

Figure 15.4 A linear stiffness element

- Linear translation:

$$F = k\Delta x = k(x_2 - x_1)$$

where k is the linear spring or stiffness constant and is measured in N/m.

- Linear rotation:

$$T = k_t\Delta\theta = k_t(\theta_2 - \theta_1)$$

where k_t is the torsional spring or stiffness constant and is measured in N·m/rad.

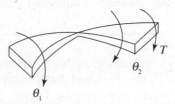

Figure 15.5 Schematic of rotational (torsional) stiffness element

Notice again the direction of spring force or torque ($x_1 - x_2$, $x_2 - x_1$ or in the case of rotation $\theta_1 - \theta_2$ or $\theta_2 - \theta_1$) is determined such that the spring force is always "resistive" to the motion.

Example 15.1

Derive the equation of motion for the following 1 DOF system. The input is the force $f(t)$ applied on the mass, and output is the mass displacement $x(t)$.

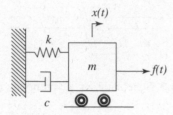

Figure 15.6 1 DOF system

Solution:

Using the FBD shown, Newton's second law for translational motion can be used to arrive at the following equations. Note that the positive acceleration is selected in the same direction as positive displacement. Also note that $\ddot{x}$ must be absolute acceleration in this expression.

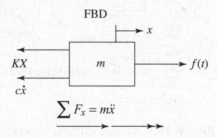

From the FBD, the equation for the balanced forces can be written as:

$$-c\dot{x}(t) - kx(t) + f(t) = m\ddot{x}(t)$$

Hence, the differential equation of motion for mass m can be easily obtained as:

$$m\ddot{x}(t) + c\dot{x}(t) + kx(t) = f(t) \qquad [15.1]$$

where $x(t)$ is the displacement of mass m and is measured from the system at equilibrium (or at rest).

Example 15.2

In many mechanical systems, a small element is used to move a much larger object through proper mechanical coupling. Figure 15.7 depicts this application in which a force f is applied to a small mass m in order to move large mass M. A combination of linear spring and damper in parallel is used to model the elastic coupling between the two masses. Derive the equations of motion governing this system, and identify input, output, and system variables.

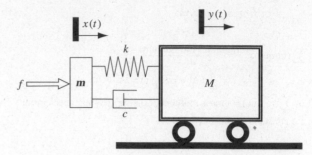

Figure 15.7 2 DOF system

Solution:

As in Example 15.1, the FBD diagram for both masses have to be drawn showing all external forces. As shown in Figure 15.8, the damping and stiffness forces are proportional to the relative velocities and displacements and always resistive to the motion.

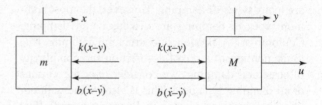

Figure 15.8 Free-body diagram of the 2 DOF system

For mass m: $\sum F_x = -k(x - y) - b(\dot{x} - \dot{y}) = m\ddot{x}$

For mass M: $\sum F_x = u + k(x - y) + b(\dot{x} - \dot{y}) = M\ddot{y}$

or

$$m\ddot{x} + b\dot{x} + kx - b\dot{y} - ky = 0$$
$$M\ddot{y} + b\dot{y} + ky - b\dot{x} - kx = u$$

Solving Differential Equations of Dynamic Systems

All lumped-parameters dynamic systems are represented by a set of ODE. Numerous techniques can be used to solve these equations. Three commonly used approaches are highlighted: the conventional solution of ODE, Laplace transform technique, and state-space representation. The first and third techniques are very briefly discussed here. The second method, the Laplace method, is more elaborated later in this chapter in the system response characteristics sections as well as feedback control sections.

Conventional solution technique for ODE: Simply stated, solving an ODE consists of two parts. The homogenous part is the solution to the initial conditions (IC). The particular solution is the solution to the external input. Example 15.3 demonstrates this technique for the 1 DOF system described in Example 15.1.

Example 15.3

Consider the DOF system of Example 15.1 where a sinusoidal force $f(t) = F_0 \sin(\omega t)$ is applied to mass m, with F_0 being the magnitude of the applied force and ω being its frequency. The differential equation of motion for mass m was easily obtained as:

$$m\ddot{x}(t) + c\dot{x}(t) + kx(t) = F_0 \sin(\omega t)$$

Obtain the solution for $x(t)$, the displacement of mass m, which is measured from the system at equilibrium or at rest.

Solution:

Utilizing the principle of superposition in equation 15.1, one can assume the following solution for displacement $x(t)$:

$$x(t) = x_c(t) + x_p(t) \qquad [15.2]$$

where $x_c(t)$, referred to as the *complimentary solution* or zero-input response, is the solution to the initial conditions (while the right-hand side of equation 15.1 is set to zero). In addition, $x_p(t)$, which is referred to as the *particular solution* or zero-state response, is the solution to the input excitation (here, f).

By considering the general solution equation 15.2, we first concentrate on the complimentary response $x_c(t)$ and assume the following solution:

$$x_c(t) = e^{st} \qquad [15.3]$$

where s is a constant (in general, a complex quantity). Substituting equation 15.3 into the equation of motion shown in equation 15.1, while also noting that the input is set to zero for this solution, yields:

$$ms^2 + cs + k \equiv CE(s) = 0 \qquad [15.4]$$

where we have utilized the fact that $e^{st} \neq 0$. Equation 15.4, or $CE(s) = 0$, is referred to as the *characteristic equation* of the system.

The roots of the characteristic equation $CE(s) = 0$ are obviously functions of the system parameters m, c, and k. Depending on the values of these parameters, three solution types can result. Two distinct real roots can occur (overdamped). Two repeated real roots can occur (critically damped). Two complex conjugate roots can occur (underdamped). We consider only the underdamped case since it is the most common. For this, we divided both sides of equation 15.4 by m to arrive at the following so-called *standard* form of the characteristic equation of 2nd-order systems:

$$s^2 + 2\zeta\omega_n s + \omega_n^2 = 0 \qquad [15.5]$$

where system natural frequency ω_n and damping ratio ζ are defined as:

$$\omega_n = \sqrt{k/m}, \quad \zeta = \frac{c}{c_r} = \frac{c}{2\sqrt{km}} \qquad [15.6]$$

The roots of quadratic equation 15.5 can be easily obtained as:

$$s_{1,2} = -\zeta\omega_n \pm \omega_n\sqrt{\zeta^2 - 1} \qquad [15.7]$$

As mentioned earlier, by restricting the discussion to only underdamped cases, we assume $\zeta < 1$ in order for equation 15.7 to result in complex conjugate roots. In this case, the roots of equation 15.7 can be recast in a more suitable form:

$$s_{1,2} = -\zeta\omega_n \pm j\omega_d, \quad j = \sqrt{-1} \qquad [15.8]$$

where $\omega_d = \omega_n\sqrt{1 - \zeta^2}$ and is referred to as the *damped natural frequency* of the system.

Hence, the general form of complimentary equation 15.3 can be written as:

$$x_c(t) = C_1 e^{s_1 t} + C_2 e^{s_2 t} = C_1 e^{(-\zeta\omega_n + j\omega_d)t} + C_2 e^{(-\zeta\omega_n - j\omega_d)t}$$

$$[15.9]$$

where C_1 and C_2 are constants and can be determined using the initial conditions. After some manipulations and using Euler's identity ($e^{j\theta} = \cos\theta + j\sin\theta$, $j = \sqrt{-1}$), exation 15.9 can be expressed in a more compact form.

$$x_c(t) = \alpha e^{-\zeta\omega_n t}\left(\sin(\omega_d t + \beta)\right) \qquad [15.10]$$

where α and β are constants and can be determined using the available initial conditions.

Now that the complimentary solution has been determined, attention can be focused on the particular solution x_p (or response to input excitation). By using an elementary differential equations background, one can assume the following general solution for x_p to comply with the type of input (or right-hand side) given in equation 15.1:

$$x_p(t) = X\left(\sin(\omega t - \phi)\right) \qquad [15.11]$$

Substituting equation 15.11 into the equation of motion (equation 15.1) and both manipulating and comparing the coefficients of sine and cosine terms on both sides of the resulting equation gives the unknowns X and ϕ as:

$$X = \frac{F_0}{k\sqrt{\left((1 - r^2)^2 + (2\zeta r)^2\right)}}, \quad \phi = \tan^{-1}\frac{2\zeta r}{1 - r^2}$$

$$[15.12]$$

where $r = \omega/\omega_n$ is the frequency ratio. The complete solution can now be obtained by superimposing the equations 15.10 and 15.11 as:

451

$$x(t) = x_c(t) + x_p(t)$$

$$= \alpha e^{-\zeta \omega_n t} \left(\sin(\omega_d t + \beta) \right) + \frac{F_0}{k \sqrt{ \left(\left(1 - r^2 \right)^2 + \left(2\zeta r \right)^2 \right) }}$$

$$\times \sin \left(\omega t - \tan^{-1} \left(\frac{2\zeta r}{1 - r^r} \right) \right)$$

[15.13]

Equation 15.13 represents the complete solution $x(t)$ for the general 1 DOF system.

The *Laplace transform technique for ODE* was briefly reviewed in Section 1.18 of this book. Here, we briefly review the definition of this transformation as well as some key properties that can be used for solving the ODE of motion of dynamic systems.

The underlying concept in this technique is to transfer the OED into a set of algebraic equations that are much easier and more convenient to handle and consequently solve. Once the solution is obtained, it could be transferred back to time-domain using inverse Laplace transform. The Laplace transform is defined as:

$$F(s) = L\left[f(t) \right] = \int_0^\infty f(t) e^{-st} dt$$

where s is a constant (typically complex quantity) and is called the Laplace variable.

The *Laplace key properties* that are useful in solving the ODE of dynamic systems are as follows:

- Superposition:

$$L\left[\alpha f_1(t) + \beta f_2(t) \right] = \alpha F_1(s) + \beta F_2(s)$$

[15.14]

- Differentiation:

$$L\left[\frac{d}{dt} f(t) \right] = sF(s) - f(0),$$

$$L\left[\frac{d^2}{dt^2} f(t) \right] = s^2 F(s) - sf(0) - \dot{f}(0)$$

[15.15]

- Integration:

$$L\left[\int f(t)\, dt \right] = \frac{F(s)}{s} - \frac{\left[\int f(t)\, dt \right]_{t=0}}{s}$$

[15.16]

- Final value theorem:

$$\lim_{t \to \infty} f(t) = \lim_{s \to 0} sF(s)$$

[15.17]

- Initial value theorem:

$$\lim_{s \to \infty} sF(s) = f(0^+)$$

[15.18]

Example 15.4 shows this technique for the same 1 DOF problem of Example 15.3.

Example 15.4

Revisit Example 15.13, and obtain the displacement of mass m, $x(t)$, in response to a unit-step function, $F(t)$.

Solution:

The equation of motion was obtained in Example 15.1:

$$m\ddot{x}(t) + c\dot{x}(t) + kx(t) = F \qquad [15.19]$$

For simplicity in the derivations, the following simple change of variable is made: $F(t) = kf(t)$. With this, the equation of motion can be put into the following standard form (as shown in Example 15.3):

$$\ddot{x}(t) + 2\zeta \omega_n \dot{x}(t) + \omega_n^2 x(t) = \frac{k}{m} f(t) = \omega_n^2 f(t)$$

[15.20]

By taking the Laplace transform of equation 15.20 and assuming a unit-step function for $f(t)$, with its Laplace being $F(s) = 1/s$, we obtain:

$$\left(s^2 + 2\zeta \omega_n s + \omega_n^2 \right) X(s) = \omega_n^2 / s$$

$$\Rightarrow X(s) = \frac{\omega_n^2}{s \left(s^2 + 2\zeta \omega_n s + \omega_n^2 \right)}$$

[15.21]

Note that for $0 < \xi < 1$, we can rewrite part of the denominator as follows:

$$s^2 + 2\zeta \omega_n s + \omega_n^2 = \left(s + \zeta \omega_n \right)^2 - \left(\zeta \omega_n \right)^2 + \omega_n^2$$

$$= \left(s + \zeta \omega_n \right)^2 + \omega_n^2 \left(1 - \zeta^2 \right)$$

$$= \left(s + \zeta \omega_n \right)^2 + \omega_d^2$$

[15.22]

where $\omega_d = \omega_n \sqrt{1 - \zeta^2}$ was defined earlier as damped natural frequency. Hence, equation 15.21 can be written as:

$$X(s) = \frac{\omega_n^2}{s \left(s^2 + 2\zeta \omega_n s + \omega_n^2 \right)} = \frac{\omega_n^2}{s \left(\left(s + \zeta \omega_n \right)^2 + \omega_d^2 \right)}$$

[15.23]

By using partial-fraction expansion (PFE), equation 15.23 reduces to:

$$X(s) = \frac{\omega_n^2}{s \left(s^2 + 2\zeta \omega_n s + \omega_n^2 \right)} = \frac{A}{s} + \frac{Bs + C}{\left(s + \zeta \omega_n \right)^2 + \omega_d^2}$$

[15.24]

By taking the common denominator and equalizing the numerators on both sides, we get:

$$\omega_n^2 = (A+B)s^2 + \left(2\zeta\omega_n A + C\right)s + A\omega_n^2$$

[15.25]

Hence,

$$\begin{cases} A\omega_n^2 = \omega_n^2 \\ A+B=0 \\ 2\zeta\omega_n A + C = 0 \end{cases} \Rightarrow \begin{cases} A=1 \\ B=-A=-1 \\ C=-2\zeta\omega_n A = -2\zeta\omega_n \end{cases}$$

[15.26]

Equation 15.23 can now be written as:

$$X(s) = \frac{\omega_n^2}{s\left(s^2 + 2\zeta\omega_n s + \omega_n^2\right)} = \frac{1}{s} - \frac{s+2\zeta\omega_n}{\left(s+\zeta\omega_n\right)^2 + \omega_d^2}$$

[15.27]

Equation 15.27 can now be used to go to any Laplace table and take the inverse Laplace of both sides. Note the following Laplace inverse equation from the table*:

$$\mathfrak{I}^{-1}\left\{\frac{Bs+C}{\left(s+a\right)^2 + \omega^2}\right\} = e^{-at}\left(B\cos(\omega t) + \frac{C-aB}{\omega}\sin(\omega t)\right)$$

[15.28]

By using this property, equation 15.27 can now be simplified as:

$$x(t) = \mathfrak{I}^{-1}\left\{X(s)\right\} = \mathfrak{I}^{-1}\left\{\frac{1}{s}\right\} - \mathfrak{I}^{-1}\left\{\frac{s+2\zeta\omega_n}{\left(s+\zeta\omega_n\right)^2 + \omega_d^2}\right\}$$

$$= 1 - e^{-\zeta\omega_n t}\left(\cos(\omega_d t) + \frac{2\zeta\omega_n - \zeta\omega_n}{\omega_d}\sin(\omega_d t)\right)$$

[15.29]

Simplifying $\dfrac{\zeta\omega_n}{\omega_d} = \dfrac{\zeta\omega_n}{\omega_n\sqrt{1-\xi^2}} = \dfrac{\zeta}{\sqrt{1-\xi^2}}$, the final displacement response, can be written as:

$$x(t) = 1 - e^{-\zeta\omega_n t}\left(\cos(\omega_d t) + \frac{\zeta}{\sqrt{1-\xi^2}}\sin(\omega_d t)\right)$$

[15.30]

which matches the result obtained from Example 15.3. One could plot the response as shown below.

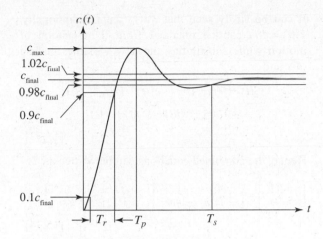

Figure 15.9 Response of a 2nd-order system to a step input

In the *state-space technique for ODE* one could represent the differential equations of motions as a set of 1st-order equations. These equations could be solved using either Laplace or standard state-space solvers (e.g., Matlab ODE solvers). This is a very powerful method and forms the basis of most modern control techniques, such as pole placement and observer techniques. This technique is briefly reviewed using the same case study as in Example 15.3. The underlying concept is to assign proper new variables to each of the states up to the order of each ODE. For example, for a 2 DOF system, we need a total of 4 new states, 2 for each of the DOF.

Example 15.5

Represent the governing equation of motion of Example 15.1 as a state-space representation in which $x(t)$ is the displacement and is considered to be the desired output of the system and $f(t)$ is the input force.

Solutions:

As shown previously, the differential equation of motion for mass m is

$$m\ddot{y}(t) + c\dot{y}(t) + ky(t) = f(t) \qquad [15.31]$$

where $y(t)$ is the displacement of mass m and is measured from system equilibrium state (or rest). Define the following new variables. Two new variables are needed as the order of the ODE is 2.

$$x_1(t) = y(t), \; x_2(t) = \dot{y}(t) \qquad [15.32]$$

*See, for example, the *Mathematical Handbook of Formulas and Tables* by M. S. Spiegel, in Schaum's Outline Series (McGraw-Hill, 1968 or later edition)

It can be easily seen that $\dot{x}_1(t) = x_2(t)$. Additionally, $\dot{x}_2(t) = \ddot{y}(t)$ can be obtained from the equation of motion while substituting the new variables instead of x and $\dot{x}$:

$$\dot{x}_2(t) = \ddot{y}(t) = 1/m\left\{-c\dot{y}(t) - ky(t) + f(t)\right\}$$
$$= 1/m\left\{-cx_2(t) - kx_1(t) + f(t)\right\}$$

[15.33]

Hence, the combined equations can be written as:

$$\left\{\begin{array}{c} \dot{x}_1(t) \\ \dot{x}_2(t) \end{array}\right\} = \left[\begin{array}{cc} 0 & 1 \\ -k/m & -c/m \end{array}\right]\left\{\begin{array}{c} x_1(t) \\ x_2(t) \end{array}\right\} + \left[\begin{array}{c} 0 \\ 1/m \end{array}\right]f(t)$$

[15.34]

By defining vector $\mathbf{x}(t) = [x_1(t)\ x_2(t)]^T$, one could rewrite equation 15.34 in the following more compact form:

$$\dot{\mathbf{x}}(t) = \mathbf{A}\mathbf{x}(t) + \mathbf{B}u(t)$$
$$\text{where } \mathbf{A} = \left[\begin{array}{cc} 0 & 1 \\ -k/m & -c/m \end{array}\right],$$
$$\mathbf{B} = \left[\begin{array}{c} 0 \\ 1/m \end{array}\right] \text{ and } u(t) = f(t)$$

[15.35]

It must be noted that $\mathbf{u}(t)$ is, in general, an R by 1 input vector (R inputs). However, for this single-input problem, $u(t)$ is scalar. The state-space equation is typically augmented with a so-called output equation in terms of state vector as well as input. This equation is given in general as:

$$y(t) = \mathbf{C}\mathbf{x}(t) + \mathbf{D}u(t)$$

where $\mathbf{C}$ and $\mathbf{D}$ are constant matrices to be determined such that the desired output is achieved. For example, for this problem, the desired output is the displacement $y(t)$, i.e., $y(t) = x(t)$. These matrices can be defined as:

$$\mathbf{C} = [\ 1\quad 0\ ],\ \mathbf{D} = 0$$
$$\Rightarrow y(t) = [\ 1\quad 0\ ]\left[\begin{array}{c} x_1 \\ x_2 \end{array}\right] + 0 \times u(t)$$
$$= x_1(t) = y(t)$$

[15.36]

In summary, the state-space representation can be given as follows:

$$\dot{\mathbf{x}}(t) = \mathbf{A}\mathbf{x}(t) + \mathbf{B}u(t)$$
$$y(t) = \mathbf{C}\mathbf{x}(t) + \mathbf{D}u(t)$$

[15.37]

where $\dot{\mathbf{x}}(t)$ is an N by 1 state vector (N state variables), $\mathbf{u}(t)$ is an R by 1 input vector (R inputs), $\mathbf{y}(t)$ is an M by 1 output vector (M outputs), $\mathbf{A}$ is an N by N system matrix, $\mathbf{B}$ is an N by R input distribution matrix, $\mathbf{C}$ is

an M by N output matrix, and finally $\mathbf{D}$ is an M by R feed-through matrix.

In the *state-space to transfer function,* the state-space representation of the dynamic system can be easily converted to the transfer function. This conversion can be numerically done using, for example, the "ss2tf" command in Matlab.

Start with the state-space equation:

$$\dot{\mathbf{x}}(t) = \mathbf{A}\mathbf{x}(t) + \mathbf{B}u(t)$$
$$y(t) = \mathbf{C}\mathbf{x}(t) + \mathbf{D}u(t)$$

[15.38]

One could take the Laplace transform of these equations as:

$$s\mathbf{X}(s) - \mathbf{x}(0) = \mathbf{A}\mathbf{X}(s) + \mathbf{B}\mathbf{U}(s)$$
$$\mathbf{Y}(s) = \mathbf{C}\mathbf{X}(s) + \mathbf{D}\mathbf{U}(s)$$

[15.39]

or

$$(s\mathbf{I} - \mathbf{A})\mathbf{X}(s) = \mathbf{B}\mathbf{U}(s) + \mathbf{x}(0)$$
$$\Rightarrow \mathbf{X}(s) = (s\mathbf{I} - \mathbf{A})^{-1}\mathbf{B}\mathbf{U}(s) + (s\mathbf{I} - \mathbf{A})^{-1}\mathbf{x}(0)$$

[15.40]

By substituting $\mathbf{X}(s)$ from equation 15.40 into equation 15.39, one could obtain the expression for the Laplace of system output as:

$$\mathbf{Y}(s) = \left[\mathbf{C}(s\mathbf{I} - \mathbf{A})^{-1}\mathbf{B} + \mathbf{D}\right]\mathbf{U}(s) + \mathbf{C}(s\mathbf{I} - \mathbf{A})^{-1}\mathbf{x}(0)$$

[15.41]

The first term in equation 15.41 is referred to as the response to the input. (Hence, the transfer function of the system can be obtained.) The second term is the response to the IC. By assuming zero IC, one can obtain the transfer function of the system for the case of single-input single-output (SISO) system as:

$$\frac{Y(s)}{U(s)} = \mathbf{C}(s\mathbf{I} - \mathbf{A})^{-1}\mathbf{B} + \mathbf{D}$$
$$\text{where } (s\mathbf{I} - \mathbf{A})^{-1} = \frac{\text{adjoint }(s\mathbf{I} - \mathbf{A})}{\det(s\mathbf{I} - \mathbf{A})}$$

[15.42]

From a mathematical background, adjoint $(s\mathbf{I} - \mathbf{A})$ is defined as $[(-1)^{i+j}\det(M_{i,j})]^T$, where $M_{i,j}$ is $s\mathbf{I} - \mathbf{A}$ with the ith and jth columns removed. For a single-input single-output system, this can simplify equation 15.42 to the following expression:

$$\frac{Y(s)}{U(s)} = \mathbf{C}(s\mathbf{I} - \mathbf{A})^{-1}\mathbf{B} + \mathbf{D} = \frac{\det\left[\begin{array}{cc} s\mathbf{I} - \mathbf{A} & \mathbf{B} \\ -\mathbf{C} & \mathbf{D} \end{array}\right]}{\det[s\mathbf{I} - \mathbf{A}]}$$

[15.43]

It is obvious from the above relationship that the characteristic equation for the system (the denominator of

$TF = 0$) is $\det[s\mathbf{I} - \mathbf{A}]$. That is, the eigenvalues of $\mathbf{A}$ are the characteristic equation roots (or poles).

Example 15.6

Given the state-space representation of Example 15.5, transfer this representation into Laplace form and obtain the transfer function of the system.

Solution:
As shown previously:

$$\frac{Y(s)}{U(s)} = \mathbf{C}(s\mathbf{I} - \mathbf{A})^{-1}\mathbf{B} + \mathbf{D}$$

$$= \frac{\det \begin{bmatrix} s\mathbf{I} - \mathbf{A} & \mathbf{B} \\ -\mathbf{C} & \mathbf{D} \end{bmatrix}}{\det[s\mathbf{I} - \mathbf{A}]}$$

$$= \frac{\det \begin{bmatrix} s & -1 & 0 \\ k/m & s+c/m & 1/m \\ -1 & 0 & 0 \end{bmatrix}}{\det \begin{bmatrix} s & -1 \\ k/m & s+c/m \end{bmatrix}}$$

$$= \frac{1/m}{s^2 + sc/m + k/m}$$

$$= \frac{1}{ms^2 + cs + k}$$

This would be the same result if one takes the Laplace transform of equation 15.1 under zero IC.

15.2 Systems Response Characteristics

The response of a dynamic system depends on the order of the system. Typically it can be qualitatively estimated using only this single parameter.

Order of a Dynamic System

The *order* of a dynamic system is defined as the difference between the highest and lowest derivatives (with respect to time) of the variable of interest in the differential equation of motion.

Example 15.7

What is the order of the following dynamic systems?

A. $m\ddot{x}(t) + c\dot{x}(t) + kx(t) = f(t)$

B. $L\dfrac{di(t)}{dt} + Ri(t) = v_a(t)$

C. $a\ddot{x}(t) + b\dot{x}(t) = f(t)$

Solution:

A. The variable of interest is obviously $x(t)$, where the differential equation of motion is written. The highest and lowest derivatives on this variable are 2 and 0, respectively. Hence, the order is $n = 2 - 0 = 2$. The system is a 2nd-order system.

B. The variable of interest is obviously $i(t)$ where the differential equation of motion is written. The highest and lowest derivatives of this variable are 1 and 0, respectively. Hence, the order is $n = 1 - 0 = 1$. The system is a 1st-order system.

C. The variable of interest is obviously $x(t)$, where the differential equation of motion is written. The highest and lowest derivatives of this variable are 2 and 1, respectively. Hence, the order is $n = 2 - 1 = 1$. The system is a 1st-order system. Although the highest-order derivative is 2, the system is 1st order since the variable $x(t)$ itself is missing. One could make a change of variable $u(t) = \dot{x}(t)$ and rewrite the equation $a\dot{u}(t) + bu(t) = f(t)$, which results in a 1st-order system as in part B.

1st-Order Dynamic Systems Response

A 1st-order dynamic system can be represented in the following general form:

$$\dot{y} + \frac{1}{\tau}y = f(t) \qquad [15.44]$$

where $f(t)$ is the input, $y(t)$ is the output, and constant τ is referred to as the time constant of the system and is a representation of how fast the system behaves. Taking the Laplace of this equation results in:

$$sY(s) - y(0) + 1/\tau Y(s) = F(s)$$

$$\Rightarrow Y(s) = \frac{y(0)}{s + 1/\tau} + \frac{F(s)}{s + 1/\tau}$$

$$[15.45]$$

The first term in the expression, $Y(s)$, is referred to as the response to IC or the zero-input response that forms the basis for internal stability. The second term refers to the response to the external input or zero-state response that forms the basis for external stability. These stability concepts are further explained in the following subsections.

By considering the input to be in the form of a step input with amplitude A (hence $F(s) = A/s$), one can write equation 15.45 as:

$$Y(s) = \frac{y(0)}{s + 1/\tau} + \frac{A}{s(s + 1/\tau)} \qquad [15.46]$$

455

The second term needs to be partitioned into simple fractions so inverse Laplace can be performed:

$$\frac{A}{s(s+1/\tau)} = \frac{a_1}{s} + \frac{a_2}{s+1/\tau} \qquad [15.47]$$

Variables a_1 and a_2 can be obtained by taking the common denominator and equating the numerator similar to the process performed in Example 15.4. With this, one could easily obtain a_1 and a_2 as $a_1 = \tau A$, $a_2 = -\tau A$. By taking the Laplace inverse of both sides of equation 15.47, we have:

$$\begin{aligned}
y(t) &= \Im^{-1}\{Y(s)\} \\
&= \Im^{-1}\left\{\frac{y(0)}{s+1/\tau}\right\} + \Im^{-1}\left\{\frac{\tau A}{s}\right\} + \Im^{-1}\left\{\frac{-\tau A}{s+1/\tau}\right\} \\
&= y(0)e^{-t/\tau} + \tau A - \tau A e^{-t/\tau}
\end{aligned}$$

Equation 15.47 can be simplified by taking the zero initial condition (i.e., $y(0) = 0$) as:

$$y(t) = \tau A\left(1 - e^{-t/\tau}\right)$$

Observe the following properties for the response:

For $(t = \tau)$, $y(\tau) = 0.632\,\tau A$

For $(t = \infty)$, $y(\infty) = \tau A \equiv y_{ss}$

Hence, $y(\tau) = 63.2\%\ y_{ss}$ $\qquad [15.48]$

$\frac{d}{dt}y(t) = Ae^{-t/\tau}$, i.e., $\dot{y}(0) = A = \tau A/A = y_{ss}/\tau$

Based on the above observations, the general response of a 1st-order dynamic system can be depicted as shown in Figure 15.10. The settling time T_s is estimated to be 4 times the time constant, τ. In other words, $T_s = 4\tau$. The rise time T_r is also defined as the time for the response to reach 90% of its final value from 10%.

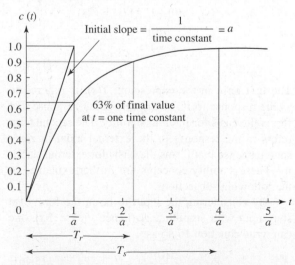

Figure 15.10 Response of a 1st-order system to a unit-step input

Example 15.8

Consider the general response of a 1st-order system $\dot{c}(t) + ac(t) = r(t)$ to unit-step input as shown in Figure 15.11 ($\tau = 1/a$ is called the *time constant*). Find an equation that relates settling time of the velocity of the mass to M.

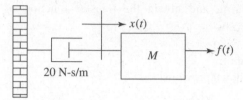

Figure 15.11 Schematic of 1 DOF

Solution:

The equation of motion can be written as (using a simple FBD):

$$M\ddot{x}(t) + c\dot{x}(t) = f(t)$$

Notice that this is a 1st-order system $(2 - 1 = 1)$. Take the new variable $\dot{x}(t) = v(t)$, and rewrite the equation as:

$$M\dot{v}(t) + cv(t) = f(t) \;\Rightarrow\; \dot{v}(t) + \frac{c}{M}v(t) = \frac{1}{M}f(t)$$
$$[15.49]$$

Comparing equation 15.49 with the standard form shown in equation 15.44 results in:

$$\tau = \frac{1}{c/M} = \frac{M}{c} = \frac{M}{20}, \quad T_s = 4\tau = 4\frac{M}{20} = \frac{M}{5}$$

2nd-Order Dynamic Systems Response

As shown in the preceding subsection, a 2nd-order dynamic system can be represented in the following general form:

$$m\ddot{x}(t) + c\dot{x}(t) + kx(t) = kf(t); \text{ where } f(t) = u(t)$$
$$[15.50]$$

It can also be shown in standard form as:

$$\ddot{x}(t) + 2\zeta\omega_n\dot{x}(t) + \omega_n^2 x(t) = \omega_n^2 u(t) \quad [15.51]$$

where

$$2\zeta\omega_n = \frac{c}{m} \text{ and } \omega_n^2 = \frac{k}{m}$$

The response to a unit-step input was shown, in Example 15.4, to be in the following form:

$$x(t) = 1 - e^{-\zeta\omega_n t}\left(\cos\omega_d t + \frac{\zeta}{\sqrt{1-\zeta^2}}\sin\omega_d t\right);$$

$$\text{where } \omega_d = \omega_n\sqrt{1-\zeta^2} \qquad [15.52]$$

This general response was depicted as shown in Figure 15.9 in the preceding section.

The properties of 2nd-order system response can be summarized based on peak time, overshoot, and settling time.

Peak time (t_p) is the time to reach the first peak or maximum value. To obtain this, one could take the derivative of the response and set it equal to zero. That is, at $t = t_p$, we have $\dot{x}(t = t_p) = 0$. By taking the derivative of $x(t)$ and setting it equal to zero, one could find the first nontrivial time as $t_p = \dfrac{\pi}{\omega_d}$.

Overshoot (M_p) is defined as the difference between the maximum of the response and its steady-state value:

$$M_p \equiv \left\{ x(t = t_p) \right\} - \left\{ x(t \to \infty) \right\}$$

$$= \left\{ 1 - e^{-\zeta \omega_n t_p} \left(\cos \omega_d t_p + \frac{\zeta}{\sqrt{1 - \zeta^2}} \sin \omega_d t_p \right) \right\} - \{1\}$$

$$= e^{-\zeta \omega_n \pi / \omega_d} = \exp\left(\frac{-\pi \zeta}{\sqrt{1 - \zeta^2}} \right)$$

$$[15.53]$$

It is interesting to note that the overshoot is only a function of the damping ratio. The percentage of overshoot ($\% M_p = M_p / x(t \to \infty)$) could be plotted as a function of the damping ratio as shown in Figure 15.12.

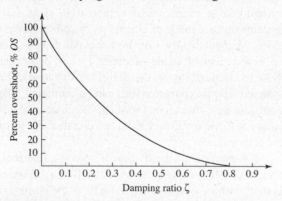

Figure 15.12 Percent overshoot vs. damping ratio (when $\xi = 0$, $M_p = 1$, or $\% M_p = 100\%$ and when $\xi = 1$, $M_p = 0$, or $\% M_p = 0\%$)

Settling time (t_s or T_S) is defined as the time for the response to reach its steady-state value within a given range. This range is typically $\pm 1\%$, $\pm 2\%$, or $\pm 5\%$. Since the dominant portion of the response is the exponential function, this time can be calculated by simply equating this portion of the response with the desired range. For example, for a $\pm 1\%$ criterion, one could write:

$$e^{-\xi \omega_n t_s} = 0.01 \implies \zeta \omega_n t_s = \ln(0.01) \implies t_s = \frac{4.6}{\zeta \omega_n}$$

$$[15.54]$$

By inspecting the equation for the characteristic roots of a 2nd-order system (see equation 15.8), one could realize that the real part of these roots is nothing other than $\sigma = \xi \omega_n$. Hence, the settling time t_s can be represented as:

$$t_s = \frac{4.6}{\sigma} \text{ based on } \pm 1\% \text{ criterion} \quad [15.55]$$

Similarly, for $\pm 2\%$ and $\pm 5\%$ criteria, this exercise can be repeated to arrive at the following settling times as:

$$t_s = \frac{4.0}{\sigma} \text{ based on } \pm 2\% \text{ criterion;}$$

$$t_s = \frac{3.0}{\sigma} \text{ based on } \pm 5\% \text{ criterion} \quad [15.56]$$

Based on this, one could conclude that the closer the (dominant) root is to the imaginary axis (i.e., the smaller the value of σ), the larger the settling time is. The following example further highlights this fact.

Example 15.9

Consider the following systems whose *pole-zero* maps are displayed. Put these systems in order from smallest to largest *settling time*. Explain your order.

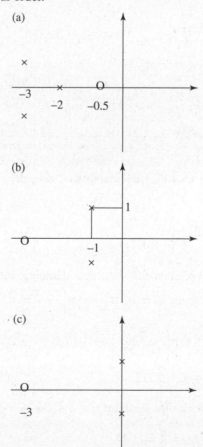

(**Note:** "x" is used to indicate a pole, while "o" is used for a zero.)

Solution:

As stated, the closer to the imaginary axis, the larger the settling time is, i.e., $t_s \propto 1/\sigma$ or $t_s = k/\sigma$ where k is a constant depending on the criterion used. Hence, we have:

$$\begin{cases} a)\ \sigma = 2 \Rightarrow t_s = k/2 \\ b)\ \sigma = 1 \Rightarrow t_s = k \\ c)\ \sigma = 0 \Rightarrow t_s = k/0 = \infty \end{cases} \rightarrow t_s^{(a)} < t_s^{(b)} < t_s^{(c)}$$

Example 15.10

Suppose you obtained the response plot of a 2nd-order system shown in Figure 15.13 either from a measured response or from a computer simulation. The steady-state value is $x_{ss} = 100$ mm. Roughly determine:

A. The percent overshoot and the damping ratio.

B. The settling time for ±5% tolerance on the steady-state error.

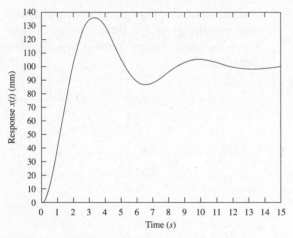

Figure 15.13 Underdamped 2nd-order system response

Solution:

A. $M_p = \dfrac{1.5 - 1.0}{1.0} \times 100 = 50\%$ or 0.5

The relationship between damping ratio and overshoot is also $M_p = \exp\left(\dfrac{-\pi\zeta}{\sqrt{1-\zeta^2}}\right)$. Hence,

$$0.5 = \exp\left(\frac{-\pi\zeta}{\sqrt{1-\zeta^2}}\right) \Rightarrow \ln(0.5) = \frac{-\pi\zeta}{\sqrt{1-\zeta^2}}$$

$$\Rightarrow \zeta = 0.212.$$

B. The peak time is about $t_p \cong 3$ s. So $t_p = \dfrac{\pi}{\omega_d}$ and

$$t_p = \omega_d = \frac{\pi}{3} = 1.047.$$

$$\omega_d = \omega_n\sqrt{1-\xi^2} \Rightarrow \omega_n = 1.047/\sqrt{1-\xi^2}$$

For the ±5% criterion,

$$t_s = \frac{3.0}{\sigma} = \frac{3.0}{\xi\omega_n} = \frac{3.0}{\xi\omega_d/\sqrt{1-\xi^2}}$$

or $t_s = \dfrac{3.0\sqrt{1-0.212^2}}{0.212(1.047)} = 17.5$ s.

To locate the characteristic roots of a 2nd-order system whose standard form is $s^2 + 2\xi\omega_n s + \omega_n^2 = 0$, one could easily obtain the relationships shown in Figure 15.14.

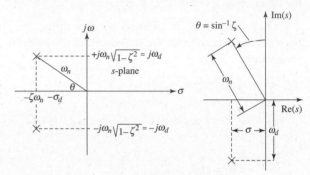

Figure 15.14 Pole location of an underdamped 2nd-order system

Higher-Order Dynamic Systems Response

The response characteristics of a higher-order dynamic system can be examined by composition of its subsystems that are made of either 1st- or 2nd-order systems. Mathematically, any higher-order system can be assumed to be made of either 1st- or 2nd-order systems. Depending on the dominant dynamics, the system response characteristics can be estimated. The dominant pole (the closest roots of the characteristics equation to the imaginary axis) can dictate the type of the response.

For example, if the dominant pole is a real root, then the system behaves dominantly like a 1st-order system with no overshoot. Similarly, if the dominant root is a complex root, then the system behaves like a 2nd-order system. Note that the overall response depends on the other roots as well and how close they are to the dominant root. For most cases, though, the behavior of the system can be clearly captured and seen from the dominant pole. Figures 15.15 and 15.16 demonstrate this concept as the response characteristics become functions of where the dominant roots are (on the complex plan) and what types (i.e., real, complex, repeated) they are.

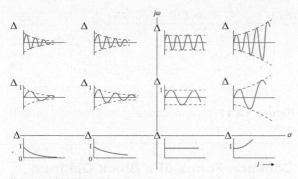

Figure 15.15 Typical response of a dynamic system as a function of the location and type of dominant characteristic poles

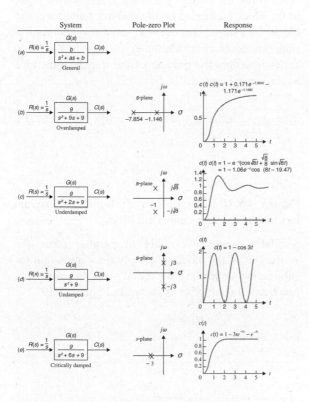

Figure 15.16 Typical response of a 2nd-order system as a function of damping ratio

Example 15.11

The characteristic roots of a dynamic system are

$$-1.7920 - j1.8160, \quad 1.7920 + j1.8160, \quad -0.4160$$

What is the order of this system? What is the settling time and damping ratio of the system?

Solution:

These poles can be plotted in the complex plan as shown below.

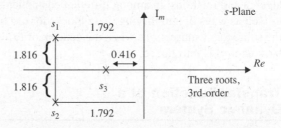

There are three poles, so the order of the system is 3. As seen from the figure, the dominant pole is at $s_3 = -0.416$. Hence, $t_s = 4.6/\sigma = 4.6/0.416 = 11.06$ s based on the ±1% criterion. Since the dominant pole is real, the damping ratio corresponding to this root will be 1 or greater than 1. Since the system is dominated by the dominant pole, it is expected that the settling time and damping ratio obtained for the dominant pole will be for the system. A numerical result (e.g., Matlab) for the step-input can show this concept.

Example 15.12

Repeat Example 15.11 when the characteristic roots of a dynamic system are given as:

$$-1.7920 - j1.8160, -1.7920 + j1.8160, -2.5$$

Solution:

In this case, the dominant pole will be $s_{1,2} = -1.7920 \pm j1.816$. Comparing this with the standard form of a complex pole of a 2nd-order system $s_{1,2} = -\xi\omega_n \pm j\omega_d$ or $s_{1,2} = -\sigma \pm j\omega_d$ reveals $\sigma = 1.7920$ and $\omega_d = 1.816$. Hence, $t_s = 4.6/\sigma = 4.6/1.7920 = 2.57$ s based on the ±1% criterion. Using Figure 15.14, one could find angle θ as $\theta = \tan^{-1}\left(\dfrac{\sigma}{\omega_d}\right) = \tan^{-1}\left(\dfrac{1.792}{1.816}\right) = 44.62°$. Hence, $\xi = \sin(\theta) = \sin(44.62°) = 0.70$.

459

15.3
Block Diagram Algebra

One method of graphically representing the relationship between system output and input is through a block diagram. In this method, the block represents a transfer function corresponding to a system's mathematical model, and the arrows represent signals (e.g., electrical voltage from a position sensor). This subsection presents a quick overview of the block diagram algebra and relationship among different components as well as ways of simplifying these blocks for use in the subsequent subsections when the concept of feedback control is introduced.

Transfer Function of a Dynamic System

Transfer function of a linear system is defined as the ratio of the Laplace transform of the output variable to the Laplace transform of the input variable under zero initial conditions (IC):

$$T(s) = \frac{\Im(\text{output})}{\Im(\text{input})}\bigg|_{\text{when all IC are set to zero}}$$

$$[15.57]$$

Let $X(s)$ be the Laplace transform of the input variable, $Y(s)$ be that of the output variable, and $G(s)$ be the transfer function. Then the relation $Y(s) = G(s) X(s)$ can be graphically shown as:

R(s) → C(s) →

Signals
(a)

R(s) → Input $G(s)$ → C(s) Output

System
(b)

Basic Components of a Block Diagram

Many systems are composed of multiple subsystems. When multiple subsystems are interconnected, a few more elements need to be added to the block diagram: *summing junctions* and *pickoff points*. The summing junction implies that the output signal is the algebraic sum of all its input signals. A pickoff point distributes the input signals to several output points. Figure 15.17 depicts examples of both summing junctions and pickoff points.

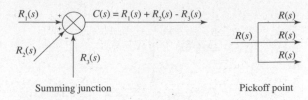

Summing junction   Pickoff point

Figure 15.17 Typical summing junction and pickoff point

Common Forms of a Block Diagram

There are three common forms in which subsystems are connected together in a block diagram: cascade (serial) form, parallel form, and feedback form. For serial form, the output of a block is simply the input of the block multiplied by the transfer function of the block. By using this concept in serial form configuration, one could conclude the equivalent of blocks in series is simply the product of the transfer functions as shown in Figure 15.18.

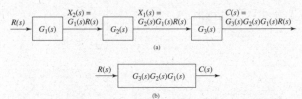

Figure 15.18 Cascade (serial) form: (a) original blocks and (b) their equivalent block

Similarly, when blocks are in parallel form, the equivalent of the blocks is simply the summation (or subtraction depending on the sign of the summing junction) of the blocks as shown in Figure 15.19.

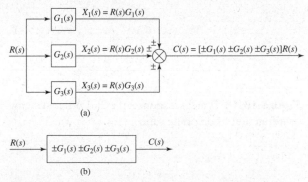

Figure 15.19 Parallel form: (a) original blocks and (b) their equivalent block

For the standard feedback form shown in Figure 15.20, let the Laplace error $e(t)$ be $E(s)$. Then from Figure 15.20 we have $E(s) = R(s) - C(s)H(s)$. Substituting this in the equation $C(s) = E(s)G(s)$ gives $C(s) = R(s)G(s) - C(s)H(s)G(s)$. Solving for $C(s) / R(s)$ from the above equation yields:

$$\frac{C(s)}{R(s)} = \frac{G(s)}{1 + G(s)H(s)} \equiv G_c(s) \qquad [15.58]$$

This is the equivalent transfer function of the feedback system. So the system can be represented by using the following block diagram.

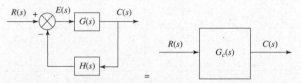

Figure 15.20 Feedback form: (left) original blocks and (right) their equivalent block

Example 15.13

Using the forms discussed so far, reduce the following multiple subsystems into one system.

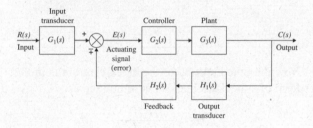

Solution:
It is clear that G_1 and G_2 are in series as are H_1 and H_2. Hence, the block diagram can be simplified as:

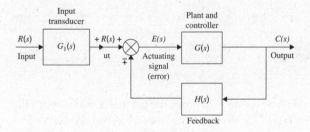

where $G(s) = G_1(s)G_2(s)$ and $H(s) = H_1(s)H_2(s)$. The above diagram represents a simple feedback form that can be further simplified as:

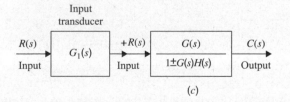

(c)

Finally, it can be simplified as:

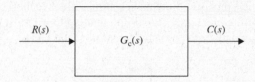

where

$$G_c(s) = \frac{C(s)}{R(s)} = \frac{G_1(s)G(s)}{1 \pm G(s)H(s)} = \frac{G_1(s)G_2(s)G_3(s)}{1 \pm G_2(s)G_3(s)H_1(s)H_2(s)}$$

Block Diagram Algebra

If the block diagram is not in any of the above forms (either common or basic forms), one could move some of the blocks around to convert the block diagram into one of these forms. This can be done using the following four simple rules.

1. **Moving forward a summing point that is behind a block:** As shown in Figure 15.21, when moving a summing point behind a block, the signal line that is moved must be multiplied by the transfer function of the block.

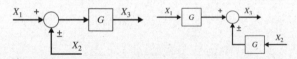

Figure 15.21 Moving forward a summing point: (left) original configuration and (right) its equivalent block

2. **Moving back a pickoff point that is in the front of a block:** As shown in Figure 15.22, when moving a pickoff point that is in the front of a block back to the front of the block, the signal line that is moved must be multiplied by the transfer function of the block.

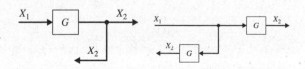

Figure 15.22 Moving back a pickoff point: (left) original configuration and (right) its equivalent block

3. **Moving forward a pickoff point that is in behind the block:** As shown in Figure 15.23, when moving a pickoff point that is behind a block to the front of the block, the signal line that is moved must be divided by the transfer function of the block.

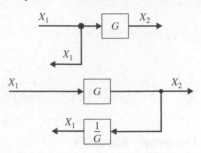

Figure 15.23 Moving forward a pickoff point: (left) original configuration and (right) its equivalent block

4. **Moving back a summing point that is in the front of a block:** As shown in Figure 15.24, when moving a summing point that is in the front of a block behind the back, the signal line that is moved must be divided by the transfer function of the block.

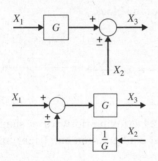

Figure 15.24 Moving back a summing point: (left) original configuration and (right) its equivalent block

Example 15.14

Using the forms discussed so far, simplify the following block diagram into one system.

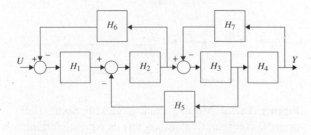

Solution:

The following steps demonstrate the simplification process used here. The steps are self-explanatory. The

first step is to move the pickoff point between the H_3 and H_4 blocks to the front of H_4. The line that carried this signal must be divided by the block TF, H_4 (see Figure 15.23).

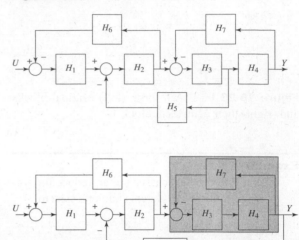

Now the tinted portion of the block diagram is in the form of a feedback and can be simplified as:

$$H_8 = \frac{H_4 H_3}{1 + H_7 H_4 H_3}$$

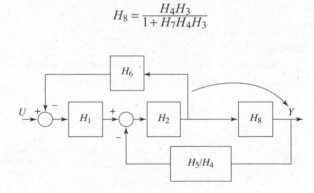

The next step is to move the pickoff point between the H_2 and H_8 blocks to the front of H_8, similar to the first step. The line that carried this signal must be divided by the block TF, H_8 (see Figure 15.23).

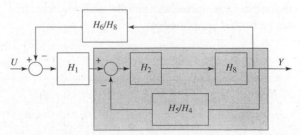

Again the tinted portion of the diagram is in the form of a feedback and can be simplified as:

$$H_9 = \frac{H_2 H_8}{1 + H_2 H_8 H_5 / H_4} = \frac{H_4 H_2 H_8}{H_4 + H_2 H_8 H_5}$$

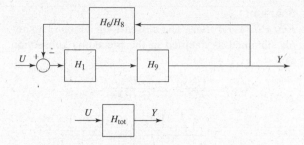

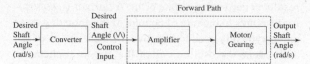

Figure 15.25 Open-loop or feedforward system

Finally, the last form of feedback can be simplified as:

$$H_{tot} = \frac{H_1 H_9}{1 + H_1 H_9 H_6 / H_8} = \frac{H_1 H_8 H_9}{H_8 + H_1 H_9 H_6}$$

One could replace all the equivalent transfer functions H_8 and H_9 from their expression into the final expression for H_{tot} and arrive at the simplified version of the closed-loop transfer function.

15.4
Basics of Feedback Control Systems

Concept of Feedback Control

The feedback control of mechanical systems involves control of one or more of several dynamic variables such as force/torque, linear/angular speeds, positions, temperatures, fluid flow, or other process variables. Feedback or feedforward control systems are used in almost every industry for process control, e.g., robots, CNC machines, continuous chemical processes, autopilots on airplanes, and nuclear reactors. In each case, the actual operations may differ, but the principle behind the use of the control system is the same. The goal is to make the system to behave desirably, e.g., faster response time with less overshoot.

Classifications of Control Systems

Control systems may be broadly classified as either open-loop (feedforward) or closed-loop (feedback). An *open-loop system* is one in which the output has no effect on the control input (Figure 15.25). Hence, the controller adjusts the system response according to a set schedule, where each operating condition is set by a specified reference input. A common example of an open-loop control system is the domestic washer. The duration of the wash cycle is independent of the cleanness of the clothing.

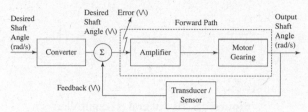

Figure 15.26 Closed-loop or feedback system

The *closed-loop or feedback control system* tends to maintain a prescribed relationship between the output and the reference input. A schematic of a feedback control system for a motor/gear unit is shown in Figure 15.26.

The output of the physical system (e.g., shaft angle) is compared with the reference input and the difference (actuating error) is input to the controller. The controller processes the error signal and produces an output that is amplified and input to the motor.

For example, in the case of the motor unit depicted in Figure 15.26, the input and output quantities could be shaft position (angles) or velocity. The transducers are used to convert these signals to voltages, which can then be compared to obtain the actuating error signal.

Three-Term PID Control

The "classical" types of control actions are *proportional, integral,* and *derivative*. The simplest type is *proportional control* in which the input (the error) is scaled up or down by a constant gain. Schematically, a proportional control action is shown in Figure 15.27.

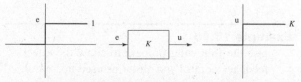

Figure 15.27 Proportional (P) controller

Using a proportional controller usually results in a steady-state error, or offset, in the response of the system. Some applications may require integral and/or derivative control action in addition to proportional. In *integral control*, the controller output is changed at a rate proportional to the actuating error signal and results in zero steady-state error. A *derivative controller* responds to the rate of change of the actuating error and is typically used to improve the transient response and reduce the oscillations of the system. Schematic representations of the integral and derivative controllers are shown in Figures 15.28 and 15.29.

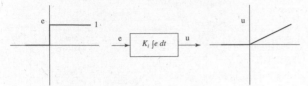

Figure 15.28 Integral (I) controller

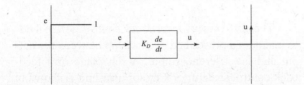

Figure 15.29 Derivative (D) controller

Generally, the output characteristics of a system can be kept within desired specifications by using proper gains. Performance characteristics of a control system are specified in terms of the transient response to a step input. The main advantage of a feedback or closed-loop system is that since it is "error driven," it can compensate for disturbances or variations from the desired input signal. Of course, the ability of the controller to compensate depends on the sophistication and the robustness of the design. However, risks are involved in using feedback. The addition of feedback to a stable system may result in the system becoming oscillatory or unstable. A *stable system* is defined as one where a bounded input results in a bounded output. Another important factor is that using the wrong signal for feedback will result in false information about the system being fed to the controller. Consequently, the controller will not operate as desired.

Example 15.15

Use the following process to show the effect of feedback control and all terms used in control.

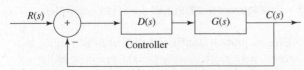

where the following transfer functions for both plant $G(s)$ and controller $D(s)$ are taken:

$$G(s) = \frac{1}{1+\tau s}, \quad D(s) = K_P + K_D s + \frac{K_I}{s}$$

(a PID controller)

Solution:

Full PID controller: The closed-loop TF can be simply obtained as detailed in the preceding subsection as:

$$G_c(s) = \frac{C(s)}{R(s)} = \frac{G(s)D(s)}{1+G(s)D(s)}$$

$$= \frac{\frac{1}{1+\tau s}\left(K_P + K_D s + \frac{K_I}{s}\right)}{1 + \frac{1}{1+\tau s}\left(K_P + K_D s + \frac{K_I}{s}\right)}$$

$$= \frac{K_P s + K_D s^2 + K_I}{(\tau + K_D)s^2 + (1 + K_P)s + K_I}$$

[15.59]

One could obtain the steady-state output of the system in response to a unit-step function as:

$$c(t \to \infty) = \lim_{s \to 0} sC(s) = \lim_{s \to 0}\left\{sR(s)G_c(s)\right\}$$

$$= \lim_{s \to 0}\left\{s\frac{1}{s}\frac{K_P s + K_D s^2 + K_I}{(\tau + K_D)s^2 + (1 + K_P)s + K_I}\right\} = 1$$

Alternatively, the steady-state error can be calculated as:

$$e_{ss} = \lim_{s \to 0}\left\{sE(s)\right\} = \lim_{s \to 0}\left\{s\left(R(s) - C(s)\right)\right\}$$

$$= \lim_{s \to 0}\left\{sR(s)\left(1 - G_c(s)\right)\right\}$$

$$= \lim_{s \to 0}\left\{s\frac{1}{s}\left(1 - \frac{K_P s + K_D s^2 + K_I}{(\tau + K_D)s^2 + (1 + K_P)s + K_I}\right)\right\} = 0$$

PD controller: If one uses only the PD controller, the closed-loop transfer function $G_c(s)$ obtained in equation 15.59 can be simplified as:

$$G_c(s) = \frac{C(s)}{R(s)} = \frac{G(s)D(s)}{1+G(s)D(s)}$$

$$= \frac{\frac{1}{1+\tau s}\left(K_P + K_D s\right)}{1 + \frac{1}{1+\tau s}\left(K_P + K_D s\right)}$$

$$= \frac{K_D s + K_P}{(\tau + K_D)s + 1 + K_P}$$

Repeating the same steps as in the case of full PID controller, one could obtain the steady-state error e_{ss} as:

$$e_{ss} = \lim_{s \to 0}\left\{sE(s)\right\}$$

$$= \lim_{s \to 0}\left\{s\left(R(s) - C(s)\right)\right\}$$

$$= \lim_{s \to 0}\left\{sR(s)\left(1 - G_c(s)\right)\right\}$$

$$= \lim_{s \to 0}\left\{s\frac{1}{s}\left(1 - \frac{K_D s + K_P}{(\tau + K_D)s + 1 + K_P}\right)\right\}$$

$$= \frac{1}{1+K_P}$$

It is obvious that we have lost the nice zero steady-state error result due to the missing I controller. If one uses only the PD controller, the proportional gain K_P must be selected relatively high in order to reduce the steady-state error.

D controller: If one uses only a D controller, the closed-loop transfer function $G_c(s)$ obtained in equation 15.59 can be written as:

$$G_c(s) = \frac{C(s)}{R(s)} = \frac{G(s)D(s)}{1 + G(s)D(s)}$$

$$= \frac{\frac{1}{1+\tau s}(K_D s)}{1 + \frac{1}{1+\tau s}(K_D s)} = \frac{K_D s}{(\tau + K_D)s + 1}$$

By repeating the same steps as in the case of full PID controller, one could obtain the steady-state error e_{ss} as:

$$e_{ss} = \lim_{s \to 0} \{ sE(s) \} = \lim_{s \to 0} \{ s(R(s) - C(s)) \}$$

$$= \lim_{s \to 0} \{ sR(s)(1 - G_c(s)) \}$$

$$= \lim_{s \to 0} \left\{ s \frac{1}{s} \left(1 - \frac{K_D s}{(\tau + K_D)s + 1} \right) \right\} = 1$$

Considering the ultimate goal is for the output to track the input, which is a unit-step function, a steady-state error of 1 equals 100% error. This is obviously not acceptable. It shows that the D controller, when used alone, leads to a permanent steady-state error that cannot be changed or improved by changing K_D. The following is a summary of important observations about the action of the P, I, and D controllers.

- A P controller is simply an amplifier with an adjustable gain. Increasing gain K_p will amplify the error that leads to better performance. However, it may lead to instability and/or actuator saturation.
- An I controller or "reset" can be used to remove the steady-state error. Sometimes, $K_I = 1/T_I$ is called the reset rate. An I controller, while removing steady-state error, may lead to oscillatory behavior.
- A D or "rate" controller improves the damping and stability. It must be used with a P controller to avoid permanent steady-state error. Increasing D will amplify the noise.
- A PD is a highly sensitive control in the sense that D tends to anticipate the actuating error, initiate an early correcting action, and increase the stability of the system.

PID Tuning Methods

As seen from the previous example, selecting controller gains is often a hit-and-miss affair. To systemati-

cally tune controller gains, several design techniques can be used. The most common techniques are the two *Ziegler–Nichols PID tuning methods*. They are the *quarter decay ratio* and the *ultimate sensitivity method*. Due to the effectiveness of the second method, we shall discuss this method.

In the *ultimate sensitivity method for PID tuning*, controller gains $D(s) = K_p(1 + 1/T_I s + T_D s)$ are tuned based on the limit of stability of the system. For this, use a proportional controller with gain K_u such that the system is put into sustained oscillations. Then this gain (K_u) and the period of the response (P_u) can be used to tune the gains as per table below.

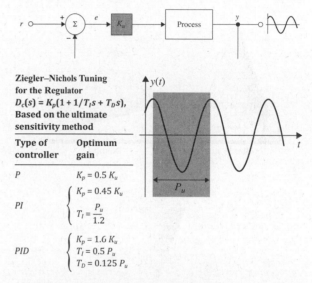

Ziegler–Nichols Tuning for the Regulator
$D_c(s) = K_p(1 + 1/T_I s + T_D s)$, Based on the ultimate sensitivity method

Type of controller	Optimum gain
P	$K_p = 0.5\,K_u$
PI	$K_p = 0.45\,K_u$ $T_I = \dfrac{P_u}{1.2}$
PID	$K_p = 1.6\,K_u$ $T_I = 0.5\,P_u$ $T_D = 0.125\,P_u$

Any feedback control system, when fully implemented, needs to be evaluated to determine the system response improvement. Although there are many characteristics to assess this improvement, two important characteristics of any feedback control system are steady-state error, and stability properties. Example 15.15 demonstrated the steady-state error characterization and relationship of the three-term PID controller on this characteristic. The following section discusses the other important characteristic, the so-called stability properties.

Stability Analysis in Control Systems

Concept of stability: There are two different viewpoints on dynamic stability: internal and external. Internal behavior includes the effect of all characteristic roots and equilibrium states. Hence, it is an inherent property of the dynamic systems. In contrast, external behavior may be affected by the types of inputs (e.g., pole-zero cancellation). A system has internal or asymptotic stability if the zero-input response (i.e., the response to initial conditions or perturbations) decays to zero as $t \to \infty$ for all possible initial conditions. This type of stability is concerned with the fact that the system will return to its equilibrium states after a disturbance has been removed. On the other hand, a system has external or bounded-input-bounded-output (BIBO) stability if the zero-state response (i.e., the response to external inputs) is bounded as $t \to \infty$ for all bounded inputs. This form of stability is concerned with whether certain types of inputs produce a bounded output. If the characteristic roots of a dynamic system are all on the left-half-plane of the complex plane, the system is stable from both viewpoints.

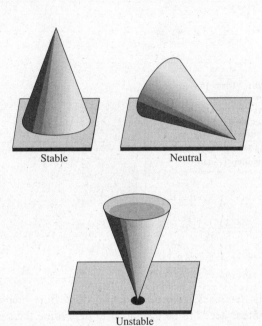

Stable   Neutral

Unstable

Figure 15.30 (Left) stable system: all poles of the closed-loop system are on LHP of the complex plane. (Middle) neutral or marginally stable system: the dominant poles of the closed-loop system are on the imaginary axis. (Right) at least one pole of the closed-loop system is on the RHP of the complex plane.

Hence, one could determine the stability by examining the *characteristic equation* of the closed-loop transfer function. This stability relation to location of the closed-loop poles can be shown by the following steps. For this, let's revisit the system output of a general linear system as:

$$Y(s) = \frac{N(s)}{D(s)} R(s) = \frac{N(s)}{(s - p_1)(s - p_2) \dots (s - p_n)} R(s)$$

$$[15.60]$$

where $p_1, p_2, \dots,$ and p_n are the closed-loop poles. By using partial fraction expansion, one can write equation 15.60 as:

$$Y(s) = \frac{k_1}{(s - p_1)} + \frac{k_2}{(s - p_2)} + \dots + \frac{k_n}{(s - p_n)}$$

$$+ \text{ the partial fraction terms from } R(s)$$

$$[15.61]$$

By taking the inverse Laplace transform of equation 15.61, one has:

$$y(t) = k_1 e^{p_1 t} + k_2 e^{p_2 t} + \dots + k_n e^{p_n t} + y_r(t)$$

As seen, if any of the poles p_i is positive, the response goes to infinity regardless of the location and types of other poles. For the system to be stable and have bounded response, all poles p_i must be negative or be located on the LHP of complex plane. This is exactly the same result observed in the previous discussion.

Example 15.16

Consider the following three system transfer functions:

$$T_1(s) = \frac{s + 2}{(s + 4)(s + 6)}, \quad T_2(s) = \frac{s - 5}{(s - 1)(s + 3)}, \text{ and}$$

$$T_3(s) = \frac{3(s - 3)}{(s + 1)(s - 3)}$$

For each system, identify whether they are internally or externally stable or unstable.

Solution:

System T_1 has two stable poles ($s_1 = -4$, $s_2 = -6$) and has no pole-zero cancellation. Hence, it is both asymptotically and BIBO stable. System T_2 has one unstable pole ($s_1 = 1$), which is not cancelled by any zero. Hence, it is both asymptotically and BIBO unstable. System T_3 has one unstable pole ($s_1 = 3$) that can be cancelled by a zero ($z_1 = 3$). Hence, it is asymptotically unstable, but BIBO stable.

Since BIBO stability predominately depends on the types of inputs, the stability analysis and assurance conditions vary for different inputs. On the other hand, the internal stability is an inherent property of a dynamic system. Therefore, a more structured and systematic analysis can be utilized.

Tools for Stability Analysis

A number of tools are available for assessing the stability of linear systems. The most commonly used are Routh–Hurwitz stability criterion, root-locus technique, and frequency-domain analysis (Nyquist and Bode plots). Leaving the details of the last two methods to the next subsection, we present here the Routh's stability criterion for its usefulness and effectiveness. This is motivated by the fact that factoring high-degree polynomials to find roots is very tedious and numerically not well conditioned. Hence, the Routh's criterion attempts to determine, without solving the characteristic roots, of how many roots are located on the RHP. This method is also useful in determining the range of parameters for which a feedback system remains stable.

Routh–Hurwitz Stability Criterion

This is a very mechanical, not intuitive, but very effective method. Three steps or tests need to be performed. Before we start the process, the polynomial needs to be ordered in the ascending order of s. Once this is done, the following steps are performed.

1. Sign agreement test: The necessary condition for the system to be stable is that all coefficients of s to have the same sign. Any sign disagreement is, indeed, an indication of system instability.
2. Missing term test: If any term is missing in the polynomial (any of the coefficients s is missing), the system is at best marginally stable. This indicates that the chance of the system being unstable is high. If one is looking for stable operation, this condition should be avoided. Alternatively, it can be used to design a proper controller or realize the type of controller that is needed to add these missing terms.
3. Routh array: If the previous two steps are checked off, then the so-called Routh's array must be formed as explained later. The number of roots of the polynomial that are on the RHP is equal to the number of sign changes in the first column of Routh's array. This implies that if all the entries in the first column are of the same sign (positive for example), the system is stable.

Routh's array: The general form of the characteristic equation can be written as:

$$CE(s) = a_n s^n + a_{n-1} s^{n-1} + \cdots a_1 s^1 + a_0$$

One would form the array as shown in Figure 15.31.

$$
\begin{array}{c|cccc}
s^n & a_n & a_{n-2} & a_{n-4} & \cdots \\
s^{n-1} & a_{n-1} & a_{n-3} & a_{n-5} & \cdots \\
s^{n-2} & b_1 & b_2 & \cdots \\
s^{n-3} & c_1 & c_2 & \cdots \\
\vdots & & & \\
s^1 & j_1 \\
s^0 & k_1
\end{array}
$$

$$b_1 = \frac{-1}{a_{n-1}} \begin{vmatrix} a_n & a_{n-2} \\ a_{n-1} & a_{n-3} \end{vmatrix}$$

$$b_2 = \frac{-1}{a_{n-1}} \begin{vmatrix} a_n & a_{n-4} \\ a_{n-1} & a_{n-5} \end{vmatrix} \cdots$$

$$c_1 = \frac{-1}{b_1} \begin{vmatrix} a_{n-1} & a_{n-3} \\ b_1 & b_2 \end{vmatrix}$$

$$c_2 = \frac{-1}{b_1} \begin{vmatrix} a_{n-1} & a_{n-5} \\ b_1 & b_3 \end{vmatrix} \cdots$$

Figure 15.31 Routh's array

Example 15.17

Determine the stability of a system whose characteristic equation is given by:

$$CE(s) = s^3 + s^2 + 2s + 8$$

Solution:

The first two steps, i.e., sign agreement and missing term, are satisfied. Hence, we need to form the Routh array:

$$
\begin{array}{c|cc}
s^3 & 1 & 2 \\
s^2 & 1 & 8 \\
s^1 & b_1 = -6 & 0 \\
s^0 & c_1 = 8 & 0
\end{array},
$$

$$b_1 = \frac{-1}{1} \begin{vmatrix} 1 & 2 \\ 1 & 8 \end{vmatrix} = -6,$$

$$c_1 = \frac{-1}{-6} \begin{vmatrix} 1 & 8 \\ -6 & 0 \end{vmatrix} = 8$$

As seen from the first column, two sign changes occur ($1 \rightarrow -6$, $-6 \rightarrow 8$). Hence, there are two roots on the RHP. A simple roots-finding method can be used to verify that there are, indeed, two roots on the RHP of the complex plane.

Example 15.18

Find the number of poles in the left half-plane, the right half-plane, and on the $j\omega$-axis for the system whose block diagram is given below.

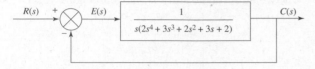

Solution:

One could obtain the closed-loop transfer function of the system as:

$$TF(s) = \frac{C(s)}{R(s)} = \frac{G(s)}{1 + G(s)}$$

$$= \frac{\dfrac{1}{s(2s^4 + 3s^3 + 2s^2 + 3s + 2)}}{1 + \dfrac{1}{s(2s^4 + 3s^3 + 2s^2 + 3s + 2)}}$$

$$= \frac{1}{2s^5 + 3s^4 + 2s^3 + 3s^2 + 2s + 1}$$

Hence, the characteristics equation can be extracted as:

$$CE(s) = 2s^5 + 3s^4 + 2s^3 + 3s^2 + 2s + 1 = 0$$

The first two steps, i.e., sign agreement and missing term, are satisfied. Hence, we need to form the Routh array:

s^5	2	2	2
s^4	3	3	1
s^3	$\emptyset\ \varepsilon$	$\frac{4}{3}$	0
s^2	$\frac{3\varepsilon-4}{\varepsilon}$	1	0
s^1	$\frac{12\varepsilon-16-3\varepsilon^2}{9\varepsilon-12}$	0	0
s^0	1	0	0

As seen, we have encountered a special case where the first entry in the first column becomes zero. Since the rows beneath this row are calculated using the inverse of this number (see the Routh array formulation), one needs to replace zero with a very small number, say ε. Note that ε can be either positive or negative. Once the zero entry in the first column is replaced with this small constant number, the process continues as before. All the remaining entries are calculated based on this new constant as shown in the table. The first column entries are then calculated considering the fact that $\varepsilon \to 0$ by either direct calculation or by taking the limit of the expression while $\varepsilon \to 0$. The first column entry in the s^2 row becomes negative when $\varepsilon \to 0$. The first column entry in the s^1 row becomes positive when $\varepsilon \to 0$. Hence, when ε is positive, there

are a total of two sign changes in the first column and 2 RHP poles. The remaining 3 (= 5 – 2) poles are on the LHP.

15.5
Classical Control Design Techniques

Any feedback control system must be properly designed and tuned before it can be implemented. Otherwise, the effect of adding a controller could result in system response characteristics that could be worse than the uncontrolled (open-loop) case. There are many classical techniques to help design the controller. Due to the limited scope of this chapter, we review only the two commonly utilized techniques: the root-locus and frequency-response techniques.

Root-locus Technique

As discussed before, the location of the characteristic roots (especially the dominant ones) dictates the response characteristics of a dynamic system. Therefore, it is very important to locate these roots or move them to a desired location for improved response characteristics.

The root locus is a plot of the path of the roots of the characteristic equation of a dynamic system (or the poles of the closed-loop transfer function) traced out on the s-plane as a system parameter (e.g., gain K) is varied. Without loss of generality, assume the following closed-loop system in which controller gain K is varied between zero and infinity. The underlying concept here is to locate the characteristic roots of the closed-loop system as parameter K changes. One could obtain the closed-loop transfer function of the system as:

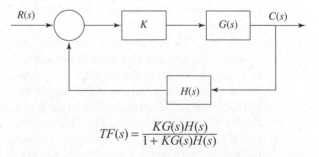

$$TF(s) = \frac{KG(s)H(s)}{1 + KG(s)H(s)}$$

The roots of the characteristic equation can be obtained from:

$$1 + KG(s)H(s) = 0 \text{ or } KG(s)H(s) = -1 \quad [15.62]$$

where $KG(s)H(s)$ is referred to as the open-loop or loop (for short) transfer function.

Properties of the root locus: From equation 15.62, we can establish the following properties for both magnitude and angle.

- Magnitude criterion: The magnitude of the open-loop transfer function must be unity for equation 15.62 to hold true. That is:

$$|KG(s)H(s)| = 1 \quad \text{or} \quad K = \frac{1}{|G(s)H(s)|}$$

[15.63]

- Angle criterion: By setting the angle condition for equation 15.62, the angle of the open-loop transfer function becomes an odd multiplier of 180°. That is:

$$\angle(KG(s)H(s)) = n180°$$
for $n = \pm1, \pm3, \pm5, \ldots$ [15.64]

By using these general properties, one can establish a step-by-step procedure for plotting the location of the characteristic roots in terms of parameter K. It can be observed from equation 15.62 that for $K = 0$, the denominator of the loop TF must vanish for the expression to hold true. This implies that when $K = 0$, we are at the poles, or the root loci start from (n) poles at the same time. Similarly, when $K = \infty$, the numerator of the loop TF must vanish, so we are at the zeros. Hence, the root loci start from n poles with n branches and end at zeros. If the number of zeros is not the same as the number of poles, then n-m branches go to infinity tangent to some asymptotes. Based on this, one could establish the step-by-step rules for constructing the root locus.

Rule 1—Poles/zeros location of $G(s)H(s)$ on the s-plane and determination of the root locus on real axis: Assume $G(s)H(s)$ has n poles and m zeros. We first locate the poles and zeros on the real axis. To the right of rightmost pole or zero, no root locus exists. A simple root (pole or zero) has asymmetry property w.r.t. root locus, i.e., the root locus exists on its either side but not both. A repeated root (or complex conjugate) has symmetry property w.r.t. root locus, i.e., if the root locus exists one side, it will exist on the other side as well.

Since the coefficients of the characteristic equations are all real (for physical systems), the characteristic roots will be complex conjugates if they are complex. Therefore, the root locus is symmetrical with respect to the real axis.

Rule 2—Determination of the number of branches and asymptotes: The number of branches is n, starting from each pole and terminating at each zero. Then n-m branches go to infinity, tangential to the (n-m) lines called *asymptotes*.

Rule 3—Intersection of asymptotes with real axis and their angles: If there are any asymptotes (i.e.,

n $m \neq 0$), the intersection of asymptotes with the real axis and their angles can be calculated according to the following:

$$\sigma_a = \frac{(\text{sum of all the finite poles}) - (\text{sum of all finite zeros})}{n - m}$$

$$\theta_a = \frac{(2k + 1)\pi}{n - m}, \quad k = 0, \pm1, \pm2, \ldots$$

Rule 4—Breakaway and the break-in points: Break-in points or breakaway points of the root locus are where two or more branches intersect. For this, one needs to rewrite the closed-loop characteristic equation in terms of gain K, i.e., $K = f(s)$, and then differentiate $f(s)$ with respect to s and solve for s. Only solutions that are on the root-locus portion of the real axis are acceptable. The rest cannot be acceptable as either break-in or breakaway points:

$$\frac{dK}{ds} = 0 \qquad [15.65]$$

Rule 5—Possible points of intersection of the root locus with the imaginary axis: For this, we set $s = j\omega$ and solve the characteristic equation $1 + KG(j\omega) = 0$. This is a complex equation that results in two real algebraic equations that can be solve simultaneously for K and ω. Since both K and ω are positive, only solutions that result in positive values are acceptable. If there are no solutions, it can be concluded that the root locus does not intersect with the imaginary axis, which is a good sign and shows that the system will remain stable for all the values of K. Any intersection indicates that the system becomes marginally stable at the intersection for that specific value of K, say $K_{crossing}$. Any K values bigger than this $K_{crossing}$ will result in an unstable situation.

Rule 6—Complete the root-locus plot: By taking a series of test points in the broad neighborhood of the origin of the s-plane, we can sketch the root loci. Then the closed-loop poles and the range of variations of gain K can be determined. The results for the stable range of K can be compared with that of Routh–Hurwitz method.

Example 15.19

Draw the root locus for the closed-loop system shown in the figure below. The purpose is to try to plot the location of the closed-loop characteristic poles as K varies from zero to infinity.

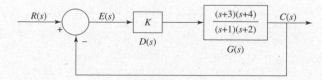

Solution:

Step 1: As shown in Figure 15.32, the poles and zeros of the open-loop TF can be shown on real axis. Subsequently, the real-axis root locus can be determined.

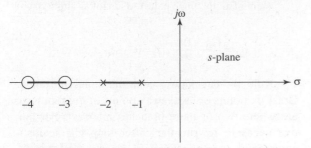

Figure 15.32 Real axis portion of the root locus

Step 2: The number of asymptotes is $n - m = 2 - 2 = 0$. So there are no asymptotes.

Step 3: Find break-in or breakaway points. The characteristic equation of motion is:

$$1 + KG(s) = 0 \quad \text{or} \quad 1 + K\frac{(s+3)(s+4)}{(s+1)(s+2)} = 0$$

Hence, we can obtain K as a function of other terms and then use the procedure explained above to obtain the possible breakaway and/or break-in points:

$$K = -\frac{(s+3)(s+4)}{(s+1)(s+2)} \Rightarrow \frac{dK}{ds} = 0$$

or

$$\frac{dK}{ds} = -\frac{[(s+2)+(s+1)][(s+3)(s+4)] - [(s+4)+(s+3)][(s+1)(s+2)]}{(s+3)^2(s+4)^2} = 0$$

$$\frac{dK}{ds} = 0 \Rightarrow 2s^2 + 10s + 11 = 0 \Rightarrow s_1 = -1.63, \; s_2 = -3.37$$

It can be verified that both of these points are on the portion of the root locus on the x-axis. So the breakaway (at $s_1 = -1.63$) and break-in (at $s_2 = -3.37$) are acceptable.

Step 4: Find the intersection with the imaginary axis. By substituting $s = j\omega$ into the characteristic equation and solving for both ω and K, one can see that there are no feasible values (i.e., positive and real) for either ω and K, so there will be no intersection with the imaginary axis. Based on this, the final plot of the root locus will be as shown in Figure 15.33.

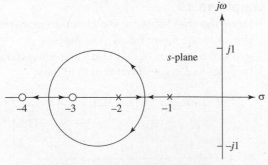

Figure 15.33 Root-locus plot

Frequency-Response Technique

Most dynamic systems can be better characterized in frequency domain since a nondimensional relationship between response characteristics and system physical parameters can be obtained. For this, the equations of motion are transferred into frequency domain by Fourier transform. In order to keep the focus, we consider only a simple case where we take a simple, single sinusoidal input to the system as follows.

Analytical expression for frequency response: Given that a linear time-invariant system is represented by a transfer function of $G(s)$:

$$\xrightarrow{R(s)} \boxed{G(s)} \xrightarrow{C(s)}$$

Let the system input be a sinusoidal motion, i.e., $r(t) = A \sin(\omega t)$, where A is the amplitude and ω is the frequency of the input signal. The Laplace transform of $r(t)$ is given by:

$$R(s) = \frac{A\omega}{s^2 + \omega^2} \qquad [15.66]$$

Then the output of the system is given as:

$$C(s) = G(s)R(s) = G(s)\frac{A\omega}{s^2 + \omega^2}$$

$$= G(s)\frac{A\omega}{(s - j\omega)(s + j\omega)} \qquad [15.67]$$

Performing a partial fraction expansion on the above equation results in:

$$C(s) = \frac{K_1}{(s - j\omega)} + \frac{K_2}{(s + j\omega)}$$

$$+ \text{ partial fraction terms from } G(s) \quad [15.68]$$

where coefficients K_1 and K_2 are calculated as:

$$K_1 = \left[(s - j\omega)C(s)\right]_{s=j\omega} = \left[\frac{A\omega}{s + j\omega}G(s)\right]_{s=j\omega}$$

$$= \frac{A}{2j}G(j\omega)$$

$$K_2 = \left[(s + j\omega)C(s)\right]_{s=-j\omega} = \left[\frac{A\omega}{s - j\omega}G(s)\right]_{s=-j\omega}$$

$$= -\frac{A}{2j}G(-j\omega)$$

$$[15.69]$$

Using Euler's formula, one can relate the complex term $G(j\omega)$ and $G(-j\omega)$ into their magnitude and angle values as:

$$G(j\omega) = |G(j\omega)|e^{j\phi}, \quad G(-j\omega) = |G(j\omega)|e^{-j\phi}$$

$$[15.70]$$

where $|G(j\omega)|$ is the magnitude of $G(j\omega)$ and ϕ is its angle. Focusing on the steady-state output (i.e., ignoring the effect of initial condition and hence zero-input response), one can perform inverse Laplace on the $C(s)$ in equation 15.68 to arrive at its time-domain expression as:

$$c(t) = K_1 e^{j\omega t} + K_1 e^{-j\omega t} \qquad [15.71]$$

By substituting the expressions for K_1 and K_2, found in equation 15.69, one could reduce equation 15.71 for $c(t)$ to:

$$c(t) = \frac{A}{2\omega} G(j\omega) e^{j\omega t} - \frac{A}{2\omega} G(-j\omega) e^{-j\omega t}$$
$$= \frac{A}{2j} |G(j\omega)| e^{j\phi} e^{j\omega t} - \frac{A}{2j} |G(j\omega)| e^{-j\phi} e^{-j\omega t}$$

$$[15.72]$$

After using the Euler identity $e^{j\theta} = \cos(\theta) + j\sin(\theta)$, one could simplify the above expression to:

$$c(t) = A |G(j\omega)| \sin(\omega t + j\phi)$$

From the above equation, it can be seen that:

- The steady-state response of the system is also in the form of a sinusoidal motion.
- The system has the same frequency as the sinusoidal input.
- The system has different amplitude and phase angle from the input. These differences are found to be functions of frequency. In particular, the amplitude is amplified by $|G(j\omega)|$, which is the magnitude of $G(j\omega)$. The phase angle is shifted by ϕ, which is the angle of $G(j\omega)$.

Frequency response technique: The frequency response of a system is defined as the frequency-dependent relation in both the amplitude and phase angle difference between the sinusoidal inputs and the resultant steady-state sinusoidal outputs. The frequency response of a system whose transfer function is $G(s)$ can be obtained *by simply replacing s in the G(s) with jω, i.e., G(jω).* Particularly, the magnitude of $G(j\omega)$ is called the magnitude frequency response and the angle of $G(j\omega)$ is called the phase frequency response. We can use either a magnitude/phase-frequency plot or a magnitude-phase (polar) plot to represent this response visually. It is clear that both the magnitude and the phase angle are functions of the input frequency and therefore vary with ω. From the frequency response, we can then obtain information about the system stability, resonance, bandwidth, etc.

1. The Bode plot is one of the frequency response techniques we study in this chapter. The magnitude and phase are plotted, respectively, as functions of frequency as the frequency varies from 0 to infinity. In the magnitude plot, the magnitude response is plotted as magnitude in decibels (dB) vs. log ω, where magnitude in a unit of dB = 20 log $|G(j\omega)|$. In the phase plot, the phase response is plotted as phase vs. log ω.

Note that depending on the information or characteristics sought from the bode plot, we can use either closed-loop TF (i.e., $G(s)/1 + GH(s)$) or loop TF (i.e., $GH(s)$).

- **Bode plot for closed-loop TF:** The information that can be obtained from the Bode plot of the closed-loop TF includes bandwidth, resonant frequency, damping ratio, and natural frequency.

- **Bode plot for loop TF:** The information that can be obtained from the Bode plot of the loop TF includes phase margin (PM), gain margin (GM), and whether or not the closed-loop system is stable (using the PM and GM).

The *resonant frequency* is the frequency at which the magnitude reaches its maximum value. A 2nd-order system can have only one resonant frequency. Higher-order systems can have many resonant frequencies. *Natural frequency* is the frequency of the natural oscillation that would show by two complex poles if the damping value were set to zero.

Based on the results obtained so far, we can conclude that the so-called *frequency response function (FRF)* or *frequency transfer function (FTF)* can be simply obtained by replacing s with $j\omega$ in the expression for transfer function:

$$FRF(\omega) \equiv T(s)\big|_{s=j\omega} \qquad [15.73]$$

For the case of the 1 DOF system (Figure 15.6) the FRF of the steady-state displacement of the system due to harmonic excitation can be expressed as:

$$G(\omega) \equiv \frac{Y(s)}{R(s)}\bigg|_{s=j\omega} = \frac{1}{m\left(\omega_n^2 - \omega^2 + 2j\zeta\omega_n\omega\right)} = |G(\omega)| e^{-j\phi}$$

$$[15.74]$$

where

$$|G(\omega)| = \frac{1}{k\sqrt{\left(\left(1-r^2\right)^2 + \left(2\zeta r\right)^2\right)}}, \quad \phi = \tan^{-1}\frac{2\zeta r}{1-r^r}$$

$$[15.75]$$

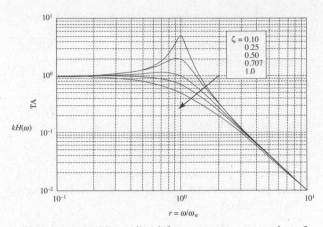

Figure 15.34 Normalized frequency response plot of system (Figure 15.6) for different damping ratio values

As mentioned earlier, the FRF and especially its magnitude can be a very helpful representative of vibration response of a dynamic system. In order to visualize this better, the plot of nondimensional magnitude of the FTF (equation 15.75) for different values of damping ratio ζ is shown in Figure 15.34. This plot can be used when designing vibration control systems. For example, for vibration attenuation in Example 15.1, one can easily see the effect of changing the system damping ratio on the steady-state response $|Y(j\omega)|$. To obtain the expression for the peak response, one could take the derivative of the amplitude $|Y(j\omega)|$ with respect to frequency ratio r and set it equal to zero. The solution to this equation will result in an expression for the resonant frequency:

$$\omega_r = \omega_n\sqrt{1 - 2\zeta^2} \quad \zeta < 0.707 \quad [15.76]$$

Substituting this into the amplitude expression $|Y(j\omega)|$ will result in the peak expression as:

$$M_{p_\omega} = |G(\omega_r)| = \left(2\zeta\sqrt{1 - \zeta^2}\right)^{-1} \quad [15.77]$$

Bandwidth is the frequency at which the frequency response has declined by 3 dB from its original value. The bandwidth of a system is a measure of the speed of response.

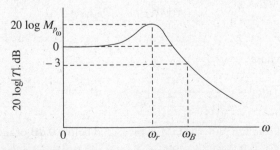

Figure 15.35 Bandwidth frequency of a dynamic system

The *frequency response technique* can be used to assess the stability of the closed-loop system using the open-loop or loop transfer functions. For this, the following are defined.

- *Gain margin (GM)* is a measure of how much the system gain would have to be increased for the open-loop transfer function so that the system enters the unstable zone. GM can be obtained from $|G(j\omega)H(j\omega)|$ when $\angle G(j\omega)H(j\omega) = -180°$.
- *Phase margin (PM)* indicates additional phase lag that is required to make the system unstable, i.e., the amount by which the phase of $G(j\omega)$ exceeds the value of $-180°$. PM can be obtained from $\varphi_m = 180 + \angle G(j\omega)H(j\omega)$ at $|G(j\omega)H(j\omega)| = 1$.
- A *stable system* has negative GM, i.e., below the horizontal axis.
- A *stable system* has also the PM above the $-180°$ line. By adding the GM into the loop TF, the system will enter the marginally stable zone.

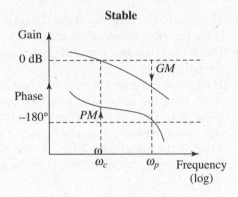

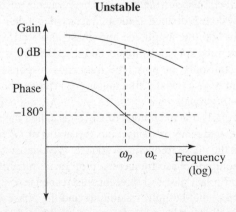

Figure 15.36 Phase margin (PM) and gain margin (GM) for stable and unstable cases

2. The Nyquist plot is another frequency response technique. In this technique, both magnitude and phase are plotted in the complex plane (called the *G*-plane). There the distance from the origin is the magnitude and the angle from the real axis is the phase angle. (Note that the counterclockwise direction is positive.)

The Nyquist diagram is a polar frequency response plot made for the *open-loop TF*. The Nyquist stability criterion determines the stability of the closed-loop system using the Nyquist plot. In this procedure, first determine the open-loop TF (i.e., the $GH(s)$) of the system, then sketch the polar plot of this complex function $G(s)H(s)$, and finally apply the Nyquist stability criterion on this polar plot of the complex function in the frequency domain.

The Nyquist stability criterion: A feedback system is stable if and only if the polar plot of $GH(s = j\omega)$ in the G-plane does not encircle the $(-1, 0)$ point when the number of poles of $GH(s)$ in the RHP of s-plane is zero. If $GH(s)$ has P poles in the RHP of s-plane, then the number of *counterclockwise* encirclements of the $(-1, 0)$ point must be equal to P for a stable system.

GM is a measure of the relative stability of a control system. It is the reciprocal of the gain $|GH(j\omega)|$ at the frequency at which the phase angle reaches $-180°$. GM is the amount of gain that can be added to the system before the closed-loop system becomes marginally stable.

PM is also a measure of the relative stability of a control system. PM is the amount of the phase shift of $G(j\omega)H(j\omega)$ at unity magnitude that will result in a marginally stable system. That is, $PM = \varphi_m = 180 + \angle G(j\omega)H(j\omega)$.

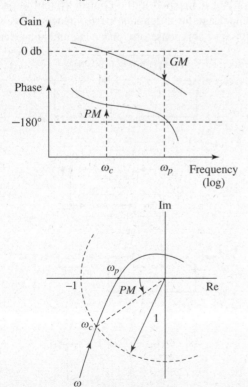

Figure 15.37 Phase margin (PM) and gain margin (GM) comparison using both Bode and Nyquist plots

The following compares different techniques:

- Transfer function and frequency response function are both complex functions.
- Transfer function is a function of the complex variable s.
- Frequency response function is a function of the input frequency ω.

In a transfer function:

1. $G(s)$ is a complex function with the complex variable s.
2. Laplace domain is a function of complex variable s.
3. Performance specifications are poles and zeros.

In a frequency response function:

1. $G(j\omega)$ is a complex function of frequency.
2. Frequency domain is a function of variable ω (frequency).
3. Performance specifications are resonant frequency and bandwidth.

Example 15.20

Draw the Bode plot of the following system whose loop TF is given, and assess the closed-loop stability.

$$G(s) = \frac{1}{s(s + 1)(s + 5)}$$

Solution:

Using the magnitude and phase plots, one can plot the response as shown in the figure below.

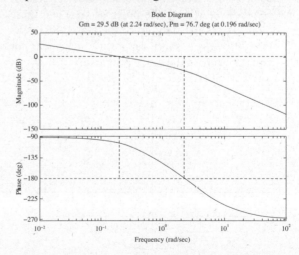

From the plot, the GM is measured below 0 dB or the PM is measured above $-180°$. So the system is stable.

Example 15.21

Draw the Bode plot of the following system whose loop TF is given, and assess the closed-loop stability.

$$G(s) = \frac{1}{(s^3 + 2s^2 + s + 0.5)(s-1)}$$

Solution:

Using the magnitude and phase plots, one can plot the response as shown in the figure below.

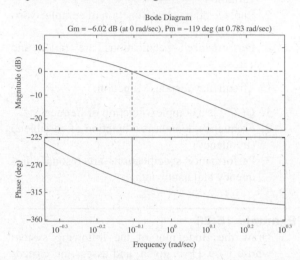

From the plot, the PM is measured below −180°. So the system is unstable.

Example 15.22

The figures below depict the Nyquist plots for a loop transfer function, $G(s)$, used in a unity feedback system with two different values of gain K. By looking at only the poles of $G(s)$ and these plots, assess the stability of the closed-loop system. Provide a short supporting statement for your answers in the space provided.

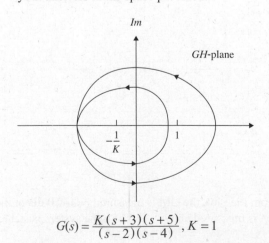

$$G(s) = \frac{K(s+3)(s+5)}{(s-2)(s-4)}, K = 1$$

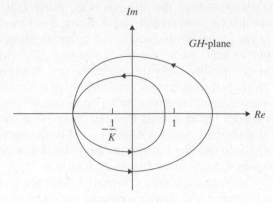

$$G(s) = \frac{K(s+3)(s+5)}{(s-2)(s-4)}, K = 0.2$$

Solution:

When $K = 1$, the $-1/K$ is located inside the circles. The loop TF of the plot on the left also has two poles on RHP (i.e., $P = 2$), and there are 2 CCW rotations around $(-1,0)$ (i.e., $N = -2$). Hence, $Z = N + P = 0$. According to the Nyquist criterion, the system is stable. When $K = 0.2$, the $-1/K$ is located outside the circle so there are no rotations around $(-1,0)$, so $N = 0$. The loop TF of the plot on the left has two poles on RHP (i.e., $P = 2$). Hence, $Z = N + P = 0 + 2 = 2$. According to the Nyquist criterion, the system is unstable.

A *compensator* is defined as an additional component that is inserted into a control system in order to compensate for a deficient performance. There are two types of compensators: phase lag and phase lead.

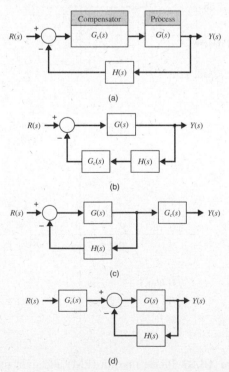

Figure 15.38 Compensator design configurations

A *phase lag compensator* possesses one zero and one pole, with the *pole closer to the origin* of the *s*-plane. This compensator reduces the steady-state tracking errors:

$$G(s) = K\alpha \frac{(1+\tau s)}{(1+\alpha\tau s)}, \alpha < 1$$

A phase lag compensator is a low-pass filter. Therefore, it filters the high-frequency noise.

- It improves the low-frequency behavior.
- It can improve the steady-state accuracy and the system's stability.
- Crossing frequency and system bandwidth are reduced by the phase lag compensator.
- Phase margin increases, and hence, it improves the system's stability.

A *phase lead compensator* possesses one zero and one pole, with the *zero closer to the origin* of the *s*-plane. This compensator increases the system's bandwidth and, hence, improves the dynamic response:

$$G(s) = K\alpha \frac{(1+\tau s)}{(1+\alpha\tau s)}, \alpha < 1$$

15.6
Data Acquisition and Processing

This section provides a brief overview of data acquisition structure, hardware architecture, and data processing. Figure 15.39 depicts an example case study of a testbed for applying impulse input (as input force) to an actively excited rubber beam and for monitoring the response in real time. Through this example, real-time control implementation, data acquisition handling, and processing are all explained (see Figure 15.40).

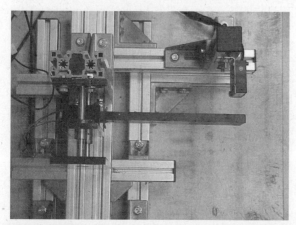

Figure 15.39 Experimental testbed for applying impulse input to an actively excited rubber beam and monitoring the response in real time

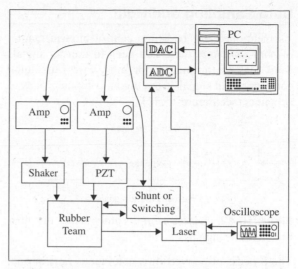

Figure 15.40 Schematic of the overall real-time control implementation and instrumentation arrangement and setup

As depicted in Figure 15.40, the process starts with designing a proper controller (e.g., developed in the Matlab/Simulink environment) and sending a controller signal out to the real world. For this, a real-time data acquisition card (see Figure 15.41) with multiple digital-to-analog (D/A) and analog-to-digital (A/D) convertors is used with a proper integration with the software used (e.g., MATLAB Simulink Real-Time Workshop) to facilitate the entire test process. More specifically, an electromagnetic shaker is used to apply the impulse force. A force sensor on the tip of the impact head measures the magnitude of the impulse force that is applied to the beam.

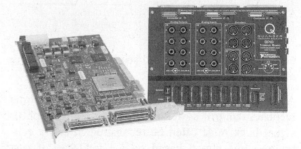

Figure 15.41 A typical data acquisition card (Quanser® MultiQ Board)

The following subsections give an overview of each of these components in a typical real-time control setting. They also briefly discuss some of the data processing and handling concepts.

Data Sampling and Hold

As mentioned, for real-time control implementation of *data-sampling,* there is a need to transfer signals from sensors (an analog for most cases) to digital computers and signals from digital computers back to actuators (see Figure 15.42).

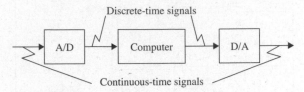

Figure15.42 Discrete-time processing of continuous-time signals

The process of converting a continuous-time or analog signal to discrete time is called "sampling operation" or "date sampling." Without loss of generality, we assume that data are sampled at equal intervals of time T, where T is referred to as the sampling period (see Figure 15.43). The inverse of this quantity, $f = 1/T$, is called the sampling rate or frequency. It is measured in Hz. Sampled data are therefore obtained at only specific instants of times or at discrete intervals. For this, a sampler is used that is basically a switch that converts a discrete function to a continuous one.

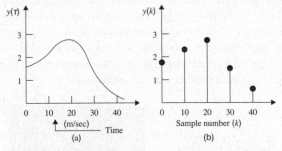

Figure 15.43 Concept of discretization or sampling of a continuous-time signal

As seen from the discussions so far in most real-time feedback control systems, sampled-data function or signal back must often be reconstructed into a continuous-time signal based on a knowledge of past samples. This process or device is referred to as an *extrapolator, data hold, or clamp.* The hold or clamp devices are classified based on the number of prior samples required for predicting the sampled data or function during the hold or waiting time (see Figure 15.44).

Zero-order hold (ZOH): In this hold operation, only one sampled value at the beginning of a sampling interval is required (see Figure 15.44).

First-order hold: For this hold, two prior samples are needed to construct the analog signal or function (see Figure 15.45).

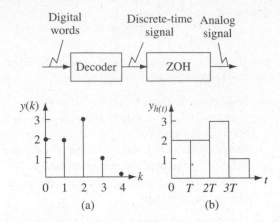

Figure 15.44 (Top) Operation performed by zero-order hold (ZOH). (Bottom) (a) Sampled data before entering ZOH, (b) analog output signal of the ZOH

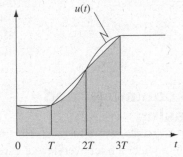

Figure 15.45 First-order hold using trapezoidal rule for integral approximation

The *Nyquist and Shannon sampling theorem* states that for the signal to be accurately constructed from the sampled data using proper hold operation, the frequency of sampling (f) must be at least twice the highest frequency component of the measured signal. That is, if we denote this highest component of the signal by f_N (also called the Nyquist frequency), then we have:

$$f > 2f_N \qquad [15.78]$$

If this condition is not met, then "signal aliasing" may occur. In other words, the ever-present signal noise may be modulated to a lower frequency by the sampling process. In a continuous-time system, the measurement noises are at a much higher frequency than the control system bandwidth. So the noises will not have significant effect on the process. However, in discrete-time and digital control environments, the frequency of these noises can be aliased down near the system bandwidth. So the effect of the noise could be significant.

Analog-to-Digital Convertor (ADC)

An ADC is required to convert analog sensor signals into a binary form suitable for use in a digital computer setting. At the most basic level, all digital signals are binary numbers consisting of many bits that are set to either 1 or 0. Due to the nature of the discretization, some analog information may be lost in the conversion process. The greater the number of bits, the more accurate the signal conversion accuracy and the longer the conversion will take.

Some ADCs have a relatively low resolution of 8 bits or less, while other convertors are more precise with the number of bits at 12 or 16. The price of ADCs generally goes up with both speed and bit size. ADCs usually have a full-scale input range of a few volts (e.g., 0–10 V for unipolar and ±5 V for bipolar). The resolution of the measured signals can be calculated as:

$$\delta R = \frac{R}{2^n} \qquad [15.79]$$

where δR denotes the resolution, R is the range of the signal (the difference between highest values, S_H, and lowest values, S_L), and n is the number of conversion bits. When ADC operates, it produces an integer number, say N, corresponding to the value of the analog signal value. Based on the number of conversion bits, the range of N would be $[0, 2^n - 1]$. Consequently, the discrete signal value can be calculated from the ADC as follows:

$$S = \delta RN + S_L \qquad [15.80]$$

with the highest measureable signal as $S_H - \delta R$ (i.e., one signal resolution less).

Digital-to-Analog Convertor (DAC)

As shown schematically in Figure 15.40, DACs are used to convert the digital signals from computer to an analog signal (typically voltage) for driving actuators or recording devices (e.g., oscilloscopes). Due to the nature of the conversions, that is reconstructing an analog signal, they are sometimes referred to as sample and hold (S/H) devices. The resolution of DACs is the same as ADCs are and calculated from equation 15.79. Since no counting or iteration is needed for DACs, they are much faster than ADCs.

Analog Signals Dynamic Range and Amplification

A sensor's dynamic range is defined as the ratio of its full-scale value to the minimum detectable signal variation. The conversion accuracy of an ADC is limited by its resolution. In order to fully utilize the resolution of the ADC, the range of signal levels must accurately match the ADC's full-scale range (typically up to 5 or 10 V). Hence, the signal levels might be scaled up to match the full-range of ADCs by means of suitable amplifying components.

Signal Filtering and Averaging

There are a number of noise sources in practical applications, such as electronic noise, electromagnetic interfaces, and noise generated in general by measurement devices. Their presence can be very undesirable, giving rise to unwanted measurement fluctuations and variations. The effect of noise can be diminished mostly by using proper passive or active analog filters. Since the frequencies obtained with most types of sensors are generally in the low range, the noise, which on the other hand occurs predominantly at one high frequency, can be separated from the measurement signals.

In order for the underlying signal to be accurately processed, the signal-to-noise ratio needs to be improved. One effective way of improving this ratio or reducing the effects of random noise is to calculate the average of the signals at several readings in quick succession. This way if the noise is really random and equally distributed around the actual signal, it should tend to average out to zero. This simple yet effective technique is referred to as averaging.

15.7
Common Sensors in Feedback Control Systems

A transducer, also known as a sensor, is referred to as a device that converts a physical quantity such as temperature, displacement, or pressure into a detectable electrical signal (typically a voltage). This section presents an overview of common sensors used in feedback control systems. Due to the limitation of this chapter, we review only a select number of sensors in different applications and discipline areas. Some less-related or seldom-used devices such as chemical and optical sensors are excluded here. Details of the operation of these devices as well as their modeling may be found in respective references. The sensitivity of each sensor is defined as the ratio of change in measured signal to the change in magnitude of the measured physical quantity. Each of the following sensors might have a detailed manufacturer's data sheet that can be used for precise details and signal-conditioning circuits. These data sheets are not given here due to the variations in the data sheets from manufacturer to manufacturer and our limited space. For the reader's convenience, we separate these sensors into two classes: solid mechanics sensors and thermofluid sensors.

Solid Mechanics Sensors

Motion sensors are typically used to measure displacement, velocity, or acceleration of a dynamic event. The following sensors are presented for measuring each of these physical quantities.

1. Linear variable differential transformers (LVDTs) are used to measure linear displacement. They are made of two coils. A primary coil in the center (coil A in Figure 15.46) is excited with a high frequency. The other type are secondary outer coils (coils B in Figure 15.46). The magnetic-flux coupling between the two coils becomes a function of the position of the core within the coils. The movement of the core along the axis of the coils results in an induced, measurable signal in the secondary coils. The LVDTs provide a relatively high sensitivity sensor and high-level voltage output. They also offer a linear response over their working range.

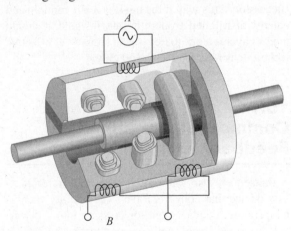

Figure 15.46 Schematic of the operation of an LVDT displacement sensor (from Wikipedia)

2. Laser displacement sensors are used to measure, through noncontact configuration, the location of a moving objective. The position or velocity of a single point on an object's surface is compared with the projected component of the incident laser beam along the direction of the laser beam. The difference in amplitude and phase of the reflected wave with the incident laser beam is a quantifiable measure of the object's out-of-plane vibration to the surface. The advantage of such an optical sensor is that it monitors system behavior remotely and requires no point of attachment to the system. Figure 15.47 depicts a commercial laser displacement sensor, the LDS 90/40, which has a measurement range of 40 mm, reference distance

of 90 mm, resolution of 0.04 mm, and output signal ranging from 0 V to 10 V. The sensitivity of the LDS device is 0.25 V/mm.

Figure 15.47 Laser displacement sensor used to measure displacement of moving objects (the LDS 90/40)

3. Potentiometer sensors are used to measure linear or angular position. A resistive wire along with a sliding component are used to measure the position of the contact. Simply, the resistance to the current flowing into the wire is a measure of the position of the contact. If used within its range and augmented with proper signal conditioning and buffering, a potentiometer sensor provides a linear output for a relatively large range of motion.

4. Accelerometers are used to measure the acceleration on moving objects. They provide an easy and inexpensive alternative to velocity (by integrating the output signal once) or displacement (by integrating the output signal twice). Figure 15.48 depicts a linear accelerometer for seismic measurements. When input acceleration displaces the base, the motion of the base is translated to the proof mass through the cantilever springs. The movement of the proof mass is captured through a simple electrical motion transducer where a variable resistor is used to produce an output voltage proportional to the acceleration of the proof mass.

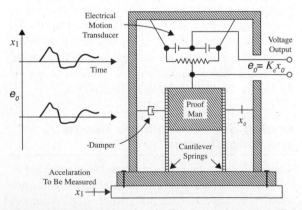

Figure 15.48 Schematic of a linear accelerometer for seismic measurements

Strain-gauge sensors belong to the class of resistance sensors in which electrical resistance varies and is measured in proportion to the mechanical strain. Since this platform of measurement is not very sensitive and is prone to environmental changes and variations, a bridge circuit is used to obtain optimum operation and precision. As shown in Figure 15.49, a fixed resistor can be replaced by a sensor (R_S in Figure 15.49). When the bridge is balanced, that is $R_s = R_1$, then $V_0 = 0$. When the sensor sees changes in its resistance due to mechanical strain, the change in the resistance can be related to the output signal voltage V_0 as:

$$V_{out} = V_{in}\left(\frac{R_2}{R_1 + R_2} - \frac{R_2}{R_S + R_2}\right) \quad [15.81]$$

$$V_{out} = V_{in}\left(\frac{R_2}{R_1 + R_2} - \frac{R_2}{R_s + R_2}\right)$$

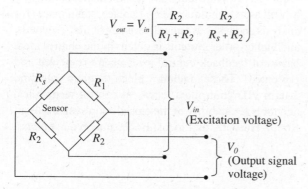

Figure 15.49 Schematic of a bridge circuit for measuring resistance changes in strain gauges

Bridges often contain two or four sensing elements by replacing the fixed resistors to enhance the overall sensitivity of the bridge. In case of nonthermal sensors, this arrangement is used to compensate for the temperature changes and variations. Most of the sensors based on strain gauge, such as load cells or pressure transducers, utilize this configuration.

When a strain-gauge sensor is used, its electrical resistance R_S (as shown in Figure 15.49) changes in response to the induced strain in the element. This relationship is expressed as:

$$\text{Gauge} - \text{Factor} = \frac{\Delta R_S / R_S}{\Delta L / L} = \frac{\Delta R_S / R_S}{\varepsilon}$$

$$[15.82]$$

where R_S is nominal resistance of the strain gauge at nominal length L, ΔR_S is the change in resistance due to the change in length ΔL, and ε is the normal strain sensed by the gauge.

Thermofluid Sensors

Temperature sensors include devices that measure temperature variations in a variety of systems. This subsection briefly presents two types of temperature sensors, the resistance temperature detectors (RTDs) and thermocouples (TC).

1. Resistance temperature detectors (RTDs) are another class of resistance sensors where a temperature-dependent resistance is utilized to relate the change in the resistance to the temperature variations. RTDs are typically constructed from a variety of metals, but platinum is the most commonly used metal. They are useful for use in the temperature ranges of –270°C to +660°C. Assuming small nonlinearities in sensor output, the linear relationship between resistance R_T at measured temperature (T) to the resistance R_0 at the reference temperature (T_0) is given by:

$$R_T = R_0[1 + \alpha(T - T_0)] \quad [15.83]$$

where α is the temperature coefficient of the RTD.

2. Thermocouples (TCs) are the simplest sensors for temperature measurements. They work based on the change in length of two dissimilar conductors when they undergo temperature variations. As shown schematically in Figure 15.50, when the two metals are heated, the difference in their thermal expansion coefficients results in different length changes. As a result, bending-induced strain occurs because the metals are bounded together. The bending can be detected by a variety of measurement sensors. Depending on the materials used, the TC output ranges from 10 to 70 μV/°C.

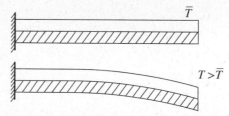

Figure 15.50 Schematic of a thermocouple under heat

Fluid pressure sensors: Applied pressure in fluids can be measured by a variety of sensor technologies such as piezoelectric, capacitive, electromagnetic, and optical. Piezoelectric effect can be used to relate the change in strain due to the applied pressure into a detectable electrical voltage (the concept of piezoelectricity). These sensors can be employed to measure absolute, gauge, vacuum, and differential pressures. Capacitive sensors can also be used to measure applied pressure by producing a variable capacitor in which the capacitance becomes a function of the

479

applied pressure. This technology is typically applicable for low-pressure measurements. Other types of pressure sensors operate similarly based on their respective transduction technologies.

Measurement Uncertainties

Most real-time feedback control implementations require precise measurements obtained by sensors used in the actual device. Although the controller might possess some levels of robustness to system parameters and especially measurement uncertainties, one needs to take into account such uncertainties in the measurement. Assume that n sensors measure n quantities $y_1, y_2, \ldots, y_n$. Moreover, assume that each sensor could be contaminated with some levels of measurement uncertainties, $w_1, w_2, \ldots, w_n$, respectively. The total output signal from each sensor is then represented as $y_1 \pm w_1, y_2 \pm w_2, \ldots, y_n \pm w_n$. Now if the calculated variable R can be made a function of the sensors output as:

$$R = f(y_1, y_2, \ldots, y_n) \qquad [15.84]$$

the uncertainties w_R in the calculated variable R can be estimated using the Kline–McClintock equations:

$$w_R = \sqrt{\left(w_1 \frac{\partial f}{\partial y_1} \right)^2 + \left(w_2 \frac{\partial f}{\partial y_2} \right)^2 + \cdots + \left(w_n \frac{\partial f}{\partial y_n} \right)^2}$$

$$[15.85]$$

15.8 Summary

This chapter provides a brief overview of modeling and analysis of dynamic systems as well as design and development of controllers, along with ways to measure their outputs to improve their response characteristics to a desired level. In the dynamic systems section, a brief introduction to mathematical modeling as well as to techniques (e.g., Laplace, state-space) for analysis and response characteristics of first-, second-, and higher-order systems is given. In the control area, basics of feedback control systems are reviewed and presented. These include block-diagram algebra, 3-term PID controllers, classical control techniques, and tools for stability of the closed-loop systems (e.g., Routh–Hurwitz, root locus, Bode, and Nyquist plots).

PRACTICE PROBLEMS

1. A damper with $B = 5$ N/(cm/sec) has one end moving with displacement $(3t + 4 \sin 10t)$ cm, while the other end has displacement $\sin 10t$ cm. What is the force generated in the damper in Newtons?
   (A) $F_b(t) = 15 + 150 \sin(10t)$
   (B) $F_b(t) = 150 \sin(10t)$
   (C) $F_b(t) = 15 + 150 \cos(10t)$
   (D) $F_b(t) = 15 - 150 \cos(10t)$

2. Which of the following equations is physically correct (i.e., represents a physical system)?
   (A)

   $m_1\ddot{x}_1 + (c_1 + c_2)\dot{x}_1 + (k_1 + k_2)x_1 - k_2x_2 - c_2\dot{x}_1 = 0$
   $m_2\ddot{x}_2 + c_2\dot{x}_2 + k_2x_2 + c_2\dot{x}_1 + k_2x_1 = F(t)$

   (B)

   $m_1\ddot{x}_1 - (c_1 - c_2)\dot{x}_1 - (k_1 - k_2)x_1 - k_2x_2 - c_2\dot{x}_2$
   $= 0;\ c_2 > c_1$
   $m_2\ddot{x}_2 + c_2\dot{x}_2 + k_2x_2 - c_2\dot{x}_1 - k_2x_1 = F(t)$

   (C)

   $m_1\ddot{x}_1 + (c_1 - c_2)\dot{x}_1 + (k_1 + k_2)x_1 - k_2x_2 - c_2\dot{x}_1 = 0$
   $m_2\ddot{x}_2 + c_2\dot{x}_2 + k_2x_2 - c_2\dot{x}_1 - k_2x_1 = F(t)$

   (D)

   $m_2\ddot{x}_1 + (c_1 + c_2)\dot{x}_1 + (k_1 + k_2)x_1 = k_2x_2 + c_2\dot{x}_2$
   $m_1\ddot{x}_2 + c_2\dot{x}_2 + k_2x_2 = F(t) + c_2\dot{x}_1 + k_2x_1$

3. What is the value of the damping ratio (ξ) for the following system?

   $$T_1(s) = \frac{12}{s^2 + 8s + 12}$$

   (A) 2.31
   (B) >1
   (C) =1
   (D) <1

4. The transfer function of a closed-loop control system is given as:

   $$F(s) = \frac{8 + 3s}{s(s+1)(s+10)}$$

   where $F(s)$ is the Laplace transform of $f(t)$. What is the value of $f(t)$ for very large t?
   (A) ∞
   (B) 0.3
   (C) 0.8
   (D) 0

Questions 5 and 6

A translational mechanical system is shown schematically in the figure below. Assume that $M = 1$ kg, $f_v = 3$ N.s/m, and $K = 4$ N/m.

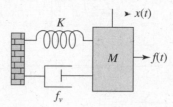

5. What is the percent overshoot of this system?
   (A) 2.50%
   (B) 2.84%
   (C) 3.84%
   (D) 0%

6. What is the settling time based on 2% criterion?
   (A) 2.667
   (B) 1
   (C) ∞
   (D) 3.667

Questions 7 and 8

For a specific value of an amplifier gain in a closed-loop control system, the following characteristic roots are obtained: $s_1 = -13.98$, $s_{2,3} = -0.51 \pm j5.96$.

7. What is the system settling time for this gain based on 2% criterion?
   (A) 7.84
   (B) 8.23
   (C) 4
   (D) 0.51

8. What is the system damping ratio for this gain?
   (A) 0.0596
   (B) 0.1
   (C) 0
   (D) 0.085

481

9. Which one of the following choices represents the simplified block of the following feedback control system?

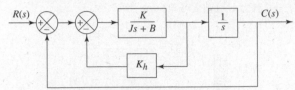

(A) $\dfrac{K}{Js^2 + Bs + K}$

(B) $\dfrac{K}{Js^2 + (B + KK_h)s + K}$

(C) $\dfrac{KK_h}{Js^2 + Bs + K}$

(D) $\dfrac{Js + B}{Js^2 + Bs + K}$

10. Consider the unity feedback control of an RC circuit whose transfer function is expressed as $G(s) = 1/(s + 1)$, as shown in the figure.

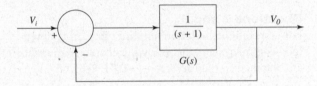

To improve the steady-state error, a suitable controller $D(s)$ must be placed on the feed forward path right before the plant $G(s)$ to increase the order of the closed-loop transfer function. Which of the following controllers will achieve this?
(A) $D(s) = as + b$
(B) $D(s) = a$
(C) $D(s) = bs$
(D) $D(s) = a + \dfrac{b}{s}$

11. Consider the unity feedback control of the 2nd-order system shown in the figure below.

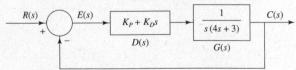

What is the value of the control gain K_D that results in a closed-loop settling time of 0.5 s. (based on 2% criterion)?
(A) $K_D = 2$
(B) $K_D = -2$
(C) $K_D = 1$
(D) $K_D = 4$

12. The closed-loop transfer function of a dynamic system is given as follows. How many characteristic roots are on the LHP?

$$T(s) = \dfrac{1}{s^5 + 6s^4 + 5s^3 + 8s^2 + s + 6}$$

(A) 2
(B) 1
(C) 0
(D) 3

13. Which of the following root loci can be a root locus?

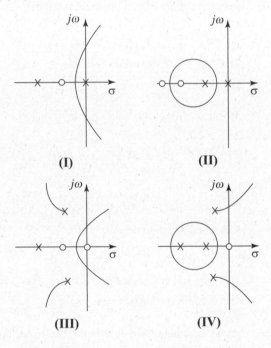

(I) (II)

(III) (IV)

(A) I
(B) II
(C) III
(D) IV

The Bode diagram of the open-loop transfer function $G(s)$ is illustrated below.

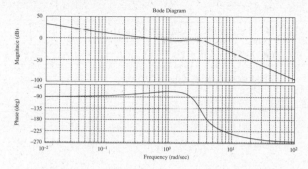

Bode Diagram

14. What is the value of gain margin (GM) for this system in dB?
(A) –10
(B) 10
(C) 5
(D) 0

15. What is the value of phase margin (PM) for this system in degrees?
(A) 90
(B) 75
(C) –105
(D) 105

16. The figure below demonstrates real-time implementation of a controlled plant. Identify the unknown blocks A and B in the order.

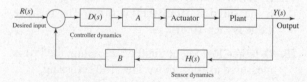

(A) S/H and Hold
(B) Filter
(C) DAC and ADC
(D) ADC and DAC

17. What is the resolution of a 12-bit ADC with a full-scale input range of 10 V?
(A) 12 mV
(B) 25 mV
(C) 2.4 mV
(D) 1 mV

18. Which of the following statements for a strain-gauge sensor in a bridge configuration is correct?
(A) When the bridge is balanced, the sensor output is not masked by the input voltage.
(B) Environmental changes do not affect the balance of the bridge.
(C) A bridge can have only one sensor resistor.
(D) None of the above.

Answer Key

1. C	**6.** A	**11.** C	**16.** C
2. D	**7.** A	**12.** D	**17.** C
3. B	**8.** D	**13.** B	**18.** A
4. C	**9.** B	**14.** B	
5. B	**10.** D	**15.** D	

Answers Explained

1. **C** We know that the damping force is given by
$F_b = B(\Delta \dot{X})$

$$\Delta \dot{X} = \frac{d}{dt}(X_1) - \frac{d}{dt}(X_2)$$
$$= 3 + 40 \cos(10t) - 10 \cos(10t)$$
$$F_b(t) = 5(3 + 30 \cos(10t)) = 15 + 150 \cos(10t)$$

2. **D**
(A) Not correct. $-c_2 \dot{x}_1$ should change to $-c_2 \dot{x}_2$ and the signs of the underlined terms in the second expression should change.
(B) Not correct. The underlined items should change to $+(c_1 + c_2)\dot{x}_1 + (k_1 + k_2)x_1$.

$$m_1 \ddot{x}_1 \underline{- (c_1 - c_2)\dot{x}_1 - (k_1 - k_2)x_1} - k_2 x_2 - c_2 \dot{x}_2$$
$$= 0; \ c_2 > c_1$$
$$m_2 \ddot{x}_2 + c_2 \dot{x}_2 + k_2 x_2 - c_2 \dot{x}_1 - k_2 x_1 = F(t)$$

(C) Not correct. The sign of $(c_1 - c_2)$ is wrong.
(D) Correct. m_1 and m_2 are just name changes.

3. **B** The roots of the characteristic equation are
$s_{1,2} = -4 \pm \sqrt{16 - 12} = -2, -6$.
Hence the system is overdamped as it possesses two distinct real roots. Hence the damping ratio is bigger than 1.

4. **C** By using FVT, we have:

$$\lim_{t \to \infty} f(t) = \lim_{s \to 0} sF(s)$$
$$= \lim_{s \to 0} s \frac{8 + 3s}{s(s+1)(s+10)}$$
$$= \lim_{s \to 0} \frac{8 + 3s}{(s+1)(s+10)} = 0.8$$

5. B (See the explanation for #6.)

6. A The EQM of this simple system can be obtained as:

$$M\ddot{x}(t) + f_v\dot{x}(t) + Kx = f(t)$$

Taking the Laplace transform of the EQM under zero ICs, one can obtain the following TF:

$$TF = \frac{X(s)}{F(s)} = \frac{1}{Ms^2 + f_vs + K} = \frac{1}{s^2 + 3s + 4}$$

Hence, $2\xi\omega_n = 3$, and $\omega_n^2 = 4$. That is, $\omega_n = 2$, $\xi = 3/4$. The setline time is therefore $t_s = 4/\xi\omega_n = 4/1.5 = 2.667$. The percent overshoot can be obtained from the following expression:

$$Mp = 100\exp\left(\frac{-\pi\xi}{\sqrt{1-\xi^2}}\right) = 2.84\%$$

7. A The dominant pole(s) are $s_{2,3} = -0.51 \pm j5.96$.

Hence, $\sigma = 0.51$ and $t_s = 4/\sigma = \frac{4}{0.51} = 7.84$ s.

8. D The dominant pole(s) are $s_{2,3} = -0.51 \pm j5.96$. Comparing this with the standard form of characteristic roots $s_{2,3} = -\xi\omega_n \pm j\omega_d = -\xi\omega_n \pm j\omega_n\sqrt{1-\xi^2}$ results in $\xi\omega_n = 0.51$ and $\omega_n\sqrt{1-\xi^2} = 5.96$.

Hence, $\frac{\xi\omega_n}{\omega_n\sqrt{1-\xi^2}} = \frac{\xi}{\sqrt{1-\xi^2}} = \frac{0.51}{5.96}$.

From this expression, ξ can be obtained as $\xi = 0.085$.

9. B Using the standard form of feedback as in equation 15.58, one could simplify the internal feedback to $G_1 = \frac{K}{s(Js + B + KK_h)}$. Then the outer feedback loop can also be simplified as:

$$G_c = \frac{\dfrac{K}{s(Js + B + KK_h)}}{1 + \dfrac{K}{s(Js + B + KK_h)}} = \frac{K}{Js^2 + (B + KK_h)s + K}$$

10. D We need to increase the order of the system to 2 to enhance the steady-state error. For this, only a controller with "I" can help. Hence, the PI controller of option (D) is selected.

11. C The closed-loop TF can be easily obtained as

$$G_c = \frac{K_p + K_Ds}{4s^2 + (K_D + 3)s + K_p}.$$ By comparing the

characteristic equation (the denominator of G_c) with the standard form, one can conclude that

$K_D + 3 = 2\xi\omega_n$. On the other hand, $t_s = 4/\xi\omega_n$. For a settling time of 0.5 s, $\xi\omega_n = 8$. Hence $K_D + 3 = 8$ or $K_D = 5$.

12. D The characteristic polynomial is:

$$Q(s) = s^5 + 6s^4 + 5s^3 + 8s^2 + s + b$$

First generate a Routh table:

s^5	1	5	1
s^4	6	8	6
s^3	$\frac{11}{3}$	0	0
s^2	8	6	0
s^1	$-\frac{11}{4}$	0	
s^0	6		

Then interpret the Routh table. Two sign changes in the first column results in two poles in the RHP. Thus, the system is unstable.

Hence, there are 3 roots of the total 5 roots on the LHP.

13. B Only (B) is a correct root locus. Choice (A) cannot have the breakaway between zero and the pole. Choice (C) cannot have the breakaway between zeros. Choice (D) cannot have the extra circle as the breakaway between pole and zero does not exist.

14. B When looking at the $-180°$ crossing, the GM is measured at 10 dB below 0 dB. The positive value is then 10 dB.

15. D When looking at the 0 dB crossing, the PM is measured at $105°$ above the $-180°$ crossing.

16. C As discussed in section 15.6, the answer is (C).

17. C The resolution is $R/2^n = 10/2^{12} = 0.00244$ or 2.44 mV.

18. A Based on the discussion in section 15.7, the correct answer is (A).

16

Safety, Health, and Environment

Taha F. Marhaba, Ph.D., P.E.

Introduction

The subject of safety, health, and environment has been recently added to the Fundamentals of Engineering (FE) General Engineering afternoon session. The chapter presents a review of the suggested topics and subtopics from an engineering perspective: hazardous properties of materials, industrial hygiene, process hazard analysis, overpressure and underpressure protection, storage and handling, waste minimization, and waste treatment. Due to the computational nature of the FE exam, emphasis has been given to equations and mathematical relationships.

16.1 Hazard Identification and Classification

A hazard can be natural, manmade, or activity related (such as the risk of driving or flying). Identifying and categorizing the hazard into a suitable kind or type is very important. Different types of hazards can have significantly different control measures.

A hazard can be classified in a number of ways. Most methods are based on the likelihood of the hazard to occur and on the seriousness of the situation it will create if the hazard were to occur. Commonly, both the likelihood and the seriousness are scaled on a numeric scale of 0 to 10. The number 0 means not likely or not serious. The number 10 means extremely likely or extremely serious. By multiplying these two scores, a comparative score can be obtained for the risk of the hazard:

$$\text{Risk} = (\text{Likelihood of occurrence score}) * (\text{Seriousness of hazard score}) \qquad [16.1]$$

Another popular method puts all hazards into either one or more of the following categories: chemical hazard, physical hazard, biological hazard, and ergonomic hazard.

Hazardous Properties of Materials

A hazard is a situation that threatens any or all of life, health, property, and environment to a noticeable extent. Great precautions are usually taken to keep the hazard dormant under all conditions. Once a hazard becomes active it can produce significant damage and create emergency situations.

Hazardous material is defined as substances that show at least one of four properties—ignitability, toxicity, corrosivity, and reactivity—and are capable of causing damage to human health. These substances include chemicals, fumes, paints, radioactive rays, dust particles, etc. The storage, handling, and disposal of hazardous materials is managed under the Resource Conservation and Recovery Act (RCRA) 240 CFR 261.

The ignitability (flammability) is how easily a material will burn, cause fire, or combust. EPA under 40 CFR 261.21 states that solid waste exhibits the property of ignitability if the representative sample of the waste has any of the following properties:

1. It is a liquid other than an aqueous solution containing less than 24% alcohol by volume and has a flash point less than 60°C.
2. It is not a liquid and is capable, under standard temperature and pressure, of causing fire through friction, absorption of moisture or spontaneous chemical changes and, when ignited, burns so vigorously and persistently that it creates a hazard.
3. It is an ignitable compressed gas.
4. It is an oxidizer.

Toxicity is the measure of damage that a material can cause to human health. Ingesting or absorbing toxic materials is injurious and lethal to human health. Toxic materials include heavy metals such as beryllium, mercury, leads, etc. Toxicity of a material is described by a laboratory test called the toxicity characteristic leaching procedure (TCLP) test method 1311. The toxicity of a substance affects human health in various ways depending upon the type of material and its effects. The effects of a toxic substance can be observed immediately or over a period of time.

Acute exposure refers to death, injury, or damage to human health caused by single exposure to a toxic substance. Acute exposure effects usually do not last longer than a day and cause immediate damage or death.

Chronic exposure refers to constant or gradual exposure to toxic substances (e.g., industrial smoke, UV rays, radioactive metals, etc.) over a period of months or even years. Chronic exposure causes irreversible side effects on human health. It may result in some health disorder or, eventually, human death.

Corrosivity refers to the capability of substances to corrode metals. Corrosive materials also can cause damage to human tissues (such as skin, flesh, muscles). EPA code 40 CFR 261.22 states that a solid waste exhibits the property of corrosivity if the representative sample of the waste has any of the following properties:

1. It is aqueous and has a pH less than or equal to 2 or greater than or equal to 12.5.
2. It is a liquid and corrodes steel (SAE 1020) at a rate greater than 6.35 mm (0.250 inch) per year at a test temperature of 55°C (130°F) .

Reactivity: EPA codes 40CFR 261.23 says a solid waste exhibits the characteristic of reactivity if a representative sample of the waste has any of the following properties:

1. It is normally unstable and readily undergoes violent change without detonating.
2. It reacts violently with water.
3. It forms potentially explosive mixtures with water.
4. When mixed with water, it generates toxic gases, vapors, or fumes in a quantity sufficient to present a danger to human health or the environment.
5. It is a cyanide- or sulfide-bearing waste that, when exposed to pH conditions between 2 and 12.5, can generate toxic gases, vapors, or fumes in a quantity sufficient to present a danger to human health or the environment.
6. It is capable of detonation or explosive reaction if it is subjected to a strong initiating source or if heated under confinement.
7. It is readily capable of detonation or explosive decomposition or reaction at standard temperature and pressure.
8. It is a forbidden explosive as defined by the U.S. Department of Transportation regulations 49 CFR 173.51, 173.53, and 173.88

Hazard Communication

The best-available preventive measure to avoid a hazard is hazard communication. All the government organizations that are in some way related to hazardous waste management have passed different regulations and norms to reduce the risk of a hazard due to hazardous waste.

The U.S. Occupational Safety and Health Administration (OSHA) mandates hazardous material handling, storage, and transport facilities to maintain material safety data sheets (MSDS) and to label the containers. MSDS is a form containing information on chemicals, chemical mixtures, and chemical compounds. The information on the form includes physical and chemical properties of the material (boiling point, flash point, melting point, etc.) along with hazards related to the material and instructions for its safe use. OSHA has also regulated standards for the handling of hazardous waste and the response to hazardous waste related incidents. OSHA's famous hazardous waste operations and emergency response (HAZWOPER) regulations (29 CFR 1910.120) have been very effective in the hazardous waste industry.

The United States Department of Transportation (DOT), as found in 49 CFR, regulates the transportation of hazardous material.

The U.S. Environmental Protection Agency (EPA) regulates the cleanup of hazardous waste sites along with the handling and disposal of hazardous waste.

Occupational Safety and Health Administration (OSHA) Standards for Worker Protection

OSHA standards as recorded in 29CFR are legally enforceable. OSHA sets standard values for allowable exposure limits in the work environment. It also mandates the use of appropriate safety equipment while handling one or more substances that are hazardous in nature. OSHA provides several guidelines to make the work environment safer for workers. Worker safety, health, and environmental control disciplines have become extremely professional since the passage of OSHA. OSHA also brought about several other changes in industrial safety regulations. Guards on all moving parts, confined space rules, hazard communication (Right to Know), and process safety management are some of the things introduced and brought into practice by OSHA.

Safe dose (human): Epidemiological and toxicological data are collected and studied on a large scale to determine a safe human dose level of a toxin. The dose-response relationship also helps in making this critical decision. The U.S. Environmental Protection Agency (EPA) recommends a stepwise method to determine the safe human dose, for noncarcinogens and for carcinogens. Carcinogens are assumed to pose a threat at all levels. The EPA model that has been built to predict the risk of exposure to carcinogens assumes a linear relationship to any exposure. Noncarcinogens, however, have a threshold value (reference dose or reference concentration) of exposure, below which there is no observed health effect. This threshold is determined by repetitive experiments and dose-response curves.

Equation 16.2 is the hazard ratio used for noncarcinogens. Equation 16.3 is used for carcinogens.

$$\text{Hazard ratio} = \frac{\text{Estimated exposure dose (mg/kg·d)}}{\text{Reference dose (mg/kg·d)}}$$

[16.2]

$$\text{Excess cancers} = (\text{risk of cancer}) * (\text{Exposed population})$$

[16.3]

The carcinogen potency factor (CPF) is defined as the probability of risk of lifetime exposure to 1.0 mg/kg·d of the carcinogen.

The probable risk of additional cancers (R) for a dose other than 1.0 mg/kg·d can be computed using the chronic daily intake (CDI) value as described in equation 16.4, where chronic daily intake is the daily dose averaged over a lifetime for an average person in mg/kg·d.

$$R = (\text{CPF}) * (\text{CDI in mg/kg·d})$$

[16.4]

Example 16.1

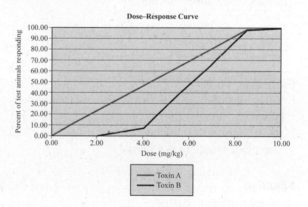

Refer to the graph above and identify the threshold value and LD_{50} for each toxin.

Solution:

A. Threshold dose is the dose level or the first point along the graph, which corresponds to positive (above zero) response.

Toxin	Approximate Threshold Dose (mg/kg)
A	0
B	2
C	4

B. LD_{50} is the point at which 50% of the test animals show a response.

Toxin	Approximate LD_{50} (mg/kg)
A	4.2
B	6.0
C	7.9

Example 16.2

The reference dose for a toxin is 15 ppm/d. What is the hazard ratio if the estimated exposure is 5 ppm/d?

Solution:

Use Equation 16.2:

$$\text{Hazard ratio} = \frac{\text{Estimated exposure dose (mg/kg·d)}}{\text{Reference dose (mg/kg·d)}}$$

Note that mg/kg is equivalent to ppm.

Estimated exposure = 5 ppm/d
Reference dose = 15 ppm/d

$$\text{Hazard ratio} = (5)/(15)$$
$$\text{Hazard ratio} = 0.33$$

Example 16.3

The risk of excess cancer for a known carcinogen is 0.01. What is the carcinogen potency factor if the daily intake averaged over a lifetime of an average recipient is 6 ppm/d?

Solution:

Use equation 16.3:

$$R = (CPF) * (CDI \text{ in mg/kg·d})$$

$R = 0.01$
$CDI = 6$ ppm/d
$CPF = ?$

$$(0.01) = (CPF)(6)$$
$$CPF = 0.00167 \text{ d/ppm}$$

16.2
Industrial Hygiene

Industrial hygiene is the study and control of environmental factors at a work place that may produce unwanted effects on the workers or the neighboring community. The American Industrial Hygiene Association (AIHA) defines industrial hygiene as that science and art devoted to the anticipation, recognition, evaluation, and control of those environmental factors or stresses— arising in or from a workplace— which may cause sickness, impaired health and well-being, or significant discomfort, and inefficiency among workers or among citizens of the community. The federal Occupational Health and Safety Act, 1970, governs industrial hygiene.

Personal Protection Equipment

Personal protective equipment (PPE) is any clothing or equipment designed to protect the wearer from injury or exposure to a toxin. Protective clothing, helmets, goggles, gloves, masks, or any other gear serving a protective purpose can be called PPE. One serious drawback of PPE, which is often overlooked, is that it does nothing to eliminate the hazard. In fact, PPE is not a hazard control method and should be used only when everything else fails. However, when there is risk of dangerous exposure or injury, PPE is the savior.

Most of the common protective equipment are made up of Nomex and Kevlar. The following describes some of the commonly used PPEs.

Respiratory protection: OSHA regulations (29 CFR 1910.134) mandate the employer to provide the worker with respiratory protective equipment whenever and wherever necessary. Additionally, a respiratory protection program is required to be implemented. As far as possible, proper engineering controls should be installed so that the use of protection would not be needed. Filter masks, gas masks, air-purifying respirators, and a self-contained breathing apparatus are some of the common types of equipment available in the market.

Protective clothing: Many materials and types of clothing are available to suit different needs and applications. Protective clothing includes electrical, thermal, and/or chemical protecting clothing such as suits, aprons, and gloves.

Head, eye, hand, and foot protection: This type of PPE is required mainly for protection against the risk of injury rather than exposure. This kind of protection should be used at all places where there is reasonable risk of injury. Glasses, goggles, and hats are some of the many types of equipment available for protection against physical injury.

Other than the types mentioned above, it is also a common practice to use ear defenders, earplugs, and sometimes other head/neck, arm/shoulder protective equipment.

Principal Entities in Industrial Hygiene

A biological hazard (biohazard) is anything that is a threat to human health. Biological hazards includes microorganisms, toxins, viruses, and medical waste.

Carcinogens: Any substance, radionuclide, or radiation that is capable of causing cancer in living tissue is called a carcinogen. Many substances, such as gamma rays or alpha particles, are considered carcinogenic because the radiation they emit is carcinogenic. In contrast, many other substances, such as asbestos and tobacco, are carcinogens by themselves when inhaled.

Cold and heat stress: Stress is the disruption of equilibrium through mental, physiological, anatomical, or physical reactions. Workers may experience stress in both cold and hot work environments. The experience of the worker can be in comfort zone, discomfort zone (where the worker is uncomfortable but there is no observable health effect), or health risk zone.

Heat stress can result in elevated heart rates. Long-term heat stress can cause dehydration, heat exhaustion, or in a rare worst-case scenario heat stroke, which can be very dangerous.

Cold stress can reduce blood circulation and can cause shivering, hypothermia, and tissue damage.

Safety measures are engineering and administrative controls, personal protective equipment, and worker training that provide effective means of avoiding cold and heat stresses.

Cumulative trauma disorder (CTD) is a condition that a person experiences from overuse of a tool or an instrument. The instrument can be anything, including a computer, knife, or guitar. Computer users and assembly-line workers are most commonly affected by CTD. Muscles, nerves, and tendons in the upper back and arms are the most commonly affected areas. The main cause of CTD is that the above-mentioned areas are kept under stress or kept tense for a very long time. Some of the popular techniques known to help prevent or reduce the effect of CTD, if it already exists, include stretches, exercises, massages, good posture, and limiting time in stressful working conditions.

Several self-explanatory names describe the CTD: repetitive strain (or stress) injury (RSI), occupational overuse syndrome, and work-related upper limb disorder (WRULD). Any of the loose group of conditions that result from overuse is usually enough for a person to be diagnosed with CTD. Table 16.1 lists some of the other disorders that share symptoms with CTD. A person suffering from any one disorder is often seen to experience several others. It is best

practice to treat the person for a single general disorder.

Table 16.1: Common Cumulative Trauma Disorders

Disorder	Brief Description
Carpal tunnel syndrome	Compressed median nerve, typically at the wrist
Reflex sympathetic dystrophy syndrome (RSDS)	Chronic progressive disease that causes severe pain
Cubital tunnel syndrome	Related to ulnar nerve entrapment
Intersection syndrome	Related to pain in the thumb side of the forearm
DeQuervain's syndrome	Inflammation of the tunnel surrounding the thumb movement control tendons
Trigger finger/ thumb	Affects the finger that pulls or pushes a trigger
Gamekeeper's thumb	Pain in or injury to the ulnar collateral ligament
Tennis elbow	Overuse of outer part of elbow, causing pain and tenderness
Golfer's elbow	Related to the pain in the flexor muscle of forearm

Noise is unwanted sound or noise pollution. Continued exposure to noise can damage the hearing permanently (long-term exposures of over 85 decibels (dB)). Additionally, long-term and sustained exposure to the noise even at lower levels (that are tolerable once in a while) can cause hearing problems. Noise in the context of industrial hygiene is traditionally considered only when it is a risk to environmental health and safety rather than a mere nuisance. However more recently, noise is being considered hazardous to the worker by an increasing variety of means. Hearing impairment is not the only concern with noise. Noise can aid or promote stress in the work environment and can raise blood pressure. Noise also acts synergistically to add to the risks of other hazards. Noise can block or mask hazards and warning signals. The concentration of a person is normally affected in the presence of noise.

Distance is the easiest way to reduce noise. Doubling the distance from a noise source will reduce the noise level sensed by 6 dB (commonly known as the rule of 6). This value, 6 dB, is twice as much as reduction (3 dB) one would achieve by shutting down one of the two identical 100 dB noise sources placed side by side. The value of 6 dB is derived from the equation of noise attenuation, which is stated in equation 16.5:

$$\text{Noise attenuation} = 10 \log_{10}\left[\left(\frac{D_2}{D_1}\right)^2\right] \qquad [16.5]$$

where D_1 and D_2 are the distances in meters over which the attenuation is to be computed.

So if the distance is doubled, i.e., $D_2 = 2D$, the attenuation will be 6.02 dB.

Particulate matter (PM) as air pollutant, or simply particulates, are suspended tiny particles (solid or liquid) in a gas. Their sizes vary from 10 nm to more than few hundred microns in diameter. The size of the particulate matter is often described by the notation PM followed by the size of the highest average diameter. For example, PM_{10} means particulate matter of size 10 µm or less. Similarly, $PM_{2.5}$ means particulate matter of size 2.5 µm or less.

Particulates originate from anthropogenic or natural sources. Naturally occurring PM is generated from volcanoes, forest and grasslands fires, and desert storms. Living vegetation and sea spray also add to the naturally occurring PM. Among the anthropogenic PM, the burning of fossil fuel gives rise to the majority of the PM in the urban environment. Roughly 10% of the total amounts of PM that floats in the atmosphere is contributed by human activities.

Inhaling PM can be very dangerous to human health. The effects of inhaling PM have been studied widely in both humans and animals. Different health effects that have been reasonably linked to inhaling PM include asthma, cardiovascular issues, lung cancer, and in some cases even premature death. The extent of damage the PM will produce depends mainly on the location in the respiratory system where it settles after being inhaled. How far the particle will go depends on the diameter of the particle. The larger the size of the particle, the higher the possibility of it getting filtered in the nose or throat, reducing the extent of damage that the particle will produce. Smaller particles (PM_{10}), however, travel farther in the respiratory system, reaching the lungs and causing serious health problems. $PM_{2.5}$ are even more dangerous as they are capable of penetrating into gas exchange parts of the lung. In general, much research has indicated the link between inhaling PM (even short-term exposure at high concentrations) and heart disease.

Ionizing radiation is energetic particles or waves capable of ionizing an atom or a molecule through interactions at an atomic level. Examples of ionizing radiation include alpha particles, beta particles, X-rays, neutrons, and nuclear radiation. The extent of ionization depends on the energy associated with the particle and not on the number of particles. Exposure to sufficient levels of such radiation can result in DNA damage in living cells and overall, can be destructive to humans. Larger doses of radiation can affect reproduction or produce mutating effects in future generations. The transfer of energy to tissue by the passing ionizing radiation causes the damage. The effects that radiation have on the body are totally different for external exposure and for internal exposure. The effect will also vary with the energy associated with the particle as well as the time of exposure to it. Major diseases due to exposure to ionizing radiation include skin and lung cancers and the shortening of life.

Nonionizing radiation is any electromagnetic radiation that does not cause ionization of atoms or molecules. There is not sufficient energy per quantum to remove an electron completely from an atom or a molecule. However, the energy is enough to excite the electron. Examples of nonionizing radiation include microwave radiation, electric and magnetic fields, radio waves, lasers, and optical radiation.

Safety measures include reducing the time of exposure, distancing the worker, and the use of protective equipment. These are all known to be effective in reducing the risk from exposure to radiation.

Example 16.4

The overall risk of a hazard is 12 on a 0–100 scale. If the seriousness of the hazard score is estimated to be 6 on a 0–10 scale, then compute the likelihood of the occurrence on a 0–10 scale.

Solution:

Use Equation 16.1:

Risk = (Likelihood of occurrence score) $*$ (Seriousness of hazard score)

Risk = 12 (0–100 scale)
Seriousness of hazard score = 6 (0–10 scale)
Likelihood of occurrence score = ?

12 = (Likelihood of occurrence score)(6)
Likelihood of occurrence = 2 (0–10 scale)

Example 16.5

For a noise pollution study, a noise reading at a distance of 8 m from the source was recorded to be 72 dB. What will be the approximate reading at a distance of 24 m from the source?

Solution:

Use Equation 16.5:

$$\text{Noise attenuation} = 10 \log_{10}\left[\left(\frac{D_2}{D_1}\right)^2\right]$$

$D_1 = 8$ m
$D_2 = 24$ m
Noise attenuation = ?

$\text{Noise attenuation} = 10 \{\log_{10}[(24/8)^2]\}$
$\qquad\qquad\qquad = 9.5$ dB
Approximate reading at 24 m = 72 − 9.5
Approximate reading at 24 m = 62.5 dB

16.3
Process Hazard Analysis

Process hazard analysis is defined as an organized process or method developed for the analysis and identification of the hazardous risk associated with the processing and treatment of the hazardous materials in industrial processes. It assists workers and employers in planning strategies that can improve safety at workplaces.

Process hazard analysis helps in the assessment of the causes and consequences of the incidents that can take place during various industrial processes, such as fire, nuclear reactions, and explosions. It also determines the factors that can potentially lead to calamities, such as method of operation, equipment, handling of materials, and human actions.

In the United States, the Occupational Safety and Health Administration (OSHA) in its Process Safety Management regulation mandated the use of process hazard analysis for the risk analysis in the design, operation, and modification of processes that involve the use of highly hazardous chemicals.

As per OSHA regulations, the employer must evaluate the hazards of a process by using one or more of the following methods, appropriate for the process being analyzed:

- What-if
- Checklist
- What-if/checklist
- Hazard and operability study (HAZOP)
- Failure mode and effects analysis (FMEA)
- Fault tree analysis
- An appropriate equivalent methodology

Fault Tree Analysis

In this method, a diagram (model) is drawn that shows the undesirable consequences that might result from a specific initiating event (for example, an explosion due to mishandling of a toxic chemical). Fault tree analysis is a quantitative method. It uses Boolean logic and symbols to associate the order of events that might result in an accident. This method is also used when investigating accidents to analyze the possible causes. There are two types of fault tree analysis, inductive and deductive.

Inductive analysis is used to analyze the outcomes of an event that might result due to specific operations.

Forward Analysis

Event ⟶ Possible Consequences

Deductive analysis is backward analysis. It is modeled to determine the possible causes of an accident that has already happened.

Backward Analysis

Event ⟶ Possible Causes

Boolean gates are used to associate the events and causes/consequences from a specific action.

- **OR gate** represents the output that can occur if an input occurs.
- **AND gate** represents the output that happens only if all input events take place (input events are independent).
- **Exclusive OR gate** represents the occurrence of an output if exactly one input event occurs.
- **Priority AND gate** represents the occurrence of an output if the input events occur in a certain sequence specified by a conditioning event.
- **Inhibit gate** represents the occurrence of an output if the input event occurs under an enabling condition specified by a conditioning event.

Both Figure 16.1 and the following list the steps involved in fault tree analysis:

- Define an undesired event that is required to be determined.
- Define the boundary of the system.
- Construct the fault tree.
- Evaluate the fault tree.
- Minimize the identified risks.

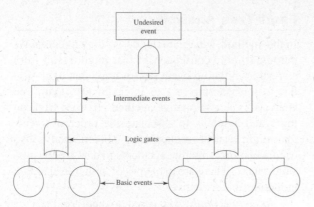

Figure 16.1 Basic fault tree structure

Fault tree analysis has many uses:

- Analyze the causes of failure of a system
- Analyze the consequences of failure of a system
- Evaluate a system upgrade
- Model system failure in the risk assessment of that system

16.4
Minimization of Waste

Waste minimization refers to the techniques adopted to minimize the amount of waste materials produced by the activities of individuals and of industrial processes. Waste minimization produces positive impacts on the environment, human health, and safety. In order to minimize the production of waste, it is necessary to acquire detailed knowledge of the process and materials used in the process. Having this knowledge promotes alternative materials or procedures that could minimize the waste and hazard risks associated with the process. The U.S. Environmental Protection Agency (EPA) has established its hierarchy of waste minimization approaches: reduce, recycle, and treatment.

Reduce (Source Reduction)

Reduction is the most desired technique. It reduces or eliminates the production of chemical waste generated at the source. It refers to the techniques that can be used to minimize the generation of waste in the first place.

For example, the use of substitute nonhazardous chemicals instead of hazardous chemicals is a simple way to minimize hazardous waste.

Recycle

Recycling is another method to minimize waste. Metals, paper, the energy value, and other useful resources in the waste can either be used for another purpose or be treated and reused in the same process. For example, high-density polyethylene (HDPE) can be recycled and used in manufacturing the bags for groceries. Low-density polyethylene (LDPE) can be recycled and used to make bags for dry cleaning or frozen food containers.

Treatment

The other minimization method is the treatment of waste. Treatment involves biological, chemical, or physical methods.

- A liquid hazardous waste stream can be detoxified and neutralized by chemical and biological treatment. The volume of waste sludge generated can be reduced by dewatering.
- Combustible hazardous waste is destroyed in special incinerators equipped with proper pollution control and monitoring systems. Ash and sludges formed are solidified/stabilized to reduce the leachability of materials.
- The remaining treated residue is disposed of in the designed landfills.

16.5
Waste Treatment

Waste can be treated in several different ways: biologically, chemically, physically/chemically, or by disposal on land.

Biological Treatment

Bio-tower or trickling filter uses the biofilm process to treat wastewater. The wastewater falls through a tower (packed bed), which is filled with permeable packing. The microorganisms attached to the packing (bed) consume the organic matter in the passing wastewater. The efficiency of the process depends on both the hydraulic loading and the organic loading. Although the predominant bacteria in the biofilm process are different than those in the activated sludge process, they function very similarly. In addition, both processes give very similar effluent quality.

Hydraulic loading

$$\text{HL} = \frac{Q + Q^r}{A_{pv}} \qquad [16.6]$$

BOD surface loading

$$SL = \frac{QS^0}{A_{pv}ha} \qquad [16.7]$$

Volumetric loading

$$VL = \frac{QS^0}{A_{pv}h} \qquad [16.8]$$

where

Q = influent flow rate (m³/d or m³/s)
Q^r = recycle flow rate (m³/d or m³/s)
A_{pv} = plan-view surface area of bio-tower (m²)
S_0 = influent BOD concentration (mg/L)
h = filter depth (m)
a = specific surface area of the medium (m²)

The activated sludge process is the most widely used of the processes for both industrial and municipal wastewater treatment. It consists of an aeration tank (reactor) and a settling tank (clarifier). Solids recycle from the clarifier to the reactor and a sludge waste line. Several enhancements to this basic process are designed and practiced to suit different site-specific or temporal changes to the raw sewage. In order to achieve this process, air (or pure oxygen) is forced continuously into a mixture of raw sewage and aggregates, or flocs, of microorganisms called activated sludge, which reduces the organic content of sewage. The mixture of raw sewage and biological mass together is termed "mixed liquor." Mixed liquor from the reactor is sent to the clarifier for settling. After sufficient settling, the supernatant is sent to treatment before final discharge, and the settled sludge is recycled to the reactor (return activated sludge or RAS) for reseeding the new raw sewage. Not all the sludge however, can be recycled. A part of it has to be wasted (waste activated sludge or WAS). This also maintains the food to microorganism (biomass) ratio (F:M in kg BOD or chemical oxygen demand (COD) applied per day per kg of total suspended solids). Though the activated sludge consists mainly of saprophytic bacteria, it also has a protozoan flora. The protozoan flora includes amoebas, spirotrichs, peritrichs, and many other filter-feeding species. Other than this, activated sludge also consists of motile and sedentary rotifers to a small extent. Figure 6.2 shows a flow process diagram of a conventional activated sludge process.

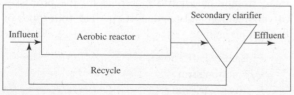

Figure 16.2 Conventional plug-flow activated sludge process

Food to microorganism ratio

$$F:M = \frac{Q^0 S^0}{VX} \qquad [16.9]$$

Solids retention time (also known as sludge age)

$$\theta_x = \frac{XV}{Q^e X^e + Q^w X^w} \qquad [16.10]$$

Volumetric organic loading rate

$$Rate = \frac{Q^0 S^0}{V} \qquad [16.11]$$

Sludge volume index

$$SVI = \frac{\text{Sludge volume after 30 minutes settling}}{\text{MLSS}} \qquad [16.12]$$

Recycle Ratio

$$R = \frac{Q^r}{Q^0} \qquad [16.13]$$

where

F:M = food to microorganism ratio (kg BOD or COD applied per day per kg of total suspended solids)
Q^0 = influent wastewater stream flow rate (m³/d or m³/s)
Q^e = effluent flow rate (m³/d or m³/s)
Q^w = waste sludge flow rate (m³/d or m³/s)
Q^r = recycle sludge flow rate (m³/d or m³/s)
S^0 = influent wastewater concentration (BOD or COD in mg/L)
V = volume of reactor (aeration tank) (m³)
X = total suspended solids concentration in the reactor (mg/L)
X^e = effluent concentration (mg/L)
X^w = waste sludge concentration (mg/L)
MLSS = mixed liquor suspended solids concentration (original) (mg/L)

Chemical Treatment

Chemical detoxification is used to reduce the hazardous effects of a particular material prior to its transportation and final disposal. The final product might not be magically nontoxic. However, insuring that the final product is less of a problem than the initial one is important.

Neutralization: A waste is considered hazardous if its pH is less than or equal to 2 or is greater than or equal to 12.5. Therefore, neutralization brings the pH of a solution into the range of 2 to 12.5. Final pH value ranges from 6 to 8 are required for good treatment processes to protect natural biota. In this process, the solution is neutralized by adding sulfuric (H_2SO_4) or hydrochloric acid (HCl) to basic solutions. Caustic soda (NaOH) or slaked lime ($Ca(OH_2)$) is added to acidic solutions.

Chemical oxidation is used to dilute a solution and is considered more expensive than biological methods. Some examples of oxidation are wet air oxidation (Zimmerman process), hydrogen peroxide, chlorine dioxide, chlorine, and ozone oxidation.

Cyanide is destroyed by oxidation under alkaline condition. Oxidation may not be desirable for high-concentration cyanide (i.e., greater than 1%) and for cyanide complexes with metals, particularly with iron and nickel.

Most frequently, chlorine is used for oxidation of cyanide. Chlorine oxidation reaction takes place in 2 steps:

$$NaCN + 2NaOH + Cl_2 \rightleftharpoons NaCNO + 2NaCl + H_2O$$

$$2NaCNO + 5NaOH + 3Cl_2 \rightleftharpoons 6NaCl + CO_2 + N_2 + NaHCO_3 + 2H_2O$$

In the first step, the pH must be maintained above 10. At low pH, highly toxic hydrogen cyanide gas may evolve. The first step in the chlorine oxidation reaction occurs quickly. The second step takes place rapidly at pH 8. However, it does not occur as quickly as the first step.

Wet air oxidation (Zimmerman process) is aqueous-phase oxidation of dissolved or suspended organic particles normally at temperatures from 175°C to 325°C. It is an efficient method and destroys most organic compounds, including pesticides. However, in some cases metal salt catalysts are added to increase the destruction efficiency. The reaction can take place at low temperature and pressure.

Precipitation reduces the solubility of metals by increasing the pH with the addition of lime and caustic. Upon reacting with lime or caustic, the hydroxide precipitates form. These precipitates are then removed at the optimum pH.

Example 16.6

A metal plating firm is installing a precipitation system to remove copper. The K_{sp} of copper hydroxide is 2.00×10^{-19}. Estimate the pH the controller should be set to in order to achieve a copper effluent concentration of 0.8 mg·L^{-1}.

Solution:

The copper hydroxide reaction is given as:

$$Cu(OH)_2 \rightleftharpoons Cu^{2+} + 2OH^-$$

The solubility product is given as :

$$K_{sp} = [Cu^{2+}][OH^-]^2$$

As defined in the problem, the concentration should not exceed 0.8 mg·L^{-1}. Therefore, calculate the moles/liter of copper:

$$[Cu^{2+}] = \frac{0.8 \text{ mg/l}}{(63.55 \text{ g/mol})\left(\frac{1000 \text{ mg}}{\text{g}}\right)}$$

$$= 1.258 \times 10^{-5} \text{ mol·L}^{-1}$$

For the concentration of hydroxide;

$$[OH^-]^2 = \frac{2.00 \times 10^{-19}}{1.258 \times 10^{-5}} = 1.58 \times 10^{-14}$$

Therefore,

$$[OH^-] = \sqrt{1.58 \times 10^{-14}} = 1.256 \times 10^{-7}$$

The pOH is

$$pOH = -\log(1.256 \times 10^{-7}) = 6.900$$

The pH set point for the controller is

$$pH = 14 - pOH = 14 - 6.9 = 7.099 \text{ or } 7.1$$

Physical/Chemical Treatment

In *carbon adsorption*, the gas molecules are absorbed, attracted to, and held on the surface of a solid by intermolecular forces. It is a mass transfer process. Activated carbon (heated charcoal), molecular sieves, silica gel, and activated alumina are the most common adsorbent materials. A properly designed carbon adsorption unit can remove gas with an efficiency exceeding 95%. Adsorption systems are configured either as stationary bed units or as moving bed units. In stationary bed adsorbers, the polluted airstream enters from the top, passes through a layer or bed of activated carbon, and exits at the bottom. In moving bed adsorbers, the activated carbon moves slowly down through channels by gravity as the air to be cleaned passes through in a cross-flow current.

In distillation, more volatile materials are separated from less volatile materials by the processes of vaporization and condensation. The four types of distillation include:

- Batch distillation
- Fractionation
- Steam stripping
- Thin film evaporation

Batch distillation is used for wastes containing high concentration solids. Fractionation is used when wastes contain minimal suspended solids and multiple constituents are required to be separated.

Air stripping is used for contaminated groundwater containing high concentrations of volatile and lower concentrations of organic compounds. Air and contaminated liquid pass countercurrently through a packed tower. The volatile compounds evaporate into the air, leaving a clean liquid stream behind. Compounds such as tetrachloroethylene, trichloroethylene, and toluene from water are removed by using this technique. See Figure 16.3.

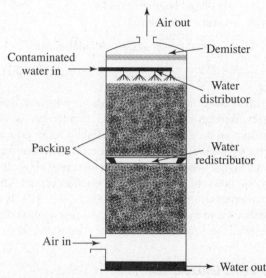

Figure 16.3 Air stripper (Source *Wikipedia*)

The air stripper design equation may be written as:

$$Z_T = \frac{L}{A} \frac{\ln\left\{(C_1/C_2) - \left(\frac{LRTg}{GH_c}\right)[(C_1/C_2) - 1]\right\}}{K_L\alpha\left[1 - \frac{LRTg}{GH_c}\right]}$$

[16.14]

where

Z_T	= packed tower depth (m)
L	= water flow (m³/min)
A	= cross-sectional area of tower (m²)
G	= air flow (m³/min)
H_C	= Henry constant
R	= universal gas constant (8.206 × 10⁻⁵ atm·m³·mol⁻¹·K⁻¹)
T_g	= temperature of air (K)
C_1, C_2	= influent and effluent organic concentration in the water (mol·m⁻³)
K_L	= liquid mass transfer coefficient
α	= effective interfacial area of packing per unit volume for mass transfer

Example 16.7

Drinking water is contaminated by trichloroethylene. The average concentration in the water is estimated to be 4,000 μg·L⁻¹. Using the following parameters, design a packed-tower stripping column to reduce water concentration to the discharge limit of 1.5 μg·L⁻¹. Note that more than one column in series may be required for reasonable tower height.

Henry law constant = 6.74 × 10⁻³
$K_L{}^a$ = 0.72 min⁻¹
Air flow rate = 60 m³·m⁻¹
G/L = 15
Temperature = 25°C
Column diameter should not exceed 4.0 m
Column height should not exceed 6.0 m

Solution:

Given that:

Henry law constant = 6.74 × 10⁻³
$K_L{}^a$ = 0.72 min⁻¹
Air flow rate = 60 m³·m⁻¹
G/L = 15
Temperature = 25°C = 298 K
Column diameter should not exceed 4.0 m
Column height should not exceed 6.0 m

Using equation 16.14:

$$Z_T = \frac{L}{A} \frac{\ln\left\{(C_1/C_2) - \left(\frac{LRTg}{GH_c}\right)[(C_1/C_2) - 1]\right\}}{K_L\alpha\left[1 - \frac{LRTg}{GH_c}\right]}$$

$$Z_T A = L \frac{\ln\left\{(C_1/C_2) - \left(\frac{LRTg}{GH_c}\right)[(C_1/C_2) - 1]\right\}}{K_L\alpha\left[1 - \frac{LRTg}{GH_c}\right]}$$

G/L = 15
$G = 60 \Rightarrow L = 4$ m³·m⁻¹
$R = 8.206 \times 10^{-5}$

$$Z_T A = 4 \times \frac{\ln\left\{\left(\frac{4,000}{1.5}\right) - \left(\frac{4 \times 8.206 \times 10^{-5} \times 298}{60 \times 6.74 \times 10^{-3}}\right)\left[\left(\frac{4,000}{1.5}\right) - 1\right]\right\}}{0.72\left[1 - \frac{4 \times 8.206 \times 10^{-5} \times 298}{60 \times 6.74 \times 10^{-3}}\right]}$$

$Z_T A = 4 \times 13.907$

$Z_T A = 55.628$ m³

The column volume is 55.628 m³. The possible solution within the boundary conditions of 4.0 m diameter and 6.0 m height could be : •

Taking height = 6 m

$\pi r^2 h = 55.628$
$r = 1.717$ m or $d = 3.43$ m

Taking height = 5 m

$\pi r^2 h = 55.628$
$r = 1.88$ m or $d = 3.76$ m

Diameter (m)	Tower Height (m)
6	3.43
5	3.76

Steam stripping is used for low volatility and highly concentrated gases (>100 ppm).The process is more likely similar to air stripping except steam is applied rather than air. Aqueous waste containing contaminants such as chlorinated hydrocarbons, xylene, acetones, methyl ethyl ketones, methanol, and pentachlorophenol are treated by this technique. The concentration treated range is 100 ppm to 10% organic chemical (US EPA 1987).

The ion exchange process helps to recover metals and ionized organic chemicals. The waste stream containing contaminants is passed through the bed of ion exchange resin. Typically zeolites, montmorillonite, clay, and soil humus are used as ion exchange resin. Ion exchangers are either cation or anion exchangers. The like charges from the resin surface remove the waste stream ions in an exchange process. As the resin bed gets saturated with the exchanged ions, the bed is shut down and the resin is regenerated.

Electrodialysis can separate a waste stream containing 1,000 to 5,000 mg/L inorganic salts into a dilute stream that contains 100 to 500 mg/L salt and a concentrated stream that contains up to 100 mg/L salt. An electrodialysis cell consists of alternate anion and cation ion exchange resin thin membrane sheets reinforced by a synthetic fabric backing. Electric potential helps in the migration of ions. As a result, positive ions pass through the cation membrane and migrate toward the cathode. Negative ions pass through the anion membrane and migrate toward the anode. Pure water is eventually left in the area between the membranes. The migration of ions is directly proportional to the applied electric potential. In *n* cells, the required current is

$$I = \frac{FQN}{n} \times \frac{E_1}{E_2}$$

where

I	=	current (ampere)
F	=	Faraday's constant = 96,487 C/g-equivalent
Q	=	flow rate (L/s)
N	=	normality of solution (g-equivalent/L)
n	=	number of cells between electrodes
E_1	=	removal efficiency (fraction)
E_2	=	current efficiency (fraction)

The voltage can be calculated as

$$E = IR \qquad [16.16]$$

where

E	=	voltage required (volts)
R	=	resistance (ohms)

The required power can be calculated as:

$$P = I^2 R \qquad [16.17]$$

Reverse osmosis is a process in which a fluid passes through a semipermeable membrane, moving from an area of lower concentration to an area of higher concentration. In reverse osmosis, the pressure on the higher concentration side is increased above the osmotic pressure and the reverse reaction occurs. The driving pressure ranges from 1,000 to 5,000 kPa. It is widely used to treat industrial waste water such as that generated from metal finishing and plating operations.

Solvent Extraction

Solvent extraction is also called liquid-liquid extraction or liquid extraction. The waste stream is mixed into a solvent to allow the mass transfer of the constituents from the waste to the solvent. The extracted solvent that contains contaminants is called extract. The waste stream that is free from contaminants is called raffinate. •

Incineration: The combustible hazardous components of waste are decomposed by oxidation at high temperature (800°C and above).The combustion of organic matter primarily produces inert ash, carbon dioxide, and water vapor. Over 90% of all incineration facilities use either liquid injection or rotary kiln incinerators.

Permitting for hazardous waste incinerators: The performance standards of incinerators are regulated under EAA rules 40CFR264.343. Hazardous waste incinerators must meet three performance standards (Theodore and Reynolds,1987)

1. **Principal organic hazardous waste constituents (POHC):** The POHC performance standard requires that hazardous waste incinerators must be designed to achieve 99.99%

or higher destruction and removal efficiency (DRE). DRE for a given POHC is defined as "the mass percentage of the POHC removed from the waste." It can be evaluated as:

$$DRE = \frac{W_{in} - W_{out}}{W_{in}} \times 100 \qquad [16.18]$$

where

W_{in} = Mass feed rate of one POHC in the waste stream

W_{out} = Mass emission rate of the same POHC present in exhaust emission prior to release to the atmosphere

2. **Hydrochloric acid (HCl):** The emission of HCl should be controlled such that it should not exceed 1.8 kg·h^{-1} or 1% of the HCl in the stack gas prior to entry into any pollution control equipment.

3. **Particulates:** The stack emission of particulate matter is limited to 180 mg per dry standard cubic meter (mg/dscm) for a stack gas corrected to 7% oxygen. This adjustment is made by calculating a corrected concentration:

$$P_c = P_m \frac{14}{21 - Y} \qquad [16.19]$$

where

P_c = corrected concentration of particulates (mg per dscm)

P_m = measured concentration of particulates (mg per dscm)

Y = percent oxygen in the dry flue gas.

Regulations for Polychlorinated Biphenyls (PCBs): Incineration of PCBs is regulated under TSCA (Wentz 1989).

1. **Time and temperature:** The regulations required to meeting either one of two conditions:

- The residence time of PCBs in the furnace must be 2 s at 1,200 ± 100°C with 2% oxygen in the stack gas, or
- The residence time of PCBs in the furnace must be 1.5 s at 1,600 ± 100°C with 2% oxygen in the stack gas.

2. **Combustion and efficiency:** The combustion efficiency should be 99.99%, which is computed as:

$$CE = \frac{C_{CO_2}}{C_{CO_2} + C_{CO}} \times 100\% \qquad [16.20]$$

3. **Monitoring and control:** The regulations require monitoring the rate and quantity of PCBs fed into the combustion system every 15 minutes. The temperature should also be checked and controlled at regular intervals. The regulations also require conducting a trial burn and monitoring the following emissions:

- Oxygen (O_2)
- Carbon monoxide (CO)
- PCBs
- Oxides of nitrogen (NO_2)
- Hydrogen chloride (HCl)
- Total chlorinated organic contents
- Total particulate matter

Example 16.8

A hazardous waste incinerator is being fed methylene chloride at a concentration of 4,858 mg·L^{-1} in an aqueous stream at a rate of 40.5 L·min^{-1}. Calculate the mass flow rate of the feed in g/min.

Solution:

Concentration of methylene chloride = 4,858 mg·L^{-1} = 4.858 g·L^{-1}

Flow rate aqueous stream = 40.5 L·min^{-1}

Mass flow rate of the feed can be calculated as:

$= 4.858$ g·L^{-1} $\times 40.5$ L·min^{-1}
$= 185.692$ g/min

The mass flow rate of the feed is 185.7 g/min.

Example 16.9

The POHs from a trial burn of POHs are shown in the table. The incinerator was operated at a temperature of 1,200°C. The stack gas flow rate was 355.9 dscm·s^{-1}. Is the unit in compliance?

Compound	Inlet (kg/h)	Outlet (kg/h)
Chlorobenzene (C_6H_6Cl)	160	0.016
Toluene (C_6H_8)	500	0.050
Xylene (C_8H_{10})	1,378.9	0.56
HCl		1.2
Particulates at 7% O_2		3.615

Solution:

Calculate the DRE first:

$$DRE = \frac{W_{in} - W_{out}}{W_{in}} \times 100$$

$$DRE_{chlorobenzene} = \frac{160 - 0.016}{160} \times 100 = 99.999\%$$

$$DRE_{toluene} = \frac{500 - 0.05}{500} \times 100 = 99.99\%$$

$$DRE_{xylene} = \frac{1{,}378.9 - 0.056}{1{,}378.9} \times 100 = 99.95\%$$

The DRE for each POH should be 99.99%. Here the DRE for xylene fails to meet the required standard.

Now check for the HCl. As per the standard, HCl emission should not exceed 1.8 kg·h^{-1}. It is given that the emission rate of HCl is 1.2 kg·h^{-1}, which is less than 1.8 kg/h.

To calculate the mass emission rate prior to control for the purpose of comparison, assume all the chlorine in the feed converted is into HCl:

$$M_{CB} = \frac{W_{CB}}{(MW)_{CB}} \times 100$$

where
M_{CB} = molar flow rate of chlorobenzene
$(MW)_{CB}$ = molecular weight of chlorobenzene

$$M_{CB} = \frac{\left(150\,\frac{kg}{h}\right)\left(1{,}000\,\frac{g}{kg}\right)}{112.5\ g/mol} \times 100$$

$$= 1{,}333.33\ mol·h^{-1}$$

Each molecule of chlorobenzene contains one atom of chlorine. Therefore,

M_{HCl} = M_{CB} = 1,333.33 mol·h^{-1}
W_{HCl} = GMW of HCl (in mol·h^{-1})
W_{HCl} = (36.5 g·mol^{-1})(1,333.33 mol·h^{-1})
$\quad$ = 48,666.545 g·h^{-1} or 48.66 kg·h^{-1}
1% of uncontrolled emission of HCl
$\quad$ = (0.01)(48.66)
$\quad$ = 0.4866 kg·h^{-1}

The emission of 1.2 kg·h^{-1} is greater than the uncontrolled emission. However, the incinerator passes the HCl limit as the emission is less than 1.8 kg·h^{-1}.

The particulate concentration was measured at 7% O_2. Therefore, the concentration does not need to be corrected. The outlet loading of the particulates is

$$W_{out} = \frac{\left(3.615\ kg·h^{-1}\right)\left(10^6\ mg·kg^{-1}\right)}{\left(355.9\ dscm·min^{-1}\right)\left(60\ min·h^{-1}\right)}$$

$$= 169.289\ mg·dscm^{-1}$$

The emission of particulate matter is 169.289 mg·dscm^{-1}, which is less than the standard 180 mg·dscm^{-1}. Therefore, the unit is in compliance with particulates.

However, the incinerator fails the DRE for xylene. So the unit is out of compliance.

Solidification and stabilization do not destroy contaminants. Instead, these processes prevent or slow the release of harmful chemicals (or leaching) from wastes, such as contaminated soil, sediment, and sludge. A waste is mixed with a binding agent such as cement or lime. Then water is added to the mixture to form a bond. The mixture is then allowed to dry and harden to form a solid block. Stabilizing agents can be modified by other additives such as silicates.

Land Disposal

The biologically, chemically/physically treated waste leaves 20% residues of the original mass. At this point, the disposal of residue in a secured landfill is the only option. The design and operation of hazardous waste landfill is regulated under EPA rules 40 CFR 264.300. It requires a minimum of the following:

1. Two or more liners. The upper component must be of composite material and thick enough to prevent/minimize the migration of hazardous components. The hydraulic conductivity of the lower component should not exceed 1×10^{-7} cm/s.
2. Leachate collection and removal (LCR) system between the liners and immediately above the bottom composite liner. In the case of multiple leachate collection and removal systems, a leak detection system is also required. The maximum permissible leachate head is 30 cm (1 foot). The LCR should be

- Constructed with a bottom slope of 1% or more
- Constructed of granular drainage materials with a hydraulic conductivity of 1×10^{-2} cm/s or more and a thickness of 12 inches (30.5 cm) or more. Alternatively, constructed of synthetic or geonet drainage materials with a transmissivity of 3×10^{-2} m^2/s or more
- Constructed of materials that are of sufficient strength and thickness to prevent collapse under the pressures exerted by overlying wastes, waste cover materials, and equipment used at the landfill
- Designed and operated to minimize clogging during the active life and postclosure care period

- Constructed with sumps and liquid removal methods (e.g., pumps) of sufficient size to collect and remove liquids from the sump and prevent liquids from backing up into the drainage layer

3. Design, construct, operate, and maintain a runoff management system to collect and control at least the water volume resulting from a 24-hour, 25-year storm.
4. Monitoring wells.
5. Landfill cap.

The migration of leachate through a clay liner can be calculated by using Darcy's law:

$$v = K \frac{dh}{dr} \qquad [16.21]$$

where

v = Darcy velocity
K = hydraulic conductivity
$\frac{dh}{dr}$ = hydraulic gradient

The seepage velocity will be

$$v' = \frac{K \frac{dh}{dr}}{\eta} \qquad [16.22]$$

where

v' = seepage velocity
η = porosity of geological material

The travel time of the contamination through a soil layer can be estimated as:

$$t = \frac{T}{v'} \qquad [16.23]$$

where

T = linear length of the flow path

Example 16.10

How long will it take for leachate to migrate through a 0.7 m clay liner if the depth of leachate above the clay layer is 30 cm? The hydraulic conductivity of the clay is 1×10^{-7} cm/s and the porosity is 45%.

Solution:

Given:
Thickness of clay liner = 0.7 m
Leachate head = 30 cm = 0.3 m
$K = 1 \times 10^{-7}$ cm/s
$\eta = 45\%$

The Darcy velocity can be calculated by:

$$v = K \frac{dh}{dr}$$

The hydraulic gradient can be estimated as:

$$\frac{dh}{dr} = \frac{0.3 + 0.7}{0.7} = 1.42$$

So

$$v = (1 \times 10^{-7}) \times (1.42) = 1.42 \times 10^{-7} \text{ cm/s}$$

Seepage velocity is given by :

$$v' = \frac{v}{\eta}$$

$$v' = \frac{1.42 \times 10^{-7}}{0.45} = 3.155 \times 10^{-7} \text{ cm/s}$$

The travel time will be

$$t = \frac{T}{v'}$$

$$t = \frac{0.7 \text{ m} \left(\frac{100 \text{ cm}}{\text{m}} \right)}{3.155 \times 10^{-7}} = t = 2.218 \times 10^{8} \text{ s}$$

or 7.03 years

The leachate will take approximately 7 years to migrate through the 0.7 m clay liner having a hydraulic conductivity of 1×10^{-7} cm/s and porosity of 45%.

16.6 Overpressure and Underpressure

Overpressure and underpressure can have several causes.

Causes of Overpressure

The major causes of overpressure include the following:

- Operating problems
- Equipment failure
- Process upset
- External fire
- Utility failures

Operating problems include mistakes such as an operator mistakenly opening or closing a valve to cause the vessel or system pressure to increase. An operator, for example, may adjust a steam regulator that results in pressures exceeding the maximum allowable working pressure (MAWP) of a steam jacket. Although the set pressure is usually at the MAWP, the design safety factors should protect the vessel for higher pressures. A vessel fails when the pressure is typically several times the MAWP.

Equipment failure can cause overpressure. For example, a heat exchanger tube rupture can increase the shell side pressure beyond the MAWP. Although the set pressure is usually the MAWP, the design safety factors should protect the vessel for higher pressures. A vessel fails when the pressure is typically several times the MAWP.

Process upset can cause overpressure. For example, a runaway reaction can cause high temperatures and pressures.

External fire: External heating, such as fire, can heat the contents of a vessel. This can result in high vapor pressure and can also cause overpressure.

Utility failures, such as the loss of cooling or the loss of agitation causing a runaway reaction, are also the cause of overpressure.

Causes of Underpressure

The major causes of underpressure include the following:

- Operating problems
- Equipment failures

Operating problems that can cause underpressure include mistakes, such as pumping liquid out of a closed system, or cooling and condensing vapors in a closed system.

Equipment failures include an instrument malfunction (e.g., vacuum gauge) or the loss of the heat input of a system that contains a material with a low vapor pressure.

Pressure Relief Devices

Pressure relief devices are added to process equipment to prevent the pressures from significantly exceeding the MAWP. (Pressures are allowed to go slightly above the MAWP during emergency reliefs.) Pressure relief devices include:

- Spring-loaded pressure relief valve
- Rupture disc
- Buckling pin
- Miscellaneous mechanical

Spring-loaded pressure relief valve: As the pressure in the vessel or pipeline at point A in Figure 16.4 exceeds the pressure created by the spring, the valve opens. The relief begins to open at the set pressure, which is usually at or below the MAWP. This pressure is usually set at the MAWP.

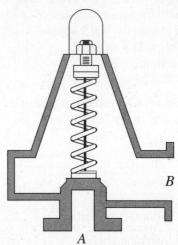

Figure 16.4 Spring-loaded pressure relief valve

Spring loaded pressure relief valves include the following several types:

1. **Safety valves:** Safety valves are designed for gases and vapors. Safety valves are further classified into:

   - Conventional
   - Balanced bellows
   - Pilot operated

2. **Relief valves:** They are designed specifically for liquids. They are further classified as:

   - Conventional
   - Balanced bellows

3. **Safety relief valves:** Safety relief valves are designed for liquids and/or gases.

Rupture disc: In Figure 16.5, the disc ruptures when the pressure at A exceeds the set pressure. Recognize, however, that it is actually the differential pressure (A – B) that ruptures the disc.

There are several types of rupture disc:

- Metal
- Graphite
- Composite
- Others

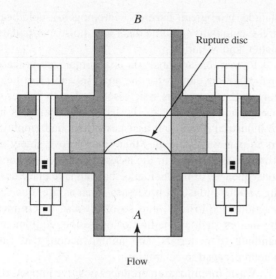

Figure 16.5 Rupture disc

Buckling pin: When the pressure exceeds the set pressure, the pin buckles. The vessel's contents exit through the open valve. The spring-operated valves close as the pressure decreases below the blowdown pressure. The blowdown pressure is the difference between the set pressure and the closing pressure, as shown in Figure 16.6.

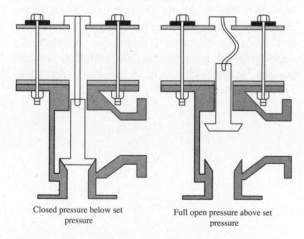

Closed pressure below set pressure

Full open pressure above set pressure

Figure 16.6 Buckling pin: The pin buckles in milliseconds at a precise, set temperature

Runaway Reaction

Overpressure can also occur due to a runaway reaction. While protecting a system from overpressure due to a runaway reaction, you must know the type and cause of a runaway reaction.

Self-heating reaction occurs due to an uncontrolled rise of temperature. Some of the reasons include:

- Loss of cooling
- Unexpected addition of heat
- Too much catalyst or reactant
- Operator mistakes
- Too fast addition of catalyst or reactant

Sleeper reaction is usually the result of operating error. For example, the addition of two immiscible reactants when the agitator is mistakenly in the off position can cause a sleeper reaction.

16.7 Storage and Handling

Storage of hazardous waste is to stock the waste for a temporary period of time prior to the waste being treated, disposed, or stored elsewhere. The storage facilities must comply with requirements listed in the RCRA regulations section 40 CFR Part 264 for permitted facilities and 40 CFR Part 265 for interim status facilities. Hazardous waste is usually stored in one of the several ways listed below.

A hazardous waste container is any portable device in which a hazardous waste is stored, transported, treated, disposed, or otherwise handled. The most common hazardous waste container is the 55-gallon drum. Other examples of containers are tanker trucks, railroad cars, buckets, bags, and even test tubes.

Tanks are constructed of nonearthen materials, including steel, plastic, fiberglass, and concrete. They are used to store or treat hazardous waste. They are stationary devices. Tanks can be opentopped or be completely enclosed.

A drip pad is a wooden drying structure used by the pressure-treated wood industry to collect excess wood preservative drippage. Drip pads are constructed of noncarthen materials with a curbed, free-draining base that is designed to convey wood preservative drippage to a collection system for proper management.

Containment buildings are completely enclosed, self-supporting structures. In other words, they have four walls, a roof, and a floor. They are used to store or treat noncontainerized hazardous waste.

501

A waste pile is an open, uncontained pile used for treating or storing waste. Hazardous waste piles must be placed on top of a double-liner system to ensure leachate from the waste does not contaminate surface or ground water supplies.

A surface impoundment is a natural topographical depression, manmade excavation, or diked area such as a holding pond, storage pit, or settling lagoon. Surface impoundments are formed primarily of earthen materials and are lined with synthetic plastic liners to prevent liquids from escaping.

16.8
Summary

Understanding hazardous properties of materials, such as ignitability, toxicity, corrosivity, and reactivity, is essential to identifying and classifying material hazards that affect health and the environment. The best available preventive measure to avoid a hazard is hazard communication. OSHA sets standard values for allowable exposure limit in the work environment. It also mandates the use of appropriate safety equipment while handling one or more substances that are hazardous in nature. Epidemiological and toxicological data are collected and studied on a large scale to determine a safe human dose level of a toxicant. Dose-response relationship also helps in making critical decisions.

The study and control of environmental factors at a workplace that may produce unwanted effects on the workers or the neighboring community is industrial hygiene. Principal entities in industrial hygiene include biological hazards, carcinogens, cold/heat stress, cumulative trauma disorders, noise, particulate matter, and radiation.

Process hazard analysis is an organized process or method developed for the analysis and identification of the hazardous risk associated with the processing and treatment of the hazardous materials in an industrial process. It assists workers and employers in planning strategies that can improve safety at workplaces. It helps in the assessment of causes and consequences of the incidents that can take place during various industrial processes, such as fire, nuclear reactions, explosions, and so on. It also determines the factors such as method of operation, equipment, handling of materials, and human action, that can potentially lead to calamities.

Waste minimization produces positive impacts on the environment, human health, and safety. In order to minimize the production of waste, it is necessary to acquire detailed knowledge of the process and materials used in the process. This means understanding the extraction of material from the earth to the impact of its disposal after processing. The purpose is that an alternative material or procedure could be suggested to minimize the waste and hazard risks associated with the process. Waste treatment through physical, chemical, and biological means is critical for reducing hazards to health and the environment. Some of the popular physical and chemical treatment processes include carbon adsorption, distillation, air stripping, steam stripping, ion exchange, electrodialysis, reverse osmosis, solvent extraction, incineration, and stabilization/solidification.

PRACTICE PROBLEMS

1. Which of the following are the major pathways for toxic agents?
   (A) Inhalation
   (B) Skin absorption
   (C) Ingestion
   (D) All of above

2. A chemical is corrosive if
   (A) it is aqueous with pH equal to or greater than 2 and less than or equal to 12.5
   (B) it is liquid with boiling point 60°C and pH less than 2.
   (C) it is aqueous with pH less than or equal to 2 and greater than or equal to 12.5
   (D) none of the above

3. Which of the following are the sources for naturally occurring particulate matter?
   1. Volcano
   2. Forest fire
   3. Fossil fuel burning
   4. Desert storm

   (A) 1 and 3 only
   (B) 1 and 4 only
   (C) 1, 2, and 3 only
   (D) 1, 2, and 4 only

4. Fire that heats the contents of a vessel giving high vapor pressures can cause
   (A) underpressure
   (B) overpressure
   (C) self-heating
   (D) sleeper reaction

5. Which of the following is the preferred method to analyze the causes of a fire explosion during work?
   (A) Boolean analysis
   (B) Inductive analysis
   (C) Deductive analysis
   (D) None of the above

6. In carbon adsorption, gas vapors are held together by:
   1. Coordinate covalent bonding
   2. Van der Waal's forces
   3. Ionic bonding
   4. Hydrogen bonding

   (A) 2 and 3 only
   (B) 2 and 4 only
   (C) 1, 2, and 3 only
   (D) 1, 2, and 4 only

7. If the stripper column volume is 40 m^3, what would be a suitable diameter if column height is limited to 5 m?
   (A) 8 m
   (B) 4.2 m
   (C) 3.19 m
   (D) 1.6 m

Answer Key

1. D
2. C
3. D
4. B
5. C
6. B
7. C

Answers Explained

1. **D** Inhalation, skin absorption, and ingestion are all major pathways for toxic agents.

2. **C** A chemical is said to be corrosive if it is aqueous with pH less than or equal to 2 and greater than or equal to 12.5

3. **D** Volcano, forest fire, and desert storm are the sources for naturally occurring particulate matter. Fossil fuel burning is not.

4. **B** External heating, such as fire that heats the contents of a vessel giving high vapor pressures, can cause overpressure.

5. **C** Deductive analysis determines the possible causes of an accident that has happened.

6. **B** In carbon adsorption, the gas molecules are absorbed, attracted to, and held on the surface of a solid by intermolecular forces, which are van der Waal's forces and hydrogen bonding.

7. **C** $Z_T A = 40 \text{ m}^3 \Rightarrow (5)(\pi r^2) = 40 \Rightarrow r = 1.596 \text{ m}$ or $d = 3.19$ m

17

Mechanical Design and Analysis

Hamid N. Hashemi, Ph.D.

Introduction

Design of a system encompasses identification of functional requirements and design parameters. Functional requirement identifies what the system is supposed to do, what the design is supposed to accomplish. For example, the system is required to lift a load, move an object, drill a hole, and so on. Design parameters specify the range of the load that is being lifted, how fast the load is lifted, the size of the machine being designed, the cost of the project, and so forth.

To achieve these goals, the designer must draw a conceptual system, develop a mock-up and sketches, provide an analysis of the project, select material, provide a failure analysis, and more. The designer may need to go through several iterations in order to satisfy functional requirements and design parameters.

Functional requirements and recognition of need
Design parameters and specifications
Conceptual design, mock-up, and sketches
Materials selections, analysis of parts, consideration of manufacturing, costs, etc.
Testing and evaluation
Presentation

Designs should be based on codes and standards. Various organizations have provided codes and standards that should be followed in the design. Some of these organizations are

Aluminum Association (AA)
American Bearing Manufacturers Association (ABMA)

American Gear Manufacturers Assocation (AGMA)
American Institute of Steel Construction (AISC)
American Iron and Steel Institute (AISI)
American National Standards Institute (ANSI)
American Society of Heating, Refrigerating and Air-Conditioning
  Engineers (ASHRAE)
American Society of Mechanical Engineers (ASME)
American Society for Testing and Materials (ASTM)
American Welding Society (AWS)
ASM International
British Standards Institution (BSI)
Industrial Fasteners Institute (IFI)
Institute of Transportation Engineers (ITE)
Institution of Mechanical Engineers (IMechE)
International Bureau of Weights and Measures (BIPM)
International Federation of Robotics (IFR)
International Standards Organization (ISO)
National Association of Power Engineers (NAPE)
National Institute for Standards and Technology (NIST)
Society of Automotive Engineers (SAE)

When designing components, you should follow these steps:

1. Identify the machine element being designed and the nature of the design calculation.
2. Draw a sketch of the element, and show all features that affect performance or stress analysis.
3. Show in a sketch the forces acting on the element (free-body diagram).
4. Identify the kind of analysis to be performed, such as bending stress, deflection, buckling, fatigue, and so on.
5. Judge the reasonableness of the results.
6. If the results are not reasonable, change the design decisions and recalculate for a different geometry or a different material.
7. Use standard sizes, convenient dimensions, readily available materials, and so on.

17.1
Materials Selection

It is the designer's responsibility to specify suitable materials for a given design. The designer must ask several questions:

- What is the function of this component?
- What load is carried by this component?
- In what environment must it operate?
- From several materials, satisfying your need, which one is readily available and can be manufactured?
- What will be the cost for each material?
- Can each material be machined?

Material properties that are considered in design could be the following:

- Tensile strength
- Yield strength
- Proportional limit
- Elastic limit
- Modulus of elasticity in tension
- Ductility and percent elongation
- Shear strength
- Fatigue properties
- Creep properties
- Fracture properties
- Corrosion properties
- Poisson's ratio
- Modulus of elasticity in shear
- Flexural strength and flexural modulus
- Hardness
- Wear in mechanical devices
- Machinability
- Density
- Coefficient of thermal expansion
- Thermal conductivity
- Electrical resistivity

Alloys of steel are identified by YYXX. YY identifies the major alloying element, and XX identifies the percentage of carbon. For example, the American Iron and Steel Institute defines steel as AISI YYXX. AISI 4340 indicates a nickel–chromium–molybdenum steel with 0.4% carbon. Table 17.28 at the end of the chapter shows some common alloy steels and their compositions. These alloys could also be heat treated to achieve the desired mechanical properties. Some of the standard heat treatments include:

- Annealing
- Normalizing
- Quenching and tempering
- Case hardening

Annealing consists of homogenizing material at a temperature around 750°C for about an hour and then cooling in the oven (slow cooling). This results in a very soft material. Normalizing is the same process except the material is cooled in air (faster cooling), resulting in a stronger but less ductile material. Quenching and tempering involve fast cooling, often in oil or water and subsequent tempering material between 200°C-400°C for about 1 hour. This results in a strong material but with a low ductility. A material could also be strengthened by performing cold or hot rolled. Cold work produces a smoother surface than hot worked. Cold-rolled steel also produces stronger material, but it is less ductile than hot-rolled steel.

Tables 17.29 through 17.33 show the mechanical properties of some alloy steels that are used in design. Nonferrous materials are also used in the design of components. Aluminum, magnesium, copper, and titanium are examples of nonferrous materials. Aluminum is identified as AYxxx, where Y identifies major alloying and xxx may define impurity. Tables 17.34 and 17.35 show the compositions of some aluminum alloys. Some aluminum alloys can be heat treated to achieve proper mechanical properties. The mechanical properties of copper and some plastics are shown in Tables 17.36 and 17.37.

17.2
Hardness Test

The hardness test is a quick method to estimate the ultimate tensile strength of a material. Brinell and Rockwell tests are common hardness test methods. Brinell hardness is found by measuring the depth of penetration of a ball with diameter D under load of P.

$$HB = \frac{P}{\pi Dt} \qquad [17.1]$$

where P is applied load in kg, D is diameter of ball, and t is depth of penetration in mm. The Brinell hardness can be used to estimate the material's ultimate tensile strength, S_u.

$$S_u = \begin{cases} 0.495 \text{ HB ksi} \\ 3.41 \text{ HB MPa} \end{cases} \qquad [17.2]$$

Example 17.1

The Brinell hardness of a cold-worked steel is found to be 120. Find its ultimate tensile strength.

Solution:

$S_u = 0.495 \times 120 = 59.4$ ksi

Rockwell hardness is also used to estimate the tensile strength of a material. Standard Rockwell tests are RA, RB, and RC. RA is based on penetration of a cone under 60 kg. RB is based on penetration of 1/16 inch ball under 100 kg. RC is based on penetration of cone under 150 kg. Tables 17.38 through 17.40 show the conversion from Rockwell data to the ultimate tensile strength.

17.3
Stress and Deformation Equations for Various Loadings

Stress σ and elongation δ of a bar under tension or compression are given as

$$\sigma = \frac{P}{A} \qquad [17.3]$$

$$\delta = \frac{PL}{EA} \qquad [17.4]$$

In equations 17.3 and 17.4, P is the applied load, L is the component length, A is the component cross section, and E is the elastic modulus of the material.

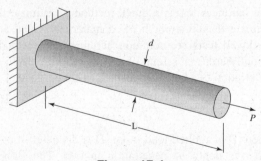

Figure 17.1

For a cylindrical component under torsion, Shear stress τ at the distance from the cylinder axis and the rotation of the cross section (angle of twist) θ are expressed as

$$\tau = \frac{Tr}{J} \qquad [17.5]$$

$$\varphi = \frac{TL}{GJ} \qquad [17.6]$$

$$J = \frac{\pi}{2}c^4 \qquad [17.7]$$

In these equations, T is the applied torque, G is the shear modulus, and J is the polar moment of inertia.

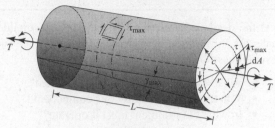

Figure 17.2

For closed, thin-walled tubes of any cross-sectional shape, shear stress in the cross section of the tube can be found from

$$\tau = \frac{T}{2At} \qquad [17.8]$$

where A is the average of the cross section and t is the tube thickness.

For components under bending, the shear and bending diagrams should be plotted to determine the magnitude of the maximum bending moment. Bending stress can be determined from

$$\sigma = \frac{Mc}{I} \qquad [17.9]$$

where c is the maximum distance to the neutral axis and I is the area of moment inertia. For the round components, $c = R$ and $I = \frac{\pi R^4}{4}$. For a rectangular cross section, $c = h/2$ and $I = \frac{bh^3}{12}$.

17.4
State of Stress in a Component

The critical location for most of the components is at the outer surface where the maximum stresses occur. The state of stress is defined as

$$\sigma_{ij} = \begin{bmatrix} \sigma_{xx} & \tau_{xy} & \tau_{xz} \\ \tau_{yx} & \sigma_{yy} & \tau_{yz} \\ \tau_{zx} & \tau_{zy} & \sigma_{zz} \end{bmatrix} \qquad [17.10]$$

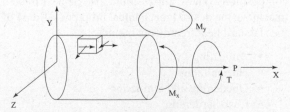

Figure 17.3

The state of stress on an element is presented as shown in Figure 17.4.

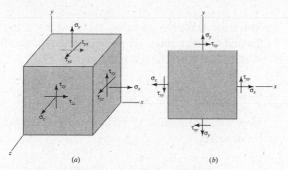

(a)   *(b)*

Figure 17.4 (a) The general 3D stress field and (b) stress field in an element located at the free surface

The principal stresses can be found from:

$$\sigma^3 - I_1\sigma^2 + I_2\sigma - I_3 = 0 \qquad [17.11]$$

where the stress invariants I_1, I_2, and I_3 are defined as:

$$I_1 = \sigma_x + \sigma_y + \sigma_z$$

$$I_2 = \sigma_x\sigma_y + \sigma_x\sigma_z + \sigma_y\sigma_z - \tau_{xy}^2 - \tau_{yz}^2 - \tau_{xz}^2$$

$$I_3 = \sigma_x\sigma_y\sigma_z + 2\tau_{xy}\tau_{yz}\tau_{xz} - \sigma_x\tau_{yz}^2 - \sigma_y\tau_{xz}^2 - \sigma_z\tau_{xy}^2$$

$$[17.12]$$

Maximum shear stress can be found by identifying the maximum and minimum roots of equation 17.11 and using equation 17.13:

$$\tau_{max} = \frac{1}{2}(\sigma_1 - \sigma_3) \qquad [17.13]$$

Example 17.2

At a critical location of a component, the stress field is given as

$$\sigma_{ij} = \begin{pmatrix} 60 & 20 & 20 \\ 20 & 0 & 40 \\ 20 & 40 & 0 \end{pmatrix} \text{ MPa}$$

Find the principal stresses and maximum shearing stress at this location.

Solution:

Substituting stress data into equation 17.11 gives:

$$\sigma^3 - 60\sigma^2 - 2{,}400\sigma + 64{,}000 = 0$$

Solving the cubic polynomial equation, the three principal stresses are

$\sigma_1 = 80$ MPa, σ_2 20 MPa, $+ \sigma_3 = -40$ MPa

Maximum shear stress at this location is then calculated using equation 17.13.

$$\tau_{max} = \frac{80 - (-40)}{2} = 60 \text{ MPa}$$

17.5 Stress Concentration

Various discontinuities in structures result in increasing the value of stress. The stress at the location of the discontinuity can be obtained from:

$$\sigma_{max} = K_t\sigma_0 \qquad [17.14]$$

Where K_t is the stress concentration obtained from Figures 17.69 through 17.80.

17.6 Failure of Components Under Static Loading and Safety Factor

There are several failure criteria for components under static loading.

Ductile Materials

For ductile materials (elongation greater than 5%), maximum shear stress theory or distortion energy criteria (DEC) can be used to predict failure load.

For the maximum shear stress theory, use equation 17.15:

$$\tau_{max} = \frac{S_y}{2\,n} \qquad [17.15]$$

where S_y is the material yield strength and n is the safety factor. Alternatively you can use equation 17.16 for distortion energy criteria (DEC):

$$\frac{S_y}{n} = \frac{1}{\sqrt{2}}\left[(\sigma_1 - \sigma_2)^2 + (\sigma_1 - \sigma_3)^2 + (\sigma_2 - \sigma_3)^2\right]^{\frac{1}{2}}$$

$$[17.16]$$

where σ_1, σ_2, and σ_3 are the principal stresses and are determined by using equation 9.25. Note that for a plane stress (2D stress) or at the outer surface of a component, $\sigma_3 = 0$.

Brittle Materials

For brittle materials (elongation < 5%), the maximum normal stress theory should be used to predict the factor of safety. For maximum normal stress theory, we can use equation 17.17:

$$S_{ut} = n\sigma_1 \qquad [17.17]$$

where S_{ut} is the ultimate tensile strength and σ_1 is the maximum tensile principal stress. For materials exhibiting both ductile and brittle behavior in tension and compression, the Coulomb-Mohr theory can be used to predict the safety factor.

The Coulomb-Mohr theory is based on the values of ultimate tensile strength (S_{ut}) as well as ultimate compressive strength (S_{uc}). Based on Figure 17.5, the theory states that fracture will occur if the largest Mohr's circle passes the failure envelope AB: BC. The factor of safety is determined by finding where the state of stress is located with respect to the failure locus.

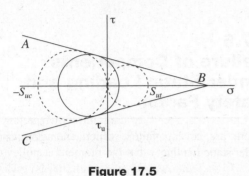

Figure 17.5

Example 17.3

The diagrams in Figure 17.6 below represent the state of stress at 2 potential points, point A and B, in a member. The material properties are listed in the table below.

S_Y	S_{UT}	% Elongation
76 MPa	186 MPa	22

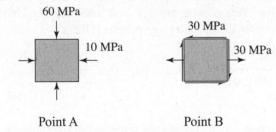

Point A   Point B

Figure 17.6

(a) Which point is the critical point? Justify your answer with calculations.
(b) What is the safety factor?

Solution:

Since the material is ductile, we can use maximum shear stress theory or DEC. For point A based on DEC, use equation 17.16. Note that in the absence of shear stress on element at point A, $\sigma_1 = 10$ MPa, $\sigma_2 = -60$ MPa, and $\sigma_3 = 0$:

$$\frac{76}{n} = \frac{1}{\sqrt{2}}\left[(10+60)^2 + (60-0)^2 + (10-0)^2\right]^{\frac{1}{2}}$$

Solve for n. We have a safety factor of $n = 1.15$. To find the maximum for point A based on maximum shear stress theory, use equation 17.13:

$$\tau_{max} = \frac{1}{2}(\sigma_1 - \sigma_3) = \frac{1}{2}(10 - (-60)) = 35 \text{ MPa}$$

Now we use equation 17.15 to determine the factor of safety.

$$35 = \frac{76}{2n} \quad \text{and } n = 1.08$$

To find point B based on DEC, we first have to determine the principal stresses by using equation 9.25. We get $\sigma_1 = 48.54$ MPa, $\sigma_2 = -18.54$ MPa, and $\sigma_3 = 0$. Now use equation 17.16 to get:

$$\frac{76}{n} = \frac{1}{\sqrt{2}}\left[(48.54+18.54)^2 + (48.54-0)^2 + (18.54-0)^2\right]^{\frac{1}{2}}$$

Solve for n.

$$n = 1.27$$

Point B based on maximum shear stress theory:

$$\tau_{max} = \frac{1}{2}(\sigma_1 - \sigma_3) = \frac{1}{2}(48.54 - (-18.54)) = 33.54 \text{ MPa}$$

Now we use equation 17.15 to determine the factor of safety:

$$33.54 = \frac{76}{2n} \quad \text{and } n = 1.13$$

We then conclude that point B is safer than point A.

Example 17.4

The member depicted in Figure 17.7 is made from 1045 hot-rolled steel with $S_y = 45$ ksi. The member is subjected to a 2,000 lb. axial load and a torque of 1,000 in·lb. Neglecting the effect of stress concentration, determine the magnitude of the following:

(a) The maximum shear stress
(b) The maximum normal stress
(c) The minimum normal stress
(d) The factor of safety for the part using an appropriate failure theory

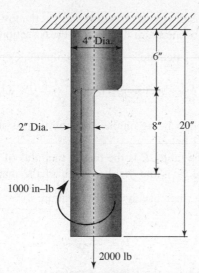

Figure 17.7

(a) Maximum shear stress

(b) Maximum energy of distortion

Assume the shaft is made of AISI 1045 hot rolled steel with $S_y = 310$ MPa.

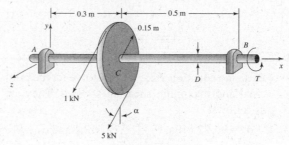

Figure 17.8

Solution:

It is observed that the critical point will be on the outside of 2-inch diameter segment of the shaft. We first solve for the normal stress at that location. Note that the bending moment at the cross section is

$$M = (2,000 \text{ lb})(1 \text{ in}) = 2,000 \text{ lb·in}$$

$$\sigma = \frac{Mc}{I} + \frac{P}{A} = \frac{2,000(1)}{\frac{\pi(2)^4}{64}} + \frac{2,000}{\frac{\pi(2)^2}{4}} = 3,185 \text{ psi}$$

$$\tau = \frac{Tc}{J} = \frac{1,000(1)}{\frac{\pi(2)^4}{32}} = 637 \text{ psi}$$

By using equation 9.25, we can solve for principal and maximum shearing stresses. They are as follows:

$$\sigma_1 = 3,307.7 \text{ psi}, \ \sigma_2 = -112.7 \text{ psi}, \ \sigma_3 = 0,$$
$$\text{and } \tau_{max} = 1,715 \text{ psi}$$

Since this material is ductile, we can use the maximum shear stress theory, equation 17.15, to determine the factor of safety:

$$1,715 = \frac{45,000}{2n} \text{ and } n = 13.1$$

Example 17.5

The bearings of the shafts described in Figure 17.8 act as simple supports. A solid steel shaft carries belt tensions (at an angle α from the y-axis in the y-z plane) at pulley C, as shown. For $\alpha = 0$ and a shaft diameter of 5.0 cm, find the factor of safety based on

Solution:

From the free-body diagram shown below, we solve for F_A and F_B, which are acting in the z-direction. Note that the moment at pulley C is about the x-axis and, as a result, has no effect on the bending moment diagram shown below.

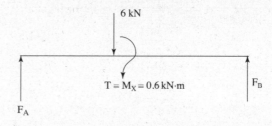

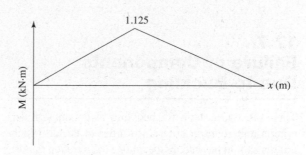

$$F_A = 3.75 \text{ kN and } F_B = 2.25 \text{ kN}$$

We now calculate the maximum bending stress and shear stress on the outside of the solid shaft:

$$\sigma = \frac{Mc}{I} = \frac{(1,125)(0.025)}{\frac{\pi(0.05)^4}{64}} = 91.7 \text{ MPa}$$

$$\tau = \frac{Tc}{J} = \frac{(600)(0.025)}{\frac{\pi(0.05)^4}{32}} = 24.46 \text{ MPa}$$

511

The stress element at pulley C on the surface of the shaft is shown in the figure below.

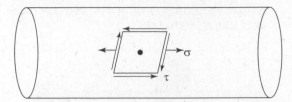

By using equation 9.25, we can solve for the principal and maximum shearing stresses. They are as follows:

$\sigma_1 = 97.82$ MPa, $\sigma_2 = -6.12$ MPa, and $\sigma_3 = 0$

By using equation 17.16, we can determine the factor of safety:

$$\frac{310}{2n} = \frac{1}{\sqrt{2}}[(97.82+6.12)^2 + (97.82-0)^2 + (-6.12-0)^2]^{\frac{1}{2}}$$

$$n = 3.07$$

Based on the maximum shear stress criteria:

$$\tau_{max} = \frac{97.82-(-6.12)}{2} = 53.49 \text{ MPa}$$

$$\frac{S_y}{2n} = \tau_{max}$$

$$\frac{310}{2n} = 53.49$$

$$n = 2.98$$

17.7
Failure of Components Due to Buckling

There are two criteria for buckling of columns: Euler criteria and Johnson criteria. In order to decide which criteria should be used, first evaluate the gyration radius:

$$k^2 A = I \qquad [17.18]$$

where k is the gyration radius and I is the lowest area moment of inertia of the column cross section.

Now calculate $\dfrac{l}{k}$ of your column. Compare it with $\left(\dfrac{l}{k}\right)_c$ which is equal to $\left(\dfrac{2\pi^2 CE}{S_y}\right)^{0.5}$ where C is constant depending on the end condition, as shown in Table 17.1 below.

In the table, E is the elastic modulus of the column, and S_y is the yield strength of the material. If $\dfrac{l}{k} > \left(\dfrac{l}{k}\right)_c$, then use the Euler equation:

$$P_c = \frac{C\pi^2 EI}{l^2} \qquad [17.19]$$

Otherwise, use the Johnson equation:

$$\frac{P_c}{A} = S_y - (\frac{S_y}{2\pi}\frac{l}{k})^2\frac{1}{CE} \qquad [17.20]$$

Example 17.6

In the design of injection molding, a large horizontal force is generated by applying a small vertical force P. Assuming a vertical force (P) of 500 lb. is applied at point A and members AB and AC are made of AISI 1045 HR steel with width of 2 inches, what should be the thickness of members AB and AC for all possible failure modes? Members are connected by pins. Consider both in plane and out of plane buckling. Use a safety factor of $n = 2$. See Figure 17.9.

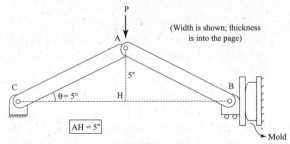

Figure 17.9

Table 17.1 End Condition Constant C for Buckling Analysis

End Condition Constant C			
Column End Conditions	**Theoretical Value**	**Conservative Value**	**Recommended Value***
Fixed-Free	¼	¼	¼
Rounded-rounded	1	1	1
Fixed-rounded	2	1	1.2
Fixed-fixed	4	1	1.2

*To be used only with liberal factors of safety when the column load is accurately known.

Solution:

AISI 1045 HR $S_y = 45$ ksi

$AB = BC = 5/(\sin 5) = 57.4$ inches

Out of plane buckling

$$I = \frac{\text{width}(t)^3}{12} = \frac{2(t)^3}{12}$$

End condition is pin-pin, $c = 1$ or $c = 1.2$.

$F_{AB} = P/(2 \sin 5) = 500/2 \sin 5 = 2,868.4$ lb

$P_c = 2,868.4 \times$ factor of safety

$$P_c = \frac{C\pi^2 EI}{l^2}$$

$$2,868.4(2) = \frac{1(\pi^2)(30 \times 10^6)\left(\dfrac{2(t)^3}{12}\right)}{(57.4)^2}$$

Solve for $t = 0.73$ inches, and check for $\dfrac{l}{k}$:

$$k_2 A = I$$

$$k^2(0.73)(2) = I = \frac{2(0.73)^3}{12}$$

$$k = 0.21$$

$$\frac{l}{k} = \frac{57.4}{0.21} = 272.4$$

$$\left(\frac{l}{k}\right)_c = \left(\frac{2\pi^2 CE}{S_y}\right)^{0.5} = \left(\frac{2\pi^2(1)(30 \times 10^6)}{45,000}\right)^{0.5}$$

$$= 114.65$$

Since $\dfrac{l}{k} > \left(\dfrac{l}{k}\right)_c$, the Euler equation is valid.

Therefore, $t = 0.73$ inches.

Example 17.7

A hydraulic actuator is used is in injection molding; see Figure 17.10. The required force to keep injection molding closed during manufacturing part is 500 lb. To activate this load, a scissor link mechanism shown below is designed. The material for arms AB and AC is 1030 HR steel. Arms AB and AC have rectangular cross sections of 0.5×2 inches.

(a) Find the required force of actuator
(b) What are the end conditions of these arms?
(c) Find the safety factor against yielding and buckling.

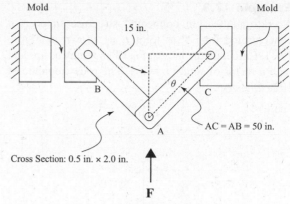

Cross Section: 0.5 in. × 2.0 in.

F

Figure 17.10

Solution:

AISI 1030 HR $S_y = 37.5$ ksi

$AB = AC = 50$ inches

$$\theta = \sin^{-1}\left(\frac{15}{50}\right) = 17.54°$$

$F_{AB} = F/(2 \sin 17.54)$

$F_{AB} = 1.67F$

$F_{AB} \cos \Theta = 500$

$F_{AB} = 524.1$ and $F = 313.8$ lb.

$k^2 A = I$

$$k^2(0.5)(2) = I = \frac{2(0.5)^3}{12}$$

$$k = 0.144$$

$$\frac{l}{k} = \frac{50}{0.144} = 346.4$$

$$c = 1.2$$

$$\left(\frac{l}{k}\right)_c = \left(\frac{2\pi^2 CE}{S_y}\right)^{0.5} = \left(\frac{2\pi^2(1.2)(30 \times 10^6)}{3,750}\right)^{0.5}$$

$$= 137.6$$

Since $\dfrac{l}{k} > \left(\dfrac{l}{k}\right)_c$, the Euler equation is valid.

$$P_c = \frac{C\pi^2 EI}{l^2} = \frac{1.2(\pi^2)(30 \times 10^6)\left(\dfrac{2(0.5)^3}{12}\right)}{(50)^2}$$

$$= 2,958 \text{ lb.}$$

Against buckling, the factor of safety is

$$n = \frac{2,958}{524.1} = 5.64.$$

Against yielding, the factor of safety is

$$n_y = \frac{37,500}{\dfrac{524.1}{2(0.5)}} = 71.6.$$

Example 17.8

A two-member pin connected structure supports a concentrated load P at joint B as shown in Figure 17.11. Calculate the largest load P that may be applied with a factor of safety $n = 2.5$. The rods are made of 1020 CD steel.

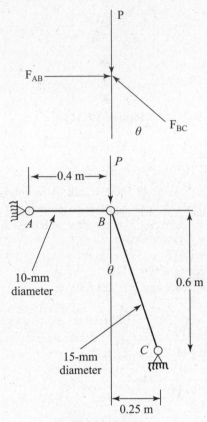

Figure 17.11

Solution:

$S_y = 390$ MPa

$$\theta = \tan^{-1}\left(\frac{0.25}{0.6}\right) = 22.62°$$

$F_{BC} \cos 22.62 = P$
$F_{BC} = 1.0833P$
$F_{AB} = 0.416P$
$c = 1$

Based on yielding: $\dfrac{1.083P}{\dfrac{\pi}{4}(0.015)^2} = \dfrac{390}{2.5}$

$P_1 = 25.4$ kN and $\dfrac{0.416\,P}{\dfrac{\pi}{4}(0.015)^2} = \dfrac{390}{2.5}$

$P_2 = 29.44$ kN

Based on buckling: $k = d/4 = 3.75$ mm

$$\left(\frac{l}{k}\right)_1 = \frac{0.65}{\dfrac{3.75}{1,000}} = 173.3$$

$$\left(\frac{l}{k}\right)_2 = \frac{0.4}{\dfrac{2.5}{1,000}} = 160$$

$$\left(\frac{l}{k}\right)_c = \left(\frac{2\pi^2 CE}{S_y}\right)^{0.5} = \left(\frac{2\pi^2(1)(200\times10^9)}{390\times10^6}\right)^{0.5} = 100.5$$

Since $\dfrac{l}{k} > \left(\dfrac{l}{k}\right)_c$, the Euler equation is valid.

$$P_{c1} = \frac{C\pi^2 EI}{l^2}\frac{1}{n_f}$$

$$= \frac{1.2\left(\pi^2\right)\left(200\times10^9\right)\dfrac{\pi}{64}(0.015)^4}{(0.65)^2}\frac{1}{2.5}$$

$$= \frac{114.3}{2.5} = 1.083P$$

$P = 42.2$ kN

$$P_{c2} = \frac{C\pi^2 EI}{l^2}\frac{1}{n_f}$$

$$= \frac{1.2\left(\pi^2\right)\left(200\times10^9\right)\dfrac{\pi}{64}(0.01)^4}{(0.4)^2}\frac{1}{2.5}$$

$$= \frac{6.046}{2.5} = 0.416P$$

$P = 5.81$ kN

Therefore, the maximum load is the smaller of the two: $P_{max} = 5.81$ kN.

Example 17.9

Link 2, shown in Figure 17.12, is 25 mm wide and is made of 1020 HR steel. The structure is subjected to 800 N of force. What should be the proper thickness of the link considering a minimum safety factor of 2 against yielding and a safety factor of 4 against buckling. The link is being supported by bearings at O and B. Consider both in plane and out of plane buckling.

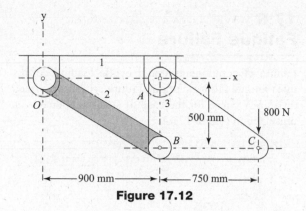

Figure 17.12

Solution:

Draw a free-body diagram of member ABC.

$$\theta = \tan^{-1}\left(\frac{500}{900}\right) = 29.05°$$

$$\sum M_A = 0$$

$F_{OB}(500\cos 29.05) - 800(750) = 0$

$F_{OB} = 1,372.7$ N

AISI 1020 HR $S_y = 210$ MPa

Based on yielding: $\dfrac{S_y}{n} = \dfrac{F_{OB}}{t(0.025)}$

$$\frac{210 \times 10^6}{2} = \frac{1,372.7}{t(0.025)}$$

$t = 0.000524$ m

For out of plane buckling:

$$l = \sqrt{900^2 + 500^2} = 1030 \text{ mm}$$

Assuming Euler buckling: $c = 1.2$

$P_c = n(F_{OB}) = 4(1,372.7) = 5,490.8$ lb

$$P_c = \frac{C\pi^2 EI}{l^2} = 5,490.8$$

$$= \frac{1.2(\pi^2)(200 \times 10^9)\left(\dfrac{0.025(t)^3}{12}\right)}{(1.03)^2}$$

$t = 10.56$ mm

Check for $\dfrac{l}{k}$:

$$k^2 A = I$$

$$k^2(0.025)(0.01056) = I$$

$$= \frac{0.025(0.01056)^3}{12}$$

$$k = 3.05 \text{ mm}$$

$$\frac{l}{k} = \frac{1,030}{3.05} = 337.7$$

$$\left(\frac{l}{k}\right)_c = \left(\frac{2\pi^2 CE}{S_y}\right)^{0.5}$$

$$= \left(\frac{2\pi^2(1.2)(200 \times 10^9)}{210 \times 10^6}\right)^{0.5} = 150$$

Since, $\dfrac{l}{k} > \left(\dfrac{l}{k}\right)_c$ the Euler equation is valid and

$t = 10.56$ mm ≈ 11 mm.

Example 17.10

To change the engine oil, a 20,000-lb. truck is lifted by a power screw jack. A recently graduated mechanical engineer recommended a double-thread Acme power screw with crest diameter of 5 inches with two threads per inch. The collar diameter of 6.0 inches and the friction coefficient for both collar and thread is 0.15. Find the torque necessary to lift the truck. If the truck is lifted 4 feet from the ground in 2 minutes, find the power required for this operation. What is the efficiency of the jack? If you consider the screws at its two end fixed and rounded conditions, is there any possibility of buckling of the screw? $E = 30$ Mpsi. Assume the screw is made of AISI 4140 steel quenched and was tempered at 600° F.

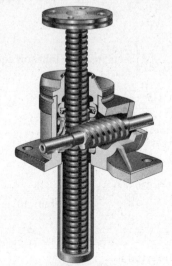

Figure 17.13

Solution:

$p = 0.5$ in

$l = 1$ in double thread

$d_m = 5 - 0.5/2 = 4.75$ inches

$\alpha = 14.5°$, thread profile angle (Acme)

sec α = sec 14.5 = 1.03

$$T_R = \frac{Fd_m}{2} \frac{l + \pi f d_m \sec\alpha}{\pi d_m - fl\sec\alpha} + \frac{f_c d_c F}{2}$$

$$= \frac{20,000(4.75)}{2}\left(\frac{1 + \pi(0.15)(4.75)(1.03)}{\pi(4.75) - 0.15(1)(1.03)}\right)$$

$$+ \frac{20,000(0.15)(6)}{2} = 19,632 \text{ lb} \cdot \text{in}$$

Each revolution lifts the truck 1 inch. $n = 4 \times 12$ = 48 revolutions in 2 min, resulting jack rpm = 24 rpm.

$$\text{hp} = \frac{Tn}{63,000} = \frac{19,632(24)}{63,000} = 7.5 \text{ hp}$$

$$e = \frac{Fl}{2\pi\, T} = \frac{20,000(1)}{2\pi(19,632)} = 0.16$$

AISI 4340 $S_y = 230$ ksi quenched and tempered at 600°F

$k = d/4 = 4.75/4 = 1.1875$

$l/k = 48/1.1875 = 40.42$

$c = 1.2$

$$\left(\frac{l}{k}\right)_c = \left(\frac{2\pi^2 CE}{S_y}\right)^{0.5}$$

$$= \left(\frac{2\pi^2(1.2)(30\times10^6)}{230\times10^3}\right)^{0.5} = 55.5$$

Since $\left(\frac{l}{k}\right) < \left(\frac{l}{k}\right)_c$, use the Johnson equation:

$$\frac{P_c}{A} = S_y - \left(\frac{S_y}{2\pi}\frac{l}{k}\right)^2 \frac{1}{C\,E}$$

$$\frac{P_c}{\frac{\pi}{4}(4.75)^2} =$$

$$230,000 - \left(\frac{230,000}{2\pi}\,40.42\right)^2 \frac{1}{1.2(30\times10^6)}$$

$P_c = 2,995,498$ lb

So the member will not buckle.

17.8 Fatigue Failure

Failure of components under cyclic loading constitutes fatigue failure. Fatigue failure criteria are based on an *S-N* curve for the case of steel components as presented in Figure 17.14.

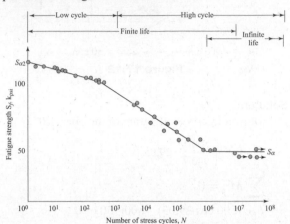

Figure 17.14

In order to have a stress life cycle greater than 10^6 cycles for a component, the stress amplitude or effective stress (Von-Mises) should be less than:

$$\sigma_a = k_a k_b k_c k_d k_e k_g\,(0.5 \times S_{ut}) \qquad [17.21]$$

where k_a is the correction factor for the surface finish and is given as:

$$k_a = aS_{ut}^{\ b} \qquad [17.22]$$

The values for a and b can be found in Table 17.2 below.

Table 17.2

Surface Finish	Factor a		Exponent b
	kpsi	MPa	
Ground	1.34	1.58	−0.085
Machined or CD	2.70	4.51	−0.265
Hot rolled	14.4	57.7	−0.718
As forged	39.9	272.0	−0.995

k_b is the correction for the size factor:

$$k_b = \begin{cases} 0.879\,d^{-0.107} & 0.11 \le d \le 2 \text{ in} \\ 0.91\,d^{-0.157} & 2 \le d \le 10 \text{ in} \\ 1.24\,d^{-0.107} & 2.79 \le d \le 51 \text{ mm} \\ 1.51\,d^{-0.157} & 51 \le d \le 254 \text{ mm} \end{cases}$$

$$[17.23]$$

For nonrotating components, the effective diameter (d_e) should be obtained prior to using Equation 17.23.

Use Table 17.3 to determine the effective diameter.

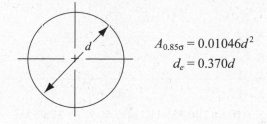

$$A_{0.85\sigma} = 0.01046d^2$$
$$d_e = 0.370d$$

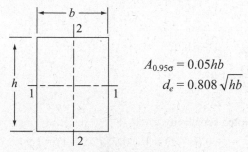

$$A_{0.95\sigma} = 0.05hb$$
$$d_e = 0.808\sqrt{hb}$$

Table 17.3

For axial loading, the size correction factor is $k_b = 1$. k_c is the load correction factor for a single applied loading:

$$k_c = \begin{cases} 1 \text{ for bending load} \\ 0.85 \text{ for axial load} \\ 0.59 \text{ for torsional load} \end{cases} \quad [17.24]$$

In the case of multiaxial loading, $k_c = 1$.

k_d is the temperature correction factor. For $T \leq 450°C$, $k_d = 1$. k_e is the reliability factor. For 50% reliability, $k_e = 1$. In addition, k_g is the miscellaneous factor used for strength reduction due to corrosion, plating, and residual stresses. In the absence of known effects, use $k_g = 1$. For a component with a finite life, the S-N curve should be constructed. The fatigue strength corresponding to 10^3 cycles is

$$S_f = fS_{ut} \quad [17.25]$$

where f can be found by using Figure 17.15

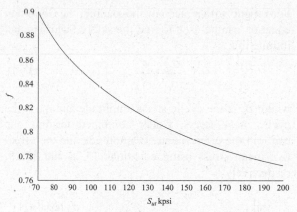

Figure 17.15

Fatigue stress concentration could be applied to applied stress before any fatigue analysis. Fatigue stress concentration is defined as:

$$K_f = 1 + q(K_t - 1) \text{ or } K_{fs} = 1 + q_{\text{shear}} q(K_{ts} - 1) \quad [17.26]$$

where q and q_{shear} are the notch sensitivity for normal and shear stress. See Figure 17.16.

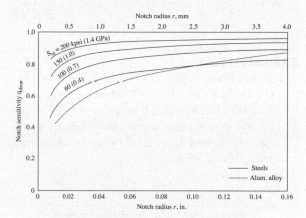

Figure 17.16 (a) Notch sensitivity for components under bending or tensile loads

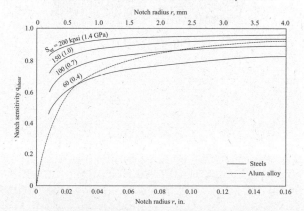

Figure 17.16 (b) Notch sensitivity for components under torsion and shear loadings

For fatigue with a mean cyclic loading, the Goodman equation can be used to find the safety factor against infinite life:

$$\frac{\sigma_a}{S_e} + \frac{\sigma_m}{S_{ut}} = \frac{1}{n} \qquad [17.27]$$

where σ_a is the cyclic stress amplitude and σ_m is the cyclic mean stress. Under combined loading one can find the effective stress amplitude and the effective mean stress using equations 17.28 and 17.29, respectively.

$$\sigma_a' = \left\{ \left[\left(K_f\right)_{\text{bending}} \left(\sigma_a\right)_{\text{bending}} + \left(K_f\right)_{\text{axial}} \frac{\left(\sigma_a\right)\text{axial}}{0.85} \right]^2 \right. \\ \left. + 3\left[\left(K_{fs}\right)_{\text{torsion}} \left(\tau_a\right)_{\text{torsion}} \right] \right\}^{1/2}$$

$$[17.28]$$

$$\sigma_m' = \left\{ \left[\left(K_f\right)_{\text{bending}} \left(\sigma_a\right)_{\text{bending}} + \left(K_f\right)_{\text{axial}} \left(\sigma_m\right)\text{axial} \right]^2 \right. \\ \left. + 3\left[\left(K_{fs}\right)_{\text{torsion}} \left(\tau_a\right)_{\text{torsion}} \right]^2 \right\}^{1/2}$$

$$[17.29]$$

Example 17.11

A machined AISI 4130 normalized steel bar shown in Figure 17.17 is subjected to an axial load which is fluctuating between –50,000 lb and 50,000 lb. Find the fatigue safety factor for infinite life as well as a life of 10^4 cycles. Assume a reliability of 50%.

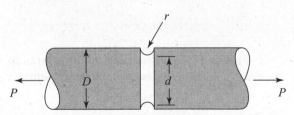

Figure 17.17

Solution:

AISI 4130 normalized, $S_{ut} = 97$ ksi

$$\sigma_a = k_a k_b k_c k_d k_e (0.5 \times S_{ut})$$

$k_a = a S_{ut}{}^b = 2.7(97)_{-0.265} = 0.8$

$k_b = 0.85$

$k_c = 1$

$k_d = 1$

$k_e = 1$

$S_e = (0.8)(1)(0.85)(1)(1)(0.5 \times 97) = 32.98$ ksi

$\sigma_a = \dfrac{P}{A} = \dfrac{50}{\dfrac{\pi 2^2}{4}} = 15.9$ ksi

$\dfrac{r}{d} = \dfrac{0.05}{2} = 0.025$

$\dfrac{D}{d} = 1.02$

$k_t = 2.2$

Notch sensitivity $q = 0.78$

$k_f = 1 + q\,(k_t - 1) = 1 + 0.78\,(2.2 - 1) = 1.94$

$\sigma = 1.94\,(15.9) = 30.8$ ksi

$n = \dfrac{s_e}{\sigma_a} = \dfrac{32.98}{30.8} = 1.07$ for infinite life.

For finite life, $n = \dfrac{s_f}{\sigma_a}$.

Construct an S–N curve.

A stress amplitude corresponding to 10^6 cycles is 32.98 ksi. A stress corresponding to 10^3 cycles is fS_{ut}. The value of f can be obtained from Figure 17.15, $f = 0.85$. Construct the S–N diagram based on these data and find the fatigue strength S_f.

$S_f = 60.01$ (fatigue strength corresponding to 10^4 cycles)

$n = \dfrac{60.01}{30.8} = 1.95$

Example 17.12

The shaft shown in Figure 17.18 rotates at high speed while the load remains fixed in space. The shaft is machined from 1080 HR steel. Identify the critical location for this shaft. Find the P necessary for the shaft to have infinite life based on a safety factor of 2.

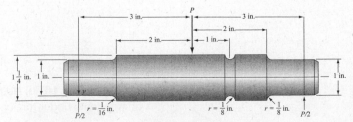

Figure 17.18

Solution:

The critical location is located at the groove.

$$\sigma = \frac{Mc}{I} = \frac{\left(\frac{P}{2}\right)(2)\left(\frac{1}{2}\right)}{\frac{\pi(1)^4}{64}} = 10.19P$$

1080 HR, $S_{ut} = 112$ ksi

$$\frac{r}{d} = \frac{1}{8} = 0.125$$

$$\frac{D}{d} = 1.25$$

$$k_t = 1.7$$

Notch sensitivity $q = 0.85$.

$$k_f = 1 + q\,(k_t - 1) = 1 + 0.85\,(1.7 - 1) = 1.6$$

$$\sigma_{max} = 1.6(10.19P) = 16.25P$$

$$S_e = k_a k_b k_c k_d k_e (0.5 \times S_{ut})$$

$$k_a = aS_{ut}^{\,b} = 2.7(112)^{-0.265} = 0.773$$

$$k_b = 0.879$$

$$k_c = 1$$

$$k_d = 1$$

$$k_e = 1$$

$$S_e = (0.773)(0.879)(1)(1)(1)(0.5 \times 112) = 38.35 \text{ ksi}$$

$$n = 2$$

$$n = \frac{S_e}{\sigma_a} = \frac{38.35}{16.25P} = 2$$

$$P = 1.18 \text{ kip}$$

Example 17.13

A round machine element, shown in Figure 17.19 is being manufactured from AISI 1050 quenched and tempered at 800°F. The part has a ground finish and is subjected to an alternating force of F.

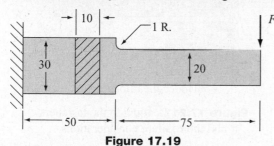

Figure 17.19

Answer the following questions:

(a) What are the material properties of this steel?
(b) What is the geometric stress concentration, K_t?
(c) What is the value of the notch sensitivity factor?
(d) What is the fatigue stress concentration factor?
(e) What is the fatigue endurance limit of the part based on 99% reliability?

(f) What is the maximum load that can be applied to the part with the safety factor of $n = 1.5$ in order for the part to survive at least 15,000 cycles? Assume force F is fluctuating between 0 and F.

Note that all dimensions are in mm.

Solution:

(a) AISI 1050 quenched and tempered at 800°F:
$S_{ut} = 1{,}090$ MPa

$S_y = 793$ MPa

% elongation = 13%

% reduction in area = 36%

Brinell hardnesss, $H_R = 444$

(b) $\dfrac{r}{d} = \dfrac{1}{20} = 0.05$

$$\frac{D}{d} = 1.5$$

$$k_t = 2.08$$

(c) Notch sensitivity $q = 0.85$

(d) $k_f = 1 + q(k_t - 1) = 1 + 0.85(2.08 - 1) = 1.918$

(e) $S_e = k_a k_b k_c k_d k_e (0.5 \times S_{ut})$

$$k_a = aS_{ut}^{\,b} = 1.58(1{,}090)^{-0.085} = 0.872$$

$$d_e = 0.37\,(20) = 7.4 \text{ mm}$$

$$k_b = 1.24\,d^{-0.107} = 1.24(7.4)^{-0.107} = 1$$

$$k_c = 1$$

$$k_d = 1$$

$$k_e = 0.814$$

$$k_g = 1$$

$$S_e = (0.872)(0.814)(1)(1)(1)(0.5 \times 1090)$$
$$\quad = 387 \text{ MPa}$$

(f) From the S-N diagram, S_f at 10^3 is equal to 861 MPa. This gives the required fatigue strength at 1,500 cycles to be 629 MPa.

$$\sigma_a = kf\left(\frac{M\,c}{I}\right)\frac{1}{2} = 0.0916F \text{ MPa}$$

$$\sigma_m = 0.0916F \text{ MPa}$$

Using the Goodman equation, force F can be found:

$$\frac{\sigma_a}{S_e} + \frac{\sigma_m}{S_{ut}} = \frac{1}{n}$$

$$\frac{0.0916F}{629} + \frac{0.0916F}{1090} = \frac{1}{1.5}$$

$$F = 2{,}903 \text{ N}$$

17.9
Screws

The helical thread screw was an important invention. It transmits angular motion to linear motion. It can also join components together. These screws can be under normal forces (separating pieces) or under shearing forces. Screw can also be used to transmit or produce large axial forces. It is always desired to reduce number of screws. Some of the important features of screws are depicted in Figure 17.20.

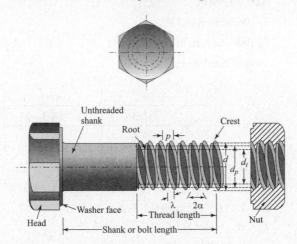

Figure 17.20

In the figure, major diameter is d, minor diameter is d_r, and mean diameter or pitch diameter is d_p. Lead l is the distance the nut moves for one turn rotation. A screw can be metric (ISO, International Standard Organization) or can be inch series (Unified National Standard, UNS). The root and crest can be either flat or round. Screws can have either fine or coarse threads. Fine threads have more threads per inch than do coarse threads. UNS screws are identified by their major diameter, fine or coarse threads, number of threads per inch, class of screw (tolerances), left or right handed, external or internal threads, as well as the grade of the screw. For example 1-12 UNRF-2A-LH grade 2.A is used for external threads while 1-12 UNRF-2A-LH grade 3.B is used for internal threads. In the International Standard Organization, screws are defined by their major diameter in millimeter and their pitch (distance between two threads) as well as grade of the screw. One example of ISO screw is M10 × 1.5 grade 5.8.

Tables 17.4 and 17.5 show dimensions of typical screws in UNF and ISO series. Tables 17.6 through 17.9 show strength and endurance limits of different grades of screws.

The most important parameters we need for the proper selection of the screws in our design are the tensile stress area, A_t, which is used in several calculations, proof strength, and their endurance limit.

The failure of components jointed together and subjected to tensile opening is shown in Figure 17.21. In addition, the figure also shows joints can also be subjected to shear loading.

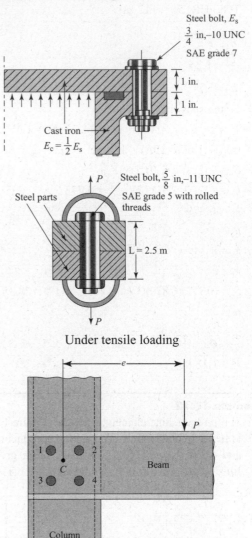

Under tensile loading

Under shear loading

Figure 17.21 Loading to the fasteners and their failure considerations

For a bolt under tensile opening, the recommended pretension is $F_i = 0.75A_t \times S_p$ if the bolt is reusable. For a permanent bolt, $F_i = 0.9A_t \times S_p$. The torque required to get this pretension can be obtained from equation 17.30.

$$T = K \times d \times F_i \qquad [17.30]$$

where K is defined as the thread engagement condition. Table 17.10 shows K values for different thread conditions.

Table 17.4 Diameters and Areas of Coarse-Pitch and Fine-Pitch Metric Threads*

Nominal		Coarse-Pitch Series			Fine-Pitch Series	
Major Diameter d (mm)	Pitch p (mm)	Tensile-Stress Area A_t (mm²)	Minor-Diameter Area A_r (mm²)	Pitch p (mm)	Tensile-Stress Area A_t (mm²)	Minor-Diameter Area A_r (mm²)
1.6	0.35	1.27	1.07			
2	0.40	2.07	1.79			
2.5	0.45	3.39	2.98			
3	0.5	5.03	4.47			
3.5	0.6	6.78	6.00			
4	0.7	8.78	7.75			
5	0.8	14.2	12.7			
6	1	20.1	17.9			
8	1.25	36.6	32.8	1	39.2	36.0
10	1.5	58.0	52.3	1.25	61.2	56.3
12	1.75	84.3	76.3	1.25	92.1	86.0
14	2	115	104	1.5	125	116
16	2	157	144	1.5	167	157
20	2.5	245	225	1.5	272	259
24	3	353	324	2	384	365
30	3.5	561	519	2	621	596
36	4	817	759	2	915	884
42	4.5	1120	1050	2	1260	1230
48	5	1470	1380	2	1670	1630
56	5.5	2030	1910	2	2300	2250
64	6	2680	2520	2	3030	2980
72	6	3460	3280	2	3860	3800
80	6	4340	4140	1.5	4850	4800
90	6	5590	5360	2	6100	6020
100	6	6990	6740	2	7560	7470
110				2	9180	9080

*The equations and data used to develop this table have been obtained from ANSB1.1-1974 and B18.3.1-1978. The minor diameter was found from the equation $d_r = d - 1.226869p$, and the pitch diameter $d_p = d - 0.649519p$. The mean of the pitch diameter and the minor diameter was used to compute the tensile-stress area.

Table 17.5 Diameters and Area of United Screw Threads UNC and UNF*

Size Designation	Nominal Major Diameter (in)	Coarse Series-UNC			Fine Series-UNF		
		Threads per inch (N)	Tensile-Stress Area A_t (in^2)	Minor-Diameter Area A_r (in^2)	Threads per Inch (N)	Tensile-Stress Area A_t (in^2)	Minor-Diameter Area A_r (in^2)
0	0.0600				80	0.00180	0.00151
1	0.0730	64	0.00263	0.00218	72	0.00278	0.00237
2	0.0860	56	0.00370	0.00310	64	0.00394	0.00339
3	0.0990	48	0.00487	0.00406	56	0.00523	0.00451
4	0.1120	40	0.00604	0.00496	48	0.00661	0.00566
5	0.1250	40	0.00796	0.00672	44	0.00880	0.00716
6	0.1380	32	0.00909	0.00745	40	0.01015	0.00874
8	0.1640	32	0.0140	0.01196	36	0.01474	0.01285
10	0.1900	24	0.0175	0.01450	32	0.0200	0.0175
12	0.2160	24	0.0242	0.0206	28	0.0258	0.0226
$\frac{1}{4}$	0.2500	20	0.0318	0.0269	28	0.0364	0.0326
$\frac{5}{16}$	0.3125	18	0.0524	0.0454	24	0.0580	0.0524
$\frac{3}{8}$	0.3750	16	0.0775	0.0678	24	0.0878	0.0809
$\frac{7}{16}$	0.4375	14	0.1063	0.0933	20	0.1187	0.1090
$\frac{1}{2}$	0.5000	13	0.1419	0.1257	20	0.1599	0.1486
$\frac{9}{16}$	0.5625	12	0.182	0.162	18	0.203	0.189
$\frac{5}{8}$	0.6250	11	0.226	0.202	18	0.256	0.240
$\frac{3}{4}$	0.7500	10	0.334	0.302	16	0.373	0.351
$\frac{7}{8}$	0.8750	9	0.462	0.419	14	0.509	0.480
1	1.0000	8	0.606	0.551	12	0.663	0.625
$1\frac{1}{4}$	1.2500	7	0.969	0.890	12	1.073	1.024
$1\frac{1}{2}$	1.5000	6	1.405	1.294	12	1.581	1.521

*This table was compiled from ANSB1.1-1974. The minor diameter was found from the equation $d_r = d - 1.299038p$ and the pitch diameter from $d_p = d - 0.649519p$. The mean of the pitch diameter and the minor diameter was used to compute the tensile-stress area.

Tablo 17.6 SAE Specifications for Steel Bolts

SAE Grade No.	Size Range Inclusive (in)	Minimum Proof Strength,* kpsi	Minimum Tensile Strength,* kpsi	Minimum Yield Strength,* kpsi	Material	Head Marking
1	$\frac{1}{4}-1\frac{1}{2}$	33	60	36	Low or medium carbon	
2	$\frac{1}{4}-\frac{3}{4}$	55	74	57	Low or medium carbon	
	$\frac{7}{8}-1\frac{1}{2}$	33	60	36		
4	$\frac{1}{4}-1\frac{1}{2}$	65	115	100	Medium carbon, cold-drawn	
5	$\frac{1}{4}-1$	85	120	92	Medium carbon, Q&T	
	$1\frac{1}{8}-1\frac{1}{2}$	74	105	81		
5.2	$\frac{1}{4}-1$	85	120	92	Low-carbon martensite, Q&T	
7	$\frac{1}{4}-1\frac{1}{2}$	105	133	115	Medium-carbon alloy, Q&T	
8	$\frac{1}{4}-1\frac{1}{2}$	120	150	130	Medium-carbon alloy, Q&T	
8.2	$\frac{1}{4}-1$	120	150	130	Low-carbon martensite, Q&T	

*Minimum strengths are strengths exceeded by 99 percent of fasteners.

Table 17.7 ASTM Specifications for Steel Bolts

ASTM Designation No.	Size Range, Inclusive, in	Minimum Proof Strength,* kpsi	Minimum Tensile Strength,* kpsi	Minimum Yield Strength,* kpsi	Material	Head Marking
A307	$\frac{1}{4} - 1\frac{1}{2}$	33	60	36	Low carbon	
A325 type 1	$\frac{1}{2} - 1$	85	120	92	Medium carbon, Q&T	
	$1\frac{1}{8} - 1\frac{1}{2}$	74	105	81		
A325 type 2	$\frac{1}{2} - 1$	85	120	92	Low-carbon martensite, Q&T	
	$1\frac{1}{8} - 1\frac{1}{2}$	74	105	81		
A325 type 3	$\frac{1}{2} - 1$	85	120	92	Weathering steel, Q&T	
	$1\frac{1}{8} - 1\frac{1}{2}$	74	105	81		
A354 grade BC	$\frac{1}{4} - 2\frac{1}{2}$	105	125	109	Alloy steel, Q&T	
	$2\frac{3}{4} - 4$	95	115	99		
A354 grade BD	$\frac{1}{4} - 4$	120	150	130	Alloy steel, Q&T	
A449	$\frac{1}{4} - 1$	85	120	92	Medium-carbon, Q&T	
	$1\frac{1}{8} - 1\frac{1}{2}$	74	105	81		
	$1\frac{3}{4} - 3$	55	90	58		
A490 type 1	$\frac{1}{2} - 1\frac{1}{2}$	120	150	130	Alloy steel, Q&T	
A490 type 3	$\frac{1}{2} - 1\frac{1}{2}$	120	150	130	Weathering steel, Q&T	

*Minimum strengths are strengths exceeded by 99 percent of fasteners.

Table 17.8 Metric Mechanical Property Classes for Steel Bolts, Screws, and Studs

Property Class	Size Range, Inclusive	Minimum Proof Strength, MPa	Minimum Tensile Strength, MPa	Minimum Yield Strength, MPa	Material	Head Marking
4.6	M5–M36	225	400	240	Low or medium carbon	4.6
4.8	M1.6–M16	310	420	340	Low or medium carbon	4.8
5.8	M5–M24	380	520	420	Low or medium carbon	5.8
8.8	M16–M36	600	830	660	Medium carbon, Q&T	8.8
9.8	M1.6–M16	650	900	720	Medium carbon, Q&T	9.8
10.9	M5–M36	830	1040	940	Low-carbon martensite, Q&T	10.9
12.9	M1.6–M36	970	1220	1100	Alloy, Q&T	12.9

Table 17.9 Fully Corrected Endurance Strengths
for Bolts and Screws with Rolled Threads*

Grade or Class	Size Range	Endurance Strength
SAE 5	$\frac{1}{4}$–1 in.	18.6 kpsi
	$\frac{1}{8}$–$1\frac{1}{2}$ in.	16.3 kpsi
SAE 7	$\frac{1}{4}$–$1\frac{1}{2}$ in.	20.6 kpsi
SAE 8	$\frac{1}{4}$–$1\frac{1}{2}$ in.	23.2 kpsi
ISO 8.8	M16–M36	129 MPa
ISO 9.8	M1.6–M16	140 MPa
ISO 10.9	M5–M36	162 MPa
ISO 12.9	M1-6-M36	190 MPa

*Repeatedly applied, axial loading, fully corrected.

Table 17.10

Bolt Condition	K
Nonplated, black finish	0.30
Zinc-plated	0.20
Lubricated	0.18
Cadmium-plated	0.16
With Bowman anti-seize	0.12
With Bowman-grip nuts	0.09

For joints subjected to external tensile force P, the components could fail due to bolt failure or part separation. The total force in the bolt can be found from:

$$F_b = F_i + C_P \qquad [17.31]$$

where C can be obtained by consideration of bolt stiffness and part stiffness.

$$C = \frac{K_b}{K_b + K_m} \qquad [17.32]$$

where k_b is the stiffness of the bolt and k_m is the stiffness of the part that can be obtained from:

$$K_b = \frac{A_d A_t E}{A_d l_t + A_t l_d} \qquad [17.33]$$

where the geometry factors can be found from Table 17.11.

Table 17.11 Suggested Procedure
for Finding Fastener Stiffness

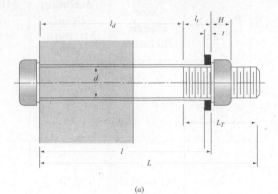

(a)

(b)

Given fastener diameter of p in mm or number of threads per inch

Grip length

For Fig. (a): l = thickness of all material squeezed between face of bolt and face of nut

For Fig. (b): $l = \begin{cases} h + t_2/2, & t_2 < d \\ h + d_2/2, & t_2 \geq d \end{cases}$

Fastener length (round up using Table A–17*)

For Fig. (a): $L > 1 + H$

For Fig. (b): $L > h + 1.5d$

Threaded length L_T Inch series:

$$L_T = \begin{cases} 2d + \frac{1}{4} \text{ in.,} & L \leq 6 \text{ in.} \\ 2d + \frac{1}{2} \text{ in.,} & L > 6 \text{ in.} \end{cases}$$

Metric series:

$$L_T = \begin{cases} 2d + 6 \text{ mm,} & L \leq 125 \text{ mm, } d \leq 48 \text{ mm} \\ 2d + 12 \text{ mm,} & 125 < L \leq 200 \text{ mm} \\ 2d + 25 \text{ mm,} & L > 200 \text{ mm} \end{cases}$$

Length of unthreaded portion in grip: $l_d = L - L_T$

Length of threaded portion in grip: $l_t = l - l_d$

Area of unthreaded portion: $A_d = \pi d^2/4$

Area of threaded portion: A_t from Table 17.4 or 17.5

Fastener stiffness: $k_b = \frac{A_d A_t E}{A_d l_t + A_t l_d}$

The stiffness of the part can be obtained by construction of frustums through the joint. Figure 17.22 shows how these frustums are constructed through the joint. Each frustum has a geometrical size of D (the smallest side of frustum: for the top and bottom frustum, $D = 1.5d$), and thickness t. The frustums are constructed from a top and bottom side of the joint by drawing two lines at 30 degrees with respect to the vertical axis at the two end of points of the line $(1.5d)$.Each frustum dimension is identified through the joint. The stiffness of each frustum is obtained from equation 17.34:

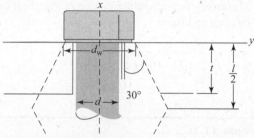

Figure 17.22 Construction of Frustum, $d_w = 1.5d$

$$K_i = \frac{0.5774\pi Ed}{\ln\frac{(1.155t + D - d)(D + d)}{(1.155t + D + d)(D - d)}} \quad [17.34]$$

where D and t are the smaller side of the frustum and thickness of the frustum, respectively, and E is the modulus of elasticity of the component frustum. The stiffness of the part is obtained from:

$$\frac{1}{K_m} = \sum \frac{1}{K_i} \quad [17.35]$$

The safety factor for bolt failure can be defined as:

$$n_L = \frac{S_p A_t - F_i}{CP} \quad [17.36]$$

The safety factor against loosening of the part can be defined as:

$$n_o = \frac{F_i}{P(1 - C)} \quad [17.37]$$

For cyclic loading, the fatigue safety factor can be defined as:

$$n_f = \frac{2S_e(S_{ut}A_t - F_i)}{CP(S_{ut} + S_e)} \quad [17.38]$$

Example 17.14

The A-frame shown in Figure 17.23 is attached to an overhead beam by 2 cap screws, each of which can be assumed carries half of the total load, F. The load, F, acts vertically downward, varying from 0 to 2,400 lb. The screws are grade 5. Despite recommended suggestions, the screws were tightened to an initial stress of $\sigma_i = 0.5S_P$. Assume that the stiffness of the part is twice the

stiffness of the screw, i.e., $K_M = 2 K_B$. Recommend the appropriate screw for a reasonable factor of safety, i.e., $2 \le n \le 4$.

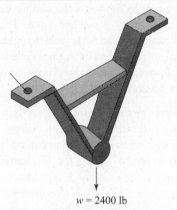

$w = 2400$ lb

Figure 17.23

Solution:

Using Tables 17.4 through 17.8:

Grade 5 $S_{ut} = 120$ ksi

$S_p = 85$ ksi

$S_e = 18.6$ ksi

$F_i = 0.5A_t S_p = 0.5A_t(85) = 42.5$ kip

Against loosening:

$$n_L = \frac{S_p A_t - F_i}{CP}$$

Find A_t from this equation:

$P = 1200$ lb $= 1.2$ kip

$$C = \frac{K_b}{K_b + K_m} = \frac{K_b}{3K_b} = \frac{1}{3}$$

$$N_L = \frac{S_p A_t - F_i}{CP} = 2 = \frac{85A_t - 42.5A_t}{\frac{1}{3}(1.2)}$$

$A_t = 0.019$ in^2

For $n = 4$, $A_t = 0.038$ in^2.

Using Tables 17.3 through 17.8

¼–20 $A_t = 0.0318$ in^2

5/16–18 $A_t = 0.0524$ in^2

Use a 5/16–18 bolt.

$F_i = 0.5A_t S_p = 0.5(0.0524)(85) = 2.23$ kip

Safety factor against fatigue:

$$n_f = \frac{2S_e(S_{ut}A_t - F_i)}{CP(S_{ut} + S_e)}$$

$$= \frac{2(18.6)(120(0.0524) - 2.23)}{\frac{1}{3}(1.2)(120 + 18.6)}$$

$$= 2.7$$

5/16–18 bolt grade 5 is fine for this application.

Example 17.15

Suppose the welded steel bracket shown in Figure 17.24 is bolted underneath a structural steel ceiling beam to support a fluctuating vertical load imposed on it by a pin and yoke. The bolts are ½-inch coarse-thread SAE grade 8, tightened to the recommended preload for nonpermanent assembly. The stiffnesses have already been computed and are $k_b = 4$ Mlb/in. and $k_m = 16$ Mlb/in.

(a) Assuming that the bolts, rather than the welds, govern the strength of this design, determine the safe repeated load that can be imposed on this assembly using the Goodman criterion with a fatigue safety design factor of 2.

(b) Compute the static overload factor to the bolt, loosening, and bolt yielding based on your answer to part (a).

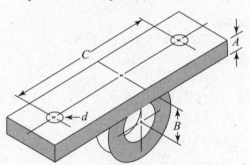

Figure 17.24

Solution:

½–UNC-13:

$A_t = 0.1419$ in^2.

$S_{ut} = 150$ ksi

$S_p = 120$ ksi

$S_e = 23.2$ ksi

$$C = \frac{K_b}{K_b + K_m} = \frac{4}{4+16} = 0.2$$

$F_i = 0.75A_t S_p = 12.8$ kip

The force on each bolt is $P/2$:

$$n_f = \frac{2S_e(S_{ut}A_t - F_i)}{CP(S_{ut} + S_e)}$$

$$= \frac{2(23.2)((150)(0.1419) - 22.8)}{0.2\left(\frac{P}{2}\right)(150 + 23.2)}$$

For $n_f = 1.5$, solve for P:

$P = 15.15$ kip

The safety factor against overload can be obtained from:

$$n_L = \frac{S_p A_t - F_i}{CP}$$

Substitute for bolt parameters:

$$n_L = \frac{120(0.1419) - 12.8}{0.2\left(\frac{15.15}{2}\right)} = 2.8$$

The safety factor against loosening can be found from:

$$n_o = \frac{F_i}{P(1 - C)}$$

$$n_o = \frac{12.8}{\frac{15.15}{2}(1 - 0.2)} = 2.1$$

The safety factor against yielding of the bolt can be obtained from:

$$n_y = \frac{S_p}{C\frac{P}{2} + F_i}{A_t} = \frac{120}{\frac{0.2\left(\frac{15.25}{2}\right) + 12.8}{(0.1419)}} = 4.18$$

Example 17.16

A 9/16-18 SAE grade 5 bolt is used to secure the bracket as shown in Figure 17.25. The bolt is well lubricated. It is preloaded to 14,000 lb and is then subjected to a cyclic load of P that varies from −5,000 lb to 10,000 lb. Assume the threaded portion of the bolt is just below the nuts. Determine:

(a) Initial torque used to tighten the bolt
(b) If the initial preload is standard based on the permanent setup

Determine the safety factor for:

(c) Static failure
(d) Loosening of the part
(e) Fatigue failure of the bolt

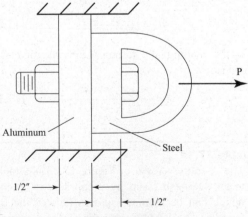

Figure 17.25

Solution:

Bolt 9/16 – 16:

$A_t = 0.203$ in^2.

$S_{ut} = 120$ ksi

$S_p = 85$ ksi

$S_e = 18.6$ ksi

$S_y = 92$ ksi

$E_s = 30 \times 10^6$ psi

$E_{A1} = 10 \times 10^6$ psi

$$k_b = \frac{AE}{l} = \frac{\frac{\pi}{4}\left(\frac{9}{16}\right)^2 (30 \times 10^6)}{1}$$

$$= 0.745 \times 10^7 \text{ lb/in.}$$

$D = 1.5 \times 9/16 = 0.8375$ in.

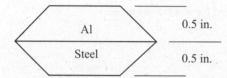

	0.5 in.
Al	
Steel	0.5 in.

Using equation 17.33, the stiffness of each frustum is

$K_1 = 3.96 \times 10^7$ lb/in and $K_2 = 1.32 \times 10^7$ lb/in

$$\frac{1}{K_m} = \sum \frac{1}{K_i}$$

$K_m = 1 \times 10^7$ lb/in.

$$C = \frac{K_b}{K_b + K_m} = 0.43$$

(a) $T = k d F_i = 0.18(9/16)(14,000) = 1,417.5$ lb·in.

(b) $F_i = 0.9 A_t S_p = 0.9\,(0.203)\,(85,000) = 15,530$ lb

So the bolt is not tightened properly.

(c)

$$n_L = \frac{S_p A_t - F_i}{CP} = \frac{0.203(85,000) - 14,000}{0.43(10,000)} = 0.76$$

(d) $$n_o = \frac{F_i}{P(1-C)} = \frac{14,000}{10,000(1-0.45)} = 2.5$$

When $P_{min} > 0$, use the equation below:

$$n_f = \frac{(S_{ut} A_t - F_i)}{C\left(P_a \dfrac{S_{ut}}{S_e} + P_m\right)} = \frac{120,000(0.203) - 14,000}{0.43\left(5,000\dfrac{120}{18.6} + 5,000\right)}$$

$$= 0.65$$

17.10 Power Screws

Figure 17.26 shows a typical power screw. It consists of a screw, a nut, and a collar as shown.

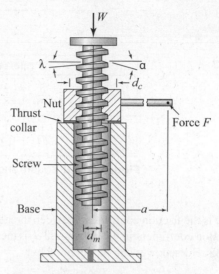

Figure 17.26 Typical power screw

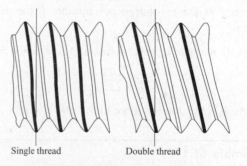

Single thread   Double thread

Figure 17.27 Single-thread and double-thread screws

Square thread or Acme thread screws are used in the design of power screws. These screws can be single thread or multiple threads. The lead (distance screw advances in one rotation) in these screws can be defined as:

$$l = np \qquad [17.39]$$

where n is the number of threads, and p is the pitch. Figure 17.27 shows single- and double-thread screws. Figure 17.28 shows square and Acme screws.

The torque required to raise and lower these screws can be obtained from:

$$T_R = \frac{F d_m}{2}\left(\frac{l + \pi f d_m \sec\alpha}{\pi d_m - fl \sec\alpha}\right) + \frac{f_c d_c F}{2} \qquad [17.40]$$

$$T_L = \frac{F d_m}{2}\left(\frac{\pi f d_m \sec\alpha - l}{\pi d_m - fl \sec\alpha}\right) + \frac{f_c d_c F}{2} \qquad [17.41]$$

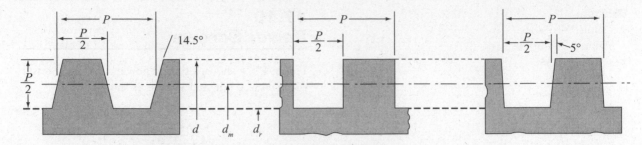

| \multicolumn{10}{c}{**Preferred Pitch for Acme Thread**} |

d_1 in	$\frac{1}{4}$	$\frac{5}{16}$	$\frac{3}{8}$	$\frac{1}{2}$	$\frac{5}{8}$	$\frac{3}{4}$	$\frac{7}{8}$	1	$1\frac{1}{4}$
p_1 in	$\frac{1}{16}$	$\frac{1}{14}$	$\frac{1}{12}$	$\frac{1}{10}$	$\frac{1}{8}$	$\frac{1}{6}$	$\frac{1}{6}$	$\frac{1}{5}$	$\frac{1}{5}$

Figure 17.28 Square and Acme threads are used for the power screw

where f is the friction coefficient between threads, f_c is the friction coefficient in the collar, and d_c is the collar diameter. Horsepower can be obtained by:

$$\text{hp} = \frac{T \times n}{63,000} \qquad [17.42]$$

where n is the revolution per minute of the power screw.

The efficiency of a power screw cab is obtained from:

$$e = \frac{Fl}{2\pi T_R} \qquad [17.43]$$

where F is the applied force to the power screw.

Example 17.17

A mechanism consists of an Acme screw with a diameter of 1 inch. The screw is attached to a spring with a stiffness of 5,000 lb/in. If the friction coefficient for the thread and collar bearing is 0.15, find the weight W required to extend the spring 0.5 inches. The collar diameter is 1.5 inch. Show the location of the collar bearing. The spring is prevented from rotation. (Figure 17.29 is not to scale.)

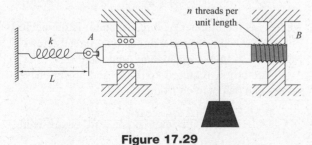

Figure 17.29

$d = 1$ in
$p = 1/5$ in
$l = p = 0.2$ in single thread
Number of threads/inch = 5
$T = W \times 0.5 = 0.5\,W$ lb·in

$$d_m = 1 - \frac{p}{2} = 1 - \frac{1}{10} = 0.9$$

$$\sec \propto = \frac{1}{\cos 14.5} = 1.033$$

$$F = 5,000 \times 0.5 = 2,500 \text{ lb}$$

$$T_R = \frac{F d_m}{2}\left(\frac{l + \pi f d_m \sec \alpha}{\pi d_m - fl \sec \alpha}\right) + \frac{f_c d_c F}{2}$$

$$= \frac{2,500(0.9)}{2}\left(\frac{0.2 + \pi(0.15)(0.9)(1.033)}{\pi(0.9 - 0.15(0.2)(1.033))}\right)$$

$$+ \frac{2,500(0.15)(1.5)}{2}$$

$$= 537.9 \text{ lb·in.}$$

$$W = T/0.5 = 1,075.8 \text{ lb}$$

Example 17.18

A machinist's clamp uses a grade 1, ½ - 20 UNF screw as shown in Figure 17.30. The collar friction radius is 0.2 inches and the coefficients of friction for the threads and collar are both 0.15. Assuming that a machinist can comfortably exert a 30 lb force on the end of the handle with an effective radius of $r = 5$ inches:

(a) Determine the clamping force developed between the jaws of the clamp
(b) Would you recommend that the spacing in the fixture, L, be smaller or larger than 18 inches? Justify your choice with calculations.

Solution:
Acme:

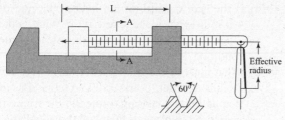

Figure 17.30

Solution:

Screw 1/2-20 UNF

$p = 1/20 = 0.05$ inch/thread

$l = p = 0.05$ single thread screw

$\alpha = 30°$

$d_m = d - 0.65p = 0.4675$ in.

$\sec \alpha = 1/\cos \alpha = 1.155$

$f = f_c = 0.15$

$r_c = 0.2$ in.

The machinist can comfortably apply $T = 30(5) = 150$ lb·in.

$$T_R = \frac{Fd_m}{2}\left(\frac{l + \pi fd_m\sec\alpha}{\pi d_m - fl\sec\alpha}\right) + \frac{Ff_cd_c}{2} = 150$$

Solve for $F = 1908$ lb

From Table 17.3: $A_t = 0.1599$ in^2

Compression stress: $\sigma_c = \dfrac{1908}{0.1599} = 11.9$ ksi

Buckling analysis:

$$\frac{\pi d^2}{4}k^2 = \frac{\pi d^4}{64}$$

$$k = \frac{d}{4}$$

$$k = \frac{0.4675}{4} = 0.1169$$

$$\frac{l}{k} = \frac{18}{0.1169} = 154$$

$$\left(\frac{1}{k}\right)_{cr} = \left(\frac{2\pi^2 CE}{S_y}\right)^{\frac{1}{2}}$$

where $C = 1$

$$E = 30 \times 10^6 \text{ psi}$$
$$S_y = 36,000 \text{ psi}$$

$$\left(\frac{1}{k}\right)_{cr} = 128$$

$$\frac{l}{k} > \left(\frac{1}{k}\right)_{cr}$$

Using the Euler equation, $\sigma_c = \dfrac{C\pi^2 E}{\left(\dfrac{l}{k}\right)^2} = 12.4$ ksi.

$$\frac{\sigma_c}{\sigma} = 1.4$$

There is a low factor of safety. So I recommend a smaller length.

Example 17.19

A load of 3,000 lb is carried by a 1.5-inch diameter double-thread Acme screw of standard size. The collar has an outside diameter of 3 inches and an inside diameter of 2.0 inches. The coefficient of friction for the threads is 0.15 and for the collar is 0.1. Determine the horsepower required by the screw to raise the load at the rate of 1 foot every 4 seconds. What is the efficiency of the power screw?

Solution:

$d = 1.5$ in.

$p = ¼$ in.

$l = np = ½$

Double-thread Acme screw:

$d_c = (3+2)/2 = 2.5$ in.

Average collar diameter:

$\sec \alpha = 1/\cos 14.5 = 1.033$

$d_m = d - p/2 = 1.5 - 1/8 = 1.375$

$$T_R = \frac{Fd_m}{2}\left(\frac{l + \pi fd_m\sec\alpha}{\pi d_m - fl\sec\alpha}\right) + \frac{Ff_cd_c}{2}$$

$$= 3,000\left[\frac{1}{2}(1.375)\left(\frac{0.5 + \pi \times 0.15 \times 1.375 \times 1.033}{\pi \times 1.375 - 0.15 \times 0.5 \times 1.033}\right)\right.$$
$$\left. + \frac{0.1 \times 2.5}{2}\right]$$

$$- 943.64 \text{ lb·in.}$$

Angular speed can be found by considering the lead in each revolution and the required movement of 1 foot in 4 seconds.

$$l \times \frac{n}{60} \times 4 = 12$$

$$0.5 \times \frac{n}{60} \times 4 = 12$$

Solve for $n = 360$ rpm:

$$\text{hp} = \frac{Tn}{63,000} = \frac{943.64(360)}{63,000} = 5.4 \text{ hp}$$

$$e = \frac{Fl}{2\pi T} = \frac{3,000(0.5)}{2\pi(943.64)} = 0.25$$

Example 17.20

A double-thread Acme screw jack with a 1-inch diameter is used to raise a load of 10,000 lb. A plain thrust collar of 2.0 inches mean diameter is used. The coefficients of friction between the threads is 0.13 and in the collar is 0.1. See Figure 17.31.

(a) Determine the screw pitch, lead, thread depth, mean diameter, and helix angle.

(b) Estimate the torque for raising the load.

(c) Estimate the efficiency of the jack when raising the load.

(d) If we desire to motorize the system and raise it 2 in./min, find the required motor horsepower.

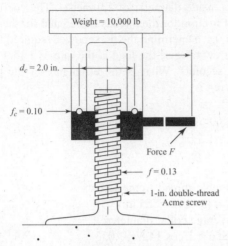

Figure 17.31

Solution:

$d = 1$ in.

$p = 1/5$ in.

Thread depth $= p/2 = 0.1$ in.

$l = np = 2(1/5) = 0.4$

$d_m = d - p/2 = 1 - 1/10 = 0.9$

$\sec \alpha = 1/\cos 14.5 = 1.033$

Helix angle $\tan(\beta) = \dfrac{l}{\pi d_m} = \dfrac{0.4}{\pi(0.9)}$ $\beta = 8.05°$

$$T_R = \frac{Fd_m}{2}\left(\frac{l + \pi fd_m\sec\alpha}{\pi d_m - fl\sec\alpha}\right) + \frac{f_c d_c F}{2}$$

$$= 2{,}263 \text{ lb} \cdot \text{in}$$

$$e = \frac{Fl}{2\pi T} = \frac{10{,}000(0.4)}{2\pi(2{,}263)} = 0.28$$

$ln = 2$ $\qquad 0.4n = 2$

Solve for n.

Angular speed is $n = 5$ rpm.

$$hp = \frac{Tn}{63{,}000} = \frac{2{,}263(5)}{63{,}000} = 0.18 \text{ hp}$$

Example 17.21

A double-threaded Acme power screw of 2-inch diameter is used in a jack having a plain thrust collar of 2.5 inches mean diameter. The coefficient of frictions between threads and collar and the base are 0.1 and 0.15, respectively.

(a) Determine the pitch, lead, thread depth, mean pitch diameter, and the helix angle of the screw.

(b) Estimate the starting torque for raising and for lowering a 5,000-lb load.

(c) If the screw is lifting the load at the rate of 4 ft/min, what is the screw rpm? What is the efficiency of the jack under this steady-state condition?

(d) What must be the friction coefficient between threads in order for the screw to unwind itself if a ball thrust bearing (of negligible friction) were used in place of the plain thrust collar?

Solution:

$d = 2$ in.

$p = 1/4$ in.

$l = np = 2(1/4) = 0.5$

$d_m = d - p/2 = 2 - 1/8 = 1.875$

$\sec \alpha = 1/\cos 14.5 = 1.033$

Helix angle $\tan(\beta) = \dfrac{l}{\pi d_m} = \dfrac{0.5}{\pi(1.875)}$ $\beta = 4.85°$

$$T_R = \frac{Fd_m}{2}\left(\frac{l + \pi fd_m\sec\alpha}{\pi d_m - fl\sec\alpha}\right) + \frac{f_c d_c F}{2} = 1{,}827 \text{ lb} \cdot \text{in}$$

$$e = \frac{Fl}{2\pi T} = \frac{5{,}000(0.5)}{2\pi(1{,}827)} = 0.22$$

$$T_L = \frac{Fd_m}{2}\left(\frac{\pi fd_m\sec\alpha - l}{\pi d_m + fl\sec\alpha}\right) + \frac{f_c d_c F}{2} = 1{,}022 \text{ lb} \cdot \text{in}$$

Angular speed is $n = 96$ rpm

$$hp = \frac{Tn}{63{,}000} = \frac{1{,}827(96)}{63{,}000} = 2.78 \text{ hp}$$

Unwind itself with no collar friction:

$$\pi fd_m\sec\alpha - l = 0 = \pi f(1.875)(1.033) - 0.5$$

$$f = 0.082$$

Since the friction coefficient between the thread is greater than 0.082, the screw will not come down by itself.

17.11
Failure of Bolts Under Shear

Figure 17.32 shows a bolt system under shear.

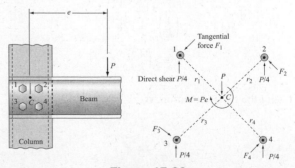

Figure 17.32

To find the dangerous bolt, first find the centroid of bolt geometry:

$$X_i = \frac{\sum x_i A_i}{\sum A_i} \qquad [17.44]$$

$$Y_i = \frac{\sum y_i A_i}{\sum A_i} \qquad [17.45]$$

Here x_i and y_i are the coordinates of the individual bolt and A_i is the individual bolt area.

Now find the applied torque to the centroid of the bolt geometry (See Figure 17.32):

$$T = F \times e$$

Next find the direct and indirect force to each bolt. The direct force to each bolt is F/n, where n is the number of bolts and its direction is the same as the applied force. The indirect force is normal to the radial line from the centroid to each bolt. The resulting torque is in the direction of the applied torque. The magnitude of these indirect forces can be obtained from:

$$F_i = \frac{T \times r_i}{\sum r_i^2} \qquad [17.46]$$

Now find the resultant force to each bolt.

Finally, the shear stress to each bolt can be found from:

$$\tau_i = \frac{F_{resultant}}{A_i} \qquad [17.47]$$

The maximum shear stress constitutes potential bolt failure.

Example 17.22

A plate is attached to a column by three identical bolts (M14 × 2) and is vertically loaded as shown in Figure 17.33. The dimensions are in millimeters. Find the maximum bolt shear force.

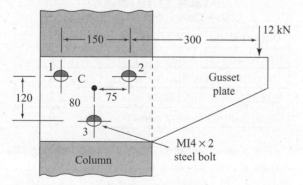

Figure 17.33

Solution:

The centroid of the bolt geometry is shown in the figure below:

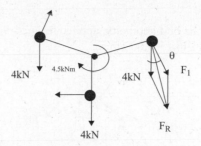

The torque at the centroid is
$$\tau = Fe = (12,000)(375) = 45,000 \text{ N·m} = 4.5 \text{ kN·m}$$
Direct force and indirect forces:

$$F_1 = \frac{4.5 \times 10^3 (85)}{85^2 + 85^2 + 80^2} = 18.35 \text{ kN}$$

$$F_2 = F_1 = 18.35 \text{ kN}$$

$$F_3 = \frac{4.5 \times 10^3 (80)}{85^2 + 85^2 + 80^2} = 17.27 \text{ kN}$$

$$\tan \Theta = 40/75 \qquad \Theta = 28.07°$$

Bolt 1 takes the maximum load.
$$(F_{resultant})_1^2 = (F_d)^2 + (F_{in})^2 + 2 F_d F_{in} \cos \Theta$$
$$(F_{resultant})_1^2 = (4)^2 + (18.35)^2 + 2(4)(18.35)$$
$$\times \cos(28.07°)$$
$$(F_{resultant})_1 = 21.96 \text{ kN}$$

$$\tau_{max} = \frac{F_{resultant}}{A} = \frac{21.96 \times 10^3}{\frac{\pi}{4}(0.014)^2} = 142.7 \text{ MPa}$$

Example 17.23

For the bracket shown in Figure 17.34, assume that the total force P is 500 lb and the distance $a = 12$ inches. If a bolt with 1/4-20 UNC grade 5 with a length of 1.5 inches is used for this application, find the safety factor against bolt failure for static loading. The thickness of the bracket is 1 inch. The bolts are placed in a square pattern, and the distance between bolts is 3 inches.

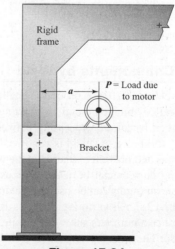

Figure 17.34

Solution:

For the bolt geometry, applied torque $\tau = (500\text{ lb}) \times (12\text{ in.}) = 6{,}000$ lb·in.

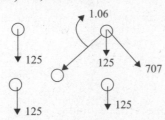

Direct force $P/4 = 500/4 = 125$ lb

$$P_{in} = \frac{Tr_i}{\sum r_i^2} = \frac{6{,}000\,(1.5\sqrt{2})}{4\,(1.5\sqrt{2})^2} = 707\text{ lb}$$

$$P_{resultant} = (125^2 + 707^2 + 2\,(125(707)\cos 45)^{1/2}$$
$$= 800\text{ lb}$$

Since $\tau = \dfrac{P_{resultant}}{A}$, we have to use either A_r or

A of the shank (no thread), which depends on whether or not the thread is between two plates. For 1/4–20 UNC grade 5:

$A_t = 0.0318$ in.^{2} $A_r = 0.0269$ in^2 (minor diameter)

Grip length is $h + d/2 = 1 + 1/8 = 1.125$ in.

$L_t = 2d + \frac{1}{4} = 0.75$ in.

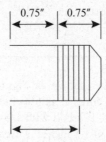

Plate thickness

So you should use A_r $\tau = \dfrac{800}{0.0269} = 29{,}750$ psi

Based on the bolt material, $S_y = 92$ ksi.

$$\tau = \frac{S_y}{2n}$$

This results in $n = 1.55$.

Joining Components by Weld

Nearly all welding is a fusion process. Metallic arc welding is also called shielded metal arc melting (SMAW). Heat is applied by an arc processing between an electrode and a workpiece. A shield is vaporized to provide a shielding gas and prevent oxidation at the weld. Direct or alternating current can be used. Various electrodes with different strengths can be used in welding.

Table 17.12 shows various electrodes used in welding. The first two numbers show the tensile strength of the electrode. The next digit refers to welding position (1 for all positions and 2 for horizontal positions). The last digit refers to welding techniques.

Table 17.12

AWS Electrode Number*	Tensile Strength kpsi (MPa)	Yield Strength kpsi (MPa)	Percent Elongation
E60xx	62 (427)	50 (345)	17–25
E70xx	70 (482)	57 (393)	22
E80xx	80 (551)	67 (462)	19
E90	90 (620)	77 (531)	14–17
E100xx	100 (689)	98 (600)	13–16
E120xx	120 (827)	107 (737)	14

*The American Welding Society (AWS) specification code numbering system for electrodes. This system uses an E prefixed to a four- or five-digit numbering system in which the first two or three digits designate the approximate tensile strength. The last digit includes variables in the welding technique, such as current supply. The next-to-last digit indicates the welding position, as, for example, flat, vertical, or overhead. The complete set of specifications may be obtained from the AWS upon request.

Stress concentration is not applied for joints under static loading. For components under cyclic loading, the endurance limit of the weld is obtained from:

$$S_e = k_a k_d k_e (0.5 S_{ut}) \qquad [17.48]$$

where you assume forged condition for the surface finish to find k_a. The alternating stress can be found from:

$$\sigma_a = k_f \times (\overline{\sigma_a}) \qquad [17.49]$$

$$\sigma_m = k_f \times (\overline{\sigma_m}) \qquad [17.50]$$

where $(\overline{\sigma_a})$ and $(\overline{\sigma_m})$ are the effective alternating stress and mean stress, respectively. k_f is the fatigue stress concentration in the weld, which can be obtained from Table 17.13.

Table 17.13 Fatigue Stress-Concentration Factors, K_{fs}

Type of Weld	K_{fs}
Reinforced butt weld	1.2
Toe of transverse fillet weld	1.5
End of parallel fillet weld	2.7
T-butt joint with sharp corners	2.0

For a weld subjected to shear and bending, the total shear stress can be found from

$$\tau = \left(\tau_d^2 + \tau_b^2\right)^{0.5} \qquad [17.51]$$

where τ_d and τ_b are the direct and bending stresses, respectively. The bending stress can be found from

$$\tau_b = \frac{Mc}{I_u \times 0.707h} \qquad [17.52]$$

l_u for weld geometry can be found from Table 17.14.

Table 17.14 Bending Properties of Fillet Welds

Weld	Throat Area	Location of G	Unit Second Moment of Area
1.	$A = 0.707hd$	$\bar{x} = 0$ $\bar{y} = d/2$	$I_u = \dfrac{d^3}{12}$
2.	$A = 1.414hd$	$\bar{x} = b/2$ $\bar{y} = d/2$	$I_u = \dfrac{d^3}{6}$
3.	$A = 1.414hb$	$\bar{x} = b/2$ $\bar{y} = d/2$	$I_u = \dfrac{bd^2}{2}$
4.	$A = 0.707h(2b + d)$	$\bar{x} = \dfrac{b^2}{2b + d}$ $\bar{y} = d/2$	$I_u = \dfrac{d^2}{12}(6b + d)$
5.	$A = 0.707h(b + 2d)$	$\bar{x} = b/2$ $\bar{y} = \dfrac{d^2}{b + 2d}$	$I_u = \dfrac{2d^3}{3} - 2d^2\bar{y} + (b + 2d)\bar{y}^2$
6.	$A = 1.414h(b + d)$	$\bar{x} = b/2$ $\bar{y} = d/2$	$I_u = \dfrac{d^2}{6}(3b + d)$
7.	$A = 0.707h(b + 2d)$	$\bar{x} = b/2$ $\bar{y} = \dfrac{d^2}{b + 2d}$	$I_u = \dfrac{2d^3}{3} - 2d^2\bar{y} + (b + 2d)\bar{y}^2$
8.	$A = 1.414h(b + d)$	$\bar{x} = b/2$ $\bar{y} = d/2$	$I_u = \dfrac{d^2}{6}(3b + d)$

Weld	Throat Area	Location of G	Unit Second Moment of Area
9.	$A = 1.414\pi hr$		$I_u = \pi r^3$

*I_u unit second moment of area, is taken about a horizontal axis through G, the centroid of the weld group, h is weld size; the plane of the bending couple is normal to the plane of the paper and parallel to the y-axis; all welds are of the same size.

Table 17.15 Torsional Properties of Fillet Welds*

Weld	Throat Area	Location of G	Unit Second Moment of Area
1.	$A = 0.707\,hd$	$\bar{x} = 0$ $\bar{y} = d/2$	$J_u = d^3/12$
2.	$A = 1.414\,hd$	$\bar{x} = b/2$ $\bar{y} = d/2$	$J_u = \dfrac{d(3b^2 + d^2)}{6}$
3.	$A = 0.707h(b + d)$	$\bar{x} = \dfrac{b^2}{2b + d}$ $\bar{y} = \dfrac{d^2}{2(b+d)}$	$J_u = \dfrac{(b+d)^4 - 6b^2d^2}{12(b+d)}$
4.	$A = 0.707h(2b + d)$	$\bar{x} = \dfrac{b^2}{2b + d}$ $\bar{y} = d/2$	$J_u = \dfrac{8b^3 + 6bd^2 + d^3}{12} - \dfrac{b^4}{2b + d}$
5.	$A = 1.414h(b + d)$	$\bar{x} = b/2$ $\bar{y} = d/2$	$J_u = \dfrac{(b+d)^3}{6}$
6.	$A = 1.414\,\pi hr$		$J_u = 2\pi r^3$

*G is centroid of weld group; h is weld size; plane of torque couple is in the plane of the paper; all welds are of unit width.

536

For a weld subjected to direct shear and torsion, you have to find the centroid of weld geometry. You also must find τ_t from

$$\tau_t = \frac{Tr}{J_u \times 0.707h} \qquad [17.53]$$

where J_u for some weld geometry is given in Table 17.15.

Example 17.24

Evaluate the design shown in Figure 17.35 with regard to stress in the welds. All parts of the assembly are made of ASTM A36 structural steel and are welded with an E60xx electrode. The load P is 2,500 lb.

(a) Find the safety factor at the connection to the **rigid surface** when the applied load is static.

(b) Find the safety factor at the same location when the applied load is fluctuating between 0 and 2,500 lb.

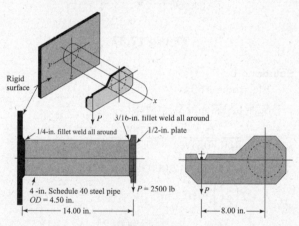

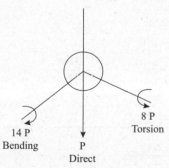

14 P
Bending

P
Direct

8 P
Torsion

Figure 17.35

Solution:

The properties of electrode E60xx are $S_{ut} = 62$ ksi and $S_y = 50$ ksi (Table 17.12). The forces and moments applied to the weld and the base are shown below.

The bending moment $M = 14P$.
The twisting moment $= T = 8P$.

$$\tau_d = \frac{P}{\pi D (0.707) h} = \frac{2,500}{\pi (4.5)(0.707)\left(\frac{1}{4}\right)} = 1,001 \text{ psi}$$

From Table 17.14 of weld geometries,

$$I_u = \pi r^3 = \pi (2.25)^3 = 35.8 \text{ in}^3.$$

$$\tau_b = \frac{Mc}{I_u (0.707) h} = \frac{14 (2,500)(2.25)}{(35.8)(0.707)\left(\frac{1}{4}\right)} = 12,445 \text{ psi}$$

From the table of weld geometries

$$J_u = 2\pi r^3 = 2\pi (2.25)^3 = 71.57 \text{ in}^3$$

$$\tau_t = \frac{Tr}{J_u (0.707) h} = \frac{8 (2,500)(2.25)}{(71.57)(0.707)\left(\frac{1}{4}\right)} = 3,548 \text{ psi}$$

Since all these shear are orthogonal to each other,

$$\tau_{\text{total}} = \sqrt{\tau_d^2 + \tau_b^2 + \tau_t^2} = 12,980 \text{ psi}$$

$$\bar{\sigma} = \sqrt{3}\tau_{\text{total}} = \frac{S_y}{n} \Rightarrow \frac{5,000}{n} = \sqrt{3}\,(12,980)$$

$n = 2.22$ static factor of safety

For fatigue:

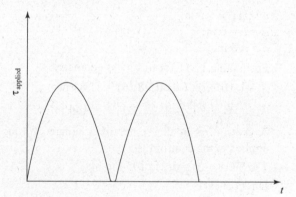

$$\frac{\bar{\sigma}_a}{S_e} + \frac{\bar{\sigma}_m}{S_{ut}} = \frac{1}{n_f}$$

From Table 17.9, the fatigue stress concentration is $k_f = 2.7$.

$$\bar{\sigma} = k_f \left(\sqrt{3}\tau_{\text{total}} \right)$$

$$\bar{\sigma}_a = \bar{\sigma}_m = 2.7 \left(\sqrt{3}\,\frac{12,980}{2} \right) = 30,351 \text{ psi}$$

$$S_e = k_a k_d k_e (0.5 S_{ut})$$

$k_d = k_e = 1$ No temperature effect based on 50% reliability

$$k_a = a(S_{ut})^b = 39.9(62)^{-0.998} = 0.657$$

$$S_e = 0.657(0.5)(62) = 20.4 \text{ ksi}$$

$$\frac{30,351}{20,400} + \frac{30,351}{62,000} = \frac{1}{n_f} \qquad n_f = 0.506$$

Example 17.25

A steel plate is welded to a column with E60xx rod along *AB* and *CD* as shown in Figure 17.36. The plate will support a steady load of $F = 5$ kips. In the figure, $m = 24$ in. and $n = 18$ in. For a 3/8 in. fillet weld with $L = 6$ in. based on the strength of the electrode:

a. Calculate the factor of safety
b. Indicate the critical point of the weld pattern

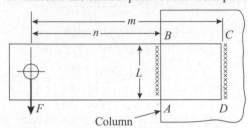

Figure 17.36

Solution:

Electrode: E60xx

$S_{ut} = 62$ ksi and $S_y = 50$ ksi

Torsional loading and direct loading from the weld geometry:

$b = m - n = 6$ in.

$d = L = 6$ in.

$h = 3/8$ in.

From table 17.11 for this weld geometry:

$A = 1.414hd = 1.414(3/8)(6) = 31.82$ in^2

$J_u \dfrac{d(3b^2 + d^2)}{6} = \dfrac{6(36^2 + 6^2)}{6} = 144$ in^3.

A centroid is at the center of a square. Torque applied at the centroid is

$T = F(18 + 3) = 5(21)\ 105$ kip·in.

Direct shear:

$\tau_d = \dfrac{F}{A} = \dfrac{5 \text{ kip}}{3.182 \text{ in}^2} = 1.57$ ksi

Indirect shear at the critical location occurs at *A* and *B*, which are critical locations.

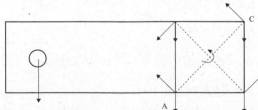

$r = \sqrt{3^2 + 3^2} = 4.24$ in

$\tau_{\text{ind}} = \dfrac{T\,r}{J_u(0.707)h} = \dfrac{105(4.24)}{(144)(0.707)\left(\frac{3}{8}\right)} = 11.67$ ksi

$\tau_{\text{total}} = \left[\tau_d^2 + \tau_{\text{ind}}^2 + 2\tau_d\tau_{\text{ind}}\cos 45\right] = 12.83$ ksi

$\bar{\sigma} = \sqrt{3}\tau = 22.22$ ksi $\quad n = \dfrac{S_y}{\bar{\sigma}} = \dfrac{50}{22.22} = 2.25$

Example 17.26

A bracket shown in Figure 17.37 supports 4,000 lb. The fillet weld extends for the full 4 in. length on both sides. What weld size is required to give a safety factor of 3.00 if an E60xx series welding rod is used? If the load is fluctuating between 0 and 4,000 lb., what should be the answer?

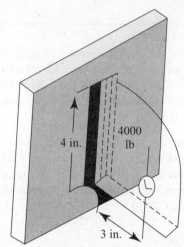

Figure 17.37

Solution:

Electrode: E60xx

$S_{ut} = 62$ ksi and $S_y = 50$ ksi

$\tau_d = \dfrac{P}{8(0.707)h} = \dfrac{4,000}{8(0.707)h} = \dfrac{707}{h}$ psi

$M = 4,000(3) = 12,000$ lb·in.

$I_u = 2\left(\dfrac{4^3}{12}\right) = 10.7$ in^3.

$\tau_{\text{ind}} = \dfrac{M\,c}{I_u(0.707)h} = \dfrac{12,000(2)}{(10.7)(0.707)(h)} = \dfrac{3,182}{h}$ psi

$\tau_R = \sqrt{\tau_d^2 + \tau_{\text{ind}}^2} = \dfrac{3,260}{h}$

$n = \dfrac{S_y}{\bar{\sigma}} = \dfrac{50,000}{\sqrt{3}\tau_R} = \dfrac{50,000}{\sqrt{3}\dfrac{3,260}{h}} = 3$

Solve for *h*: $h = 0.34$ in.

For fatigue:

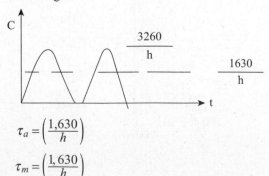

$\tau_a = \left(\dfrac{1,630}{h}\right)$

$\tau_m = \left(\dfrac{1,630}{h}\right)$

$$S_e = k_a(0.5S_{ut})$$

$$k_a = a(S_{ut})^b = 39.9(62)^{-0.998} = 0.657$$

$$k_f = 2.7$$

$$S_e = 0.657(0.5)(62) = 20.4 \text{ ksi}$$

$$\frac{2.7\sqrt{3}\left(\dfrac{1630}{h}\right)}{20,400} + \frac{2.7\sqrt{3}\left(\dfrac{1630}{h}\right)}{62,000} = \frac{1}{n} = \frac{1}{3}$$

Solve for h weld size:

$$h = 1.5 \text{ in.}$$

Since this is a big weld size, change to a higher-strength electrode.

Example 17.27

Two pieces arc being welded together using electrode E6010. Considering $a = 5$ inches and $l_1 = 8$ inches, find the critical location of the weld and the safety factor against both static loading and fatigue loading considering P is fluctuating between 0 and 20 kip.

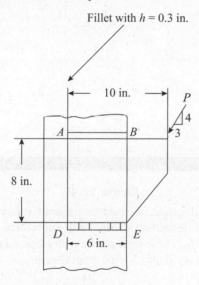

Fillet with $h = 0.3$ in.

10 in.

A B P

4

3

8 in.

D E

6 in.

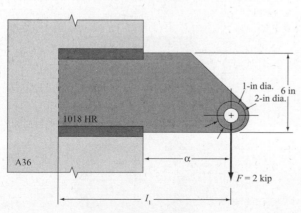

Figure 17.38

Solution:

$$T = 2,000(6.5) = 13,000 \text{ lb·in.}$$

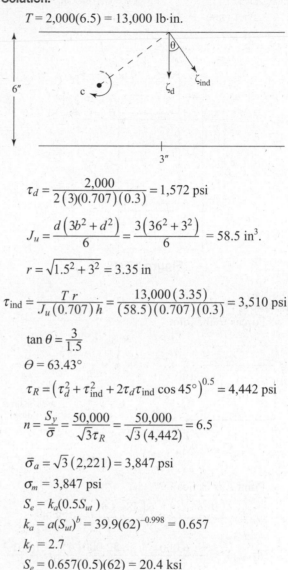

6"

θ

c τ_d τ_{ind}

3"

$$\tau_d = \frac{2,000}{2(3)(0.707)(0.3)} = 1,572 \text{ psi}$$

$$J_u = \frac{d\left(3b^2 + d^2\right)}{6} = \frac{3\left(36^2 + 3^2\right)}{6} = 58.5 \text{ in}^3.$$

$$r = \sqrt{1.5^2 + 3^2} = 3.35 \text{ in}$$

$$\tau_{ind} = \frac{T\,r}{J_u(0.707)\,h} = \frac{13,000(3.35)}{(58.5)(0.707)(0.3)} = 3,510 \text{ psi}$$

$$\tan\theta = \frac{3}{1.5}$$

$$\theta = 63.43°$$

$$\tau_R = \left(\tau_d^2 + \tau_{ind}^2 + 2\tau_d\tau_{ind}\cos 45°\right)^{0.5} = 4,442 \text{ psi}$$

$$n = \frac{S_y}{\bar{\sigma}} = \frac{50,000}{\sqrt{3}\tau_R} = \frac{50,000}{\sqrt{3}(4,442)} = 6.5$$

$$\bar{\sigma}_a = \sqrt{3}(2,221) = 3,847 \text{ psi}$$

$$\sigma_m = 3,847 \text{ psi}$$

$$S_e = k_a(0.5S_{ut})$$

$$k_a = a(S_{ut})^b = 39.9(62)^{-0.998} = 0.657$$

$$k_f = 2.7$$

$$S_e = 0.657(0.5)(62) = 20.4 \text{ ksi}$$

$$\frac{2.7\sqrt{3}(3,847)}{20,400} + \frac{2.7\sqrt{3}(3,847)}{62,000} = \frac{1}{n}$$

Solve for the factor of safety n.

$$n = 1.5$$

Example 17.28

Calculate the proper size of the fillet weld for attaching a plate to a frame. The plate supports an inclined force $P = 10$ kips. Assume we are using electrodes E60 for this job, and use safety factor of 2. If the load is fluctuating between 0 and 10 kips, what is the proper size of the weld?

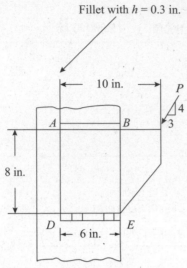

Fillet with $h = 0.3$ in.

10 in.

A B

8 in.

P

4

3

D E

6 in.

Figure 17.39

Solution:

Forces to the joint:

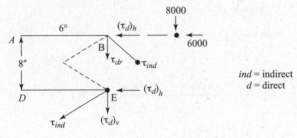

ind = indirect
d = direct

Point E is critical.

$$J_u = \frac{d\left(3b^2 + d^2\right)}{6} = \frac{6\left(38^2 + 6^2\right)}{6} = 228 \text{ in.}^3$$

$$\left(\tau_{total}\right)_h = \frac{6,000}{12\left(0.707\right)h} + \frac{32,000\left(4\right)}{228\left(0.707\right)h} = \frac{1,501.3}{h}$$

$$\left(\tau_{total}\right)_v = \frac{8,000}{12\left(0.707\right)h} + \frac{32,000\left(3\right)}{228\left(0.707\right)h} = \frac{1,538.5}{h}$$

$$\tau_{resultant} = \frac{2,149.6}{h} \text{ ksi}$$

Electrode: E60xx

$S_y = 50$ ksi

$$\bar{\sigma} = \sqrt{3}\tau_{resultant} = \frac{S_y}{n} = \frac{50,000}{n} = \sqrt{3}\,\frac{2149.6}{h}$$

For $n = 2$, we get $h = 0.15$ in.

17.12
Gears

You need to know the following 4 facts about gears:

1. Type of gears
2. Terminologies and nomenclatures
3. Forces transmitted
4. Design of a gearbox

Types of Gears: Spurs, Helical, Bevel, and Worm

Spur gears have teeth parallel to the shaft. In contrast, helical gears have teeth at an angle to the transmitting shaft. Spur gears are used to transfer power between two parallel shafts. Helical gears can transmit power between parallel as well as between nonparallel shafts. Helical gears are less noisy compared to spur gears. Figure 17.40 shows a typical spur or helical gear set.

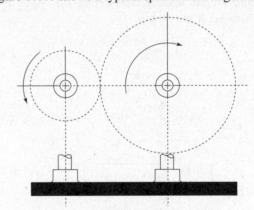

Figure 17.40

Bevel gears are used to transmit rotary motion between intersecting shafts. Teeth are formed on conical surfaces. These teeth can be either straight or spiral. Figure 17.41 shows a typical bevel gear.

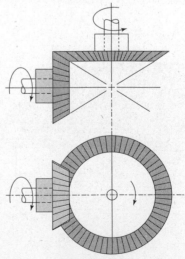

Figure 17.41

Worm gears are used for transmitting motion between nonparallel and non-intersecting shafts, Depending on the number of teeth engaged, these are either called single or double gears. Worm gears are mostly used when the speed ratio is quite high, 3 or more. Figure 17.42 shows a typical worm gear set.

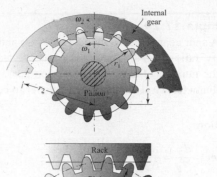

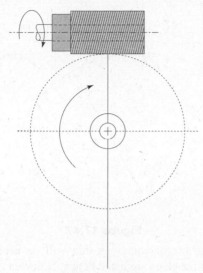

Figure 17.42

Spur Gear Terminologies

Consider two frictional cylinders that can be used to transmit power from one cylinder to the other. The teeth could be built on these cylinders as shown in Figure 17.43. The perimeters of the cylinders are called pitch circles. So the pitch circles corresponding to each gear are tangent to each other.

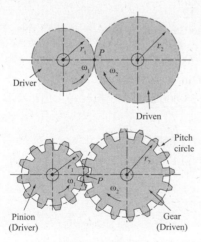

Figure 17.43

The small gear is called the pinion, while the larger one is called the gear. All calculations are based on the pitch circle diameter. The teeth could be also internal. Figure 17.44 shows an internal spur gear system.

Figure 17.44

Gear geometry and associated geometry nomenclature are shown in Figure 17.45.

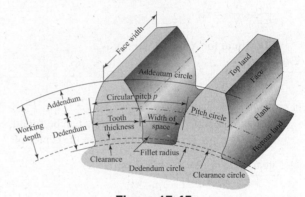

Figure 17.45

Circular pitch, p, is the the distance from one tooth to the next along the pitch circle: $p = \pi d/N$.

m is the module and is equal to d/N, which is the pitch circle diameter/number of teeth:

$$p = \pi m$$

P is the diametral pitch and is equal to $P = N/d$:

$$pP = \pi$$

Two matching gears should have the same circular pitch, diametral pitch, or module. The velocity ratio between two gears is shown by the equation:

$$\frac{\omega_1}{\omega_2} = \frac{d_2}{d_1} \qquad [17.54]$$

where d_1 and d_2 are pitch circle diameters and ω is the angular velocity. The torque transmitted between shafts is shown by:

$$\frac{T_1}{T_2} = \frac{d_1}{d_2} \qquad [17.55]$$

Example 17.29

A spur gear set has a module of 4 mm and a velocity ratio of 2.8. The pinion has 20 teeth. Find the number of teeth on the driven gear, the pitch diameters, and the theoretical center to center.

Solution:

$$m = \frac{d_p}{N_p} = \frac{d_g}{N_g} = 4$$

$$m = \frac{d_p}{20} = \frac{d_g}{N_g} = 4$$

Solving for d_p = 80 mm results in 80 mm. From the speed ratio:

$$\frac{n_p}{n_g} = \frac{N_g}{N_p} = 2.8$$

Substitute for N_p and solve for N_g:

$$\frac{N_g}{20} = 2.8$$

$$Ng = 56$$

The pitch diameter of the gear now can be found: d_g = 224 mm. The distance between the center of gears is $c = \frac{d_g + d_p}{2} = 152$ mm.

Gear Train and Transmitted Forces

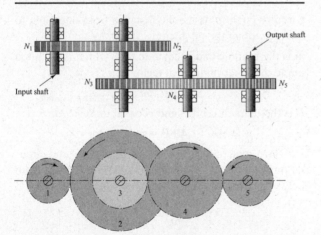

Figure 17.46

The speed ratio in the gear train shown in Figure 17.46 is given as

$$\frac{n_5}{n_1} = \left(-\frac{N_1}{N_2}\right)\left(-\frac{N_3}{N_4}\right)\left(-\frac{N_4}{N_5}\right) \quad [17.56]$$

Transmitted Load

With a pair of gears or gear sets, power is transmitted by the force developed between contacting teeth. See Figure 17.47.

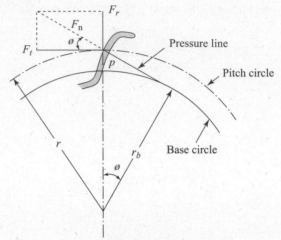

Figure 17.47

Some useful equations that will be used in the design of the gears are listed below. Note that F_t and F_r are tangential to the gear and a normal force, V is pitch circle speed in ft/min, hp is the torque transmitted, and n is the gear speed in rpm. In the metric system, power is in KW, T is $n \cdot m$, and pitch circle velocity is in m/s.

$$F_t = F_n \cos \phi$$

$$F_r = F_n \sin \phi$$

$$V = d/2\omega = d \times \frac{2\pi \text{ rpm}}{60}$$

$$V = \frac{\pi d n}{12}$$

$$d \cdot \text{inch}, n \cdot \text{rpm}, V \cdot \text{fpm}$$

$$hp = \frac{Tn}{63,000}$$

$$T, lb \cdot \text{in.}$$

$$F_t = \frac{33,000 \text{ hp}}{V}$$

$$KW = \frac{F_t V}{1,000} = \frac{Tn}{9,549}$$

$$T, N \cdot m, V \cdot m / s, F, \text{Newton}$$

Bending Stress in Gears

Bending stress in gears can be found using AGMA formulations. Design for the bending strength of gear tooth—the AGMA method:

$$\sigma = F_t k_o k_v \frac{P}{b} \frac{k_s k_m}{J} \quad \text{U.S. Customary}$$

$$\sigma = F_t k_o k_v \frac{1.0}{bm} \frac{k_s k_m}{J} \quad \text{SI units}$$

$\sigma =$ Bending stress at the root of the tooth

$F_t =$ Transmitted tangential load

$K_0 =$ Overload factor

$K_v =$ Velocity factor

$P =$ Diameterial pitch, P

$b =$ Face width

$m =$ Metric module

$K_s =$ Size factor

$K_m =$ Mounting factor

$J =$ Geometry factor

The factors needed to evaluate bending stress can be obtained from the various graphs presented next. Most of these factors should be obtained by considering a gear's quality, whether the gear is subjected to shock, and how precisely it is mounted. When calculating these factors, the pitch circle velocity should be evaluated in units of fpm (foot/min). The safety factor is obtained by comparing bending stress with that of the manufacturer's recommended allowable stress.

Table 17.16 Overload Correction Factor K_0

| Source of power | Load on driven machine | | |
	Uniform	Moderate shock	Heavy Shock
Uniform	1.00	1.25	1.75
Light shock	1.25	1.50	2.00
Medium shock	1.50	1.75	2.25

The stress should not exceed allowable stress.

$$\sigma_{all} = \frac{S_t K_L}{K_T K_R}$$

$\sigma_{all} =$ Allowable bending stress

$S_t =$ Bending Strength

$K_L =$ Life factor

$K_T =$ Temperature factor

$K_R =$ Reliability factor

The velocity factor, K_v, used in the AGMA equation depends on the gear quality.

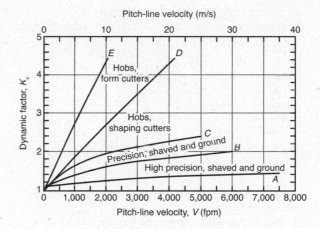

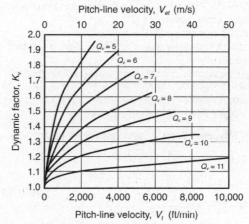

–Even with steady loads tooth impact can cause shock loading

–Impact strength depends on quality of the gear and the speed of gear teeth (pitch line velocity)

–Gears are classified with respect to manufacturing tolerances:

  –Q_V 3 – 7, commercial quality

  –Q_V 8 – 12, precision

–Graphs are available which chart K_y for different quality factors.

Figure 17.48 Dynamic factor – K_v

Table 17.17 Reliability Factor K_R

Reliability (%)	90	99	99.9	99.99
Factor K_R	0.85	1.00	1.25	1.50

SOURCE: The AGMA

–Adjusts for reliability other than 99%

  –$K_R = 0.658 - 0.0759 \ln (1\text{-R})$ $0.5 < R < 0.99$

  –$K_R = 0.50 - 0.109 \ln (1\text{-R})$ $0.99 < R < 0.9999$

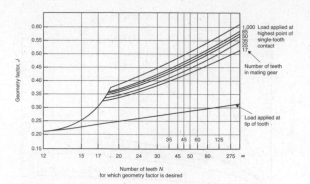

–Updated Lewis Form Factor includes effect of stress concentration at fillet
  –Different charts for different pressure angles
–Available for Precision Gears where we can assume load sharing (upper curves)
  –HPSTC-highest point of single tooth contract
  –Account for meshing gear and load sharing (contact ratio > 1)
–Single tooth contact conservative assumption (bottom curve)
  –J = 0.311 lnN + 0.15 (20 degree)
  –J = 0.367 lnN + 0.2016 (25 degree)

Figure 17.49 AGMA geometry factor – *J*

Table 17.18 Mounting Correction Factor K_M

Condition of support	Face Width (in.)			
	0 to 2	6	9	16 up
Accurate mounting, low bearing clearances, maximum deflection, precision gears	1.3	1.4	1.5	1.8
Less rigid mountings, less accurate gears, contact across the full face	1.6	1.7	1.8	2.2
Accuracy and mounting such that less than full-face contact exists	Over 2.2			

–Failure greatly depends on how load is distributed across face
  –Accurate mounting helps ensure even distribution
–For larger face widths even distribution is difficult to attain
–Note formula depends on face width which has to be estimated for initial iteration
  –Form goal: b < D_D 6 < b ×P < 16

Table 17.19 Bending Strength

Material	Heat treatment	Minimum hardness or tensile strength	S_t	
			ksi	(MPa)
Steel	Normalized	140 Bhn	19–25	(131–172)
	Q & T	180 Bhn	25–33	(172–223)
	Q & T	300 Bhn	36–47	(248–324)
	Q & T	400 Bhn	42–56	(290–386)
	Case carburized	55 R_C	55–65	(380–448)
		60 R_C	55–70	(379–483)
	Nitrided AISI-4140	48 R_C case	34–45	(234–310)
		300 Bhn core		
Cast iron				
AGMA Grade 30		175 Bhn	8.5	(58.6)
AGMA Grade 40		200 Bhn	13	(89.6)
Nodular iron ASTM Grade:				
60-40-18			15	(103)
80-55-06	Annealed		20	(138)
100-70-18	Normalized		26	(179)
120-90-02	Q & T		30	(207)
Bronze, AGMA 2C	Sand cast	40 ksi (276 MPa)	5.7	(39.3)

SOURCE: AGMA 218.01.
Q & T = Quenched and tempered.
–Tabulated Data similar to fatigue strength
–Range given because value depends on Grade
–Based on life of 10^7 cycles and 99% reliability

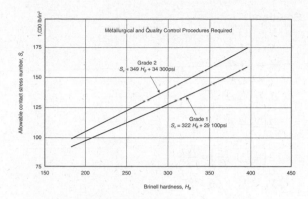

–Through hardened steel gears
–Different charts for different manufacturing methods
—Grade 1 – good quality
  $S_t = 77.3\,H_B + 12{,}800$
– Grade 2 – premium quality
  $S_t = 102\,H_B + 16{,}400$

Figure 17.50 S_t – Analytical estimate

Example 17.30

A gear set is transmitting power from a pinion gear to a gear. The pinion width is 5/8 in. and is made of 180 Bhn steel with 24 teeth and a diameteral pitch of 12/in. The gear is made of AGMA

30 cast iron and corresponds to a quality factor curve D. Is this gear system safe if the transmitted power is 1.2 hp and the pinion is rotating at 1600 rpm and the gear has 60 teeth? Assume a reliability of 99.9%.

Solution:

$$d_P = \frac{N_p}{P} = \frac{24}{12} = 2 \text{ in.}$$

$$d_G = \frac{60}{12} = 5 \text{ in.}$$

$$hp = \frac{T\,n}{63{,}000} = \frac{T\,(1{,}600)}{63{,}000} = 1.2$$

$$T = 47.25 \text{ lb} \cdot \text{in.}$$

$$F_t = \frac{T}{r} = \frac{47.25}{1} = 47.25 \text{ lb}$$

$$V = \frac{\pi d n}{12} = \frac{\pi\,(2)\,(1{,}600)}{12} = 837.8 \text{ fpm}$$

From Figure 17.49, $J = 0.36$.
From Figure 17.48, and Tables 17.6 through 17.8:
  $k_v = 1.698$
  $k_m = 1.6$
  $k_s = 1$

Number of cycles	160 Bhn	250 Bhn	450 Bhn	Case carburized (55–63 R_c)
10^3	1.6	2.4	3.4	2.7–4.6
10^4	1.4	1.9	2.4	2.0–3.1
10^5	1.2	1.4	1.7	1.5–2.1
10^6	1.1	1.1	1.2	1.1–1.4
10^7	1.0	1.0	1.0	1.0

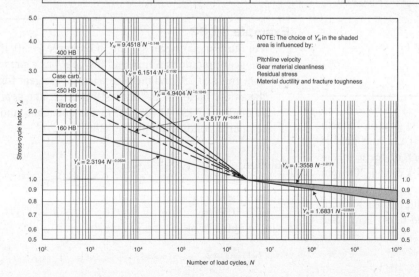

–Adjusts for life goals other than 10^7 cycles
–Fatigue effects vary with material properties and surface finishes
–$K_L = 1.6831^{-0.0323}$ N > 3E6
Note: @ 2000 rpm reach 3 million cycles in 1 day of service

Figure 17.51 Bending strength life factor - K_L

AGMA:

$$\sigma = F_t k_o k_v \frac{P}{b} \frac{k_s k_m}{J}$$

$$\sigma = 6.846 \text{ ksi}$$

For 10^7 cycles, the life factor $K_L = 1.0$. The reliability factor for 99.9% is $K_R = 1.25$. There is no temperature effect, $K_T = 1.0$.

$$\sigma_{all} = \frac{S_t k_L}{k_T k_R} = \frac{29(1)}{1(1.25)} = 23.2 \text{ ksi}$$

Since $\sigma_{all} > \sigma$, it is safe.
The safety factor $n = (23.2/6.846) = 3.4$.

Planetary Gear System

A planetary gear system consists of a sun gear, planet gears, a ring gear, and an arm. Figure 17.52 shows a typical planetary gear system. A planetary gear system has two inputs. This is in contrast to a gear train, which has one input. The gear ratio is defined as:

$$e = \frac{n_L - n_A}{n_F - n_A} \qquad [17.57]$$

where n_L is the speed of the last gear, n_A is the speed of the arm, and n_F is the speed of the first gear.

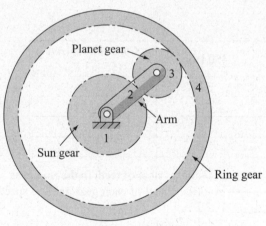

Planet gear

Arm

Sun gear

Ring gear

Figure 17.52

Example 17.31

A planetary gear system has a sun gear of 21 teeth and a planet gear of 33 teeth. If the ring gear is held stationary, find the number of teeth in the ring. If the system transmits 83 kW at 1,200 rpm (at the sun attachment), find the speed of the arm and planet gear. Find the torque on each shaft. How much torque is transmitted to the arm. See Figure 17.48.

Solution:

$N_p = 33$
$N_s = 21$
$d_p = 33/P$
$d_s = 21/P$
$d_R = N_R/P$

$$2\left(\frac{d_s}{2} + d_p\right) = d_R$$

$$2\left(\frac{21}{2P} + \frac{33}{P}\right) = \frac{N_R}{P}$$

$$N_R = 21 + 66 = 87$$

$$kW = \frac{Tn}{9,549}$$

$$T = \frac{(83)\,9,549}{1,200} = 660.5 \text{ N} \cdot \text{m}$$

$$e = \frac{n_L - n_A}{n_F - n_A} = -\left(\frac{N_s}{N_P}\right)\left(\frac{N_P}{N_R}\right) = -\frac{21}{87}$$

$$\frac{0 - n_A}{1,200 - n_A} = -\frac{21}{87}$$

$$n_A = 381.8 \text{ rpm}$$

The arm rotates in the same direction as the sun gear. For the planet gear:

$$e = \frac{n_p - n_A}{n_F - n_A} = -\frac{N_s}{N_p} = -\frac{21}{33}$$

$$\frac{n_p - 381.8}{1,200 - 381.8} = -\frac{21}{33}$$

$$n_p = -138.87 \text{ rpm}$$

The planet rotates in the opposite direction of the sun gear.

$$T_s \cdot n_s = T_A \cdot n_A \qquad \text{Power is conserved.}$$

$$(660.5)(1,200) = T_A (381.8)$$

$$T_A = 2,076 \text{ N} \cdot \text{m}$$

Example 17.32

For the planetary gear system in Example 17.30, if the gear module is 5 mm and the gear has a pressure angle of 25, find the face width of the gear for the safety factor of 2.8 against bending. The gear is made of AISI 4140 steel in the quenched and tempered condition with 180 Bhn.

Solution:

$m = d_s/N$
$m = 5 = d_s/21$
$d_s = 105 \text{ mm}$
$T = 660.5 \text{ N} \cdot \text{m}$

$$F_t\left(\frac{d_s}{2}\right) = T$$

$$F_t = \frac{660.5}{\frac{0.105}{2}} = 12,580 \text{ N}$$

For 4140 steel, $S_t = 197.5$ MPa.

$$\sigma_{all} = \left(\frac{S_t k_L}{k_T k_R}\right)\frac{1}{SF}$$

$k_L = 1$
$k_T = 1$
$k_R(50\% \text{ reliability}) = 1$
$SF = 2.8$

$$\sigma_{all} = \left(\frac{S_t k_L}{k_m k_R}\right)\frac{1}{SF} = \frac{197.5(1)}{(1)(1)}\frac{1}{2.8} = 70.54 \text{ MPa}$$

$$\sigma_{all} = F_t k_o k_v \frac{1}{m}\frac{1}{b}\frac{k_s k_m}{J}$$

$F_t = 12{,}580 \text{ N·m} = 5$
$b = ?$

Based on moderate shock, $k_o = 1.75$.
Interface speed is $V = r\omega$

$$V = \frac{0.105}{2}\left(\frac{2\pi(1{,}200)}{60}\right) = 6.6 \frac{\text{m}}{\text{s}}$$

For an average quality gear, using Figure 17.48 and curve C:
$k_v = 1.6$
$k_m = 1.6$
$k_s = 1$
J, which is the geometry factor from Figure 17.49, is $J = 3.3$.

$$\sigma_{all} = F_t k_o k_v \frac{1}{m}\frac{1}{b}\frac{k_s k_m}{J}$$

$$70.54 = 12{,}580(1.75)(1.6)\frac{1(1.6)}{b(5)(3.3)}$$

$b = 48.42$ mm

Example 17.33

For the gear train shown:

(a) Find the tangential and normal forces acting on each gear.
(b) Find the reaction to the shaft at C.

The gear system has a diametrical pitch of 3 teeth/in. Gear 1 is transmitting 30 hp at 4,000 rpm through the idler gear 2.

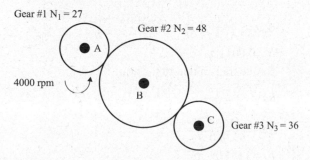

Gear #1 $N_1 = 27$
Gear #2 $N_2 = 48$
4000 rpm
A
B
C
Gear #3 $N_3 = 36$

Solution:

$$P = \frac{N_1}{d_1} = \frac{N_2}{d_2} = \frac{N_3}{d_3} = 3$$

$d_1 = 9$ in.
$d_2 = 16$ in.
$d_3 = 12$ in.

$$hp = \frac{Tn}{63{,}000} = \frac{\frac{d_1}{2}(F_t)n}{63{,}000}$$

$F_t = 105$ lb

$F_n = F_t \tan(25°) = 105(\tan(25°)) = 48.96$ lb

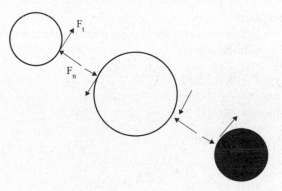

$$R_c = \sqrt{105^2 + 48.96^2} = 115.9 \text{ lb}$$

Example 17.34

A planetary gear system is designed with one sun, 6 planets, and 1 ring gear as shown in the figure. The ring gear is stationary. A 2 hp motor rotating at 1,600 rpm is attached to the sun. The sun has a 4-inch circular pitch and 18 teeth. The first planet has 45 teeth. The second planet has 90 teeth.

(a) Find the number of teeth in the ring gear.
(b) Find the speed of each gear and the speed of the arm.
(c) Show a detail of this design in terms of bearings and mounting.
(d) If the gear is made of steel with a Bhn of 180, find the proper face width based on average quality, medium shock and not accurate mounting, and infinite life with a safety factor of 2.
(e) Find the torque transmitted to the arm.

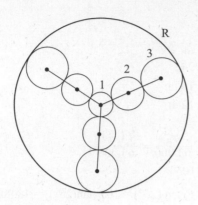

Solution:

In this problem, the circular pitch is given, not the diametral pitch.

$$p = \frac{\pi d_1}{N_1} = \frac{\pi d_2}{N_2} = \frac{\pi d_3}{N_3} = \frac{\pi d_R}{N_R}$$

$$4 = \frac{\pi d_s}{18}$$

$$d_s = 22.93 \text{ in.}$$

$$4 = \frac{\pi d_{p_1}}{45}$$

$$d_{p_1} = 57.32 \text{ in.}$$

$$4 = \frac{\pi d_{p_2}}{90}$$

$$d_{p_2} = 114.65 \text{ in.}$$

$$d_R = 2\left(\frac{d_s}{2} + d_{p_1} + d_{p_2}\right)$$

$$= 2\left(\frac{22.93}{2} + 57.32 + 114.65\right)$$

$$= 366.87 \text{ in.}$$

$$p = \frac{\pi d_R}{N_R} = \frac{\pi(366.87)}{N_R}$$

$$N_R = 288$$

$$e = \frac{n_L - n_A}{n_F - n_A}$$

Apply this equation between the sun and ring gears:

$$\left(-\frac{18}{45}\right)\left(\frac{-45}{90}\right)\left(\frac{90}{288}\right) = \frac{18}{288}$$

$$\frac{0 - n_A}{1600 - n_A} = \frac{18}{288}$$

$$n_A = -106.7 \text{ rpm}$$

The arm is rotating opposite the sun gear rotation. Apply the equation between sun gear and planet 1 gear:

$$\frac{-18}{45} = \frac{n_{p_1} + 106.7}{1600 + 106.7}$$

$$n_{p_1} = -789.4 \text{ rpm}$$

The planet rotates in the opposite direction to the sun gear. Apply the equation between the sun gear and planet 2 gear:

$$\left(\frac{-18}{45}\right)\left(\frac{-45}{90}\right) = \frac{n_{p_2} + 106.7}{1600 + 106.7}$$

$$n_{p_2} = 234.67 \text{ rpm}$$

The planet rotates in the same direction as the sun gear.

$$\sigma_{\text{all}} = F_t k_o k_v \frac{P}{b} \frac{k_s k_m}{J}$$

Based on uniform moderate shock, $k_o = 1.25$.

$$V = \frac{\pi \, d_n}{12} = \frac{\pi(22.93)(1,600)}{12} = 9,600 \text{ fpm}$$

For the average quality of the gear from Table 17.16 and from Figures 17.48 to 17.51:

$$k_v = 1.25$$
$$k_s = 1$$
$$k_m = 1.6$$

Also from the graph for the sun gear with $N_s = 18$ and $N_{P_1} = 45$:

$$J = 0.35$$

$$hp = \frac{Tn}{63,000} = \frac{\frac{d_s}{2}(F_t)n}{63,000}$$

$$2 = \frac{F_t \frac{22.93}{2}(1,600)}{63,000}$$

$$F_t = 2.289 \text{ lb}$$

$$\sigma_{\text{all}} = F_t k_o k_v \frac{P}{b} \frac{k_s k_m}{J}$$

$$= 2.289(1.25)(1.6)\frac{\frac{18}{22.93}}{b}\frac{1(1.6)}{0.35}$$

$$\sigma = \frac{16.43}{b}$$

$$S_t = 29,000 \text{ psi}$$
$$k_L = 1$$
$$k_T = 1$$
$$k_R = 1$$
$$SF = 2$$

$$\sigma_{\text{all}} = \left(\frac{S_t \, k_L}{k_T \, k_R}\right)\frac{1}{SF} = \frac{29,000}{2} = \frac{16.43}{b}$$

$$b = 0.0011 \text{ in.}$$

Torque transmitted to the arm:

$$hp = \frac{T_A \, n_A}{63,000}$$

$$2 = \frac{T_A(106.7)}{63,000}$$

$$T_A = 1181 \text{ lb} \cdot \text{in}$$

Example 17.35

A conveyer drive involving heavy shock torsional loading is to be operated by an electric motor turning at a speed of 1,600 rpm. The speed ratio of the spur gears connecting to the motor and conveyor or speed reducer is to be 1:2. Determine the maximum horsepower that the gear set can transmit based on bending strength and applying AGMA. The gear set has dimetrial pitch of 10/in., rotating at 1,600 rpm, face width of 1.5 in., and Bhn of 300.

Solution:

$$d_p = \frac{N}{P} = \frac{18}{10} = 1.8 \text{ in}.$$

Gear ratio:

$$\frac{N_1}{N_2} = \frac{1}{2} = \frac{18}{N_2}$$

$N_2 = 36$ teeth

Pitch circle velocity:

$$V = \frac{\pi d_n}{12} = \frac{\pi (1.8)(1,600)}{12} = 754 \text{ fpm}$$

Based on the given information and using Table 17.17 and Figure 17.51,

$k_L = 1$
$K_T = 1$ (oil temperature < 160°F)
$k_R = 1.25$ (99.9% reliability)
$S_t = 41.5$ ksi (average strength of 300 Bhn)

$$\sigma_{\text{all}} = \left(\frac{S_t \, k_L}{k_T \, k_R} \right) = \frac{41.5(1)}{1(1.25)} = 33.2 \text{ ksi}$$

Forces to the teeth can be calculated based on allowable stress:

$$\sigma_{\text{all}} = F_t k_o k_v \frac{P}{b} \frac{k_s k_m}{J}$$

For a gear with $P = 10$ in. and $b = 1.5$ in, $k_v = 1.55$ from curve C on Figure 17.48. The gear geometry factor from Figure 17.49 is $J = 0.235$.

$$F_t = \frac{33,200(1.5)(0.235)}{1.75(1.55)(10)(1.0)(1.6)} = 270 \text{ lb}$$

$$hp = \frac{F_t \, V}{33,000} = \frac{270(754)}{33,000} = 6.2 \text{ hp}$$

Design of Gears Based on Wear and Contact Stress

The contact stress in a gear system can be obtained from the concept of Hertz contact stress. This can be expressed as:

$$\sigma_c = C_p \left(F_t k_o k_v \frac{k_s}{bd} \frac{k_m C_f}{I} \right)^{1/2} \quad [17.58]$$

Equation 17.58 applies when equation 17.59 is satisfied:

$$C_p = 0.564 \left[\frac{1}{\frac{1 - \vartheta_p^2}{E_p}} + \frac{1}{\frac{1 - \vartheta_g^2}{E_g}} \right]^{1/2} \quad [17.59]$$

I = AGMA elastic coefficients C_p for spur gears, in $\sqrt{\text{psi}}$ and $(\sqrt{\text{MPa}})$ $\frac{\sin \Phi \cos \Phi}{2 \, m_N} \frac{m_G}{m_G + 1}$

[17.60]

where C_p is the elastic coefficient and can be obtained from Table 17.20.

Table 17.20

Pinion Material	E ksi (GPa)	Gear Material			
		Steel	Cast Iron	Aluminum Bronze	Tin Bronze
Steel	30,000 (207)	2300 (191)	2000 (166)	1950 (162)	1900 (158)
Cast iron	19,000 (131)	2000 (166)	1800 (149)	1800 (149)	1750 (145)
Aluminum bronze	17,500 (121)	1950 (162)	1800 (149)	1750 (145)	1700 (141)
Tin bronze	16,000 (110)	1900 (158)	1750 (145)	1700 (141)	1650 (137)

Remember the following:

k_v is the velocity factor and is obtained from Figure 17.48.

k_s is the size factor

b is the face width

d is the pitch circle diameter

k_m is the mounting factor, also known as the load distribution factor

C_f is the surface condition factor and varies from 1.0 to 1.5, depending on the surface condition

I is the geometry factor

m_G is the gear ratio, which is $\frac{d_g}{d_p} = \frac{N_g}{N_p}$ and for internal gear m_G is negative

m_N is the load sharing ratio and is equal to 1 for spur gears

E is modulus of elasticity

v is Poisson's ratio

φ is the pressure angle

Allowable contact stress can be defined as:

$$\sigma_{c, \text{all}} = \frac{S_c \, C_L \, C_H}{k_T \, k_R} \quad [17.61]$$

where S_c is surface strength and can be obtained from Table 17.21.

Table 17.21 Surface Fatigue Strength or Allowable Contact Stress S_c

Material	Minimum hardness or tensile strength	S_c	
		ksi	(MPa)
	Through hardened		
Steel	180 Bhn	85–95	(586–655)
	240 Bhn	105–115	(724–793)
	300 Bhn	120–135	(827–931)
	360 Bhn	145–160	(1000–1103)
	400 Bhn	155–170	(1069–1172)
	Case caburized		
	55 R_c	180–200	(1241–1379)
	60 R_c	200–225	(1379–1551)
	Flame or induction hardened		
	50 R_c	170–190	(1172–1310)
Cast iron			
AGMA grade 20		50–60	(345–414)
AGMA grade 30	175 Bhn	65–75	(448–517)
AGMA grade 40	200 Bhn	75–85	(517–586)
Nodular (ductile) iron			
Annealed	165 Bhn	90–100% of the S_c value of steel with the same hardness	
Normalized	210 Bhn		
OQ&T	255 Bhn		
Tin bronze			
AGMA 2C(10–12% tin)	40 ksi (276 MPa)	30	(207)
Aluminum bronze			
ASTM B 148-52 (alloy (9C-H.T.)	90 ksi (621 MPa)	65	(448)

C_L is life factor and is obtained from Figure 17.53

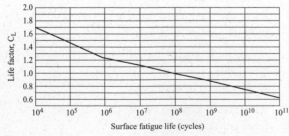

Figure 17.53

C_H is the hardness ratio. It is determined by $C_H = 1 + A$

$\times \left(\dfrac{N_g}{N_p} - 1 \right)$, where $A = 8.98 \times 10^{-3} \left(\dfrac{HBp}{HBg} \right)$ and where

HBp and HBg are the Brinell hardness of the pinion and gear, respectively. Additionally, k_T is the temperature factor and k_R is the reliability factor.

Example 17.36

A conveyer system is shown in the figure. The pinion and gear are made of steel with 300 Bhn. Both gears have a face width of $b = 1.5$ in. The pinion is rotating at 1,600 rpm. The gears have a diametrial pitch of 10/in. and a gear ratio of 1: 2. The number of teeth on the pinion is 18. Based on the contact stress, find the maximum hp that can be transmitted. Assume a safety factor of 2.

Solution:

$N_p = 18$

$N_g = 36$

Based on the gear ratio:

$d_p = 18/10 = 1.8$ in. and $d_g = 3.6$ in.

Allowable contact stress can be evaluated from:

$\sigma_{c,\,all} = \dfrac{S_C\,C_L\,C_H}{k_T\,k_R}\dfrac{1}{SF}$ where $k_T = k_R = 1$ and

$C_L = 1$ for infinite life. The parameter C_H, which is the hardness ratio, can be found from

$C_H = 1 + A\left(\dfrac{N_g}{N_p} - 1\right)$ and

$A = 8.98 \times 10^{-3}\left(\dfrac{HBp}{HBg}\right) - 8.29 \times 10^{-3} = 0.00069$

Therefore $C_H \approx 1$. $S_c = 127.5$ ksi from Table 17.17. The allowable contact stress is therefore found as:

$\sigma_{c,\,all} = \dfrac{S_C\,C_L\,C_H}{k_T\,k_R}\dfrac{1}{SF} = \left(\dfrac{127.5(1)(1)}{1(1)}\right)\left(\dfrac{1}{2}\right) = 63.7$ ksi

The contact force F_t can be obtained by considering the contact stress:

$\sigma_c = C_p\left(F_t\,k_o\,k_v\,\dfrac{k_s}{b\,d}\dfrac{k_m\,C_f}{I}\right)^{1/2}$

$k_o = 1.75$ for moderate shock

$V = \dfrac{\pi\,d_n}{12} = \dfrac{\pi(1.8)(1600)}{12} = 754$ rpm

From Figure 17.47, $k_v = 1.55$, $k_s = 1$, and $k_m = 1.6$. The pressure angle is 25°, and $m_N = 1$, $m_G = (N_G/N_P) = 2$.

$I = \dfrac{\sin\theta\cos\theta}{2\,m_N}\dfrac{m_G}{m_G + 1} = \dfrac{\sin 25°\cos 25°}{2(1)}\dfrac{2}{2+1} = 0.127$

$C_f = 1$ for a smooth surface. $C_p = 2300\sqrt{\text{psi}}$ from Table 17.20.

$P = N/d_p$

$10 = 18/d_p$

$d_p = 1.8$ in.

Now substitute in equation 17.58 to solve for F_t:

$\sigma_c = C_p\left(F_t\,k_o\,k_v\,\dfrac{k_s}{b\,d}\dfrac{k_m\,C_f}{I}\right)^{1/2}$

63.75×10^3

$= 2{,}300\left(F_t(1.75)(1.55)\dfrac{1}{1.5(1.8)}\dfrac{1.6(1)}{0.127}\right)^{1/2}$

$F_t = 60.7$ lb

$hp = \dfrac{F_t V}{33{,}000} = \dfrac{(60.7)(754)}{33{,}000} = 1.4$ hp

17.13
Belts, Clutches, and Brakes

Belts, clutches, and ropes are used to transmit power over a long distance. They are often used as a replacement for gears. They are less noisy and absorb shocks and vibration. In contrast to other systems where friction is not helpful, here we rely on the friction to transmit power.

Brakes and clutches are essentially the same devices. Each is associated with rotation. Brakes absorb the kinetic energy of the moving bodies and convert it to heat. Clutches transmit power between two shafts.

	Size range	Center distance
	T = 0.75 to 5 mm	No upper limit
	d = 3 to 19 mm	No upper limit
	A = 13 to 38 mm B = 8 to 23 mm 2β = 34° to 40°	Limited
	p = 2 mm and up	Limited

Figure 17.54 Four types of belts and their center distance

Belts are made of fiber-reinforced polyurethane or rubber-impregnated fabric reinforced with steel or nylon. Flat belts have to operate at a higher tension than the V-belt. The V-belt speed should be in the range of 7,000 fpm. V-belts are slightly less efficient than flat belts. However, V-belts can transmit more power.

Timing belts transmit power at a constant angular velocity ratio. They are used when a precise speed ratio is desired.

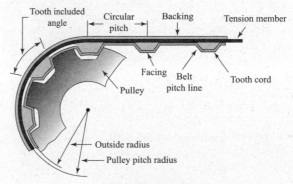

Figure 17.55

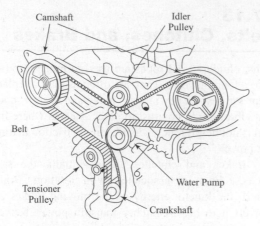

Figure 17.56 Typical V6 Timing Belt

Belt Drives and Their Dimensions

Figure 17.57 shows a typical belt arrangement and its dimensions.

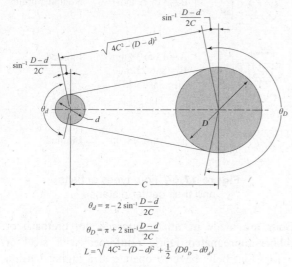

$$\theta_d = \pi - 2\sin^{-1}\frac{D-d}{2C}$$

$$\theta_D = \pi + 2\sin^{-1}\frac{D-d}{2C}$$

$$L = \sqrt{4C^2 - (D-d)^2} + \frac{1}{2}(D\theta_D - d\theta_d)$$

Figure 17.57(a)

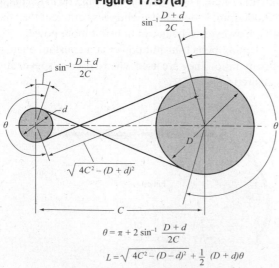

$$\theta = \pi + 2\sin^{-1}\frac{D+d}{2C}$$

$$L = \sqrt{4C^2 - (D-d)^2} + \frac{1}{2}(D+d)\theta$$

Figure 17.57(b)

Torque Transmitted by Belts

Figure 17.58 shows the tension in the tight and slack sections of the belt. The transmitted torque can be obtained from:

$$T = (F_1 - F_2) \times r \qquad [17.62]$$

The initial tension in the belt is set at $F_i = \frac{1}{2}(F_1 + F_2)$.

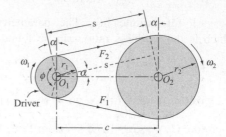

Figure 17.58(a)

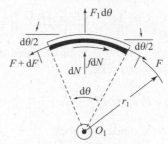

Figure 17.58(b)

The transmitted horsepower can be found from:

$$hp = \frac{Tn}{63,000} \qquad [17.63]$$

The speed ratio between two pulleys can be found from

$$\frac{n_1}{n_2} = \frac{r_2}{r_1}$$

The relationship between the tight and slack forces in a belt can be obtained by considering a free body diagram of the belt section.

To find the tension in a flat belt, use equation 17.64.

$$\frac{F_1 - F_c}{F_2 - F_c} = e^{f\varphi} \qquad [17.64]$$

To find the tension in a V-belt, use equation 17.65.

$$\frac{F_1 - F_c}{F_2 - F_c} = e^{\frac{f\varphi}{\sin(\beta)}} \qquad [17.65]$$

where f is the friction coefficient between the belt and the pulley, φ is the belt contact angle, and F_c is the centrifugal force obtained from equation 17.66.

$$F_c = \frac{w}{g}V^2 \qquad [17.66]$$

where w is the belt weight per unit belt length and V is the belt speed. β is the belt angle for a V-belt as shown in Figure 17.59.

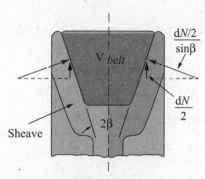

Figure 17.59

Belts also experience bending. The maximum tension in a belt, considering the effects of bending, is determined by:

$$F_{max} = K_s F_1 \qquad [17.67]$$

where K_s is a service factor and is obtained from Table 17.22.

Table 17.22

Driven machine	Driver (motor or engine)	
	Normal torque characteristic	High or nonuniform torque
Uniform	1.0 to 1.2	1.1 to 1.3
Light shock	1.1 to 1.3	1.2 to 1.4
Medium shock	1.2 to 1.2	1.4 to 1.6
Heavy shock	1.3 to 1.5	1.5 to 1.8

Belts are also subjected to fatigue loading as shown in Figure 17.60.

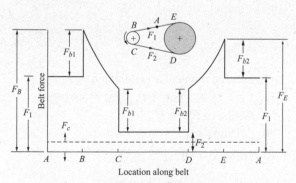

Figure 17.60

Example 17.37

A plastic, flat belt 60 mm wide and 0.5mm thick belt transmits 10 kW. The input pulley has a diameter of 300 mm and rotates at 2,800 rpm. The output pulley speed is 1,600 rpm. The pulleys are 700 mm apart. The coefficient of friction is 0.2, and the belt-specific weight is 25 kN/m³.

(a) Find the input torque at the small pulley.
(b) Find the contact angles.
(c) Find the maximum tension in the belt, assuming both pulleys are under medium shocks.

Solution:

To find torque:

$$kW = \frac{Tn}{9{,}549}$$

$$10 = \frac{T(2{,}800)}{9{,}549}$$

$$T = 34.1 \, \text{N} \cdot \text{m}$$

$$\frac{n_1}{n_2} = \frac{D}{d_1}$$

$$\frac{2{,}800}{1{,}600} = \frac{D}{d_1}$$

$$D = 525 \, \text{mm}$$

The length of the belt:

$$L = \sqrt{4C^2 - (D-d)^2} + \frac{1}{2}\left(D\theta_D + d\theta_d\right)$$

where

$$\theta_D = \pi + 2\sin^{-1}\left(\frac{D-d}{2C}\right)$$

$$\theta_D = \pi + 2\sin^{-1}\left(\frac{525-300}{2(700)}\right) = 3.46 \, \text{rad}$$

$$\theta_d = \pi - 2\sin^{-1}\left(\frac{D-d}{2C}\right)$$

$$\theta_D = \pi - 2\sin^{-1}\left(\frac{525-300}{2(700)}\right) = 2.82 \, \text{rad}$$

$$L = \sqrt{4C^2 - (D-d)^2} + \frac{1}{2}\left(D\,\theta_D + d\,\theta_d\right)$$

$$= \sqrt{4(700)^2 - (525-300)^2}$$

$$+ \frac{1}{2}\left(525(3.46) + 300(2.82)\right) = 2{,}708 \, \text{mm}$$

$$\frac{F_1 - F_c}{F_2 - F_c} = e^{f\varphi}$$

$$T = (F_1 - F_2) \times r = (F_1 - F_2)\frac{d_1}{2} = 34.1$$

$$F_c = \frac{w}{g}V^2$$

First find w, which is the weight per unit length of the belt: w = specific weight × belt cross section.

$$w = (25{,}000)(0.06)(0.0005) = 0.75 \, \text{N/m}$$

$$V = r\omega = \frac{d}{2}\left(\frac{2\pi n}{60}\right) = \frac{\pi d n}{60}$$

$$= \frac{\pi(0.3)(2{,}800)}{60} = 43.98 \, \text{m/s}$$

$$F_c = \frac{w}{g} V^2 = \frac{0.75}{9.81} 43.98^2 = 148 \text{ N}$$

$$\frac{F_1 - F_c}{F_2 - F_c} = e^{f\varphi}$$

$$\frac{F_1 - 148}{F_2 - 148} = e^{0.2(2.82)}$$

$$(F_1 - F_2)\frac{d_1}{2} = 34.1$$

Solve for F_1 and F_2 from these two equations:

$F_1 = 675$ N
$F_2 = 448$ N
$F_{max} = k_s F_1 = 1.2(675) = 810$ N

Example 17.38

The tension in a flat belt is given by the motor weight as shown. The motor mass is 80 kg and is assumed to be concentrated at the motor shaft position. The motor speed is 1,405 rpm. Pulley diameters are 400 mm and 200 mm. Calculate the belt width when the allowable belt stress is 6.0 MPa. The coefficient of friction between the belt and pulley is 0.5. The belt thickness is 5 mm, and its density is 200 kg/m³. Find the power transmitted by this pulley system.

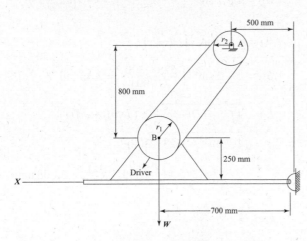

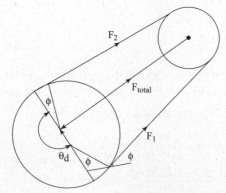

Solution:

Coordinates of axles A and B are as follows: A (500, 1050) and B (700, 250)

The distance between pulleys is

$$C = \sqrt{200^2 + 800^2} = 824.62 \text{ mm}$$

The contact angle is

$$\theta_d = \pi - 2 \sin^{-1}\left(\frac{D-d}{2C}\right)$$

$$\theta_D = \pi - 2 \sin^{-1}\left(\frac{400-200}{2(824.62)}\right) = 3.384 \text{ rad}$$

$$\varphi = \sin^{-1}\left(\frac{D-d}{2C}\right)$$

$$= \sin^{-1}\left(\frac{400-200}{2(824.62)}\right)$$

$$= 14 \text{ degrees}$$

$$F_1 \cos\varphi + F_2 \cos\varphi = (F_1 + F_2)\cos 14$$
$$= 0.97(F_1 + F_2)$$
$$= F_{total}$$

The free-body diagram of the supports is now drawn.

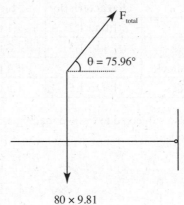

To take the moment about the pin:

$F_{total} \cos 75.96 (250) + F_{total} \sin 75.96 (700)$
$- (80)(9.81)(700) = 0$

$F_{total} = 742.64$ N

$0.97(F_1 + F_2) = F_{total}$

$$F_1 + F_2 = \frac{742.64}{0.97} = 765.6 \text{ N (1)}$$

$$F_c = \frac{w}{g} V^2 \quad w = 200(9.81)(0.005)(\text{width})$$

$$V = r\omega = \frac{d}{2}\left(\frac{2\pi n}{60}\right) = \frac{\pi d n}{60}$$

$$= \frac{\pi (0.4)(1405)}{60} = 30.37 \text{ m/s}$$

$$F_c = \frac{w}{g} V^2 = \frac{200(9.81)(0.005)(\text{width})}{9.81} 30.37^2$$

$$= 865 (\text{width}) \text{ N}$$

$$\frac{F_1 - F_c}{F_2 - F_c} = e^{f\varphi}$$

$$\frac{F_1 - F_c}{F_2 - F_c} = e^{0.5(3.384)} \quad (2)$$

From Equations (1) and (2), solve for F_1 and F_2 in terms of F_c.

$F_2 = 119.067 + 4.43$

$F_c = 119.067 + 4.43(865w)$
$\quad = 119.067 + 3831.95w \quad (3)$

$F_1 = 646.53 - 4.43$

$F_c = 646.53 - 4.43(865w) = 646.53 - 3831.95w \quad (4)$

$F_{max} = k_s F_1 = 1.2\,(646.53 - 4.43\,F_c)$
$\quad = 775.84 - 5.32\,F_c \quad (5)$

The allowable stress in the belt is 6 MPa. Use equation (5):

$$\frac{775.84 - 5.32(865w)}{0.005w} = 6\left(10^6\right)$$

Solve for $w = 0.00224$ m $= 22.4$ mm.

Now find F_1 and F_2. From equations (3) and (4):

$F_1 = 560.5$ N

$F_2 = 204.98$ N

$T = (F_1 - F_2) \times r = (560.5 - 204.98)(0.2) = 71.12$ N·m

$$kW = \frac{T\,n}{9{,}549}$$

$$kW = \frac{71.12\,(1{,}405)}{9{,}549} = 10.47$$

Example 17.39

A flat belt is 6 inches wide and is 1/3 inch thick. It transmits 15 hp. The center distance is 8 feet. The driving pulley has a 6-inch diameter and rotates at 2,000 rpm. The driven pulley has an 18-inch diameter. The belt weighs 0.035 lb/in³. If the friction coefficient for this successful transmission is 0.3:

(a) Find F_1 and F_2. What is the required pretension? What is the maximum stress if the system is under moderate shock?
(b) What is the belt length?
(c) If oil spilled between the pulley and belt reduces the friction coefficient to 0.2, what are F_1 and F_2? Does the belt slip?

Solution:

Belt width $b = 6$ in.

Small pulley diameter $= 6$ in.

Belt thickness $t = 1/8$ in.

Large pulley diameter $= 18$ in.

Distance between pulleys $C = 96$ in.

Friction coefficient $f = 0.3$.

Weight/length $= 0.035(6)(1/3) = 0.07$ lb/in.

$$\theta_d = \pi - 2\sin^{-1}\left(\frac{D - d}{2\,C}\right)$$

$$\theta_D = \pi - 2\sin^{-1}\left(\frac{18 - 6}{2(96)}\right) = 3.015 \text{ rad}$$

$$\theta_D = \pi + 2\sin^{-1}\left(\frac{D - d}{2\,C}\right)$$

$$\theta_D = \pi + 2\sin^{-1}\left(\frac{18 - 6}{2(96)}\right) = 3.265 \text{ rad}$$

$$F_c = \frac{w}{g}V^2 = \frac{(0.07)}{32.2\,(12)}\left(\frac{(3)(2\pi)(2{,}000)}{60}\right)^2$$
$$= 71.5 \text{ lb}$$

$$T = \frac{hp\,(63{,}000)}{2{,}000} = \frac{15\,(63{,}000)}{2{,}000} = 472.5 \text{ lb·in}$$

$$T = (F_1 - F_2) \times r$$
$$472.5 = (F_1 - F_2)(3)$$

$$\frac{F_1 - F_c}{F_2 - F_c} = e^{f\varphi}$$

$$\frac{F_1 - 71.5}{F_2 - 71.5} = e^{0.3(3.302)}$$

Solve for F_1 and F_2:

$F_1 = 319.92$ lb

$F_2 = 162.4$ lb

$$F_i = \frac{1}{2}(F_1 + F_2) = 241.2 \text{ lb}$$

$$\sigma_{max} = \frac{k_s F_1}{b\,t} = \frac{1.2\,(319.92)}{6\left(\frac{1}{3}\right)} = 191.9 \text{ psi}$$

$$L = \sqrt{4\,C^2 - (D - d)^2} + \frac{1}{2}(D\,\theta_D + d\,\theta_d)$$
$$= \sqrt{4\,(96)^2 - (18 - 6)^2} + \frac{1}{2}(18\,(3.265) + 6\,(3.015))$$
$$= 230 \text{ in}.$$

The oil slips and $f = 0.2$.

$$\frac{F_1 - 71.5}{F_2 - 71.5} = e^{0.2(3.015)}$$

$F_1 + F_2 = 482.2$ based on initial tension.

Solve for F_1 and F_2

$F_1 = 290.46$ lb

$F_2 = 191.9$ lb

$T = (F_1 - F_2) \times (290.46 - 191.9)(3) = 295.68$ lb·in.
Since 295.68 lb·in. < 472.5 lb·in., which is generated by the motor, the belt will slip.

Brakes

The belt can also be used as a braking device as shown in Figure 17.60. Forces in the belt can be evaluated using all the equations described, except for $F_c = 0$ since the belt is not moving.

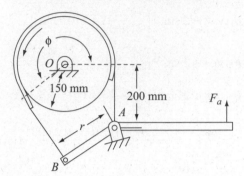

Figure 17.61

Example 17.40

The differential brake depicted in Figure 17.62 is rotating at 1,200 rpm. The mass moment of inertia of the drum is 100 N·m s². We would like to stop the drum in 2 seconds by applying force F_a, which is at a distance of 500 mm from point A. The friction coefficient between the belt and drum is 0.3. The design specifies that the angle between the belt at A and arm AB is 120°. From the geometry given in the figure, determine

(a) The angle of wrap
(b) The length of arm s from the geometry of the brake
(c) The force F_a necessary to stop the drum

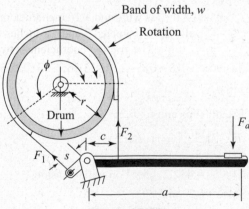

Figure 17.62

Solution:

Here we are interested in finding the force required to stop the drum in a specified amount of time.

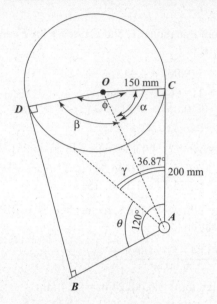

From the geometry of $ABDOC$:

$\Phi = 540 - 3(90) - 120 = 150°$

From triangle ACO:

$AO = \sqrt{200^2 + 150^2} = 250$ mm

Angle α can be obtained from triangle COA:

$\tan \alpha = \dfrac{200}{150}$

$\alpha = 53.13°$

Angle $\beta = 150° - 53.13° = 96.87°$:

Using the law of cosines for triangle ODA:

$DA = 306.55$ mm

Using the law of sines:

$\dfrac{306.55}{\sin 96.87°} = \dfrac{150}{\sin \gamma}$

$\gamma = 29.34°$

$\Theta = 120° - 36.87° - 29.34° = 53.79°$

$s = AD \cos 53.79° = 181.1$ mm

Angular deceleration:

$\alpha = \dfrac{\omega_0}{t} = \dfrac{2\pi (1,200)}{60(2)} = 62.83 \ \dfrac{\text{rad}}{\text{s}^2}$

Torque:

$T = I\alpha = (10)(62.83) = 6,283$ N·m

As shown in the figure, torque is acting opposite to the rotation.

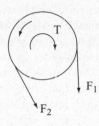

$$T = (F_1 - F_2) \times r = (F_1 - F_2)(0.15) = 6,283$$

$$\frac{F_1 - F_c}{F_2 - F_c} = e^{f\varphi}$$

$$\frac{F_1 - 0}{F_2 - 0} = e^{0.3(3.67)}$$

Solve for F_1 and F_2:

$$F_1 = 62,725.8 \text{ N}$$

$$F_2 = 20,839 \text{ N}$$

To find F_a, we take the moment about point A and set it equal to zero:

$$F_a(500) - F_2(181.1) = 0$$

$$F_a = 7,548 \text{ N}$$

Example 17.41

A drum weighing 50 lb is rotating at 2,000 rpm. The drum is stopped by a belt system shown in Figure 17.63. The radius of the drum is 5 inches. This drum should be stopped in 2 seconds. The friction coefficient between the belt and drum is 0.3.

(a) Based on the brake system geometry, find the belt contact angle.

(b) Find the force required to stop the drum.

(c) If the belt width is 3 inches and its thickness is 0.125 inches, find the maximum pressure on the belt.

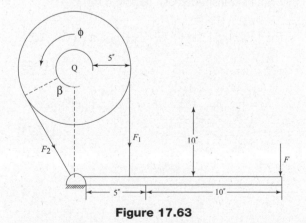

Figure 17.63

Solution:

Angular deceleration:

$$\alpha = \frac{\omega_0}{t} = \frac{2\pi(2,000)}{60(2)} = 104.72 \ \frac{\text{rad}}{\text{s}^2}$$

$$I = \frac{1}{2}mr^2 = \frac{1}{2}\frac{50}{32.2}\left(\frac{5}{12}\right)^2 = 0.1348 \ \text{lb} \cdot \text{ft} \cdot \text{s}^2$$

Torque:

$$T = I\alpha = (0.1348)(104.72)$$
$$= 14.11 \ \text{lb} \cdot \text{ft} = 169.32 \ \text{lb} \cdot \text{in.}$$

$$\varphi = 360 - 90 - \beta = 360 - 90 - \cos^{-1}(5/10)$$
$$= 210° = 3.665 \ \text{rad}$$

$$T = (F_1 - F_2) \times r = (F_1 - F_2)(5) = 169.32$$

$$\frac{F_1 - F_c}{F_2 - F_c} = e^{f\varphi}$$

$$\frac{F_1 - 0}{F_2 - 0} = e^{0.3(3.665)}$$

Solve for F_1 and F_2:

$$F_1 = 50.78 \text{ N}$$

$$F_2 = 16.9 \text{ N}$$

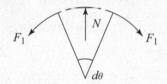

$$P_{max} = \frac{N}{(Rd\theta)\,\text{width}} = \frac{F_1 d\theta}{(Rd\theta)(\text{width})}$$
$$= \frac{50.78}{3(0.125)} = 135.41 \ \text{psi}$$

Example 17.42

A flat belt 4 inches wide and 3/16 inch thick operates on pulleys of diameter 5 and 15 inches and transmits 10 hp. The friction coefficient between the belt and pulleys is 0.3. The belt specific weight is 0.04 lb/in³. The pulleys are rotating at 1,500 rpm and they are 5 feet apart. Determine:

(a) The required belt tension

(b) The length of the belt

Solution:

$$W = (3/16)(4)(0.04)(12) = 0.36 \ \text{lb/ft}$$

$$T = \frac{63,000(10)}{1,500} = 420 \ \text{lb} \cdot \text{in.}$$

$$V = \frac{\pi d n}{12} = \frac{\pi(5)(1,500)}{12} = 1963.5 \ \text{ft/min}$$

$$F_1 - F_2 = \frac{33,000 \ \text{hp}}{V} = \frac{33,000(10)}{1963.5} = 168.1 \ \text{lb}$$

$$F_c = \frac{w}{g} V^2 = \frac{0.36}{32.2} \left(\frac{1963.5}{60} \right)^2 = 11.97 \text{ lb}$$

$$\theta_d = \pi - 2 \sin^{-1} \left(\frac{D-d}{2C} \right)$$

$$\theta_D = \pi - 2 \sin^{-1} \left(\frac{15-5}{2(5 \times 12)} \right) = 2.975 \text{ rad}$$

$$\theta_D = \pi + 2 \sin^{-1} \left(\frac{D-d}{2C} \right)$$

$$\theta_D = \pi + 2 \sin^{-1} \left(\frac{15-5}{2(5 \times 12)} \right) = 3.31 \text{ rad}$$

$$\frac{F_1 - F_c}{F_2 - F_c} = e^{f\theta_d}$$

$$\frac{F_1 - 11.97}{F_2 - 11.97} = e^{0.3(2.975)}$$

Solve for F_1 and F_2

$F_1 = 296.22 \text{ lb}$
$F_2 = 128.5 \text{ lb}$

$$L = \sqrt{4C^2 - (D-d)^2} + \frac{1}{2}(D \theta_D + d \theta_d)$$

$$= \sqrt{4(60)^2 - (15-5)^2} + \frac{1}{2}(15(2.975) + 6(3.31))$$

$$= 151.8 \text{ in.}$$

17.14 Clutches

Clutches are used to transmit power between two shafts. Typical disk clutches are shown in Figure 17.64.

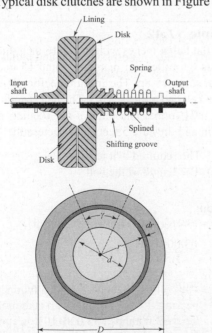

Figure 17.64

Here the force required to couple two disks is desired without causing slippage between two disks. The analysis of disks is based on assuming either uniform pressure or uniform wear in order to find the torque that can be transmitted. Uniform wear is more conservative in estimating the torque that can be transmitted. Based on uniform wear, the pressure between two disks is assumed to be

$$p = \frac{p_{max}d}{2r} \qquad [17.68]$$

where p_{max} is the maximum pressure that will develop at the surface of the disk, r is the radial distance to any location on the disk, and d is the disk's internal diameter. The torque transmitted can be found from:

$$T = \int_{\frac{d}{2}}^{\frac{D}{2}} \frac{p_{max}d}{2r} \left(2\pi r^2 \right) fdr = \frac{1}{8} p_{max}d\pi f \left(D^2 - d^2 \right) \qquad [17.69]$$

The axial force required for coupling disks can be obtained from:

$$F_a = \int_{\frac{d}{2}}^{\frac{D}{2}} \frac{p_{max}d}{2r} \left(2\pi r \right) dr = \frac{1}{2} p_{max}d\pi (D - d) \qquad [17.70]$$

An expression relating the torque to the applied axial force can be found from:

$$T = \frac{1}{4} F_a (D + d) f \qquad [17.71]$$

The maximum torque capacity occurs for the disk with:

$$d = \frac{D}{\sqrt{3}} \qquad [17.72]$$

In design the disk size is usually $0.45D < d < 0.8D$. If a design is based on uniform pressure and if the clutch disk lining is flexible, the pressure distribution between disks is fairly uniform. In this case, the torque that can be transmitted can be found from:

$$T = \int_{\frac{d}{2}}^{\frac{D}{2}} p_{max} \left(2\pi r^2 \right) fdr = \frac{1}{12} p_{max}\pi f \left(D^3 - d^3 \right) \qquad [17.73]$$

$$F_a = \int_{\frac{d}{2}}^{\frac{D}{2}} p_{max} \left(2\pi r \right) dr = \frac{1}{4} p_{max}\pi \left(D^2 - d^2 \right) \qquad [17.74]$$

$$T = \frac{1}{3} F_a f \frac{D^3 - d^3}{D^2 - d^2} \qquad [17.75]$$

Example 17.43

A disk clutch with a lining friction coefficient of 0.3, an outside diameter of 10 inches, and an inside diameter of 6 inches is used to transmit 5 hp. The shaft is rotating at 2,000 rpm. Find the axial force that must be applied for successful transmission of the power.

Solution:

$$hp = \frac{Tn}{63,000}$$

$$5 = \frac{T(2,000)}{63,000}$$

$$T = 157.5 \text{ lb} \cdot \text{in.}$$

For uniform wear:

$$T = \frac{1}{4} F_a (D + d) f$$

$$157.5 = \frac{1}{4} F_a (10 + 6) \times 0.3$$

$$F_a = 209.7 \text{ lb}$$

Based on uniform pressure:

$$T = \frac{1}{3} F_a f \frac{D^3 - d^3}{D^2 - d^2}$$

$$157.5 = \frac{1}{3} F_a (0.3) \frac{10^3 - 6^3}{10^2 - 6^2}$$

$$F_a = 128.57 \text{ lb}$$

Example 17.44

A clutch system is going to transmit 40 hp between two shafts. One of the shafts is not rotating at the time of coupling. The speed of the rotating shaft is 1,000 rpm. The clutch diameters of the disk are $d = 2$ inches and $D = 8$ inches. The total mass moment of inertia of the system is 20 lb·in·s². Find the axial force for successful coupling and the time needed to stop the slippage. The friction coefficient between linings is 0.3.

Solution:

$$hp = \frac{Tn}{63,000}$$

$$T = \frac{hp(63,000)}{n}$$

$$T = \frac{40(63,000)}{1,000} = 2,520 \text{ lb} \cdot \text{in.}$$

Based on uniform wear:

$$T = \frac{1}{4} F_a (D + d) f$$

$$2,520 = \frac{1}{4} F_a (8 + 2) \times 0.3$$

$$F_a = 3,360 \text{ lb}$$

$$T = I\alpha$$

$$2,520 = 20\alpha$$

$$\alpha = 126 \frac{\text{rad}}{\text{s}^2}$$

$$\alpha = \frac{\omega}{t}$$

$$t = \frac{\omega}{\alpha} = \frac{\frac{2\pi(1,000)}{60}}{126} = 0.83 \text{ s}$$

Based on uniform pressure:

$$T = \frac{1}{3} F_a f \frac{D^3 - d^3}{D^2 - d^2}$$

$$2,520 = \frac{1}{3} F_a (0.3) \frac{8^3 - 2^3}{8^2 - 2^2}$$

$$F_a = 3,000 \text{ lb}$$

In addition, $t = 0.83$, which is the same as above.

17.15
Rolling Contact Bearings

Roller bearings are referred to by different names in many design books. The most common names are rolling-contact bearing, antifriction bearing, and roller bearing. These bearings are used to reduce friction by transferring the load through the contact rolling elements. Friction in the roller bearing is governed by the contact normal load, lubricant, and speed of the shaft. The friction coefficient in these types of bearings is between 0.001–0.002. Some of the advantages of roller bearings include:

- Ease of lubrication
- Ability to support both axial and radial forces
- Can be easily replaced
- Will generate noise when about to fail

Some disadvantages of roller bearings include:

- More noisy as compared with journal bearings
- Complex failure mechanisms such as fatigue, pitting, spalling, and chipping on the contact surfaces
- Poor damping behavior

Figure 17.65 shows a general roller bearing and its significant components. Roller bearings are categorized in terms of their load-carrying conditions. The bearings could be designed to carry normal loads, axial loads, or both. The rolling elements can also be ball bearing or roller bearing. Roller bearings can take more radial loads compared with ball bearings. However, roller bearings are less flexible for any misalignments. In ball bearings, the balls are running in grooves called raceways. The radius of a groove is slightly bigger than the radius of the ball. Deeper grooves are used for bearings designed to support axial loads as well. They are many types of ball bearing and roller bearings. Figures 17.65 and 17.66 are typical ball and roller bearings.

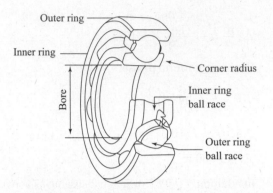

Figure 17.65 General roller bearing

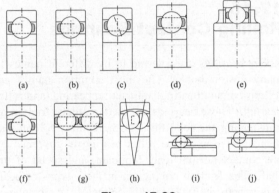

Figure 17.66

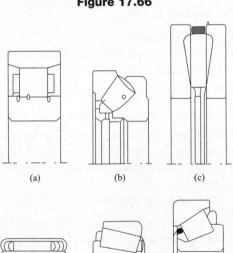

Figure 17.67

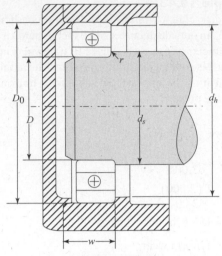

Figure 17.68

The Anti-Friction Bearing Manufacturing Association, AFMBA, has identified bearings by a two-digit number, called the dimension series code, such as 02 and 03. Tabulation by AFBMA should be referred to in order to find the bearing dimensions and load-carrying capacity. Figure 17.67 shows a ball bearing, shaft, and housing dimensions. The dimension of a 02 ball series and both a 02 series and 03 series straight roller bearings are shown in Tables 17.23 and 17.24.

The life of the bearing is defined as the number of hours the bearing can operate at a uniform speed under an operating load. Rating life L_{10} refers to number of revolutions that 90% of the bearings survive under operating conditions. Median life is defined as the time when 50% of the bearings survive under operating conditions. Median life is also used to rate bearing life. Median life is often 5 times greater than L_{10}. The loading rate for series 02 and series 03 for getting 10^6 revolutions is defined as C and is tabulated in Tables 17.23 and 17.24. The parameter C_s is defined as the maximum static load on the bearing without impairing its running operation.

For bearings subjected to both the axial and radial loads the equivalent radial load is defined as:

$$P = K_s(XVF_r + YF_a) \qquad [17.76]$$
$$P = K_sVF_r \qquad [17.77]$$

The greater value of these two equations should be used to obtain the life of the bearing.

Here P is the equivalent radial load, F_r is the applied radial load, F_a is the applied axial load, V is the rotation factor (1 for inner ring rotating, 1.2 for the outer ring rotating), X and Y are the radial and axial factors, and K_s is the shock or service factor (overloading) obtained from Table 17.19. The X and Y loading factors can be found from Tables 17.25 and 17.26.

Table 17.23 Dimensions and Basic Load Ratings for 02-Series Ball Bearings

Bore D (mm)	OD D_o (mm)	Width W (mm)	Fillet radius r (mm)	Load ratings (kN)			
				Deep groove		Angular contact	
				C	C_s	C	C_s
10	30	9	0.6	5.07	2.24	4.94	2.12
12	32	10	0.6	6.89	3.10	7.02	3.05
15	35	11	0.6	7.80	3.55	8.06	3.65
17	40	12	0.6	9.56	4.50	9.95	4.75
20	47	14	1.0	12.7	6.20	13.3	6.55
25	52	15	1.0	14.0	6.95	14.8	7.65
30	62	16	1.0	19.5	10.0	20.3	11.0
35	72	17	1.0	25.5	13.7	27.0	15.0
40	80	18	1.0	30.7	16.6	31.9	18.6
45	85	19	1.0	33.2	18.6	35.8	21.2
50	90	20	1.0	35.1	19.6	37.7	22.8
55	100	21	1.5	43.6	25.0	46.2	28.5
60	110	22	1.5	47.5	28.0	55.9	35.5
65	120	23	1.5	55.5	34.0	63.7	41.5
70	125	24	1.5	61.8	37.5	68.9	45.5
75	130	25	1.5	66.3	40.5	71.5	49.0
80	140	26	2.0	70.2	45.0	80.6	55.0
85	150	28	2.0	83.2	53.0	90.4	63.0
90	160	30	2.0	95.6	62.0	106	73.5
95	170	32	2.0	108	69.5	121	85.0

Note: Bearing life capacities, C, for 10^6 revolution life with 90% reliability.

Table 17.24 Dimensions and Basic Load Ratings for Straight Cylindrical Bearings

Bore D (mm)	02 series			03 series		
	OD D_o (mm)	Width w (mm)	Load rating C (kN)	OD D_o (mm)	Width w (mm)	Load rating C (kN)
25	52	15	16.8	62	17	28.6
30	62	16	22.4	72	19	36.9
35	72	17	31.9	80	21	44.6
40	80	18	41.8	90	23	56.1
45	85	19	44.0	100	25	72.1
50	90	20	45.7	110	27	88.0
55	100	21	56.1	120	29	102
60	110	22	64.4	130	31	123
65	120	23	76.5	140	33	138
70	125	24	79.2	150	35	151
75	130	25	91.3	160	37	183
80	140	26	106	170	39	190
85	150	28	119	180	41	212
90	160	30	142	190	43	242
95	170	32	165	200	45	264

Note: Bearing life capacities, C, for 10^6 revolution life with 90% reliability.

Table 17.25 Factors for Deep-Groove Ball Bearings

F_d/C_s	e	$F_a/VF_r \leq e$ X	Y	$F_a/VF_r > e$ X	Y
0.014*	0.19				2.30
0.21	0.21				2.15
0.028	0.22				1.99
0.042	0.24				1.85
0.056	0.26				1.71
0.070	0.27	1.0	0	0.56	1.63
0.084	0.28				1.55
0.110	0.30				1.45
0.17	0.34				1.31
0.28	0.38				1.15
0.42	0.42				1.04
0.56	0.44				1.00

* Use 0.014 if $F_d/C_s < 0.014$

Table 17.26 Factors for Commonly Used Angular-Contact Ball Bearings

Contact angle (α)	e	$\dfrac{i\,F_a^*}{C_s}$	Single-row bearing $F_a/VF_r > e$ X	Y	Double-row bearing $F_a/VF_r \leq e$ X	Y	$F_a/VF_r > e$ X	Y
	0.38	0.015		1.47		1.65		2.39
	0.40	0.029		1.40		1.57		2.28
	0.43	0.058		1.30		1.46		2.11
	0.46	0.087		1.23		1.38		2.00
15°	0.47	0.12	0.44	1.19	1.0	1.34	0.72	1.93
	0.50	0.17		1.12		1.26		1.82
	0.55	0.29		1.02		1.14		1.66
	0.56	0.44		1.00		1.12		1.63
	0.56	0.58		1.00		1.12		1.63
25°	0.68		0.41	0.87	1.0	0.92	0.67	1.41
35°	0.95		0.37	0.66	1.0	0.66	0.60	1.07

*i is the number of rows of balls

Table 17.27 Shock or Service Factors K_s

Type of load	Ball bearing	Roller bearing
Constant or steady	1.0	1.0
Light shocks	1.5	1.0
Moderate shocks	2.0	1.3
Heavy shocks	2.5	1.7
Extreme shocks	3.0	2.0

The bearing life in million revolutions can be obtained from:

$$L = K_r \left(\frac{C}{P}\right)^a \qquad [17.78]$$

where K_r is the reliability factor for a reliability of more than 90%, which is obtained from Figure 17.68, C is the basic loading rate given by the bearing series in Tables 17.22 and 17.24, and $a = 3$ for a ball bearing and $a = 3.333$ for a roller bearing.

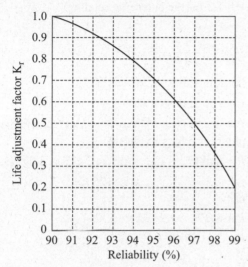

Figure 17.69

Example 17.45

A 60 mm bore 9 series 02 deep groove ball bearing is operating under combined radial and axial loads of 10 kN and 5 kN, respectively. The shaft is rotating at 1,000 rpm and under moderate shocks. Find:

(a) The equivalent radial load to the bearing
(b) The life of this bearing for 90% reliability and for 95% reliability and based on a median life of 50%

Solution:

The equivalent loading is determined by:

$P = K_s(XVF_r + YF_a)$ or $P = K_sVF_r$, whichever produces a larger value of P.

For moderate shock, $k_s = 2$.

The shaft is rotating, so $v = 1.0$.

The values of x and y are obtained by using Tables 17.17 and 17.18.

From Table 17.23, for a 60 mm deep groove, $C = 47.5$ kN and $C_s = 28$ kN.

From Table 17.25:

$$\frac{F_a}{C_s} = \frac{5}{28} = 0.1785$$

$$e = 0.34$$

$$\frac{F_a}{V F_r} = \frac{5}{10} = 0.2$$

$$\frac{F_a}{V F_r} < 0.34$$

$$x = 1.0 \text{ and } y = 0$$

$$P = K_s(XVF_r + YF_a) = 2(1)(1)(10) = 20 \text{ kN}$$

With life based on 90% reliability: $k_r = 1$.

$$L_{10} = K_r \left(\frac{C}{P}\right)^a = 1\left(\frac{47.5}{20}\right)^3 = 13.39$$

Life expectancy is 13.39×10^6 revolutions.

Based on 95% reliability, $k_r = 0.62$.

$$L = K_r \left(\frac{C}{P}\right)^a = 0.62\left(\frac{47.5}{20}\right)^3 = 8.3$$

Life expectancy is 8.3×10^6 revolutions.

Median life is $5 \times L_{10} = 5 \times 13.39 \times 10^6 = 66.98 \times 10^6$ revolutions.

Example 17.46

A ball bearing is used in the design of a rotating shaft with diameter of 20 mm. The engineer wants to increase the life of this system by 30%. How much should the radial load be changed? Assume the axial load remains at 2.5 kN and that a 02 ball bearing series is used in this design.

Solution:

For series 02 ball bearings with a deep groove, use Table 17.23:

$C = 12.7$ kN
$C_s = 6.2$ kN

From Table 17.25:

$$\frac{F_a}{C_s} = \frac{2.5}{6.2} = 0.4$$

$$e = 0.42$$

$$V = 1.0$$

$\dfrac{F_a}{V F_r}$ is assumed to be less than 0.42 and will be checked later.

$$x = 1.0 \text{ and } y = 0$$

$$L = K_r \left(\frac{C}{P}\right)^a$$

$$L = \left(\frac{C}{P}\right)^3$$

$$L_1 = \left(\frac{C}{P_1}\right)^3$$

For 30%, more life $= 1.3 \times L_1$

$L_2 = 1.3 L_1$

$$\left(\frac{P_1}{P_2}\right)^3 = 1.3$$

$$\frac{P_1}{P_2} = 1.091$$

$P_2 = 0.916 P_1$

$P = K_s(XVF_r + YF_a)$ for $x = 1$ and $y = 0$

$P_1 = K_s(XVF_{r1})$

$P_1 = K_s F_{r1}$ and $P_2 = K_s(XVF_{r2})$

$$\frac{F_{r2}}{F_{r1}} = \frac{P_2}{P_1} = 0.916$$

$F_{r2} = 0.916 F_{r1}$

So the radial load should be decreased by 8%.

For $\frac{F_a}{V F_r} > e$, $x = 0.56$ and $y = 1.04$.

$P_1 = K_s(0.56 F_{r1} + 1.04\,(2.5))$ and

$P_2 = K_s(0.56 F_{r2} + 1.04\,(2.5))$

$$\frac{P_2}{P_1} = 0.916 = \frac{0.56\,F_{r2} + 2.6}{0.56\,F_{r1} + 2.6}$$

$0.56\,F_{r2} + 2.6 = 0.513\,F_{r1} + 2.38$

$F_{r2} = 0.916\,F_{r1} - 0.393$ kN

This is the relationship between radial forces when the life is increased by 30%.

Example 17.47

A 30 mm (02 series) deep groove ball bearing carries a radial load of 2.5 kN and an axial load of 3.5 kN at 1,800 rpm. The outer ring rotates, and the load is steady. Determine the life of this bearing in hours.

Solution:

For 30 mm bearings from Table 17.23:

$C = 19.5$ kN

$C_s = 10\ 4N$

From Table 17.22:

$$\frac{F_a}{C_s} = \frac{3.5}{10} = 0.35$$

$e = 0.4$

$V = 1.2$ for the outer ring rotating:

$$\frac{F_a}{VF_r} = \frac{3.5}{1.2\,(2.5)} = 1.16$$

So from Table 17.25:

$x = 0.56$ and $y = 1.095$

$P = K_s(XVF_r + YF_a)$ or $P = K_s(VF_r)$, whichever produces a larger value of P.

$P = K_s(XVF_r + YF_a) = (0.56(1.2)(2.5) + 1.095\,(3.5))$

$\qquad\qquad\qquad = 5.5125$ kN

or

$P = K_s VF_r = 1.2\,(2.5) = 3$ kN

$P = 5.5125$ kN

$$L = K_r\left(\frac{C}{P}\right)^a$$

$$L = \left(\frac{19.5}{5.5125}\right)^3 = 44.264$$

$$\frac{44.264 \times 10^6}{1,800\,(60)} = 409.85 \text{ hours}$$

Table 17.28 Steel Alloys and Their Compositions

AISI or SAE number	Composition
10xx	Plain carbon steels
11xx	Plain carbon (resulfurized for machinability)
13xx	Manganese (1.5–2.0%)
23xx	Nickel (3.25–3.75%)
25xx	Nickel (4.75–5.25%)
31xx	Nickel (1.10–1.40%), chromium (0.55–0.90%)
33xx	Nickel (3.25–3.75%), chromium (1.40–1.75%)
40xx	Molybdenum (0.20–0.30%)
41xx	Chromium (0.40–1.20%), molybdenum (0.08–0.25%)
43xx	Nickel (1.65–2.00%), chromium (0.40–0.90%), molybdenum (0.20–0.30%)
46xx	Nickel (1.40–2.00%), molybdenum (0.15–0.30%)
48xx	Nickel (3.25–3.75%), molybdenum (0.20–0.30%)
51xx	Chromium (0.70–1.20%)
52xx	Chromium (1.30–1.60%)
61xx	Chromium (0.70–1.10%), vanadium (0.10%)
81xx	Nickel (0.20–0.40%), chromium (0.30–0.55%), molybdenum (0.08–0.25%)
86xx	Nickel (0.30–0.70%), chromium (0.40–0.85%), molybdenum (0.08–0.25%)
87xx	Nickel (0.40–0.70%), chromium (0.40–0.60%), molybdenum (0.20–0.30%)
92xx	Silicon (1.80–2.20%)

xx Carbon content, 0.xx w/o.

All steels have $0.50 \pm$ w/o manganese, unless stated otherwise.

Table 17.29 Mechanical Properties of Hot-rolled (HR) and Cold-rolled (CD) Steels

UNS number	AISI/ SAE number	Processing	Ultimate strength* S_u, MPa	Yield strength* S_y, MPa	Elongation in 50 mm, %	Reduction in area, %	Brinell hardness H_B
G10060	1006	HR	300	170	30	55	86
		CD	330	280	20	45	95
G10100	1010	HR	320	180	28	50	95
		CD	370	300	20	40	105
G10150	1015	HR	340	190	28	50	101
		CD	390	320	18	40	111
G10200	1020	HR	380	210	25	50	111
		CD	470	390	15	40	131
G10300	1030	HR	470	260	20	42	137
		CD	520	440	12	35	149
G10350	1035	HR	500	270	18	40	143
		CD	550	460	12	35	163
G10400	1040	HR	520	290	18	40	149
		CD	590	490	12	35	170
G10450	1045	HR	570	310	16	40	163
		CD	630	530	12	35	179
G10500	1050	HR	620	340	15	35	179
		CD	690	580	10	30	197
G10600	1060	HR	680	370	12	30	201
G10800	1080	HR	770	420	10	25	229
G10950	1095	HR	830	460	10	25	248

SOURCE: 1986 SAE Handbook, p.215.

*Values listed are estimated ASTM minimum values in the size range 18 to 32 mm.

Note: To convert from MPa to ksi, divide given values by 6.895.

Table 17.30 Mechanical Properties of Heat-Treated Steels

AISI number	Treatment	Temperature, °C	Ultimate strength S_u, MPa	Yield strength S_y, MPa	Elongation in 50 mm, %	Reduction in area, %	Brinell hardness H_B
1030	WQ&T	205	848	648	17	47	495
	WQ&T	425	731	579	23	60	302
	WQ&T	650	586	441	32	70	207
	Normalized	925	521	345	32	61	149
	Annealed	870	430	317	35	64	137
1040	OQ&T	205	779	593	19	48	262
	OQ&T	425	758	552	21	54	241
	OQ&T	650	634	434	29	65	192
	Normalized	900	590	374	28	55	170
	Annealed	790	519	353	30	57	149
1050	WQ&T	205	1120	807	9	27	514
	WQ&T	425	1090	793	13	36	444
	WQ&T	650	717	538	28	65	235
	Normalized	900	748	427	20	39	217
	Annealed	790	636	365	24	40	187
1060	OQ&T	425	1080	765	14	41	311
	OQ&T	540	965	669	17	45	277
	OQ&T	650	800	524	23	54	229
	Normalized	900	776	421	18	37	229
	Annealed	790	626	372	22	38	179
1095	OQ&T	315	1260	813	10	30	375
	OQ&T	425	1210	772	12	32	363
	OQ&T	650	896	552	21	47	269
	Normalized	900	1010	500	9	13	293
	Annealed	790	658	380	13	21	192
4130	WQ&T	205	1630	1460	10	41	467
	WQ&T	425	1280	1190	13	49	380
	WQ&T	650	814	703	22	64	245
	Normalized	870	670	436	25	59	197
	Annealed	865	560	361	28	56	156
4140	OQ&T	205	1770	1640	8	38	510
	OQ&T	425	1250	1140	13	49	370
	OQ&T	650	758	655	22	63	230
	Normalized	870	870	1020	18	47	302
	Annealed	815	655	417	26	57	197

SOURCE: ASM Metals Reference Book, 2nd ed. Metals Park, OH: American Society for Metals, 1983
Notes: To convert from MPa to ksi, divide given values by 6.895.
Values tabulated for 25-mm round sections and of gage length 50 mm. The properties for quenched and tempered steel are from a single heat. OQ&T = oil-quenched and tempered; WQ&T = water-quenched and tempered.

Table 17.31 Mechanical Properties of Gray Cast Iron

ASTM class*	Ultimate strength S_u, MPa	Compressive strength S_{uc}, MPa	Modulus of elasticity, GPa		Brinell hardness	Fatigue stress concentration factor
			Tension	Torsion	H_B	K_f
20	150	575	66–97	27–39	156	1.00
25	180	670	79–102	32–41	174	1.05
30	215	755	90–113	36–45	201	1.10
35	250	860	100–120	40–48	212	1.15
40	295	970	110–138	44–54	235	1.25
50	365	1135	130–157	50–54	262	1.35
60	435	1295	141–162	54–59	302	1.50

*Minimum values of S_u (in ksi) are given by the class number.
Note: To convert from MPa to ksi, divide given values by 6.895.

Table 17.32 Mechanical Properties of Some Stainless Steels

AISI type	Ultimate strength S_u (MPa)		Yield strength S_y (MPa)		Elongation in 50 mm, %		Izod impact J (N·m)	
	An.	CW	An.	CW	An.	CW	An.	CW
Austentic								
302	586	758	241	517	60	35	149	122
303	620	758	241	552	50	22	115	47
304	586	758	241	517	60	55	149	122
347, 348	620	758	241	448	50	40	149	—
Martensitic								
410	517	724	276	586	35	17	122	102
414	793	896*	620*	862	20	15*	68	—
431	862	896*	655*	862*	20	15*	68	—
440 A,B,C	724	796*	414	620*	14	7*	3	3*
Ferritic								
430, 430F	517	572	296	434	27	20	—	—
446	572	586	365	483	23	20	3	—

SOURCES: *Metal Progress Databook 1980*, Vol 118, no.1, Metals Park, OH: American Society for Metals (June 1980); *ASME Handbook Metal Properties*, New York: McGraw-Hill, 1954.
Note: To convert from MPa to ksi, divide given values by 6.895.
* Annealed and cold drawn.

Table 17.33 Average Properties of Common Engineering Materials*

Material	Density Mg/m³	Ultimate strength, MPa			Yield strength,† MPa		Modulus of elasticity, GPa	Modulus of rigidity, GPa	Coefficient of thermal expansion, $10^{-6}/°C$	Elongation in 50 mm, %	Poisson's ratio
		Tension	Compression**	Shear	Tension	Shear					
SI Units											
Steel											0.27–0.3
Structural, ASTM-A36	7.86	400	—	—	250	145	200	79	11.7	30	
High Strength, ASTM-A242	7.86	480	—	—	345	210	200	79	11.7	21	
Stainless (302), cold rolled	7.92	860	—	—	520	—	190	73	17.3	12	
Cast iron											0.2–0.3
Gray, ASTM A 48	7.2	170	650	240	—	—	70	28	12.1	0.5	
Malleable, ASTM A-47	7.3	340	620	330	230	—	165	64	12.1	10	
Wrought iron	7.7	350	—	240	210	130	190	70	12.1	35	0.3
Aluminum											0.33
Alloy 2014-T6	2.8	480	—	290	410	220	72	28	23	13	
Alloy 6061-T6	2.71	300	—	185	260	140	70	26	23.6	17	
Brass, yellow											0.34
Cold rolled	8.47	540	—	300	435	250	105	39	20	8	
Annealed	8.47	330	—	220	105	65	105	39	20	60	
Bronze, cold rolled (510)	8.86	560	—	—	520	275	110	41	17.8	10	0.34
Copper, hard drawn	8.86	380	—	—	260	160	120	40	16.8	4	0.33
Magnesium alloys	1.8	140–340	—	165	80–280	—	45	17	27	2–20	0.35
Nickel	8.08	310–760	—	—	140–620	—	210	80	13	2–50	0.31
Titanium alloys	4.4	900–970	—	—	760–900	—	100–120	39–44	8–10	10	0.33
Zinc alloys	6.6	280–390	—	—	210–320	—	83	31	27	1–10	0.33
Concrete											0.1–0.2
Medium strength	2.32	—	28	—	—	—	24	—	10	—	
High strengh	2.32	—	40	—	—	—	30	—	10	—	
Timber‡ (air dry)											
Douglas fir	0.54	—	55	7.6	—	—	12	—	4	—	
Southern pine	0.58	—	60	10	—	—	11	—	4	—	
Glass. 98% silica	2.19	—	50	—	—	—	65	28	80	—	0.2–0.27
Graphite	0.77	20	240	35	—	—	70	—	7	—	
Rubber	0.91	14	—	—	—	—	—	—	162	600	0.45–0.5

*Properties may vary widely with changes in composition, heat treatment, and method of manufacture.

** For ductile metals the compression strengths is assumed to be the same as that in tension.

†Offset of 0.2%

‡Loaded parallel to the grain.

Table 17.34 Composition of Some Aluminum Alloys

Alumium Alloy Identification Codes	
Composition designations	
1xyy	Unalloyed aluminum (>99% Al)
2xxx	Al + Cu as principal alloying element
3xxx	Al + Mn as principal alloying element
4xxx	Al + Si as principal alloying element
5xxx	Al + Mg as principal alloying element
6xxx	Al + Mg + Si as principal alloying elements
7xxx	Al + Zn as principal alloying element
8xxx	Al + other elements
yy = points of purity. Thus 1060 = 99.60% Al; 1090 = 99.90% Al; etc.	

Temper designations (suffixes to composition designations)	
—F	As fabricated
—O	As annealed
—H	Strain hardened by a cold-working process —H1X Hardened only, with X representing fraction hardness (8 = fully hard) —H2X Hardened and partially annealed —H3X Hardened and stabilized
—T	Heat treated —T2 Annealed (cast alloys) —T3 Solution heat-treated and cold worked —T4 Solution heat-treated and aged naturally —T5 Artificially aged only —T6 Solution heat-treated and artificially aged —T7 Solution heat-treated and stabilized —T8 Solution heat-treated, cold—worked, and aged —T9 Solution heat-treated, aged, and cold-worked —T10 Same as —T5, followed by cold-working

Table 17.35 Mechanical Properties of Aluminum Alloys

Alloy	Ultimate strength S_u		Yield strength S_y		Elongation in 50 mm, %	Brinell hardness H_R
	MPa	(ksi)	MPa	(ksi)		
Wrought:						
1100-H14	125	(18)	115	(17)	20	32
2011-T3	380	(55)	295	(43)	15	95
2014-T4	425	(62)	290	(42)	20	105
2024-T4	470	(68)	325	(47)	19	120
6061-T6	310	(45)	275	(40)	17	95
6063-T6	240	(35)	215	(31)	12	73
7075-T6	570	(83)	505	(73)	11	150
Cast:						
201-T4*	365	(53)	215	(31)	20	—
295-T6*	250	(36)	165	(24)	5	—
355-T6*	240	(35)	175	(25)	3	—
-T6**	290	(42)	190	(27)	4	—
356-T6*	230	(33)	165	(24)	2	—
-T6**	265	(38)	185	(27)	5	—
520-T4*	330	(48)	180	(26)	16	—

SOURCES: *ASM Metals Reference Book*, Metals Park, OH: American Society for Metals, 1981; *1981 Materials Selector*, vol. 92, no.6, Cleveland: Materials Engineering, Penton/IPC (December 1980).
*Sand casting
**Permanent-mold casting

Table 17.36 Mechanical Properties of Some Common Copper Alloys

Alloy	UNS number	Ultimate strength S_u, MPa	Yield strength S_y, MPa	Elongation in 50 mm, %
Wrought:				
Leaded				
Beryllium copper	C17300	469–1379	172-1227	43–3
Phos bronze	C54400	469–517	393–434	20–15
Aluminum				
Silicon-bronze	C64200	517–703	241–469	32–22
Silicon-bronze	C65500	400–745	152–414	60–13
Manganese bronze	C67500	448–579	207–414	33–19
Cast:				
Leaded				
Red brass	C83600	255	117	30
Yellow brass	C85200	262	90	35
Manganese bronze	C86200	655	331	20
Bearing bronze	C93200	241	124	20
Aluminum bronze	C95400	586–724	241–372	18–8
Copper nickel	C96200	310	172	20

SOURCES: 1981 materials reference issue, Machine Design, 53, no. 6 (March 19, 1981).
Note: To convert from MPa to ksi, divide given values by 6.895.

Table 17.37 Mechanical Properties of Some Common Plastics

Plastic	Ultimate strength S_u		Elongation in 50 mm, %	Izod impact strength	
	MPa	(ksi)		J	(ft·lb)
Acrylic	72	(10.5)	6	0.5	(0.4)
Cellulose acetate	14–18	(2–7)	—	1.4–9.5	(1–7)
Epoxy (glass-filled)	69–138	(10–20)	4	2.7–4.1	(3)
Fluorocarbon	23	(3.4)	300	4.1	(3)
Nylon (6/6)	83	(12)	60	1.4	(1)
Phenolic (wood-flour filled)	48	(7)	0.4–0.8	0.4	(0.3)
Polycarbonate	62–72	(9–10.5)	110–125	16–22	(12–16)
Polyester (25% glass filled)	110–90	(16–23)	1–3	1.4–2.6	(1.0–1.9)
Polypropylene	34	(5)	10–20	0.7–3.0	(0.5–2.2)

SOURCES: 1981 materials reference issue, Machine Design, 53, no. 6 (March 19, 1981);
1981 materials selector issue, Materials Engineering, 92, no.6 (December 1980).

Table 17.38 Hardness Conversion Chart (Higher Hardness)

Brinell Hardness Tungsten Carbide Ball	Rockwell Hardness			Approximate Tensile Strength	
3000 kg	A Scale 60 kg	B Scale 100 kg	C Scale 150 kg	(ksi)	(MPa)
—	85.6	—	68.0	—	
—	85.3	—	67.5	—	
—	85.0	—	67.0	—	
767	84.7	—	66.4	—	
757	84.4	—	65.9	—	
745	84.1	—	65.3	—	
733	83.8	—	64.7	—	
722	83.4	—	64.0	—	
710	83.0	—	63.3	—	
698	82.6	—	62.5	—	
684	82.2	—	61.8	—	
682	82.2	—	61.7	—	
670	81.8	—	61.0	—	
656	81.3	—	60.1	—	
653	81.2	—	60.0	—	
647	81.1	—	59.7	—	
638	80.8	—	59.2	329	2267
630	80.6	—	58.8	324	2232
627	80.5	—	58.7	323	2225
601	79.8	—	57.3	309	2129
578	79.1	—	56.0	297	2046
555	78.4	—	54.7	285	1964
534	77.8	—	53.5	274	1888
514	76.9	—	52.1	263	1812
495	76.3	—	51.0	253	1743

Table 17.39 Hardness Conversion Chart (Intermediate Hardness)

Brinell Hardness Tungsten Carbide Ball	Rockwell Hardness			Approximate Tensile Strength	
3000 kg	A Scale 60 kg	B Scale 100 kg	C Scale 150 kg	(ksi)	(MPa)
477	75.6	—	49.6	243	1674
461	74.9	—	48.5	235	1619
444	74.2	—	47.1	225	1550
429	73.4	—	45.7	217	1495
415	72.8	—	44.5	210	1447
401	72.0	—	43.1	202	1378
388	71.4	—	41.8	195	1343
375	70.6	—	40.4	188	1295
363	70.0	—	39.1	182	1254
352	69.3	—	37.9	176	1213
341	68.7	—	36.6	170	1171
331	68.1	—	35.5	166	1144
321	67.5	—	34.3	160	1102
311	66.9	—	33.1	155	1068
302	66.3	—	32.1	150	1033
293	65.7	—	30.9	145	999
285	65.3	—	29.9	141	971
277	64.6	—	28.8	137	944
269	64.1	—	27.6	133	916
262	63.6	—	26.6	129	889
255	63.0	—	25.4	126	868
248	62.5	—	24.2	122	840
241	61.8	100.0	22.8	118	813

Table 17.40 Hardness Conversion Chart (Lower Hardness)

Brinell Hardness Tungsten Carbide Ball	Rockwell Hardness			Approximate Tensile Strength	
	A Scale 60 kg	B Scale 100 kg	C Scale 150 kg	(ksi)	(MPa)
3000 kg					
235	61.4	99.0	21.7	115	792
229	60.2	98.2	20.5	111	765
223	59.6	97.3	20.0	109	751
217	59.0	96.4	18.0	105	723
212	58.6	95.5	17.0	102	703
207	58.3	94.6	16.0	100	689
201	57.6	93.8	15.0	98	675
197	56.9	92.8	—	95	655
192	56.4	91.9	—	93	641
187	55.7	90.7	—	90	620
183	55.2	90.0	—	89	613
179	54.6	89.0	—	87	599
174	53.9	87.8	—	85	586
170	53.2	86.8	—	83	572
167	52.8	86.0	—	81	558
163	52.3	85.0	—	79	544
156	51.0	82.9	—	76	523
149	49.9	80.8	—	73	503
143	49.3	78.7	—	71	489
137	47.6	76.4	—	67	461
131	46.3	74.0	—	65	448
126	45.3	72.0	—	63	434
121	44.2	69.8	—	60	413
116	43.1	67.6	—	58	400
111	42.1	65.7	—	56	386

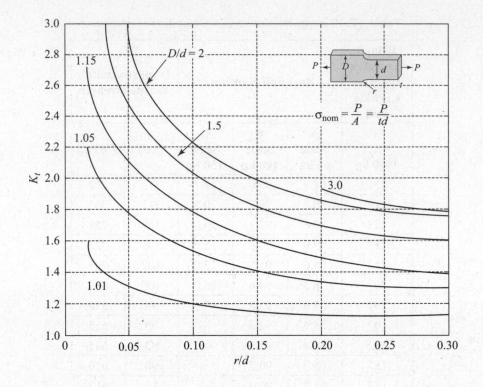

Approximate formula, $K_t \approx B\left(\dfrac{r}{d}\right)^a$, where:		
D/d	**B**	**a**
2.00	1.100	−0.321
1.50	1.077	−0.296
1.15	1.014	−0.239
1.05	0.998	−0.138
1.01	0.977	−0.107

Figure 17.70 Theoretical stress-concentration factor K_t for a filleted bar in axial tension

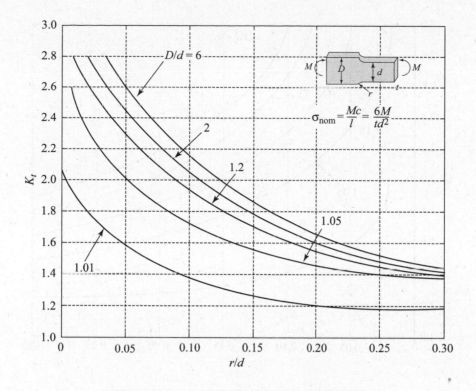

$$\sigma_{nom} = \frac{Mc}{l} = \frac{6M}{td^2}$$

Approximate formula, $K_t \approx B\left(\dfrac{r}{d}\right)^a$, where:		
D/d	**B**	**a**
6.00	0.896	−0.358
2.00	0.932	−0.303
1.20	0.996	−0.238
1.05	1.023	−0.192
1.01	0.967	−0.154

Figure 17.71 Theoretical stress-concentration factor K_t for a filleted bar in bending

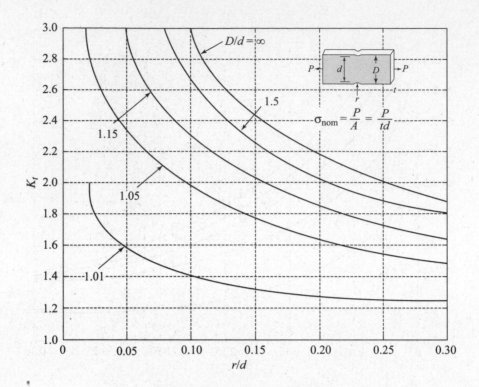

Approximate formula, $K_t \approx B \left(\dfrac{r}{d} \right)^a$, where:		
D/d	**B**	**a**
∞	1.110	−0.417
1.50	1.133	−0.366
1.15	1.095	−0.325
1.05	1.091	−0.242
1.01	1.043	−0.142

Figure 17.72 Theoretical stress-concentration K_t for a notched bar in axial tension

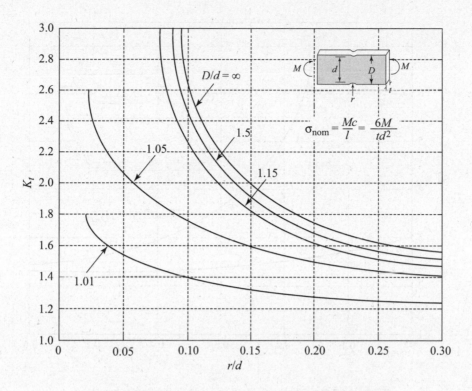

$$\sigma_{nom} = \frac{Mc}{l} = \frac{6M}{td^2}$$

Approximate formula, $K_t \approx B\left(\dfrac{r}{d}\right)^a$, where:		
D/d	**B**	**a**
∞	0.971	−0.357
1.50	0.983	−0.334
1.15	0.993	−0.303
1.05	1.025	−0.240
1.01	1.061	−0.134

Figure 17.73 Theoretical stress-concentration factor K_t for a notched bar in bending

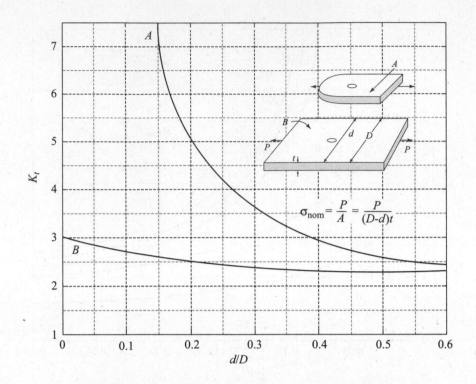

Figure 17.74 Theoretical stress-concentration factor K_t:
A—for a flat bar loaded in tension by a pin through the transverse hole
B—for a flat bar with a transverse hole in axial tension

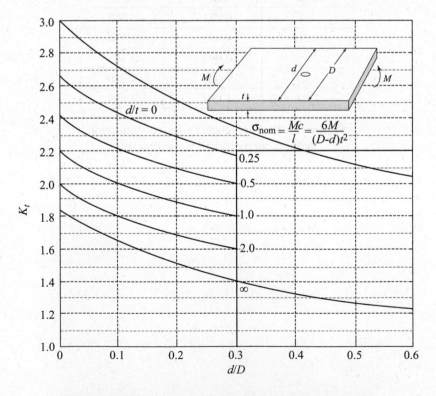

Figure 17.75 Theoretical stress-concentration factor K_t
for a flat bar with a transverse hole in bending

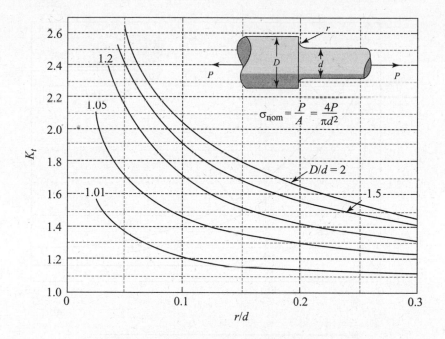

$$\sigma_{nom} = \frac{P}{A} = \frac{4P}{\pi d^2}$$

Approximate formula, $K_t \approx B\left(\frac{r}{d}\right)^a$, where:		
D/d	**B**	**a**
2.00	1.015	−0.300
1.50	1.000	−0.282
1.20	0.963	−0.255
1.05	1.005	−0.171
1.01	0.984	−0.105

Figure 17.76 Theoretical stress-concentration factor K_t for a shaft with shoulder fillet in axial tension

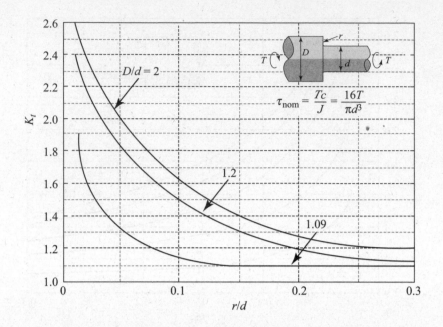

$$\tau_{nom} = \frac{Tc}{J} = \frac{16T}{\pi d^3}$$

Approximate formula, $K_t \approx B \left(\dfrac{r}{d} \right)^a$, where:

D/d	B	a
2.00	0.863	−0.239
1.20	0.833	−0.216
1.09	0.903	−0.127

Figure 17.77 Theoretical stress-concentration factor K_t for a shaft with a shoulder fillet in torsion

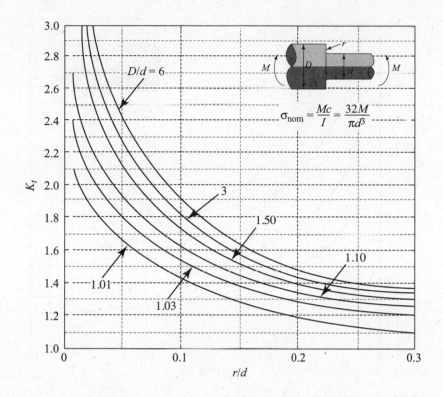

Approximate formula, $K_t \approx B\left(\dfrac{r}{d}\right)^a$, where:		
D/d	**B**	**a**
6.00	0.879	−0.332
3.00	0.893	−0.309
1.50	0.938	−0.258
1.10	0.951	−0.238
1.03	0.981	−0.184
1.01	0.919	−0.170

Figure 17.78 Theoretical stress-concentration factor K_t for a shaft with a shoulder fillet in bending

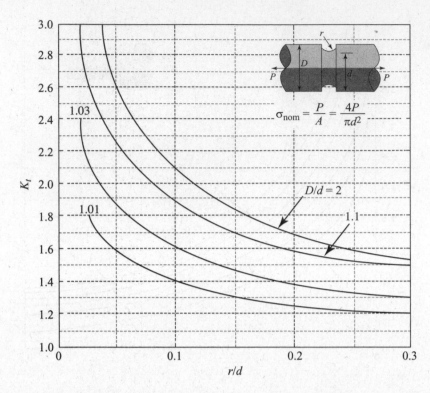

Figure 17.79 Theoretical stress-concentration factor K_t for a grooved shaft in axial tension

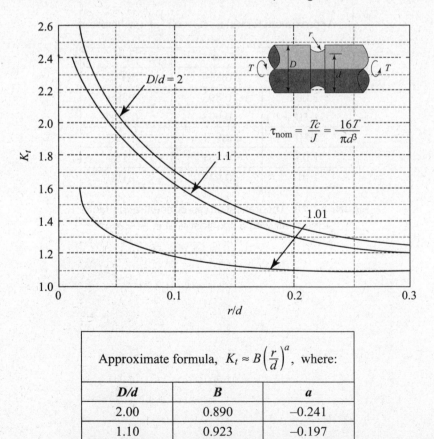

Approximate formula, $K_t \approx B\left(\dfrac{r}{d}\right)^a$, where:		
D/d	**B**	**a**
2.00	0.890	−0.241
1.10	0.923	−0.197
1.01	0.972	−0.102

Figure 17.80 Theoretical stress-concentration factor K_t for a grooved shaft in torsion

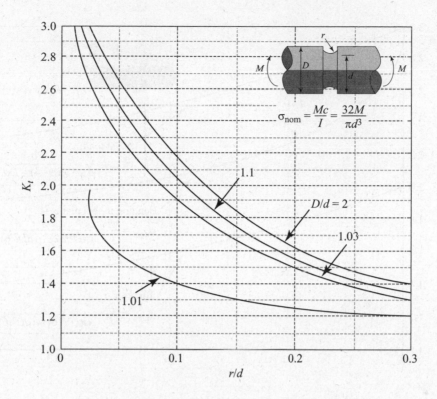

$$\sigma_{nom} = \frac{Mc}{I} = \frac{32M}{\pi d^3}$$

Approximate formula, $K_t \approx B\left(\dfrac{r}{d}\right)^a$, where:		
D/d	**B**	**a**
2.00	0.936	−0.331
1.10	0.955	−0.283
1.03	0.990	−0.215
1.01	0.994	0.152

Figure 17.81 Theoretical stress-concentration factor K_t for a grooved shaft in bending

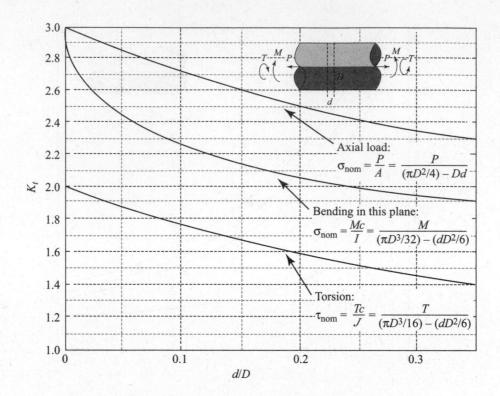

Figure 17.82 Theoretical stress-concentration factor K_t for a shaft with a transverse hole in axial tension, bending, and torsion

Summary

This chapter provided the procedures, concepts, and decision skills necessary to design machine elements. Design of a system encompasses identification of functional requirements as well as design parameters.

The concepts of materials selection, stress analysis, and different failure criteria for different loading situations (e.g., static and cyclic loadings) were discussed, along with the design of power screws and fasteners under various loadings, the attachments of components by weld, and the design of a gear system against teeth bending and contact stresses. Example problems were presented. Bearings, belts, and clutches were also covered in this chapter.

PRACTICE PROBLEMS

1. A component shown in the figure is subjected to a tensile force of 1,000 N. Find the maximum tensile stress. $D = 6$ cm, $d = 4$ cm, and $r = 0.5$ cm.

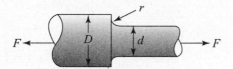

(A) 0.79 MPa
(B) 1.4 MPa
(C) 2.1 MPa
(D) 3.0 MPa

2. The bar shown in the figure has a diameter of 1 in. and the properties listed in the table. If the bar is subjected to a tensile force of 2,000 lb, determine the factor of safety.

S_Y (ksi)	S_{UT} (ksi)	S_e (ksi)	q
45	82	28	10%

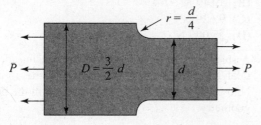

(A) 2.2
(B) 1.6
(C) 11.8
(D) 3.6

3. The bar shown below is made of ASTM class 50 gray cast iron and has dimensions $d = 50$ mm and $D = 56$ mm. If it is subjected to a static torque of 2,000 N·m and a static axial load of 225 kN, what is the safety factor?

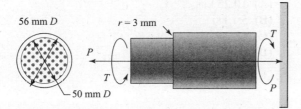

(A) 1.37
(B) 2.9
(C) 1.8
(D) 10.4

4. A car jack is designed as shown in the figure. Members AB and AC form a rectangular cross section and are made of 1045 HR steel. The width of the plate AB and AC is 2 inches. Find the thickness of these two plates based on both static failure as well as in plane and out of plane buckling. All connections are pinhole connections. Neglect stress concentration. $P = 1000$ lb. and the safety factor is $n = 2$.

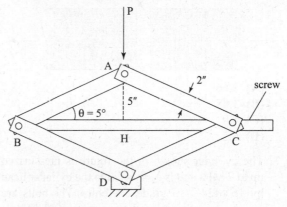

(A) 0.23 in.
(B) 0.127 in.
(C) 1.6 in.
(D) 0.05 in.

5. A structure that consists of the beam AB and the column CD is supported and loaded as shown in the figure. The structure is designed using AISI 1030 HR. What is the largest load F that may be applied with a safety factor of $n = 1.5$? Consider CD in your failure analysis.

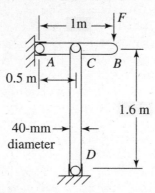

(A) 108.8 kN
(B) 165.2 kN
(C) 62.5 kN
(D) 32.25 kN

6. A machined AISI 4130 normalized steel bar shown in the figure is subjected to an axial load that is fluctuating between 20 kip and 5 kip while at the same time is subjected to a steady-state torque of 5000 lb·in. Use the Goodman diagram to find the safety factor.

Given: $D = 2\frac{1}{8}$ in., $d = 2$ in., $r = 0.05$ in.

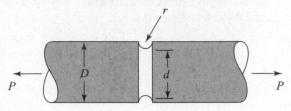

(A) 3.66
(B) 3.88
(C) 1.9
(D) 6.2

7. The cylinder shown in the figure is pressurized up to 7 MPa and is connected to the cylinder head by 16 M24 × 3 grade 8.8 bolts. The bolts are evenly spaced around the perimeters of the two circles with diameters of 1.2 and 1.5 m, respectively. The cylinder is made of cast iron, and its head is made of high-carbon steel. Assume that the force in each bolt is inversely related to its radial distance from the center of the cylinder head. Calculate the safety factor against failure due to static failure if the pressure is fluctuating between 0 and 7 MPa. Assume the design is made with replacement consideration.

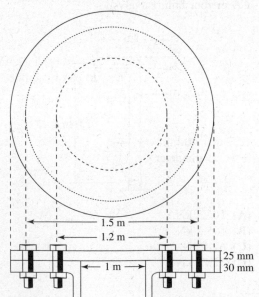

(A) 0.84
(B) 0.49
(C) 1.2
(D) 2.6

8. A valve for high-pressure air is shown in the figure below. The spindle has thread M14 × 2. After relating torque and axial thrust force, determine the axial force against seating when the applied torque is 10 N·m during the tightening. The coefficient of friction in the thread and against the seat is 0.12.

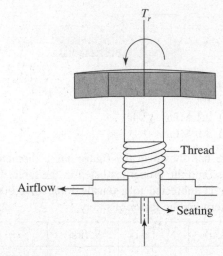

(A) 10,280 N
(B) 5,140 N
(C) 9,629 N
(D) 25,000 N

9. A steel plate is attached to a column using identical bolts (UNC 1-8) as shown in the figure below. Find the maximum shear stress in the bolt geometry.

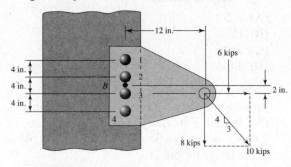

(A) 10.25 ksi
(B) 5.1 ksi
(C) 16.2 ksi
(D) 2.65 ksi

10. Two medium-carbon steel plates (AISI 1040) are attached by parallel-loaded fillet welds, as shown using electrode 6010. Each weld is 3 inches long. What is the minimum fillet leg length (h) that must be used if a static load of 1,000 lb is applied?

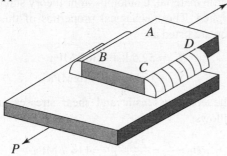

(A) 0.01 in.
(B) 0.024 in.
(C) 0.015 in.
(D) 0.2 in.

11. Three gears have a module of 5 mm and a 20° pressure angle. Driving gear 1 transmits 40 kW at 2,000 rpm to the idler gear 2 on shaft B. The output gear is mounted on shaft C, which drives a machine. Find the reaction on shaft B.

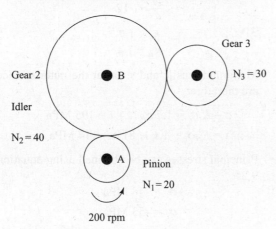

(A) 7.37 kN
(B) 8.66 kN
(C) 1.25 kN
(D) 12.0 kN

12. A pulley drive system uses a V-belt to transmit torque. The pulley radius is 100 mm, and the friction coefficient between the pulley and belt is 0.3. The belt weighs 2.25 N/m. Its angle $2\beta - 36°$ and contact angle $\varphi = 153°$. Based on 10 kW power transmission at a speed of 1,800 rpm, find the maximum tension in the belt. Assume uniform operation.
(A) 1,200 N
(B) 525.6 N
(C) 781.3 N
(D) 256 N

13. A V-belt drive pulley with a 200 mm diameter, 170° contact angle, 38° sheave angle, and coefficient of friction of 0.15 is rotating at 1,600 rpm. The belt weighs 8 N/m, and tight side tension is 3 kN. Determine the power capacity of this drive.
(A) 34 kW
(B) 22 kW
(C) 9 kW
(D) 12.6 kW

14. A disk clutch with an outsider diameter four times the inside diameter is used to transmit 40 hp at 1,000 rpm. The maximum pressure is found to be 20 psi, and the friction coefficient between the lining is 0.3. Find the clutch diameter.
(A) 3.6 in.
(B) 4.8 in.
(C) 14.8 in.
(D) 16.2 in.

15. A series 02 ball bearing is installed on a shaft with a diameter of 15 mm. The shaft is rotating at 1,800 rpm and is expected to last at least 25 khr. Assuming there is no axial force and that the system is operating under a moderate shock, find the maximum radial load that can be applied to this bearing.
(A) 280 N
(B) 560 N
(C) 140 N
(D) 100 N

Answer Key

1. B	6. A	11. A
2. C	7. B	12. C
3. A	8. C	13. A
4. B	9. A	14. C
5. D	10. B	15. A

Answers Explained

1. B In order to find the stress concentration factor, we need the following:

$$\frac{r}{d} = \frac{0.5}{4} = 0.125$$

$$\frac{D}{d} = \frac{6}{4} = 1.5$$

Using the figure, we determine $K_t = 1.75$:

$$\sigma_0 = \frac{F}{A} = \frac{1,000}{\frac{\pi d^2}{4}} = 0.79 \text{ MPa}$$

$$\sigma_{max} = 1.75 \times 0.79 = 1.4 \text{ MPa}$$

2. C Using the figure from question 1, we first determine the stress concentration factor k_t:

$$\frac{r}{d} = \frac{\frac{d}{4}}{d} = 0.25$$

$$\frac{D}{d} = \frac{3}{2}$$

$$k_t = 1.5$$

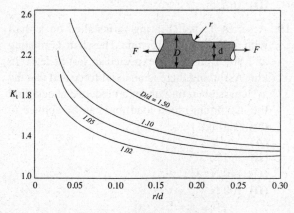

$$\sigma_0 = \frac{P}{A} = \frac{2,000}{\frac{\pi 1^2}{4}} = 2,546.7 \text{ psi}$$

$$\sigma_{max} = 1.5 \times 2546.7 = 3820 \text{ psi}$$

Since this material is ductile, we can use the maximum shear stress theory shown in equation 17.15 to determine the safety factor.

Note that $\tau_{max} = \frac{1}{2}(3820) = 1,910$ psi.

$$1,910 = \frac{45,000}{2n}$$

$$n = 11.8$$

3. A Since ASTM class 50 gray cast iron is ductile, brittle material, Coulomb-Mohr theory should be applied. The mechanical properties of this cast iron are listed as:

$$S_{ut} = 52.5 \text{ ksi} = 362 \text{ MPa}$$

$$S_{uc} = 164 \text{ ksi} = 1136 \text{ MPa}$$

The nominal tensile and shear stresses are as follows:

$$\sigma_0 = \frac{P}{A} = \frac{225,000}{\frac{\pi}{4} 50^2} = 114.7 \text{ MPa}$$

$$\tau_0 = \frac{TR}{J} = \frac{2000 \times 10^3 \times 25}{\frac{\pi}{32} 50^4} = 81.4 \text{ MPa}$$

The stress concentration factors can be obtained from charts:

$$\frac{r}{d} = \frac{3}{50} = 0.06$$

$$\frac{D}{d} = 1.12$$

$$K_t = 1.7$$

$$K_{ts} = 1.4$$

Maximum tensile and shear at the outer surface are therefore:

$$\sigma = K_t \sigma_0 = 1.7 \times 114.7 = 195 \text{ MPa}$$

$$\tau = K_{ts} \sigma_0 = 1.4 \times 81.4 = 114 \text{ MPa}$$

Principal stresses can be obtained using equation 9.25:

$$\sigma_1 = 248 \text{ MPa}$$

$$\sigma_2 = -53 \text{ MPa}$$

$$\sigma_3 = 0$$

Using Coulomb-Mohr theory, the safety factor can be obtained:

$$\frac{\sigma_1}{S_{ut}} - \frac{\sigma_2}{S_{uc}} = \frac{1}{n}$$

$$\frac{248}{362} - \frac{-53}{1,136} = \frac{1}{n}$$

$$n = 1.37$$

4. B AISI 1045 HR $S_y = 45$ ksi

$AB = 5/(\sin 5) = 57.4$ in.

$F_{AB} = P/(2 \sin 5) = 1,000/2 \sin 5 = 5,737$ lb

For out of plane buckling and an end condition of fixed-fixed, $c = 1.2$.

$P_e = 5737 \times$ factor of safety

$$P_c = \frac{C\pi^2 EI}{l^2}$$

$$5,737(2) = \frac{1(\pi^2)(30 \times 10^6)\left(\frac{2(t)^3}{12}\right)}{(57.4)^2}$$

$$t = 0.86 \text{ in. or } \frac{7}{8} \text{ in.}$$

Check for $\frac{l}{k}$:

$k^2 A = I$

$$k^2 (0.86)(2) = I = \frac{2(0.86)^3}{12}$$

$k = 0.248$ in.

$$\frac{l}{k} = \frac{57.4}{0.248} = 231.2$$

$$\left(\frac{l}{k}\right)_c = \left(\frac{2\pi^2 C E}{S_y}\right)^{0.5} = \left(\frac{2\pi^2 (1.2)(30 \times 10^6)}{45,000}\right)^{0.5}$$
$$= 125$$

Since $\left(\frac{l}{k}\right) > \left(\frac{l}{k}\right)_c$, the Euler equation is valid.

Based on yielding, $\dfrac{S_y}{n} = \dfrac{5,737}{2t}$

$t = 0.127$

So $t = 7/8$ in. satisfies both.

5. D AISI 1030 HR $S_y = 260$ MPa

$k^2 A = 1$

$$k^2 \frac{\pi}{4} d^2 = \frac{\pi}{64} d^4$$

$k = 10$ mm

$$\frac{l}{k} = \frac{1,600}{10} = 160$$

$c = 1$

$$\left(\frac{l}{k}\right)_c = \left(\frac{2\pi^2 CE}{S_y}\right)^{0.5} = \left(\frac{2\pi^2 (1)(200 \times 10^9)}{260 \times 10^6}\right)^{0.5}$$
$$= 123.2$$

Since $\left(\frac{l}{k}\right) > \left(\frac{l}{k}\right)_c$, the Euler equation is valid.

$P_c = 2F$

$$= \frac{C\pi^2 EI}{l^2} \frac{1}{n_f}$$

$$= \frac{1.2\left(\pi^2\right)\left(200 \times 10^9\right)\frac{\pi}{64}(0.04)^4}{(1.6)^2} \frac{1}{1.5}$$

$F = 32.25$ kN

Against yielding, $2F = \dfrac{S_y A}{n} = \dfrac{(260 \times 10^6)\frac{\pi}{64}(0.04)^4}{1.5}$

$F = 10.8$ kN

Therefore, $F_c = 32.25$ kN.

6. A AISI 4130 normalized, $S_{ut} = 97$ ksi

$\sigma_a = k_a k_b k_c k_d k_e k_g (0.5 \times S_{ut})$

$k_a = a S_{ut}{}^b = 2.7(97)^{-0.265} = 0.8$

$k_b = 0.879 (2)^{-0.107} = 0.8162$

$k_c = 1$

$k_d = 1$

$k_e = 1$

$k_g = 1$

$S_e = (0.8)(0.8162)(1)(1)(1)(0.5 \times 97) = 32$ ksi

$$\sigma_a = \frac{P}{A} = \frac{50}{\frac{\pi 2^2}{4}} = 15.9 \text{ ksi}$$

$$\frac{r}{d} = \frac{0.05}{2} = 0.025$$

$$\frac{D}{d} = 1.02$$

$k_t = 2.2$

Notch sensitivity $q = 0.78$.

$k_f = 1 + q(k_t - 1) = 1 + 0.78(2.2 - 1) = 1.94$

$k_{ts} = 1.55$

$q_s = 0.9$

$$\sigma_a = \frac{P_a}{A} = \frac{7.5}{\frac{\pi 2^2}{4}} = 2.39 \text{ ksi}$$

$$\sigma_m = \frac{P_m}{A} = \frac{12.5}{\frac{\pi 2^2}{4}} = 3.98 \text{ ksi}$$

$$\tau_m = \frac{T c}{J} = \frac{(5)(1)}{\frac{\pi(1)^4}{32}} = 3.18 \text{ ksi}$$

Using equations 17.28 and 17.29:

$$\sigma'_a = \sqrt{\left(1.94\left(\frac{2.39}{0.85}\right)\right)^2} = 5.45 \text{ ksi}$$

$$\sigma'_m = \sqrt{((1.94)(3.98))^2 + 3((1.5)(3.18))^2} = 9.95 \text{ ksi}$$

Goodman:

$$\frac{\sigma_a}{S_e} + \frac{\sigma_m}{S_{ut}} = \frac{1}{n}$$

$$\frac{5.45}{32} + \frac{9.95}{97} = \frac{1}{n}$$

$$n = 3.66$$

7. **B** M24 × 3 grade 8.8

$A_t = 353 \text{ mm}^2$

$S_{ut} = 830 \text{ MPa}$

$S_y = 660 \text{ MPa}$

$S_p = 600 \text{ MPa}$

$S_e = 129 \text{ MPa}$

Forces to inner bolts:

$$8F + 8\left(\frac{1.2}{1.5}\right)F = (7 \times 10^6)\frac{\pi}{4} \quad (1)$$

$F = 3.82 \times 10^5$ N external force to each inner bolt.

Stiffness of the bolt:

$$K_b = \frac{A_d A_t E}{A_d l_t + A_t l_d} = \frac{\frac{\pi}{4}(24^2)(353)E}{\frac{\pi}{4}(24^2)(30) + 353(25)}$$

$$= 7.13E \text{ N/mm}$$

To find K_m, construct the frustum:

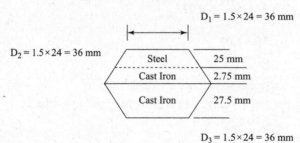

D₁ = 1.5 × 24 = 36 mm

D₂ = 1.5 × 24 = 36 mm

Steel	25 mm
Cast Iron	2.75 mm
Cast Iron	27.5 mm

D₃ = 1.5 × 24 = 36 mm

Using equation 17.34, the stiffness of each frustum can be found:

$K_1 = 52.28E$

$K_2 = 547.87E$

$K_3 = 25.04E$

Elastic modulus of cast iron is $E_c = E/2$.

$$\frac{1}{K_m} = \sum \frac{1}{K_i}$$

$$K_m = 16.42E$$

$$C = \frac{K_b}{K_b + K_m} = 0.281$$

The safety factor against bolt failure:

$$F_i = 0.75A_t S_p = 0.75(353)(600) = 158,850 \text{ N}$$

$$n_L = \frac{S_p A_t - F_i}{CP} = \frac{600(353) - 158,850}{0.281(3.82 \times 10^5)} = 0.49$$

8. **C** M12 screws From the table for this screw:

$p = 1.75$ mm

$2\acute{\alpha} = 60°$

$\acute{\alpha} = 30°$

Assuming no friction in the seating:

$$T_R = \frac{Fd_m}{2}\left(\frac{l + \pi f d_m \sec\alpha}{\pi d_m - f l \sec\alpha}\right)$$

$$d_m = d - 0.649p$$

From Tables 17.3 and 17.4:

$d_m = 10.864$ mm

$$T = F\left(\frac{10.864}{2}\right)\left(\frac{1.75 + \pi(0.12)(10.864)(\sec 30)}{\pi(10.864) - 0.12(1.75)(\sec 30)}\right)$$

$$T = 1.0386F \text{ N·mm}$$

$$1.0386F = 10,000$$

$$F = 9,628.5 \text{ N}$$

9. **A** The centroid of the bolt geometry is located in the middle of the bolt geometry shown in the figure below.

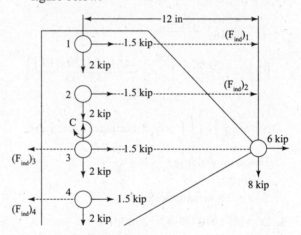

Torque at the centroid:

$$T = 8(12) - 6(2) = 84 \text{ kip·in}$$

The indirect forces are $(F_{ind})_1$, $(F_{ind})_2$, $(F_{ind})_3$, and $(F_{ind})_4$.

From the figure, bolt #1 is the critical bolt:

$$(F_{ind})_1 = \frac{T(6)}{(6^2 + 2^2)2} = 6.3 \text{ kip}$$

$(1.5 + 6.3) = 7.8$ kip

2 kip $F_{\text{Resultant}} = 8.052$ kip

$$\tau = \frac{8.052}{\frac{\pi}{4}(1)^2} = 10.25 \text{ ksi}$$

10. B Electrode: E6010 $S_{ut} = 62$ ksi

$S_y = 50$ ksi

$$\tau = \frac{P}{6(0.707)h} = \frac{1{,}000}{(0.707)h} = \frac{236}{h}$$

$$\bar{\sigma} = \sqrt{3}\,\tau = \sqrt{3}\left(\frac{236}{h}\right) = \frac{S_y}{n} = \frac{50{,}000}{3}$$

$h = 0.024$ in.

11. A For these gears based on the module of 5 mm:

$d_1 = N_1 m_1 = 20(5) = 100$ mm
$d_2 = N_2 m_2 = 40(5) = 200$ mm
$d_3 = N_3 m_3 = 30(5) = 150$ mm

$$T = \frac{9{,}549 \text{ kW}}{n} = \frac{9{,}549(40)}{2{,}000} = 191 \text{ N·m}$$

$$F_t\left(\frac{d_1}{2}\right) = 191$$

$$F_t = \frac{191}{0.05} = 3.82 \text{ kN}$$

$$F_n = F_t \tan(20) = 3.82 \tan(20) = 1.39 \text{ kN}$$

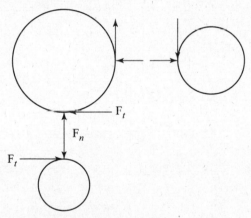

The reaction to shaft B:

$$F = \sqrt{5.21^2 + 5.21^2} = 7.37 \text{ kN}$$

12. C $\varphi = 153° = 2.76$ radians

$\beta = 18°$

$$V = r\omega = \frac{d}{2}\left(\frac{2\pi n}{60}\right) = \frac{\pi d n}{60}$$

$$F_c = \frac{w}{g}V^2 = \frac{2.25}{9.81}\left(\frac{\pi(0.2)(1{,}800)}{60}\right)^2 = 8.15 \text{ N}$$

$$\frac{F_1 - F_c}{F_2 - F_c} = e^{\frac{f\varphi}{\sin(\beta)}} = \frac{F_1 - 81.5}{F_2 - 81.5} = e^{\frac{0.3(2.76)}{\sin(18)}} - 14.57$$

$$\frac{[(F_1 - F_2) \times 0.1]\,n}{9{,}549} = kW$$

$$\frac{[(F_1 - F_2) \times 0.1]\,(1{,}800)}{9{,}549} = 10$$

$$F_1 - F_2 = 530.5$$

Solve for F_1 and F_2:

$F_1 = 651.1$ N
$F_2 = 120.6$ N

Pretension $F_i = \frac{1}{2}(F_1 + F_2) = 385.85$ N
$F_{max} = k_s F_1 = 1.2(651.1) = 781.3$ N

13. A $V = \dfrac{\pi d n}{60} = \dfrac{\pi(0.2)(1{,}600)}{60} = 16.75 \dfrac{\text{m}}{\text{s}}$

$$F_c = \frac{w}{g}V^2 = \frac{8}{9.81}(16.75)^2 = 228.9 \text{ N}$$

$\varphi = 170° = 2.967$ rad

$\beta = 19°$

$$\frac{F_1 - F_c}{F_2 - F_c} = e^{f\varphi/\sec\beta}$$

$$\frac{3{,}000 - 228.9}{F_2 - 228.9} = e^{0.3(3.665)/\sec 19} = 3.924$$

$F_2 = 935.1$ N

$T = (F_1 - F_2) \times r = (3{,}000 - 935.1)(0.1) = 206.5$ N

$$kW = \frac{Tn}{9{,}549} = \frac{206.5(1600)}{9{,}549} = 34 \text{ kW}$$

14. C Based on the uniform wear:

$$hp = \frac{Tn}{63{,}000}$$

$$40 = \frac{T(1{,}000)}{63{,}000}$$

$T = 252$ lb · in

$$T = \frac{1}{8}\,p_{max}d\pi f\left(D^2 - d^2\right)$$

$$252 = \frac{1}{8}(20)\,d\pi(0.3)\left((4d)^2 - d^2\right)$$

Solve for $d = 1.9$ in ≈ 2 in.

$D = 4d = 8$ in.

Based on uniform pressure:

$$T = \frac{1}{12}\,p_{max}\pi f\left(D^3 - d^3\right)$$

$$252 = \frac{1}{12}(20)\pi(0.3)\left((4d)^3 - d^3\right)$$

Solve for $d = 3.7$ in.

$D = 4d = 14.8$ in.

15. A For the ball bearing with $D = 15$ mm, and using Table 17.23:

$C = 7.8$ kN

$C_s = 3.35$ kN

$L = (25,000)(60)(1,800)/10^6 = 2,700$

$$L = K_r \left(\frac{C}{P}\right)^a$$

$$2,700 = 1\left(\frac{7.8}{P}\right)^3$$

$P = 0.56$ kN

From Table 17.25, for no axial loading:

$$\frac{F_a}{C_s} = 0$$

$e = 0.19$

Since $\dfrac{F_a}{V\,F_r} < 0.19$, $x = 1.0$ and $y = 0$.

Also for moderate shock, $k_s = 2$ and for shaft rotating, $V = 1.0$.

$P = K_s x V F_r$

$0.56 = 2\,(1 \times 1)\,F_r$

$F_r = 0.28$ kN is the maximum radial force.

PRACTICE EXAMINATIONS

General

Taking the practice exams included in this book is an important part of your preparation for the FE/EIT exam. Take the practice exams only after you have thoroughly reviewed the concepts and worked the sample problems presented in this book.

Included are two full-length practice exams. The first practice exam covers the material presented in the Morning Session exam. As on the actual exam, you have four hours to complete the practice test. For examinees not electing to take a discipline-specific afternoon exam, the second practice exam covers the material in the afternoon General Exam option. Again, as on the actual exam, you have four hours to complete the practice test.

Training is the best way to increase your test-taking stamina! For this reason it is strongly recommended that you take these practice exams in two back-to-back, four-hour sessions. Simulate test conditions; use only your calculator and your copy of the "NCEES FE Reference Handbook" while taking the practice exams. Read each question carefully, study the diagrams, and eliminate obviously incorrect answers. Since there is no penalty for wrong answers, make your best guess when you do not know the answer to a question.

Allow enough time in your study program not only to take the practice exams but also to review any problem areas that become evident when you analyze your incorrect answers. Fill in the charts at the end of each exam as you score your results. An analysis of incorrect answers is a useful self-teaching tool and will help you to fine-tune your final study sessions.

ANSWER SHEET
Practice Exam—Mechanical Discipline

1. Ⓐ Ⓑ Ⓒ Ⓓ
2. Ⓐ Ⓑ Ⓒ Ⓓ
3. Ⓐ Ⓑ Ⓒ Ⓓ
4. Ⓐ Ⓑ Ⓒ Ⓓ
5. Ⓐ Ⓑ Ⓒ Ⓓ
6. Ⓐ Ⓑ Ⓒ Ⓓ
7. Ⓐ Ⓑ Ⓒ Ⓓ
8. Ⓐ Ⓑ Ⓒ Ⓓ
9. Ⓐ Ⓑ Ⓒ Ⓓ
10. Ⓐ Ⓑ Ⓒ Ⓓ
11. Ⓐ Ⓑ Ⓒ Ⓓ
12. Ⓐ Ⓑ Ⓒ Ⓓ
13. Ⓐ Ⓑ Ⓒ Ⓓ
14. Ⓐ Ⓑ Ⓒ Ⓓ
15. Ⓐ Ⓑ Ⓒ Ⓓ
16. Ⓐ Ⓑ Ⓒ Ⓓ
17. Ⓐ Ⓑ Ⓒ Ⓓ
18. Ⓐ Ⓑ Ⓒ Ⓓ
19. Ⓐ Ⓑ Ⓒ Ⓓ
20. Ⓐ Ⓑ Ⓒ Ⓓ
21. Ⓐ Ⓑ Ⓒ Ⓓ
22. Ⓐ Ⓑ Ⓒ Ⓓ
23. Ⓐ Ⓑ Ⓒ Ⓓ
24. Ⓐ Ⓑ Ⓒ Ⓓ
25. Ⓐ Ⓑ Ⓒ Ⓓ
26. Ⓐ Ⓑ Ⓒ Ⓓ
27. Ⓐ Ⓑ Ⓒ Ⓓ
28. Ⓐ Ⓑ Ⓒ Ⓓ

29. Ⓐ Ⓑ Ⓒ Ⓓ
30. Ⓐ Ⓑ Ⓒ Ⓓ
31. Ⓐ Ⓑ Ⓒ Ⓓ
32. Ⓐ Ⓑ Ⓒ Ⓓ
33. Ⓐ Ⓑ Ⓒ Ⓓ
34. Ⓐ Ⓑ Ⓒ Ⓓ
35. Ⓐ Ⓑ Ⓒ Ⓓ
36. Ⓐ Ⓑ Ⓒ Ⓓ
37. Ⓐ Ⓑ Ⓒ Ⓓ
38. Ⓐ Ⓑ Ⓒ Ⓓ
39. Ⓐ Ⓑ Ⓒ Ⓓ
40. Ⓐ Ⓑ Ⓒ Ⓓ
41. Ⓐ Ⓑ Ⓒ Ⓓ
42. Ⓐ Ⓑ Ⓒ Ⓓ
43. Ⓐ Ⓑ Ⓒ Ⓓ
44. Ⓐ Ⓑ Ⓒ Ⓓ
45. Ⓐ Ⓑ Ⓒ Ⓓ
46. Ⓐ Ⓑ Ⓒ Ⓓ
47. Ⓐ Ⓑ Ⓒ Ⓓ
48. Ⓐ Ⓑ Ⓒ Ⓓ
49. Ⓐ Ⓑ Ⓒ Ⓓ
50. Ⓐ Ⓑ Ⓒ Ⓓ
51. Ⓐ Ⓑ Ⓒ Ⓓ
52. Ⓐ Ⓑ Ⓒ Ⓓ
53. Ⓐ Ⓑ Ⓒ Ⓓ
54. Ⓐ Ⓑ Ⓒ Ⓓ
55. Ⓐ Ⓑ Ⓒ Ⓓ
56. Ⓐ Ⓑ Ⓒ Ⓓ

57. Ⓐ Ⓑ Ⓒ Ⓓ
58. Ⓐ Ⓑ Ⓒ Ⓓ
59. Ⓐ Ⓑ Ⓒ Ⓓ
60. Ⓐ Ⓑ Ⓒ Ⓓ
61. Ⓐ Ⓑ Ⓒ Ⓓ
62. Ⓐ Ⓑ Ⓒ Ⓓ
63. Ⓐ Ⓑ Ⓒ Ⓓ
64. Ⓐ Ⓑ Ⓒ Ⓓ
65. Ⓐ Ⓑ Ⓒ Ⓓ
66. Ⓐ Ⓑ Ⓒ Ⓓ
67. Ⓐ Ⓑ Ⓒ Ⓓ
68. Ⓐ Ⓑ Ⓒ Ⓓ
69. Ⓐ Ⓑ Ⓒ Ⓓ
70. Ⓐ Ⓑ Ⓒ Ⓓ
71. Ⓐ Ⓑ Ⓒ Ⓓ
72. Ⓐ Ⓑ Ⓒ Ⓓ
73. Ⓐ Ⓑ Ⓒ Ⓓ
74. Ⓐ Ⓑ Ⓒ Ⓓ
75. Ⓐ Ⓑ Ⓒ Ⓓ
76. Ⓐ Ⓑ Ⓒ Ⓓ
77. Ⓐ Ⓑ Ⓒ Ⓓ
78. Ⓐ Ⓑ Ⓒ Ⓓ
79. Ⓐ Ⓑ Ⓒ Ⓓ
80. Ⓐ Ⓑ Ⓒ Ⓓ
81. Ⓐ Ⓑ Ⓒ Ⓓ
82. Ⓐ Ⓑ Ⓒ Ⓓ
83. Ⓐ Ⓑ Ⓒ Ⓓ
84. Ⓐ Ⓑ Ⓒ Ⓓ

85. Ⓐ Ⓑ Ⓒ Ⓓ
86. Ⓐ Ⓑ Ⓒ Ⓓ
87. Ⓐ Ⓑ Ⓒ Ⓓ
88. Ⓐ Ⓑ Ⓒ Ⓓ
89. Ⓐ Ⓑ Ⓒ Ⓓ
90. Ⓐ Ⓑ Ⓒ Ⓓ
91. Ⓐ Ⓑ Ⓒ Ⓓ
92. Ⓐ Ⓑ Ⓒ Ⓓ
93. Ⓐ Ⓑ Ⓒ Ⓓ
94. Ⓐ Ⓑ Ⓒ Ⓓ
95. Ⓐ Ⓑ Ⓒ Ⓓ
96. Ⓐ Ⓑ Ⓒ Ⓓ
97. Ⓐ Ⓑ Ⓒ Ⓓ
98. Ⓐ Ⓑ Ⓒ Ⓓ
99. Ⓐ Ⓑ Ⓒ Ⓓ
100. Ⓐ Ⓑ Ⓒ Ⓓ
101. Ⓐ Ⓑ Ⓒ Ⓓ
102. Ⓐ Ⓑ Ⓒ Ⓓ
103. Ⓐ Ⓑ Ⓒ Ⓓ
104. Ⓐ Ⓑ Ⓒ Ⓓ
105. Ⓐ Ⓑ Ⓒ Ⓓ
106. Ⓐ Ⓑ Ⓒ Ⓓ
107. Ⓐ Ⓑ Ⓒ Ⓓ
108. Ⓐ Ⓑ Ⓒ Ⓓ
109. Ⓐ Ⓑ Ⓒ Ⓓ
110. Ⓐ Ⓑ Ⓒ Ⓓ

Practice Exam— Mechanical Discipline

1. A particle travels in a straight line in such a way that its distance s in meters from a given point on the line after time t in seconds is given by the equation $s = 10t^3 - 2t^5$. What is the instantaneous velocity of the particle at $t = 1$ second?
   (A) -2 m/s
   (B) 15 m/s
   (C) 50 m/s
   (D) 20 m/s

2. The slope of the line that passes through points (a,b) and (b,a) is
   (A) 0
   (B) 1
   (C) -1
   (D) undefined

3. What are the coordinates of the center of a sphere whose equation is $x^2 + y^2 + z^2 - 4y + 6z = 0$?
   (A) $(0, -2, 3)$
   (B) $(0, 2, -3)$
   (C) $(1, 4, 9)$
   (D) $(1, -2, -3)$

4. The volume of the solid generated by revolving the area bounded by $y = \sqrt{x}$, $y = 2$, and the y-axis about the y-axis is
   (A) 8π cubic units
   (B) $\dfrac{16\pi}{3}$ cubic units
   (C) $\dfrac{32\pi}{5}$ cubic units
   (D) $\dfrac{8\pi}{3}$ cubic units

5. For $0 < x < \dfrac{\pi}{2}$, the trigonometric expression $\tan x \csc x \cos x$ is equal to
   (A) 1
   (B) 0
   (C) $\sin x$
   (D) $\cos 2x$

6. For $0 < a < b$, the value of $\int_{\ln a}^{\ln b} e^{-x}\, dx$ is
   (A) $a - b$
   (B) $b - a$
   (C) $\dfrac{1}{a} - \dfrac{1}{b}$
   (D) $\dfrac{1}{b} - \dfrac{1}{a}$

7. The polishing time for a sample of eight bowls is 6, 12, 14.2, 10, 13, 15.2, 11, and 8 minutes. What is the standard deviation of the polishing time of this sample?
   (A) 2.9076
   (B) 3.1084
   (C) 9.6621
   (D) 11.175

8. A fair coin is flipped 5 times. What is the probability of the coin showing 5 tails given that at least 4 tails are showing?
   (A) 0.0313
   (B) 0.2000
   (C) 0.1563
   (D) 0.1667

9. The latitude, degrees north of the equator, and the January minimum temperature for a sample of five U.S. cities is given in the table below. Use a least squares regression line to predict the January minimum temperature for a city of latitude 42 degrees.

Latitude (°N)	31.2	32.9	33.6	35.4	40.7	31
Temperature (°C)	44	38	35	31	15	45

   (A) 7
   (B) 9
   (C) 11
   (D) 13

10. Suppose the mass of a foot-long metal rod can be approximated by the density function $f(x) = 3x^2$ for $0 \leq x \leq 1$ and by $f(x) = 0$ elsewhere. Where is the center of gravity for the metal rod?
    (A) 5 inches
    (B) 9 inches
    (C) 10 inches
    (D) 12 inches

11. The following example is a typical flow chart. The end result of the chart is

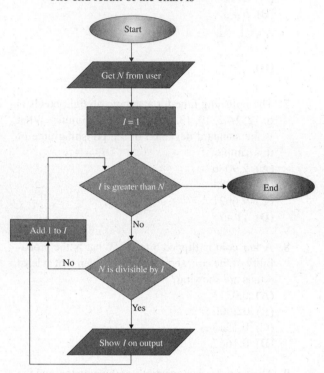

    (A) Sums a series of N's.
    (B) Checks if 3 numbers are sides of triangle.
    (C) Calculates all divisors of N.
    (D) Checks if number is prime.

Questions 12 and 13

The following is a segment of a spreadsheet. Use the data in the cells to answer questions 42 and 43.

	A	B	C	D
1	23	24	25	26
2	7	A2^2	B2*A$1	
3	8	A3^2	B3*B$1	
4	9	A4^2	B4*C$1	
5	10	A5^2	B5*D$1	

12. What will be the top to bottom values in Column B?
    (A) 12, 14, 16, 18
    (B) 16, 18, 20, 24
    (C) 14, 16, 18, 20
    (D) 49, 64, 81, 100

13. What will be the top to bottom values in Column C?
    (A) 1127, 1536, 2025, 2600
    (B) 340, 390, 460, 530
    (C) 70, 96, 126, 160
    (D) 160, 240, 320, 450

14. An unlicensed engineer interviews for a position as Director of Plant Operations at a university. On his resume, he states that he is a member of the National Society of Professional Engineers (NSPE) and includes his membership number. He is, in fact, an associate member of NSPE. Does his claim that he is a member violate the engineers' code of ethics?
    (A) Yes. It implies that he is a licensed engineer.
    (B) No. He was implying membership, not licensure.
    (C) Yes. A member number could be construed as a PE registration number.
    (D) No. He was indicating professional society involvement.

15. A forensic engineer accepts a retainer from an attorney for a plaintiff and receives selected documentation from the attorney. Subsequently the engineer bills for work that includes a review of the case documentation. Still later, because of a disagreement with the attorney, the engineer ceases to perform the work without delivering a report to the attorney or receiving any additional payment for her services. Thereafter, the engineer returns the full retainer fee to the attorney and all of the file documentation provided earlier. Several months later, the engineer is approached by the defense attorneys in the same case and accepts an assignment to function as an expert for the defense in the same legal dispute. Was it ethical for the engineer to accept the new assignment?
    (A) No. She drew no conclusions during the first review of the material.
    (B) Yes. She returned all moneys paid to her.
    (C) No. She did not have the consent of all concerned parties.
    (D) Yes. Several months passed before she was approached by the defense attorneys.

16. Snow and ice heavily damaged a building that was designed by Registered Engineers and Architects. The owner was not able to recover damages because of
    (A) the insolvency of the bond holder
    (B) the statute of limitations
    (C) a hold-harmless clause between architect and engineer
    (D) an Act of God clause in the insurance policy

17. A company can manufacture parts for a manufacturing cost of $4.00 per unit plus $4,000 for tooling. A semiautomated system has been developed as an alternative. The costs for the proposed method are $40,000 for equipment and $1.00 per unit manufacturing cost per unit. What is the break-even number of units per year?
    (A) 12,000
    (B) 11,000
    (C) 4,000
    (D) 15,000

18. A $50,000 municipal bond is being sold. The interest rate is 8% with $2,000 to the bond holder every 6 months. The bond life is 10 years. At that time, the owner will receive the $50,000 back plus the last $2,000 interest payment. If your minimum interest rate is 12% compounded semiannually, what would you pay for the bond?
    (A) $23,520
    (B) $25,630
    (C) $38,530
    (D) $77,160

19. Two machines are being considered for purchase. The first machine costs $10,000 with a life of 10 years and a salvage value of $1,000. The second machine costs twice as much as the first with the same life and a salvage value of $4,000. The net annual revenues of the first machine are $1,500. If interest is 8%, what would be the annual revenues of the second machine for the two machines to be equally desirable?
    (A) $7,460
    (B) $5,370
    (C) $2,920
    (D) $2,780

20.

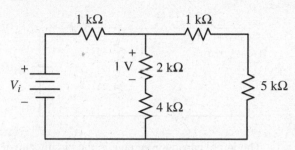

The value of the source voltage, V_i, for the circuit shown above is
    (A) 4 V   (C) 5.5 V
    (B) 7 V   (D) 9.5 V

21.

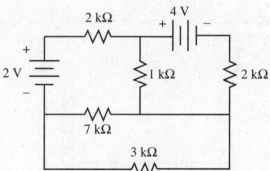

What is the voltage drop across the 1-kiloohm resistor shown in the diagram above?
    (A) 0.294 V   (C) 3.24 V
    (B) 1.43 V   (D) 4.56 V

22. The voltage between the terminals of a 0.01microfarad capacitor is 50 volts. What is the charge on the plates of the capacitor?
    (A) 50 μC
    (B) 5.0 μC
    (C) 0.5 μC
    (D) 10 μC

23.

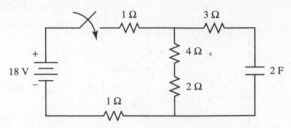

The initial capacitor voltage in the circuit shown above is 0 volt. What voltage will be measured across the capacitor 2 seconds after the switch is closed?

(A) 8.65 V
(B) 1.23 V
(C) 10.64 V
(D) 2.18 V

24. What is the moment of the 200-N force about point *A*?

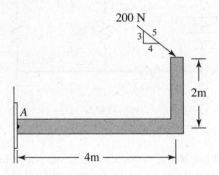

(A) 160 N·m
(B) 800 N·m
(C) 984.4 N·m
(D) 480 N·m

25. Determine distance *d* so that the resultant couple is 200 N·m clockwise.

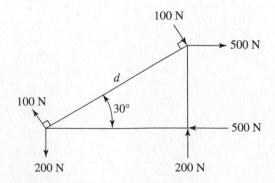

(A) 1.13 m
(B) 0.57 m
(C) 1.33 m
(D) 1.0 m

26. Determine the vertical reaction at smooth surface at point *B*, due to the 10-lb force.

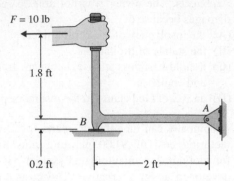

(A) 5.55 lb
(B) 5.0 lb
(C) 10 lb
(D) 20 lb

27. Determine the force in member *FE* of the truss shown.

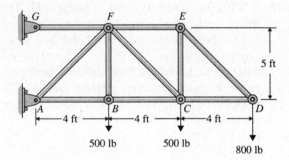

(A) 800 lb
(B) 1000 lb
(C) 1800 lb
(D) 640 lb

602

28. Determine the reaction at the rocker (point F) on the frame shown.

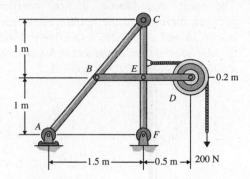

(A) 266.67 N
(B) 293.33 N
(C) 200 N
(D) 100 N

29. Determine the maximum bending moment in the cantilevered beam shown.

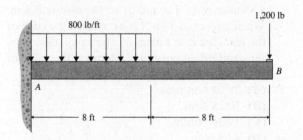

(A) 70,400 lb·ft
(B) 19,200 lb·ft
(C) 25,600 lb·ft
(D) 44,800 lb·ft

30. Determine the minimum force P required to hold the 60-lb uniform bar AB from sliding. The surface at B is smooth and the coefficient of friction at A is 0.3.

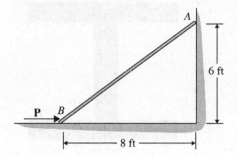

(A) 40 lb
(B) 23.1 lb
(C) 66.67 lb
(D) 60 lb

31. Determine the y-coordinate of the centroid of the shaded area.

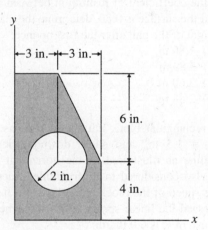

(A) 6.25 in
(B) 4.2 in
(C) 6.4 in
(D) 5.85 in

32. Determine the moment of inertia with respect to the horizontal axis through the centroid of the I-beam shown. The web thickness is 1.0 inch.

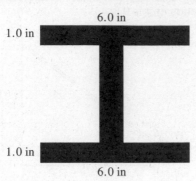

6.0 in

1.0 in

1.0 in

6.0 in

(A) 256 in^4
(B) 148 in^4
(C) 166 in^4
(D) 200 in^4

33. A compressed rubber ball at rest is dropped from a height of 4.5 m above a solid surface. If the coefficient of restitution between the ball and the surface is 0.80, determine the maximum height of the ball after the first bounce.

(A) 1.80 m
(B) 3.89 m
(C) 3.60 m
(D) 2.88 m

34. A rectangular block initially at rest has a mass $m_B = 1.5$ lb$_m$ and slides down a frictionless incline as illustrated. If the horizontal surface is also considered to be frictionless, determine the speed of the block at the instant it has compressed the linear spring 2.0 inches. The spring constant k_s is 6.0 lb$_{in}$/in.

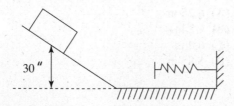

30 "

(A) 41.68 ft/s
(B) 14.29 ft/s
(C) 10.87 ft/s
(D) 30.84 ft/s

35. The slender rod shown in the figure has a mass of 120 kg and a moment of inertia $I_g = 80$ kg·m^2. The static and kinetic friction coefficients between the rod and the horizontal surface are $\mu_s = 0.35$ and $\mu_k = 0.25$, respectively. Find the angular acceleration of the rod at the instant that a 500 N horizontal force is applied.

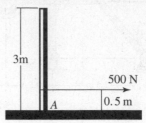

3m

500 N

A

0.5 m

(A) –0.73 rad/s^2 counterclockwise
(B) 1.48 rad/s^2 clockwise
(C) –11.77 rad/s^2 counterclockwise
(D) 0.73 rad/s^2 clockwise

36. A torsional pendulum is composed of a thin disc having a mass of 7.0 kg that is attached to a slender rod. The radius of the disc is 0.8 m. When a force of 38.0 N is exerted on the edge of the disc, the disc rotates ¼ of a revolution from the equilibrium position. Determine the value of the torsional spring constant.

(A) 38.72 Nm/rad
(B) 30.25 N/m
(C) 19.36 Nm/rad
(D) 47.5 N/m

37. An inelastic cord is attached at stationary point P as shown in the figure.

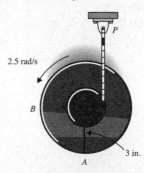

P

2.5 rad/s

B

3 in.

A

The cord unravels from the inner hub of a wheel, which has a radius of 2.0 in. Points A and B are located on the outer edge of the wheel. At the instant when the wheel is rotating at angular speed $\omega = 2.5$ rad/s, determine the magnitude of the speed at point A with respect to point P.

(A) 12.4 in/s
(B) 17.5 in/s
(C) 9.0 in/s
(D) 13.5 in/s

38. The missile illustrated is projected from a cannon with an initial velocity $v_0 = 480$ m/s. If the mass of the missile is 5 kg, determine the angular momentum about point O at the instant when it has attained its maximum height along the trajectory.

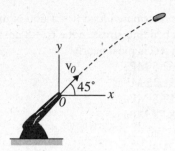

(A) 2.82×10^7 kg·m^2/s
(B) 9.96×10^6 kg·m^2/s
(C) 1.99×10^7 kg·m^2/s
(D) 3.98×10^6 kg·m^2/s

39. A velocity vs. time graph for an airplane that travels from rest along a horizontal runway is given in the figure. Determine the displacement of the plane during the time interval $2 \le t \le 30$ s.

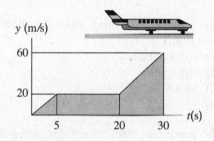

(A) 1542.0 m
(B) 492.0 m
(C) 742.0 m
(D) 692.0 m

40. A 3 kg masonry block has an initial velocity of 4.0 m/s along a horizontal surface. Due to the application of a constant force **F** as illustrated below, the block changes direction and has a velocity of -3.0 m/s at the instant when **F** has been acting for a period of 6.0 seconds. Determine the required magnitude of **F**.

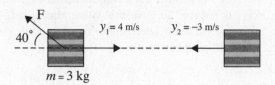

(A) 5.4 N
(B) 3.5 N
(C) 4.6 N
(D) The required applied force magnitude will cause the block to lift off of the surface.

41. A 7.0 lb collar slides down a frictionless vertical rod as shown. The rod is semicircular with a radius of curvature of 2.0 ft. Assuming that the collar is released from rest at point A, determine the normal reaction force at the instant when the collar has traveled to position B.

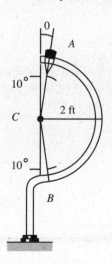

(A) 34.5 lb$_f$
(B) 27.6 lb$_f$
(C) 14.3 lb$_f$
(D) 33.8 lb$_f$

42. An underdamped shock absorber is designed for a prototype motorcycle that has a mass of 200 kg. When the motorcycle is tested for performance, a road bump causes a vertical velocity that results in a damped vibration motion as illustrated by the displacement vs. time graph shown below. If the damped period of vibration is 2.0 s and the initial amplitude X_o is reduced to ¼ of this value after a half-cycle of vibration, determine the required viscous damping constant c.

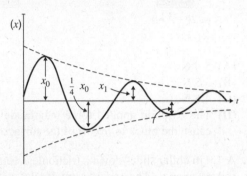

    (A) 235.6 N·s/m
    (B) 10.7 kg·rad/s
    (C) 554.3 N·s/m
    (D) 100.2 kg·rad/s

Questions 43–45

Given the diameter of rod AB is equal to 4 cm and the diameter of pin A is equal to 3 cm and note that the pin at A is in double shear.

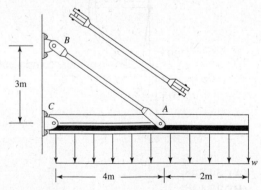

43. If the distributed load $w = 1,000$ N/m, the average normal stress in rod AB is most nearly
    (A) 5.97 MPa
    (B) 1.49 MPa
    (C) 12 MPa
    (D) 6 MPa

44. If the distributed load $w = 1000$ N/m, the average shear stress in pin A is most nearly
    (A) 2.6 MPa
    (B) 10.6 MPa
    (C) 5.3 MPa
    (D) 11.2 MPa

45. If the distributed load $w = 1,000$ N/m, the maximum bending stress in the 6×8 cm rectangular beam AC is most nearly
    (A) 62.5 MPa
    (B) 31.25 MPa
    (C) 17.5 MPa
    (D) 8.78 MPa

Questions 46 and 47

Shaft AB has a diameter of 5 cm.

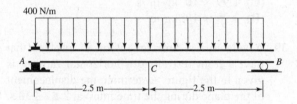

46. The maximum bending stress in shaft AB is most nearly
    (A) 101.8 MPa
    (B) 203.6 MPa
    (C) 50.9 MPa
    (D) 25.6 MPa

47. The maximum transverse shear stress in shaft AB is most nearly
    (A) 10 kPa
    (B) 0
    (C) 13.58 kPa
    (D) 679 kPa

Questions 48 and 49

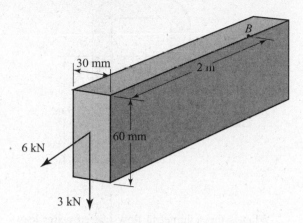

48. The normal stress at point B is most nearly
    (A) 3.33 MPa
    (B) 333.33 MPa
    (C) 336.67 MPa
    (D) 330 MPa

49. The shear stress at point B is most nearly
    (A) 0
    (B) 3.33 MPa
    (C) 5.0 MPa
    (D) 8.0 MPa

50. The 5 m wooden column has a 10×10 cm square cross section with fixed ends and a modulus of elasticity $E = 9$ MPa. The critical load due to bucking is most nearly

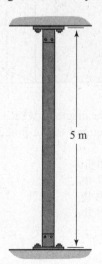

    (A) 118.4 N
    (B) 29.6 N
    (C) 59.2 N
    (D) 60.4 N

51. A solid shaft is rotating at a speed of 1,800 RPM while transmitting 5 KW of power. If the allowable angle of twist for a 2 m length is 4 degrees, determine the minimum diameter of the shaft. Use $G = 60$ GPa.
    (A) 12.6 mm
    (B) 17.4 mm
    (C) 17.9 mm
    (D) 12.0 mm

52. When carbon is dissolved in iron to a concentration of 0.5 percent carbon, what phase(s) is(are) present at 750°C.
    (A) Ferrite
    (B) Ferrite and pearlite
    (C) Austenite
    (D) Martensite

53. A metal alloy is to be selected for an application as a hanger. It is to be 1.0 mm in diameter and must support a static load of 800 N. What is the BHN of eligible alloys?
    (A) 295
    (B) 116
    (C) 428
    (D) 550

54. A copper hook is attached to a plain carbon steel chain below the surface in a saltwater bay. If galvanic corrosion occurs, what will be the anodic reaction?
    (A) $Fe \rightleftarrows Fe^{2+} + 2e^-$
    (B) $Cu \rightleftarrows Cu^{2+} + 2e^-$
    (C) $Cu^{2+} + 2e^- \rightleftarrows Cu$
    (D) $NaCl \rightleftarrows Na^+ + Cl^-$

55. A solution is required to be
    (A) a single phase
    (B) a liquid composed of two or more components
    (C) homogeneous
    (D) comprised partly of water

56. A tensile specimen with a diameter of 10 millimeters is subjected to a load of 500 newtons. What is the engineering stress?
    (A) 50 N/mm
    (B) 30 ksi
    (C) 5.2 N/mm^2
    (D) 6.36 MN/m^2

57. In a normalized plain carbon steel containing 0.5 weight percent carbon, approximately what percentage of the alloy is pearlite?
(A) 37%
(B) 63%
(C) 91%
(D) 7.4%

58. At room temperature, a plain carbon steel containing 0.2 weight percent carbon is
(A) hypereutectoid
(B) ferromagnetic
(C) brittle
(D) highly alloyed

59. A bar is rolled to a reduction in area of 30%. What is the true strain?
(A) 0.357
(B) 1.43
(C) 0.154
(D) 0.30

60. Which of the following is (are) used for corrosion protection?
(A) annealing
(B) tempering
(C) all of the above
(D) none of the above

61.

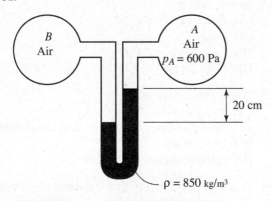

The pressure in vessel *B* shown in the diagram above is most nearly
(A) 2,268 Pa
(B) 684 Pa
(C) 433 Pa
(D) 767 Pa

62.

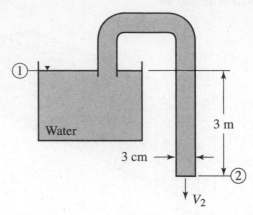

What is the volumetric flow rate of water leaving the siphon in the above diagram? (Assume that the diameter of the tank is much larger than the diameter of the siphon tube.)
(A) 0.32 m³/s
(B) 0.00542 m³/s
(C) 1.28 m³/s
(D) 0.65 m³/s

63.

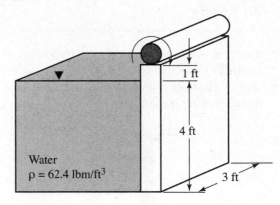

In the above diagram, what is the approximate moment required at the hinge to keep the gate closed?
(A) 3,000 ft·lb$_f$
(B) 4,500 ft·lb$_f$
(C) 5,500 ft·lb$_f$
(D) 4,000 ft·lb$_f$

64.

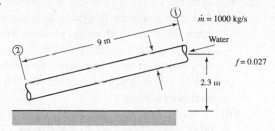

For the pipe of 0.2 m diameter shown above, what is the pressure drop $(p_1 - p_2)$?
(A) 2 kPa
(B) −4 kPa
(C) 2.06 kPa
(D) −2 kPa

65.

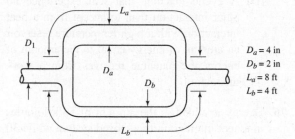

What is the approximate velocity through tube a in the above diagram if the total volume flow rate is 260 $\dfrac{\text{ft}^3}{\text{min}}$?

(A) 40 ft/s
(B) 20 ft/s
(C) 50 ft/s
(D) 30 ft/s

66.

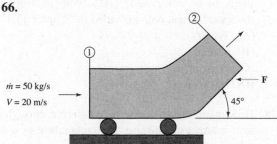

What force, **F**, is required to keep the cart shown above from moving? (Assume in the cross-sectional area that $Area_1 = Area_2$)
(A) 1,707 N
(B) 720 N
(C) 707 N
(D) 293 N

Questions 67–70

Water flows from the tank shown in the diagram through a cast iron pipe that is 75 meters long and has a 10-centimeter inside diameter. The jet of water leaving the pipe is expected to turn a turbine. The turbine operates most efficiently when the horizontal force (F_x) on the blade is at least 100 newtons. (The density of water is 1,000 kg/m^3, and its viscosity is 8×10^{-4} N·s/m^2. The roughness for cast iron pipe is 0.25 mm.)

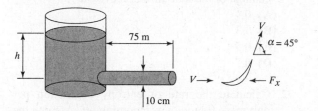

67. What should the velocity of the water leaving the pipe be in order to produce a horizontal force of 100 newtons?
(A) 6.59 m/s
(B) 8.00 m/s
(C) 8.50 m/s
(D) 5.44 m/s

68. What is the Reynolds number of the water at this velocity?
(A) 1.00×10^6
(B) 6.80×10^5
(C) 8.24×10^5
(D) 1.06×10^5

69. What is the friction factor for this water velocity and this pipe?
(A) 0.25
(B) 0.026
(C) 0.05
(D) 0.018

70. What is the head loss in the pipe due to friction?
(A) 43.16 m
(B) 4.05 m
(C) 34.36 m
(D) 12.37 m

Questions 71–73

An inventor claims to have developed a device that does not require any energy input yet is able to produce hot and cold streams of air from a single stream of air. Air comes in at 75° F and 5 bar pressure. It exits at 150°F and at 0°F, both at 1 bar pressure. Air properties are C_p = 0.24 BTU/lbm·R, R = 0.0686 BTU/lbm·R.

71. Assuming the claim is correct, calculate the percentage of mass of air that leaves at 150°F.
   (A) 0.1
   (B) 0.3
   (C) 0.5
   (D) 0.7

72. Calculate the rate of entropy production per unit mass flow rate of incoming air. Assume the incoming air is divided into two equal flows at the exits.
   (A) 0 BTU/R
   (B) 0.03 BTU/R
   (C) −0.03 BTU/R
   (D) −0.06 BTU/R

73. If entropy production term per unit mass flow of air is −0.06 BTU/R, then which of the following is true?
   (A) The claim does not violate the second law of thermodynamics.
   (B) The claim does violate the second law of thermodynamics.

74. Which of the following statements is NOT true regarding the state of a system?
   (A) It remains unchanged for a steady-state process.
   (B) It is determined by the values of any two independent properties in a macroscopically homogeneous substance.
   (C) It remains unchanged for a work interaction with a closed system.
   (D) It may in some instances change spontaneously until the entropy of the system reaches a maximum value.

75. Which of the following devices is possible?
   (A) A cyclic machine that will experience no other interaction than to produce energy through a work interaction while transferring energy from a high-temperature reservoir to a low-temperature reservoir through heat interactions
   (B) A cyclic machine that will experience no other interaction than to transfer to a thermal reservoir an amount of energy equal to the amount of energy the machine receives from a work interaction
   (C) A device that will change the thermodynamic state of a material from one equilibrium state to another without experiencing a change in the amount of energy contained in the material, in the amount of material, or in the external forces placed on the material
   (D) A cyclic machine that will experience no other interaction than to accept from a heat interaction with a high-temperature reservoir an amount of energy equal to the amount of energy the machine receives from a work interaction

76. Energy is added in the amount of 50 kilojoules in a heat interaction to a closed system while 30 kilojoules of work is done by the system. The change in the internal energy of the system is
   (A) 80 kJ
   (B) 20 kJ
   (C) −20 kJ
   (D) −80 kJ

77. If a sample experiencing a change in temperature from 23°C to 46°C also experiences a change in specific enthalpy of 120 kilojoules per kilogram, of what material is the sample most likely to be composed? (Base your answer on the specific heat data provided in Chapter 13.)
   (A) Water
   (B) Helium
   (C) Ammonia
   (D) Hydrogen

78. How much energy must be transferred through a heat interaction to raise the temperature of a 4-kilogram sample of methane in a closed system from 15°C to 35°C?
   (A) 34.8 kJ
   (B) 45 kJ
   (C) 139 kJ
   (D) 180 kJ

79. For an isentropic process of an ideal gas ($k = 1.40$), with an initial pressure of 50 pounds per square inch absolute, an initial specific volume of 8.2 cubic feet per pound mass, and a final pressure of 120 psia, what is the final value of the specific volume?
(A) 8.2 ft³/lbm
(B) 3.42 ft³/lbm
(C) 19.7 ft³/lbm
(D) 4.39 ft³/lbm

80. One kilogram of air undergoes an isometric process, starting from 200°C and 600 kilopascals, in which the temperature drops to 110°C. The final pressure is
(A) 1,091 kPa
(B) 741 kPa
(C) 330 kPa
(D) 486 kPa

81. An ideal gas ($k = 1.4$), with an initial specific volume of 30 cubic meters per kilogram, experiences an isothermal compression from 200 kilopascals to 800 kilopascals. What is the final specific volume of the gas?
(A) 7.5 m³/kg
(B) 11.14 m³/kg
(C) 120 m³/kg
(D) 75 m³/kg

Questions 82–84

Steam enters an adiabatic turbine at 1 megapascal and 300°C and expands to atmospheric pressure. The turbine efficiency is 80%.

82. What would the temperature of the water exiting the turbine be if the turbine were ideal?
(A) 60°C
(B) 80°C
(C) 100°C
(D) 120°C

83. What would be the approximate work output of an ideal turbine with the same inlet and exit conditions of water?
(A) 375 kJ/kg
(B) 461 kJ/kg
(C) 789 kJ/kg
(D) 2,632 kJ/kg

84. What is the approximate actual work of this turbine?
(A) 369 kJ/kg
(B) 461 kJ/kg
(C) 576 kJ/kg
(D) 2,100 kJ/kg

85. An insulated aluminum rod ($k = 302$ W/(mK)) with a 3 cm diameter is connected between two systems of 150°C and 30°C. How long does it take for 10 kJ of energy to be transferred from the high-temperature system to the cold one if the rod is 50 cm in length?
(A) 1.5 min
(B) 3.2 min
(C) 4.5 min
(D) 10 min

86. A stone with emissivity of 0.9 in Arizona reaches 52°C in the summer sun. What is the radiant heat energy transfer from the stone if its surface area is 200 cm²?
(A) 6 W
(B) 8 W
(C) 10 W
(D) 12 W

87. A 2m by 2m plate with an emissivity of 0.12 and a temperature of 400°C is suspended vertically in a large 20°C room. What is the net heat transfer rate from the plate?
(A) 5.4 kW
(B) 6.2 kW
(C) 8.0 kW
(D) 9.2 kW

88. What is the radius of the insulation surrounding a copper wire ($k = 0.08$ W/(mK)) on a calm day with $h = 2$ W/(m²k)?
(A) 2 cm
(B) 4 cm
(C) 6 cm
(D) 8 cm

89. A spherical thermocouple junction ($k = 20$ W/mK, $c = 400$ J/kg, and $\rho = 8,500$ kg/m³) is used to measure gas temperatures where $h = 400$ W/m²K. Calculate the junction diameter if the thermocouple has a time constant of 1 second.
(A) 0.3 mm
(B) 0.7 mm
(C) 1.0 mm
(D) 7.0 mm

90. A spherical thermocouple ($k = 20$ W/mK, $c = 400$ J/kg, $\rho = 8,500$ kg/m³, $D = 0.5$ mm, $T = 25°C$) is placed in a gas stream of $200°C$. How long will it take for the junction to reach $199°C$ if $h = 400$ W/m²K?
    (A) 2 seconds
    (B) 3 seconds
    (C) 3.7 seconds
    (D) 4.4 seconds

Questions 91–94

Wind chill is related to heat transfer on a windy day. Consider a tissue ($k = 0.2$ W/m·K) with a thickness of 4-mm and an interior temperature of $37°C$. Convection heat transfer coefficient is 20 W/m²·K for a calm day and 70 W/m²·K for a windy day. In both cases, ambient air is $-10°C$.

91. What is the heat flux on a calm day?
    (A) 600 W/m²
    (B) 625 W/m²
    (C) 650 W/m²
    (D) 670 W/m²

92. What is the heat flux on a windy day?
    (A) 1240 W/m²
    (B) 1260 W/m²
    (C) 1280 W/m²
    (D) 1300 W/m²

93. What will be the skin outer surface temperature on a calm day?
    (A) 20°C
    (B) 24°C
    (C) 26°C
    (D) 29°C

94. What will be the skin outer surface temperature on a windy day?
    (A) 10°C
    (B) 12°C
    (C) 14°C
    (D) 16°C

Questions 95–96

For a specific value of a controller gain, the following closed-loop system characteristic roots are obtained:

$$s_1 = -13.98, \; s_{2,3} = -0.51 \pm j\,5.96$$

95. What is the system settling time (in seconds) for this gain based on 2% criterion?
    (A) 0.286
    (B) 7.84
    (C) 0.51
    (D) 5.96

96. What is the system damping ratio for this gain?
    (A) 8.5%
    (B) 10%
    (C) 1%
    (D) Overdamped

97. Given the control system in the figure below, what is the value of K so that there is 10% error in the steady-state response to a unit ramp (i.e., $R(s) = 1/s^2$)?

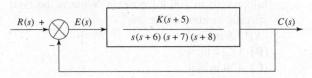

    (A) 762
    (B) 572
    (C) 672
    (D) 752

98. What is the simplified transfer function of the following block diagram?

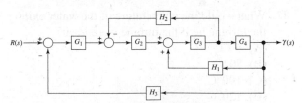

    (A) $\dfrac{G_1 G_2 G_3 G_4}{1 + G_3 G_4 H_1 + G_2 G_3 H_2 + G_1 G_2 G_3 G_4 H_3}$

    (B) $\dfrac{G_1 G_2 G_3 G_4 H_3}{1 - G_3 G_4 H_1 - G_2 G_3 H_2 + G_1 G_2 G_3 G_4 H_3}$

    (C) $\dfrac{G_1 G_2 G_3 G_4 H_3}{1 - G_3 G_4 H_1 + G_2 G_3 H_2 + G_1 G_2 G_3 G_4 H_3}$

    (D) $\dfrac{G_1 G_2 G_3 G_4}{1 - G_3 G_4 H_1 + G_2 G_3 H_2 + G_1 G_2 G_3 G_4 H_3}$

99. An open-loop system has a transfer function of $G(s) = \dfrac{5}{s(s^2 + 2s + 10)}$ and is to be controlled by a proportional controller with gain K, using a unity negative feedback loop. What is the range of K that guarantees the stability of the system?
(A) $K < 4$
(B) $0 < K < 4$
(C) $K > 0$
(D) $0 < K < 2$

100. A feedback system has the following state-space representation:

$$\dot{\mathbf{x}}(t) = \begin{bmatrix} -4 & 0 \\ 1 & -1 \end{bmatrix} \mathbf{x}(t) + \begin{bmatrix} 1 \\ 0 \end{bmatrix} u(t)$$

$$y(t) = \begin{bmatrix} 0 & 1 \end{bmatrix} \mathbf{x}(t) + \begin{bmatrix} 0 \end{bmatrix} u(t)$$

What is the system's settling time (based on 2% criterion)?
(A) 1 s
(B) 4 s
(C) 2 s
(D) 5 s

101. A joint is welded together using electrode 6,010 with a tensile strength of 60 ksi. If the joint is subjected to a fluctuating tensile load of −10,000 and 10,000, find the fatigue safety factor. Assume the stress concentration is 2.7.

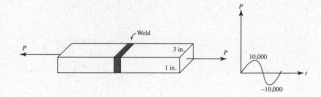

(A) 1.54
(B) 2.26
(C) 1.78
(D) 3.26

102. A pinion and gear are used to transmit a torque. The pinion is attached to a motor with ½ hp and rotating at 1,800 rpm. The pinion has 20 teeth, and the gear has 50 teeth. Find the torque transmitted to the gear.
(A) 18.75 lb·in
(B) 43.75 lb·in
(C) 52.75 lb·in
(D) 21.75 lb·in

103. If the gear system in question 102 is made of cast iron with a Bhn of 200 and a tensile strength of 13 ksi, what is the factor of safety against bending? Assume this gear system has a diametral pitch of 2/in, average quality, moderate shocks, and not accurate mounting. The gear is 0.1-inch thick.
(A) 5.1
(B) 10.1
(C) 7.01
(D) 8.10

104. A belt is used to transfer power from one pulley to a second pulley. The diameter of pulleys arc 10 inches and 30 inches. The pulleys are located 100 inches apart. What is the length of the belt?
(A) 250.1 in
(B) 262.8 in
(C) 278.1 in
(D) 285.7 in

105. If we are transmitting 5 hp from the small pulley to the large pulley in problem 104 and the small pulley is rotating at 1,800 rpm, what is the value to belt tension in the tight segment? The belt weighs 0.1 lb/inch. The friction coefficient between the belt and pulleys is 0.15.
(A) 250 lb
(B) 325 lb
(C) 275 lb
(D) 300 lb

106. A 60 mm bore (02 series) deep-groove ball bearing is subjected to the radial force of 10 kN and no axial force. If the shaft is rotating at 1,500 rpm, find the expected life of this bearing in hours.
(A) 1,000 hrs
(B) 1,191 hrs
(C) 1,250 hrs
(D) 1,395 hrs

107. Find the bearing life if the bearing described in question 106 is also subjected to an axial force of 2 kN. This bearing is subjected to a moderate shock.
(A) 129 hrs
(B) 139 hrs
(C) 149 hrs
(D) 159 hrs

108. A planetary gear system consists of a sun, planet, and ring gears. The sun has a diametral pitch of 2/in and has 20 teeth. The planet has 50 teeth. Find the number of teeth in the ring gear. The ring gear is stationary.

(A) 120
(B) 140
(C) 100
(D) 70

109. In the planetary gear system described in question 108, if the sun is rotating at 1,500 rpm, find the speed of the arm.

(A) 1,500 rpm
(B) 1,286 rpm
(C) 1,352 rpm
(D) 1,326 rpm

110. A horizontal component is connected to a column by using steel rivets with a diameter of ½ inch and a yield strength of 40 ksi. The system is subjected to a force $P = 100$ lb, as shown in the figure. Find the safety factor against rivet shearing.

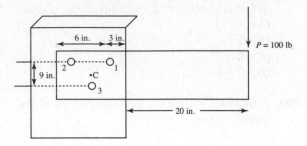

(A) 1.5
(B) 1.8
(C) 2.2
(D) 2.7

Answer Key

1.	D	29.	D	57.	B	85.	B
2.	D	30.	B	58.	B	86.	D
3.	B	31.	A	59.	A	87.	A
4.	C	32.	C	60.	D	88.	B
5.	A	33.	D	61.	A	89.	B
6.	C	34.	C	62.	B	90.	C
7.	B	35.	A	63.	C	91.	D
8.	D	36.	C	64.	C	92.	C
9.	C	37.	D	65.	A	93.	B
10.	B	38.	B	66.	D	94.	A
11.	C	39.	C	67.	A	95.	B
12.	D	40.	C	68.	C	96.	A
13.	A	41.	A	69.	B	97.	C
14.	A	42.	C	70.	A	98.	D
15.	C	43.	A	71.	C	99.	B
16.	C	44.	C	72.	D	100.	B
17.	A	45.	B	73.	B	101.	B
18.	C	46.	A	74.	C	102.	B
19.	D	47.	D	75.	A	103.	B
20.	A	48.	C	76.	B	104.	B
21.	B	49.	A	77.	B	105.	B
22.	C	50.	A	78.	C	106.	B
23.	D	51.	C	79.	D	107.	C
24.	B	52.	B	80.	D	108.	A
25.	A	53.	A	81.	A	109.	B
26.	C	54.	A	82.	C	110.	C
27.	D	55.	A	83.	B		
28.	B	56.	D	84.	A		

Answers Explained

1. D $v = \dfrac{ds}{dt} = 30t^2 - 10t^4$. Therefore at $t = 1$ s,

$$v = 20 \text{ m/s}$$

2. D slope $m = \dfrac{a-b}{b-a} = -1$

3. B Complete the squares, and rewrite the equation as $(x-0)^2 + (y-2)^2 + (z+3)^2 = 13$. Hence, the center of the sphere is at $(0, 2, -3)$.

4. C Volume $= \pi \displaystyle\int_0^2 (y^2)^2 \ dy = \pi \left(\dfrac{y^5}{5}\right)\Big|_0^2 = \dfrac{32\pi}{2}$ cubic units.

5. A Rewrite the expression as

$$\dfrac{\sin x}{\cos x} \cdot \dfrac{1}{\sin x} \cdot \dfrac{\cos x}{1}$$

If $\sin x \neq \cos x \neq 0$, the expression is equal to 1.

6. C $\displaystyle\int_{\ln a}^{\ln b} e^{-x}\, dx = -e^{-x} = -\left(\dfrac{1}{b} - \dfrac{1}{a}\right) = \dfrac{1}{a} - \dfrac{1}{b}$

7. B We first calculate the sample mean to get

$$\bar{x} = \dfrac{6 + 12 + 14.2 + 10 + 13 + 15.2 + 11 + 8}{8} = 11.175$$

The variance is

$$s^2 = \dfrac{1}{8-1}\sum_{i=1}^{8}(x_i - 11.175)^2 = \dfrac{67.635}{7} = 9.6621$$

Taking the square root gives $s = 3.1084$.

8. D Let X be the number of tails showing on the coin. Note that X is binomial with $n = 5$ and $p = 0.5$. We want to calculate

$$P(X = 5 \mid X \geq 4) = \dfrac{P(X = 5, X \geq 4)}{P(X \geq 4)} = \dfrac{P(X = 5)}{P(X \geq 4)}$$

The last statement follows since if the coin shows 5 tails, it satisfies showing at least 4. By properties of the binomial random variable, we have

$$P(X = 5) = C(5, 5)(0.05)^5(1 - 0.05)^0 = 0.03125$$

and

$$P(X \geq 4) = P(X = 4) + P(X = 5)$$
$$= C(5, 4)(0.5)^4(1 - 0.5)^1 + C(5, 5)(0.05)^5(1 - 0.05)^0$$
$$= 0.1875$$

Finally, $P(X = 5 \mid X \geq 4) = \dfrac{0.03125}{0.1875} = 0.1667$

9. C Calculating we find, $\bar{x} = 34.13, \bar{y} = 34.67$,

$$S_{xy} = \sum_{i=1}^{6} x_i y_i - (1/6)\left(\sum_{i=1}^{6} x_i\right)\left(\sum_{i=1}^{6} x_i\right)$$
$$= 6901.9 - (1/6)(204.8)(208)$$
$$= -197.83$$

$$S_{xx} = \sum_{i=1}^{6} x_i^2 - (1/6)\left(\sum_{i=1}^{6} x_i\right)^2$$
$$= 7055.46 - (1/6)(204.8)^2$$
$$= 64.95$$

Now $\hat{b} = S_{xy}/S_{xx} = -3.046$, and $\hat{a} = \bar{y} = \hat{b}\bar{x} = 138.63$. Thus $\hat{y}(42) = 138.63 - 3.046(42) = 10.698$.

10. B Let X represent the random variable given by the density function. The expected value is the center of gravity. Hence

$$E(X) = \int_0^1 xf(x)\,dx = \int_0^1 3x^3 = 3/4$$

and the center of gravity is at 9 inches.

11. C Start at the beginning of the chart. Follow the arrows from the beginning to the end, noting decision points (diamond shape). In this example, N is tested for its magnitude and whether it is divisible by I. When I is greater, Is are listed and the program ends.

12. D

A	B	C		D
1 23	24	25		26
2 7	**49**	**49*23=1127**		
3 8	**64**	**64*24=1536**		
4 9	**81**	**81*25=2025**		
5 10	**100**	**100*26=2600**		

13. A Refer to the solution for problem 12 and look at Column C.

14. A Choices (B) and (D) are the least reasonable responses. A registered engineer would state membership and a number, making (C) a possible selection. However, (A) is the best choice because the implication is that the applicant is licensed, which is unethical.

616

15. C Not writing written conclusions, returning moneys paid, and passing time do not make it ethical to accept the other assignment. If all parties had agreed and thus had been informed, the engineer could have accepted the assignment. Choices (A), (B), and (D) are not acceptable.

16. C Insolvency of the bond holder would not release a party. There is no mention of a time constraint; therefore, (A) and (B) are incorrect. Ice recurs based on temperature, not an Act of God, and should be a design consideration. Therefore, (C) is the correct answer.

17. A Use the following equation:

$$4.00x + 4,000 = 1.00x + 40,000$$
$$3x = 36,000$$
$$x = 12,000$$

18. C Since interest is paid every 6 months, the number of periods is 20. Your MARR is 12% or 6% per 6 months.

Bond value
= PW of all receipts over life of the bond
= 2,000(P/A, 6%, 20) + 50,000)(P/F, 6%, 20)
= 2,000(11.470) + 50,000(.3118) = $38,530

19. D Since the lives are the same, PW analysis can be used by setting the alternatives equal to each other.

−10,000 + $1,500(P/A, 8%, 10) + 1,000(P/F, 8%, 10) =
−$20,000 + Rev(2)(P/A, 8%, 10) + $4,000(P/F, 8%, 10)

Solve for Rev(2)

Rev(2)= ($18,147.2 + 528.2)/6.71 = $2,780

20. A Solve for the current through the center branch:

$$I_2 = \frac{1\ V}{2\ k\Omega} = 0.5\ mA$$

The voltage across the 2 kΩ and 4 kΩ resistors is:

$$V_2 = (0.5\ mA)(2\ k\Omega + 4\ k\Omega) = 3\ V$$

The voltage across the 1 kΩ and 5 kΩ resistors is also 3 V. Therefore, the current through this combination is

$$I_3 = \frac{3\ V}{1\ k\Omega + 5\ k\Omega} = 0.5\ mA$$

The total current is

$$I_1 = I_2 + I_3 = 0.5\ mA + 0.5\ mA = 1.0\ mA$$

The voltage across R_1 (the first 1 kΩ resistor) is

$$V_1 = (1.0\ mA)(1\ k\Omega) = 1\ V$$

Apply Kirchhoff's voltage law around the input loop:

$$V_i = V_1 + V_2 = 1\ V + 3\ V = 4\ V$$

21. B Select the loop currents so that only one (I_A for this problem) flows through the 1 kΩ resistor. Therefore, only loop current I_A will have to be calculated before applying Ohm's law to solve for the voltage.

Loop A path includes the 2 V source, resistors 2 kΩ, 1 kΩ, and 7 kΩ.
Loop B path includes the 2 V and 4 V sources, resistors 2 kΩ, 2 kΩ, and 3 kΩ.
Loop C path includes resistors 7 kΩ and 3 kΩ.
Assume all loop currents are chosen in a clockwise direction.
Loop A: 10 kΩ + 2 kΩ − 7 kΩ = 2 V
Loop B: 2 kΩ + 7 kΩ + 3 kΩ = −4 V + 2 V
Loop C: −7 kΩ + 3 kΩ + 10 kΩ = 0
Solve for I_A (note the answer will be in mA):

$$I_A \frac{\begin{vmatrix} 2 & 2 & -7 \\ -2 & 7 & 3 \\ 0 & 3 & 10 \end{vmatrix}}{\begin{vmatrix} 10 & 2 & -7 \\ 2 & 7 & 3 \\ -7 & 3 & 10 \end{vmatrix}} = \frac{204}{143} = 1.43\ mA$$

$$V = (1.43\ mA)(1\ k\Omega) = 1.43\ V$$

22. C Apply equation 12.26:

$$Q = C \times V = (0.01 \times 10^{-6}\ F) \times 50\ V$$
$$= 0.5\ \mu C$$

23. D To determine the maximum voltage the capacitor can charge to and the time constant, first obtain the Thévenin equivalent circuit. The Thévenin equivalent resistance is found by replacing the voltage source with 0 Ω. Then the 1 Ω and 2 Ω resistors are in series, and this series combination is in parallel with the series combination of 4 Ω and 2 Ω.

Let $R_a = 1\ \Omega + 2\ \Omega = 3\ \Omega$
and $R_b = 4\ \Omega + 2\ \Omega = 6\ \Omega$. Then,

$$R_{Th} = \frac{3\ \Omega \times 6\ \Omega}{3\ \Omega + 6\ \Omega} + 3\ \Omega = 5\ \Omega$$

The Thévenin equivalent voltage is found by removing the capacitor and measuring the voltage across R_b (the voltage across the resistor combination of 4 Ω and 2 Ω). This Thévenin voltage may be calculated by using the voltage division law:

$$V_{Th} = \frac{R_b}{R_a + R_b} \times 18 \text{ V}$$

$$= \frac{6 \text{ Ω}}{3 \text{ Ω} + 6 \text{ Ω}} \times 18 \text{ V} = 12 \text{ V}$$

The time constant is $\tau = 5 \text{ Ω} \times 2 \text{ F} = 10$ s. Apply equation 12.33:

$$v_c(t) = 12 \text{ V} \times (1 - e^{-2/10}) = 2.18 \text{ V}$$

24. B Clockwise moment is "+."

$$(200/5)(2) + 200 (3/5) (4) = 800 \text{ N·m}$$

25. A Clockwise moment is "+."

$$(100)(d) + (500)(d \sin 30°) - (200)(d \cos 30°) = 200$$

$$d = 1.13 \text{ m}$$

26. C $\sum M_A = 0$

$$10(2) - F_B(2) = 0$$

$$F_B = 10 \text{ lb}$$

27. D Using method of sections, cut the truss vertically through members EF, FC, and BC.

$$\sum M_C = 0$$

$$F_{EF}(5) - 800(4) = 0$$

$$F_{EF} = 640 \text{ lb}$$

28. B Draw a free-body diagram of the frame.

$$\sum M_A = 0$$

$$F_F(1.5) - 200 (2.2) = 0$$

$$F_F = 293.3 \text{ N}$$

29. D

$$\sum M_A = 1200 (16) + 6400 (4) = 44,800 \text{ lb·ft}$$

30. B Draw a free-body diagram for bar AB.

$$\sum M_A = 0$$

$$(0.3 \, N_A)(8) + P (6) - 60 (4) = 0$$

$$\sum F_x = 0 \quad N_A = P$$

$$P = 23.1 \text{ N}$$

31. A

$$\bar{Y} = \frac{\sum yA}{A} = \frac{(2)(24) + (7)(18) + 6(9) - (4)(4\pi)}{24 + 18 + 9 - 4\pi} = 6.25 \text{ in}$$

32. C

$$\bar{I} = 1/12 [6 (8)^3 - 5 (6)^3] = 166 \text{ in}^4$$

33. D Use the conservation of energy principle to determine the velocity of the ball immediately before impact with the surface.

$\frac{1}{2} mv_1^2 = mgh_1$, where $h_1 = 4.5$ m initial height.

$$v_1 = \sqrt{2gh_1}$$

$$v_1 = \sqrt{2(9.81)(4.5)} = 9.40 \text{ m/s}$$

The conservation of linear momentum principle is now used for partially elastic collisions to calculate the upward velocity of the ball v_2 immediately after impact. Since the surface is assumed not to move, we assign $V_s = 0$ before and after the collision.

$$e(v_1 - v_s) = v_s - v_2$$

$$0.80 (9.40 - 0) = -v_2$$

Solving -7.52 m/s $= v_2$ denotes the velocity of the ball has changed direction upward. The conservation of energy principle is again used to solve the maximum height h_2 of the ball.

$$\frac{1}{2} mv_2^2 = mgh_2$$

Since the mass m cancels,

$$h_2 = \frac{v_2^2}{2g} = \frac{(7.52)^2}{2(9.81)} = 2.88 \text{ m}$$

34. C The conservation of energy principle is used due to the frictionless motion. From the initial position 0 at the top of the incline to the horizontal surface just before the spring is compressed at position 1:

$$KE_0 + PE_0 = KE_1 + PE_1$$

$$\frac{1}{2}mv_0^2 + mgh_0 = \frac{1}{2}mv_1^2 + mgh_1$$

Since $v_0 = 0$ and $h_1 = 0$ and using the horizontal surface as a zero PE datum,

$$v_1 = \sqrt{2gh_0} = \sqrt{2(32.2)(2.5 \text{ ft})} = 12.69 \text{ ft/s}$$

Conservation of energy is employed again from position 1 to where the spring is compressed 2.0 inches at position 2:

$$KE_1 + PE_1 = KE_2 + PE_2$$

PE_1 and PE_2 represent the potential energy stored in the spring.

Since $PE_1 = 0$ due to no initial deformation of the spring,

$$KE_1 - KE_2 = PE_2$$

$$\frac{1}{2}m(v_1^2 - v_2^2) = \frac{1}{2}kx_2^2$$

$$\frac{1}{2}(1.5 \text{ lbm})((12.69 \text{ ft/s})^2 - v_2^2)$$

$$= \frac{1}{2}(6 \text{ lbf/in})(2.0 \text{ in})^2$$

$$120.7 \text{ lbm ft}_2/\text{s}^2 - 0.75 \text{ lbm } v_2^2 = 12.0 \text{ in·lbf}$$
$$= 1.0 \text{ ft·lbf}$$

$$120.7/32.2 \text{ ft·lbf} - 0.75 \text{ lbm } v_2^2 = 1.0 \text{ ft·lbf}$$

$$-v_2^2 = (1.0 \text{ ft·lbf} - 3.75 \text{ ft·lbf})/0.75 \text{ lbm}$$

$$v_2 = \sqrt{(2.75)(32.2)/0.75} = 10.87 \text{ ft/s}$$

35. A Draw the FBD of the rod as illustrated below.

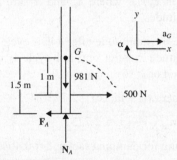

Equations of motion for the rod:

$$\sum F_x = m(a_x) \rightarrow 500 - F_f = 120(a_x)$$

$$500 - \mu_k N_A = 120a_x \qquad (1)$$

$$\sum F_y = m(a_y) \rightarrow a_y = 0 \text{ due to vertical}$$
equilibrium

$$N_A - W = 0 \rightarrow N_A - 120(9.81) = 0$$

therefore $N_A = 1{,}177.2$ N reaction. $\qquad (2)$

Sum moments about the center of gravity G of the rod:

$$\nearrow \sum M_G = I_g\alpha \rightarrow F_f(1.5) - 500(1.0) = 80\alpha \quad (3)$$

If no slipping occurs at A, then necessarily static friction $F_{fs} = \mu_s N_A = 0.35(1{,}177.2)$

$$F_{fs} = 0.35(1{,}177.2) = 412.0 \text{ N}$$

Since the applied force F_A is > 412.0 N, slipping occurs.

Kinetic friction force $F_f = 0.25(1{,}177.2) = 294.3$ N

Solving equation 3 for the angular acceleration:

$$(294.3)(1.5) - 500(1.0) = 80\alpha$$

$$-58.55/80 = \alpha$$

$$\rightarrow \alpha = -0.73 \text{ rad/s}^2 \nwarrow \text{counterclockwise}$$

36. C Use Hooke's law for a torsional pendulum.

$T = -k\Theta$, where T is the applied torque and Θ is the angular displacement.

$$T = F \cdot R = (38.0)(0.80) = 30.4 \text{ N·m}$$

and $\Theta = \frac{1}{4}(-2\pi) = -\pi/2$ rad

$$\rightarrow k = -T/\Theta = \frac{-30.4}{-\pi/2} = 19.36 \text{ N·m/rad}$$

37. D The absolute speed of point A is determined by finding the velocity of center point O and using the relative velocity of A with respect to P.

Note that the direction of the velocity of O is always downward due to unraveling and that the velocity direction of A is tangential relative to center O due to the circular motion of points on the wheel. Relative velocity equation for point A with respect to point O:

$$V_A = V_O + V_{A/O} = (2.0)(2.5) \downarrow + (5.0)(2.5) \rightarrow$$

Resolve velocity vectors for the magnitude of absolute speed at A:

$$V_A = \sqrt{(5)^2 + (12.5)^2} = 13.5 \text{ in/s}$$

38. B By using the equations of motion for a projectile that is affected by only gravity and by assuming negligible wind resistance, the maximum height h is calculated. An x-y coordinate frame is attached to the C.G. of the missile and the vertical component v_{oy} of the initial velocity is employed. At the instant of maximum height, all of the velocity is horizontal only.

$$v_{oy} = (480)\sin 45° = 339.4 \text{ m/s}$$

$$\frac{1}{2}m v_{oy}^2 = mgh \rightarrow h = v_{oy}^2/2g$$

$$= (339.4)^2/2(9.81)$$

$$= 115{,}192.4/19.62 = 5{,}871.2 \text{ m}$$

The maximum height of the missile is also the radius of curvature r at the top of the missile's trajectory at this point M. Also note that the horizontal component of the missile's velocity v_{ox} remains constant.

$$v_{ox} = v_o\cos 45° = 339.4 \text{ m/s}$$

The angular momentum of the missile at point M about point O is $H_o = rmv_{ox}$ since v_{ox} is the tangential velocity at the instant of maximum height.

$H_o = (5{,}871.2 \text{ m})(5 \text{ kg})(339.4 \text{ m/s}) = 9{,}963{,}426.4$ $\text{kg·m}^2/\text{s} = 9.96 \times 10^6 \text{ kg·m}^2/\text{s}$

39. C The displacement of the plane is the change of position Δx of the plane from $t = 2$ to $t = 30$ s. Displacement is determined from calculating the area of the velocity vs. time graph for the given time interval. During the first 5 seconds, the linear increase in velocity has a rate of change of 4 since $v = 20$ at $t = 5$ s. To find the area of the graph from $t = 2$ s to $t = 5$ s, the velocity at $t = 2$ is needed.

$v(2) = 8$ m/s due to the slope of 4.

The displacement Δx of the plane is calculated using the sum of areas of three sections of the graph.

$\Delta x = A_{2\to5} + A_{5\to20} + A_{20\to30} = \frac{1}{2}(8 + 20)(3) +$ $(20)(15) + \frac{1}{2}(20 + 60)(10) = 742.0$ m

40. C The principle of impulse and momentum is used to solve for the constant applied force **F**.

$\text{Imp} = \int \mathbf{F}dt = mv_2 - mv_1$

Since the velocity directions are entirely in a horizontal x-direction, the scalar components of the impulse-momentum equation are employed.

$\text{Imp}_x = F_x\Delta t = m(v_2 - v_1) = (3 \text{ kg})(-3 - 4) \text{ m/s} =$ -21 kg·m/s.

$F_x\Delta t = (F \cos 40°)(6.0 \text{ s}) = F(0.766)(6.0) =$ -21 kg·m/s

Solving for magnitude F

$F = \dfrac{-21 \text{ kg·m/s}}{4.60} = -4.57 \text{ kg·m/s}^2 \approx 4.6 \text{ N}$

41. A Since there is no friction on the collar, the conservation of energy principle is used to calculate the speed V_B at point B.

$KE_A + PE_A = KE_B + PE_B$

Let point B represent a zero PE datum. The height h_A of point A relative to B is calculated by doubling the height of A above point C.

$h_A = 2 \cdot (2 \cos 10°) = 3.94$ ft

$\frac{1}{2}mv_A^2 + mgh_A = \frac{1}{2}mv_B^2 + 0$

By canceling the mass m, $g(3.94 \text{ ft}) = \frac{1}{2}v_B^2$

$\sqrt{2(32.2 \text{ ft/s}^2)(3.94 \text{ ft})} = v_B$

$v_B = 15.93$ ft/s

Draw a FBD of the collar at B and then attach a tangential and normal reference frame. For circular motion, the reaction force N occurs in the normal direction toward the center of curvature C.

$\sum F_n = mv_B^2/R$

$-w \cos 10° + \text{N} = (7 \text{ lb})(15.93 \text{ ft/s})^2/2.0 \text{ ft}$

$-(7 \text{ lb}_f)(0.985) + \text{N} = 888.2 \text{ lb·ft/s}^2$

$\text{N} = 888.2/32.2 + 6.9 = 34.5 \text{ lb}_f$ normal reaction force.

42. C The motorcyle's shock absorber is modeled as a damped free vibration. The equations for a linear spring-mass system with viscous damping are employed. Since the damped period of vibration is given $T_d = 2.0$ s, we first calculate the damped frequency $\omega_d = 2\pi/T_d = 2\pi/2 = \pi$ rad/s.

$\omega d = \omega_n\sqrt{1 - \zeta^2}$ and

$\omega_n = \omega_d/\sqrt{1 - \zeta^2} = \pi/\sqrt{1 - \zeta^2}$

The damping factor ζ is calculated using the logarithmic decrement for the vibration.

$\partial = \ln(x_0/x_1)$, where x_0 and x_1 are successive amplitudes.

Since the amplitude after a half cycle is $\frac{1}{4}x_0$ the amplitude x_1 after one full cycle is necessarily $\frac{1}{4}(\frac{1}{4}x_0)$ or $1/16x_0$.

$\delta = \ln(x_0/x_1) = \ln(16) = 2.773$

$\delta = 2\pi\zeta/\sqrt{1 - \zeta^2} = 2.773$

Solving for damping factor, $\zeta = 0.404$.

Substituting above, $\omega_n = \pi/\sqrt{1 - (0.404)^2} = 3.43$ rad/s.

The damping factor c is now found using $\zeta = c/c_c$.

$C = \zeta c_c = \zeta(2m\omega_n) = 0.404(2)(200 \text{ kg})(3.43 \text{ rad/s}) =$ $554.3 \text{ kg·rad/s} = 554.3 \text{ N·s/m}$

43. A Draw the free-body diagram of the beam and

$$\sum M_C = 0$$

$(3/5 \, F_{AB})(4) - (1{,}000)(6)(3) = 0$
$F_{AB} = 7{,}500$ N

$$\sigma_{AB} = \frac{F_{AB}}{A_{AB}} = \frac{7{,}500}{\frac{\pi}{4}(0.04)^2} = 5.97 \text{ MPa}$$

44. C Since the pin at A is in double shear, then

$$V = \frac{F_{AB}}{2} = 3,750 \text{ N}$$

$$\tau_{ave} = \frac{V}{A} = \frac{3,750}{\frac{\pi}{4}(0.03)^2} = 5.3 \text{ MPa}$$

45. B The vertical pin reaction at the pin C is equal to 1,500 N. Draw the shear and the bending moment diagrams. Note that the maximum moment is 2,000 N·m at point A, then

$$I = \frac{1}{12}(0.06)(0.08)^3 = 2.56 \times 10^{-6} \text{ m}^4$$

$$\sigma = \frac{M\ c}{I} = \frac{(12,000)(0.04)}{2.56 \times 10^{-6}} = 31.25 \text{ MPa}$$

46. A The maximum moment occurs at the center of the shaft point C and is equal to

$$M_{max} = \frac{(400)(5)^2}{8} = 1,250 \text{ N·m}$$

$$I = \frac{\pi}{4} r^4 = \frac{\pi}{64} d^4 = \frac{\pi}{4}(0.05)^4 = 3.07 \times 10^{-7} \text{ m}^4$$

$$\sigma_{max} = \frac{M\ c}{I} = \frac{(1,250)(0.025)}{3.07 \times 10^{-7}} = 101.8 \text{ MPa}$$

47. D For a solid circular shaft, the maximum transverse shear stress can be determined by using the following equation:

$$\tau_{max} = \frac{4 V_{max}}{3 A} = \frac{4(1,000)}{3\left(\frac{\pi}{4}(0.05)^2\right)} = 679 \text{ kPa}$$

Note that the maximum shear load is equal to the reaction at the supports A or B.

48. C

$$V = 3 \text{ kN} \quad N = 6 \text{ kN} \quad M_B = (3)(2) = 6 \text{ kN·m}$$

$$I = \frac{1}{12}(0.03)(0.06)^3 = 5.4 \times 10^{-7} \text{ m}^4$$

$$\sigma_B = \frac{N}{A} + \frac{M_B c}{I} = \frac{(6,000)(0.03)}{5.4 \times 10^{-7}} = 336.67 \text{ MPa}$$

49. A Since point B is located at the top fiber of the beam, the first moment of the area is zero. As a result, the transverse shear stress at point B is equal to zero.

50. A Since both ends are fixed, the effective length is half of the actual length and

$$L_{eff} = 0.5 \ \ell = 0.5(5) = 2.5 \text{ m}$$

$$I = \frac{1}{12}(0.01)(0.01)^3 = 8.33 \times 10^{-6} \text{ m}^4$$

$$P_{cr} = \frac{\pi^2 EI}{l_{eff}^2} = \frac{\pi^2(9 \times 10^6)(8.33 \times 10^{-6})}{(2.5)^2} = 118.4 \text{ N}$$

51. C

$$P = 5 \text{ KW} = 5,000 \text{ W}$$

$$\omega = 1,800 \text{ RPM} = 188.49 \text{ rad/sec}$$

$$P = T\omega \quad T = P/\omega \quad T = \frac{5,000}{188.49} = 26.52 \text{ N·m}$$

$$\phi = 5 \text{ degrees} = 0.08727 \text{ radians}$$

$$\varphi = \frac{T\ l}{G\ J} \text{ where } J = \frac{\pi}{32} d^4$$

$$d = \sqrt[4]{\frac{32\ T\ l}{\pi\ G\ \varphi}} = \sqrt[4]{\frac{32(26.52)(2)}{\pi(60 \times 10^9)(0.08727)}}$$

$$= 0.0179 \text{ m} = 17.9 \text{ mm}$$

52. B Refer to Figure 10.8, on iron-carbon phase diagram.

53. A In order to be eligible for service, the UTS of the alloy must not be exceeded by the application. The UTS is related to the BHN by the factor 3.45:

$$1 \frac{MN}{m^2} = 1 \frac{N}{mm^2}$$

$$\text{Area} = \frac{\pi}{4}(1 \text{ mm}^2) = 0.785 \text{ mm}^2$$

$$\text{UTS} = \frac{800\ N}{0.785 \text{ mm}^2} = 1,019 \frac{N}{mm^2}$$

$$\text{BHN} = \frac{1,019}{3.45} = 295$$

54. A Use the standard emf series in Table 10.4 as an approximation of the behavior of dissimilar metals in salt water. Iron is anodic with respect to copper. Therefore, the anodic oxidation half-cell reaction for iron will occur.

55. A Although the solute and solvent may be in different phases (e.g., when the solute is solid table salt, and the solvent is liquid water), the solution has only a single phase (in this example, liquid).

56. D Engineering stress, σ, is defined as load per area. Here the load is 500 N and the area is

$$\pi r^2 = 7.85 \text{ mm}^2 = 7.85 \times 10^{-6} \text{ m}^2$$

Then the stress is

$$\sigma = \frac{500 \text{ N}}{7.85 \times 10^{-6}} = 6.36 \times 10^6 \frac{\text{N}}{\text{m}^2} = 6.36 \text{ MN/m}^2$$

57. B For a plain carbon steel containing 0.5 weight percent carbon, the lever law for the amount of pearlite should be drawn in the austenite plus ferrite region, just above the eutectoid temperature. The amount of austenite at this temperature is the amount of pearlite when the austenite is cooled to room temperature:

$$f_\gamma = \frac{0.5 - 0.02}{0.5 - 0.02} \cong \frac{5}{8} \cong 63\%$$

58. B Refer to Figure 10.8, a phase diagram for iron carbon.

59. A A reduction in area can be expressed as a fraction:

$$RA = \frac{A_0 - A}{A_0}$$

True strain ϵ is $\ln \frac{1}{l_0} = \ln \frac{A_0}{A}$. Therefore,

$$\epsilon = \ln \frac{1}{1 - RA} = \ln \frac{1}{0.7} = 0.357$$

60. D Annealing and tempering affect the microstructure and hardness. They do not protect against corrosion.

61. A Because the density of air has a much smaller value than that of the manometer fluid, pressure differences within the air sections are negligible compared with pressure differences within the manometer fluid. Therefore, p_A is very nearly the pressure at the top of the manometer fluid exposed to vessel A, and p_B is very nearly the pressure at the surface of the manometer fluid exposed to vessel B. The two pressures can be related as follows:

$$p_B - p_A = \rho_m g h_m$$

This equation is rearranged as:

$$p_B = p_A + \rho_m g h_m$$
$$= 600 \text{ Pa} + \left(850 \frac{\text{kg}}{\text{m}^3}\right)\left(9.81 \frac{\text{m}}{\text{s}^2}\right)(0.20 \text{ m})$$
$$\times \left(\frac{\frac{\text{N}}{\text{kg} \cdot \text{m}}}{\text{s}^2}\right)\left(\frac{\text{Pa}}{\frac{\text{N}}{\text{m}^2}}\right)$$
$$= 2{,}268 \text{ Pa}$$

62. B This problem is solved using the Bernoulli equation:

$$\frac{p_1}{\gamma} + \frac{\overline{V}_1^2}{2g} + z_1 = \frac{p_2}{\gamma} + \frac{\overline{V}_2^2}{2g} + z_2$$

In this case, because the diameter of the tank is assumed to be much larger than the diameter of the siphon tube, the average velocity at section 1 is nearly 0. Also, the pressures of the fluid at section 1 and section 2 are equal to the atmospheric pressure. Using these facts, simplify the Bernoulli equation and solve for $\overline{V}_2$:

$$\frac{\overline{V}_2^2 - \overset{=0}{\overline{V}_1^2}}{2g} = \frac{\overset{=0}{p_1 - p_2}}{\gamma} + (z_1 - z_2)$$
$$\overline{V}_2 = \sqrt{2g(z_1 - z_2)}$$
$$= \sqrt{2\left(9.81 \frac{\text{m}}{\text{s}^2}\right)(3 \text{ m})}$$
$$= 7.67 \text{ m/s}$$

The volumetric flow rate is determined from the average velocity using $Q = \overline{V}A$, where the area A is given by the equation

$$A = \frac{\pi D^2}{4}$$

and where D is the diameter of the siphon tube.

$$Q = \overline{V}\left(\frac{\pi D^2}{4}\right)$$
$$= \left(7.67 \frac{\text{m}}{\text{s}}\right)\left[\frac{\pi (0.03 \text{ m})^2}{4}\right]\left(\frac{3{,}600 \text{ s}}{\text{h}}\right)$$
$$= 0.00542 \frac{\text{m}^3}{\text{s}}$$

63. C The total force acting on the gate, applied by the water pressure, can be modeled as a point force located at the center of pressure. Using the magnitude of this point force and its distance from the axis of the hinge, the required moment is determined. (Refer to the figure below.)

To determine the magnitude of the point force, use these equations:

$$F = \frac{p_1 + p_2}{2} A_v$$

$$p_1 = \overbrace{\rho g z_1}^{= 0}$$

$$= 0 \frac{\text{lbf}}{\text{ft}^2}$$

$$p_2 = \rho g z_2$$

$$= \left(62.4 \frac{\text{lbm}}{\text{ft}^3}\right)\left(32.2 \frac{\text{ft}}{\text{s}^2}\right)(4 \text{ ft})\left(\frac{\text{lbf}}{32.2 \frac{\text{lbm}\cdot\text{ft}}{\text{s}^2}}\right)$$

$$= 249.6 \frac{\text{lbf}}{\text{ft}^2}$$

$$A_v = (3 \text{ ft})(4 \text{ ft})$$
$$= 12 \text{ ft}^2$$

$$F = \frac{0 \frac{\text{lbf}}{\text{ft}^2} + 249.6 \frac{\text{lbf}}{\text{ft}^2}}{2}(12 \text{ ft}^2)$$

$$= 1{,}497.6 \text{ lbf}$$

To determine the center of pressure, use these equations:

$$z^* = \frac{\rho g I_{y_c} \overbrace{\sin(90°)}^{= 1}}{F}$$

$$I_{y_c} = \frac{wh^3}{12}$$

$$= \frac{(3 \text{ ft})(4 \text{ ft})^3}{12}$$

$$= 16 \text{ ft}^4$$

$$z^* = \frac{\left(62.4 \frac{\text{lbm}}{\text{ft}^3}\right)\left(32.2 \frac{\text{ft}}{\text{s}^2}\right)(16 \text{ ft}^4)}{1{,}497.6 \text{ lbf}}\left(\frac{\text{lbf}}{32.2 \frac{\text{lbm}\cdot\text{ft}}{\text{s}^2}}\right)$$

$$= 0.667 \text{ ft}$$

To determine the length of the moment arm, use this equation:

$$z_m = 1 \text{ ft} + z_c + z^*$$
$$= 1 \text{ ft} + 2 \text{ ft} + 0.667 \text{ ft}$$
$$= 3.667 \text{ ft}$$

And the total moment is:

$$M = F z_m$$
$$= (1{,}497.6 \text{ lbf})(3.667 \text{ ft})$$
$$= 5{,}496 \text{ lbf}\cdot\text{ft}$$

64. C Solve this problem using the Bernoulli equation, modified to include the frictional head loss:

$$\frac{p_1}{\gamma} + \frac{\overline{V}_1^2}{2g} + z_1 = \frac{p_2}{\gamma} + \frac{\overline{V}_2^2}{2g} + z_2 + h_f$$

In this case, the average velocity at the outlet is equal to the average velocity at the inlet. So the pressure loss is given by the equation:

$$p_1 - p_2 = \gamma(z_2 - z_1 + h_f)$$

The frictional head loss is expressed as:

$$h_f = f\left(\frac{L}{D}\right)\frac{\overline{V}^2}{2g}$$

Also:

$$V = \frac{\dot{m}}{\rho A}$$

where

$$A = \frac{\pi(0.02 \text{ m})^2}{4}$$

$$= 0.157 \text{ m}^2$$

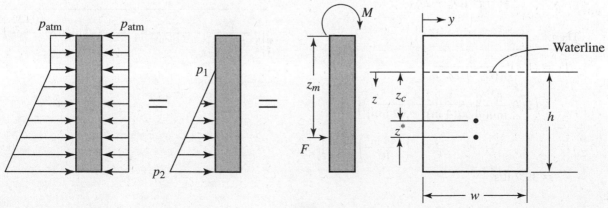

Then

$$\overline{V} = \frac{\left(1,000\,\frac{kg}{s}\right)}{\left(1,000\,\frac{kg}{m^3}\right)(0.157\,m^2)}$$

$$= 6.37\ m/s$$

and

$$h_f = (0.027)\left(\frac{9\,m}{0.20\,m}\right)\left(\frac{\left(6.37\,\frac{m}{s}\right)^2}{2\left(9.81\,\frac{m}{s^2}\right)}\right)$$

$$= 2.51\ m$$

Finally,

$$p_1 - p_2 = \left(9,810\,\frac{N}{m^3}\right)(-2.3\,m + 2.51\,m)\left(\frac{Pa}{\frac{N}{m^2}}\right)$$

$$= 2,060\ Pa$$

$$= 2.06\ kPa$$

65. A Since the head loss through each pipe must be equal:

$$f_a\,\frac{L_a}{D_a}\,\frac{\overline{V}_a^2}{2g} = f_b\,\frac{L_b}{D_b}\,\frac{\overline{V}_b^2}{2g}$$

With $f_a = f_b$:

$$\frac{V_a}{V_b} = \sqrt{\frac{D_a}{D_b}\frac{L_b}{L_a}}$$

$$= \sqrt{\frac{4\,in}{2\,in}\frac{4\,ft}{8\,ft}}$$

$$= 1$$

$$(\overline{V}_a = \overline{V}_b)$$

Use this result in the continuity equation:

$$\overline{V}A = \overline{V}_a A_a + \overline{V}_b A_b$$
$$= V_a(A_a + A_b)$$
$$= V_a\left(\frac{\pi D_a^2}{4} + \frac{\pi D_b^2}{4}\right)$$
$$= \frac{\pi V_a}{4}(D_a^2 + D_b^2)$$

Then:

$$\overline{V}_a = (\overline{V}A)\frac{4}{\pi(D_a^2 + D_a^2)}$$

$$= \left(260\,\frac{ft^3}{min}\right)\left[\frac{4}{\pi[(4\,in)^2 + (2\,in)^2]}\right]$$

$$\left(\frac{144\,in^2}{ft^2}\right)\left(\frac{min}{60\,s}\right)$$

$$= 39.7\ ft/s$$

66. D Use the impulse-momentum principle to solve this problem:

$$\overrightarrow{F} = \dot{m}\overrightarrow{V}_2 - \dot{m}\overrightarrow{V}_1$$

In this case, only the horizontal, or x-direction, component is important. The x-direction equation for the impulse momentum principle is as follows:

$$|\overrightarrow{F}_x| = \dot{m}|\overrightarrow{V}_2|\cos\alpha - \dot{m}|\overrightarrow{V}_1|$$

Since $A_1 = A_2$, $|\overrightarrow{V}_1| = |\overrightarrow{V}_2| = 20$ m/s, so

$$|\overrightarrow{F}_x| = \dot{m}|\overrightarrow{V}_1|(\cos\alpha - 1)$$

$$= \left(50\,\frac{kg}{s}\right)\left(20\,\frac{m}{s}\right)(\cos 45° - 1)\left(\frac{N}{\frac{kg\cdot m}{s^2}}\right)$$

$$= -293\ N$$

67. A The force on a fixed blade is given by the equation

$$|\mathbf{F}_x| = \dot{m}(|\mathbf{V}_1| - |\mathbf{V}_2|\cdot\cos\alpha)$$

Because friction along the face of the blade can be neglected, $|\mathbf{V}_1| = |\mathbf{V}_2|$, and

$$|\mathbf{F}_x| = \dot{m}|\mathbf{V}_1|(1 - \cos\alpha)$$
$$\dot{m} = \rho Q$$
$$= \rho|\mathbf{V}_1|A$$
$$= \rho|\mathbf{V}_1|\left(\frac{\pi D^2}{4}\right)$$

and

$$|\mathbf{F}_x| = \rho\left(\frac{\pi D^2}{4}\right)|\mathbf{V}_1|^2(1 - \cos\alpha)$$

Then

$$|\mathbf{V}_1| = \sqrt{\frac{1}{\rho}\left(\frac{4}{\pi D^2}\right)\left(\frac{1}{1-\cos\alpha}\right)|\mathbf{F}_x|}$$

$$= \sqrt{\left[\frac{1}{\left(1,000\,\frac{kg}{m^3}\right)}\left(\frac{4}{\pi(0.10\,m)^2}\right)\left(\frac{1}{1-\cos 45°}\right)\right](100\,N)\left(\frac{\frac{kg\cdot m}{s^2}}{N}\right)}$$

$$= 6.59\ m/s$$

68. C Find the Reynolds number as follows:

$$\text{Re} = \frac{\rho \bar{V} D}{\mu}$$

$$= \frac{\left(1{,}000 \ \frac{kg}{m^3}\right)\left(6.59 \ \frac{m}{s}\right)(0.10 \ m)}{8 \times 10^{-4} \ \frac{N \cdot s}{m^2}}\left(\frac{N}{\frac{kg \cdot m}{s^2}}\right)$$

$$= 8.24 \times 10^5$$

69. B The friction factor is determined using the Moody diagram at the end of Chapter 11. The parameters required are the Reynolds number and the relative roughness. The Reynolds number was determined for Problem 68. The relative roughness is found as follows:

$$\frac{e}{D} = \frac{0.25 \ mm}{10 \ cm}$$

$$= \frac{0.00025 \ m}{0.10 \ m}$$

$$= 0.0025$$

From the Moody diagram, the friction factor is found to be 0.026.

70. A The head loss in the pipe due to friction is found as follows:

$$h_f = f\left(\frac{L}{D}\right)\left(\frac{\bar{V}^2}{2g}\right)$$

$$= 0.026\left(\frac{9 \ m}{0.20 \ m}\right)\left(\frac{\left(6.37 \ \frac{m}{s}\right)^2}{2\left(9.81 \ \frac{m}{s^2}\right)}\right)$$

$$= 43.16 \ m$$

71. C

$$\dot{m}_1 = \dot{m}_2 + \dot{m}_3$$
$$\dot{m}_1 h_1 = \dot{m}_2 h_2 + \dot{m}_3 h_3$$
$$\dot{m}_3 = \dot{m}_1 - \dot{m}_2$$
$$\dot{m}_1 h_1 = \dot{m}_2 h_2 + (\dot{m}_1 - \dot{m}_2)h_3 = \dot{m}_2 h_2 + \dot{m}_1 h_3 - \dot{m}_2 h_3$$
$$\dot{m}_1 = (h_1 - h_3) = \dot{m}_2(h_2 - h_3)$$
$$\frac{\dot{m}_2}{\dot{m}_1} = \frac{h_1 - h_3}{h_2 - h_3} = \frac{c_p(T_1 - T_3)}{c_p(T_2 - T_3)} = \frac{(460 + 75) - (460 + 0)}{(460 + 150) - (460 + 0)}$$
$$= 0.5$$

72. D

$$\frac{d\cancel{s}}{\cancel{dt}} = \int \frac{\delta \cancel{Q}}{T} + \sum \dot{m}_i s_i - \sum \dot{m}_e s_e + \dot{\sigma}$$

$$\dot{\sigma} = \dot{m}_1 s_1 - \dot{m}_2 s_2 - \dot{m}_3 s_3$$
$$= 0.5 \dot{m}_1 s_1 + 0.5 \dot{m}_1 s_1 - 0.5 \dot{m}_1 s_2 - 0.5 \dot{m}_1 s_3$$
$$= 0.5 \dot{m}_1 (s_1 - s_2) + 0.5 \dot{m}_1 (s_1 - s_3)$$

$$\frac{\dot{\sigma}}{\dot{m}} = 0.5\left[(s_1 - s_2) + (s_1 - s_3)\right]$$

$$= 0.5\left[\left(c_p \ln \frac{T_1}{T_2} - R \ln \frac{p_1}{p_2}\right)\right]$$
$$+ \left(c_p \ln \frac{T_1}{T_3} - R \ln \frac{p_1}{p_3}\right)$$

$$= 0.5\left[\left(0.24 \ln \frac{610}{535} - 0.0686 \ln \frac{5}{1}\right)\right]$$
$$+ \left(0.24 \ln \frac{610}{460} - 0.0686 \ln \frac{5}{1}\right)$$

$$= -0.06$$

73. B

74. C A steady-state system is defined as a system in which the state remains constant with time. Therefore, the state of a system remains unchanged for a steady-state process, and statement (A) is true.

It can be shown that the state of a system that is comprised of a macroscopically homogeneous substance is determined by the values of any two independent properties. So statement (B) is true.

Because during a work interaction, energy is exchanged between the system and its surroundings, the state of the system must change. Therefore, statement (C) is false.

One statement of the second law of thermodynamics is that there exists one stable equilibrium state for which the value of entropy is a maximum for a given value of energy. If the value of entropy is lower than this maximum value, it is possible that the entropy will increase spontaneously. So statement (D) is true.

75. A If an energy balance and an entropy balance are applied to the system described in statement (A), it is possible to configure the system in such a way that the values of energy and entropy do not change with time. Device (A) is possible.

By applying an entropy balance to the system described in statement (B), it can be shown that the value of entropy must be continuously decreasing. Therefore, device (B) is not possible.

An equilibrium state is determined by the amount of energy contained in the material, the amount of material, and the external forces placed on the material. Therefore, there cannot be two equilibrium states under these conditions. Device (C) is not possible.

By applying an energy balance to statement (D), it can be shown that the value of energy must be continuously increasing. Therefore, device (D) is not possible.

76. B When the energy balance equation, equation 13.14, is applied to this system, it can be simplified as follows:

$$\Delta E = Q - W$$

where Q = the amount of energy transferred to a system during a heat interaction
W = the amount of energy transferred from a system during a work interaction.

Substitute the given values:

$$\Delta E = 50 \text{ kJ} - 30 \text{ kJ}$$
$$= 20 \text{ kJ}$$

77. B The relationship between a change in specific enthalpy and the temperature of a system is given by equation 13.38:

$$\Delta h = c_p(\Delta T)$$

Then

$$c_p = \frac{\Delta h}{\Delta T}$$

$$= \frac{120 \frac{\text{kJ}}{\text{kg}}}{46°C - 23°C}$$

$$= 5.22 \frac{\text{kJ}}{\text{kg·°C}} \text{ or } 5.22 \frac{\text{kJ}}{\text{kg·K}}$$

Therefore, based on the property data tables at the end of Chapter 13, the substance is most likely helium, which has a constant-pressure specific heat of 5.19 kJ/kg·K.

78. C The change in internal energy for a closed system is equal to:

$$\Delta U = m \, \Delta u$$

The change in specific internal energy is related to the change in temperature by equation 13.38:

$$\Delta u = c_v \, \Delta T$$

so that

$$\Delta U = m \, c_v \, \Delta T$$

$$= (4 \text{ kg})\left(1.74 \frac{\text{kJ}}{\text{kg·K}}\right)(35°C - 15°C)$$

$$= 139.2 \text{ kJ}$$

79. D For an isentropic process in a system composed of an ideal gas, from equation 13.41:

$$p_1 v_1{}^k = p_2 v_2{}^k$$

$$v_2 = v_1\left(\frac{p_1}{p_2}\right)^{1/k}$$

$$= \left(8.2 \frac{\text{ft}^3}{\text{lbm}}\right)\left(\frac{50 \text{ psia}}{120 \text{ psia}}\right)^{\frac{1}{1.40}}$$

$$= 4.39 \text{ ft}^3/\text{lbm}$$

80. D From equation 13.47, for an isometric process of a system composed of an ideal gas:

$$\frac{T_1}{p_1} = \frac{T_2}{p_2}$$

$$p_2 = p_1 \frac{T_2}{T_1}$$

$$= (600 \text{ kPa})\frac{(110 + 273)\text{K}}{(200 + 273)\text{K}}$$

$$= 486 \text{ kPa}$$

81. A One relationship that describes an isothermal process of an ideal gas is Boyle's law, equation 13.49:

$$p_1 v_1 = p_2 v_2$$

$$v_2 = v_1 \frac{p_1}{p_2}$$

$$= \left(30 \frac{\text{m}^3}{\text{kg}}\right)\frac{(200 \text{ kPa})}{(800 \text{ kPa})}$$

$$= 7.5 \text{ m}^3/\text{kg}$$

82. C From the superheated steam tables at the end of Chapter 13, the value of entropy at the entrance state of the turbine is 7.1229 kJ/kg·K. Since normal atmospheric pressure is 101.35 kPa, the table for superheated steam at 100 kPa will provide values very close to those expected at atmospheric pressure. Since the value of entropy for saturated vapor at this pressure is greater than the value at the turbine inlet, some condensation will occur. Therefore, the temperature of the exiting stream of water will be equal to the saturation temperature of water at atmospheric pressure, or 100°C.

83. **B** Turbine work is determined using equation 13.77:

$$\dot{W} = \dot{m}(h_i - h_e) \quad \text{or} \quad w = (h_i - h_e)$$

The value of enthalpy in the inlet is taken from the superheated steam table:

$$h_i = 3051.2 \text{ kJ/kg}$$

To find the value of enthalpy in the outlet stream of an ideal turbine, the quality of the exit stream for the ideal turbine must be determined. Use equation 13.66d:

$$s_{mix} = s_f + x(s_g - s_f) \quad \text{or} \quad x = \frac{(s_{mix} - s_f)}{(s_g - s_f)}$$

Find the values of s_f and s_g using the saturated water table:

$$s_f = 1.3069 \text{ kJ/kg·K}, \qquad s_g = 7.3549 \text{ kJ/kg·K}$$

In this case, the value of s_{mix} is equal to the value of entropy in the inlet stream of the turbine:

$$s_{mix} = 7.1229 \text{ kJ/kg·K}$$

Substitute known values:

$$x = \frac{s_{mix} - s_f}{s_g - s_f}$$

$$= \frac{\left(7.1229 \dfrac{\text{kJ}}{\text{kg·K}} - 1.3069 \dfrac{\text{kJ}}{\text{kg·K}}\right)}{\left(7.3549 \dfrac{\text{kJ}}{\text{kg·K}} - 1.3069 \dfrac{\text{kJ}}{\text{kg·K}}\right)}$$

$$= 0.962$$

Using this value for quality, determine the value of enthalpy for the exit stream of the ideal turbine by applying equation 13.66c, where $h_e = h_{mix}$:

$$h_e = h_f + x(h_g - h_f)$$

Find the values of h_f and h_g using the saturated water table:

$$h_f = 419.04 \text{ kJ/kg}, \qquad h_g = 2{,}676.1 \text{ kJ/kg}$$

Then

$$h_e = 419.04 \frac{\text{kJ}}{\text{kg}} + 0.962\left(2{,}676.1 \frac{\text{kJ}}{\text{kg}} - 419.04 \frac{\text{kJ}}{\text{kg}}\right)$$

$$= 2590.3 \text{ kJ/kg}$$

The work produced in an ideal turbine under these conditions is:

$$\dot{w} = (h_I - h_e)$$

$$= \left(3{,}051.2 \frac{\text{kJ}}{\text{kg}} - 2{,}590.3 \frac{\text{kJ}}{\text{kg}}\right)$$

$$= 460.9 \text{ kJ/kg}$$

$$= 461 \text{ kJ/kg}$$

84. **A** Turbine efficiency as given is defined as the amount of work a turbine produces, divided by the amount of work produced in an ideal turbine operating at the same conditions:

$$\eta_{turb} = \frac{w_{real}}{w_{ideal}}$$

Then

$$w_{real} = (\eta_{turb})(w_{ideal})$$

$$= (0.80)\left(461 \frac{\text{kJ}}{\text{kg}}\right)$$

$$= 368.8 \text{ kJ/kg}$$

$$= 369 \text{ kJ/kg}$$

85. **B**

$$\dot{Q} = \frac{tkA\Delta T}{L} \rightarrow t = \frac{QL}{kA\Delta T}$$

$$= 195.18 \text{ s}$$

$$= 3.25 \text{ min}$$

$$= 3.2 \text{ min}$$

86. **D**

$$\dot{Q} = \in \sigma A T^4 = 12 \text{ W}$$

87. **A**

$$\dot{Q} = \in \sigma A(T_p^4 - T_R^4)$$

$$= 5{,}382 \text{ W}$$

$$= 5.38 \text{ kW}$$

$$= 5.4 \text{ kW}$$

88. **B**

$$r_{cr} = \frac{k_{insu}}{h_\infty} = 0.04 \text{ m} = 4 \text{ cm}$$

89. **B**

$$\tau_c = \frac{\rho c V}{hA_s} = \frac{\rho c}{h} \cdot \frac{\frac{\pi}{6}D^3}{\pi D^2} = \frac{\rho c D}{6h}$$

$$= 0.0007 \text{ m}$$

$$= 0.7 \text{ mm}$$

90. C Calculate the Biot number to verify the use of lumped capacitance.

$$B_i = \frac{hL_c}{k} = \frac{h}{k} \cdot \left(\frac{V}{A_s}\right) = 0.0017 < 0.1$$

Use lumped capacitance.

Now calculate the time it will take to reach 199°C using LC.

$$\frac{T - T_\infty}{T_i - T_\infty} = e^{-\frac{hA_st}{\rho cV}} \rightarrow t = \frac{\rho cV}{hA_s} \cdot \ln\left(\frac{T_i - T_\infty}{T - T_\infty}\right)$$

$$= 3.7 \text{ seconds}$$

91. D

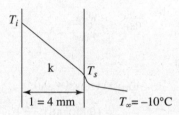

$$q'' = \frac{q}{A} = \frac{T_i - T_\infty}{\frac{L}{k} + \frac{1}{h}} = \frac{37 - (-10)}{\frac{0.009}{0.2} + \frac{1}{20}}$$

$$= 671 \text{ W/m}^2$$

$$= 670 \text{ W/m}^2$$

92. C

$$q'' = \frac{37 - (-10)}{\frac{0.004}{0.2} + \frac{1}{70}} = 1{,}282 \text{ W/m}^2 = 1{,}280 \text{ W/m}^2$$

93. B

$$\frac{T_i - T_\infty}{\frac{1}{k} + \frac{1}{h}} = q'' = \frac{T_i - T_s}{\frac{1}{k}}$$

$$T_s = T_i - \frac{\frac{1}{k}(T_i - T_\infty)}{\frac{1}{k} + \frac{1}{h}}$$

$$T_s = 37 - \frac{\frac{0.004}{0.2}(37 - (-10))}{\frac{0.004}{0.2} + \frac{1}{20}} = 23.6°C = 24°C$$

94. A

$$T_s = 37 - \frac{\frac{0.004}{0.2}(37 - (-10))}{\frac{0.004}{0.2} + \frac{1}{70}} = 9.6°C = 10°C$$

95. B The dominant pole is $s_{2,3} = -0.51 \pm j5.96$. So $\sigma = 0.51$. Based on the 2% setting time criterion, we get $t_s \cong 4/0.51 = 7.84$.

96. A As discussed and depicted in Figure 15.14, by comparing the dominant roots with the standard form we have:

$$s_{2,3} = -0.51 \pm j5.96 = -\xi\omega_n \pm j\omega_d$$

$$\tan\theta = 5.96/0.51 = 11.69 \rightarrow \theta = \tan^{-1}(11.69)$$
$$= 85.11$$

Hence, $\xi = \cos(\theta) = \cos(85.11) = 0.085$ or $\xi = 8.5\%$.

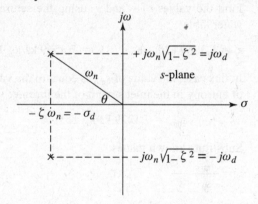

97. C

$$R(s) = \frac{1}{s^2} \text{ (unit ramp input)}$$

$$E(s) = R(s) - C(s), \ C(s) = G(s) \cdot E(s)$$

$$\Rightarrow E(s) = R(s) - G(s) \cdot E(s) \Rightarrow E(s) = \frac{R(s)}{1 + G(s)}$$

$$e_{ss} = \lim_{t \to \infty} = e(t) = \lim_{s \to 0} SE(s) = \lim_{s \to 0} \frac{SR(s)}{1 + G(s)}$$

$$R(s) = \frac{1}{s^2} \Rightarrow e_s = \lim_{s \to 0} \frac{1}{s + sG(s)}$$

$$e_{ss} = \lim_{s \to 0} \frac{1}{s + s\frac{K(s+s)}{s(s+6)(s+7)(s+8)}} = \frac{6 \times 7 \times 8}{K \times 5}$$

$$e_{ss} = 0.1 = \frac{6 \times 7 \times 8}{5K} \Rightarrow K = \frac{6 \times 7 \times 8}{0.5}$$

$$K = 672$$

98. D The following shows the detailed steps on simplifying this block diagram.

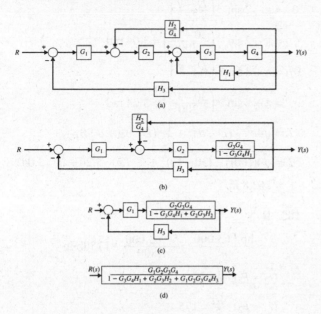

(a)

(b)

(c)

(d)

99. B The closed-loop TF is

$$G_C = \frac{KG(s)}{1 + KG(s)}$$

$$= \frac{\dfrac{5K}{s(s^2 + 2s + 10)}}{1 + \dfrac{5K}{s(s^2 + 2s + 10)}}$$

$$= \frac{5K}{s(s^2 + 2s + 10) + 5K}$$

$$= \frac{5K}{s^3 + 2s^2 + 10s + 5K}$$

The $CE(s)$ is then the denominator G_C:

$$CE(s) = s^3 + 2s^2 + 10s + 5K = 0$$

From our experience with the root-locus method, one approach to find the stable range for parameter K is to check the possible intersection of the root locus with the imaginary axis. If there is an intersection, it implies that the system will become marginally stable for this value and unstable for all the values greater than this value. Since we consider positive only the gain root locus, the minimum value for K to result in a stable system is zero, that is $K > 0$. The upper bound can then be obtained by this intersection. This intersection can simply be obtained by replacing s in the $CE(s)$ with $j\omega$. This results in:

$$CE(s) = (j\omega)^3 + 2(j\omega)^2 + 10j\omega + 5K = 0$$

$$-j\omega^3 - 2\omega^2 + 10j\omega + 5K = 0$$

$$(5K - 2\omega^2) + j\omega(10 - \omega^2) = 0$$

or

$$(5K - 2\omega^2) = 0 \text{ and } \omega(10 - \omega^2) = 0$$

Since we are looking for a nontrivial intersection (that is, $\omega \neq 0$), then the 2nd equation in the above expression results in $\omega^2 = 10$. Upon substituting this value into the 1st equation, one can obtain the value of K as:

$$(5K - 2\omega^2) = 0 \Rightarrow 5K - 20 \Rightarrow K = 4$$

Hence for this system to be stable, the range of K should be

$$0 < K < 4$$

100. B

$$CE(s) = \det(s\mathbf{I} - \mathbf{A}) = 0$$

Hence,

$$CE(s) = \det(s\mathbf{I} - \mathbf{A})$$

$$= \det\left(s\begin{bmatrix} 1 & 0 \\ 0 & 1 \end{bmatrix} - \begin{bmatrix} -4 & 0 \\ 1 & -1 \end{bmatrix} \right)$$

$$= \det\begin{pmatrix} s+4 & 0 \\ -1 & s+1 \end{pmatrix}$$

$$= (s+4)(s+1)$$

$$= 0$$

This results in two stable characteristic roots -1 and -4. So the open-loop system is stable. There are two real roots. So the system is definitely overdamped ($\xi > 1$). The dominant root is at -1, so $\sigma = 1$, and hence $t_s = 4/\sigma = 4$ s.

101. B The finished condition for the weld is considered a forged condition in the fatigue analysis.

$$k_a = a(S_{ut})^b = 39.9(60)^{-0.995} = 0.678$$

$$S_e = k_a k_b k_c(0.5\,S_{ut}) = 0.678 \cdot 1 \cdot 1 \cdot (0.5 \cdot 60)$$

$$= 20.34 \text{ ksi}$$

$$\sigma = K_f \frac{P}{A} = 2.7\frac{1000}{3} = 9000 \text{ psi} = 9 \text{ ksi}$$

$$n = \frac{S_e}{\sigma_a} = \frac{20.34}{9} = 2.26$$

102. B

$$\frac{n_p}{n_g} = \frac{N_g}{N_p}$$

$$\frac{1,800}{n_g} = \frac{50}{20}$$

Speed of the gear $n_g = 720$ rpm

$$\text{hp} = \frac{T n_g}{63,000}$$

$$\frac{1}{2} = \frac{T \cdot 720}{63,000}$$

$$T = 43.75 \text{ lb·in}$$

103. B

$$P = \frac{N_p}{d_p} = \frac{N_g}{d_g}$$

$$2 = \frac{20}{d_p} = \frac{50}{d_g}$$

$d_p = 10$ in, $d_g = 25$ in

$$T_{\text{input}} = \frac{\text{hp} \cdot 63,000}{\text{rpm}} = \frac{\frac{1}{2} \cdot 63,000}{1,800} = 17.5 \text{ lb·in}$$

From the AGMA equation for bending and Tables 17.16–17.19:

$$\sigma_b = F_t K_o K_v \frac{P}{b} \frac{K_s K_m}{J}$$

The overload factor for moderate shock $K_o = 1.75$.

The size factor $K_s = 1.0$.

The mounting factor, less rigid, $K_m = 1.6$.

To find the K_v, the pitch circle velocity should be obtained:

$$v = \frac{\pi d n}{12} = \frac{\pi \cdot 10 \cdot 1,800}{12} = 4,712 \text{ fpm}$$

From the chart for average quality of the gear, $K_v = 2.3$.

The geometry factor, J is obtained from Table 17.10 for pinion with number of teeth 20 and gear with 50 teeth, $J = 0.35$.

Substitute in the AGMA equation results:

$$\sigma_b = 3.5 \cdot 1.75 \cdot 2.3 \frac{2}{0.1} \frac{1 \cdot 1.6}{0.35} = 1,288 \text{ psi}$$

The allowable stress is obtained from

$$\sigma_{\text{all}} = \frac{S}{K_T} \frac{K_L}{K_R}$$

$$K_L = K_T = K_R = 1$$

$$\sigma_{\text{all}} = \frac{13 \cdot 1}{1 \cdot 1} = 13 \text{ ksi}$$

$$n = \frac{13,000}{1,288} = 10.09$$

104. B

$$\theta_d = \pi - 2 \sin^{-1}\left(\frac{D-d}{2c}\right)$$

$$= \pi - 2 \sin^{-1}\left(\frac{30-10}{200}\right) = 3.04 \text{ rad}$$

$$\theta_D = \pi + 2 \sin^{-1}\left(\frac{D-d}{2c}\right)$$

$$= \pi + 2 \sin^{-1}\left(\frac{30-10}{200}\right) = 3.24 \text{ rad}$$

$$L = \sqrt{4c^2 - (D-d)^2} + \frac{1}{2} \cdot (D \cdot \theta_D + d \cdot \theta_D)$$

$$L = \sqrt{4(100)^2 - (30-10)^2} + \frac{1}{2} \cdot (30 \cdot 3.24 + 10 \cdot 3.04)$$

$$= 262.8 \text{ in}$$

105. B

$$T = \frac{\text{hp} \cdot 63,000}{n} = \frac{5 \cdot 63,000}{1,800} = 175 \text{ lb·in}$$

$$(F_1 - F_2) \cdot \frac{d}{2} = T$$

$$(F_1 - F_2) \cdot \frac{10}{2} = 175$$

$$(F_1 - F_2) = 35 \text{ lb}$$

$$F_c = \frac{w}{g} v^2 = \frac{0.1}{386.4}\left(\frac{\pi}{60} \cdot 10 \cdot 1,800\right)^2 = 230 \text{ lb}$$

$$\frac{F_1 - F_c}{F_2 - F_c} = e^{f\theta}$$

$$\frac{F_1 - 230}{F_2 - 230} = e^{(0.15 \cdot 3.04)}$$

Solve for F_1 and F_2:

$$F_1 = 325.34 \text{ lb}, F_2 = 290.34 \text{ lb}$$

106. B For a 60 mm bore ball bearing, use Table 17.23, $C = 47.5$ kN, $C_s = 28$ kN

The equivalent force is

$$P = WVF_r$$

$$\frac{F_a}{C_s} = 0, e = 0.19, X = 1.0, V = 1.0$$

$$P = 1 \cdot 1 \cdot F_r = 10 \text{ kN}$$

$$L_{10} = \left(\frac{C}{P}\right)^a = \left(\frac{47.5}{10}\right)^3 = 107.17$$

$$\text{hrs} = \frac{107.17 \cdot 10^6}{1,500 \cdot 60} = 1,190.8$$

107. C

$$P = K_s(XVF_r + YF_a)$$

For moderate shock, $K_s = 2.0$.

$$\frac{F_a}{C_s} = \frac{2}{28} = 0.071$$

$$e = 0.19$$

$$\frac{F_a}{VF_r} = \frac{2}{1 \cdot 10} = 0.2$$

$$X = 1.0, Y = 0$$

$$P = 2(1 \cdot 1 \cdot 10 + 0 \cdot 2) = 20 \text{ kN}$$

$$L_{10} = \left(\frac{C}{P}\right)^a = \left(\frac{4.75}{20}\right)^3 = 13.396$$

$$\text{hours} = \frac{13.396 \cdot 10^6}{1{,}500 \cdot 60} = 148.84 \sim 149$$

108. A

$$P = \frac{N_s}{d_s} = \frac{N_p}{d_p} = \frac{N_R}{d_R}$$

$$2 = \frac{20}{d_s} = \frac{50}{d_p} = \frac{N_R}{d_R}$$

$$d_s = 10 \text{ in}, d_p = 25 \text{ in}$$

$$d_R = 2\left(\frac{d_s}{2} + d_p\right) = 60 \text{ in}$$

$$2 = \frac{N_R}{d_R} = \frac{N_R}{60}$$

$$N_R = 120$$

109. B

$$e = \frac{n_F - n_A}{n_L - n_A}$$

$$\left(-\frac{N_s}{N_p}\right) = \left(\frac{N_p}{N_R}\right) = \frac{n_F - n_A}{n_L - n_A}$$

$$\left(-\frac{20}{50}\right)\left(\frac{50}{120}\right) = \frac{1{,}500 - n_A}{0 - n_A}$$

$$n_A = 1{,}285.72 \sim 1{,}286 \text{ rpm}$$

110. C

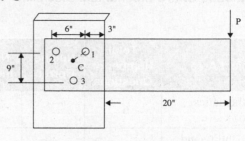

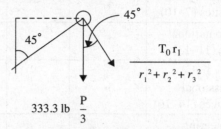

$$333.3 \text{ lb} \quad \frac{P}{3}$$

$$T = P \cdot L_c = 1{,}000(26) = 26{,}000 \text{ lb·in}$$

Bolt 1 is a dangerous bolt.

Direct force $= F_d$

$$= \frac{P}{3} = 333.3 \text{ lb}$$

Indirect force $= F_{id}$

$$= \frac{Tr_1}{r_1^2 + r_2^2 + r_3^2}$$

$$= \frac{26{,}000 \cdot 4.24}{4.24^2 + 4.24^2 + 6^2} = 1{,}532 \text{ lb}$$

Using the cosine rule,

$$F_R^2 = F_d^2 = F_{id}^2 + 2F_d F_{id} 1 \cos(\theta)$$

$$\theta = 45°$$

Substituting for forces, $F_R = 1783$ lb.

$$\tau_R = \frac{1{,}783}{\frac{\pi}{4} \cdot 0.5^2} = 9{,}085.9 \text{ psi}$$

$$\tau_R = \frac{S_y}{2n}$$

$$n = \frac{40{,}000}{2(9{,}085.6)} = 2.2$$

Practice Exam—Mechanical Discipline

Diagnostic Chart: Mechanical Discipline

Subject Area	Total Number of Questions	Number Correct	Number Incorrect	Reason for Incorrect Answer			
				Lack of Knowledge	Misread Problem	Careless or Mathematical Error	Wrong Guess
Mathematics (1–6)	6						
Probability and Statistics (7–10)	4						
Computational Tools (11–13)	3						
Ethics and Professional Practices (14–16)	3						
Engineering Economics (17–19)	3						
Electricity and Magnetism (20–23)	4						
Statics (24–32)	9						
Dynamics, Kinematics, and Vibrations (33–42)	10						
Mechanics of Materials (43–51)	9						
Materials Properties and Processing (52–60)	9						
Fluid Mechanics (61–70)	10						
Thermodynamics (71–84)	14						
Heat Transfer (85–94)	10						
Measurement, Instrumentations, and Controls (95–100)	6						
Mechanical Design and Analysis (101–110)	10						
Total	**110**						

ANSWER SHEET
Practice Exam—Other Discipline

1. Ⓐ Ⓑ Ⓒ Ⓓ	29. Ⓐ Ⓑ Ⓒ Ⓓ	57. Ⓐ Ⓑ Ⓒ Ⓓ	85. Ⓐ Ⓑ Ⓒ Ⓓ
2. Ⓐ Ⓑ Ⓒ Ⓓ	30. Ⓐ Ⓑ Ⓒ Ⓓ	58. Ⓐ Ⓑ Ⓒ Ⓓ	86. Ⓐ Ⓑ Ⓒ Ⓓ
3. Ⓐ Ⓑ Ⓒ Ⓓ	31. Ⓐ Ⓑ Ⓒ Ⓓ	59. Ⓐ Ⓑ Ⓒ Ⓓ	87. Ⓐ Ⓑ Ⓒ Ⓓ
4. Ⓐ Ⓑ Ⓒ Ⓓ	32. Ⓐ Ⓑ Ⓒ Ⓓ	60. Ⓐ Ⓑ Ⓒ Ⓓ	88. Ⓐ Ⓑ Ⓒ Ⓓ
5. Ⓐ Ⓑ Ⓒ Ⓓ	33. Ⓐ Ⓑ Ⓒ Ⓓ	61. Ⓐ Ⓑ Ⓒ Ⓓ	89. Ⓐ Ⓑ Ⓒ Ⓓ
6. Ⓐ Ⓑ Ⓒ Ⓓ	34. Ⓐ Ⓑ Ⓒ Ⓓ	62. Ⓐ Ⓑ Ⓒ Ⓓ	90. Ⓐ Ⓑ Ⓒ Ⓓ
7. Ⓐ Ⓑ Ⓒ Ⓓ	35. Ⓐ Ⓑ Ⓒ Ⓓ	63. Ⓐ Ⓑ Ⓒ Ⓓ	91. Ⓐ Ⓑ Ⓒ Ⓓ
8. Ⓐ Ⓑ Ⓒ Ⓓ	36. Ⓐ Ⓑ Ⓒ Ⓓ	64. Ⓐ Ⓑ Ⓒ Ⓓ	92. Ⓐ Ⓑ Ⓒ Ⓓ
9. Ⓐ Ⓑ Ⓒ Ⓓ	37. Ⓐ Ⓑ Ⓒ Ⓓ	65. Ⓐ Ⓑ Ⓒ Ⓓ	93. Ⓐ Ⓑ Ⓒ Ⓓ
10. Ⓐ Ⓑ Ⓒ Ⓓ	38. Ⓐ Ⓑ Ⓒ Ⓓ	66. Ⓐ Ⓑ Ⓒ Ⓓ	94. Ⓐ Ⓑ Ⓒ Ⓓ
11. Ⓐ Ⓑ Ⓒ Ⓓ	39. Ⓐ Ⓑ Ⓒ Ⓓ	67. Ⓐ Ⓑ Ⓒ Ⓓ	95. Ⓐ Ⓑ Ⓒ Ⓓ
12. Ⓐ Ⓑ Ⓒ Ⓓ	40. Ⓐ Ⓑ Ⓒ Ⓓ	68. Ⓐ Ⓑ Ⓒ Ⓓ	96. Ⓐ Ⓑ Ⓒ Ⓓ
13. Ⓐ Ⓑ Ⓒ Ⓓ	41. Ⓐ Ⓑ Ⓒ Ⓓ	69. Ⓐ Ⓑ Ⓒ Ⓓ	97. Ⓐ Ⓑ Ⓒ Ⓓ
14. Ⓐ Ⓑ Ⓒ Ⓓ	42. Ⓐ Ⓑ Ⓒ Ⓓ	70. Ⓐ Ⓑ Ⓒ Ⓓ	98. Ⓐ Ⓑ Ⓒ Ⓓ
15. Ⓐ Ⓑ Ⓒ Ⓓ	43. Ⓐ Ⓑ Ⓒ Ⓓ	71. Ⓐ Ⓑ Ⓒ Ⓓ	99. Ⓐ Ⓑ Ⓒ Ⓓ
16. Ⓐ Ⓑ Ⓒ Ⓓ	44. Ⓐ Ⓑ Ⓒ Ⓓ	72. Ⓐ Ⓑ Ⓒ Ⓓ	100. Ⓐ Ⓑ Ⓒ Ⓓ
17. Ⓐ Ⓑ Ⓒ Ⓓ	45. Ⓐ Ⓑ Ⓒ Ⓓ	73. Ⓐ Ⓑ Ⓒ Ⓓ	101. Ⓐ Ⓑ Ⓒ Ⓓ
18. Ⓐ Ⓑ Ⓒ Ⓓ	46. Ⓐ Ⓑ Ⓒ Ⓓ	74. Ⓐ Ⓑ Ⓒ Ⓓ	102. Ⓐ Ⓑ Ⓒ Ⓓ
19. Ⓐ Ⓑ Ⓒ Ⓓ	47. Ⓐ Ⓑ Ⓒ Ⓓ	75. Ⓐ Ⓑ Ⓒ Ⓓ	103. Ⓐ Ⓑ Ⓒ Ⓓ
20. Ⓐ Ⓑ Ⓒ Ⓓ	48. Ⓐ Ⓑ Ⓒ Ⓓ	76. Ⓐ Ⓑ Ⓒ Ⓓ	104. Ⓐ Ⓑ Ⓒ Ⓓ
21. Ⓐ Ⓑ Ⓒ Ⓓ	49. Ⓐ Ⓑ Ⓒ Ⓓ	77. Ⓐ Ⓑ Ⓒ Ⓓ	105. Ⓐ Ⓑ Ⓒ Ⓓ
22. Ⓐ Ⓑ Ⓒ Ⓓ	50. Ⓐ Ⓑ Ⓒ Ⓓ	78. Ⓐ Ⓑ Ⓒ Ⓓ	106. Ⓐ Ⓑ Ⓒ Ⓓ
23. Ⓐ Ⓑ Ⓒ Ⓓ	51. Ⓐ Ⓑ Ⓒ Ⓓ	79. Ⓐ Ⓑ Ⓒ Ⓓ	107. Ⓐ Ⓑ Ⓒ Ⓓ
24. Ⓐ Ⓑ Ⓒ Ⓓ	52. Ⓐ Ⓑ Ⓒ Ⓓ	80. Ⓐ Ⓑ Ⓒ Ⓓ	108. Ⓐ Ⓑ Ⓒ Ⓓ
25. Ⓐ Ⓑ Ⓒ Ⓓ	53. Ⓐ Ⓑ Ⓒ Ⓓ	81. Ⓐ Ⓑ Ⓒ Ⓓ	109. Ⓐ Ⓑ Ⓒ Ⓓ
26. Ⓐ Ⓑ Ⓒ Ⓓ	54. Ⓐ Ⓑ Ⓒ Ⓓ	82. Ⓐ Ⓑ Ⓒ Ⓓ	110. Ⓐ Ⓑ Ⓒ Ⓓ
27. Ⓐ Ⓑ Ⓒ Ⓓ	55. Ⓐ Ⓑ Ⓒ Ⓓ	83. Ⓐ Ⓑ Ⓒ Ⓓ	
28. Ⓐ Ⓑ Ⓒ Ⓓ	56. Ⓐ Ⓑ Ⓒ Ⓓ	84. Ⓐ Ⓑ Ⓒ Ⓓ	

Practice Exam— Other Discipline

1. The partial derivative $\frac{\partial z}{\partial x}$ of

   $z = 2x^2y - 2(x + 3y)$ is
   (A) $2x^2 + 3$
   (B) $4xy - 2$
   (C) $2x^2y$
   (D) $4xy + 3$

2. The value of the determinant $\begin{vmatrix} -1 & 4 & -2 \\ 6 & 3 & -4 \\ 0 & -2 & 1 \end{vmatrix}$ is
   (A) 5
   (B) 0
   (C) -1
   (D) 6

3. The graph of the general quadratic equation
   $3x^2 - 10xy + 3y^2 + 8 = 0$ is
   (A) an ellipse
   (B) a parabola
   (C) a hyperbola
   (D) a circle

4. The general solution of the differential equation
   $\frac{dy}{dx} + 3y = 0$ with $y(0) = 1$ is
   (A) $y = e^{-3x}$
   (B) $y = e^{3x}$
   (C) $y = \ln 3x$
   (D) $y = xe^{3x}$

5. The general solution of the differential equation
   $y'' + 4y' + 4 = 0$ is
   (A) $y = c_1e^{2x}$
   (B) $y = e^{-2x}(c_1 + c_2x)$
   (C) $y = c_1e^{2x} + c_2xe^{2x}$
   (D) $y = c_1e^{-2x} + c_2$

6. A unit vector perpendicular to the vector
   $\mathbf{A} = 3\mathbf{i} - 4\mathbf{j}$ is
   (A) $4\mathbf{i} - 3\mathbf{j}$
   (B) $\mathbf{i}$
   (C) $\frac{3}{5}\mathbf{i} - \frac{4}{5}\mathbf{j}$
   (D) $\frac{4}{5}\mathbf{i} + \frac{3}{5}\mathbf{j}$

7. The area of Quadrant I that is bounded by
   $y = x^2$, $x = 3$, and the x-axis is
   (A) 9 square units
   (B) 27 square units
   (C) 6 square units
   (D) 10 square units

8. What is the maximum value of the function f
   defined by $f(x) = x^3 - 3x$?
   (A) -2
   (B) 2
   (C) 0
   (D) 9

9. What is the inverse of the matrix $\begin{bmatrix} 3 & 1 \\ -5 & -2 \end{bmatrix}$?
   (A) $\begin{bmatrix} 2 & 1 \\ -5 & -3 \end{bmatrix}$
   (B) $\frac{1}{11}\begin{bmatrix} 2 & -1 \\ 5 & 3 \end{bmatrix}$
   (C) $\frac{1}{11}\begin{bmatrix} -2 & 1 \\ 5 & 3 \end{bmatrix}$
   (D) undefined

10. The complex number $3 - 3i$ is equivalent to
    (A) $\sqrt{2}e^{(\pi/4)i}$
    (B) $3\sqrt{2}e^{(\pi/4)i}$
    (C) $3\sqrt{2}e^{(5\pi/4)i}$
    (D) $3\sqrt{2}e^{(7\pi/4)i}$

11. The measure of the smallest angle in a triangle with sides that are 4, 5, and 8 meters in length is
    (A) 30.8°
    (B) 24.1°
    (C) 22.0°
    (D) 29.7°

12. What is the sum of the infinite geometric series $\sum_{i=1}^{\infty} (0.3)^i$?
    (A) $\dfrac{3}{10}$
    (B) $\dfrac{1}{2}$
    (C) $\dfrac{2}{5}$
    (D) $\dfrac{3}{7}$

13. The value of the definite integral $\int_{2}^{3} \dfrac{1}{x-1}dx$ is
    (A) ln 2
    (B) ln 3
    (C) 1
    (D) 0

14. The equation of the tangent to the curve $x^2 + y^2 = 25$ at point (3,4) is
    (A) $4x - 3y = 0$
    (B) $3x - 4y = -7$
    (C) $4x + 3y = 24$
    (D) $3x + 4y = 25$

15. In order to be hired for a product assembly position at a factory, a prospective employee must show that he or she can complete 30 assembly tasks on average in less than 2.5 minutes. It is assumed that the standard deviation of the completion time is $\sigma = 0.15$ minutes. If a prospective employee completes the 30 tasks with $\bar{x} = 2.45$ minutes, what can we conclude?
    (A) At both the $\alpha = 0.05$ and $\alpha = 0.01$ level of significance, sufficient evidence shows that the prospective employee can complete the task in less than 2.5 minutes.
    (B) At the $\alpha = 0.05$ but not at the $\alpha = 0.01$ level of significance, sufficient evidence shows that the prospective employee can complete the task in less than 2.5 minutes.
    (C) At the $\alpha = 0.01$ but not at the $\alpha = 0.05$ level of significance, sufficient evidence shows that the prospective employee can complete the task in less than 2.5 minutes.
    (D) At both the $\alpha = 0.05$ and $\alpha = 0.01$ level of significance, sufficient evidence does not show that the prospective employee can complete the task in less than 2.5 minutes.

16. Let X be a random variable whose density function is $f(x) = kx$ for $0 \le x \le 2$ and $f(x) = 0$ elsewhere. Find $P(X > 1)$.
    (A) 0.2500
    (B) 0.4375
    (C) 0.5625
    (D) 0.7500

17. Currently the population of weights of men has a mean 191 pounds and a standard deviation of 30 pounds. A 40-year-old elevator is limited to 32 occupants and built when the average weight of a male was 170 pounds. For safety reasons, it was designed for 200-pound males, so it will be overloaded if the weight of the occupants exceeds 6,400 pounds. If 32 men get on this elevator, what is the probability the elevator will be overloaded?
    (A) 0.0446
    (B) 0.1587
    (C) 0.3821
    (D) 0.9554

18. A researcher is planning to do a hypothesis test with $H_1: \mu \ne \mu_0$. She would like the probability of a type I error to be 0.05 and the probability of a type II error to be 0.10. Moreover, she would like to be able to detect a difference between μ_0 and μ_1, the actual population mean, of 0.1. Assume that the population standard deviation is 0.3. Approximately what sample size is necessary?
    (A) 78
    (B) 79
    (C) 95
    (D) 96

19. A county has recently unveiled plans to reconstruct one of its roads. The current road has 10-foot-wide lanes and the proposed road has 12-foot-wide lanes. A total of 45 area residents are asked what lane widths they prefer. The responses are given in the table below. The standard deviation is $s = 1.16$ feet. Based on inputs from residents, the county changed the plans to 11-foot-wide lanes. The residents responded by stating that the data suggest that the area residents would actually prefer a road with lanes less than 11 feet wide. Based on the information given,

Preferred Width	Frequency
9 feet	9
10 feet	14
11 feet	13
12 feet	6
13 feet	3

(A) we can reject the residents' claim at both the 0.01 and 0.005 level of significance
(B) we can reject the residents' claim at the 0.01 but not at the 0.005 level of significance
(C) we can reject the residents' claim at the 0.005 but not at the 0.01 level of significance
(D) we cannot reject the residents' claim at both the 0.01 and 0.005 level of significance

20. An owner of four gas stations in a county knows, based on historical data, that of all customers that put gas into their cars, station A handled 20% of all customers, station B handled 30%, station C served 15%, and station D served 35%. Recently, a new road was constructed that may have changed traffic patterns in the county and, hence, the distribution of customers using the four stations. A recent count of customers putting gas into their cars at one of the four stations on a particular day showed that station A had 505 cars, station B had 775 cars, station C had 360 cars, and station D had 860 cars. Based on this data, what can we conclude?
(A) At the 0.05 and 0.01 level of significance, we conclude that the distribution has not changed.
(B) At the 0.05 but not at the 0.01 level of significance, we conclude that the distribution has changed.
(C) At the 0.01 but not at the 0.05 level of significance, we conclude that the distribution has changed.
(D) At the 0.05 and 0.01 level of significance, we conclude that the distribution has changed.

21. A college is trying to determine if there is a difference between verbal SAT scores for females and males on campus. Information from a random sample is given in the table below. Based on a 98% confidence interval for the difference of the two means, what can we conclude?

	n	$\bar{x}$	s
Female	33	584.50	77.00
Male	56	571.25	66.64

(A) The females have significantly higher scores than the males.
(B) The males have significantly higher scores than the females.
(C) There is no significant difference in the scores.
(D) We cannot conclude anything based on this information.

22. What is the valence or oxidation state of sulfur in sodium thiosulfate, $Na_2S_2O_3$?
(A) $+1$
(B) $+2$
(C) $+3$
(D) -2

23. How many grams of hydrochloric acid (HCl) are present in 500 mL of 0.015 M HCl solution?
(A) 2.738 grams
(B) 0.2738 grams
(C) 0.0075 grams
(D) 0.075 grams

24. What is the approximate freezing point of an aqueous solution of 111 grams of $CaCl_2$ dissolved in 1 kilogram of H_2O? $[K_f(H_2O) = 1.86]$
(A) $-5.5°C$
(B) $-3.7°C$
(C) $-1.8°C$
(D) $+5.5°C$

25. What is the correct expression for the equilibrium constant K_c for the chemical reaction shown below?

$$N_2(g) + 3H_2(g) \Leftrightarrow 2NH_3(g)$$

(A) $K_c = \dfrac{[NH_3]}{[H_2][N_2]}$

(B) $K_c = \dfrac{2[NH_3]}{3[H_2][N_2]}$

(C) $K_c = \dfrac{[NH_3]^2}{[N_2][H_2]^3}$

(D) $K_c = \dfrac{[N_2][H_2]^3}{[NH_3]^2}$

26. In the following reaction:

$$CuSO_4(aq) + Zn(s) \rightarrow Cu(s) + ZnSO_4(aq)$$

the oxidizing agent is
(A) Zn
(B) Cu
(C) $CuSO_4$
(D) $ZnSO_4$

Questions 27 and 28

Ethane gas burns according to this reaction:

$$2C_2H_6 + 7O_2 \rightarrow 4CO_2 + 6H_2O$$

27. How many grams of CO_2 gas are produced from 300 grams of C_2H_6?
(A) 300 g
(B) 440 g
(C) 600 g
(D) 880 g

28. How many liters of CO_2 gas are produced, assuming ideal gas behavior, at STP?
(A) 224 L
(B) 448 L
(C) 660 L
(D) 880 L

29. What body has jurisdiction over an alleged ethics violation by a Registered Professional Engineer?
(A) National Society of Professional Engineers
(B) State Board of Registration for Engineers
(C) National professional society (e.g., ASME, IEEE) of the engineer involved
(D) Local district court

30. An engineer working for a consulting firm has decided to go into private practice. One of her first steps should be to
(A) submit her letter of resignation
(B) tell colleagues about her plans
(C) send e-mail to the firm regarding her plans
(D) discuss her plans with her supervisor

31. As an engineer, your ethical responsibility, in order from highest to lowest importance, is to
(A) client, society, yourself, profession
(B) yourself, profession, client, society
(C) profession, yourself, society, client
(D) society, client, profession, yourself

32. The annual payment into a fund is $800, and payments will be made for 15 years. Each payment is made at the end of the year. How much will the fund be worth when it matures if interest is 10% per year?
(A) $31,215
(B) $25,418
(C) $35,600
(D) $26,400

33. A residential building lot is purchased for $400,000 cash. If the lot is held for 5 years and the return is 18% before taxes, what is the selling price closest to if there is a 6% annual inflation rate?
(A) $1,855,000
(B) $1,225,000
(C) $1,195,000
(D) $890,000

34. A self-employed engineer deposits $2,000 into a savings account at the end of every year for 8 years. She makes no deposits for the next 4 years. How much will be in the account at the end of 12 years if interest is 6% per year?
(A) $28,570
(B) $26,960
(C) $24,990
(D) $22,750

35. Superior Air Flights is considering the purchase of a helicopter to connect service between its base airport and a new field being built about 25 miles away. The choppers are assumed to be needed for only 6 years until a rapid transit service is completed. Estimates for the two craft under consideration are as follows:

	Whirl 2B	ROT8
First cost	$95,000	$120,000
Annual maintenance	$3,000	$9,000
Salvage value	$12,000	$25,000
Life in years	3	6

What is the annual cost advantage of selecting ROT8?
(A) $0
(B) $700
(C) $2,100
(D) $4,216

638

36. It costs Superior Air Flights $1,200 to run a particular flight, empty or full. Additionally, each passenger generates costs of $40. Regular tickets sell for $90. A plane holds 65 people, but each flight carries only about 35 passengers. The marketing manager proposes to sell special tickets for $50 to people not normally flying this route. He thinks he can sell 15 additional seats for each flight without cutting into the regular business. If he is correct, how much will the total profit be on this flight?

(A) $200
(B) $500
(C) $700
(D) $1,000

37. Two alternative buildings are being considered. The first has an initial cost of $550,000 and annual costs of $162,000 per year. The second building is estimated to cost $750,000 with annual costs of $124,500. With interest at 10% per year, what is the useful life of these two buildings to have the same equivalent annual cost?

(A) 6.5 years
(B) 7 years
(C) 8 years
(D) 8.5 years

38. New equipment is being considered for purchase. The initial investment is $40,000 with an estimated salvage value of $10,000 at the end of the equipment's useful life of 20 years. The annual operating costs are estimated at $9,000 per year with an annual material savings of $12,850. The before tax rate of return is closest to what rate?

(A) 6%
(B) 8%
(C) 10%
(D) 12%

39.

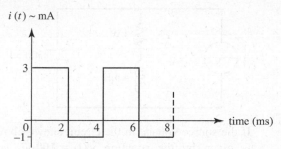

The waveform shown in the above diagram repeats every 4 milliseconds. The average and rms values of the waveform are

(A) 2.0 V, 2.82 V
(B) 1.0 V, 2.24 V
(C) 0.5 V, 1.0 V
(D) 1.25 V, 2.82 V

40.

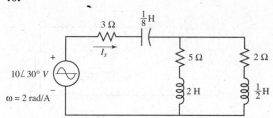

What is the source current, I_s, in the circuit shown above?

(A) $1.84\angle 66°$ A
(B) $2.56\angle -42°$ A
(C) $0.56\angle 28°$ A
(D) $3.31\angle -25°$ A

41. A series-tuned circuit is designed to have a resonant frequency of 31.8 kilohertz and a bandwidth of 2 kilohertz. If the inductor value is 10 millihenrys, the values of the capacitor and the resistor are

(A) 7.5 μF, 5 Ω
(B) 5.5 μF, 2.5 Ω
(C) 2.5 μF, 125.7 Ω
(D) 4.5 μF, 20 Ω

42.

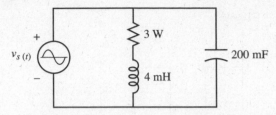

If the source voltage in the circuit shown above is given by the equation $v_s(t) = 100\sqrt{2}$ cos $1,000t$ volts, what is the complex power delivered to the load?
(A) $565\angle 52°$ VA
(B) $12.65\angle 18.4°$ VA
(C) $720\angle -22.5°$ VA
(D) $975\angle -65°$ VA

Questions 43 and 44 relate to the figure shown below.

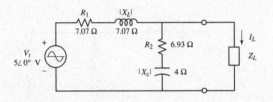

43. Determine the Thévenin equivalent circuit, V_{Th} and Z_{Th}, for the load impedance, Z_L, in the circuit.
(A) $5.51 \angle 32.2°$ V; $3.67 \angle -17.4°$ Ω
(B) $2.79 \angle -42.4°$ V; $5.58 \angle 2.63°$ Ω
(C) $1.9 \angle 63.4°$ V; $4.12 \angle -31.5°$ Ω
(D) $3.47 \angle -17.8°$ V; $7.83 \angle 8.9°$ Ω

44. What is the load current, I_L, if the load impedance is $(5 - j7)$ Ω?
(A) $0.223 \angle -9.87°$ A
(B) $(1.24 + j0.84)$ A
(C) $0.517 \angle -21.3°$ A
(D) $(3.13 + j1.87)$ A

Questions 45 and 46 relate to the figure shown below.

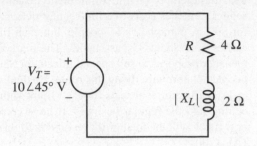

45. The total volt amperes (VA) for the circuit are
(A) $(15.3 + j12)$ VA
(B) $18.6 \angle -16.4°$ VA
(C) $22.4 \angle -26.6°$ VA
(D) $(19 - j4.3)$ VA

46. If the frequency of the source is 400 hertz, what capacitor value must be connected in series with R and L so that the circuit has unity power factor?
(A) 87.3 μF
(B) 26.6 μF
(C) 142 μF
(D) 199 μF

47.

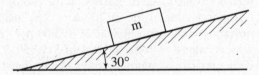

The minimum coefficient of static friction for the block shown in the diagram to be in equilibrium is most nearly
(A) 0.58
(B) 0.5
(C) 1
(D) 0.87

48.

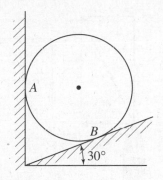

A cylinder weighing 100 newtons rests between two smooth walls, as shown in the diagram. The reaction at point *B* is most nearly

(A) 100 N
(B) 115.5 N
(C) 50 N
(D) 86.6 N

49.

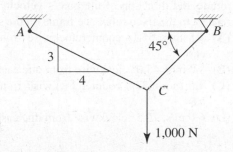

The tension in cable *AC,* shown in the diagram above, is most nearly

(A) 806.5 N
(B) 714 N
(C) 1,414 N
(D) 600 N

50.

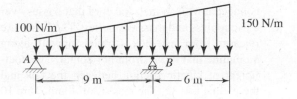

The reaction at point *A* in the diagram is most nearly

(A) 208 N
(B) 1,625 N
(C) 500 N
(D) 937 N

51.

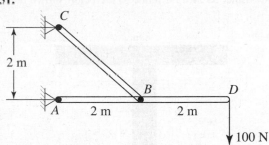

The resultant pin reaction at point *B* in the above diagram (neglecting the weights of *BC* and *AD*) is most nearly

(A) 283 N
(B) 141.5 N
(C) 100 N
(D) 200 N

52.

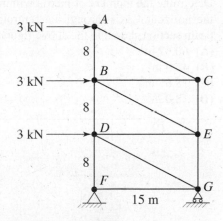

In the above diagram, the force in member *DE* is most nearly

(A) 9 kN
(B) 24 kN
(C) 6 kN
(D) 9.6 kN

641

Questions 53 and 54 relate to the beam shown below.

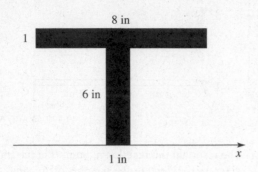

8 in

1

6 in

1 in

x

53. Determine the y-coordinate of the centroid with respect to the x-axis.
   (A) 1.57 in
   (B) 5.0 in
   (C) 6.2 in
   (D) 3.0 in

54. Determine the moment of inertia with respect to the horizontal axis through its centroid for the beam section shown in the above figure.
   (A) 60.67 in^4
   (B) 42.0 in^4
   (C) 18.67 in^4
   (D) 18.0 in^4

55.

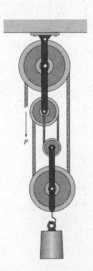

P

Determine the force P needed for equilibrium if the weight of the cylinder is 60 lb.
   (A) 60 lb
   (B) 20 lb
   (C) 30 lb
   (D) 15 lb

56.

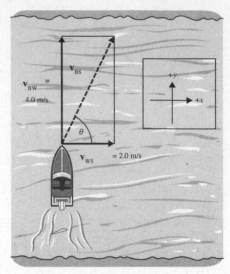

$\mathbf{v}_{BS}$

$\mathbf{v}_{BW} = 4.0$ m/s

θ

$\mathbf{v}_{WS}$ = 2.0 m/s

$+y$

$+x$

A riverboat is crossing a stream. The velocity of the stream current is 2.0 m/s due east relative to the shore. The boat's direction of motion is perpendicular to the stream. The boat travels at 4.0 m/s relative to the water. Determine the magnitude and direction of the boat's velocity with respect to the fixed reference frame on the shore.
   (A) 3.5 m/s, 55.1° counterclockwise from due east
   (B) 2.0 m/s, 45.0° clockwise from due east
   (C) 4.5 m/s, 63.6° counterclockwise from due east
   (D) 6.0 m/s, 70.5° clockwise from due east

57.

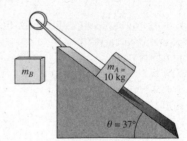

m_B

$m_A = 10$ kg

$\theta = 37°$

Two boxes illustrated in the figure above are connected with an inelastic cord that passes over a frictionless and massless pulley. The boxes are released from rest. Box A travels upward along an incline that is 37° from horizontal. The coefficient of kinetic friction between the box and the incline is 0.32. The masses of A and B are 10 kg and 13 kg, respectively. Determine the acceleration of box A along the incline if the tension in the cord is known to be 103.0 N.
   (A) 4.1 m/s^2
   (B) 1.3 m/s^2
   (C) 7.0 m/s^2
   (D) 1.9 m/s

58.

A child swings a rigid ball that is attached to an inelastic string as shown in the figure above. The mass of the ball is 0.16 kg. The ball travels in a horizontal, circular path that has a diameter of 1.20 m. If the ball travels at a rate of 2.0 revolutions per second, find the force $\mathbf{F}_T$ that the child exerts on the string. The mass of the string is considered negligible.

(A) 1.07 N
(B) 15.16 N
(C) 16.73 N
(D) 1.54 N

59.

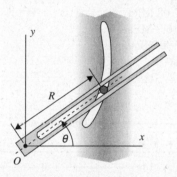

A rigid, smooth, steel ball has curvilinear plane motion due to the rotation of the slotted link mechanism about point O as illustrated in the figure above. The equations of motion are as follows:

Radial distance from O:
$R = 0.8\theta$ units of meters

Constant angular velocity of the link:
$\omega = d\theta/dt = 3.5$ rad/s counterclockwise.

Determine the acceleration component of the ball that is perpendicular to R at the instant when $\theta = 10.5$ radians.

(A) 102.9 m/s^2
(B) 19.6 m/s^2
(C) 1.4 m/s^2
(D) Not enough information is given.

60.

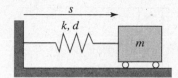

The cart shown in the figure above is initially at rest and is connected to a linear spring. Due to an initial spring deformation of 3.0 inches, the cart rolls along a frictionless horizontal surface and experiences oscillating motion. The spring constant is 0.2 lbf/in. The mass of the cart is 8.0 lbm. Determine the velocity of the cart after 4.5 s of motion.

(A) –9.1 in/s
(B) 0.9 in/s
(C) –2.3 in/s
(D) 1.1 in/s

61. A 1.4 lbm basketball rolls toward a wall in a gymnasium at a velocity of 9.0 ft/s. The ball makes contact with the wall for 0.2 s and then reverses its direction with a velocity of 7.5 ft/s immediately after the collision. Determine the impact force of the wall on the ball.

(A) 17.5 lbf
(B) 115.5 lbf
(C) 0.5 lbf
(D) 3.6 lbf

62.

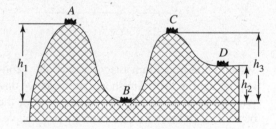

A roller coaster car carrying several passengers at a theme park has a total mass of 2,500 kg and travels from left to right as illustrated by the figure above. The speed of the car at the highest point A is known to be 1.5 m/s. The dimensions of the track are as follows: $h_1 = 35$ m , $h_2 = 16$ m, $h_3 = 28$ m, radius of curvature at C is $R_c = 22$ m. Assuming negligible friction between the car and the tracks, determine the apparent weight of the car w/passengers at point C.

(A) 2,318.3 N
(B) 37.7 kN
(C) 15.8 kN
(D) 8,675.4 N

63.

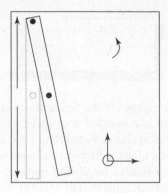

The uniform steel rod shown above pivots about point O and has a length D of 1.4 m and a mass of 9.0 kg. The rod oscillates as a pendulum about the pivot point without friction. Determine the angular momentum H_o of the rod about pivot O at the instant when the angular velocity of the rod is 3.6 rad/s.

(A) 10.6 kg·m²/s
(B) 1.3 kg·m²/s
(C) 21.2 kg·m²/s
(D) 5.3 kg·m²/s

64.

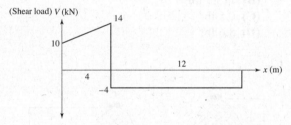

What is the maximum magnitude of the bending moment in the beam shown in the above diagram? The bending moment at each end of the beam is 0, and there are no concentrated couples along the beam.

(A) 96 kN·m
(B) 48 kN·m
(C) 24 kN·m
(D) 40 kN·m

65. If the allowable shearing stress is 900 kilopascals, the torque transmitted by a 150-millimeter diameter solid shaft is most nearly

(A) 1,193 N·m
(B) 298 N·m
(C) 596 N·m
(D) 1,590 N·m

66.

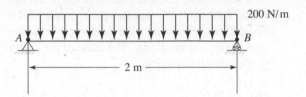

The maximum bending stress for the 4 × 6 centimeter beam shown in the diagram above is most nearly

(A) 4.17 MPa
(B) 2.08 MPa
(C) 16.7 MPa
(D) 100 MPa

67.

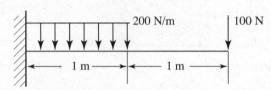

The maximum transverse shearing stress for the 10-centimeter diameter shaft beam shown in the above diagram is most nearly

(A) 57.3 kPa
(B) 38.2 kPa
(C) 25.5 kPa
(D) 50.9 kPa

68.

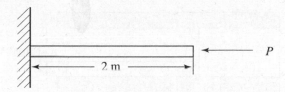

The critical load, P, for the 2-meter long cantilever beam shown in the diagram is most nearly
(A) 2.468 EI
(B) 1.234 EI
(C) 0.617 EI
(D) 0.5 EI

69.

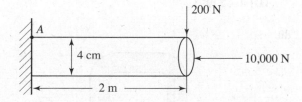

In the above diagram, the normal stress at point A is most nearly
(A) 71.6 MPa
(B) 55.71 MPa
(C) 7.95 MPa
(D) 63.6 MPa

70.

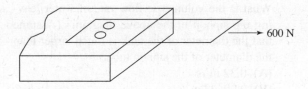

The average shearing stress for each 3-centimeter diameter bolt shown above is most nearly
(A) 0.424 MPa
(B) 1.2 MPa
(C) 0.848 MPa
(D) 0.56 MPa

71.

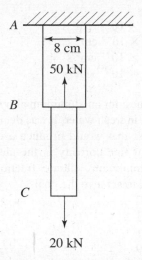

The normal stress in section AB of the above diagram is most nearly
(A) 9.95 MPa
(B) 3.97 MPa
(C) 13.92 MPa
(D) 5.97 MPa

72.

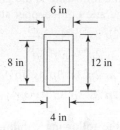

Determine the maximum stress in (psi) due to bending for the beam shown.
(A) 304.6 psi
(B) 35,200 psi
(C) 7,310.8 psi
(D) 3,655.4 psi

73. The diffusion coefficient of carbon in iron at 720°C is:
    (A) 4.4×10^{-13} cm²/s
    (B) 2.4×10^{-7} cm²/s
    (C) 2.7×10^{-9} cm²/s
    (D) 8.5×10^{-11} cm²/s

74. In the quest for an aluminum alloy that would be buoyant in fresh water, it was decided to attempt a process that would produce a uniform distribution of fine porosity in the aluminum. What is the minimum volume fraction of porosity required to achieve the goal?
    (A) 0.37
    (B) 0.63
    (C) 0.51
    (D) 0.49

75. Which of the four polymers listed in Table 10.6 can be the most suitable choice for use in a squeeze bottle application?
    (A) Polyethylene
    (B) Polypropylene
    (C) Polystyrene
    (D) Polyvinylchloride

76. A material whose electrons cannot be excited across the band gap is considered
    (A) a metal
    (B) a semiconductor
    (C) an insulator
    (D) a superconductor

77. Which of the following is the ground-state electron configuration of a copper atom?
    (A) $1s^2 2s^2 2p^6 3s^2 3p^6 4s^1 3d^{10}$
    (B) $1s^2 1p^6 2s^2 2p^6 3s^2\ 3p^6 3d^5$
    (C) $1s^2 1p^2 2s^2 2p^6 3s^2\ 3p^6 4s^2 4p^6 4d^1$
    (D) $1s^2 2s^2 2p^6 3s^2 3p^6 3d^6 4s^2 4p^3$

78. Which of the following exhibits an increase in resistivity with increasing temperature?
    (A) A metal
    (B) A semiconductor
    (C) A superconductor
    (D) An insulator

79.

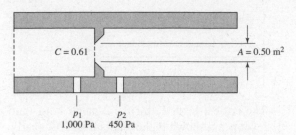

In the above diagram, where C is the discharge coefficient, what is the volumetric flow rate through the orifice?
    (A) 0.32 m³/s
    (B) 0.28 m³/s
    (C) 1.28 m³/s
    (D) 0.65 m³/s

80.

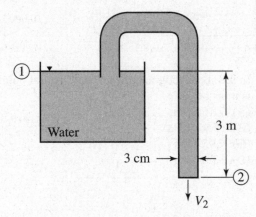

What is the volumetric flow rate of water leaving the siphon in the above diagram? (Assume that the diameter of the tank is much larger than the diameter of the siphon tube.)
    (A) 0.32 m³/s
    (B) 0.00542 m³/s
    (C) 1.28 m³/s
    (D) 0.65 m³/s

81.

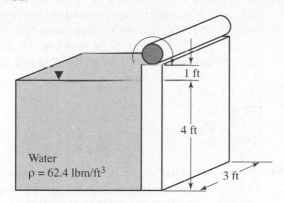

Water
$\rho = 62.4$ lbm/ft³

In the above diagram, what is the approximate moment required at the hinge to keep the gate closed?
(A) 3,000 ft·lbf
(B) 4,500 ft·lbf
(C) 5,500 ft·lbf
(D) 4,000 ft·lbf

82.

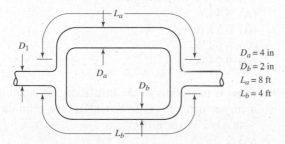

$D_a = 4$ in
$D_b = 2$ in
$L_a = 8$ ft
$L_b = 4$ ft

What is the approximate velocity through tube a in the above diagram if the total volume flow rate is

$260 \dfrac{\text{ft}^3}{\text{min}}$?

(A) 40 ft/s
(B) 20 ft/s
(C) 50 ft/s
(D) 30 ft/s

83.

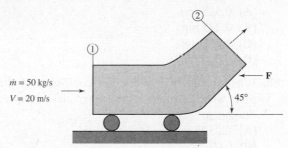

$\dot{m} = 50$ kg/s
$V = 20$ m/s

What force, F, is required to keep the cart shown above from moving? (Assume in the cross-sectional area that Area₁ = Area₂.)
(A) 1,707 N
(B) 720 N
(C) 707 N
(D) 293 N

84. A mechanical pressure gauge attached to a closed reservoir tank of an air compressor indicates a gauge pressure of 827 kPa on a day when the barometer height is 724 mmHg. What is the absolute pressure in the tank?
(A) 726 kPa
(B) 928 kPa
(C) 731 kPa
(D) 923 kPa

Questions 85–88

Data were obtained from measurements on a vertical section of old, corroded, galvanized iron pipe of 2.5 cm inside diameter. At one section, the pressure was $P_1 = 690$ kPa; at a second section, 6 meters lower, the pressure was $P_2 = 520$ kPa. The volumetric flow rate of 20°C water was 11.2 m³/hr.

85. What is the major head loss?
(A) 12.4 m
(B) 19.0 m
(C) 23.3 m
(D) 7.4 m

86. What is the friction coefficient of the pipe?
(A) 0.05
(B) 0.24
(C) 0.08
(D) 0.12

87. What is the relative roughness of the pipe?
    (A) 0.525 mm
    (B) 0.243 mm
    (C) 1.203 mm
    (D) 0.850 mm

88. What would be the friction coefficient if the pipe were restored to its new, clean relative roughness?
    (A) 0.0525
    (B) 0.0325
    (C) 0.0234
    (D) 0.0153

Questions 89–92

Steam enters an adiabatic turbine at 1 megapascal and 300°C and expands to atmospheric pressure. The turbine efficiency is 80%.

89. What would the temperature of the water exiting the turbine be if the turbine were ideal?
    (A) 60°C
    (B) 80°C
    (C) 100°C
    (D) 120°C

90. What would be the approximate work output of an ideal turbine with the same inlet and exit conditions of water?
    (A) 375 kJ/kg
    (B) 461 kJ/kg
    (C) 789 kJ/kg
    (D) 2,632 kJ/kg

91. What is the approximate actual work of this turbine?
    (A) 369 kJ/kg
    (B) 461 kJ/kg
    (C) 576 kJ/kg
    (D) 2,100 kJ/kg

92. What is the value of enthalpy at the exit of this turbine?
    (A) 419 kJ/kg
    (B) 1,859 kJ/kg
    (C) 2,257 kJ/kg
    (D) 2,682 kJ/kg

93. An insulated aluminum rod ($k = 302$ W/(mK)) with a 3 cm diameter is connected between two systems of 150°C and 30°C. How long does it take for 10 kJ of energy to be transferred from the high-temperature system to the cold one if the rod is 50 cm in length?
    (A) 1.5 min
    (B) 3.2 min
    (C) 4.5 min
    (D) 10 min

94. A stone with emissivity of 0.9 in Arizona reaches 52°C in the summer sun. What is the radiant heat energy transfer from the stone if its surface area is 200 cm^2?
    (A) 6 W
    (B) 8 W
    (C) 10 W
    (D) 12 W

95. A 2 m by 2 m plate with an emissivity of 0.12 and a temperature of 400°C is suspended vertically in a large 20°C room. What is the net heat transfer rate from the plate?
    (A) 5.4 kW
    (B) 6.2 kW
    (C) 8.0 kW
    (D) 9.2 kW

96. What is the radius of the insulation surrounding a copper wire ($k = 0.08$ W/(mK)) on a calm day with $h = 2$ W/(m^{2}K)?
    (A) 2 cm
    (B) 4 cm
    (C) 6 cm
    (D) 8 cm

97. A spherical thermocouple junction ($k = 20$ W/mK, $c = 400$ j/kg, and $\rho = 8,500$ kg/m^3) is used to measure gas temperatures where $h = 400$ W/m^{2}K. Calculate the junction diameter if the thermocouple has a time constant of 1 second.
    (A) 0.3 mm
    (B) 0.7 mm
    (C) 1.0 mm
    (D) 7.0 mm

98. A spherical thermocouple ($k = 20$ W/mK, $c = 400$ j/kg, and $\rho = 8,500$ kg/m^3, $D = 0.5$ mm, $T = 25°C$) is placed into a gas stream of 200°C. How long will it take for the junction to reach 199°C if $h = 400$ W/m^{2}K?
(A) 2 seconds
(B) 3 seconds
(C) 3.7 seconds
(D) 4.4 seconds

Questions 99–102

Wind chill is related to heat transfer on a windy day. Consider a 4 mm thick tissue ($k = 0.2$ W/m·K) with an interior temperature of 37°C. The convection heat transfer coefficient is 20 W/m^2·K for a calm day and 70 W/m^2·K for a windy day. In both cases, ambient air is –10°C.

99. What is the heat flux on a calm day?
(A) 600 W/m^2
(B) 625 W/m^2
(C) 650 W/m^2
(D) 670 W/m^2

100. What is the heat flux on a windy day?
(A) 1,240 W/m^2
(B) 1,260 W/m^2
(C) 1,280 W/m^2
(D) 1,300 W/m^2

101. What will be the skin outer surface temperature on a calm day?
(A) 20°C
(B) 24°C
(C) 26°C
(D) 29°C

102. What will be the skin outer surface temperature on a windy day?
(A) 10°C
(B) 12°C
(C) 14°C
(D) 16°C

103. The frequency components of the measured signal from an accelerometer used in a car airbag deploying system are 100 Hz, 180 Hz, 300 Hz, and 400 Hz. What should be the maximum sampling time for the signal to be accurately constructed from the sampled data?
(A) 2.5 ms
(B) 10 ms
(C) 5 ms
(D) 1.25 ms

104.

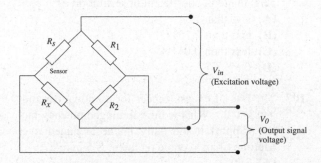

The Wheatstone bridge shown in the figure has the following arm values; $R_1 = 1.6$ kΩ, $R_2 = 500$ Ω, and $R_s = 2.4$ kΩ. What should be the value of the unknown resistor R_x for the bridge to be balanced (or null condition)?
(A) 500 Ω
(B) 1,600 Ω
(C) 750 Ω
(D) 2,400 Ω

105. A hot-wire sensor is used to measure the temperature within a jet of hot gas. If the coefficient of thermal coefficient is $\alpha = 0.00395/°C$, what is the percentage of increase in the resistance of this sensor if the temperature changes from 0°C to 50°C?
(A) 20%
(B) 15%
(C) 10%
(D) 5%

106. The coefficient of restitution of a ball can be determined by dropping the ball from a known height, h_1, onto a surface. If the ball bounces back to the height h_2, this coefficient can be calculated as:

$$e = \sqrt{\frac{h_2}{h_1}}$$

For an experimental case study, these heights are measured as $h_1 = 0.5$ m and $h_2 = 0.4$ m. If the uncertainty in the height measurement is 1 mm, what is the combined uncertainty measurement percentage in coefficient of restitution e?

(A) less than 1%
(B) less than 0.15%
(C) less than 0.015%
(D) 0.5%

107. The risk of excess cancer for a known carcinogen is 0.02. What is the carcinogen potency factor (d/ppm) if the daily intake averaged over lifetime of an average recipient is 5 ppm/d?

(A) 250
(B) 0.1
(C) 0.02
(D) 0.04

108. How long will it take for leachate to migrate through a 0.5 m clay liner if the depth of leachate above the clay layer is 50 cm. The hydraulic conductivity of the clay is 1×10^{-7} cm/s and porosity is 50%.

(A) 1 year
(B) 4 years
(C) 8 years
(D) 12 years

109. For a noise pollution study, the noise reading at a distance of 14 m from the source was recorded to be 70 dB. What will be the approximate reading at a distance of 280 m from the source?

(A) 26 dB
(B) 46 dB
(C) 62 dB
(D) 700 dB

110. The reference dose for a toxicant is 12 ppm/d. What is the hazard ratio if the estimated exposure is 3 ppm/d?

(A) 4
(B) 36
(C) 0.33
(D) 0.25

Answer Key

1.	B	29.	B	57.	D	85.	C
2.	A	30.	D	58.	B	86.	A
3.	C	31.	D	59.	B	87.	A
4.	A	32.	B	60.	A	88.	B
5.	B	33.	B	61.	D	89.	C
6.	D	34.	C	62.	D	90.	B
7.	A	35.	D	63.	C	91.	A
8.	B	36.	C	64.	B	92.	D
9.	A	37.	C	65.	C	93.	B
10.	D	38.	B	66.	A	94.	D
11.	B	39.	B	67.	D	95.	A
12.	D	40.	A	68.	C	96.	B
13.	A	41.	C	69.	B	97.	B
14.	D	42.	B	70.	A	98.	C
15.	B	43.	B	71.	D	99.	D
16.	D	44.	A	72.	D	100.	C
17.	A	45.	C	73.	B	101.	B
18.	D	46.	D	74.	B	102.	A
19.	B	47.	A	75.	A	103.	D
20.	A	48.	B	76.	C	104.	C
21.	C	49.	B	77.	A	105.	A
22.	B	50.	A	78.	A	106.	B
23.	B	51.	A	79.	A	107.	D
24.	A	52.	C	80.	B	108.	B
25.	C	53.	B	81.	C	109.	B
26.	C	54.	A	82.	A	110.	D
27.	D	55.	D	83.	D		
28.	B	56.	C	84.	D		

Answers Explained

1. B Treat y as a constant to obtain $\frac{\partial z}{\partial x} = 4xy - 2$.

2. A Expand about the third row to obtain

$$2\begin{vmatrix} -1 & -2 \\ 6 & -4 \end{vmatrix} + 1\begin{vmatrix} -1 & 4 \\ 6 & 3 \end{vmatrix} = 2(16) + 1(-27) = 5$$

3. C The discriminant

$$B^2 - 4AC = (-10)^2 - 4(3)(3) = 64 > 0$$

Hence, the graph is a hyperbola.

4. A The equation is linear in y with integrating factor $e^{\int 3dx} = e^{3x}$. Hence, the general solution is $ye^{3x} = c$, or $y = ce^{-3x}$. For $y(0) = 1$, $c = 1$. Therefore, $y = e^{-3x}$.

5. B The solution of the auxiliary equation $m^2 + 4m + 4 = 0$ is the repeated root $m = -2$. Hence, the general solution is $y = c_1e^{-2x} + c_2xe^{-2x}$ or $y = e^{-2x}(c_1 + c_2x)$.

6. D $\mathbf{B} = 4\mathbf{i} + 3\mathbf{j}$ is perpendicular to $\mathbf{A} = 3\mathbf{i} - 4\mathbf{j}$ since $\mathbf{A} \cdot \mathbf{B} = 0$. Since $|\mathbf{B}| = 5$, a unit vector perpendicular to $\mathbf{A}$ is $\frac{4}{5}\mathbf{i} + \frac{3}{5}\mathbf{j}$.

7. A Area $= \int_0^3 x^2\,dx = \frac{x^3}{3}\Big|_0^3 = 9$ square units.

8. B $f'(x) = 3x^2 - 3$ and $f''(x) = 6x$. Solving $f'(x) = 0$ yields $x = \pm 1$, with $f(1) = -2$ and $f(-1) = 2$. Since $f''(-1) = -6 < 0$, the maximum value of f is 2.

9. A $|A| = \begin{vmatrix} 3 & 1 \\ -5 & -2 \end{vmatrix} = -1$

$$A^T = \begin{bmatrix} 3 & -5 \\ 1 & -2 \end{bmatrix}$$

implies $\text{adj}(A) = \begin{bmatrix} -2 & -1 \\ 5 & 3 \end{bmatrix}$

Hence,

$$A^{-1} = \frac{\text{adj}(A)}{|A|} = \begin{bmatrix} 2 & 1 \\ -5 & -3 \end{bmatrix}$$

10. D $r = \sqrt{3^2 + (-3)^2} = 3\sqrt{2}$
Also, $\theta = \tan^{-1}(-1) = -\pi/4$, which is equivalent to $7\pi/4$. Hence, $3 - 3i = 3\sqrt{2}e^{(7\pi/4)i}$.

11. B The smallest angle is opposite the smallest side. Thus, by the law of cosines,

$$4^2 = 5^2 + 8^2 - 2(5)(8)\cos\theta,$$

which implies that $\theta = 24.1°$.

12. D $a_1 = 0.3$ and $r = 0.3$. Hence,

$$S = \frac{0.3}{1 - 0.3} = \frac{3}{7}$$

13. A $\int_2^3 \frac{1}{x-1}\,dx = \ln|x-1|\Big|_2^3 = \ln 2 - \ln 1$
$$= \ln 2$$

14. D Differentiate implicitly to obtain $2x + 2y\frac{dy}{dx} = 0$, which implies $\frac{dy}{dx} = -\frac{x}{y}$. At $(3, 4)$, slope $m = -3/4$. Hence, the equation of the tangent is $(y - 4) = -\frac{3}{4}(x - 3)$ or $3x + 4y = 25$.

15. B We perform a hypothesis test with a null hypothesis of $\mu = 2.5$ and an alternative of $\mu < 2.5$. Note that this is a one-tailed test. The test statistic is $z = \frac{\bar{x} - \mu_0}{\sigma/\sqrt{n}} = \frac{2.45 - 2.5}{0.15/\sqrt{30}} = -1.83$. By using the unit normal distribution table, we find that the critical value for a level of significance of $\alpha = 0.05$ is about -1.65 whereas the critical value for a level of significance of $\alpha = 0.01$ is about -2.3. Thus at $\alpha = 0.05$, we can reject the null hypothesis and the employee can be hired. However, at $\alpha = 0.01$, we cannot reject the null hypothesis and the employee cannot be hired.

16. D We first need to find the value of k. Since the area under a density curve must equal 1, we have $1 = \int_0^2 kx\,dx = 2k$, and so $k = 0.5$. Now

$$P(X > 1) = \int_1^2 0.5x\,dx = 1 - 0.25 = 0.75.$$

17. A The elevator will be overloaded if the average weight of 32 randomly selected males is over 200 pounds. So we want to calculate $P(\bar{x} > 200)$. Using the central limit we know that $\bar{x}$ is approximately normally distributed with a mean of 191 and a standard deviation of $30/\sqrt{32}$. Converting to unit normal and then using the unit normal distribution table, we have:

$$P(\bar{x} > 200) = P\left(\frac{\bar{x} - \mu}{\sigma/\sqrt{n}} > \frac{200 - 191}{30/\sqrt{32}}\right) = P(z > 1.7) = 0.0446$$

18. D We use the formula $n \cong \dfrac{(Z_{\alpha/2} + Z_\beta)^2 \sigma^2}{(\mu_1 - \mu_2)^2}$

$\alpha = 0.05$ and $\beta = 0.10$. From the unit normal distribution table, we find $Z_{\alpha/2} \cong 1.95$, and $Z_\beta \cong 1.3$. Hence,

$$n \cong \frac{(Z_{\alpha/2} + Z_\beta)^2 \sigma^2}{(\mu_1 - \mu_2)^2} \cong \frac{(1.95 + 1.3)^2 (0.3)^2}{(0.1)^2} \cong 95.0625.$$

In this case, we need to round up since increasing the sample size will only improve accuracy. So $n \cong 96$.

19. B We need to do a hypothesis test with H_1: $\mu < 11$. Since the population standard deviation is unknown, we perform a t-test. By using the data we find $\bar{x} = 10.56$. So the test statistic is

$$t = \frac{\bar{x} - \mu_0}{s/\sqrt{n}} = \frac{10.56 - 11.00}{1.16/\sqrt{45}} = -2.544.$$ According

to the t-distribution table, the critical value associated with a 0.01 level of significance is -2.326 and the critical value associated with a 0.005 level of significance is -2.576. Thus, we can reject the null hypothesis at the 0.01 level but not at the 0.005 level.

20. A We perform a goodness of fit hypothesis test. The null hypothesis states that all proportions have stayed the same, while the alternative is that at least one is different. Creating a table to organize our calculations we have:

Location	Claim	Observed (*o*)	Expected (*e*)	$(o-e)^2/e$
A	0.20	505	500	0.050
B	0.30	775	750	0.833
C	0.15	360	375	0.600
D	0.35	860	875	0.257
Totals	1.00	2,500	2,500	1.740

The test statistic is given by

$$\chi^2 = \sum \frac{(Observed - Expected)^2}{Expected}$$

We have $\chi^2 = 1.740$, which is the value in the bottom right of the table. The critical values of interest here are $\chi^2_{0.05,3} = 7.81473$ and $\chi^2_{0.01,3} = 11.3449$. We reject the null hypothesis if $\chi^2 > \chi^2_{\alpha,3}$. In both cases, we fail to reject the null hypothesis.

21. C Let μ_1 be the mean score of females and μ_2 be the mean score of males. We construct a t-interval for $\mu_1 - \mu_2$. For a 98% confidence interval with $n_1 + n_2 - 2 = 87$ degrees of freedom, we have $t_{0.02/2,87} = 2.326$. For the confidence interval:

$$\sqrt{\frac{(1/n_1 + 1/n_2)\left[(n_1 - 1)s_1^2 + (n_2 - 1)s_2^2\right]}{n_1 + n_2 - 2}}$$

$$= \sqrt{\frac{(1/33 + 1/56)\left[(33-1)77^2 + (56-1)66.64^2\right]}{33 + 56 - 2}}$$

$$= 15.500$$

So,

$$584.50 - 571.25 - 2.326(15.500) \le \mu_1 - \mu_2$$
$$\le 584.50 - 571.25 + 2.326(15.500)$$
$$-22.803 \le \mu_1 - \mu_2 \le 49.303$$

Thus, the difference in the population mean SAT scores is between -22.803 and 49.303. Since the interval contains 0, we cannot say that either females or males have higher scores.

22. B See Example 3.4. Sulfur may have more than one valence or oxidation state. Set up a simple algebraic equation, letting x equal the unknown valence of sulfur. Sodium has a valence of $+1$, and oxygen has a valence of -2. The net charge on $Na_2S_2O_3$ is 0, since it's uncharged. Sum up the individual atomic charges, and set them equal to 0:

$$2(+1) + 2(x) + 3(-2) = 0$$
$$x = +2$$

23. B First, find the number of moles of HCl present in the solution by using the relationship #moles HCl $= (V)(M)$, where V = volume of solution and M = molarity of solution.

$$\#moles\ HCl = (0.500\ L)(0.015\ M)$$
$$= 0.0075\ moles\ HCl$$

Next, convert this into grams by multiplying by HCl's molar mass of 36.5 grams/mole:

(0.0075 moles HCl)(36.5 grams/mole)
= 0.2738 grams HCl

To be precise, this answer should be rounded off to three significant figures, i.e., 0.274 grams.

24. **A** This is a freezing-point-depression problem, so use equation 3.13. The solute is $CaCl_2$, and the solvent is water. $CaCl_2$ is a strong electrolyte, so $i = 3$.

The K_f of water is given as 1.86. The molality m of the solution is

$$m = \frac{111 \text{ g}/111 \text{ g/mol}}{1.0 \text{ kg}} = 1.0 \text{ molal}$$

Substitute these numbers into equation 3.13, and solve for ΔT. Thus $\Delta T = 5.5°C$. Hence the freezing point T_f of the solution is $-5.5°C$.

25. **C** The correct expression for the equilibrium constant K_c is given by equation 3.30. Of the choices given, only $K_c = \dfrac{[NH_3]^2}{[N_2][H_2]^3}$ is in this form.

26. **C** The oxidizing agent is the reactant species that removes electrons from another reactant. Since copper is in the $+2$ oxidation state in $CuSO_4$ and goes to the zero or uncharged state on the product side, it must gain 2 electrons by removing them from zinc. Thus, $CuSO_4$ is the oxidizing agent.

27. **D** The reaction is already balanced; assume O_2 is in excess. Use the factor-label method:

$$300 \text{ g } C_2H_6 \frac{(1.0 \text{ mol } C_2H_6)}{(30.0 \text{ g } C_2H_6)} \frac{(4.0 \text{ mol } CO_2)}{(2.0 \text{ mol } C_2H_6)}$$
$$\frac{(44.0 \text{ g } CO_2)}{(1.0 \text{ mol } CO_2)} = 880 \text{ g}$$

28. **B** Since CO_2 can be regarded as an ideal gas, convert 880 g to liters using the ideal gas law:

$$PV = nRT$$

where $n = m/MM = \dfrac{880 \text{ g}}{44.0 \text{ g/mol}} = 20.0 \text{ mol}$.

STP means that $T = 273$ K and $P = 1.0$ atm. Also $R = 0.0821$ L·atm/K·mol. Substitute the numbers and solve for V: $V = 448$ L.

29. **B** Choices (A) and (C) are professional societies representing PEs and specific engineering disciplines, including PEs. They do not have the power to settle disputes of any nature. Ethics violations are civil suits and could be settled in court; however, the Board of Registration in each state is empowered to handle ethics violations. Choice (B) is the best response.

30. **D** Choices (B) and (C) are unethical at this point in time. Formal resignation is required but, again, not now. A discussion with her supervisor is the first step in the process. Choice (D) is the correct response.

31. **D** The hierarchy of ethical responsibility is stated in the Code of Ethics. Choice (D) is the correct response.

32. **B**

$F = 800(F/A, 10, 15)$
 $= 800(31.7725)$
 $= \$25,418$

33. **B** Calculate the future selling price (F).

$F = \$400,000(F/P, 18\%, 5)(F/P, 6\%, 5)$

 $= \$400,000(2.288)(1.338) = \$1,225,000$

34. **C** Calculate the amount in the savings account.

Total amount at end of 12 years $= FW = \$2,000$ $(F/A, 6, 8)(F/P, 6, 4)$

$FW = \$2,000(9.8975)(1.2625) = \$24,991$

35. **D** $AC(2B) = 95K(A/P, 10, 3) + 3K - 12K$
 $(A/F, 10, 3)$
 $= 95K(0.421) + 3K - 12K(0.3021)$
 $= \$37,566$

and

$AC(ROT) = 120K(A/P, 10, 6) + 9K - 25K (A/F, 10, 6)$
 $= 120K(0.2296) + 9K - 25K(0.1296)$
 $= \$33,350$

Difference in annual cost $= \$37,566 - \$33,350$
 $= \$4,216$

36. C Use this equation:

Profit = Revenues − Costs
= [(35·$90) + (15·$50)] − [(35+15)·$40] − $1,200
= [($3,150) + ($750)] − [(50)($40)] − $1,200
= $700

37. C Set the PW equal to each other since the alternatives have the same life.

PW(1) = PW(2)

$550,000 + 162,000(P/A, 10\%, n)$
$= \$750,000 + \$124,500(P/A, 10\%, n)$
$(P/A, 10\%, n)$
$= (750,000 - 550,000)/(162,999 - 124,500)$
$= 5.33$

Go to the 10% interest table at the end of Chapter 6, and find $n = 8$ years

38. B Solve by trial and error.

At 6%:
$\$40,000 = (\$12,850 - \$9,000)(P/A, i, 20)$
$\qquad + \$10,000(P/F, i, 20)$
$\$40,000 = (\$3,850)(11.4699)$
$\qquad + \$10,000(0.3118)$
$\$40,000 = \$47,277 + \$3,118$
$\$40,000 = \$50,395$

Try a higher interest rate.

At 10%:
$\$40,000 = (\$12,850 - \$9,000)(P/A, i, 20)$
$\qquad + \$10,000(P/F, i, 20)$
$\$40,000 = (\$3,850)(8.5136)$
$\qquad + \$10,000(0.1486)$
$\$40,000 = \$32,777 + \$1,486$
$\$40,000 = \$34,263$

39. B For the average value of this waveform, apply equation 12.36(b):

$$I_{\text{avg}} = \frac{(3\,\text{V} \times 2\,\text{ms}) + (-1\,\text{V} \times 2\,\text{ms})}{4\,\text{ms}}$$

$$= 1.0\,\text{V}$$

For the rms value of this waveform, apply equation 12.37 or use:

$$I_{\text{rms}} = \sqrt{\frac{(9\,\text{V}^2 \times 2\,\text{ms}) + (1\,\text{V}^2 \times 2\,\text{ms})}{4\,\text{ms}}}$$

$$= 2.24\,\text{V}$$

40. A Solve for the impedance of each branch:

$$Z_1 = 3\,\Omega - j\,\frac{1}{\left(2\,\text{rad/s} \times \frac{1}{8}\,\text{H}\right)}$$

$$= (3 - j4)\,\Omega$$
$$Z_2 = 5\,\Omega + j(2\,\text{rad/s} \times 2\,\text{H}) = (5 + j4)\,\Omega$$
$$Z_3 = 2\,\Omega + j\left(2\,\text{rad/s} \times \frac{1}{2}\,\text{H}\right) = (2 + j1)\,\Omega$$

Find the total impedance:

$$Z_T = Z_1 + \frac{Z_2 \times Z_3}{Z_2 + Z_3}$$

$$= (3 - j4)\,\Omega + \frac{(5 + j4)\,\Omega \times (2 + j1)\,\Omega}{(5 + j4)\,\Omega + (2 + j1)\,\Omega}$$

$$= (4.45 - j3.18)\,\Omega$$

Apply Ohm's law:

$$I_s = \frac{10\angle 30°\,\text{V}}{(4.45 - j3.18)\,\Omega} = 1.84\angle 66°\,\text{A}$$

41. C Rearrange equation 12.48 to solve for C:

$$C = \frac{1}{(2\pi \times 31.8 \times 10^3\,\text{Hz})^2 \times (10 \times 10^{-3}\,\text{H})}$$

$$= 2.5 \times 10^{-9}\,\text{F} = 2.5\,\mu\text{F}$$

Rearrange equation 12.52(b) to solve for R:

$$R = 2\pi \times BW \times L$$
$$= 2\pi \times (2 \times 10^3\,\text{Hz}) \times (10 \times 10^{-3}\,\text{H})$$
$$= 125.7\,\Omega$$

42. B Determine the impedance of each branch.

Let $\quad Z_a = R + j\omega L = (3 + j(1,000)(4\,\text{mH}))$
$$= (3 + j4)\,\Omega$$

and $\quad Z_b = 0 - j\,\frac{1}{(1,000\,\text{rad/s})(200\,\mu\text{F})}$

$$= -j5\,\Omega$$

$$Z_T = \frac{Z_a Z_b}{Z_a + Z_b} = \frac{(3 + j4)\,\Omega \times (-j5)\,\Omega}{(3 + j4)\,\Omega + (-j5)\,\Omega}$$

$$= (7.5 - j2.5)\,\Omega$$

The input voltage phasor is $10.0\angle 0°$ V. Apply Ohm's law:

$$I = \frac{10.0\angle 0°\,\text{V}}{(7.5 - j2.5)\,\Omega} = 1.265\angle 18.4°\,\text{A}$$

Complex power = $10.0\angle 0°$ V × $1.265\angle 18.4°$
A = $12.65 \angle 18.4°$ VA

43. B The Thévenin equivalent voltage is determined by using the voltage division law:

$$V_{Th} = \frac{Z_2}{Z_1 + Z_2} \times V_T$$

$$= \frac{(6.93 - j4)\,\Omega}{(7.07 + j7.07)\,\Omega + (6.93 - j4)\,\Omega}$$

$$\times 5\angle 0°\,V = 2.79\angle -42.4°\,V$$

The Thévenin equivalent impedance is determined as follows:

$$Z_{Th} = \frac{Z_1 \times Z_2}{Z_1 + Z_2}$$

$$= \frac{(7.07 + j7.07)\,\Omega \times (6.93 - j4)\,\Omega}{(7.07 + j7.07)\,\Omega + (6.93 - j4)\,\Omega}$$

$$= (5.57 + j0.256)\,\Omega = 5.58\angle 2.63°\,\Omega$$

44. A Use the Thévenin equivalent values to obtain the load current:

$$I_L = \frac{V_{Th}}{Z_{Th} + Z_L}$$

$$= \frac{2.79\angle -42.4°\,V}{(5.57 + j0.256)\,\Omega + (5 - j7)\,\Omega}$$

$$= 0.223\angle -9.87°\,A$$

45. C Solve for the load current:

$$I_L = \frac{10\angle 45°\,V}{(4 + j2)\,\Omega} = 2.24\angle 18.4°\,A$$

The phase angle from voltage to current is $26.6°$ ($45° - 18.4°$).

Apply equations 12.45 and 12.46:

$P = 10\,V \times 2.24\,V \times \cos(26.6°) = 20\,W$

$Q = 10\,V \times 2.24\,A \times \sin(26.6°) = 10\,VAR$

For an inductive circuit, the power factor is lagging. So apply equation 12.47:

$S = (20 - j10)\,VA = 22.4\angle -26.6°\,VA$

46. D $X_L = X_C = \dfrac{1}{\omega C} = 2\,\Omega$

Solve for C:

$$C = \frac{1}{(2\,\Omega)(2\pi)(400\,Hz)} = 199\,\mu F$$

47. A $\mu_s = \tan\theta = \tan(30°) = 0.577 \cong 0.58$

48. B $\Sigma F_y = 0 \qquad B\cos(30°) = 100$

$B = 115.5\,N$

49. B $\Sigma F_x = 0 \quad \dfrac{4}{5}T_{AC} = T_{BC}\cos 45°$

$\Sigma F_y = 0 \quad \dfrac{3}{5}T_{AC} + T_{BC}\sin 45° - 1{,}000 = 0$

Solve for T_{AC}:

$$T_{AC} = 714\,N$$

50. A $\Sigma M_B = 0 \quad A(9) - 1{,}500(1.5) + 375(1) = 0$

$A = 208.33\,N$

51. A $\Sigma M_A = 0 \qquad (F_{BC}\cos 45°)(2) - 100(4) = 0$

$$F_{BC} = 283\,N$$

52. C Cut the truss through members *BD*, *DE*, and *EG*. Use the FBD of the upper section of the truss:

$$\Sigma F_x = 0 \qquad F_{DE} = 6\,kN$$

53. B $\bar{Y} = \dfrac{\Sigma yA}{A} = \dfrac{(6.5)(8) + (3)(6)}{8 + 6} = 5\,in.$

54. A

$$I = I_{web} + I_{flange}$$

$$I_{web} = \frac{1}{12}(1)(6)^3 + (6)(2)^2 = 42\,in^4$$

$$I_{flange} = \frac{1}{12}(8)(1)^3 + (8)(1.5)^2 = 18.67\,in^4$$

$$I = I_{web} + I_{flange} = 42 + 18.67 = 60.67\,in^4$$

55. D Draw the free-body diagram of the cylinder:

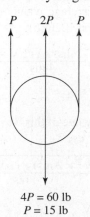

$4P = 60\,lb$

$P = 15\,lb$

56. C To solve for the velocity of the boat relative to the fixed shore reference, let V_{BS} represent the desired resultant vector. The velocity of the boat relative to the water is given by V_{BW}. Assign V_{WS} to represent the given water velocity relative to the shore. Since $V_{BS} = V_{BW} + V_{WS}$, these vectors are resolved into the desired resultant.

The magnitude of V_{BS} is determined using the Pythagorean theorem.

$$|V_{BS}| = \sqrt{4^2 + 2^2} = 4.47 \approx 4.5 \text{ m/s}$$

The direction of V_{BS} is found using

$$\Theta = \cos^{-1}(2.0/4.5) = 63.6°$$

counterclockwise from due east.

57. D

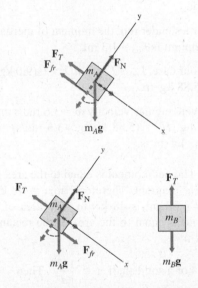

The FBD of box A displays the forces acting along the inclined direction and perpendicular to the normal direction. Since the block is assumed to have no motion in the normal direction, then $\sum F_n = 0$ and the normal reaction F_n is determined.

$$F_n - W_n = 0 \rightarrow F_n = mg \cos 37°$$

$$= (10 \text{ kg})(9.81)(0.80) = 78.5 \text{ N}$$

Since the cord is inelastic, the velocity of box B in a vertical direction must equal the velocity of box A along the incline. The accelerations of A and B are necessarily equal for the same reason.

Motion of A upward along the incline using case ii FBD: $\sum F_A = m_A a_A$

$$F_T - mg \sin 37° - F_{fr} = (10 \text{ kg})a_A \rightarrow$$
$$103.0 \text{ N} - 98.1 \text{N}(0.60) - 0.32(78.5 \text{ N}) = (10 \text{ kg})a_A$$

$$19.0 \text{ N} = (10 \text{ kg})a_A \text{ and } a_A = 19.0 \text{ N}/10 \text{ kg} = 1.9 \text{ m/s}^2$$

Please note that this problem also could have been solved by using the FBD of B, and solving for the acceleration of B, which equals the desired acceleration of A.

58. B The FBD of the ball will contain only two external forces—the ball's weight mg and the desired tension F_T exerted by the string. The circular motion of the ball creates a centripetal force toward the center of the path that is effectively the value of the tension since the weight acts in a direction perpendicular to vertical. Using a tangential and normal reference frame for circular motion:

$\sum F_n = mv_B^2/r$, where v_B is the constant speed of the ball tangent to the circular path.

$$v_B = r\omega = (0.6 \text{ m})(2 \text{ rev/s} \cdot 2\pi \text{ rad/rev})$$

$$= 2.4\pi \text{ m/s} \approx 7.54 \text{ m/s}$$

$$\sum F_n = F_T = (0.16 \text{ kg})(7.54 \text{ m/s})^2/0.6 \text{ m}$$

$$= 15.16 \text{ N}$$

59. B The curvilinear plane motion of the steel ball due to the slotted link is best described by using the radial and transverse equations of motion for a rigid body. The component of acceleration perpendicular to R is the transverse acceleration a_θ.

$$a_\theta = R(d^2\theta/dt^2) + 2(dR/dt)(d\theta/dt)$$

Given the constant angular velocity $\omega = d\theta/dt = 3.5, \rightarrow \int d\theta = \int 3.5 dt$ and $\theta = 3.5t$.

By substitution, the radial distance R at any time t is $R = 0.8(3.5t) = 2.8t$.

Necessarily, $dR/dt = 2.8 \text{ m/s} = $ constant at any time t.

$$d^2\theta/dt^2 = d\omega/dt = 0 \text{ since } \omega = \text{constant.}$$

Although the transverse acceleration is needed when $\theta = 10.5$ radians, note that the above derivatives are constant or 0 for any angle θ or t.

Therefore $a_\theta = $ constant.

$$a_\theta = 0 + 2(2.8 \text{ m/s})(3.5 \text{ rad/s}) = 19.6 \text{ m/s}^2$$

60. A The equations of motion for an undamped free vibration of a spring/mass system having no initial velocity are

Displacement: $x(t) = x_0 \cos(\omega_n t)$, where x_0 is the initial displacement of the spring.

Velocity: $v(t) = dx(t)/dt = -x_0 \omega_n \sin(\omega_n t)$

Natural circular frequency

$$\omega_n = \sqrt{k/m}$$
$$= \sqrt{(0.2 \text{ lbf/in})/8.0 \text{ lbm}}$$
$$= \sqrt{(2.41 \text{ lbf/ft})(32.2)/8.0 \text{ lbm}}$$

$\omega_n = 3.1$ rad/s and substituting $t = 4.5$ s into the velocity equation yields

$$v(2) = -(3.0 \text{ in})(3.1 \text{ rad/s}) \sin(3.1 \cdot 4.5)$$
$$= -(9.3 \text{ in/s})\sin(13.95 \text{ rad})$$
$$= -(9.3)(0.98)$$
$$= -9.1 \text{ in/s}$$

61. D The principle of impulse and momentum is employed to calculate the average impulsive force **F** of the wall on the wall.

$$\text{Imp}_{1\rightarrow2} = \int \mathbf{F} dt = m(\mathbf{v}_2 - \mathbf{v}_1)$$

Since the wall force is assumed constant during the impact, $\mathbf{F}\Delta t = m(\mathbf{v}_2 - \mathbf{v}_1)$. Note that $\mathbf{v}_2 = -7.5$ and $\mathbf{v}_1 = 9.0$ due to the opposite velocity directions.

$$\mathbf{F}(0.2 \text{ s}) = (1.4 \text{ lbm})(-7.5 - 9.0) \text{ ft/s}$$
$$= -23.1 \text{ lbm ft/s} \rightarrow \mathbf{F} = -23.1/0.2$$
$$= -115.5 \text{ lbm ft/s}$$

Converting to lb force,

$$\mathbf{F} = -(115.5/32.2) = -3.59 \approx -3.6 \text{ lbf}$$

Note that the negative sign attached to the force indicates the opposing direction of the force with respect to the positive velocity assigned to the initial direction of the ball.

62. D The apparent weight of a rigid object is the reaction force N_R, which the track exerts on the car at point C. The speed of the car at point C is needed. The conservation of energy principle is used due to the "frictionless" condition of the roller coaster track.

$$KE_A + PE_A = KE_C + PE_C$$

$$\tfrac{1}{2}mv_A{}^2 + mgh_1 = \tfrac{1}{2}mv_C{}^2 + mgh^3$$

$$\tfrac{1}{2}(2,500 \text{ kg})(1.5 \text{ m/s})^2 + (2,500 \text{ kg})(9.81 \text{ m/s}^2)(35 \text{ m})$$
$$= \tfrac{1}{2}(2.500)v_C{}^2 + (2,500)(9.81)(28 \text{ m})$$

$$2812.5 \text{ N} \cdot \text{m} + 858,375 \text{ N} \cdot \text{m}$$
$$= 1,250 \text{ kg } v_3{}^2 + 686,700 \text{ N} \cdot \text{m}$$

$$174,487.5 \text{ N} \cdot \text{m} = 1,250 \text{ kg } v_C{}^2 \rightarrow v_C$$
$$= \sqrt{(174,487.5/1250)} = 11.81 \text{ m/s}$$

At point C, sum forces in the normal direction toward the center of curvature.

$$\sum F_n = mv_3{}^2/R$$
$$= 2,500 \text{ kg}(11.81 \text{ m/s})^2/22 \text{ m}$$
$$= 15,849.6 \text{ N}$$

$$\sum F_n = W - N_R$$
$$= 15,849.6 \text{ N} \rightarrow N_R = W - 15,849.6 \text{ N}$$
$$= (2,500)(9.81) - 15,849.6 = 8,675.4 \text{ N}$$

The apparent weight is 8,675.4 N.

63. C The angular momentum H_o of a rigid body about a given point is given by $H_o = I_o\omega$, where I_o represents the mass moment of inertia of the rod.

For a slender rod, the moment of inertia about its endpoint is $I_{\text{rod}} = 1/3 \, mL^2$

In our case, $I_{\text{rod}} = I_o = \tfrac{1}{3}mD^2 = \tfrac{1}{3}(9.0 \text{ kg})(1.4 \text{ m})^2$ $= 5.88 \text{ kg} \cdot \text{m}^2$.

Given angular velocity $\omega = 3.6$ rad/s and substituting, $H_o = (5.88 \text{ kg} \cdot \text{m}^2)(3.6 \text{ rad/s}) = 21.16 \approx 21.2 \text{ kg} \cdot \text{m}^2/\text{s}$.

64. B The net moment is equal to the area under the shear diagram. Therefore, area = 48 kN·m, the net moment is 48 kN·m. The area of the trapezoid is equal to the area of the rectangle = 48 kN·m.

65. C For a solid shaft, $\tau = \dfrac{16T}{\pi d^3}$. Then

$$T = \frac{\tau \pi d^3}{16} = 596.4 \text{ N} \cdot \text{m}$$

66. A Here, $M_{\text{max}} = 100$ N·m. Then:

$$\sigma_{\text{max}} = \frac{M_{\text{max}}C}{I} = \frac{100(0.03)}{\dfrac{1}{12}(0.04)(0.06)^3}$$

$$= 4.166 \text{ MPa}$$

67. D Here, $V_{\text{max}} = 300$ N. For a circular cross section,

$$\tau = \frac{4}{3}\frac{V}{A} = \frac{4(300)}{3\dfrac{\pi}{4}(0.1)^2} = 50.9 \text{ kPa}$$

68. C The critical load is given by the equation

$$P_{cr} = \frac{\pi^2 EI}{L_e^2} = \frac{\pi^2 EI}{4^2} = 0.617 \text{ EI}$$

Note that $L_e = 2L = 2(2) = 4$ m.

69. B The normal stress at point A is given by the equation

$$\sigma_A = \frac{Mc}{I} - \frac{P}{A}$$

$$= \frac{(400)(0.02)}{\frac{\pi}{64}(0.04)^4} - \frac{10,000}{\frac{\pi}{4}(0.04)^2} = 55.7 \text{ MPa}$$

70. A For each 3-cm diameter bolt:

$$\tau = \frac{V}{A_s} = \frac{300}{\frac{\pi}{4}(0.03)^2} = 0.424 \text{ MPa}$$

71. D The normal stress in section AB is given by the equation

$$\sigma_{AB} = \frac{P_{AB}}{A_{AB}} = \frac{30,000}{\frac{\pi}{4}(0.08)^2} = 5.968 \text{ MPa}$$

72. D Maximum moment occurs at point A and is:

$$M_{max} = 1,000\,(16) + 600(8)(4)$$

$$= 35,200 \text{ lb·ft} = 422,400 \text{ lb·in}$$

The moment of inertia with respect to the neutral axis is

$$\bar{I} = \frac{1}{12}\,bh^3 = \frac{1}{12}\,[6(12)^3 - 4(8)^3] = 693.33 \text{ in}^4$$

$$\sigma = \frac{Mc}{I} = \frac{(422,400)(6)}{693.33} = 3,655.4 \text{ psi}$$

73. B At 720°C, iron is body centered cubic. According to Table 10.3, D_0 is 3.9×10^{-7} m^2/s and Q is 80 kJ/mol. Therefore:

$$D = D_0 e^{-\frac{Q}{RT}} = 3.9 \times 10^{-7} e^{-\frac{80,000}{(8.31)(993)}} \text{ m}^2/\text{s}$$
$$= 2.4 \times 10^{-11} \text{ m}^2/\text{s} = 2.4 \times 10^{-7} \text{ cm}^2/\text{s}$$

74. B Using the rule of mixtures:

$$\rho_c = f_1(Al) + f_2(air)$$
$$1 = f_1(2.7) + f_2(0)$$
$$f_1 = \frac{1}{2.7} = 0.37$$
$$f_1 + f_2 = 1$$
$$f_2 = 0.63$$

75. A At ambient temperatures, both polyethylene and polypropylene would be suitable since their glass transition temperatures are below room temperature. However, the very low T_g of polyethylene makes it more suitable over a wider range of low temperatures.

76. C Metal is a conductor; free electrons can easily move through it. A semiconductor allows some electrons to move through it. A superconductor has no electrical resistance.

77. A In choices (B) and (C), the $1p$ level does not exist. In choice (D), the $3d$ level is a higher energy level than the $4s$ level.

78. A This is one of the definitions of a metal.

79. A The flow rate of fluid through an orifice flow meter is given by the equation

$$Q = CA\sqrt{2g\left[\left(\frac{p_1}{\gamma} + z_1\right) - \left(\frac{p_2}{\gamma} + z_2\right)\right]}$$

Because z_1 is equal to z_2, this equation can be simplified to

$$Q = CA\sqrt{2g\left(\frac{p_1 - p_2}{\gamma}\right)}$$

Substitute known values:

$$Q = (0.61)(0.50 \text{ m}^2) \times$$

$$\sqrt{2\left(9.81\,\frac{\text{m}}{\text{s}^2}\right)\left(\frac{1,000 \text{ Pa} - 450 \text{ Pa}}{9,810\,\frac{\text{N}}{\text{m}^3}}\right)\left(\frac{\frac{\text{N}}{\text{m}^2}}{\text{Pa}}\right)}$$

$$= 0.32 \text{ m}^3/\text{s}$$

80. B This problem is solved using the Bernoulli equation:

$$\frac{p_1}{\gamma} + \frac{\bar{V}_1^2}{2g} + z_1 = \frac{p_2}{\gamma} + \frac{\bar{V}_2^2}{2g} + z_2$$

In this case, because the diameter of the tank is assumed to be much larger than the diameter of the siphon tube, the average velocity at section 1 is nearly 0. Also, the pressures of the fluid at section 1 and at section 2 are equal to the atmospheric pressure. Using these facts, simplify the Bernoulli equation and solve for $\bar{V}_2$:

659

$$\frac{\overline{V}_2^{\,2} - \overset{=0}{\overline{V}_1^{\,2}}}{2g} = \overset{=0}{\frac{p_1 - p_2}{\gamma}} + (z_1 - z_2)$$

$$\overline{V}_2 = \sqrt{2g(z_1 - z_2)}$$

$$= \sqrt{2\left(9.81\,\frac{m}{s^2}\right)(3\,m)}$$

$$= 7.67\,\text{m/s}$$

The volumetric flow rate is determined from the average velocity using $Q = \overline{V}A$, where the area A is given by the equation

$$A = \frac{\pi D^2}{4}$$

and where D is the diameter of the siphon tube.

$$Q = \overline{V}\left(\frac{\pi D^2}{4}\right)$$

$$= \left(7.67\,\frac{m}{s}\right)\left[\frac{\pi(0.03\,m)^2}{4}\right]\left(\frac{3,600\,s}{h}\right)$$

$$= 0.00542\,\frac{m^3}{s}$$

81. C The total force acting on the gate, applied by the water pressure, can be modeled as a point force located at the center of pressure. Using the magnitude of this point force and its distance from the axis of the hinge, the required moment is determined. (Refer to the figure below.)
To determine the magnitude of the point force, use these equations:

$$F = \frac{p_1 + p_2}{2}A_v$$

$$p_1 = \overset{=0}{\rho g z_1}$$

$$= 0\,\frac{\text{lbf}}{\text{ft}^2}$$

$$p_2 = \rho g z_2$$

$$= \left(62.4\,\frac{\text{lbm}}{\text{ft}^3}\right)\left(32.2\,\frac{\text{ft}}{s^2}\right)(4\,\text{ft})\left(\frac{\text{lbf}}{32.2\,\frac{\text{lbm}\cdot\text{ft}}{s^2}}\right)$$

$$= 249.6\,\frac{\text{lbf}}{\text{ft}^2}$$

$$A_v = (3\,\text{ft})(4\,\text{ft})$$

$$= 12\,\text{ft}^2$$

$$F = \frac{0\,\dfrac{\text{lbf}}{\text{ft}^2} + 249.6\,\dfrac{\text{lbf}}{\text{ft}^2}}{2}(12\,\text{ft}^2)$$

$$= 1,497.6\,\text{lbf}$$

To determine the center of pressure, use these equations:

$$z^* = \frac{\rho g I_{y_c}\overset{=1}{\sin(90°)}}{F}$$

$$I_{y_c} = \frac{wh^3}{12}$$

$$= \frac{(3\,\text{ft})(4\,\text{ft})^3}{12}$$

$$= 16\,\text{ft}^4$$

$$z^* = \frac{\left(62.4\,\dfrac{\text{lbm}}{\text{ft}^3}\right)\left(32.2\,\dfrac{\text{ft}}{s^2}\right)(16\,\text{ft}^4)}{1,497.6\,\text{lbf}}\left(\frac{\text{lbf}}{32.2\,\frac{\text{lbm}\cdot\text{ft}}{s^2}}\right)$$

$$= 0.667\,\text{ft}$$

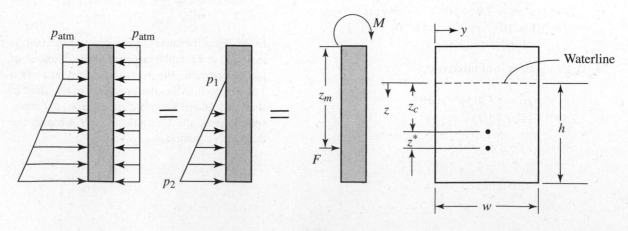

To determine the length of the moment arm, use this equation:

$$z_m = 1 \text{ ft} + z_c + z^*$$
$$= 1 \text{ ft} + 2 \text{ ft} + 0.667 \text{ ft}$$
$$= 3.667 \text{ ft}$$

And the total moment is:

$$M = Fz_m$$
$$= (1{,}497.6 \text{ lbf})(3.667 \text{ ft})$$
$$= 5{,}496 \text{ lbf·ft}$$

82. A Since the head loss through each pipe must be equal:

$$f_a \frac{L_a}{D_a} \frac{\overline{V}_a^{\,2}}{2g} = f_b \frac{L_b}{D_b} \frac{\overline{V}_b^{\,2}}{2g}$$

With $f_a = f_b$:

$$\frac{V_a}{V_b} = \sqrt{\frac{D_a}{D_b} \frac{L_b}{L_a}}$$
$$= \sqrt{\frac{4 \text{ in}}{2 \text{ in}} \frac{4 \text{ ft}}{8 \text{ ft}}}$$
$$= 1$$
$$(\overline{V}_a = \overline{V}_b)$$

Use this result in the continuity equation:

$$\overline{V}A = \overline{V}_a A_a + \overline{V}_b A_b$$
$$= V_a(A_a + A_b)$$
$$= V_a\left(\frac{\pi D_a^2}{4} + \frac{\pi D_b^2}{4}\right)$$
$$= \frac{\pi V_a}{4}(D_a^2 + D_b^2)$$

Then:

$$\overline{V}_a = (\overline{V}A)\frac{4}{\pi(D_a^2 + D_a^2)}$$
$$= \left(260 \frac{\text{ft}^3}{\text{min}}\right)\left[\frac{4}{\pi[(4 \text{ in})^2 + (2 \text{ in})^2]}\right]$$
$$\times \left(\frac{144 \text{ in}^2}{\text{ft}^2}\right)\left(\frac{\text{min}}{60 \text{ s}}\right)$$
$$= 39.7 \text{ ft/s}$$
$$= 40 \text{ ft/s}$$

83. D Use the impulse-momentum principle to solve this problem:

$$\overrightarrow{F} = \dot{m}\overrightarrow{V}_2 - \dot{m}\overrightarrow{V}_1$$

In this case, only the horizontal, or x-direction, component is important. The x-direction equation for the impulse momentum principle is as follows:

$$|\overrightarrow{F_x}| = \dot{m}|\overrightarrow{V}_2| \cos \alpha - \dot{m}|\overrightarrow{V}_1|$$

Since $A_1 = A_2, |\overrightarrow{V}_1| = |\overrightarrow{V}_2| = 20 \text{ m/s}$, so

$$|\overrightarrow{F_x}| = \dot{m}|\overrightarrow{V}_1|(\cos \alpha - 1)$$
$$= \left(50 \frac{\text{kg}}{\text{s}}\right)\left(20 \frac{\text{m}}{\text{s}}\right)(\cos 45° - 1)\left(\frac{\text{N}}{\frac{\text{kg·m}}{\text{s}^2}}\right)$$
$$= -293 \text{ N}$$

84. D

$$P_a = P_{\text{gauge}} + P_{\text{atm}} = P_{\text{gauge}} + \rho_{\text{Hg}}g(0.724)$$
$$P_a = 827 \text{ kPa} + (13.560 \text{ kg/m}^3)(9.81 \text{ m/s}^2)$$
$$(0.724 \text{ m})$$
$$P_a = 827 \text{ kPa} + 96.3 \text{ kPa} = 923 \text{ kPa}$$

85. C Water at 20°C has $\rho = 1{,}000 \text{ kg/m}^3$ and $\mu = 10^{-3}$ kg/ms. Use the Bernoulli equation:

$$\frac{P_1 - P_2}{\rho g} + \frac{\overline{V}_1^2 - \overline{V}_2^2}{2g} + z_1 - z_2 = h_f$$

$$\frac{690 \text{ kPa} - 520 \text{ kPa}}{(1{,}000 \text{ kg/m}^3)(9.81 \text{ m/s}^2)} + 0 + 6 = h_f$$

Note: $V_1 = V_2$

Therefore, $h_f = 23.3 \text{ m}$

86. A The velocity of water in a pipe can be calculated using the volumetric flow rate:

$$\overline{V} = \frac{4Q}{\pi D^2} = \frac{4(11.2 \text{ m}^3/\text{h.1 h}/3{,}600 \text{ s})}{\pi(0.025 \text{ m})^2}$$
$$= 6.34 \text{ m/s}$$

With that, the Reynold's number for the flow in a pipe is

$$\text{Re}_D = \frac{\rho \overline{V} D}{\mu} = \frac{(1{,}000 \text{ kg/m}^3)(6.34 \text{ m/s})(0.025 \text{ m})}{(10^{-3} \text{ kg/ms})}$$
$$= 158{,}500$$

Finally, the friction coefficient is

$$f = h_f\left(\frac{D}{L}\right)\left(\frac{2g}{\overline{V}^2}\right)$$
$$= (23.3 \text{ m})\left(\frac{0.025 \text{ m}}{6 \text{ m}}\right)\left(\frac{2(9.81 \text{ m/s}^2)}{(6.34 \text{ m/s})^2}\right)$$
$$= 0.047 \approx 0.05$$

87. **A** Knowing $f = 0.05$ and from the Moody diagram (Chapter 11):

$\dfrac{\varepsilon}{D} \approx 0.021$, which results in:

$\varepsilon = (0.021)(0.025 \text{ m}) = 0.525 \text{ mm}$

88. **B** For new, galvanized iron pipe, one can find in the table (Chapter 13): $\varepsilon = 0.15 \text{ mm}$

Then from the Moody diagram (Chapter 11), knowing

$\text{Re}_D = 158{,}500$ and $\dfrac{\varepsilon}{D} = \dfrac{0.15 \times 10^{-3}}{0.025 \text{ m}} = 0.006$,

we can find

$f = 0.0325$

89. **C** From the superheated steam tables at the end of Chapter 13, the value of entropy at the entrance state of the turbine is 7.1229 kJ/kg·K. Since normal atmospheric pressure is 101.35 kPa, the table for superheated steam at 100 kPa will provide values very close to those expected at atmospheric pressure. Since the value of entropy for saturated vapor at this pressure is greater than the value at the turbine inlet, some condensation will occur. Therefore, the temperature of the exiting stream of water will be equal to the saturation temperature of water at atmospheric pressure, or 100°C.

90. **B** Turbine work is determined using equation 13.77:

$\dot{W} = \dot{m}(h_i - h_e)$ or $w = (h_i - h_e)$

The value of enthalpy in the inlet is taken from the superheated steam table in Chapter 13:

$h_i = 3{,}051.2 \text{ kJ/kg}$

To find the value of enthalpy in the outlet stream of an ideal turbine, the quality of the exit stream for the ideal turbine must be determined. Use equation 13.66d:

$s_{\text{mix}} = s_f + x(s_g - s_f)$ or $x = \dfrac{(s_{\text{mix}} - s_f)}{(s_g - s_f)}$

Find the values of s_f and s_g using the saturated water table in Chapter 13:

$s_f = 1.3069 \text{ kJ/kg·K}$, $s_g = 7.3549 \text{ kJ/kg·K}$

In this case, the value of s_{mix} is equal to the value of entropy in the inlet stream of the turbine:

$s_{\text{mix}} = 7.1229 \text{ kJ/kg·K}$

Substitute known values:

$$x = \dfrac{s_{\text{mix}} - s_f}{s_g - s_f}$$

$$= \dfrac{\left(7.1229 \dfrac{\text{kJ}}{\text{kg·K}} - 1.3069 \dfrac{\text{kJ}}{\text{kg·K}}\right)}{\left(7.3549 \dfrac{\text{kJ}}{\text{kg·K}} - 1.3069 \dfrac{\text{kJ}}{\text{kg·K}}\right)}$$

$$= 0.962$$

Using this value for quality, determine the value of enthalpy for the exit stream of the ideal turbine by applying equation 13.66c, where $h_e = h_{\text{mix}}$:

$$h_e = h_f + x(h_g - h_f)$$

Find the values of h_f and h_g using the saturated water table:

$h_f = 419.04 \text{ kJ/kg}$, $h_g = 2{,}676.1 \text{ kJ/kg}$

Then

$$h_e = 419.04 \dfrac{\text{kJ}}{\text{kg}} + 0.962\left(2{,}676.1 \dfrac{\text{kJ}}{\text{kg}} - 419.04 \dfrac{\text{kJ}}{\text{kg}}\right)$$

$$= 2{,}590.3 \text{ kJ/kg}$$

The work produced in an ideal turbine under these conditions is:

$$\dot{w} = (h_I - h_e)$$

$$= \left(3{,}051.2 \dfrac{\text{kJ}}{\text{kg}} - 2{,}590.3 \dfrac{\text{kJ}}{\text{kg}}\right)$$

$$= 460.9 \text{ kJ/kg}$$

$$= 461 \text{ kJ/kg}$$

91. **A** Turbine efficiency as given is defined as the amount of work a turbine produces divided by the amount of work produced in an ideal turbine operating at the same conditions:

$$\eta_{\text{turb}} = \dfrac{w_{\text{real}}}{w_{\text{ideal}}}$$

Then:

$$w_{\text{real}} = (\eta_{\text{turb}})(w_{\text{ideal}})$$

$$= (0.80)\left(461 \dfrac{\text{kJ}}{\text{kg}}\right)$$

$$= 368.8 \text{ kJ/kg}$$

$$= 369 \text{ kJ/kg}$$

92. D Determine the value of enthalpy for the exit state of the real turbine as follows:

$$w = (h_i - h_e)$$
$$h_e = h_i - w$$
$$= 3{,}051.2 \, \frac{kJ}{kg} - 368.8 \, \frac{kJ}{kg}$$
$$= 2{,}682.4 \, kJ/kg$$

93. B

$$\dot{Q} = \frac{tkA\Delta T}{L} \rightarrow t = \frac{QL}{kA\Delta T} = 195.18 \, s = 3.25 \, min$$

94. D

$$\dot{Q} = \in \sigma A T^{\,4} = 12 \, W$$

95. A

$$\dot{Q} = \in \sigma A (T_p^4 - T_R^4) = 5{,}382 \, W = 5.38 \, kW$$

96. B

$$r_{cr} = \frac{k_{insu}}{h_\infty} = 0.04 \, m = 4 \, cm$$

97. B

$$\tau_c = \frac{\rho c V}{h A_s} = \frac{\rho c}{h} \cdot \frac{\frac{\pi}{6} D^3}{\pi D^2} = \frac{\rho c D}{6h}$$
$$= 0.0007 \, m$$
$$= 0.7 \, mm$$

98. C Calculate the Biot number to verify the use of lumped capacitance.

$$B_i = \frac{hL_c}{k} = \frac{h}{k} \cdot \left(\frac{V}{A_s} \right) = 0.0017 < 0.1$$

Use lumped capacitance.

Now calculate the time it will take to reach 199°C using LC:

$$\frac{T - T_\infty}{T_i - T_\infty} = e^{-\frac{h A_s t}{\rho c V}} \rightarrow t = \frac{\rho c V}{h A_s} \cdot \ln \left(\frac{T_i - T_\infty}{T - T_\infty} \right)$$
$$= 3.7 \, seconds$$

99. D

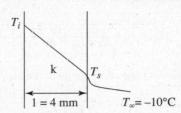

$$q'' = \frac{q}{A} = \frac{T_i - T_\infty}{\frac{L}{k} + \frac{1}{h}} = \frac{37 - (-10)}{\frac{0.009}{0.2} + \frac{1}{20}}$$
$$= 671 \, W/m^2$$

100. C

$$q'' = \frac{37 - (-10)}{\frac{0.004}{0.2} + \frac{1}{70}} = 1{,}282 \, W/m^2 = 1{,}280 \, W/m^2$$

101. B

$$\frac{T_i - T_\infty}{\frac{1}{k} + \frac{1}{h}} = q'' = \frac{T_i - T_s}{\frac{1}{k}}$$

$$T_s = T_i - \frac{\frac{1}{k}(T_i - T_\infty)}{\frac{1}{k} + \frac{1}{h}}$$

$$T_s = 37 - \frac{\frac{0.004}{0.2}(37 - (-10))}{\frac{0.004}{0.2} + \frac{1}{20}} = 23.6°C = 24°C$$

102. A

$$T_s = 37 - \frac{\frac{0.004}{0.2}(37 - (-10))}{\frac{0.004}{0.2} + \frac{1}{70}} = 9.6°C = 10°C$$

103. D For the signal to be accurately constructed from the sampled data using proper hold operation, the frequency of sampling (f) must be at least twice the highest frequency component of the measured signal. That is, $f = 2 \times 400$ Hz $= 800$ Hz. Hence, $\Delta t = 1/800 = 0.00125$ s or 1.25 ms.

104. C As described in equation 15.81, the output of the bridge is obtained as:

$$V_{out} = V_{in}\left(\frac{R_x}{R_s + R_x} - \frac{R_2}{R_1 + R_2}\right)$$

For the bridge to be balanced, the output voltage must be zero. That is, $\frac{R_x}{R_s + R_x} - \frac{R_2}{R_1 + R_2} = 0$.

This results in $\frac{R_x}{R_s} = \frac{R_1}{R_2}$.

Hence, $R_x = \frac{R_2 R_s}{R_1} = \frac{500 \times 2{,}400}{1{,}600} = 750\ \Omega$.

105. A As seen from the thermal sensor discussions, the resistance of the thermal sensor is a function of the temperature as $R_T = R_0[1 + \alpha(T - T_0)]$.

Hence, $\frac{R_T - R_0}{R_0} = \alpha(T - T_0) = 0.00395 \times 50 = 0.1975$ or $\approx 20\%$.

106. B If the measurement quantities $y_1, y_2, \ldots, y_n$ are contaminated with some levels of measurement uncertainties, say, $w_1, w_2, \ldots, w_n$ respectively, the uncertainties w_R in the calculated variable R $(R = f(y_1, y_2, \ldots, y_n))$ can be estimated using the Kline-McClintock equation as:

$$w_R = \sqrt{\left(w_1 \frac{\partial f}{\partial y_1}\right)^2 + \left(w_2 \frac{\partial f}{\partial y_2}\right)^2 + \cdots + \left(w_n \frac{\partial f}{\partial y_n}\right)^2}$$

Here, $e_R = \sqrt{\left(w_{h1} \frac{\partial e}{\partial h_1}\right)^2 + \left(w_{h2} \frac{\partial e}{\partial h_2}\right)^2}$.

Now

$$\frac{\partial e}{\partial h_1} = \frac{-h_2}{2h_1^2 \sqrt{h_2/h_1}}$$

$$= \frac{-0.4}{2(0.5)^2 \sqrt{0.4/0.5}}$$

$$= -0.8944,$$

$$\frac{\partial e}{\partial h_2} = \frac{1}{2h_1 \sqrt{h_2/h_1}}$$

$$= \frac{1}{2(0.5)\sqrt{0.4/0.5}}$$

$$= 1.1180.$$

Hence,

$$e_R = \sqrt{(0.001 \times (-0.8944))^2 + (0.001 \times 1.1180)^2}$$
$$= 0.0014317 \text{ or } 0.14\%.$$

107. D

$$R = (CPF) \cdot (\text{Chronic daily intake (CDI)(mg/kg} \cdot \text{d)})$$

Given, $R = 0.02$

$\qquad$ CDI = 5 ppm/d

$\qquad$ CPF = ?

$\qquad$ $(0.02) = (CPF)(5)$

$\qquad$ CPF = 0.004 d/ppm

108. B Given that:

Thickness of clay liner = 0.5 m
Leachate head = 50 cm = 0.5 m
$K = 1 \times 10^{-7}$ cm/s
$\eta = 50\%$

The Darcy velocity can be calculated by

$$v = K\frac{dh}{dr}$$

The hydraulic gradient can be estimated as

$$\frac{dh}{dr} = \frac{0.5 + 0.5}{0.5} = 2.0$$

So $v = (1 \times 10^{-7}) \times (2.0) = 2 \times 10^{-7}$ cm/s

Seepage velocity is given by $v' = \frac{v}{\eta}$

$$v' = \frac{2 \times 10^{-7}}{0.50} = 4 \times 10^{-7} \text{ cm/s}$$

The travel time will be $t = \frac{T}{v'}$

$$t = \frac{0.5\text{ m} \frac{100\text{ cm}}{\text{m}}}{4 \times 10^{-7}} = t = 1.25 \times 10^8 \text{ s}$$

or 3.96 years

It will take approximately 4 years to migrate through a 0.5 m clay liner having hydraulic conductivity of 1×10^{-7} cm/s and porosity 50%.

109. B Noise attenuation $= 10 \log_{10}\left[\left(\frac{D_2}{D_1}\right)^2\right]$

Given $\quad D_1 = 14$ m

$\qquad\quad D_2 = 280$ m

Noise attenuation = ?

Noise attenuation $= 10 \log 10\ [(280/14)^2]$

$$= 26 \text{ dB}$$

Approximate reading at 280 m = 72 − 26 = 46 dB.

110. D

$$\text{Hazard ratio} = \frac{\text{Estimated exposure dose (mg/kg·d)}}{\text{Reference dose (mg/kg·d)}}$$

Note that mg/kg is equivalent to ppm.

Given, estimated exposure = 3 ppm/d

reference dose = 12 ppm/d

hazard ratio = (3)/(12)

hazard ratio = 0.25

Diagnostic Chart: Other Discipline

Subject Area	Total Number of Questions	Number Correct	Number Incorrect	Reason for Incorrect Answer			
				Lack of Knowledge	Misread Problem	Careless or Mathematical Error	Wrong Guess
Mathematics (1–14)	14						
Probability and Statistics (15–21)	7						
Chemistry (22–28)	7						
Ethics and Professional Practices (29–31)	3						
Engineering Economics (32–38)	7						
Electricity and Magnetism (39–46)	8						
Statics (47–55)	9						
Dynamics, Kinematics, and Vibrations (56–63)	8						
Mechanics of Materials (64–72)	9						
Materials Properties and Processing (73–78)	6						
Fluid Mechanics of Liquids and Gases (79–82)	14						
Heat Transfer (93–102)	10						
Instrumentation and Data Acquisitions (103–106)	4						
Safety, Health, and Environment (107–110)	4						
Total	**110**						

Index